SIGMA-DELTA CONVERTERS

SIGMA-DELTA CONVERTERS

Practical Design Guide

SECOND EDITION

JOSÉ M. DE LA ROSA

Institute of Microelectronics of Seville
IMSE-CNM (CSIC, University of Seville)
Spain

IEEE PRESS

WILEY

Edition History

This edition first published in English copyright year 2013

Registered Offices
John Wiley & Sons, Inc., 111 River Street, Hoboken, NJ 07030, USA
John Wiley & Sons Ltd, The Atrium, Southern Gate, Chichester, West Sussex, PO19 8SQ, UK

Editorial Office
John Wiley & Sons Ltd, The Atrium, Southern Gate, Chichester, West Sussex, PO19 8SQ, UK

For details of our global editorial offices, customer services, and more information about Wiley products visit us at www.wiley.com.

Wiley also publishes its books in a variety of electronic formats and by print-on-demand. Some content that appears in standard print versions of this book may not be available in other formats.

Library of Congress Cataloging-in-Publication Data

Names: Rosa, José M. de la, author.
Title: Sigma-delta converters : practical design guide / José M. de la Rosa, University of Seville, Spain.
Other titles: CMOS sigma-delta converters
Description: Second edition. | Hoboken, NJ, USA : Wiley-IEEE Press, 2018. | Revised edition of: CMOS sigma-delta converters. | Includes bibliographical references and index. |
Identifiers: LCCN 2018023691 (print) | LCCN 2018026688 (ebook) | ISBN 9781119275756 (Adobe PDF) | ISBN 9781119275763 (ePub) | ISBN 9781119275787 (hardcover)
Subjects: LCSH: Metal oxide semiconductors, Complementary–Design and construction. | Analog-to-digital converters–Design and construction.
Classification: LCC TK7871.99.M44 (ebook) | LCC TK7871.99.M44 R668 2019 (print) | DDC 621.3815/9–dc23
LC record available at https://lccn.loc.gov/2018023691

Cover design by Wiley
Cover image: Courtesy of José M. de la Rosa

Set in 9.5/11.5pt NimbusRomNo9L by SPi Global, Chennai, India

M WEP319113 241024

This book is dedicated to the memory of two colleagues and friends who have been key persons in my life:

Mariano Marcos Bárcena, professor of professors, and my first teacher of Physics, who inspired my passion for science and technology.

Fernando Medeiro Hidalgo, one of the best analog and mixed-signal IC designers I have known, with whom I was very fortunate to take my first steps in this fascinating world of ΣΔ converters.

God bless you, Mariano, Fernando.

To my wife Visi, my daughter María, my son Jaime
and to the memory of my little son José Manuel

"Engineering isn't about perfect solutions; it's about doing the best you can with limited resources."

Randy Pausch, Professor, Carnegie Mellon (*The Last Lecture*, Hyperion 2008)

CONTENTS

PREFACE

Sigma-Delta modulators ($\Sigma\Delta$Ms) have become one of the best choices for the implementation of analog/digital interfaces of electronic systems integrated in CMOS technologies. Compared to other kinds of analog-to-digital converters (ADCs), $\Sigma\Delta$Ms cover the widest conversion region of the resolution-versus-bandwidth plane. They are the most efficient way to digitize very diverse types of signal in an increasing number of application scenarios, from high-resolution low-bandwidth data conversions for digital audio, sensor interfaces, and instrumentation, to ultra-low-power biomedical systems and medium-resolution broadband wireless communications. This versatility, together with their robustness and their simplicity in many practical situations, has made more and more engineers today consider $\Sigma\Delta$Ms as the first choice for their research projects and their industrial products.

The idea underlying the operation of $\Sigma\Delta$Ms was patented by Cutler in 1960 [1], although its application to the construction of data converters was first reported in the published literature by Inose *et al.* in 1962 [2]. The operation of $\Sigma\Delta$Ms is relatively simple to describe, although sometimes very difficult to analyze. Essentially, the fundamental principle behind $\Sigma\Delta$Ms is based on the combination of two signal processing techniques, namely: *oversampling* and quantization *noise shaping*. The former consists of taking the signal samples at a higher rate than the one dictated by the Nyquist sampling theorem. These samples are commonly quantized with a large error using a low-resolution quantizer. The resulting oversampled quantization error is filtered in the modulator feedback loop, so that its frequency spectrum is *shaped* in such a way that a large portion of its power is pushed out of the signal band, where it is removed by a digital filter. The outcome of the combined action of oversampling and noise shaping allows $\Sigma\Delta$Ms to achieve high-precision digitization using a low-resolution coarse quantizer. Therefore, unlike other kinds of ADC architectures that require high-precision analog circuits, $\Sigma\Delta$Ms trade the accuracy of their analog circuitry for speed of digital signal processing, thus achieving a higher degree of insensitivity to circuit error mechanisms and potentially benefiting from CMOS technology's evolution towards the nanometer scale.

Prompted by these benefits and fueled by technology downscaling and industry trends in consumer digital electronics, the original concept of noise shaping described above has evolved over the last five decades through many ΣΔM generations, giving rise to a plethora of architectures, circuit- and system-design techniques, and a number of integrated circuits (ICs), which have pushed the state of the art on ΣΔMs forward, yielding innovative research results and successful industry products.

All these advances and research studies have led (and continue to do so) to a vast amount of technical literature. Indeed, since the publication of pioneering works such as the widely cited papers written by Candy [3, 4] and Boser and Wooley [5], the number of publications has increased significantly, now including hundreds of patents, thousands of research papers, some tutorial papers [6–8], as well as dozens of introductory and specialized monographs [9–31]. However, with so much material and such an abundance of technical information published, many designers – particularly novel designers, but also some experienced designers focused on specific subtopics of ΣΔMs – may become sometimes disoriented and lose their way. This has motivated some authors to put all these pieces of information together in a comprehensive and systematic way.

Apart from earlier books aiming to catalogue the existing publications on ΣΔMs [9], one of the first attempts to present a guide for ΣΔM designers is the book edited by Norsworthy *et al.* in 1997 [10], also known as "the yellow book" by the ΣΔM community. This book, which deals with a number of important subjects in ΣΔMs, had contributions by a number of experts in the field, thus making it hard to present its contents in a coherent and consistent way. With this objective in mind, some authors have put their efforts into writing tutorial monographs dealing with the systematic design of ΣΔMs.

Among others, the book written by Schreier and Temes, published in 2005 [21], often referred to as "the green book", has become one of the most popular books on ΣΔ converters. This book provides an excellent and comprehensive treatment of ΣΔMs, their operating principles, and main architectures, presenting several design examples constructed using the well-known Schreier's MATLAB toolbox [32]. A revised second edition of this book was written by Pavan, Schreier and Temes, and was published in 2017 [33]. This new edition expanded the contents of the first edition with more sections dealing with continuous-time (CT) circuit implementations and circuit design considerations, without losing the main intention of the first edition, namely to give a basic understanding of the operation of ΣΔ converters.

Some other remarkable and pioneering examples are the book written by Medeiro *et al.* in 1999 [13] – focused on the systematic design of SC ΣΔMs – and the book of Ortmanns and Gerfers [22], published in 2006, which is still one of the most complete monographs on CT ΣΔMs to date. All of these books, as well as other monographs reported in the technical literature, give incomplete views of ΣΔMs, paying more attention to particular aspects of their design, and/or a type of architecture, circuit technique, or application.

This being the case, and following the spirit of the first edition of this book, this second edition attempts to cover some of these knowledge gaps in the ΣΔ literature, by providing a comprehensive and systematic description of the *universe* of ΣΔMs, their diverse architectures, circuit techniques, analysis and synthesis methods and CAD tools, as well as their practical design considerations. As in the first edition, one of the main purposes of this book is to be an educational and reference textbook for undergraduate and graduate students. With this goal in mind, and based on the courses already given by the author and the feedback received from readers and course attendees, the contents of the second edition of the book have been updated, completed and structured to address a large audience: from senior designers who want to acquire a deeper and up-to-date insight into ΣΔMs, to inexperienced engineers who are looking for a uniform and self-contained reference into this hot topic. The new contents and materials make this new edition a unique monograph, a result of the compiling and updating of the enormous number of technical and research studies reported to date on the topic of ΣΔMs. It presents the results of this compilation in a didactical, pedagogical, and intuitive style.

Another key feature of this book (as mentioned in the title) is that it can be used as a *practical guide for designers*, emphasizing explanations of the multiple trade-offs involved in the whole design flow of ΣΔMs – from specifications to chip implementation and characterization. To this end, a *top-down* approach is followed, presenting the contents in a hierarchical way; in other words, going from the theoretical fundamentals, system-level design equations, and behavioral models to circuit, transistor-level, and physical implementations, in order to provide readers with the necessary understanding and insights into the recent advances, trends, and challenges involved in the design of state-of-the-art ΣΔM ICs.

This second edition emphasizes two key points, which were not covered in such depth in the first edition. The first is to include more detailed explanation of ΣΔMs implemented using CT circuits, going from system-level synthesis to practical circuit/physical limitations. The second point is to include more practical case studies and applications, as well giving a deeper description of the synthesis methodologies and CAD tools employed in the design of ΣΔ converters. Due to the quantity of all these new materials, the table of contents of the first edition has been re-organized and expanded, going from five chapters and two appendixes to ten chapters and three appendixes in this second edition.

The top-down approach adopted in this book inspires the hierarchical way in which the contents are structured. Thus, Chapter 1 begins from the top, giving an introduction to data converters and explaining the basic concepts and fundamentals behind ΣΔ modulation, its main building blocks, the signal processing involved, its performance metrics and basic examples to illustrate the concepts of noise shaping and oversampling – the main ingredients of ΣΔ converters. Chapter 2 gives a taxonomical description of the diverse variety of ΣΔM architectures, the nature of signals (low-pass and band-pass), as well as the dynamics involved (either discrete-time or continuous-time). In this chapter, ΣΔMs are considered ideal systems, except for their inherent quantization error; CT synthesis methods and architectures will be explained in more detail than in the first edition of the book.

Chapters 3 and 4 descend one level in the modulator hierarchy to analyze the effect of the main circuit error mechanisms, as well as architectural and timing nonidealities. The mathematical models, analytical procedures and design guidelines described in this chapter provide an understanding of the main practical issues affecting the performance of ΣΔMs. Chapter 3 mostly focuses on SC circuit error mechanisms, while Chapter 4 updates and expands on the contents of the first edition, dealing with CT circuit nonideal effects and compensation techniques. This is complemented by the state-space analysis of clock jitter in CT-ΣΔMs given in Appendix A.

The knowledge derived from the first four chapters is an essential part of the systematic *top-down/bottom-up* synthesis methodology of ΣΔMs, which is described in Chapters 5 and 6. Thus, Chapter 5 deals with the synthesis methodology of ΣΔ converters, focusing on high-level behavioural modelling and simulation techniques, and giving an introduction to the SIMSIDES simulator. The updates of the new version of this simulator are also included in this second edition and distributed through the book's companion website. Chapter 6 focuses on optimization techniques, illustrating different approaches to combine simulation and optimization for the high-level design of ΣΔ modulators. The contents of this chapter are extended and complemented in Appendices B and C. Appendix B includes a user guide to SIMSIDES and Appendix C provides an overview of all behavioral models and libraries included in this simulator.

The circuit- and physical-level considerations are presented in Chapters 7 and 8, which update and expand on the contents of the first edition. This update and re-organization is motivated by the fact that the previous version (presented in a single chapter) was too long and, according to readers' feedback, needed to be restructured into two chapters. These chapters give a number of necessary design recommendations and practical recipes to complete the design flow of a ΣΔM, showing the step-by-step methodology to transform a system-level model into an electrical schematic – initially based on macromodels, and then implemented with transistors – and, finally, concluding the design cycle with the layout and chip implementation. Plenty of examples, case studies, and simulation test benches are provided to illustrate the practical issues and design considerations addressed in the chapter, which cover everything from electrical analysis and simulation using SPICE-like simulators, to layout design considerations, chip prototyping, and experimental measurements of ΣΔMs in the laboratory. These chapters are complemented by Chapter 9, where some more content dealing with practical chip implementation, case studies and experimental characterization are provided.

To conclude the book, Chapter 10 gives an overview of state-of-the-art ΣΔM ICs, comparing their performance with Nyquist-rate ADCs. Overall, more than 500 state-of-the-art IC references have been studied in detail and considered in this review, including papers published up until August 2017. Therefore, following the practical philosophy that inspires this book, the diverse families of state-of-the-art ΣΔM architectures and circuit techniques are exhaustively analyzed and compared to extract practical and empirical design guidelines from the statistical data, trying to identify the coming trends, design challenges, as well as the solutions proposed by cutting-edge ICs at the frontiers of ΣΔMs.

As stated above, the book contents are addressed and structured for a wide audience: from senior designers to students starting to work in the area of ΣΔMs. Bearing this in mind, the style and main purpose of the book is to serve also as an educational and reference textbook for undergraduate and graduate students. Indeed, the book is based on several graduate courses given by the author, including masters and doctorate degree programs, invited lectures, and IEEE conference tutorials, distinguished lectures, and courses. All these materials have been adapted and updated so that a large portion of the book can also be used (and indeed it has been used) in both undergraduate and graduate courses.

However, in spite of the encyclopaedic nature of the book, it is impossible to give an exhaustive description of all the topics contained in the thousands of publications dealing with ΣΔMs. Instead, the book tries to cover the main subtopics, providing sufficient insight to allow the reader to understand the others, which are only reviewed in brief, or sometimes even omitted. In order to try to palliate these unavoidable deficiencies, a list of references is included at the end of each chapter. Overall, the book contains a list of references in order to guide readers to an increase in their understanding of the diverse research topics in the ΣΔ world.

The huge quantity of information contained in the book is complemented and updated by a number of electronic resources, which enlarge on the resources provided with the first edition. They are freely available on the Web. To this end, all the data analyzed in the state-of-the-art survey presented in Chapter 10 have been collected in a spreadsheet, which is available at `http://www.imse-cnm.csic.es/~jrosa/CMOS-SDMs-Survey-IMSE-JMdelaRosa.xlsx`. This database is periodically kept up to date, and aims to be a complement to the popular Murmann's ADC survey data collection [34]. In addition, a fully functional version of the time-domain behavioral simulator SIMSIDES is freely available on demand at `http://www.imse-cnm.csic.es/simsides`. The simulator incorporates many examples, including the case studies presented in the book and many more examples. Apart from the SIMSIDES software, the majority of examples and test benches of different CAD tools used throughout the book are also available on the Web at `http://www.imse-cnm.csic.es/~jrosa`.

Last but not least, although the author has done his best to cover the many new and hot topics in the world of ΣΔ converters, there is some content that cannot yet be covered in detail while other areas had to be omitted to allow the book to be completed in a reasonable time and in a reasonable number of pages.

I hope that readers understand these limitations, find this edition of the book useful and practical, and enjoy reading (and using) it as much as I have enjoyed revising and writing it. As in the first edition, your feedback is very important and very welcome!

José M. de la Rosa
Sevilla, January 2018

References

[1] C. C. Cutler, "Transmission System Employing Quantization," *U.S. Patent No. 2,927,962*, 1960.

[2] H. Inose, Y. Yasuda, and J. Murakami, "A Telemetering System by Code Modulation – $\Delta - \Sigma$ Modulation," *IRE Trans. on Space Electronics and Telemetry*, vol. 8, pp. 204–209, September 1962.

[3] J. Candy and O. J. Benjamin, "The Structure of Quantization Noise from Sigma-Delta Modulation," *IEEE Transactions on Communications*, pp. 1316–1323, 1981.

[4] J. Candy, "A Use of Double Integration in Sigma-Delta Modulation," *IEEE Transactions on Communications*, vol. 33, pp. 249–258, March 1985.

[5] B. E. Boser and B. A. Wooley, "The Design of Sigma-Delta Modulation Analog-to-Digital Converters," *IEEE J. of Solid-State Circuits*, vol. 23, pp. 1298–1308, December 1988.

[6] P. M. Aziz *et al.*, "An Overview of Sigma-Delta Converters," *IEEE Signal Processing Magazine*, vol. 13, pp. 61–84, January 1996.

[7] I. Galton, "Delta-Sigma Data Conversion in Wireless Transceivers," *IEEE Trans. on Microwave Theory and Techniques*, vol. 50, pp. 302–315, January 2002.

[8] J. M. de la Rosa, "Sigma-Delta Modulators: Tutorial Overview, Design Guide, and State-of-the-Art Survey," *IEEE Transactions on Circuits and Systems I: Regular Papers*, vol. 58, pp. 1–21, January 2011.

[9] J. Candy and G. Temes, *Oversampling Delta-Sigma Data Converters: Theory, Design and Simulation*. IEEE Press, 1991.

[10] S. R. Norsworthy, R. Schreier, and G. C. Temes, *Delta-Sigma Data Converters: Theory, Design and Simulation*. IEEE Press, 1997.

[11] J. Cherry and W. Snelgrove, *Continuous-Time Delta-Sigma Modulators for High-Speed A/D Conversion*. Kluwer Academic Publishers, 1999.

[12] J. V. Engelen and R. van de Plassche, *BandPass Sigma-Delta Modulators: Stability Analysis, Performance and Design Aspects*. Kluwer Academic Publishers, 1999.

[13] F. Medeiro, B. Pérez-Verdú, and A. Rodríguez-Vázquez, *Top-Down Design of High-Performance Sigma-Delta Modulators*. Kluwer Academic Publishers, 1999.

[14] V. Peluso, M. Steyaert, and W. Sansen, *Design of Low-Voltage Low-Power CMOS Delta-Sigma A/D Converters*. Kluwer Academic Publishers, 1999.

[15] S. Rabii and B. A. Wooley, *The Design of Low-Voltage, Low-Power Sigma-Delta Modulators*. Kluwer Academic Publishers, 1999.

[16] L. Breems and J. H. Huijsing, *Continuous-Time Sigma-Delta Modulation for A/D Conversion in Radio Receivers*. Kluwer Academic Publishers, 2001.

[17] Y. Geerts, M. Steyaert, and W. Sansen, *Design of Multi-bit Delta-Sigma A/D Converters*. Kluwer Academic Publishers, 2002.

[18] J. M. de la Rosa, B. Pérez-Verdú, and A. Rodríguez-Vázquez, *Systematic Design of CMOS Switched-Current Bandpass Sigma-Delta Modulators for Digital Communication Chips*. Kluwer Academic Publishers, 2002.

[19] M. Kozak and I. Kale, *Oversampling Delta-Sigma Modulators*. Springer, 2003.

[20] O. Bajdechi and J. Huising, *Systematic Design of Sigma-Delta Analog-to-Digital Converters*. Kluwer Academic Publishers, 2004.

[21] R. Schreier and G. C. Temes, *Understanding Delta-Sigma Data Converters*. IEEE Press, 2005.

[22] M. Ortmanns and F. Gerfers, *Continuous-Time Sigma-Delta A/D Conversion: Fundamentals, Performance Limits and Robust Implementations*. Springer, 2006.

[23] K. Philips and A. H. M. van Roermund, *Sigma Delta A/D Conversion for Signal Conditioning*. Springer, 2006.

[24] R. del Río, F. Medeiro, B. Pérez-Verdú, J. M. de la Rosa, and A. Rodríguez-Vázquez, *CMOS Cascade ΣΔ Modulators for Sensors and Telecom: Error Analysis and Practical Design*. Springer, 2006.

[25] L. Yao, M. Steyaert, and W. Sansen, *Low-Power Low-Voltage Sigma-Delta Modulators in Nanometer CMOS*. Springer, 2006.

[26] P. G. R. Silva and J. H. Huijsing, *High Resolution IF-to-Baseband ΣΔ ADC for Car Radios*. Springer, 2008.

[27] R. H. van Veldhoven and A. H. M. van Roermund, *Robust Sigma Delta Converters*. Springer, 2011.

[28] A. Morgado, R. del Río, and J. M. de la Rosa, *Nanometer CMOS Sigma-Delta Modulators for Software Defined Radio*. Springer, 2012.

[29] E. Janssens and A. van Roermund, *Look-Ahead Based Sigma-Delta Modulation*. Springer, 2011.

[30] R. Gaggl, *Delta-Sigma A/D Converters: Practical Design for Communication Systems*. Springer, 2013.

[31] M. Bolatkale, L. Breems, and K. Makinwa, *High Speed and Wide Bandwidth Delta-Sigma ADCs*. Springer, 2015.

[32] R. Schreier, *The Delta-Sigma Toolbox*. [Online]. Available: http://www.mathworks.com/matlabcentral, 2017.

[33] S. Pavan, R. Schreier, and G. C. Temes, *Understanding Delta-Sigma Data Converters*. Wiley-IEEE Press, 2nd ed., 2017.

[34] B. Murmann, *ADC Performance Survey 1997–2018*. [Online]. Available: http://www.stanford.edu/~murmann/adcsurvey.html.

ACKNOWLEDGEMENTS

I would like to express my deepest gratitude to my colleagues at the Institute of Microelectronics of Sevilla and the Department of Electronics and Electromagnetism of the University of Sevilla. I am especially indebted to Prof. Bélen Pérez-Verdú and Prof. Angel Rodríguez-Vázquez for so many things I have learned from them, not only related to the topic of this book, but also the other values and skills I have put into practice throughout my professional career. I am also particularly grateful to my friends and colleagues, Dr. Rocío del Río – who coauthored the first edition of this book – and Dr. Fernando Medeiro, with whom I took my first steps in the fascinating world of ΣΔ converters, working side by side in so many research and industrial projects.

I would also like to acknowledge my former students, especially Jesús Ruíz-Amaya, Rafael Romay, Ramón Tortosa, Alonso Morgado, Edwin Becerra, Gerardo García, Gerardo Molina, Luis Guerrero-Linares, Mohammad Honarparvar and Sohail Asghar. Some of the material used for the preparation of this book was adapted from their theses.

I would like to give my thanks to the staff at Wiley, a great team of editors led by Peter Mitchell, as well as the anonymous reviewers of the book proposal for their constructive and valuable comments and suggestions, which became very useful for me in making the second version of this book a reality. I also appreciate the discussions I have had over the years with so many experts in the field of ΣΔ converters, who I met at numerous conferences and technical meetings, most of them organized by the IEEE Circuits and Systems Society and the Solid-State Circuits Society. I was very honoured to coauthor papers with some of the recognized experts in analog and ΣΔ IC design, like Prof. Shanthi Pavan, Dr. Richard Schreier, Prof. Maurits Ortmanns and Prof. Mohamad Sawan.

In addition to all these great people, my work has been supported by a number of national and international research and industrial projects. More recently, it was funded by the Spanish Ministry of Economy, Industry and Competitiveness (with support from the European Regional Development Fund) under contract TEC2016-75151-C3-3-R.

Finally, my most grateful thanks go to my family. Without their love, patience and support, none of this would have been possible. Thank you so much for everything!

LIST OF ABBREVIATIONS

ΣΔ	Sigma-Delta
ΣΔM	Sigma-Delta modulator
AAF	Antialiasing filter
A/D	Analog-to-digital
ADC	Analog-to-digital converter
BB	Baseband
BGA	Ball grid array
BP	Band-pass
CAD	Circuit-aided design
CDS	Correlated double sampling
CIC	Cascaded integrator comb
CIFF	Cascade of integrators in feedforward form
CIFB	Cascade of integrators in feedback form
CLA	Clocked averaging
CMFB	Common-mode feedback
CMOS	Complementary MOSFET
CRFF	Cascade of resonators in feedforward form
CRFB	Cascade of resonators in feedback form
CS	Current-steering
CT	Continuous-time
D/A	Digital-to-analog
DAC	Digital-to-analog converter
DC	Direct current
DCL	Digital cancelation logic
DCR	Direct conversion receiver

DEM	Dynamic element matching
DMT	Discrete multi-tone
DNL	Differential nonlinearity
DR	Dynamic range
DRC	Design rule checker
DSP	Digital signal processor
DT	Discrete-time
DWA	Data weighted averaging
EA	Evolutionary algorithm
ELD	Excess loop delay
ENOB	Effective number of bits
ESD	Electrostatic discharge
FE	Forward-Euler
FIR	Finite impulse response
FFT	Fast Fourier transform
FOM	Figure of merit
FPGA	Field programmable gate array
FS	Full scale
GB	Gain-bandwidth product
GPU	Graphics processing unit
GRO	Gated switched-ring oscillator
GUI	Graphic user interface
HDL	Hardware description language
HPF	High pass filter
HRZ	Half-delay return to zero
IBN	In-band noise power
IC	Integrated circuit
IF	Intermediate frequency
IIR	Infinite impulse response
IIT	Impulse-invariant transformation
ILA	Individual level averaging
INL	Integral nonlinearity
I/O	Input–output
ISI	Intersymbol interference
ITF	Integrator transfer function
LDI	Lossless discrete integrator
LP	Low-pass
LPE	Layout parasitic extractor
LSB	Least significant bit
LTCC	Low-temperature cofired ceramic
LTI	Linear time-invariant
LVS	Layout versus schematic
MASH	Multi-stage noise shaping
MDT	Modulator dependent term
MEM	Micro electromechanical
MICTOR	Matched impedance connector
MiM	Metal-insulator-metal
MOEA	Multi-objective evolutionary algorithm

MoM	Metal-oxide-metal
MOOP	Multi-objective optimization problem
MOS	Metal-oxide-semiconductor
MOSFET	MOS field-effect transistor
MOST	MOS transistor
MR	Multi-rate
MTPR	Multitone power ratio
nMOS	n-channel MOSFET
NSGA	Nondominated sorting genetic algorithm
NSIQ	Noise-shaped integrated quantizer
NRZ	Non-return-to-zero
NTF	Noise transfer function
OBG	Out of band gain
OL	Overload level
OSR	Oversampling ratio
OTA	Operational transconductance amplifier
PCB	Printed circuit board
PDF	Probability density function
PDM	Pulse-density modulation
PLL	Phase-locked loop
pMOS	p-channel MOSFET
PoF	Pareto-optimal front
PSD	Power spectral density
PWL	Piecewise linear
PWM	Pulse-width modulation
QFP	Quad flat package
QPLL	Quadrature phase locked loop
RF	Radio frequency
RTF	Resonator transfer function
RTO	Return to open
ROM	Read only memory
RZ	Return-to-zero
SAR	Successive approximation register
SC	Switched-capacitor
SDR	Software-defined radio
SDT	Signal dependent term
SFDR	Spurious-free dynamic range
S/H, S&H	Sample and hold
SI	Switched-current
SMASH	Sturdy multi-stage noise shaping
SMD	Surface-mount device
SNR	Signal-to-noise ratio
SNDR	Signal-to-noise-plus-distortion ratio
SPR	Serial-to-parallel register
SQNR	Signal-to-quantization-noise ratio
SoC	System on chip
SR	Slew rate
TDC	Time-to-digital converter

TEQ	Time-encoding quantizer
THD	Total harmonic distortion
TI	Time interleaved
TIA	Transimpedance amplifier
USTF	Unity signal transfer function
VCO	Voltage-controlled oscillator
VCRO	Voltage-controlled ring oscillator
ZOH	Zero-order hold

CHAPTER 1

INTRODUCTION TO ΣΔ MODULATORS: FUNDAMENTALS, BASIC ARCHITECTURE AND PERFORMANCE METRICS

This chapter is conceived as an introduction to ΣΔ data converters. Their operating principle consists in combining oversampling, quantization error processing, and negative feedback for improving the effective resolution of a coarse quantizer. These basic concepts are presented in Section 1.1, putting especial emphasis on the two main processes involved in the analog-to-digital conversion, namely sampling and quantization. The errors associated with their inherent continuous-to-discrete transformations are analyzed and the way in which they can be mitigated by combining oversampling and noise shaping is presented. On the basis of these ingredients, the performance of ΣΔ converters is compared with Nyquist-rate converters in order to illustrate how to achieve the same specifications by trading circuit element accuracy for signal processing.

Section 1.2 shows the basic architecture, ideal behavior, and performance metrics of ΣΔ modulators and the simplest way to implement such an architecture is presented in Section 1.3, where the so-called first-order ΣΔ modulator is analyzed in detail. This simple architecture is used as an illustration to show the principles of operation behind ΣΔMs and to highlight the main drawbacks of the correlation between the quantization error and the input signal. System-level design parameters and strategies to enhance the performance of ΣΔ converters are discussed in Section 1.4, and the different types of ΣΔ topologies and their implementations are outlined. Everything is put together in Sections 1.5 and 1.6, where the different building blocks required to implement ADCs and DACs are described.

Sigma-Delta Converters: Practical Design Guide, Second Edition. José M. de la Rosa.

1.1 Basics of Analog-to-Digital Conversion

ADCs are electronic systems that perform the transformation of analog signals–which are continuous in time and in amplitude–into digital signals–which are discrete in both time and amplitude. Figure 1.1a illustrates the general block diagram of an ADC intended for the conversion of low-pass (LP) signals, It essentially consists of an *antialiasing filter* (AAF), a *sampler*, a *quantizer* and a *coder*.

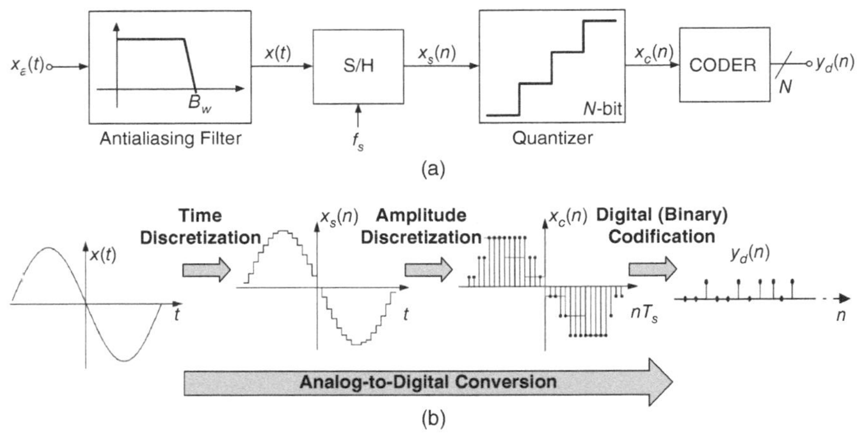

Figure 1.1 Analog-to-digital conversion: (a) conceptual block diagram; (b) signal processing. A Nyquist-rate ADC is assumed.

The operation of these blocks is illustrated in Figure 1.1b. First, the analog input signal $x_a(t)$ of the ADC passes through the AAF: an LP analog filter that prevents out-of-band components from folding over the signal bandwidth B_w during the subsequent sampling, which would corrupt the signal information according to the Nyquist sampling theorem. The resulting band-limited signal $x(t)$ is sampled at a rate f_s by the sampling and hold (S/H) circuit, thus yielding a discrete-time signal $x_s(n) = x(nT_s)$, where $T_s = 1/f_s$ stands for the sampling period. The continuous range of amplitudes of $x_s(n)$ are quantized using N bits, so that each continuous-valued input sample of the quantizer is mapped onto the closer discrete-valued level out of the 2^N that cover the input range. Finally, the digitization process is completed by assigning a unique digital code to each output level of the quantizer, normally using binary coding, which yields the N-bit digital output $y_d(n)$.

As conceptually shown in Figure 1.1b, the fundamental processes involved in A/D conversion are sampling and quantization. Both processes implement a continuous-to-discrete transformation, the former in time and the latter in amplitude. These two continuous-to-discrete transformations limit the performance of ADCs, defining their main specifications, in terms of the speed and accuracy, the latter also referred to as *resolution*[1] and measured in bits. It is therefore very common to compare the performance of different types of ADCs in the resolution-versus-speed plane, as illustrated in Figure 1.2. In this picture, the state-of-the-art performance of different digitization techniques – ΣΔ, Flash, two-step, folding, pipeline and SAR – are depicted. It can be seen that the so-called ΣΔ ADCs – the topic of this book – have the widest conversion region. This is one of the motivations for considering ΣΔ techniques for the implementation of ADCs, but it is not the only one. To understand

[1] As will be explained later in this book, the accuracy of an ADC is not only limited in practice by the quantization process, but also by a number of nonideal effects caused by the circuit and the physical implementation of the chip.

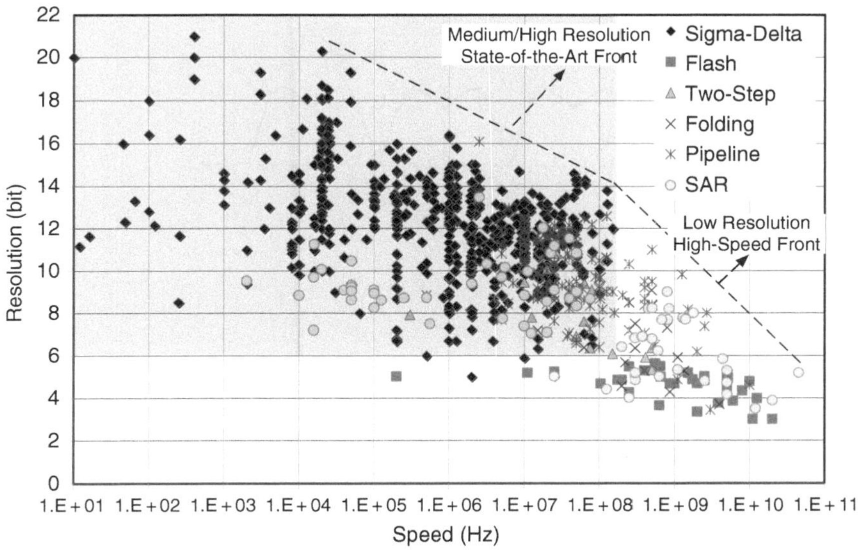

Figure 1.2 Resolution (accuracy) versus speed achieved by state-of-the-art ADCs.

the main benefits of ΣΔ ADCs it is important first to analyze in detail the processes involved in A/D conversion, namely sampling and quantization.

1.1.1 Sampling

As stated above, the sampling process performs the continuous-to-discrete transformation of the input signal in time and imposes a limit on the bandwidth of the analog input signal. According to the Nyquist theorem, to prevent information loss, $x(t)$ must be sampled at a minimum rate of $f_N = 2B_w$, often referred to as the Nyquist frequency. On the basis of this criterion, ADCs in which the analog input signal is sampled at the minimum rate ($f_s = f_N$) are called *Nyquist-rate ADCs*. Conversely, ADCs in which $f_s > f_N$ are called *oversampling ADCs*. How much faster than required the input signal is sampled is expressed in terms of the *oversampling ratio* (OSR), defined as:

$$\mathrm{OSR} = \frac{f_s}{2B_w} \tag{1.1}$$

Whether oversampling is used or not in an ADC has a noticeable influence on the requirements of its antialiasing filter. Since in Nyquist-rate ADCs the input signal bandwidth B_w coincides with $f_s/2$, aliasing will occur if $x_a(t)$ (as shown in Figure 1.1) contains frequency components above $f_s/2$. Higher-order analog AAFs are thus required to implement sharp transition bands capable of removing out-of-band components with no significant attenuation of the signal band, as illustrated in Figure 1.3a for the LP case. Conversely, since $f_s/2 > B_w$ in oversampling ADCs, the replicas of the input signal spectrum that are created by the sampling process are farther apart than in Nyquist-rate ADCs. As illustrated in Figure 1.3b, frequency components of the input signal in the range $[B_w, f_s - B_w]$ do not alias within the signal band, so that the filter transition band can be smoother; this greatly reduces the order required for the AAF and simplifies its design.

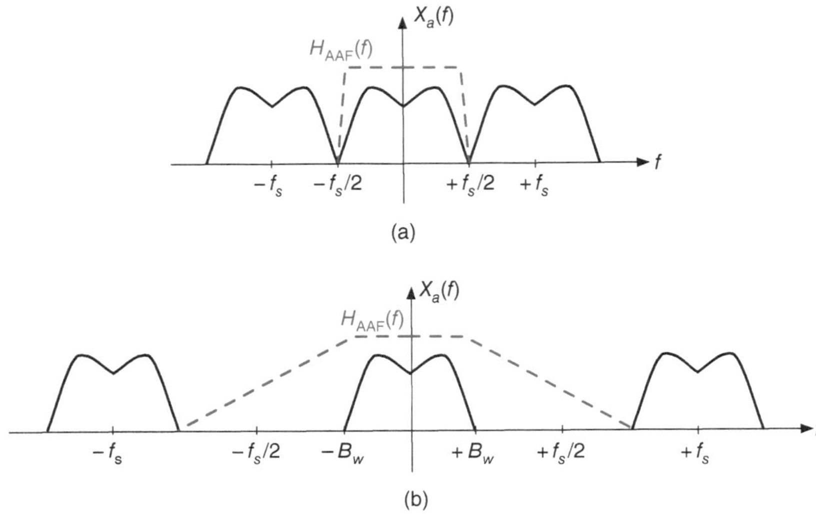

Figure 1.3 Antialiasing filter for: (a) Nyquist-rate ADCs; (b) oversampling ADCs.

1.1.2 Quantization

The quantization process also introduces a limitation on the performance of an ideal ADC, since an error is generated while performing the continuous-to-discrete transformation of the input signal in amplitude. This is commonly referred to as *quantization error*. The basic operation of quantizers is illustrated in Figure 1.4.

As an example, Figure 1.4c depicts the I/O characteristic of a quantizer with $N = 2$, although the results also apply to a generic N-bit quantizer. Input amplitudes within the full-scale input range $[-X_{FS}/2, +X_{FS}/2]$ are *rounded* to one out of the 2^N different output levels, which are usually encoded into a binary digital representation. If these levels are equally spaced, the quantizer is said to be *uniform* and the separation between adjacent output levels is defined as the *quantization step*,

$$\Delta = \frac{Y_{FS}}{2^N - 1} \tag{1.2}$$

where Y_{FS} stands for the full-scale output range. Since X_{FS} and Y_{FS} are not necessarily equal, the quantizer may exhibit a gain different from unity, as indicated in Figure 1.4c by the slope k_q. As shown in Figure 1.4e, the quantizer operation thus inherently generates a *rounding* error that is a nonlinear function of the input. Note that, if $q(n)$ is kept within the range $[-X_{FS}/2, +X_{FS}/2]$, the quantization error $e(n)$ is bounded within $[-\Delta/2, +\Delta/2]$. The former input range is known as the *nonoverload region* of the quantizer, as opposed to ranges with $|q(n)| > \Delta/2$, for which the magnitude of $e(n)$ grows monotonously. Figure 1.4 also shows the operation of a single-bit quantizer ($N = 1$). Note from Figure 1.4d that, compared to the multi-bit case, the output of a single-bit quantizer is determined by the input sign only, regardless its magnitude. Therefore, the gain k_q is undefined and can be arbitrarily chosen.

The quantization characteristics shown in Figure 1.4a correspond to those of so-called *midriser* quantizers. This term comes from what happens in the region around the zero value of the I/O characteristics, considering the analogy with a *staircase*. As a result, the number of levels is always an even number. Alternatively, there is another type of quantizer block called a *midtread* quantizer,

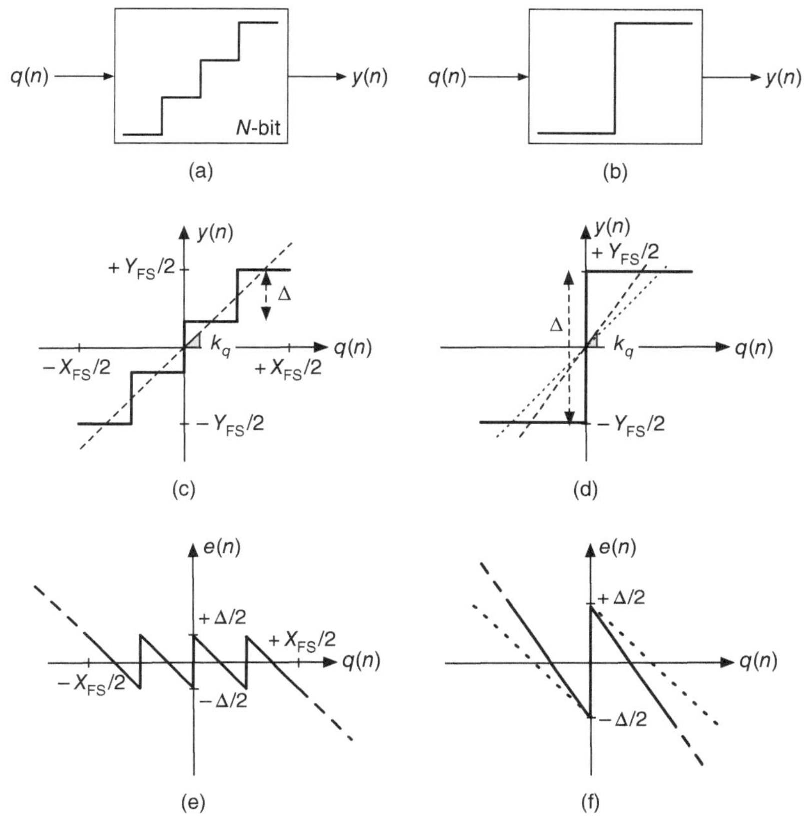

Figure 1.4 Quantization process: (a) multi-bit quantizer block; (b) single-bit quantizer block; (c) I/O characteristic of a multi-bit quantizer; (d) I/O characteristic of a single-bit quantizer; (e) multi-bit quantization error; (f) single-bit quantization error.

with a conceptual I/O characteristic like that shown in Figure 1.5a, in which the level corresponding to zero – like the tread of a staircase – is also considered a valid level for quantization purposes. These quantization I/O characteristics are usually exploited by fully-differential circuits, in which the zero value can be obtained by simply subtracting the characteristics of the two branches implementing the differential circuitry. For instance, Figure 1.5b illustrates how to obtain a three-level quantizer from two single-level quantizers; that is, comparator functions. In what follows, a midriser quantization will be considered, but without loss of generality.

1.1.3 Quantization White Noise Model

In practice, an ideal quantizer such as that shown in Figure 1.6a can often be modeled using the linear scheme in Figure 1.6b if several assumptions are made about the statistical properties of the quantization error [1–3]. As already shown in Figure 1.4e, the quantization error $e(n)$ is systematically determined by the quantizer input signal $q(n)$. Nevertheless, if $q(n)$ is assumed to change randomly from sample to sample within the range $[-\Delta/2, +\Delta/2]$, $e(n)$ will also be uncorrelated from sample to sample. This behavior is also seen if the number of quantizer levels increases. This is illustrated in

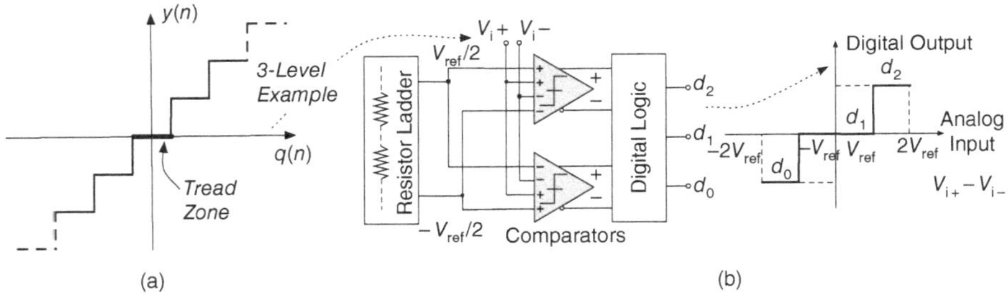

Figure 1.5 Midtread quantization: (a) conceptual I/O characteristic; (b) illustration of a three-level ADC made up of two single-bit quantizers (comparators).

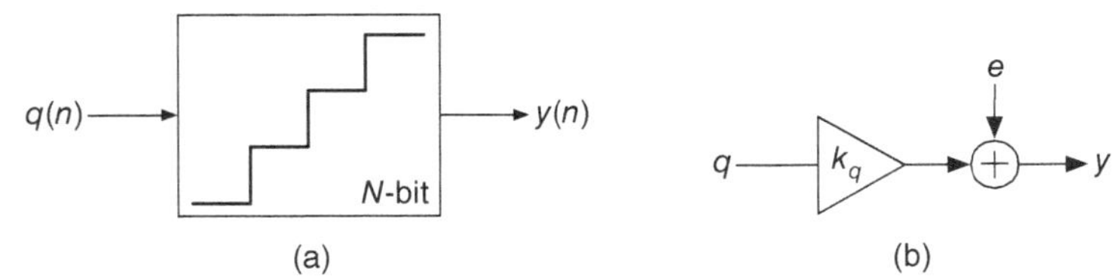

Figure 1.6 Quantization linear model: (a) multi-bit quantizer block; (b) equivalent model with additive white noise.

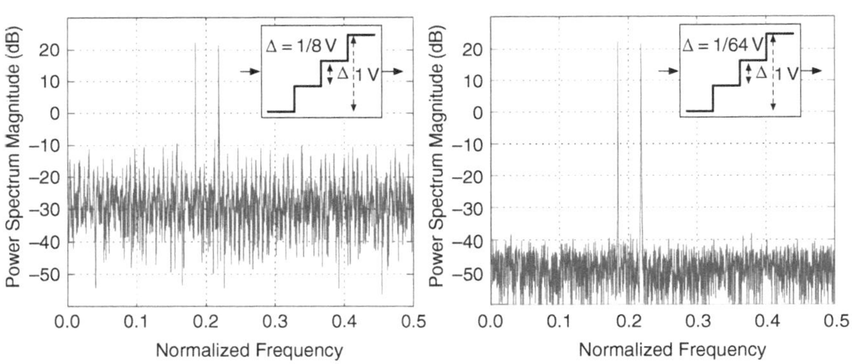

Figure 1.7 Illustrating the validity of the white-noise model for a quantizer when the number of levels of the quantizer is increased when a two-tone input signal is applied.

Figure 1.7, where a quantizer is excited by a two-tone signal. Note that a number of intermodulation tones can be observed in the output spectrum of the quantizer. These tones are caused by the strong correlation between the quantizer input and the quantization error, as conceptually depicted in Figure 1.4e. However, as the number of quantizer levels is increased (by decreasing Δ), the number of tones is drastically reduced, apart from the obvious reduction of the noise power.

Under these requirements, the quantization error can be viewed as a *random process with a uniform probability distribution* in the range $[-\Delta/2, +\Delta/2]$, as illustrated in Figure 1.8a. The power

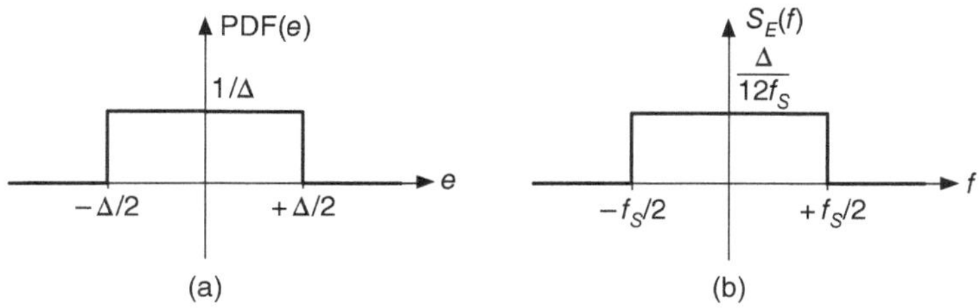

Figure 1.8 Quantization white noise: (a) probability density function (PDF); (b) power spectral density.

associated with the quantization error can thus be computed as:

$$\overline{e^2} = \sigma_e^2 = \int_{-\infty}^{+\infty} e^2 \mathrm{PDF}(e)\mathrm{d}e = \frac{1}{\Delta}\int_{-\Delta/2}^{+\Delta/2} e^2 \mathrm{d}e = \frac{\Delta^2}{12} \tag{1.3}$$

The former assumption implies that, as illustrated in Figure 1.8b, the power of the quantization error will be also *uniformly* distributed in the range $[-f_s/2, +f_s/2]$, yielding

$$\overline{e^2} = \int_{-\infty}^{+\infty} S_E(f)\mathrm{d}f = S_E \int_{-f_s/2}^{+f_s/2} \mathrm{d}f = \frac{\Delta^2}{12} \tag{1.4}$$

so that the power spectral density (PSD) of the quantization error in that range is:

$$S_E = \frac{\overline{e^2}}{f_s} = \frac{\Delta^2}{12f_s} \tag{1.5}$$

These assumptions are collectively known to as the *additive white noise approximation* of the quantization error and allow the representation of a quantizer that is deterministic and nonlinear with the random linear model in Figure 1.6b. Here $y(n) = k_q q(n) + e(n)$, with $e(n)$ being *quantization noise*.[2]

With the approximation of the quantization error as white noise, the performance of ideal ADCs can be easily evaluated. For a Nyquist ADC in which $f_s = 2B_w$, all the quantization noise power falls inside the signal band and passes to the ADC output as a part of the input signal itself, as illustrated in Figure 1.9a. Conversely, if an oversampled signal is quantized, since $f_s > 2B_w$, only a fraction of the total quantization noise power lies within the signal band, as illustrated in Figure 1.9b. The *in-band noise power* (IBN) caused by the quantization process in an ideal oversampling ADC is thus,

$$\mathrm{IBN} = \int_{-B_w}^{+B_w} S_E(f)\mathrm{d}f = \int_{-B_w}^{+B_w} \frac{\Delta^2}{12f_s}\mathrm{d}f = \frac{\Delta^2}{12\mathrm{OSR}} \tag{1.6}$$

so that the larger the oversampling ratio, the smaller the IBN.[3]

[2]Although the assumptions underlying the additive white noise approximation are rarely met in practice and are not strictly valid, it is commonly used in ADC design and usually yields good results – and better for a larger number of bits in the quantizer, as illustrated in Figure 1.7. Even though strictly speaking it is not valid for stand-alone single-bit quantizers, the assumption is also employed in the design of single-bit ΣΔ modulators [4].

[3]Note that Equation (1.6) for the IBN of oversampling ADCs also holds true for Nyquist ADCs, just considering OSR = 1. The same applies for subsequent expressions derived from Equation (1.6).

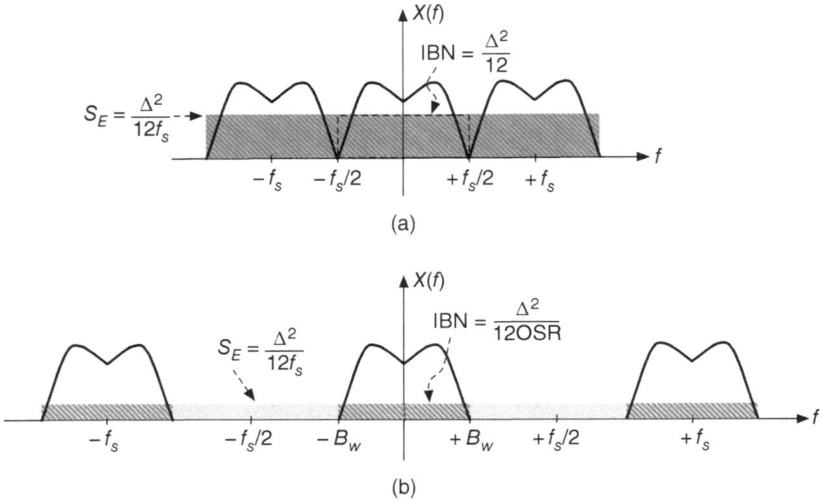

Figure 1.9 Quantization noise in: (a) Nyquist-rate ADCs; (b) oversampling ADCs.

The *dynamic range* (DR) of an ideal ADC can be determined as the ratio of the output power at the frequency of an input sinusoid with maximum amplitude to the in-band quantization noise power:

$$\mathrm{DR\ (dB)} = 10\log_{10}\left(\frac{P_{\mathrm{sig,out,max}}}{\mathrm{IBN}}\right) \tag{1.7}$$

From Figure 1.4c, the maximum input amplitude in the nonoverload region of an N-bit quantizer is $X_{\mathrm{FS}}/2$ and its corresponding output power can be approximated to [5]:

$$P_{\mathrm{sig,out,max}} \approx \frac{(2^N\Delta/2)^2}{2} = 2^{2N-3}\Delta^2 \tag{1.8}$$

so that, using Equations (1.6) and (1.8), the DR of an ideal oversampling ADC yields:

$$\mathrm{DR\ (dB)} \approx 6.02N + 1.76 + 10\log_{10}(\mathrm{OSR}) \tag{1.9}$$

Note that, for a Nyquist ADC – that is, OSR $= 1$ in Equation (1.9) – each additional bit in the quantizer results in a DR increase of approximately 6 dB. For an oversampling ADC, the DR further increases with OSR by approximately 3 dB/octave, so that using, for instance, an OSR of 4 is similar to having one extra bit in the N-bit quantizer.

1.1.4 Noise Shaping

An approach to further increase the accuracy of an oversampling ADC is *shaping* the quantization white noise in the frequency domain – that is, filtering it – in such a way that most of its power lies outside the signal band. This is illustrated in Figure 1.10a, where the quantization noise is conceptually obtained by subtracting the quantizer input signal $q(n)$ from its output $y(n)$. It then passes through a filter transfer function, usually called *noise transfer function* (NTF).

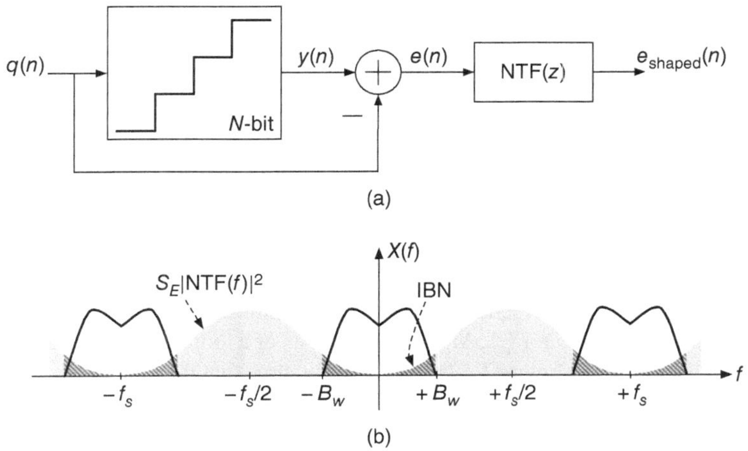

Figure 1.10 Quantization noise shaping: (a) conceptual block diagram; (b) effect on the in-band noise of an oversampling noise-shaping ADC.

For quantizers working on LP signals, NTF is of the high-pass type and can be easily obtained from a differentiator filter, with a Z-domain transfer function given by,

$$\mathrm{NTF}(z) = (1 - z^{-1})^L \tag{1.10}$$

where L stands for the filter order. Taking into account that $z = e^{sT_s} = e^{j2\pi f/f_s}$, the magnitude of the pure-differentiator NTF in Equation (1.10) can be approximated for low frequencies to

$$|\mathrm{NTF}(f)| = |1 - e^{-j2\pi f/f_s}|^L = \left[2\sin\left(\frac{\pi f}{f_s}\right)\right]^L \approx \left(\frac{2\pi f}{f_s}\right)^L, \quad \text{for } f \ll f_s \tag{1.11}$$

so that the power due to the shaped quantization noise that lies within the signal band yields:

$$\mathrm{IBN} = \int_{-B_w}^{+B_w} S_E(f)|\mathrm{NTF}(f)|^2 \mathrm{d}f \approx \frac{\Delta^2}{12} \frac{\pi^{2L}}{(2L+1)\mathrm{OSR}^{(2L+1)}} \tag{1.12}$$

Using Equations (1.8) and (1.12), the DR of an ideal oversampling noise-shaping ADC can be obtained as:

$$\mathrm{DR\ (dB)} \approx 6.02N + 1.76 + 10\log_{10}\left(\frac{2L+1}{\pi^{2L}}\right) + (2L+1)10\log_{10}(\mathrm{OSR}) \tag{1.13}$$

Note that, in comparison with Equation (1.9), if oversampling is used in combination with noise shaping, the DR increases with OSR by approximately $3(2L+1)$ dB/octave.

1.2 Sigma-Delta Modulation

In contrast to the ADCs discussed so far, which are open-loop systems from a control perspective, ΣΔ ADCs rely on a feedback path to achieve closed-loop control of the quantization error. The fundamentals on how the shaping of quantization noise is implemented in practice, as well as the basic architecture, performance metrics, and ideal behavior of oversampling noise-shaping ADCs is presented in this section.

1.2.1 From Noise-shaped Systems to ΣΔ Modulators

The conceptual block diagram shown in Figure 1.10 only processes the quantization noise. In order to be used to digitize signals, the sampled input signal, $x_s(n)$, should be processed in parallel with the quantization noise, as conceptually depicted in Figure 1.11, where a *signal transfer function* (STF) is applied to the quantized version of $x_s(n)$. This, together with $e_{\text{shaped}}(n)$, yields a digital representation of the input signal, $y(n)$, which can be represented in the Z-domain as:

$$Y(z) = \text{STF}(z)X(z) + \text{NTF}(z)E(z) \tag{1.14}$$

There are many ways of implementing the system of Figure 1.11 in practice. One of the simplest forms is the so-called ΣΔ modulator (ΣΔM), which consists of a loop filter $H(z)$ and a B-bit quantizer in a feedback loop, as shown in Figure 1.12a [6]. Let us assume that the gain of the loop filter is large within the signal band and small outside it. Due to the action of the negative feedback, the analog input signal x and the analog version of the ΣΔM output y will coincide within the signal band, so that the error signal $x - y$ in this closed-loop system is very small within the signal band. Since the B-bit quantizer is uniform, most of the differences between the input and the output of the ΣΔM will be placed at higher frequencies, so that the quantization noise is shaped in the frequency domain and most of its power is pushed outside of the signal band.

Using the linear additive white noise model in Figure 1.6b for the embedded quantizer, the ΣΔM in Figure 1.12a can be modeled as the two-input (x and e) one-output (y) linear system shown in

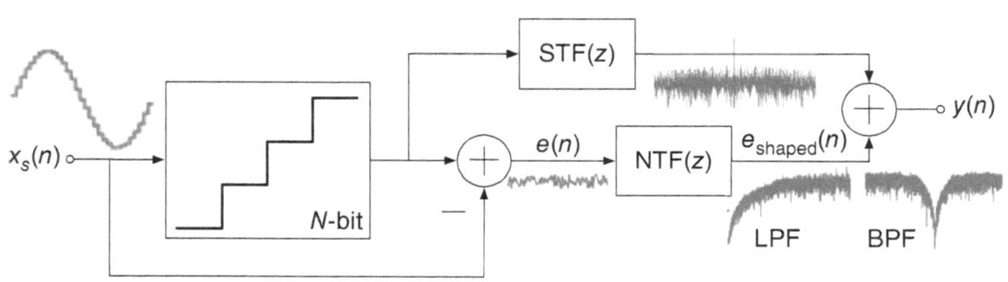

Figure 1.11 Conceptual block diagram of a noise-shaped ADC. Two different shaped noise output spectra are illustrated, for when the NTF is either a band-pass or a low-pass filter.

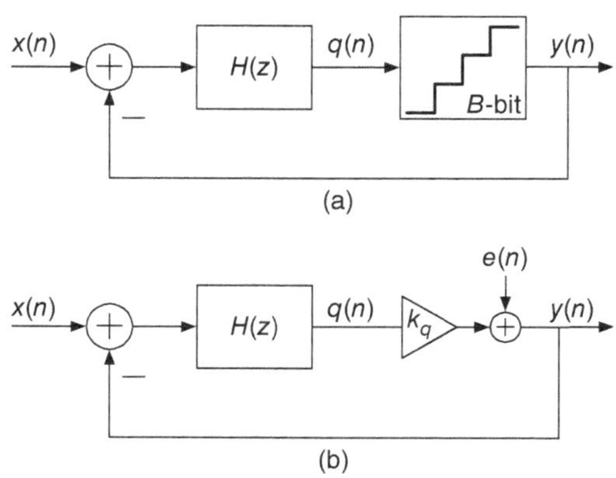

Figure 1.12 ΣΔ modulator: (a) block diagram; (b) ideal linear model.

Figure 1.12b, which is described in the Z-domain by Equation (1.14), where the STF and NTF are given by:

$$\mathrm{STF}(z) = \frac{k_q H(z)}{1 + k_q H(z)}, \quad \mathrm{NTF}(z) = \frac{1}{1 + k_q H(z)} \tag{1.15}$$

Note that, if the loop filter is designed such that $|H(f)| \gg 1$ within the signal band, then $|\mathrm{STF}(f)| \approx 1$ and $|\mathrm{NTF}(f)| \ll 1$; in other words, the quantization noise is ideally canceled while the input signal is perfectly transferred to the output.

1.2.2 Performance Metrics of ΣΔMs

In contrast to Nyquist-rate ADCs, whose performance is mainly characterized by static performance metrics – monotonicity, gain and offset errors, differential nonlinearity (DNL), and integral nonlinearity (INL) [5] – the characteristics of ΣΔ ADCs are typically measured using dynamic performance metrics, which are obtained from the frequency-domain representation of the time-domain digital output sequence. This therefore requires the computation of the fast Fourier transform (FFT) of a finite-length output sequence with a specific *windowing function*, as will be discussed in Chapter 5. From that power spectrum representation of a ΣΔM output sequence, various spectral metrics are directly measured and other noise and power metrics are derived.

Figure 1.13 illustrates an exemplary spectrum of a ΣΔM output sequence when a sinusoidal signal with frequency f_{in} is applied at the input. The main characteristics of the spectrum are highlighted: for example, the length of the digital sequence from which the FFT has been computed, the output signal peak at the frequency f_{in} corresponding to the converted signal, and so on. As will be discussed in Chapters 3 and 4, due to nonidealities of the circuitry used for implementing the ΣΔM, the output spectrum differs in practice from that from a *purely* shaped quantization noise. On the one hand, linear errors give rise to a noise floor, as well as to a degradation of the shaping order. On the other, nonlinear errors generate distortion, which is typically noticeable for large input amplitudes, but submerged under the noise floor for small input signal amplitudes. Spectral metrics such as the *spurious-free dynamic range* (SFDR) – the ratio of the signal power to the strongest spectral tone [5] – can be directly measured from the output modulator spectrum, as shown in Figure 1.13.

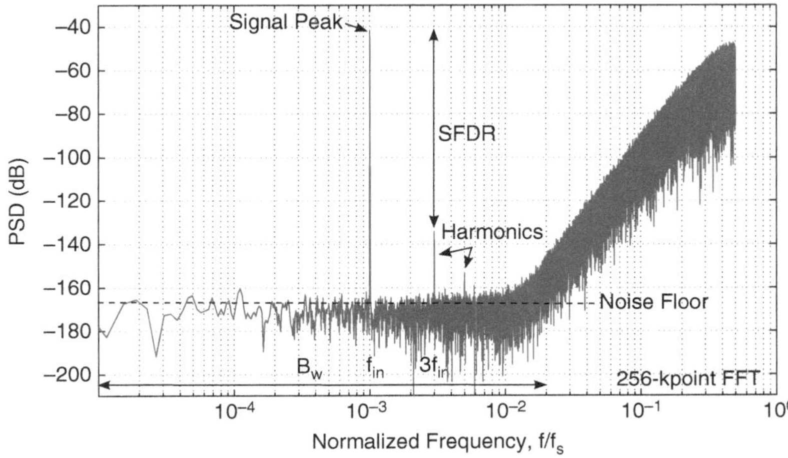

Figure 1.13 Illustration of a typical experimental output spectrum of a ΣΔ modulator and its main characteristics. An LP ΣΔM is assumed.

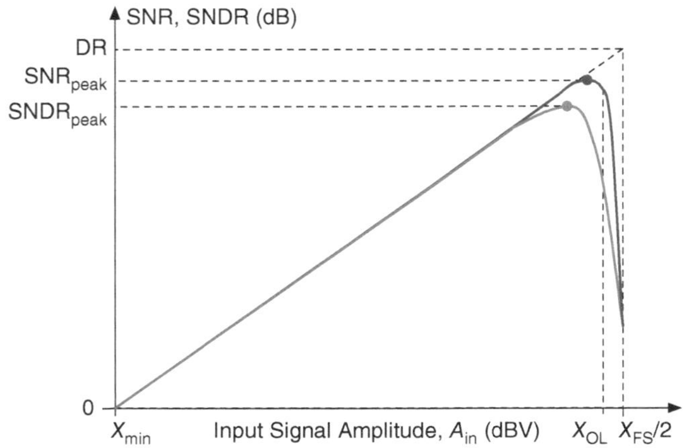

Figure 1.14 Illustration of the performance metrics of a ΣΔM on a typical SNR curve.

Noise and power metrics are derived from the ΣΔM output spectra by integration over the signal bandwidth. They are typically collected in a single plot, as shown in Figure 1.14. These metrics are usually the most important measures and comprise:

- **Signal-to-noise ratio (SNR)** is the ratio of the output power at the frequency of an input sinusoid to the *uncorrelated* in-band noise power:

$$\text{SNR (dB)} = 10\log_{10}\left(\frac{P_{\text{sig,out}}}{\text{IBN}}\right) \tag{1.16}$$

  This accounts for the modulator linear performance only, so that the in-band power associated with the harmonics of the input signal is not considered part of the IBN for SNR computations. If an ideal ΣΔM is considered and only the in-band quantization noise is accounted for in the IBN computation, the term *signal-to-quantization-noise ratio* (SQNR) is often employed.

- **Signal-to-noise-plus-distortion ratio (SNDR)** is defined as the ratio of the output power at the frequency of an input sinusoid to the total IBN power, also accounting for possible harmonics at the ΣΔM output. As illustrated in Figure 1.14, this makes a typical SNDR curve deviate from the SNR curve only for large input amplitudes, for which the generated distortion is noticeable. Therefore, the output spectra from which the SNDR curve is computed are typically obtained by applying an input signal at $f_{\text{in}} \leq B_w/3$ (for LP ΣΔMs), so that at least the second and third harmonics lie within the signal band.

- **Dynamic range (DR)** can be defined as the ratio of the output power at the frequency of an input sinusoid with maximum amplitude to the output power for a small input amplitude for which SNR = 0 dB; i.e., so it cannot be distinguished from the error. Ideally, a sinusoid with maximum amplitude at the modulator input will provide an output sinusoid sweeping the full-scale range Y_{FS} of the embedded quantizer, so that

$$\text{DR (dB)} = 10\log_{10}\left(\frac{P_{\text{sig,out,max}}}{\text{IBN}}\right) = 10\log_{10}\left[\frac{(Y_{\text{FS}}/2)^2}{2\text{IBN}}\right] \tag{1.17}$$

- **Effective number of bits (ENOB).** Since the DR of an ideal N-bit Nyquist-rate converter is given by Equation (1.9) with OSR = 1, a similar expression can be established for ΣΔMs,

$$\text{ENOB(bit)} = \frac{\text{DR (dB)} - 1.76}{6.02} \tag{1.18}$$

  where ENOB can be defined as the number of bits needed for an ideal Nyquist-rate ADC to achieve the same DR as the ΣΔ ADC. The performance of oversampled ΣΔ converters and Nyquist-rate ADCs can thus be compared in simple way [7]. Instead of the DR, the peak SNDR is also often used in Equation (1.18) to express the accuracy of the A/D conversion in a ΣΔ modulator in bits.

- **Overload level (OL).** As illustrated in Figure 1.14, the SNR of a ΣΔ modulator increases monotonously with the input signal amplitude (A_{in}), but sharply drops for input amplitudes close to half of the full-scale input range of the embedded quantizer ($X_{\text{FS}}/2$) due to its overload and the associated IBN increase. The overload level is considered to define the maximum input amplitude for which the ΣΔM still operates correctly, and can almost be arbitrarily defined, but it is typically chosen as the amplitude for which the SNR drops by 6 dB below the peak SNR [8].

1.3 The First-order ΣΔ Modulator

For the conversion of LP signals, the simplest loop filter $H(z)$ that exhibits the desired frequency performance defined in Equation 1.15 is an integrator, which has a Z-domain transfer function given by:

$$\text{ITF}(z) = \frac{1}{z-1} = \frac{z^{-1}}{1-z^{-1}} \tag{1.19}$$

The above transfer function can be realized using SC circuits, as illustrated in Figure 1.15. This circuit can be combined with an embedded quantizer, to implement a first-order ΣΔM. Note that the quantizer block is made up of an ADC in the feedforward path and a DAC in the feedback path. In this example, a single-comparator is used to implement the 1-bit ADC, whose output Y is the output of the modulator. This signal controls the feedback connection of reference voltages $V_{\text{ref+}}$ and $V_{\text{ref-}}$ to the integrator through an SC branch, which implements the required feedback 1-bit DAC. In this simple circuit example, both the modulator input and the DAC feedback signals are processed through the same sampling capacitor, C, in the integrator. Assuming a linear model for the quantizer with $k_q = 1$, the Z-transform of the modulator is given by:

$$Y(z) = z^{-1}X(z) + (1 - z^{-1})E(z) \tag{1.20}$$

This builds up a first-order high-pass shaping of the quantization noise – see Equation (1.10).

For the sake of illustration, Figure 1.16a shows the output waveform of a first-order ΣΔM with an embedded 3-bit (8-level) quantizer for a sinusoidal input signal. Note that, due to the combined action of oversampling and negative feedback, the modulator output is a pulse-density modulated (PDM) signal whose local average tracks the input signal value within adjacent code transitions.

Figure 1.16b plots the output *pulse stream*, often referred to as the *bitstream*, of a first-order ΣΔM with a 1-bit (2-level) quantizer, when a stair-waveform signal is applied at the input of the

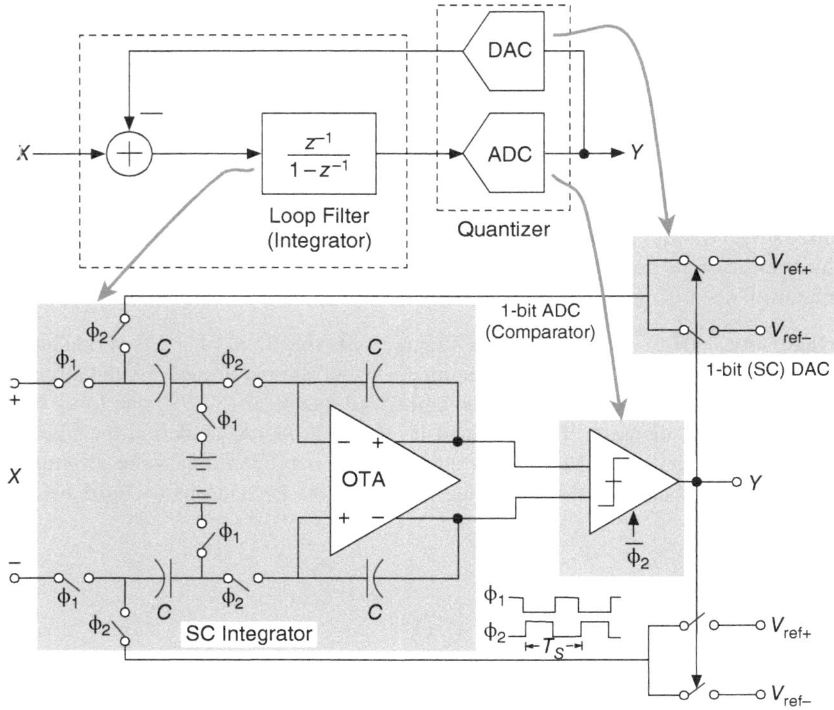

Figure 1.15 Block diagram of a first-order ΣΔM and its corresponding fully-differential SC circuit implementation with 1-bit quantization.

modulator. Note that the ΣΔM output pulse density is different for each input level, thus making the average of the feedback signal coincident with the corresponding stair step. The latter is better shown by considering an input DC signal and counting the number of positive (logic one) and negative (logic zero) pulses obtained at the output of the modulator. Table 1.1 illustrates the values of the input and output of a 1-bit quantizer in a first-order modulator by considering different cases of an input DC signal level measured by reference to the quantizer full-scale (FS) range, namely: 0, 1/3 and 2/3. It can be seen that there is always a repetitive pattern of output pulses – a digital representation of the input value – given by [9]:

$$\overline{Y} = \frac{P_{+1} - P_{-1}}{P_{+1} + P_{-1}} \cdot Y_{\rm FS} \tag{1.21}$$

where P_{+1} and P_{-1} represent respectively the number of logic ones and logic zeroes included in the repetitive pattern (highlighted in Table 1.1). This behavior is the result of the action of the feedback in a ΣΔM, which forces the difference between the input and the output to be zero.

In spite of its simplicity, a negative effect of using a 1-bit quantizer in a first-order ΣΔM is the strong correlation between the input signal and the quantization error. This effect is illustrated in the so-called *noise pattern* [10] – depicted in Figure 1.17 – which represents the in-band quantization error power versus the DC value of the modulator input for OSR = 64. This strongly nonlinear behavior

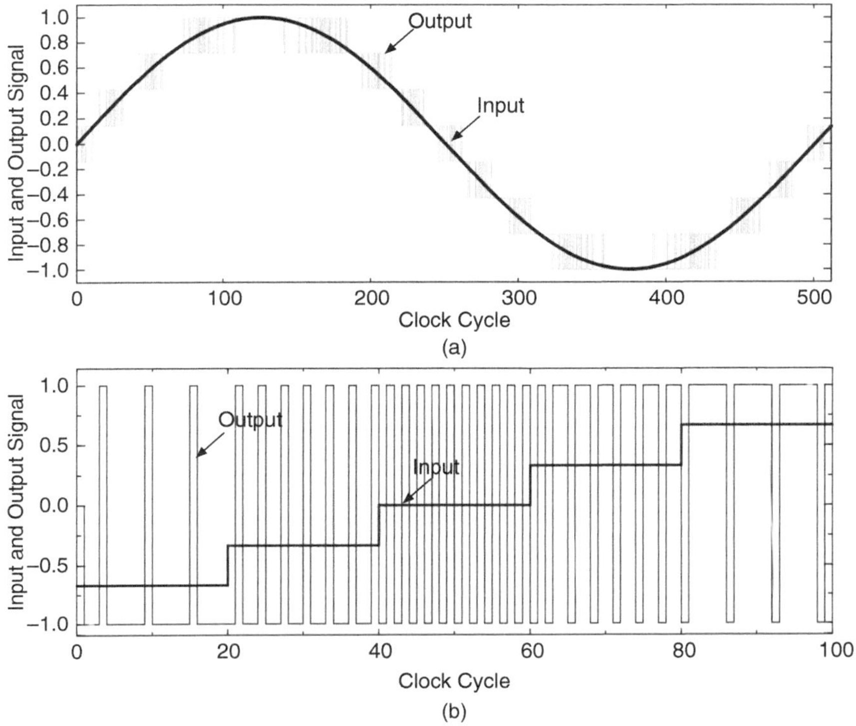

Figure 1.16 PDM output signal of a first-order ΣΔ modulator with a embedded (a) 3-bit quantizer for an input sinusoid and (b) 1-bit quantizer and an input stair waveform.

Table 1.1 Repetitive patterns in 1-bit ΣΔMs with DC inputs.

n	$q(n), y(n)$ $[\boldsymbol{x(n)} = \mathbf{0}]$	$q(n), y(n)$ $[\boldsymbol{x(n)} = \mathbf{1/3}]$	$q(n), y(n)$ $[\boldsymbol{x(n)} = \mathbf{1/2}]$
0	$0, \mathbf{1}$	$0, \mathbf{1}$	$0, \mathbf{1}$
1	$-1, \mathbf{-1}$	$-2/3, \mathbf{-1}$	$-1/2, \mathbf{-1}$
2	$0, 1$	$2/3, \mathbf{1}$	$1, \mathbf{1}$
3	$-1, -1$	$0, 1$	$1/2, \mathbf{1}$
4	$0, 1$	$-2/3, -1$	$0, 1$
…	…	…	…

causes the linear model to be less valid and produces a number of discrete tones (often referred to as *idle tones* [9]) in the output spectrum, as illustrated in Figure 1.18. This tonal behavior disappears if the order of the loop filter, L, and/or the number of bits of the embedded quantizer, B, are increased, resulting in the white noise model being a better approximation for the quantizer. As an illustration, Figure 1.18 also shows the output spectrum of a first-order modulator with $B = 5$. It can be seen how, in addition a reduciton in the quantization noise power, the output spectrum tonality is drastically reduced as well.

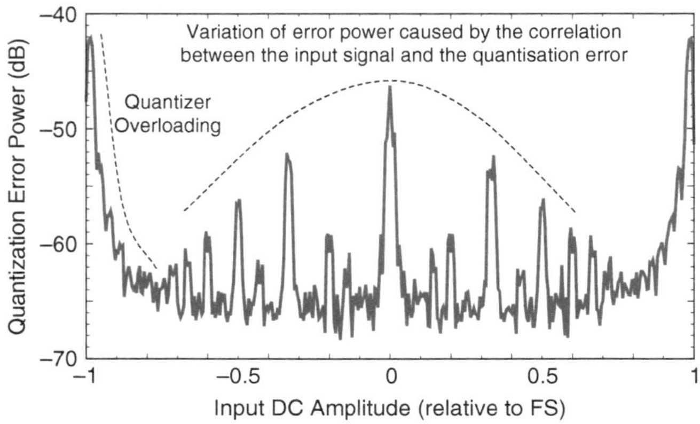

Figure 1.17 Noise pattern of a first-order ΣΔM with 1-bit quantizer.

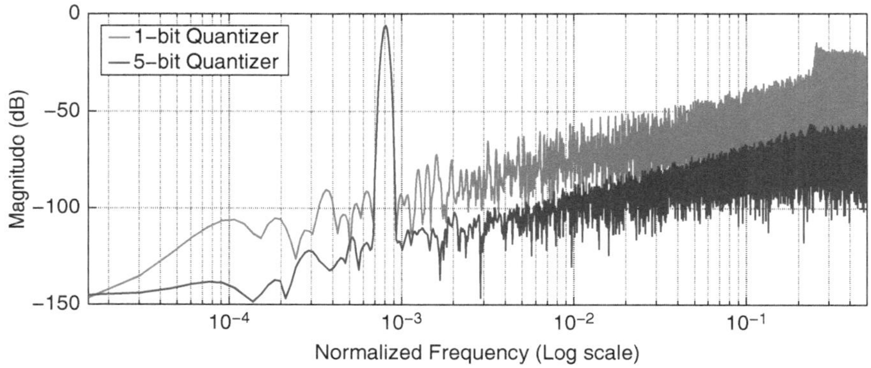

Figure 1.18 Output spectra of a first-order ΣΔM for a 1-bit and a 5-bit quantizer.

1.4 Performance Enhancement and Taxonomy of ΣΔMs

According to Equation 1.14, the output of an ideal LP Lth-order ΣΔM in the Z-domain can be considered to be

$$Y(z) = \text{STF}(z)X(z) + \text{NTF}(z)E(z) = z^{-L}X(z) + (1 - z^{-1})^L E(z) \tag{1.22}$$

where $|\text{STF}(f)| = 1$ and the NTF builds up an Lth-order high-pass shaping of the quantization noise of the embedded quantizer. If a B-bit quantizer is employed, the dynamic range of the ΣΔM can be obtained from Equations (1.12) and (1.17) to ideally yield,

$$\text{DR (dB)} = 10 \log_{10}\left(\frac{P_{\text{sig,out,max}}}{\text{IBN}}\right) \approx 10 \log_{10}\left[\frac{3}{2}(2^B - 1)^2 \frac{(2L+1)\text{OSR}^{(2L+1)}}{\pi^{2L}}\right] \tag{1.23}$$

taking into account that $Y_{\text{FS}} = (2^B - 1)\Delta$ – see Equation (1.2) – and considering quantization noise as the only contribution to the IBN.

1.4.1 ΣΔM System-level Design Parameters and Strategies

Note from Equation (1.23) that the dynamic range of a ΣΔ modulator is ideally determined by the values of L, OSR, and B, which can thus be considered as the three key parameters that define the ΣΔM at the top level. The pros and cons of increasing the DR of a ΣΔ modulator by increasing each of these parameters are briefly discussed next and will be analyzed in more detail in Chapter 2:

- **High-order ΣΔ modulators:** The accuracy of the A/D conversion can be considerably improved by increasing the noise-shaping order, since a larger fraction of the total quantization noise power will be pushed out of the signal band. Figure 1.19 illustrates the ideal noise-shaping functions of orders ranging from 1 to 5. The case $L = 0$ – no shaping – is also included for comparison purposes. The DR enhancement if L is increased by one for a given OSR can be obtained from Equation (1.23):

$$\Delta\text{DR (dB)} \approx 10\log_{10}\left[\frac{2L+3}{2L+1}\left(\frac{\text{OSR}}{\pi}\right)^2\right] \tag{1.24}$$

  This means that, for instance, the DR of a fourth-order ΣΔM with OSR = 32 is ideally 21.3 dB (3.5 bit) larger than that of a third-order ΣΔM. However, the use of high-order ($L > 2$) loop filters gives rise to stability problems in a ΣΔM. Although these problems can be circumvented, the dynamic range of a high-order ΣΔM will in practice be smaller than predicted by Equation (1.23).

- **High-OSR ΣΔ modulators:** Figure 1.20 shows the ideal dynamic range as a function of OSR for noise-shaping orders ranging from 0 (no shaping) to 5 and assuming a single-bit embedded quantizer ($B = 1$). As illustrated, the combination of oversampling and noise-shaping considerably enhances the ΣΔM performance for OSR > 4. Note from Equation (1.23) that the DR of an ideal L-order ΣΔM increases with OSR by $3(2L + 1)$ dB/octave. However, for a given conversion bandwidth B_w, the oversampling ratio cannot be arbitrarily increased, since it leads to a higher sampling frequency f_s for the operation of the ΣΔ circuitry. The latter, if achievable in practice for a given technological process, leads to larger power consumption.

- **Multibit ΣΔ modulators:** An increase in B leads to a decrease of the quantization step Δ and thus to a reduction of the quantization noise power. Each additional bit in the embedded quantizer of a ΣΔM is considered to typically yield a 6-dB (1-bit) improvement in the DR [11].

  However, a multi-bit embedded quantizer requires a multi-level DAC to close the negative feedback loop in the ΣΔM. In contrast to a two-level feedback DAC ($B = 1$), which is inherently

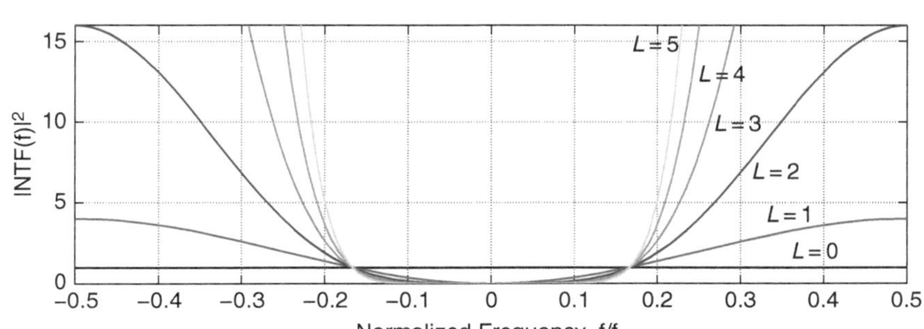

Figure 1.19 Illustration of the shaping of quantization noise as a function of frequency in a ΣΔM. NTF is given by Equation (1.10) and L is the noise-shaping order.

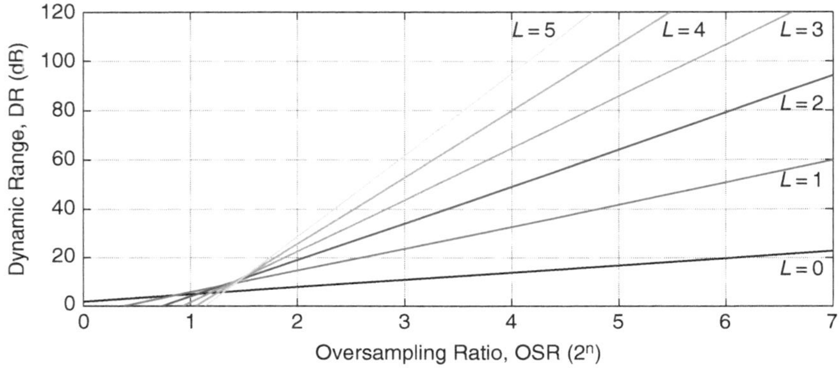

Figure 1.20 Ideal dynamic range of a ΣΔM as a function of the oversampling ratio for different noise-shaping orders (L). A single-bit internal quantizer ($B = 1$) is assumed.

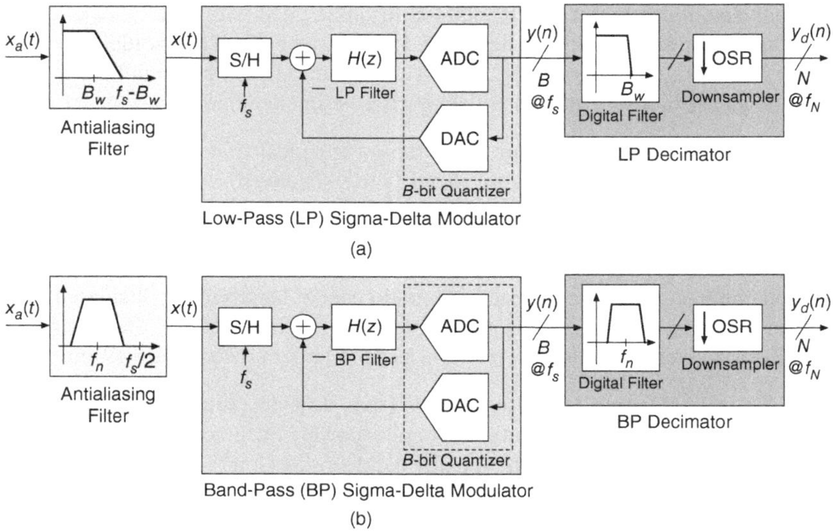

Figure 1.21 General block diagram of a ΣΔ ADC with: (a) LP signal; (b) BP signal.

linear, a multi-level DAC will in practice be nonlinear to some extent. As noticeable from Figure 1.21, the DAC nonlinearity will be directly added to the ΣΔM input and will thus appear at the output, since $|\mathrm{STF}(f)| \approx 1$ within the signal band. Therefore, the linearity required in a multi-bit DAC equals in practice that wanted for the ΣΔ modulator.

1.4.2 Classification of ΣΔMs

The strategies discussed above for improving the dynamic range of a ΣΔM may be combined in many different ways, giving rise to the huge number of ΣΔM topologies reported in literature, which can be grouped according to different classification criteria [12]:

- **Single-bit versus multi-bit ΣΔMs**, depending on the number of bits in the embedded quantizer.
- **Single-loop versus cascade ΣΔMs**, depending on the number of quantizers employed: ΣΔMs employing only one quantizer are called *single-loop* topologies, whereas those using several quantizers are often referred to as *cascade* or MASH ΣΔMs.
- **Low-pass versus band-pass ΣΔMs**, depending on the nature of the signals being converted. The A/D conversion of LP signals has been assumed in previous sections, but band-pass (BP) ΣΔMs can also be built.
- **Discrete-time versus continuous-time ΣΔMs**, depending on the nature of loop filter dynamics. The use of a discrete-time (DT) loop filter in the ΣΔM has been assumed in previous sections. However, continuous-time (CT) ΣΔMs can be also implemented in practice. According to this classification criteria, *hybrid* CT/DT ΣΔMs take advantage of the benefits of both DT and CT implementations.

Describing all possible ΣΔM architectures derived from these classification criteria goes beyond the scope of this book. A detailed analysis of them can be found in the vast quantity of papers and books available in the ΣΔ literature [4, 10, 11, 13–24]. Instead, this book will hereafter focus on the most representative families of ΣΔMs, and their benefits and drawbacks will be further discussed in Chapter 2. In the next sections, the different components required for use of ΣΔMs for the implementation of ADCs and DACs will be analyzed.

1.5 Putting All The Pieces Together: From ΣΔMs to ΣΔ ADCs

In order to use ΣΔMs for the implementation of ADCs, some additional building blocks are required to properly filter the input signal and remove the out-of-band quantization noise. Figure 1.21 illustrates the basic block diagram of a ΣΔ ADC, considering an LP signal (Figure 1.21a), and a BP signal (Figure 1.21b). In both cases, there are three main building blocks required to implement the ΣΔ ADC:

- **An antialiasing filter (AAF)** band limits the analog input signal to prevent aliasing during its subsequent sampling. As discussed in Section 1.1.1, oversampling can considerably relax the attenuation requirements of the AAF, so that smoother transition bands are usually sufficient, compared to Nyquist-rate ADCs. Typically, low-order (either first-order or second-order) filters are enough in the majority of applications. Moreover, in the case of CT-ΣΔMs, the ΣΔM implements an implicit AAF, as will be explained later in this book. This greatly simplifies this block in practice.
- **A sigma-delta modulator (ΣΔM)** is where the oversampling and quantization of the band-limited analog signal take place. The quantization noise of the embedded B-bit quantizer is shaped in the frequency domain by placing an appropriate loop filter $H(z)$ before it and closing a negative feedback loop around them both. As stated above, low-resolution quantizers, with B typically in the range 1–5 bits, are sufficient for obtaining small IBN and high accuracy in the A/D conversion.
- **A decimation filter** uses a high-selectivity digital filter to sharply remove the out-of-band spectral content of the ΣΔM output and thus most of the shaped quantization noise. The decimator also reduces the data rate from f_s down to the Nyquist frequency, while increasing the word length from B to N bits to preserve resolution.

The ΣΔ modulator is the block that has most influence on the performance metrics of the ADC, in essence because it is responsible for the sampling and quantization processes that ultimately limit the accuracy of the A/D conversion. This is the main reason why the majority of design efforts focus on the design of the modulator. In this book we will mainly focus on this block, although the reader should bear in mind that in order to build a complete ΣΔ ADC, the other blocks are always required.

1.5.1 Some Words about ΣΔ Decimators

Figure 1.22 depicts the signal processing of ΣΔ decimators considering both an LP-ΣΔ ADC and a BP-ΣΔ ADC. In this example, an single-tone input signal is used as a test signal, and hence the AAF does not have any action over the signal since it is band limited. The influence of the decimator is illustrated by firstly removing the out-of-band quantization noise and afterwards by downsampling the signal to the Nyquist rate. Note that the conceptual scheme shown in Figure 1.21b may require a highly selective digital BPF to implement decimation in BP-ΣΔ ADCs. However, as will be discussed in Section 2.5, this problem can be solved by mixing down the digital signal to baseband, where an LP decimator can be used.

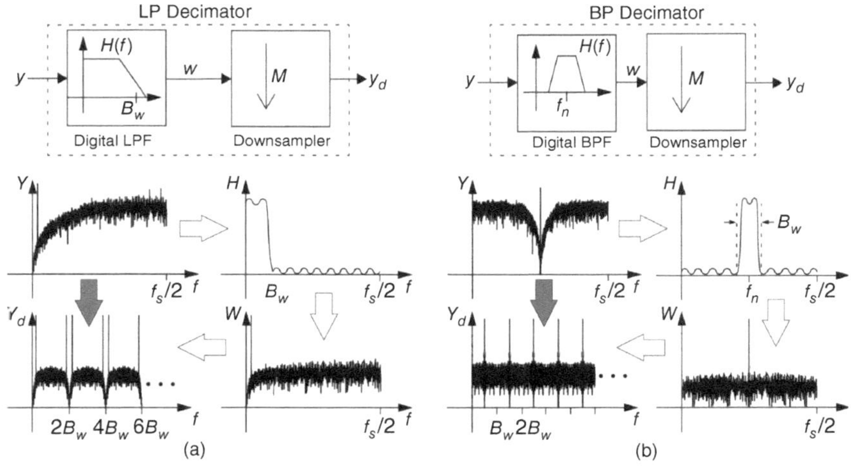

Figure 1.22 Illustrating the signal processing in ΣΔ decimators for: (a) LP ΣΔ ADCs, (b) BP ΣΔ ADCs.

As stated above, decimation filters reduce the oversampling frequency used by ΣΔMs down to the Nyquist rate, $f_s/2$, in order to process the signal more efficiently in the digital domain. The so-called cascaded-integrator-comb (CIC) structure – conceptually depicted in Figure 1.23 – is a compact way to implement decimation filters in ΣΔ ADCs, only requiring adders and delay elements for its implementation [25].

In the majority of practical situations, a multi-stage architecture is usually an efficient way to implement the filter and decimation stages in order to obtain a k-order decimation filter, as illustrated in Figure 1.23 [4]. The Z-domain transfer function of each stage is given by:

$$H_{\mathrm{CIC}}(z) = \frac{1}{M} \cdot \frac{1 - z^{-M}}{1 - z^{-1}} \tag{1.25}$$

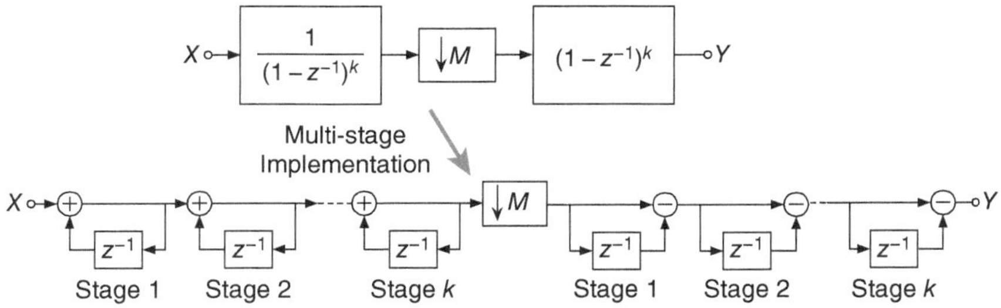

Figure 1.23 Block diagram of a k-order (sinc^k) CIC decimation filter.

which is also referred to as a *sinc* filter, whose frequency response is:

$$H_{\text{CIC}}(e^{j2\pi f}) = \frac{\text{sinc}(M\pi f)}{\text{sinc}(\pi f)} \tag{1.26}$$

where M is the decimation factor and $\text{sinc}(x) \equiv sin(x)/x$ stands for the *sinc* function. This function can be implemented at the logic level, as illustrated in Figure 1.24 [4].

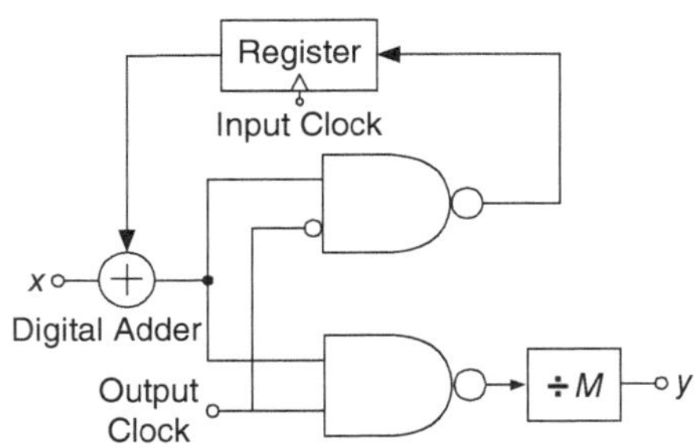

Figure 1.24 Decimation stage filter implementation [4].

It can be shown that a K-order CIC decimator produces a digital output with a word-length, b_{out}, given by [26]:

$$b_{\text{out}} = b_{\text{in}} + K \cdot \log_2 M \tag{1.27}$$

where b_{in} is the number of bits of the decimator input, i.e. the ΣΔM output. Note that if M is assumed to be a power of 2 – that is, $M = 2^p$ – then $b_{\text{out}} = b_{\text{in}} + p \cdot K$. For example, if a second-order decimator ($K = 2$), is used together with a 1-bit first-order ΣΔM with $M = 32$, an 11-bit word-length digital output is produced by the ΣΔ ADC.

It is important to note that a CIC decimator introduces a droop in the signal band, thus penalizing the resolution achieved by the ΣΔ ADC. Of course, a more accurate response can be obtained by increasing the decimator filter order, K, but at the price of increasing the circuit complexity and power consumption. As a rule of thumb, $K = L + 1$ is usually sufficient in most practical applications [19]. As an illustration, Figure 1.25 shows the time response of a ΣΔ ADC made up of a second-order ΣΔM and a third-order CIC decimator with $M = \text{OSR} = 128$ and $B = 1$, considering a half-scale input sine wave. It can be seen how, although the ΣΔM output is a PDM (single-bit) signal, the decimator output – the ADC output – is a multi-bit digital representation of the analog input signal (a sine wave in this example).

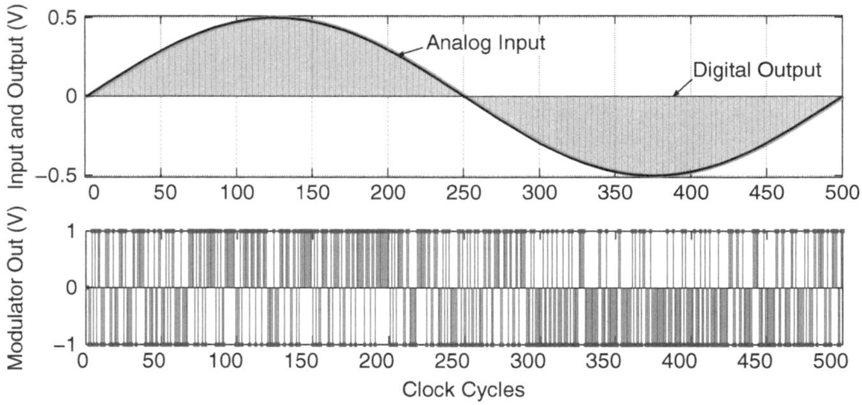

Figure 1.25 Signal waveforms in a ΣΔ ADC made up of a 1-bit second-order ΣΔM and a third-order CIC decimator.

Apart from CIC structures, other decimator topologies – such as recursive, multi-rate topologies, and so on – can be used to improve the efficiency in terms of silicon area and power consumption. This is especially so for high values of OSR, where the power increase is mostly due to the digital integrators, in which the adders operate at the highest sampling rate and with a full bit-width. A detailed analysis of different alternative decimation topologies is beyond the scope of this book and the interested reader can find a number of interesting publications related to this topic in the literature [26, 27].

1.6 ΣΔ DACs

DACs can take advantage of ΣΔ techniques to increase their performance by trading the accuracy of their analog components by digital signal processing. Figure 1.26 shows the block diagram of a ΣΔ DAC, which is made up of four main blocks: an interpolator, a digital ΣΔM, a DAC and an analog *reconstruction* filter. In this case, the ΣΔM is fully implemented in the digital domain, and combines oversampling and feedback to reduce the truncation error resulted from transforming a $N_{\rm in}$-bit input signal into a $N_{\rm out}$-bit output signal, with $N_{\rm in} \ll N_{\rm out}$. The ΣΔM input is provided by an interpolator filter, which generates the required oversampled $N_{\rm in}$-bit data sequence sampled at $\mathrm{OSR} \cdot f_N$. The truncation error that results from this transformation is shaped by the ΣΔM feedback loop, so that most of its power is pushed out of the signal band, where the ΣΔM digital output is converted into an analog signal by using a low-resolution highly-linear (typically $N_{\rm in} = 1, 2$) DAC[4], and the out-of-band truncation noise can be subsequently removed using an analog filter.

1.6.1 System Design Trade-offs and Signal Processing in ΣΔ DACs

As illustrated in Figure 1.26, ΣΔ DACs are mostly-digital systems, except for the low-resolution DAC block and the reconstruction filter. However, the design of these blocks can be notably simplified

[4]The DAC building blocks used in ΣΔ DACs are typically implemented using current steering or SC circuit techniques, depending on the performance required in terms of speed and resolution. If high speed is required, current-steering circuits are usuaully the best approach, as will be discussed in Chapter 8.

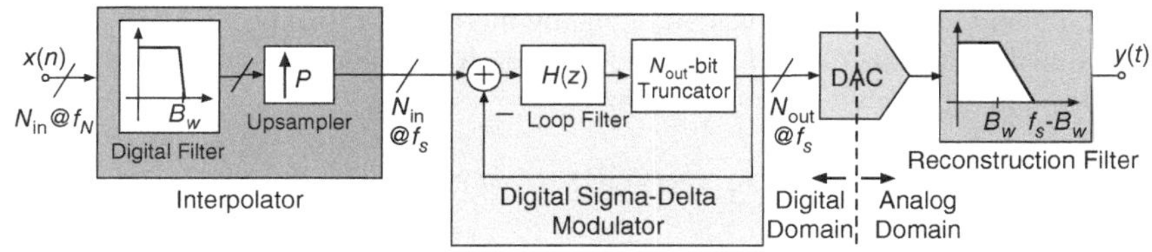

Figure 1.26 Conceptual block diagram of a ΣΔ DAC.

if OSR is high and $N_{out} = 1$. In this case, perfect linearity is guaranteed in the DAC and the filter specifications can be relaxed due to the action of OSR, in a similar way to what happens to the AAF in ΣΔ ADCs. Indeed, there are several design trade-offs involving the system-level parameters characterizing the performance of ΣΔ DACs, namely: N_{in}, N_{out}, OSR and L, with L being the ΣΔM loop-filter order.

On the one hand, using a low value of N_{out}, for instance $N_{out} = 1$, simplifies the design of the DAC circuit, and makes it more linear and robust against nonidealities. However, the stability of 1-bit ΣΔMs is more difficult to guarantee, specially for high values of L. Moreover, the high slew-rate of 1-bit DAC output (analog) signals and the large amount of out-of-band truncation noise, makes the design of the analog filter more complex. These design issues can be relaxed by using multi-bit ΣΔMs, but at the price of reducing the linearity of the DAC circuit [4].

Figure 1.27 shows the signal processing in ΣΔ DACs, where the operation of each building block is illustrated by the signal that results at its output. The role of the interpolator filter is the opposite of

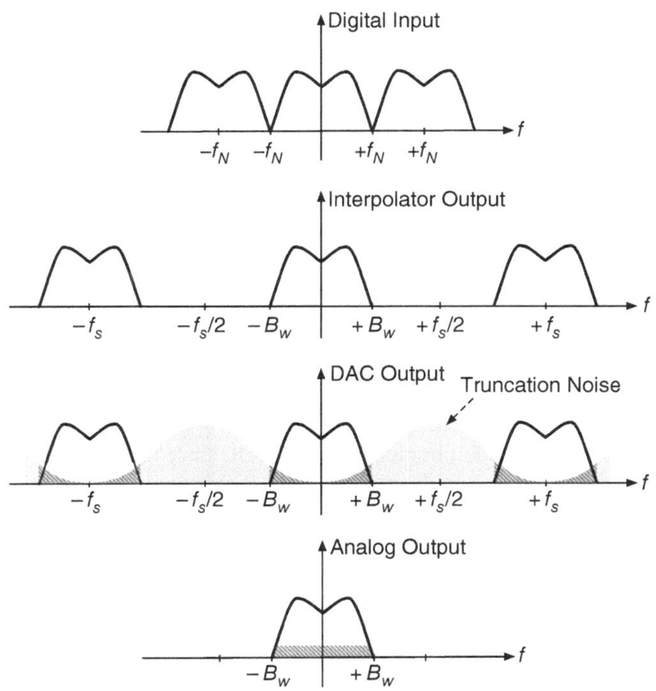

Figure 1.27 Illustration of signal processing in ΣΔ DACs.

a decimator: an interpolator oversamples the signal by increasing f_s from f_N to OSR $\cdot f_N$. One way of interpolating signals is to add OSR − 1 zeroes between each sample of the input signal. This operation can be formulated in the discrete-time domain as:

$$y(n) = \begin{cases} x(n/\text{OSR}), & n = m \cdot \text{OSR} \\ 0, & \text{Otherwise} \end{cases} \tag{1.28}$$

where m is an integer number. Taking the Z-transform gives:

$$Y(z) = X(z^{\text{OSR}}) \tag{1.29}$$

The above function introduces spectral images or replicas at multiples of f_N, which must be suppressed by the interpolator filter, as depicted in Figure 1.27. To this end, a tap-based FIR filter can be used, although with this kind of filter topology, the chip area and the power consumption increase along with the number of taps included. A more efficient way of implementing interpolator filters is based on the use of multi-stage topologies – similar to those used in decimators – where the interpolation factor, $P \equiv \text{OSR}$, is factorized in multiple steps, for example $P = P_1 \cdot P_2 \cdot P_3 \ldots$, where P_i is the interpolation factor of the ith stage.

An alternative interpolator implementation consists of repeating the signal samples at the oversampling rate; that is, without introducing zeroes between two consecutive input samples. In this case, the discrete-time response of the interpolator can be written as:

$$y(n) = x(n/P) \tag{1.30}$$

which can be represented in the frequency domain as a sinc function,

$$\frac{Y(f)}{X(f)} = \frac{\text{sinc}(\pi f)}{\text{sinc}(\pi f/P)} \tag{1.31}$$

As might be expected, the above transfer function is the inverse of that produced by decimator filters, as given in Equation (1.26). There are many ways of implementing interpolators. However, a detailed description is beyond the scope of this book and can be read in a number of references in the literature [28].

1.6.2 Implementation of Digital ΣΔMs used in DACs

As stated above, the operation of the ΣΔMs used to build DACs is similar to those used in ADCs. The main difference is related to the circuit implementation itself, since the ΣΔMs used in DACs are fully implemented in the digital domain, and hence, because all signals involved in the feedback systems are digital, no data conversion is required. Although their principle of operation is the same, and in theory similar architectures and system-level strategies can be applied to both ΣΔ ADCs and DACs, the digital implementation allows loop-filter solutions to be used, which are more robust against the effects of circuit nonidealities than the analog ones. Therefore, the loop-filter is implemented by digital accumulators – instead of analog integrators – and digital adders, delays and multipliers. They can be notably simplified in terms of hardware if the loop-filter coefficients are integer numbers, and even more so if they are powers of two.

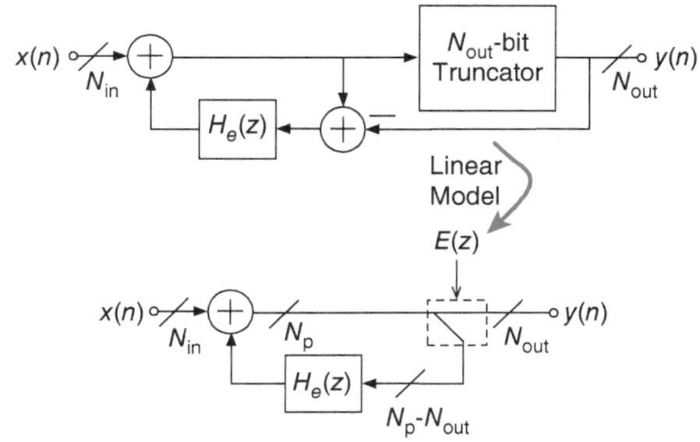

Figure 1.28 Conceptual block diagram of a digital ΣΔM typically used in ΣΔ DACs.

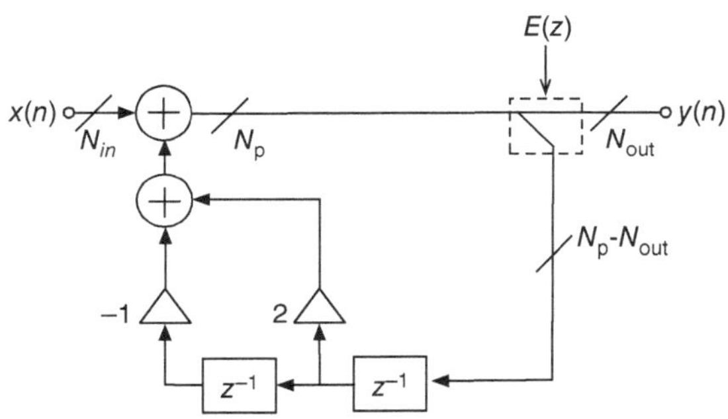

Figure 1.29 Block diagram of a second-order digital ΣΔM.

Figure 1.28 shows a system-level diagram of a highly efficient digital ΣΔM. Assuming a linear model for the truncator, it can be shown that the Z-domain transfer function of the ΣΔM output is given by:

$$Y(z) = X(z) + [1 - H_e(z)] \cdot E(z) \tag{1.32}$$

where $H_e(z)$ is the loop-filter transfer function, which – in contrast to analog ΣΔMs – is placed in the feedback path of the modulator. In the case of a first-order ΣΔM, $H_e(z) = z^{-1}$, and hence $STF(z) = 1$ and $NTF(z) = 1 - z^{-1}$. As an illustration, Figure 1.29 shows a second-order digital ΣΔM. In this case, $H_e(z) = 2z^{-1} - z^{-2}$, and hence $NTF(z) = (1 - z^{-1})^2$. As stated above, higher-order digital ΣΔMs can be implemented following the same philosophy as for their analog counterparts. Moreover, some state-of-the-art ΣΔ DACs use time-interleaving techniques in order to relax the speed specifications. This enables them to be used to handle signals in telecom applications operating in the gigahertz range, while keeping low values of the OSR [29].

1.7 Summary

This chapter has presented an introduction to ΣΔ converters. The benefits of employing oversampling and quantization noise shaping in the digitization of signals have been analyzed and compared to the performance of Nyquist-rate ADCs. Among the blocks that build up a ΣΔ ADC, the chapter has focussed on the ΣΔ modulator, including its general architecture, ideal operation, and performance metrics.

The simplest topology of a ΣΔM – based on a first-order loop filter – was analyzed in detail to illustrate the principles of operation behind ΣΔ techniques that achieve high accuracy with low-resolution embedded quantization. The main limitations of first-order ΣΔMs – caused by the strong correlation between the quantization error and the input signal – were discussed and existing methods for increasing the effective resolution of ΣΔMs were highlighted as an introduction to the taxonomy of ΣΔM architectures; this will be discussed in more detail in Chapter 2.

The chapter concluded by showing how to build both ADCs and DACs using ΣΔMs. In both data-conversion systems, the core building block is the ΣΔM, which is implemented in the analog domain in ADCs and in the digital domain in DACs. The majority of system-level strategies – as will be discussed in Chapter 2 – can be applied to both analog and digital ΣΔMs. However, the former are more sensitive to nonideal-circuit effects, thus requiring more careful design and being in many cases the design bottleneck in very diverse electronic systems. Therefore, the rest of the book is mostly devoted to the practical design issues related to analog ΣΔMs, starting in next chapter with an analysis of their most representative system architectures.

References

[1] W. Bennett, "Spectra of Quantized Signals," *Bell System Technical J.*, vol. 27, pp. 446–472, July 1948.

[2] B. Widrow, "Statistical Analysis of Amplitude-Quantized Sampled-Data Systems," *Trans. AIEE - Part II: Applications and Industry*, pp. 555–568, January 1960.

[3] A. B. Sripad and D. L. Snyder, "A Necessary and Sufficient Condition for Quantization Errors to be Uniform and White," *IEEE Trans. on Acoustics, Speech and Signal Processing*, vol. 25, pp. 442–448, October 1977.

[4] S. R. Norsworthy, R. Schreier, and G. C. Temes, *Delta-Sigma Data Converters: Theory, Design and Simulation.* IEEE Press, 1997.

[5] R. van de Plassche, *CMOS Integrated Analog-to-Digital and Digital-to-Analog Converters.* Springer, 2003.

[6] H. Inose, Y. Yasuda, and J. Murakami, "A Telemetering System by Code Modulation – Δ – Σ Modulation," *IRE Trans. on Space Electronics and Telemetry*, vol. 8, pp. 204–209, September 1962.

[7] B. E. Boser and B. A. Wooley, "The Design of Sigma-Delta Modulation Analog-to-Digital Converters," *IEEE J. of Solid-State Circuits*, vol. 23, pp. 1298–1308, December 1988.

[8] A. Marques, V. Peluso, M. S. Steyaert, and W. M. Sansen, "Optimal Parameters for ΔΣ Modulator Topologies," *IEEE Trans. on Circuits and Systems II: Analog and Digital Signal Processing*, vol. 45, pp. 1232–1241, September 1998.

[9] J. Candy and O. J. Benjamin, "The Structure of Quantization Noise from Sigma-Delta Modulation," *IEEE Transactions on Communications*, pp. 1316–1323, 1981.

[10] J. Candy and G. Temes, *Oversampling Delta-Sigma Data Converters: Theory, Design and Simulation.* IEEE Press, 1991.

[11] Y. Geerts, M. Steyaert, and W. Sansen, *Design of Multi-bit Delta-Sigma A/D Converters.* Kluwer Academic Publishers, 2002.

[12] A. Rodríguez-Vázquez, F. Medeiro, J. M. de la Rosa, R. del Río, R. Tortosa, and B. Pérez-Verdú, "Sigma-Delta CMOS ADCs: An Overview of the State-of-the-Art," in *CMOS Telecom Data Converters* (A. Rodríguez-Vázquez, F. Medeiro, and E. Janssens, editors), Kluwer Academic Publishers, 2003.

[13] F. Medeiro, B. Pérez-Verdú, and A. Rodríguez-Vázquez, *Top-Down Design of High-Performance Sigma-Delta Modulators.* Kluwer Academic Publishers, 1999.

[14] J. Cherry and W. Snelgrove, *Continuous-Time Delta-Sigma Modulators for High-Speed A/D Conversion.* Kluwer Academic Publishers, 1999.

[15] V. Peluso, M. Steyaert, and W. Sansen, *Design of Low-Voltage Low-Power CMOS Delta-Sigma A/D Converters.* Kluwer Academic Publishers, 1999.

[16] S. Rabii and B. A. Wooley, *The Design of Low-Voltage, Low-Power Sigma-Delta Modulators.* Kluwer Academic Publishers, 1999.

[17] L. Breems and J. H. Huijsing, *Continuous-Time Sigma-Delta Modulation for A/D Conversion in Radio Receivers.* Kluwer Academic Publishers, 2001.

[18] J. M. de la Rosa, B. Pérez-Verdú, and A. Rodríguez-Vázquez, *Systematic Design of CMOS Switched-Current Bandpass Sigma-Delta Modulators for Digital Communication Chips.* Kluwer Academic Publishers, 2002.

[19] R. Schreier and G. C. Temes, *Understanding Delta-Sigma Data Converters.* IEEE Press, 2005.

[20] R. del Río, F. Medeiro, B. Pérez-Verdú, J. M. de la Rosa, and A. Rodríguez-Vázquez, *CMOS Cascade ΣΔ Modulators for Sensors and Telecom: Error Analysis and Practical Design.* Springer, 2006.

[21] M. Ortmanns and F. Gerfers, *Continuous-Time Sigma-Delta A/D Conversion: Fundamentals, Performance Limits and Robust Implementations.* Springer, 2006.

[22] L. Yao, M. Steyaert, and W. Sansen, *Low-Power Low-Voltage Sigma-Delta Modulators in Nanometer CMOS.* Springer, 2006.

[23] P. G. R. Silva and J. H. Huijsing, *High Resolution IF-to-Baseband* ΣΔ ADC for Car Radios. Springer, 2008.

[24] R. H. van Veldhoven and A. H. M. van Roermund, *Robust Sigma Delta Converters*. Springer, 2011.

[25] J. Candy, "A Use of Double Integration in Sigma-Delta Modulation," *IEEE Transactions on Communications*, vol. 33, pp. 249–258, March 1985.

[26] E. Hogenauer, "An Economical Class of Digital Filters for Decimation and Interpolation," *IEEE Trans. on Acoustic, Speech and Signal Processing*, vol. 29, pp. 155–162, 1981.

[27] H. Aboushady *et al.*, "Efficient Polyphase Decomposition of Comb Decimation Filters in ΣΔ Analog-to-Digital Converters," *IEEE Trans. on Circuits and Systems II: Analog and Digital Signal Processing*, vol. 48, pp. 898–903, 2001.

[28] S. Pavan, R. Schreier, and G. C. Temes, *Understanding Delta-Sigma Data Converters.* Wiley-IEEE Press, 2nd ed., 2017.

[29] A. Bhide and A. Alvandpour, "An 11 GS/s 1.1 GHz Bandwidth Interleaved ΔΣ DAC for 60 GHz Radio in 65nm CMOS," *IEEE J. of Solid-State Circuits*, vol. 50, pp. 2306–2318, 2015.

CHAPTER 2

TAXONOMY OF ΣΔ ARCHITECTURES

This chapter presents an overview of practical topologies for the implementation of ΣΔ modulation. The large variety of existing ΣΔ realizations is classified according to the type of modulator architecture (single-loop or cascade), the circuit techniques employed (discrete-time or continuous-time), and the nature of the signals being converted (low-pass or band-pass). Starting with discrete-time low-pass single-bit ΣΔ modulators, the implications of these different alternatives are then presented in an incremental way.

Section 2.1 is dedicated to the second-order ΣΔ modulator. Higher-order single-loop ΣΔ architectures are considered in Section 2.2, taking into account issues related to their practical implementation and problems not addressed by linear models, such as instabilities. Cascade ΣΔ topologies are covered in Section 2.3. In Section 2.4, the topological study is extended to ΣΔ modulators using multi-bit embedded quantizers, analyzing their pros and cons. Techniques to circumvent the disadvantages, such as dynamic element matching or dual-quantization, are revised.

The conversion of band-pass signals is covered in Section 2.5, taking into account its application in wireless receivers. Section 2.6 addresses the realization of continuous-time ΣΔ modulators, discussing their advantages compared to discrete-time ones and the existing alternatives for the loop filter and the feedback implementation. Finally, the chapter concludes with the main synthesis methods for designing CT-ΣΔMs, based on a discrete-to-continuous time transformation (Section 2.7) or a direct synthesis of the CT-ΣΔM loop filter (Section 2.8).

Sigma-Delta Converters: Practical Design Guide, Second Edition. José M. de la Rosa.

2.1 Second-order ΣΔ Modulators

ΣΔ modulators that make use of only one embedded quantizer are usually referred to as single-loop topologies, as stated in Chapter 1. To get familiar with these architectures, their performance, their circuit-level implementation, as well as other practical aspects, we first address second-order ΣΔMs. Afterwards, the problem of stability in higher-order ΣΔMs is considered, together with architectural alternatives to circumvent it.

Let us consider the block diagram of a first-order ΣΔM, as shown in Figure 2.1a, where coefficient $a_{1,2}$ stands for the loop-filter integrator *scaling* or *weight*. If the embedded quantizer is replaced by a first-order ΣΔM, the block diagram depicted in Figure 2.1b is obtained. This ΣΔM topology – commonly referred to as a second-order DT-ΣΔM – can be built by cascading two DT

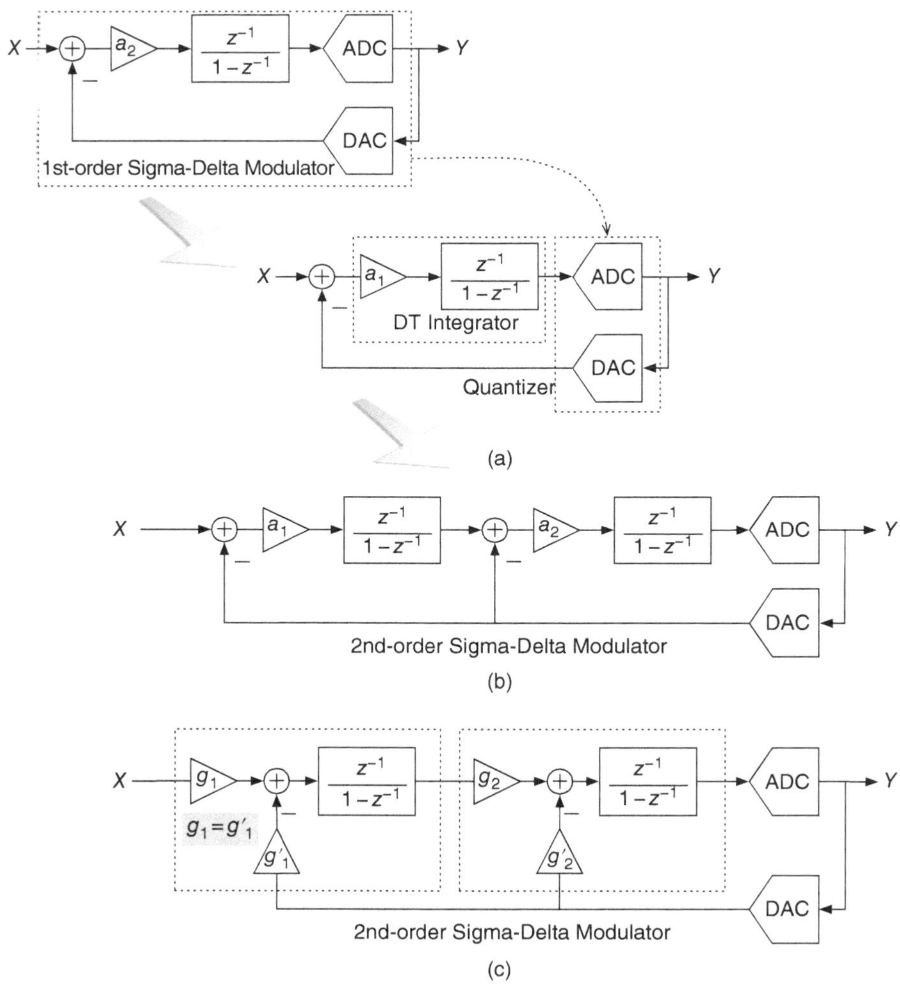

Figure 2.1 Block diagram of second-order ΣΔMs and different notations: (a) synthesis of the architecture from a first-order ΣΔM; (b) general representation of the second-order DT-ΣΔM using the loop-filter notation in [1, 2]; (c) alternative representation of the second-order DT-ΣΔM using the notation in [3, 4].

integrators [5], with each integrator receiving a weighted feedback path from the DAC, implemented by coefficients a_i [5].

Under linear analysis, the modulator output in the Z-domain yields,

$$Y(z) = \frac{k_q a_1 a_2 \frac{z^{-2}}{(1-z^{-1})^2} X(z) + E(z)}{1 + k_q a_1 a_2 \frac{z^{-2}}{(1-z^{-1})^2} + k_q a_2 \frac{z^{-1}}{(1-z^{-1})}} \tag{2.1}$$

where k_q stands for the gain of the quantizer. For a pure second-order shaping, Equation (2.1) needs to be simplified to:

$$Y(z) = z^{-2}X(z) + (1 - z^{-1})^2 E(z) \tag{2.2}$$

so that the following expressions for the integrator coefficients need to be fulfilled:

$$\begin{aligned} k_q a_1 a_2 &= 1 \\ k_q a_2 &= 2 \end{aligned} \tag{2.3}$$

Figure 2.2a compares the output spectrum of a first- and a second-order ΣΔM with a 1-bit quantizer and a half-scale input sine wave. In addition to the higher noise-shaping filtering, the second-order ΣΔM output spectrum barely shows tonal behavior compared to the first-order ΣΔM – as already discussed in Section 1.3. This is better illustrated in Figure 2.2b, which shows the quantization noise pattern corresponding to a second-order ΣΔM with 1-bit quantizer and OSR = 64. It is clear how the in-band quantization error power is better predicted by the linear (white-noise) model for the quantizer, except for those input signal amplitudes closer to the full-scale range (Δ = 1 V in this example). In these amplitude ranges, the quantizer overloads and the quantization error power increases monotonously and the SNR of the modulator drops out drastically.

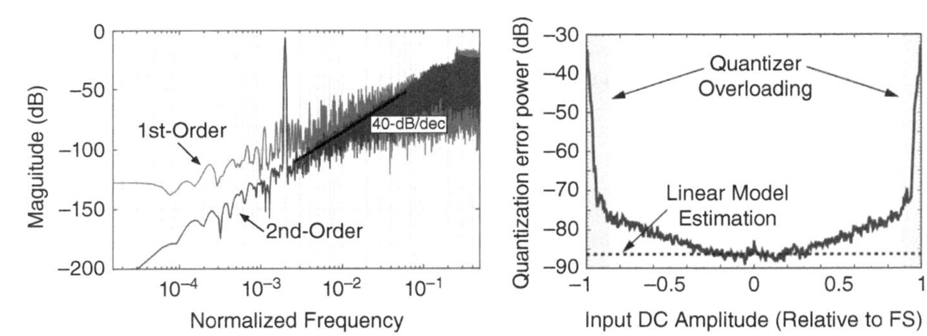

Figure 2.2 Illustrating the noise-shaping of second-order ΣΔMs: (a) half-scale output spectrum; (b) noise pattern.

2.1.1 Alternative Representations of Second-order ΣΔMs

Figure 2.1c shows an alternative representation of the second-order ΣΔM, using a coefficients notation [3, 4] that allows different weights to be allocated in the forward and feedback paths of each integrator,

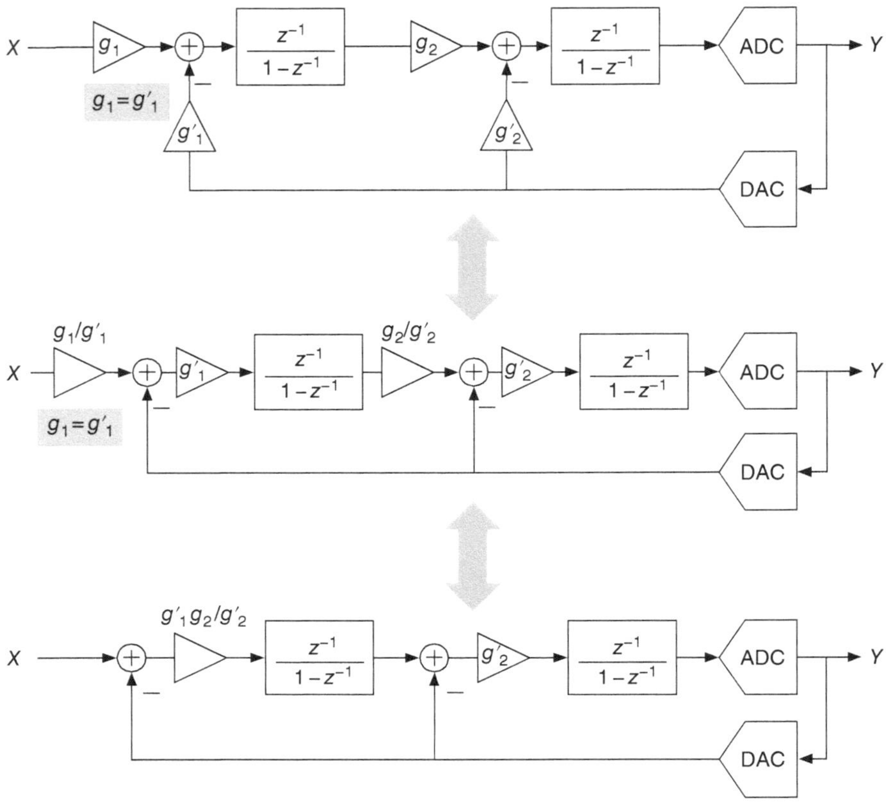

Figure 2.3 Equivalence between the DT representations in Figures 2.1b and 2.1c.

using coefficients g_i and g_i', respectively. As illustrated in Figure 2.3, the notations in Figures 2.1b and 2.1c can be easily connected with the equalities:

$$\begin{aligned} a_1 &= \frac{g_1' g_2}{g_2'} \\ a_2 &= g_2' \end{aligned} \tag{2.4}$$

Both nomenclatures for the integrator scaling coefficients of ΣΔ modulators will be used throughout this book. The notation in [1, 2] is closer to the modulator architectural level, whereas the notation in [3, 4] is closer to the actual circuit-level implementation, in which integrators with more than one SC input branch are usually employed. The latter is thus useful to accurately account for some nonidealities of practical ΣΔM implementations, which will be covered in Chapter 3.

For the sake of illustration, Figure 2.4 shows a possible implementation of the second-order ΣΔM in Figure 2.1 using fully-differential SC circuitry and assuming single-bit quantization. Similar to the circuit in Figure 1.15, the modulator differential input signal is denoted by X and the modulator digital output Y controls the feedback connection of references voltages $V_{\text{ref}+}$ and $V_{\text{ref}-}$ to the integrators. The modulator's full scale range Δ thus equals $2V_{\text{ref}}$, with $V_{\text{ref}} = V_{\text{ref}+} - V_{\text{ref}-}$. Note from the first SC integrator in Figure 2.4 that both the modulator input signal and the DAC feedback signal are

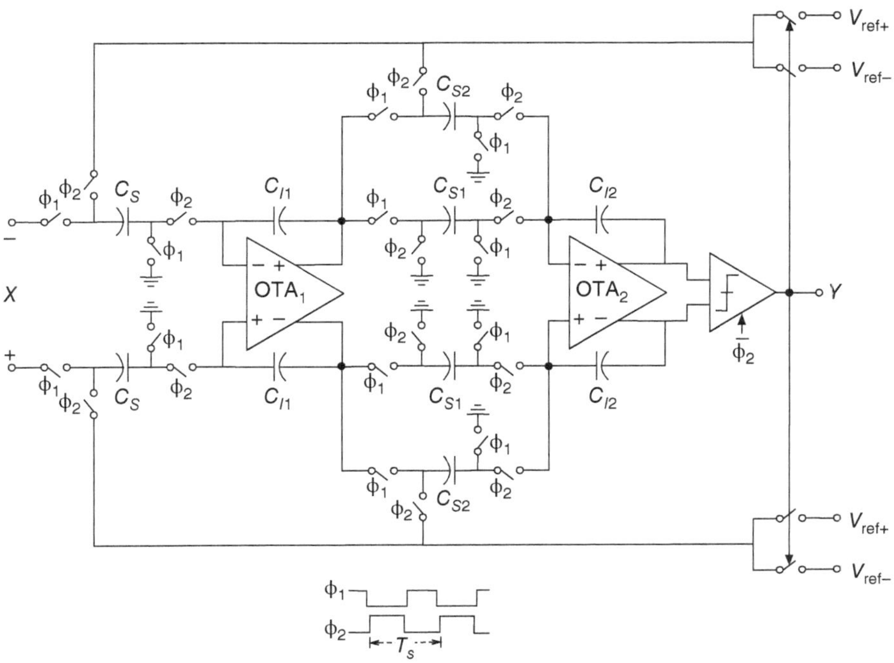

Figure 2.4 Fully-differential SC implementation of a second-order ΣΔ modulator.

processed through the same sampling capacitor, C_S. For the second integrator, the output of the first integrator is processed through both C_{S1} and C_{S2}, whereas the DAC feedback signal is processed only through C_{S2}. The modulator scaling coefficients are thus implemented as the capacitor ratios shown in Figure 2.1c.

$$g_1 = g_1' = \frac{C_S}{C_{I1}}$$
$$g_2 = \frac{C_{S1} + C_{S2}}{C_{I2}}, \quad g_2' = \frac{C_{S2}}{C_{I2}} \tag{2.5}$$

In practice the value of the integrator weights are selected to fulfill the relations in Equation (2.3), also taking into account their effects on some aspects of the modulator performance, such as:

- Keeping the state variables (integrator outputs) bounded to ensure modulator stability. The second-order ΣΔM is stable for inputs in the range $[-0.9\Delta/2, +0.9\Delta/2]$ if $g_2' > 1.25 g_1' g_2$, regardless of the quantizer gain k_q [5]. This condition is already met if $g_2' = 2g_1' g_2$ – see Equations (2.3) and (2.4).
- Keeping the modulator overload level as close as possible to the full scale to ensure a high peak SNR – see Figure (1.14).
- Minimizing the required signal range at the integrator outputs; in other words, the integrator output swing demands must be attainable with the intended voltage supply and as low as possible to reduce power consumption and to facilitate circuit design.
- Simplifying the practical implementation of the integrator weights as ratios of unit capacitors.

Generally speaking, the selection of the scaling coefficients of a ΣΔ modulator involves trade-offs between architectural-, circuit-, and technological-level aspects of the practical implementation; as a result, the optimum selection for a given application may not apply elsewhere. For the sake of illustration, Table 2.1 shows several sets of weights reported for the second-order single-bit ΣΔM in Figure 2.1. All sets exhibit an overload level $X_{OL} \approx -4$ dBFS (i.e., −4 dB below the full-scale amplitude $V_{ref} = \Delta/2$). The required integrator output swing and the minimum number of unit capacitors are also included. Capacitor sharing between weights in the same integrator has been considered.

Table 2.1 Comparison of some sets of coefficients reported for the second-order single-bit ΣΔM.

	[6]	[7]	[8]	[9]
g_1, g_1'	1/2, 1/2	1/4, 1/4	1/2, 1/2	1/3, 1/3
g_2, g_2'	1/2, 1/2	1/2, 1/4	1, 1/2	3/5, 2/5
a_1, a_2	0.5, 0.5	0.5, 0.25	0.5, 0.5	0.5, 0.4
Overload level	−4 dBFS	−4 dBFS	−4 dBFS	−4 dBFS
Integrator output swing	$\pm 1.5\ V_{ref}$	$\pm 0.75\ V_{ref}$	$\pm 1.25\ V_{ref}$	$\pm 1.0\ V_{ref}$
Unit capacitors (2× in fully-diff)	6(= 3 + 3)	11(= 5 + 6)	9(= 5 + 4)	12(= 4 + 8)

2.1.2 Second-Order ΣΔM with Unity STF

Figure 2.5 shows an alternative second-order ΣΔM topology, which makes use of feedforward paths to implement a so-called *unity* STF [10, 11].

Under linear analysis, the modulator output in the Z-domain yields,

$$Y(z) = X(z) + (1 - z^{-1})^2 E(z) \tag{2.6}$$

so that STF(z) = 1 – in other words, the signal transfer function equals 1 at all frequencies – whereas NTF is unaffected.

One of the most appealing features of the unity-STF (USTF) ΣΔM in Figure 2.5 is that ideally there is no input signal trace processed by the integrators. Indeed, the integrator inputs in the Z-domain

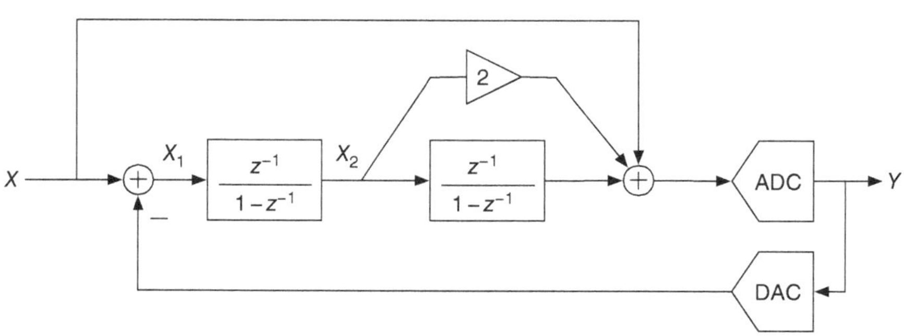

Figure 2.5 Second-order ΣΔ modulator with unity STF.

can be obtained as,

$$
\begin{aligned}
X_1(z) &= -(1 - z^{-1})^2 E(z) \\
X_2(z) &= -z^{-1}(1 - z^{-1})E(z)
\end{aligned}
\tag{2.7}
$$

showing that they depend only on the quantization error. This is illustrated in Figure 2.6a, which compares the magnitude spectrum of the integrators' output nodes for the modulator in Figure 2.5 and that obtained in Figure 2.1b with $a_1 = 1/2$ and $a_2 = 1/4$. In all cases, a 3-bit quantizer is considered. Note that the modulator with USTF shows only shaped quantization noise at the output of the integrators, thus allowing their output swing to be reduced, as shown in Figure 2.6b.

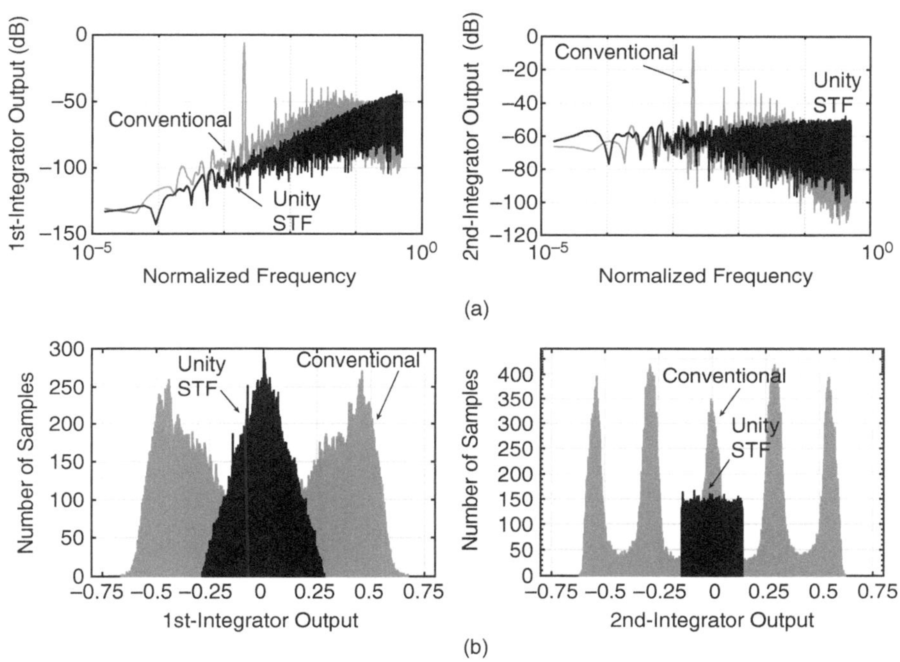

Figure 2.6 Illustrating the use of unity-STF in second-order ΣΔMs with 3-bit quantization: (a) magnitude spectrum of integrators' outputs; (b) output swings of integrators.

In practice, there will be some residual component of the modulator input signal at the integrator inputs, but it is normally negligible. This means that, if nonlinearities of the circuit implementation are accounted for, the generated distortion will be considerable lower for the USTF ΣΔM in Figure 2.5 than for the traditional ΣΔM in Figure 2.1. Moreover, the technique is effective for any oversampling ratio, which makes USTF ΣΔMs especially suited for reducing sensitivity to circuit imperfections in wideband applications, in which low OSR values are required.

The concept of the USTF can be extended to noise shaping of any order. The only requirement is to ideally make STF(z) = 1, without changing the modulator NTF. In recent years, USTF has been often applied in ΣΔMs for wideband and multimode applications [12–16].

2.2 High-order Single-loop ΣΔMs

The simplest way to extend a ΣΔM to an arbitrary Lth-order shaping involves including L integrators before the quantizer. Extending the second-order ΣΔM in Figure 2.1a, the topology in Figure 2.7 is obtained. This is known as an Lth-order single-loop ΣΔM with *distributed feedback* [17].[1] Ideally, its NTF can be derived from linear analysis and equated to Equation (1.10) to derive a set of relations between the integrator scaling coefficients to be fulfilled for obtaining a pure-differentiator noise shaping – similar to what was done in Equation (2.3) for the second-order ΣΔM. The in-band quantization noise and the modulator dynamic range would thus ideally be as given by Equations (1.12) and (1.13), respectively.

However, this modulator performance cannot be produced in practice because ΣΔMs with pure-differentiator FIR NTFs are prone to instability if $L > 2$, exhibiting unbounded states and poor SNR compared to that predicted by linear analysis. In general, instability appears at the modulator output as a large-amplitude low-frequency oscillation, leading to long bitstreams of alternating values of +1 and −1. This tendency to instability can be explained as follows [18]. For a ΣΔM to be stable, the quantizer input must not be allowed to become too large. Since the quantizer input is obtained as (see Figure 2.7):

$$Q(z) = \mathrm{STF}(z)X(z) + [\mathrm{NTF}(z) - 1]E(z) \tag{2.8}$$

the gain of NTF(z) − 1, or simply NTF(z), must not be too large. However, as clearly visible from Figure 1.19, the out-of-band gain of FIR NTFs of the form $(1 - z^{-1})^L$ rapidly increases for $L > 2$, yielding $\max[|\mathrm{NTF}(f)|^2] = 2^{2L}$ at $f = f_s/2$. Consequently, it starts to overload the quantizer, which yields a significant decrease of the modulator SNR.

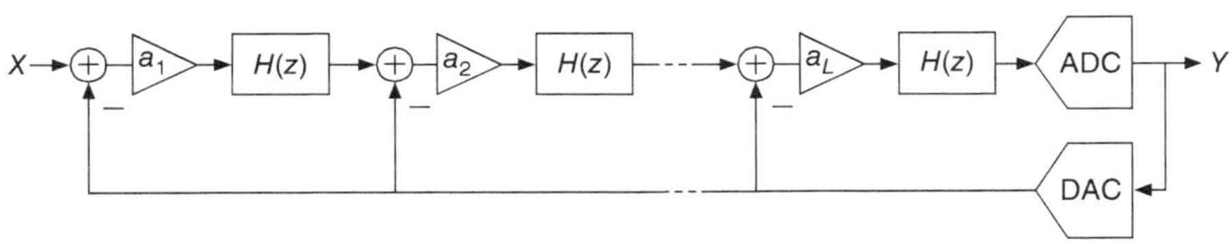

Figure 2.7 High-order single-loop ΣΔM with distributed feedback.

This problem can be circumvented by resorting to single-loop ΣΔMs with IIR NTFs of the form $\mathrm{NTF}(z) = (z - 1)^L/D(z)$, with $D(z)$ being a polynomial determined by the modulator scaling coefficients; this helps to limit the out-of-band gain of NTF. Note that the scaling coefficients can be designed to build a high-pass Butterwoth or Chebyshev filter for the NTF, with cutoff frequency beyond the signal band and approximately flat gain in the pass-band. Thus, for the ΣΔM to remain stable while maximizing SNR, this gain must be properly adjusted. Based on empirical conditions derived from a number of stable high-order single-loop designs, there is a rule of thumb that the optimal out-of-band gain of the NTF is given by:

$$||\mathrm{NTF}(z)||_\infty = \max[\mathrm{NTF}(z)] \simeq 1.5 \tag{2.9}$$

This is widely used as an approximate stability criterion, and is commonly known as the (modified) Lee's rule [19]. However, unlike second-order ΣΔMs, for which a stability condition have been extracted [5], determining exact conditions that guarantee the stability of higher-order single-loop ΣΔMs is still an open question. It has been shown, using behavioral simulations, that high-order ΣΔMs are *conditionally* stable; that is, with proper selection of the scaling coefficients, stable operation can be obtained for inputs restricted to within a certain range and for certain initial conditions of the state variables [1, 20]. However, despite the absence of general stability conditions, successful high-order

[1]The discrete-time filters $H(z)$ in Figure 2.7 are assumed to be integrators with the transfer function in Equation (1.19).

ΣΔMs have been designed since the late 1980s [19]. Indeed, the topology in Figure 2.7 has been widely used, and optimal coefficients for it have been presented in the literature [1]. Its STF and NTF can be calculated under linear analysis as:[2]

$$\mathrm{STF}(z) = \frac{k_q \prod_{i=1}^{L} a_i H^L(z)}{1 + k_q \sum_{i=1}^{L} \left[\left(\prod_{j=i}^{L} a_j \right) H^{L-i+1}(z) \right]} \tag{2.11}$$

$$\mathrm{NTF}(z) = \frac{1}{1 + k_q \sum_{i=1}^{L} \left[\left(\prod_{j=i}^{L} a_j \right) H^{L-i+1}(z) \right]} \tag{2.12}$$

If the integrators described by Equation (1.19) are used as filters – that is, $H(z) = \mathrm{ITF}(z)$ – the NTF can be approximated for low frequencies to [1]:

$$|\mathrm{NTF}| \approx \frac{|1 - z^{-1}|^L}{k_q \prod_{i=1}^{L} a_i} \tag{2.13}$$

Here the complete scaling of the outermost feedback branch of the ΣΔM dominates the noise-shaping behavior. Similar to Equations (1.11) and (1.12), the IBN of an Lth-order single-loop ΣΔM with distributed feedback thus yields,

$$\mathrm{IBN}_L \approx \frac{\Delta^2}{12} \frac{1}{(k_q \prod_{i=1}^{L} a_i)^2} \frac{\pi^{2L}}{(2L+1)\mathrm{OSR}^{2L+1}} \tag{2.14}$$

so that the IBN increases by a factor $1/(k_q \prod_{i=1}^{L} a_i)^2$ compared to an ideal ΣΔM.

A major drawback of this loop filter implementation in Figure 2.7 is that the integrator outputs contain a significant amount of the input signal [24], so that the integrators require significant swing capabilities and/or the scaling coefficients need to be low. This problem can be circumvented using the loop filter topology in Figure 2.8, which is a chain of integrators with *feedforward summation* [25].Under linear analysis, the corresponding STF and NTF can be calculated as:

$$\mathrm{STF}(z) = \frac{k_q \sum_{i=1}^{L} a_{L-i+1} H^{L-i+1}(z)}{1 + k_q \sum_{i=1}^{L} a_{L-i+1} H^{L-i+1}(z)} \tag{2.15}$$

$$\mathrm{NTF}(z) = \frac{1}{1 + k_q \sum_{i=1}^{L} a_{L-i+1} H^{L-i+1}(z)} \tag{2.16}$$

[2]The quantizer gain k_q of the linear quantizer model is explicitly considered hereinafter. As discussed in Section 1.1.2, the gain of a multi-bit quantizer is clearly defined – say $k_q = 1$ if its input and output full-scale ranges are the same (see Figure 1.6) – but that of a single-bit quantizer can be arbitrarily chosen. Nevertheless, the effective value of k_q needs to be estimated to quantitatively analyze the performance of ΣΔMs. There are many different approaches to finding a good approximation [1, 21, 22]. Among them, the one adopted here for the case of ΣΔMs with distributed feedback corresponds to that of Ortmanns and Gerfers [23],

$$k_q = \begin{cases} 1/a_1, & \text{single-bit first-order } \Sigma\Delta\text{M} \\ 2/a_L, & \text{single-bit } L\text{th-order } \Sigma\Delta\text{M} \\ 1, & \text{multi-bit } \Sigma\Delta\text{M} \end{cases} \tag{2.10}$$

because of its simplicity and its good agreement with simulations.

Note that NTF structure is the same as in Equation (2.12), so the IBN of this ΣΔM topology yields an expression similar to that in Equation (2.14).

However, for the case of the distributed feedback topology in Figure 2.7 and the feedforward summation topology in Figure 2.8, the loop filter for the STF and the NTF are essentially identical – compare Equation (2.11) with (2.12) and Equation (2.15) with (2.16). This means that, if the NTF is designed for the desired noise-shaping behavior, both topologies also fix the STF function, since $STF(z) = 1 - NTF(z)$. In addition, the STF of the feedforward summation architecture exhibits some peaking at high frequencies, and this may jeopardize the modulator stability if precautions are not taken. Alternatively, NTFs with inverse Chebyshev filtering characteristics can also be designed [24].

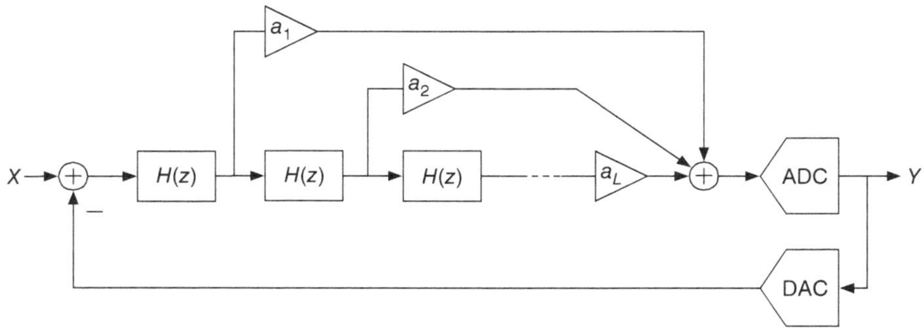

Figure 2.8 High-order single-loop ΣΔM with feedforward summation.

If a certain degree of freedom is desired in designing both the modulator NTF and STF, the topology in Figure 2.9 can be used. This is a chain of integrators with *distributed feedback and distributed feedforward input paths* [26].

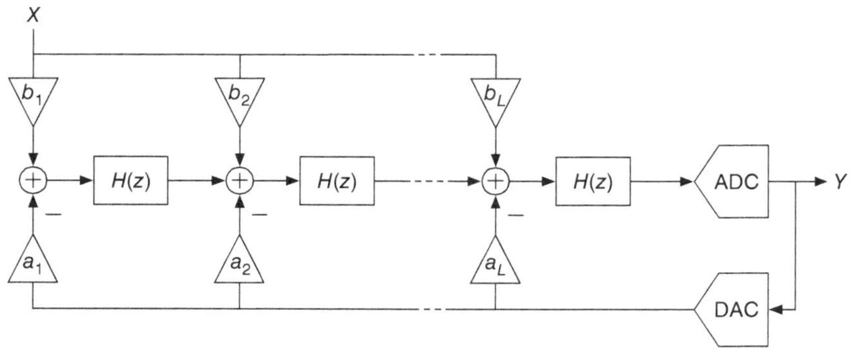

Figure 2.9 High-order single-loop ΣΔM with distributed feedback and distributed feedforward input paths.

In this architecture, the zeros of the STF can be fixed with coefficients b_i without affecting the pole placement, so both the STF and NTF can be separately optimized to some extent [24].

Local feedback loops – illustrated in Figure 2.10 for a single-loop ΣΔM with feedforward summation – can be included in both Figures 2.8 and 2.9, for creating the notches in the NTF, thus moving its zeros away from DC, and optimally spreading them over the signal band to minimize

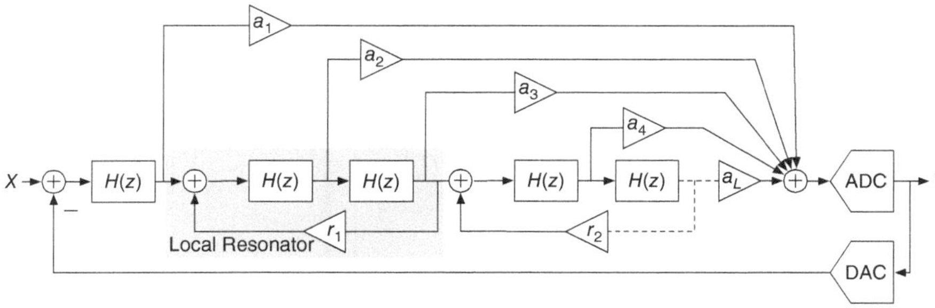

Figure 2.10 High-order single-loop ΣΔM with feedforward summation and local feedback loops.

the IBN [27]. Schreier has shown that the optimal placement of the complex-conjugate zeros for minimizing the total in-band error power is [27]:

$$\min\left[\int_0^{B_w} |\mathrm{NTF}(f)|^2 df\right] \tag{2.17}$$

This is illustrated in Figure 2.11, where alternatives for the NTF of a fourth-order single-loop ΣΔM are depicted. In this case, two notches are introduced in the signal band, corresponding to the two complex-conjugate zeros obtained from solving Equation (2.17) for OSR = 64. It can be shown that this approach minimizes the NTF more than when using either IIR or FIR NTFs, with their zeros placed at DC.

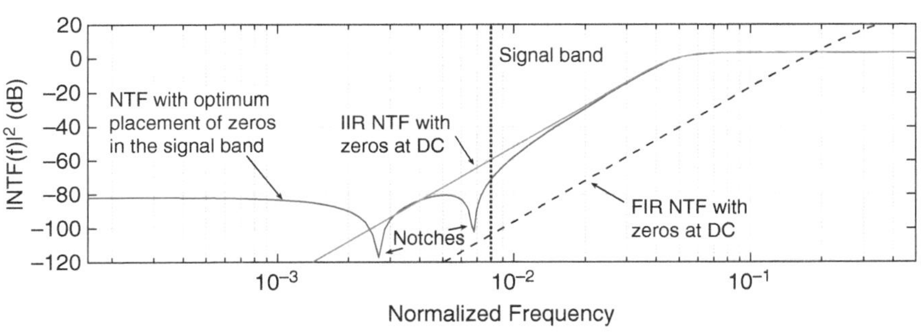

Figure 2.11 Illustrating different implementations of a fourth-order NTF with OSR = 64.

2.3 Cascade ΣΔ Modulators

As discussed in Section 2.2, stability problems arising from implementing a high-order NTF with a single-loop ΣΔM can be circumvented with adequate scaling coefficients, but this results in a significant decrease of the DR compared with an ideal ΣΔM. An alternative approach to obtaining a high-order noise-shaping while avoiding instabilities is found in the so-called *cascade* ΣΔMs, also known as *multiloop* ΣΔMs or *multi-stage noise-shaping* (MASH) ΣΔMs [28–31]. Their conceptual architecture is illustrated in Figure 2.12 and consists of N stages of ΣΔ modulators, in which each stage remodulates a scaled version of the quantization error generated in the preceding one. The outputs of the cascaded stages y_i are conveniently processed in the digital domain to cancel out, at

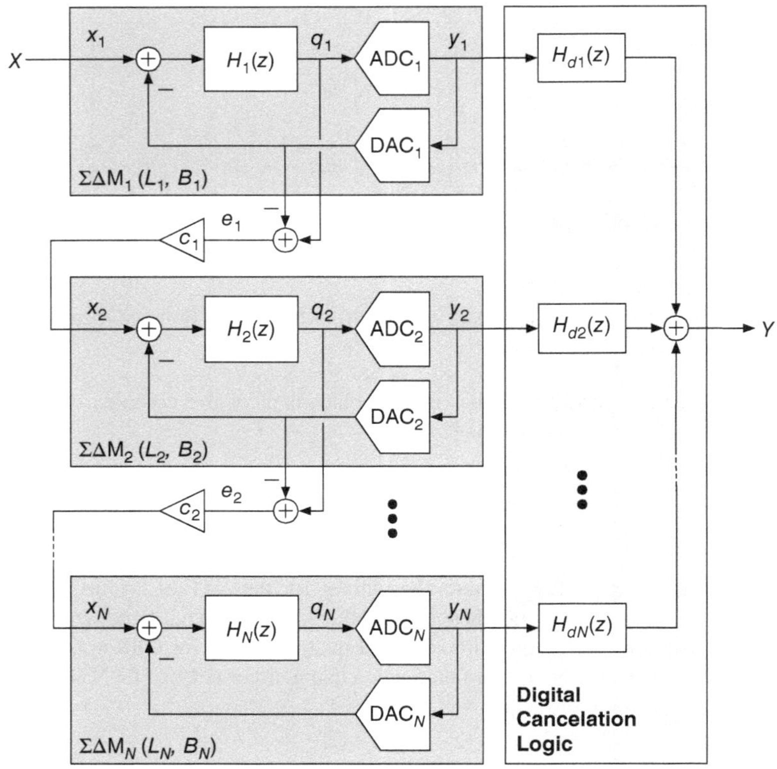

Figure 2.12 General topology of an N-stage cascade ΣΔ modulator.

the overall ΣΔM output y, all the quantization errors except that of the back-end stage. In addition, the latter quantization error appears at the cascade output, shaped with an order L equal to the sum of those of the cascaded stages ($L = L_1 + L_2 + \ \dots \ + L_N$). Unconditionally stable high-order shaping can thus be obtained if only first- and second-order ΣΔMs are cascaded ($L_i \leq 2$), since all feedback loops are local to the low-order ΣΔ stages and there is no interstage feedback. Therefore, the performance of a multi-stage ΣΔM is similar to that of an ideal high-order single-loop one, and there are no stability issues.[3]

The operating principle of cascade ΣΔMs can be easily understood by considering an example two-stage cascade. Using linear analysis, the stage outputs can be expressed in the Z-domain as,

$$
\begin{aligned}
Y_1(z) &= \mathrm{STF}_1(z)X_1(z) + \mathrm{NTF}_1(z)E_1(z) \\
Y_2(z) &= \mathrm{STF}_2(z)X_2(z) + \mathrm{NTF}_2(z)E_2(z)
\end{aligned}
\tag{2.18}
$$

[3]Cascade ΣΔMs can be *ideally* extended to an arbitrary number of stages. However, as will be shown in detail in Chapter 3, the effectiveness of cascading a large number of ΣΔ stages to achieve an arbitrary high-order noise shaping is limited in practice by circuit nonidealities, which preclude the complete cancelation of low-order-shaped quantization errors of the front-end ΣΔ stages at the modulator output. This effect is known as *noise leakage*.

where the inputs of the ΣΔ stages are given by $X_1(z) = X(z)$ and $X_2(z) = -c_1E_1(z)$, and $E_1(z)$ and $E_2(z)$ stand for the quantization errors of the respective stages. The overall modulator output after the digital cancelation logic (DCL) can thus be obtained from the expressions above as,

$$\begin{aligned} Y(z) &= H_{d1}(z)Y_1(z) + H_{d2}(z)Y_2(z) \\ &= \mathrm{STF}_{\mathrm{casc}}(z)X(z) + \mathrm{NTF}_{1,\mathrm{casc}}(z)E_1(z) + \mathrm{NTF}_{2,\mathrm{casc}}(z)E_2(z) \end{aligned} \tag{2.19}$$

where $\mathrm{STF}_{\mathrm{casc}}(z)$, $\mathrm{NTF}_{1,\mathrm{casc}}(z)$, and $\mathrm{NTF}_{2,\mathrm{casc}}(z)$ stand for the overall cascade transfer functions for the input signal and the quantization errors, respectively:

$$\begin{aligned} \mathrm{STF}_{\mathrm{casc}}(z) &= H_{d1}(z)\mathrm{STF}_1(z) \\ \mathrm{NTF}_{1,\mathrm{casc}}(z) &= H_{d1}(z)\mathrm{NTF}_1(z) - c_1H_{d2}(z)\mathrm{STF}_2(z) \\ \mathrm{NTF}_{2,\mathrm{casc}}(z) &= H_{d2}(z)\mathrm{NTF}_2(z) \end{aligned} \tag{2.20}$$

Note that, if the signal processing in the DCL matches part of the signal processing in the analog side, Equation (2.20) yields,

$$\left.\begin{aligned} H_{d1}(z) &= \mathrm{STF}_2(z) \\ H_{d2}(z) &= \frac{1}{c_1}\mathrm{NTF}_1(z) \end{aligned}\right\} \Rightarrow \left\{\begin{aligned} \mathrm{STF}_{\mathrm{casc}}(z) &= \mathrm{STF}_1(z)\mathrm{STF}_2(z) \\ \mathrm{NTF}_{1,\mathrm{casc}}(z) &= 0 \\ \mathrm{NTF}_{2,\mathrm{casc}}(z) &= \frac{1}{c_1}\mathrm{NTF}_1(z)\mathrm{NTF}_2(z) \end{aligned}\right. \tag{2.21}$$

so that the first-stage quantization error is canceled at the overall output and that of the second stage is attenuated, with an order equal to the sum of the orders of the two stages ($L = L_1 + L_2$).

As an illustration, Figure 2.13 shows the output spectrum of a two-stage cascade ΣΔM made up of two first-order single-loop ΣΔMs, such as the one shown in Figure 1.15; in other words, $L_1 = L_2 = 1$. Note that the noise-shaping of the cascade modulator is equivalent to that obtained by a second-order single-loop ΣΔM, while the tonal behavior of the front-end stage quantization error is cancelled at the output of the cascade. In practice, however, this cancellation is not perfect, thus leading to a more nonlinear output spectrum because of the strong correlation between the input signal and the quantization error in the front-end stage.

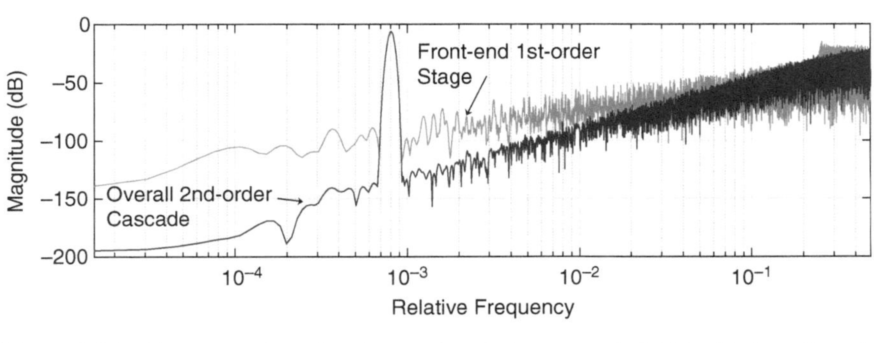

Figure 2.13 Illustrating the noise-shaping in a 2-stage cascade ΣΔM with $L_1 = L_2 = 1$.

For the generic N-stage cascade ΣΔM in Figure 2.12, assuming $\text{STF}_i(z) = z^{-L_i}$ and $\text{NTF}_i(z) = (1 - z^{-1})^{L_i}$ in the individual stages, the overall modulator output thus yields,

$$Y(z) = z^{-L}X(z) + \frac{1}{\prod_{i=1}^{N-1} c_i}(1 - z^{-1})^L E_N(z), \text{ with } L = \sum_{i=1}^{N} L_i \tag{2.22}$$

under the required matching between the analog processing in the ΣΔ stages and the DCL. The in-band quantization noise of the cascade is then given by,

$$\text{IBN}_{\text{casc}} = \frac{\Delta_N^2}{12} \frac{1}{\prod_{i=1}^{N-1} c_i^2} \frac{\pi^{2L}}{(2L+1)\text{OSR}^{2L+1}} \tag{2.23}$$

where Δ_N stands for the quantization step employed in the last ΣΔ stage. Note from Equation (2.23) that only the interstage scaling coefficients c_i, which prevent premature overload of the subsequent stages, decrease the performance below that of an ideal Lth order ΣΔM. Typical values of $1/\prod_{i=1}^{N-1} c_i$ are between 2 and 4 if single-bit quantization is employed. This leads to a reduction in the ideally attainable DR of only 6–12 dB (1–2 bit). These performance losses are inherent to cascade ΣΔMs, but they are significantly smaller than those resulting from optimized high-order single loops. In addition, they are independent of the OSR.

The aforementioned benefits of cascade ΣΔMs have favored the development of a large number of different topologies:

- 2-1 ΣΔM, a third-order two-stage ΣΔM built up with a second-order stage followed by a first-order one [29] – it is also known as a *SOFO cascade*, standing for second-order first-order;
- 2-2 ΣΔM, which represents a fourth-order cascade built up with two second-order stages [32, 33];
- 2-1-1 ΣΔM, a fourth-order three-stage cascade [7];
- 2-2-1 ΣΔM [34];
- 2-1-1-1 ΣΔM [35];
- 2-2-2 ΣΔM [36];
- and others.

Note from the cascade topologies above that the first stage is usually a second-order ΣΔM and that first-order ΣΔMs are avoided at the front-end [31]. The reason for this is twofold. First, the quantization error from the first stage would be only first-order shaped, and noise leakages would be larger. Second, the tonal behavior of a first-order first stage would degrade the performance of the cascade, as discussed earlier. In addition, although it is low-order ΣΔ stages that are most frequently cascaded, 3-1 and 3-2 ΣΔMs have also been reported [37, 38].

Figure 2.14 illustrates the block diagram of a 2-1-1 cascade, in which coefficients a_i represent the in-loop integrator scaling factors, and coefficients b_i and c_i determine the interstage scaling factors.

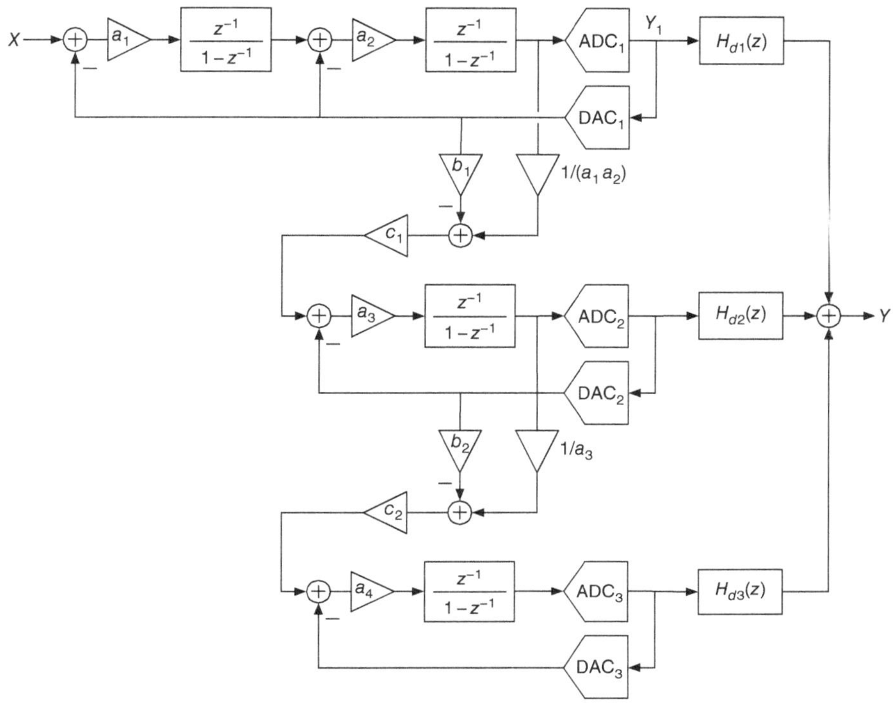

Figure 2.14 Block diagram of a 2-1-1 cascade ΣΔ modulator using the notation in [1, 2]. The relations in Equations (2.24) and (2.25) must be fulfilled for the correct operation of the cascade.

The first stage performs as an ideal second-order ΣΔM and the second and third stages as an ideal first-order ΣΔM under the assumptions,

$$\begin{aligned} &k_{q1}a_1a_2 = 1, \; k_{q1}a_2 = 2 \\ &k_{q2}a_3 = 1 \\ &k_{q3}a_4 = 1 \end{aligned} \tag{2.24}$$

where k_{qi} stands for the gain of the quantizer in the ith ΣΔ stage. The matching required between the signal processing in the ΣΔ stages and the digital processing in the DCL leads to,

$$\begin{aligned} H_{d1}(z) &= z^{-2}[1 + (b_1 - 1)(1 - z^{-1})^2][1 + (b_2 - 1)(1 - z^{-1})^3] \\ H_{d2}(z) &= \frac{1}{c_1}z^{-1}(1 - z^{-1})^2[1 + (b_2 - 1)(1 - z^{-1})^3] \\ H_{d3}(z) &= \frac{1}{c_1c_2}(1 - z^{-1})^3 \end{aligned} \tag{2.25}$$

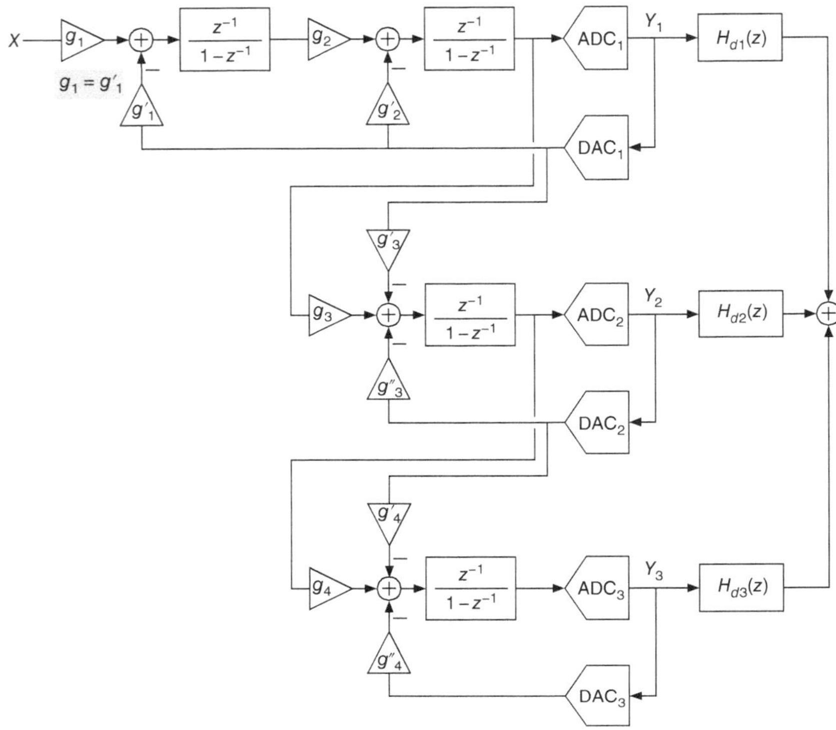

Figure 2.15 Alternative representation of the 2-1-1 ΣΔM in Figure 2.14 using the notation in [3, 4]. Modulator coefficients (a_i, b_i, c_i) are mapped onto integrator input coefficients (g_i, g_i', g_i''), which are closer to the circuit-level implementation.

for a complete cancelation of the first- and second-stage quantization errors at the cascade output. For the sake of completeness, Figure 2.15 shows an alternative representation of the 2-1-1 ΣΔM according to the notation in [3, 4]. Proceeding in a similar way as in Figure 2.3 for the second-order ΣΔM, the two notations are easily connected with the equalities:

$$
\begin{aligned}
a_1 &= \frac{g_1' g_2}{g_2'}, \ a_2 = g_2', \ a_3 = g_3'', \ a_4 = g_4'' \\
b_1 &= \frac{g_3'}{g_1' g_2 g_3}, \ b_2 = \frac{g_4'}{g_3'' g_4} \\
c_1 &= \frac{g_1' g_2 g_3}{g_3''}, \ c_2 = \frac{g_3'' g_4}{g_4''}
\end{aligned}
\tag{2.26}
$$

The value of the analog coefficients in the 2-1-1 ΣΔM need to be selected to fulfill the relations in Equation (2.24), but many different sets of values can do the work. In addition, several extra considerations are taken into account in practice:

- The resulting loss of performance is minimized in comparison to an ideal ΣΔM.
- The modulator overload level is maximized to achieve high peak SNR.

- The circuit-level implementation of the set of analog coefficients is simplified: for SC-ΣΔMs this means taking into account practical capacitor ratios using unit elements, enabling capacitor sharing within the integrators, reducing the total number of unit capacitors for area saving, and so on.
- The integrator output swing requirements are minimized, especially in case of a low-voltage supply.
- The implementation of the DCL is simplified; to that end, power-of-two coefficients are often preferred for b_1, b_2, $1/c_1$ and $1/c_2$ – see Equation (2.25) – in order to use shift registers only.

Table 2.2 illustrates some sets of analog coefficients reported for the 2-1-1 single-bit ΣΔM, together with the main resulting features. Optimized coefficients for some other cascade topologies can be found in the literature [1].

Figure 2.16 illustrates the noise shaping in the different stages of Figure 2.15, for the optimized coefficients of del Rio *et al.* [39]. All stages in the cascade use a 1-bit embedded quantizer, except for the last-stage, which includes a 3-bit quantizer, thus further reducing the in-band noise power, as will be discussed in Section 2.4.

Table 2.2 Comparison of some sets of coefficients reported for the 2-1-1 single-bit ΣΔM.

	[39]	[7]	[9]
g_1, g_1'	$1/4, 1/4$	$1/4, 1/4$	$1/3, 1/3$
g_2, g_2'	$1, 1/2$	$1/2, 1/4$	$3/5, 2/5$
g_3, g_3', g_3''	$1, 1/2, 1/2$	$1, 3/8, 2/8$	$5/6, 3/6, 2/6$
g_4, g_4', g_4''	$1, 1/2, 1/2$	$1, 1/4, 1/4$	$1, 1/3, 1/3$
a_1, a_2, a_3, a_4	$0.5, 0.5, 0.5, 0.5$	$0.5, 0.25, 0.25, 0.25$	$0.5, 0.4, 0.33, 0.0.33$
b_1, c_1, b_2, c_2	$2, 0.5, 1, 1$	$3, 0.5, 1, 1$	$3, 0.5, 1, 1$
$1/(c_1c_2)$	2	2	2
Loss of DR	−6 dB	−6 dB	−6 dB
Overload level	−3 dBFS	−2.5 dBFS	−2 dBFS
Integ. output swing	$\pm(0.75, 1, 1, 1)V_{\text{ref}}$	$\pm(0.75, 0.7, 0.6, 0.6)V_{\text{ref}}$	$\pm(1, 1, 0.9, 0.8)V_{\text{ref}}$
Unit capacitors (2×)	$17(= 5 + 4 + 4 + 4)$	$35(= 5 + 6 + 16 + 8)$	$29(= 4 + 8 + 11 + 6)$

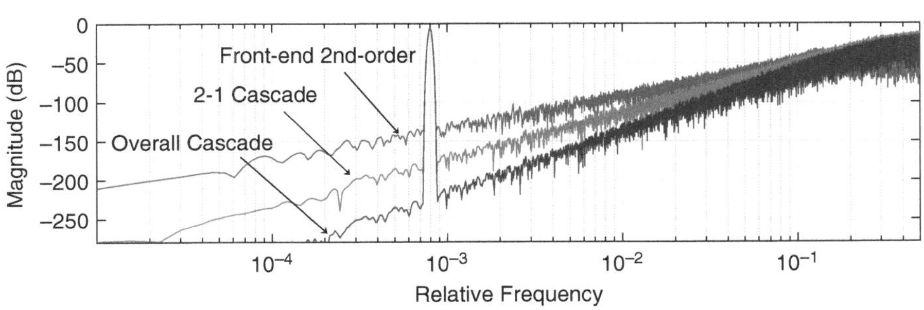

Figure 2.16 Illustrating the noise-shaping in a 2-1-1 cascade ΣΔM with 3-bit quantization in the back-end stage and the loop-filter coefficients of del Rio *et al.* [39].

2.3.1 SMASH ΣΔM Architectures

One of the main limitations of cascade ΣΔMs is their higher sensitivity to mismatch than their single-loop counterparts. This limitation has motivated ΣΔ designers to look for alternative implementations of cascade ΣΔMs that reduce sensitivity to noise leakages. This is the case of the so-called *sturdy* MASH (SMASH) architectures [40, 41]. The main idea behind this kind of cascade ΣΔMs consists of replacing the DCL with the analog signal processing provided by the loop-filter integrators themselves, together with an additional inter-stage digital feedback path. As a result, the output of the modulator can be obtained from a direct digital addition (or subtraction) operation of the different stage outputs in the cascade – with no need for a DCL as in conventional cascades – and the subsequent elimination of matching requirements between the analog and the digital filtering.

Figure 2.17 shows the conceptual block diagram of a two-stage SMASH ΣΔM.[4] Assuming a linear model for the embedded quantizers, the Z-transform of the modulator output can be written as [41],

$$Y(z) = \mathrm{STF}_1(z)X(z) + \mathrm{NTF}_1(z)[1 - \mathrm{STF}_2(z)]E_1(z) - \mathrm{NTF}_1(z)\mathrm{NTF}_2(z)E_2(z) \tag{2.27}$$

where NTF_i and STF_i stand for the NTF and STF of the ith stage in the cascade, respectively. Note that the overall NTF of the cascade modulator – that is, the NTF filtering the second-stage quantization error $E_2(z)$ – is the same as that of a conventional cascade ΣΔM. However, the first-stage quantization error is not completely cancelled. Instead, it is filtered by $\mathrm{NTF}_1(z)[1 - \mathrm{STF}_2(z)]$.

Figure 2.18 illustrates how to implement the concept of SMASH ΣΔMs in a fourth-order 2-2 topology [41]. Note that the DCL is replaced by direct feedback paths from the output of the second stage to the input of the first stage, although an additional feedback DAC (DAC_3 in Figure 2.18) is required. Analyzing Figure 2.18 with linear models for the embedded quantizers, it is easy to show

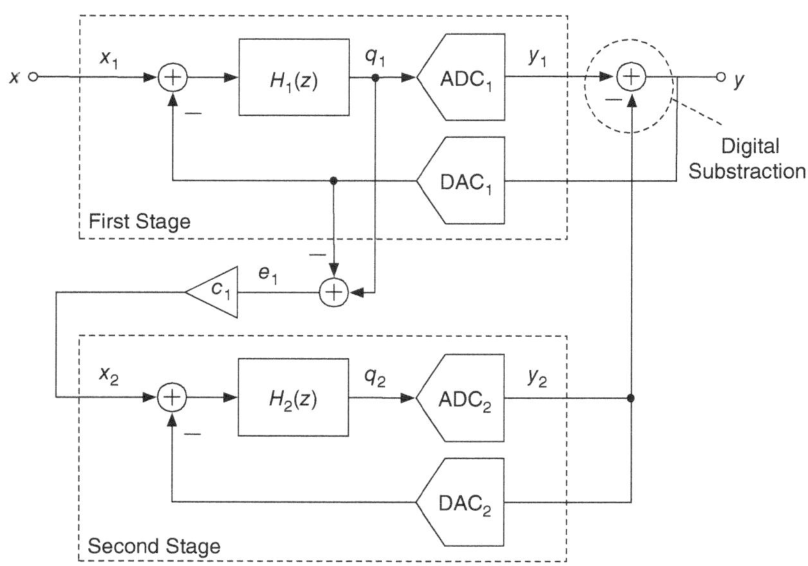

Figure 2.17 Conceptual block diagram of a two-stage SMASH ΣΔM [40, 41].

[4]The concept of SMASH ΣΔMs has also been applied to multi-stage cascade architectures [42].

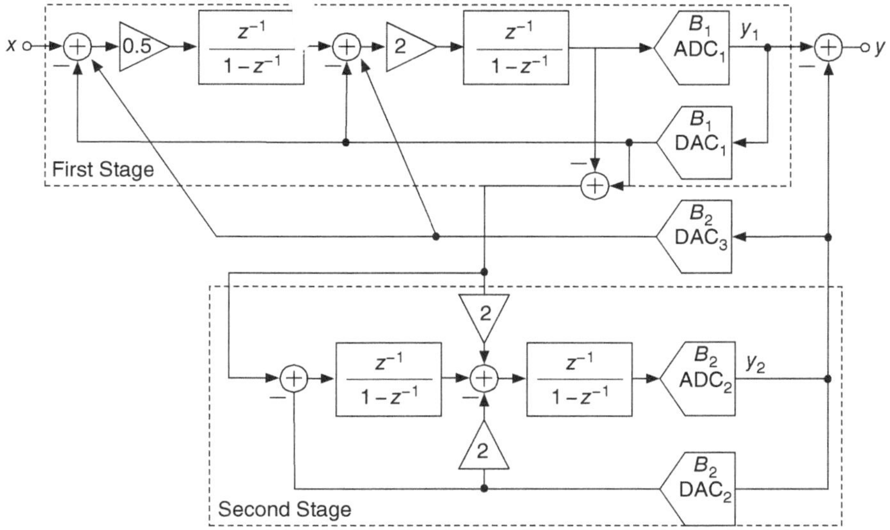

Figure 2.18 Block diagram of a 2-2 SMASH ΣΔM [41].

that the Z-transform of the modulator output is given by:

$$Y(z) = z^{-2}X(z) + (1 - z^{-1})^4 E_1(z) - (1 - z^{-1})^4 E_2(z) \tag{2.28}$$

Note from the above expression that fourth-order high-pass filtering is obtained for both $E_1(z)$ and $E_2(z)$. In this case, $\text{STF}_2(z) = 2z^{-1} - z^{-2}$, which according to Equation (2.27), does not give a complete cancellation of $E_1(z)$. However, $E_1(z)$ can be ideally cancelled if $\text{STF}_2 = 1$; in other words, if USTF is used in the second stage of the SMASH topology. Indeed, the performance of SMASH ΣΔMs can be markedly improved if a USTF is implemented in all stages of the modulator [43]. As stated in Section 2.1.2, the main advantage of using a USTF is that integrators ideally process quantization error only and so their output swing can be reduced and the tolerance to some signal-dependent errors – such as amplifier nonlinearities – is increased.[5]

Figure 2.19 shows a ΣΔM architecture that extends the underlying principle of SMASH ΣΔMs to the implementation of USTFs in both stages, thus taking advantage of both strategies (SMASH and USTF) to achieve higher resolution in wideband applications with lower values of OSR. In this example, the Z-transform of the modulator output is given by:

$$Y(z) = X(z) - \frac{1}{d}(1 - z^{-1})^4 E_2(z) \tag{2.29}$$

Note that, contrary to conventional SMASH ΣΔMs, the quantization error in the front-end stage is ideally cancelled, while avoiding any digital filtering. In addition, the use of a scaling factor d (which is usually implemented as a power of 2) will further reduce the in-band noise power of the second-stage quantization error at the output. Moreover, using USTFs in both stages of the modulator allows for relaxation of the output swing and gain demands in the amplifiers used to build the

[5]The use of USTFs in continuous-time (CT) ΣΔMs can degrade the implicit antialias filter implemented by these kinds of ΣΔMs – as discussed in Section 2.6.

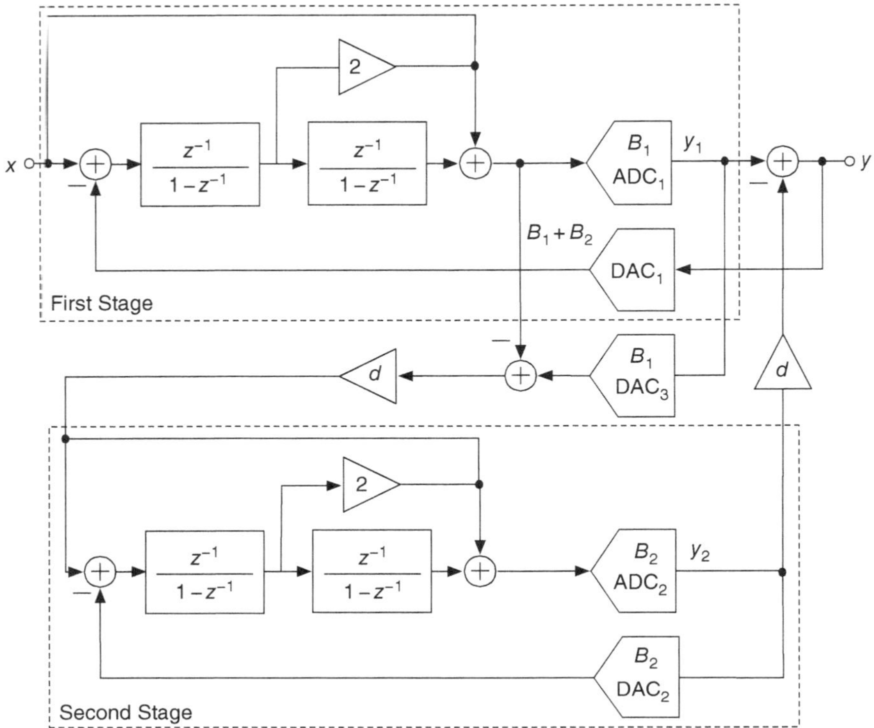

Figure 2.19 SMASH ΣΔM with USTF [43].

loop-filter integrators, while having better sensitivity to mismatch than simple SMASH and cascade topologies. However, the price to pay is that a front-end DAC with a full scale larger than that of the quantizers in the cascade (with resolutions B_1 and B_2) is required to account for the summation of the digital outputs of the stages. Nevertheless, the location of the digital adder helps to increase considerably the robustness to mismatches in the loop-filter coefficients – implemented in practice by capacitor ratios. As an illustration, Figure 2.20 shows a Monte Carlo simulation of Figure 2.19, considering $d = 4$, OSR= 16 and a standard deviation of 0.1% in all loop-filter coefficients. It can be seen that the architecture in Figure 2.18 is more robust against mismatch than the conventional SMASH ΣΔM shown in Figure 2.17 [43].

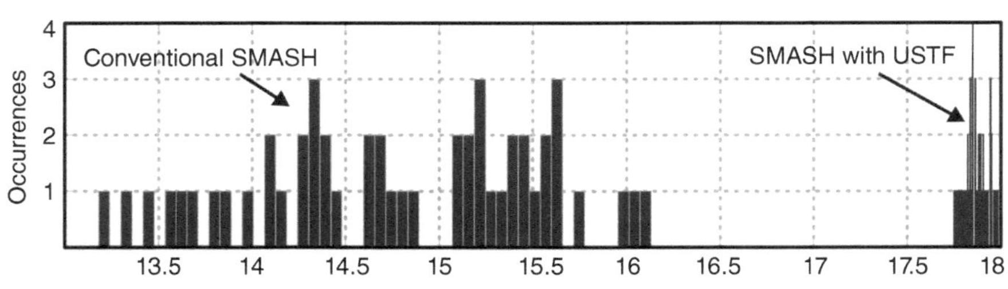

Figure 2.20 Effect of mismatch on the effective resolution of SMASH ΣΔMs [43].

2.4 Multi-bit ΣΔ Modulators

As discussed in Chapter 1, the dynamic range of a ΣΔM can be enhanced if the resolution of the embedded quantizer is increased. The main advantages of resorting to multi-bit ΣΔ modulators are:

- The in-band quantization noise power is reduced by approximately 6 dB per additional bit in the embedded quantizer, thanks to the smaller quantization step Δ.
- Internal nonlinearities are weaker in multi-bit ΣΔMs than in their single-bit counterparts. The quantizer operation better fits the additive white noise model in Section 1.1.3, and phenomena caused by nonlinear dynamics are less evident.
- For a given order in the loop filter, the stability properties of multi-bit ΣΔMs are better than for single-bit ΣΔMs [44].

These benefits suggest that, for targeted performance of the ΣΔM, multi-bit quantization can be traded for noise shaping and/or oversampling. Indeed, multi-bit ΣΔMs are often employed in broadband applications to compensate for the limited OSR. Nevertheless, multi-bit quantizers also have important drawbacks that may counter the advantages:

- They require more analog circuitry and are more difficult to design than single-bit ones.
- In contrast to 1-bit quantizers, which are *intrinsically* linear because only two levels are used for quantization, multi-bit quantizers exhibit in practice some nonlinearities in their transfer characteristics, mostly due to device *mismatching*, which significantly influences the ΣΔM performance.

2.4.1 Influence of Multi-bit DAC Errors

Figure 2.21 illustrates an enhanced version of the linear model of the multi-bit ΣΔM in Figure 1.12b. Errors related to the multi-bit conversion are added to the quantization error e that has been considered so far; namely, an error e_{ADC} associated with the A/D conversion and an error e_{DAC} in the subsequent D/A conversion required to reconstruct the analog feedback signal. Note that e_{ADC} is injected in the same path as the quantization error e and therefore it too is attenuated within the signal band by noise shaping. However, DAC errors are injected in the feedback path and therefore they directly add to the ΣΔM input signal and pass to the ΣΔM output as part of the input signal itself. Consequently, the linearity of a multi-bit ΣΔM will be *no better* than that of a multi-bit embedded DAC and the latter must be designed to achieve the linearity targeted for the whole ΣΔ ADC, which may be challenging under the influence of component mismatching.

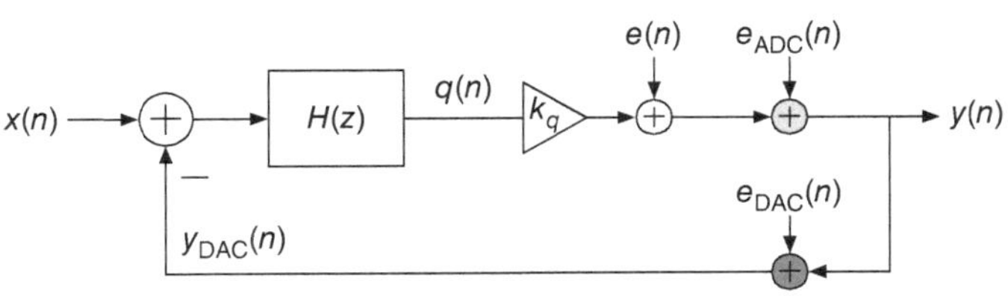

Figure 2.21 Linear model of a multi-bit ΣΔM including errors in the embedded ADC and DAC.

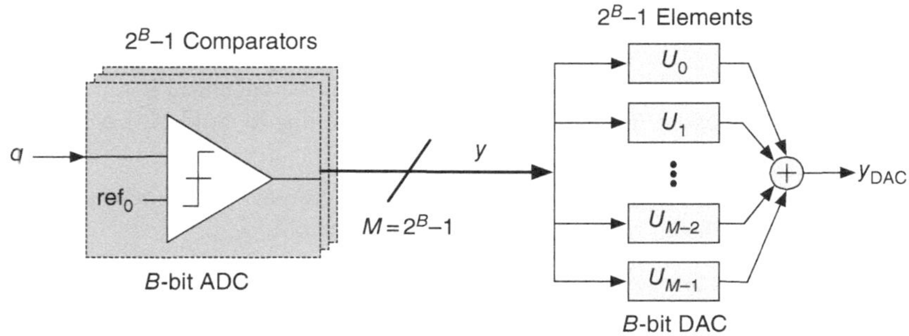

Figure 2.22 Parallel topology of a typical multi-bit quantizer embedded in a ΣΔM.

Figure 2.22 conceptually illustrates the parallel architecture that is typically used for the multi-bit quantizer in ΣΔMs, in which the resolution is usually low ($B \leq 5$). The B-bit ADC consists of a bank of $2^B - 1$ comparators; this digitizes the loop filter output into thermometer code, which will be subsequently coded into binary. The DAC employs $2^B - 1$ unit elements (capacitors, resistors, current sources, and so on, depending on the circuit implementation) to reconstruct the analog feedback signal using 2^B levels (numbered from 0 to $M = 2^B - 1$). The ith analog output level is generated by activating i unit elements and adding their outputs (charges or currents). DAC errors are caused by the mismatch between its unit elements, which makes the DAC output levels deviate from their nominal values. Assuming that the actual value of each unit element follows a Gaussian distribution, the worst-case relative error in the DAC output y_{DAC} can be estimated as,

$$\sigma\left(\frac{\Delta y_{\mathrm{DAC}}}{y_{\mathrm{DAC}}}\right) \approx \frac{1}{2\sqrt{2^B}}\,\sigma\left(\frac{\Delta U}{U}\right) \tag{2.30}$$

where $\sigma(\Delta U/U)$ stands for the relative error in the value of the unit element. Obviously, the DAC accuracy increases with the number of unit elements thanks to the parallel topology. However, for a ΣΔM with 4-bit embedded quantization to achieve 16-bit linearity, the required matching for the DAC unit elements should be better than 0.01% (13 bits). Device matching achieved in most nanometer CMOS processes is nevertheless in the range of 0.1% (10 bits) and the required accuracy in the elements can only be obtained through the parallel connection of many more of them (~ 64). This means that achieving linearities better than 12 or 13 bits in multi-bit ΣΔMs by means of relying only on *standard* device matching usually leads to prohibitive area occupation.

A direct method to improve the standard device matching is laser trimming, which can sometimes be done at the foundry, but at the expense of additional fabrication and/or measurement steps and increased cost. Calibration and correction schemes have also been proposed, either in the analog or digital domain [45], but these are often expensive to implement in terms of system design complexity, hardware requirements, and power consumption.

Among the different alternatives that have been developed through the years for achieving highly linear multi-bit ΣΔMs, two approaches clearly prevail because of the modest component matching required and the reduced circuit complexity involved. These approaches are discussed next.

2.4.2 Dynamic Element Matching Techniques

As previously discussed, mismatches between the unit elements cause DAC nonlinearities that generate harmonic distortion in the ΣΔM. For a multi-bit DAC with the topology in Figure 2.22, there is a univocal correspondence between the thermometric input code (y) and the respective error of the DAC output (y_{DAC}), because the *same* unit elements are always used for generating a given DAC output level. The operating principle of dynamic element matching (DEM) is to break this direct correspondence by varying over time the set of elements that are employed for generating a given DAC level, thus transforming its fixed error into a *time-varying* one. To that end, as conceptually illustrated in Figure 2.23, a digital block is incorporated. This controls the selection of unit elements at each clock cycle according to an algorithm that tries to null the average error in each DAC level over time. In this way, the part of the DAC error power that was in the low-frequency range is pushed to higher frequencies and removed by the decimator.

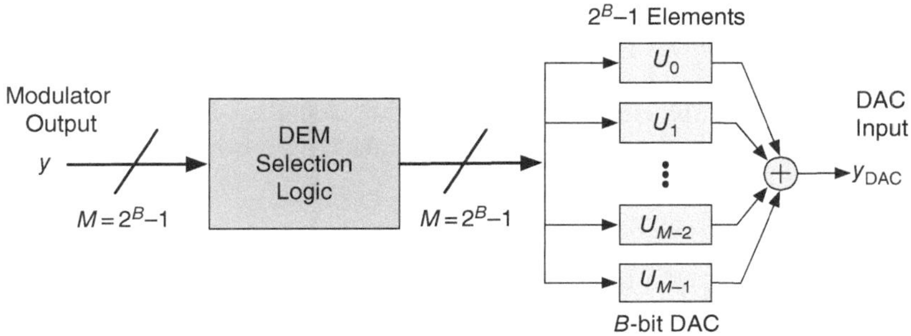

Figure 2.23 Incorporating an element selection logic for applying DEM to a multi-bit DAC.

The incorporation of DEM techniques in multi-bit ΣΔMs is facilitated by digitally-oriented CAD tools and represents only a little area overhead in the modern CMOS process, which explains the large number of algorithms that have been developed in this area over the years. A detailed overview of many of them can be found in the literature [2, 45, 46]. They can be categorized as follows:

- **Randomization algorithms**, in which the DAC unit elements are selected according to pseudo-randomly configured networks (e.g., butterfly structures) [47]. Harmonic distortion induced by the DAC is transformed into white noise, whose out-of-band power will be removed by the decimation filter. The DAC error power lying within the signal band will nevertheless increase the noise floor of the ΣΔM.
- **Rotation algorithms**, in which the DAC unit elements are selected in a periodic manner for shifting harmonic distortion out of the signal band. The ΣΔM noise floor is not increased, but the signal processing may cause mixed frequency components that fold over the modulator passband. *Clocked averaging* (CLA) [48] is an example of this kind of DEM technique.
- **Mismatch-shaping algorithms**, in which the DAC unit elements are selected according to algorithms that make the mismatching error push most of its power to higher frequencies. The order of the mismatch shaping is normally limited to one or two. *Individual level averaging* (ILA) [49] and *data weighted averaging* (DWA) [50] – and its many modifications – are examples of this kind of algorithm.
- **Vector-quantizer structures**, in which a digital ΣΔ converter in error-feedback configuration is incorporated to achieve high-order shaping [51].

Among the many different DEM alternatives, DWA [50] – or a modified version of DWA – dominates in high-speed high-resolution ΣΔ modulators with multi-bit quantization. In such high-speed ΣΔMs, the complexity of the DEM algorithm becomes a concern because the delay introduced by the DEM selection logic in the ΣΔ feedback loop can limit the maximum achievable clock frequency. In this regard, DWA results in highly practical implementations, especially when the number of DAC elements is large [46].

Consider an M-element DAC, as shown in Figure 2.23, with input code $y(n)$. In conventional DWA [50], the DAC unit elements are selected sequentially from the DAC array, beginning with the next available unused element. An index pointer ptr stores the address of that element in a digital register and manages the rotational element selection process, so that the DAC elements selected at time n are those from $ptr(n)$ to $[ptr(n) + y(n) - 1] \bmod M$, in increasing order. Every clock cycle, the index pointer is incremented by modulus M by the DAC input code $y(n)$ according to

$$ptr(n+1) = [ptr(n) + y(n)] \bmod M, \quad 0 \leq ptr(n) \leq M-1 \tag{2.31}$$

Figure 2.24 illustrates how the elements of a 4-bit DAC would be selected following the DWA algorithm above, which achieves a *first-order shaping* of the DAC mismatch errors.

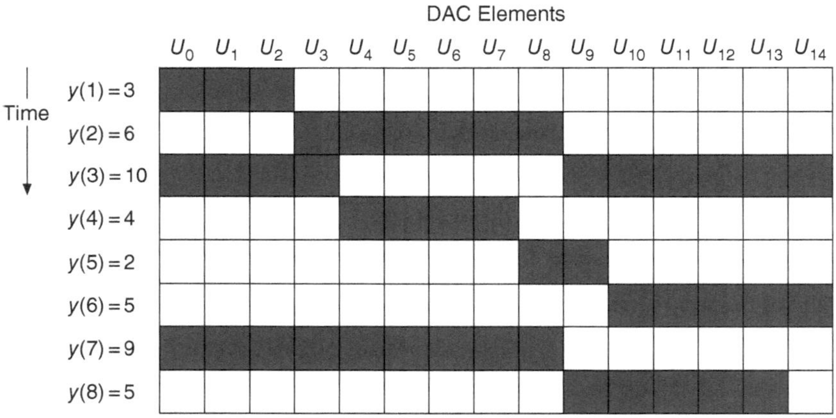

Figure 2.24 Selection of unit elements in a 15-element DAC according to the DWA algorithm in Equation (2.31). The shaded boxes indicate the elements that contribute to generate the DAC output level $y_{\mathrm{DAC}}(n)$ for the corresponding input code $y(n)$.

However, if the ΣΔM input is DC or varies slowly, and there is a rational relationship between its value and the number of DAC elements, out-of-band tones are generated that may fold back to the signal band due to the modulation by the ΣΔM's output waveform [50]. Several modifications of the conventional DWA algorithm have been proposed to reduce its *tonal behavior*, such as

- *rotated* DWA (R-DWA) [52]
- *randomized* DWA (Rn-DWA) [53]
- *bidirectional* DWA (Bi-DWA) [54]
- *partitioned* DWA (P-DWA) [34]
- *pseudo* DWA [55].

Of these, pseudo DWA significantly decreases the tonal power with a minor modification of the DWA algorithm. It modifies the DWA scheme by inverting with periodicity n_{inv} the least-significant bit (LSB) of the DAC input code $y(n)$ used to update the index pointer in Equation (2.31). The element-selection process is thus essentially the same as in conventional DWA except that, every n_{inv} clock cycles, a DAC element is either reselected or skipped depending on whether the previous DAC input code was odd or even, respectively. This modification of the DWA algorithm breaks the cyclic nature of the element-selection process and so reduces the tonal behavior [55]. The choice of n_{inv} is, however, a compromise between linearity and resolution. If n_{inv} is too large – note that $n_{\text{inv}} = \infty$ corresponds to conventional DWA – the signal-dependent tones will not be eliminated. If n_{inv} is too small, different DAC elements will be used at significantly different rates, resulting in an increase of the in-band noise. Deriving an analytical expression for the optimum value of n_{inv} is rather complex, but simple behavioral simulations can be used to find an appropriate value for a given multi-bit ΣΔM [55].

As an illustration, let us consider a second-order feedforward ΣΔM with USTF and 4-bit quantization. Assuming that the loop-filter is implemented by SC circuits and considering a capacitor mismatch of 1%,[6] the performance of the modulator can be severely degraded, as illustrated in Figure 2.25, if a DWA algorithm is not embedded in the modulator feedback DAC.

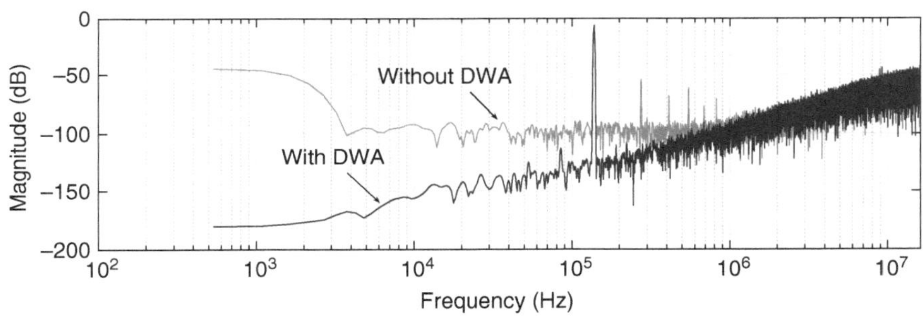

Figure 2.25 Illustrating the action of DWA in the noise-shaping performance of a second-order feedforward SC-ΣΔM with USTF and 4-bit quantization and $f_s = 35.2$MHz.

2.4.3 Dual Quantization

Rather than reducing the DAC mismatch errors that are injected in the feedback of a multi-bit ΣΔM – as DEM techniques do – dual-quantization basically involves injecting them at a *different point*, where their influence on the overall ΣΔM linearity is not severe. The operation principle of dual-quantization is to employ both single- and multi-bit quantization at a time in a ΣΔM: two-level quantization for its intrinsic linearity and multi-bit quantization for its reduced error power. The concept is applicable to both single-loop and cascade ΣΔMs.

2.4.3.1 Dual-quantization Single-loop ΣΔMs Figure 2.26 shows a third-order single-loop ΣΔM with dual quantization [56], in which the first two integrators are fed by a two-level DAC, whereas the third one is fed by a multi-bit DAC. The outputs of the corresponding ADCs are digitally processed to cancel the *coarse* 1-bit quantization error at the overall output, in which only the *fine* multi-bit quantization error ideally remains. The modulator linearity is not threatened, since DAC

[6]This is an exaggerated value in practice, but it is used here for illustration purposes.

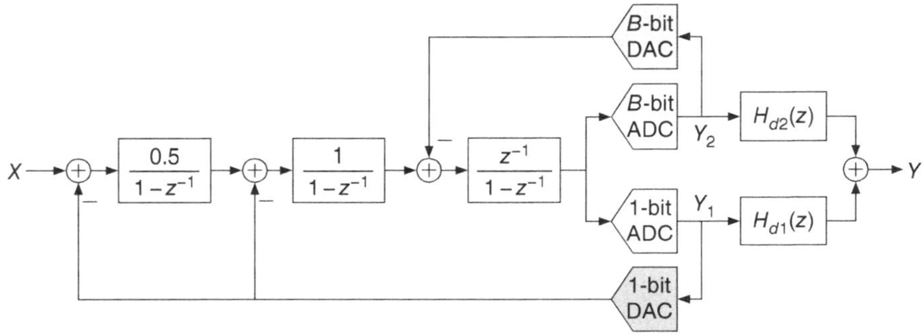

Figure 2.26 A third-order single-loop ΣΔM employing dual quantization [56].

mismatch errors are suppressed by the gain of the first two integrators and are thus second-order high-pass shaped. The third-order ΣΔM also benefits from improved stability thanks to the multi-bit feedback in the back-end integrator. Note, however, that the topology requires a digital cancelation logic and thus suffers from noise leakage.

The concept in Figure 2.26 can be generalized to higher-order ΣΔMs. As the order of the loop filter increases, the number of back-end integrators with multi-bit feedback is a trade-off between aggressive noise shaping (to improve stability) and linearity requirements of the multi-bit DAC.

2.4.3.2 Dual-quantization Cascade ΣΔMs

Dual-quantization schemes are *naturally* incorporated in cascade ΣΔMs [57]. As shown in Section 2.3, the output of a cascade ΣΔM ideally contains only the input signal and the last-stage quantization error, whereas the quantization errors from the remaining stages are removed by the DCL. The dynamic range of the ΣΔM can be thus easily increased using multi-bit quantization only in the back-end stage. The remaining quantizers can be single-bit, so as to retain linear feedback in the front-end stages. Therefore, the resulting topology is as illustrated in Figure 2.12, with $B_i = 1$ $(i = 1, \dots, N-1)$ and $B_N = B$. In this way, nonlinearities in the multi-bit DAC will appear at the overall cascade output with a shaping of order equal to the sum of the order of the preceding ΣΔ stages, so that Equation (2.22) yields

$$Y(z) = z^{-L}X(z) + \frac{1}{\prod_{i=1}^{N-1} c_i}(1 - z^{-1})^L E_N(z) + \frac{1}{\prod_{i=1}^{N-1} c_i}(1 - z^{-1})^{(L-L_N)} E_{\mathrm{DAC}}(z), \text{ with } L = \sum_{i=1}^{N} L_i \tag{2.32}$$

if the error in the back-end multi-bit DAC $E_{\mathrm{DAC}}(z)$ is taken into account in the linear analysis. Many cascade ΣΔM integrated circuits using this dual-quantization scheme can be found in the literature [58–60].

Many cascade ΣΔMs employing multi-bit quantization in all stages have also been reported [34, 54]. Note that, under ideal conditions, the quantization errors of all but the last stage are canceled at the overall cascade output by the DCL. Use of multi-bit quantization in these stages has a twofold purpose:

- For reducing the power of the respective quantization errors that will in practice leak to the output. DEM techniques can be employed in the front-end ΣΔ stage for achieving the targeted modulator linearity, whereas the remaining multi-bit stages often rely just on the shaping provided by the preceding integrators [34].

- For increasing the value of the interstage scaling coefficients c_i in the ΣΔ cascade – see Figure 2.12 – in comparison with a 1-bit approach, while avoiding overloading [54]. In this way, the factor $1/\prod_{i=1}^{N-1} c_i$ that amplifies the last-stage quantization error in Equation (2.32) can be made smaller than unity, which increases the dynamic range over that ideally attainable with an Lth-order B-bit ΣΔM.

Cascade ΣΔMs using trilevel (1.5-bit) quantizers [36] are also often employed, since they yield a 3-dB reduction of the quantization error power compared to 1-bit quantization. Although trilevel coding is not inherently linear, it is often used in fully-differential SC ΣΔMs, since highly-linear trilevel DACs can be easily implemented using just one extra switch [61].

2.5 Band-pass ΣΔ Modulators

The operating principle of low-pass ΣΔ modulators (LP-ΣΔMs) – in which the quantization noise is high-pass shaped in order to give suppression around DC – can be extended to the more general case of stop-band filtering of the quantization noise in order to give suppression around a *nonzero* frequency in the so-called *band-pass* ΣΔ modulators (BP-ΣΔMs) [62, 63]. Direct application of this band-pass approach can be found in the A/D conversion of many wireless receiver systems, a BP-ΣΔM being used for the digitization of *intermediate frequency* (IF) signals [60, 64]. The application of BP-ΣΔMs has also been extended to the direct digitization of *radio frequency* (RF) signals. Although RF ΣΔ ADCs were initially implemented in SiGe processes [65, 66], they are already a reality in CMOS technologies too [67–75].[7]

Although a broadband Nyquist-rate ADC could a priori be an alternative to BP-ΣΔMs for IF/RF digitization in wireless receivers, the bandwidth of RF signals is typically much smaller than the carrier frequency and the reduction of the quantization noise over the entire Nyquist band becomes thus very inefficient, as illustrated in Figure 2.27. Instead, if a BP-ΣΔM is used, the quantization noise is attenuated only in a narrow band around the carrier frequency, thus taking advantage of a high OSR to meet a high DR requirement, even in the presence of strong interfering signals.

Figure 2.28 illustrates a typical block diagram of the IF-to-baseband section in a digital receiver based on a BP-ΣΔM. As shown, the BP-ΣΔM contains a band-pass loop filter $H(z)$ in order to

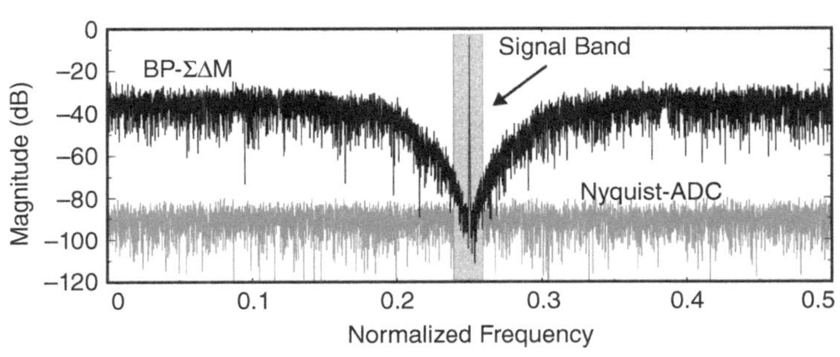

Figure 2.27 Output spectrum of a BP-ΣΔM and a Nyquist-rate ADC.

[7] Details of ΣΔMs for RF-to-digital conversion can be found in Section 10.4.

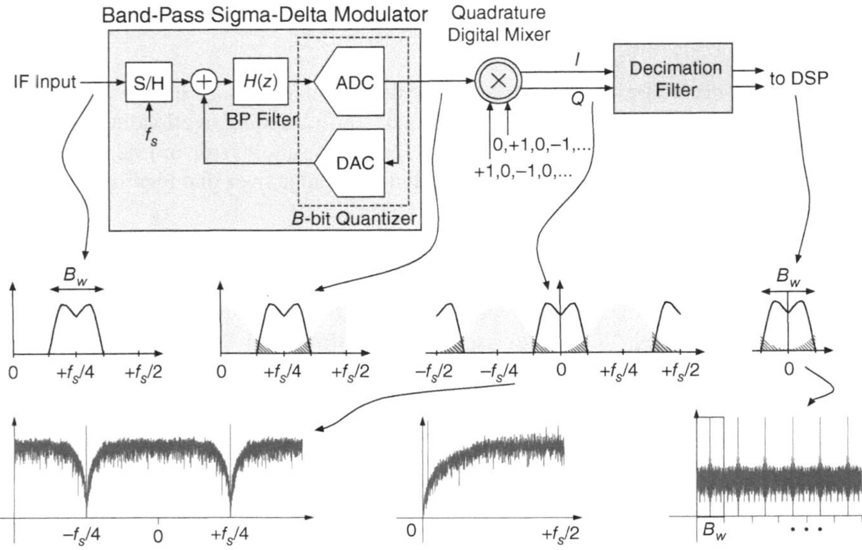

Figure 2.28 Typical section of an IF-conversion receiver based on a band-pass ΣΔ ADC. For the sake of illustration, the input IF frequency f_{IF} is assumed to be $f_s/4$ and a notch frequency $f_n = f_s/4$ is assumed in the BP-ΣΔM and in the quadrature digital mixer.

give a stop-band NTF. This has zeros placed at a nonzero frequency, often referred to as the *notch frequency*, f_n. The digital output of the modulator is mixed to DC by a digital quadrature mixer and is then low-pass filtered and decimated by a quadrature decimation filter to remove out-of-band spectral components and quantization noise. Finally, the resulting baseband digital data is processed in a DSP.

Note that a LP-ΣΔM has its NTF zeros at or near $z = 1$ (which corresponds to DC), whereas those of a BP-ΣΔM are placed elsewhere on the unit circle; i.e., $z = e^{j2\pi f_n/f_s}$ (which corresponds to f_n, with $0 < f_n < f_s/2$). Since the NTF zeros occur in complex conjugate pairs, realizing L zeros of the NTF in the passband of a BP-ΣΔM thus requires a $2L$th-order band-pass loop filter. In other words, the shaping performed on the quantization noise by a $2L$th-order BP-ΣΔM is equivalent to that performed by an Lth-order LP-ΣΔM.

2.5.1 Quadrature BP-ΣΔMs

The early IF/RF digitization in wireless receivers illustrated in Figure 2.28 can also be modified in such a way that the BP ΣΔ ADC directly digitizes quadrature input signals, using a so-called *quadrature* BP-ΣΔM [76, 77], as conceptually depicted in Figure 2.29a. Note that the main difference with respect to a conventional BP-ΣΔM is that a complex BP filter is embedded in the modulator loop. This filter can be implemented by cross-coupling conventional filters, as illustrated in Figure 2.29b. As a result, the resulting NTF has complex zeros that are not necessarily conditioned to be placed symmetrically with respect to DC. In that case, the quantization noise needs to be stop-band filtered only for positive (or negative) frequencies. Quadrature BP-ΣΔMs can thus be more efficient than conventional BP-ΣΔMs, since no power is dedicated to digitizing the negative-frequency content of the input [60]. This is illustrated in Figure 2.30, which represents the quantization noise shaped by a fourth-order quadrature BP-ΣΔM with three of its zeros located at $f_n = f_s/4$, and the other one

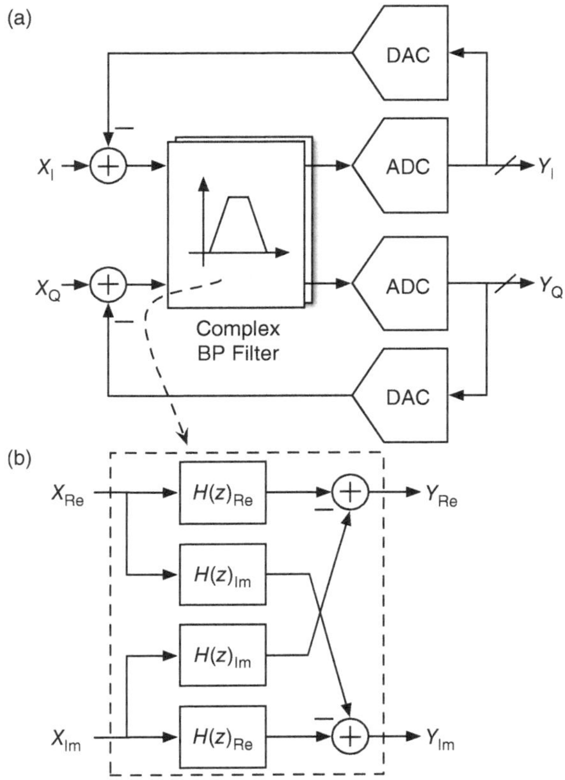

Figure 2.29 Quadrature BP-ΣΔMs: (a) conceptual block diagram; (b) example of complex filter realization.

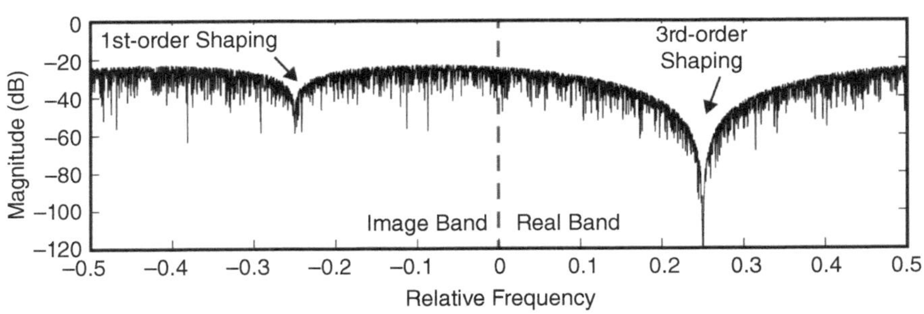

Figure 2.30 Illustrating the noise shaping of a fourth-order quadrature BP-ΣΔM with asymmetric NTF.

at $f_n = -f_s/4$. The latter can be used to further attenuate the image signal component caused by mismatches between the real and the imaginary channels of the quadrature loop filter [76].

2.5.2 The $z \rightarrow -z^2$ LP–BP Transformation

In spite of the aforementioned benefits of quadrature BP-ΣΔMs, one of the most common design choices for the notch frequency is $f_n = f_s/4$, since this location optimizes the trade-off between antialiasing filtering and image-rejection filtering in digital radio receivers [64]. It also helps to simplify the quadrature digital mixer to baseband, since the general digital cosine and sine sequences ($e^{j2\pi n f_n/f_s}$) reduce to a quadrature data series of +1s, 0s, and −1s, as illustrated in Figure 2.28. More importantly, the NTF zeros of a so-called $f_s/4$ BP-ΣΔM are placed at $z = \pm j$, so that the synthesis of the band-pass ΣΔ loop filter can be easily derived from an initial low-pass prototype by applying the transformation,

$$z \xrightarrow[0\,-\,f_s/4]{\text{LP–BP}} -z^2 \tag{2.33}$$

which is often referred to as the $z \rightarrow -z^2$ LP–BP transformation of ΣΔ modulators.

As an example, in Figure 2.31 this transformation is applied to a second-order LP-ΣΔM to obtain a fourth-order $f_s/4$ BP-ΣΔM, in which the integrators are replaced by *resonators*. The resulting BP architecture preserves all the properties of its LP original, as reguards dynamics, stability, resolution, and so on. Indeed, it can be shown that the main performance figures (IBN, SNR, DR, and so on) of $f_s/4$ BP-ΣΔMs have the *same* expressions as the ones derived in previous sections for the low-pass case. For the sake of illustration, if the $z \rightarrow -z^2$ transformation is applied to the NTF of an ideal Lth-order low-pass ΣΔM in Equation (1.22), that resulting for an ideal $2L$th-order band-pass ΣΔM yields:

$$\text{NTF}(z) = (1 + z^{-2})^L \tag{2.34}$$

By integrating the shaped quantization noise over the signal band, the resulting IBN is found to be,

$$\text{IBN} = \int_{f_n-\frac{B_w}{2}}^{f_n+\frac{B_w}{2}} S_E(f)|\text{NTF}(f)|^2 \text{d}f \approx \frac{\Delta^2}{12}\frac{\pi^{2L}}{(2L+1)\text{OSR}^{(2L+1)}} \tag{2.35}$$

so that the expression equals that of the LP case in Equation (1.12).

However, centering the signal passband at $f_s/4$ has some disadvantages. On the one hand, in the presence of nonlinearities in the analog circuitry of the ΣΔM, any intermodulation distortion products resulting from the mixing of tones at $f_s/2$ with the input signal will fall inside the modulator passband and will thus corrupt the signal information. On the other, for a given IF input, the clock rate demands are more restrictive than placing f_n between $f_s/4$ and $f_s/2$.[8]

2.5.3 BP-ΣΔMs with Optimized NTF

In practice, the $z \rightarrow -z^2$ transformation is widely used, but there are many other custom-designed loop filters in which the synthesis of the optimized STF and NTF is directly addressed through suitable

[8]The $z \rightarrow -z^2$ transformation also offers the possibility of centering the IF input at $3f_s/4$ [78] instead of $f_s/4$, given that the spectrum is symmetrical with respect to $f_s/2$. Making $f_{\text{IF}} = 3f_s/4$ preserves the requirements of the antialiasing filter and the image-rejection filter specifications can be relaxed. In addition, it allows for either the clock rate to be reduced to 1/3 or the signal processing to be three times faster. The only disadvantage is that OSR is reduced by a factor of three.

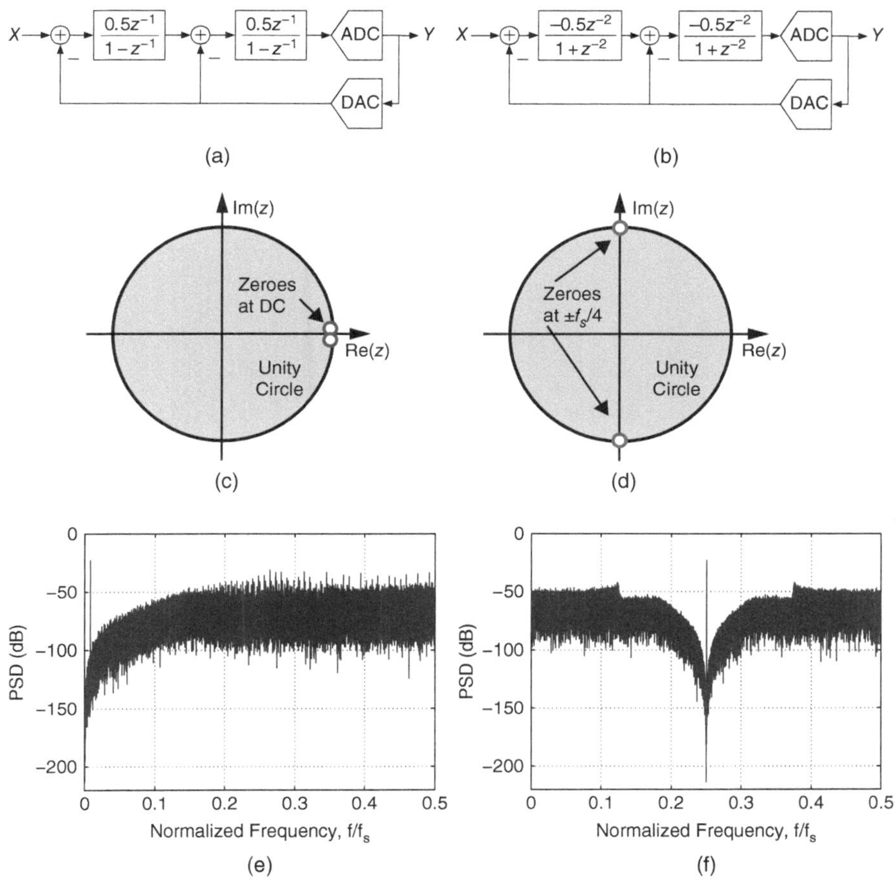

Figure 2.31 Illustration of the $z \rightarrow -z^2$ LP–BP transformation of ΣΔMs: (a) block diagram of a second-order LP-ΣΔM; (b) block diagram of the resulting fourth-order $f_s/4$ BP-ΣΔM; (c) zero-pole plot of the NTF of the LP-ΣΔM; (d) zero-pole plot of the NTF of the BP-ΣΔM; (e) output spectrum of the LP-ΣΔM ($f_{\text{in}} \approx f_s/128$); (f) output spectrum of the BP-ΣΔM ($f_{\text{in}} \approx f_s/4$).

placement of the poles and zeros [79–81]. This is especially the case if the loop filter has to have special characteristics, such as adjacent-channel suppression or partial mixing inside the ΣΔM [82–84]. In this respect, BP-ΣΔMs exhibit the same architectural variety as LP-ΣΔMs and the trade-offs between the different structures are also essentially the same [60]. Bandpass modulators can be implemented using either single-loop or cascade topologies, with a similar trade-off between improved stability and increased sensitivity to noise leakages due to nonidealities in the analog circuitry. Similarly, the loop filter of BP-ΣΔMs can be implemented using any of the topologies commonly employed for LP-ΣΔMs, such as the feedback, feedforward, and hybrid topologies shown in Figures 2.7, 2.8, and 2.9, respectively.

Figure 2.32 shows two architectural alternatives for the implementation of a fourth-order BP-ΣΔM, both of which allow the optimization of both STF and NTF. The band-pass architecture in Figure 2.32a is based on the LP-ΣΔM with distributed feedback in Figure 2.7, whereas that in Figure 2.32b is based on the LP-ΣΔM with feedforward summation of Figure 2.8. Both band-pass

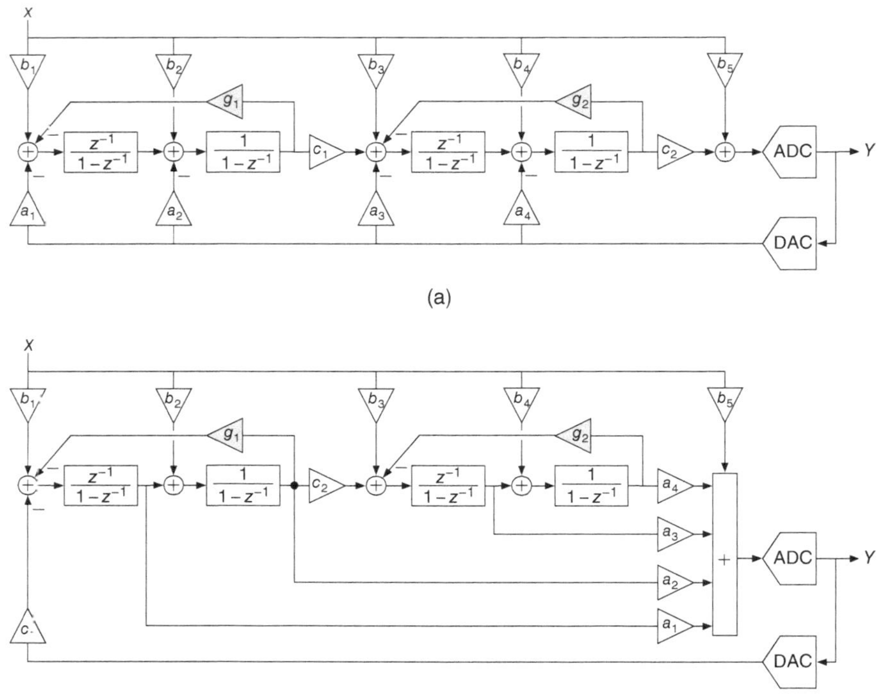

Figure 2.32 Illustration of fourth-order band-pass ΣΔ topologies which allow the NTF optimization: (a) cascade of resonators with feedback; (b) cascade of resonators with feedforward summation.

topologies include input feedforward for eliminating the input-signal content from the loop filter and thus relaxing the requirements of the analog circuitry [11]. In addition, if designed properly, input feedforward flattens the STF to a constant without any peaking, which prevents undesirable amplification of out-of-band interferers [60].

Note for Figure 2.32 that the band-pass loop filters are obtained by shifting the poles from DC to nonzero frequencies by adding internal feedback paths (coefficients g_i) in the initial low-pass topologies. Figure 2.33a depicts the *lossless discrete integrator* (LDI) loop that builds the resonator.[9] This can be implemented using SC techniques, as shown in Figure 2.33b. The transfer function of the LDI resonator is given by,

$$\mathrm{RTF}(z) = \frac{z^{-1}}{1 - (2 - g)z^{-1} + z^{-2}} \tag{2.36}$$

so that the poles are $z_p = (1 - g/2) \pm j\sqrt{1 - (1 - g/2)^2}$. Note that, for $0 < g < 4$, the poles of the RTF lie on the unit circle, and thus the quality factor Q of the resonator is ideally infinite.[10] The resonant

[9]Note that one of the DT integrators has a z^{-1} delay in the numerator, whereas the other does not. A slightly less effective resonator would be implemented if both integrators had a z^{-1} delay, since the poles would be moved on a vertical line from the $(1, j0)$ point, away from the real axis, and would not exhibit infinite gain at the resonance frequency [58].

[10]The resonator Q is limited in practice by the finite DC gain of the amplifiers. However, finite resonator Q is not usually a limiting nonideality, since $Q > 100$ is easily achieved.

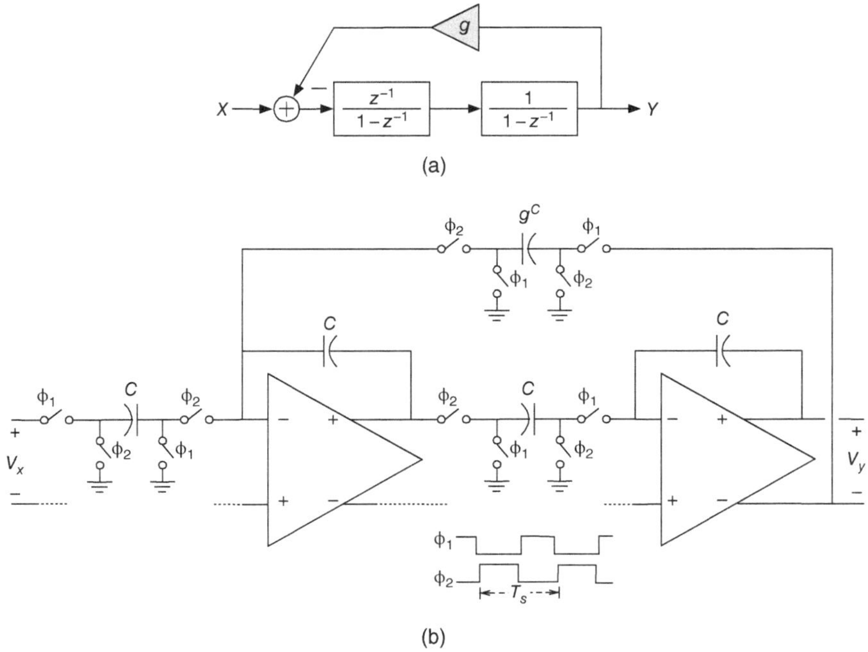

Figure 2.33 Illustration of an LDI resonator: (a) block diagram; (b) SC implementation. Only half of the differential circuit is explicitly shown for the sake of simplicity.

frequency of a resonator corresponds to a notch in the NTF of the BP-ΣΔM at a frequency,

$$\frac{2\pi f_n}{f_s} = \cos^{-1}\left(1 - \frac{g}{2}\right) \tag{2.37}$$

which can thus be placed at any arbitrary position from 0 to $f_s/2$ depending on the value of g – which is determined in an SC implementation by a capacitor ratio (see Figure 2.33b).

Therefore, the center frequency of a BP-ΣΔM can be made *programmable* by changing the resonant frequency of its resonators [85], for example by using programmable banks of capacitors for the SC implementation of coefficients g_i. This feature is commonly exploited in digital radios, since the *same* programmable BP-ΣΔM can be used for the digitization of different IF/RF frequencies.[11]

2.5.4 Time-interleaved and Polyphase BP-ΣΔMs

Another approach to increase the flexibility of BP-ΣΔMs in terms of programmable f_n with reduced f_s is to use time-interleaved (TI) or *P*-path loop filters [87]. Essentially the idea involves partitioning a BP-ΣΔM clocked at f_s in a number of TI BP-ΣΔMs, each one operating at f_s/P, with P being the

[11]However, this approach is only suitable when varying the modulator passband over a narrow frequency range. To maintain stability and SNR over a wide programmability range, other parameters, such as the NTF poles, the STF, and the resonator signal swings, need to be controlled in addition to the resonant frequencies by programming more modulator coefficients. This has been done using SC techniques [86].

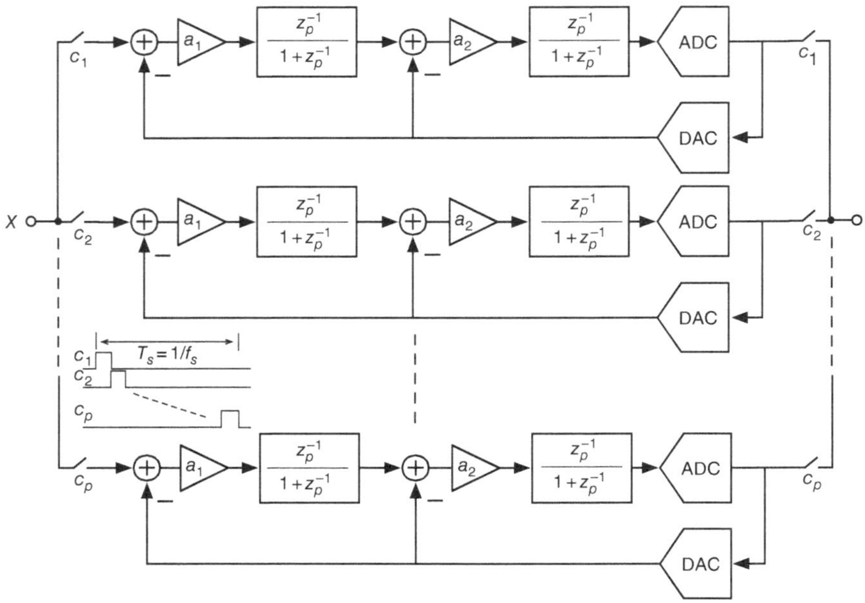

Figure 2.34 Conceptual block diagram of a TI BP-ΣΔM.

number of paths, as illustrated in Figure 2.34. In this way, the transfer function of the modulator can be expressed as:

$$H(z) = H(z_p)|_{z_p=z^P} \tag{2.38}$$

For example, a resonator transfer function can be partitioned into two ($P = 2$) TI high-pass filters, each one sampled at $f_s/2$,

$$\frac{z^{-2}}{1+z^{-2}} = \left.\frac{z_p^{-1}}{1+z_p^{-1}}\right|_{z_p=z^2} \tag{2.39}$$

The main problem of TI BP-ΣΔMs is due to the offset, gain, and timing mismatch caused by circuit imperfections, which cause spur tones to appear at the output spectrum, thus degrading the noise-shaping [88].

An alternative architecture of TI BP-ΣΔM is based on the so-called *polyphase* descomposition of the NTF of BP-ΣΔMs [89]. In this case, the overall STF is a delay that ranges from z^{-1} to z^{-P}, depending on the particular implementation. The NTF is given by:

$$\text{NTF}_{\text{PF}} = \sum_{0}^{P-1} z^{-j} \cdot N_j(z^P) \tag{2.40}$$

where $N_j(z^P)$ stands for the transfer functions of the jth path modulator resulting from the polyphase decomposition, which can be generically expressed as:

$$N_j = \sum_{m=0}^{k} \alpha_{mP+j} \cdot z^{-Pj} \tag{2.41}$$

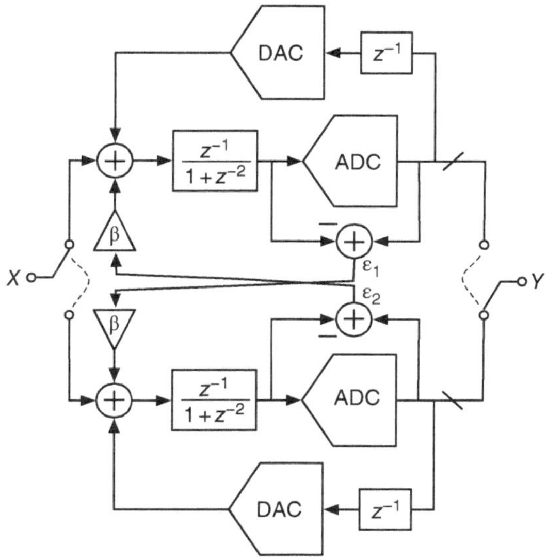

Figure 2.35 Block diagram of a second-order polyphase BP-ΣΔM [89].

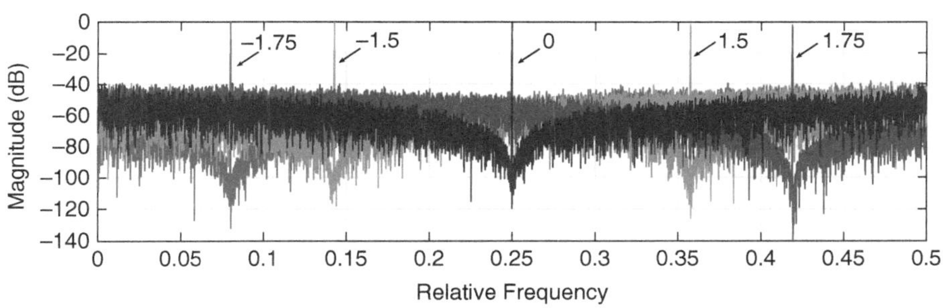

Figure 2.36 Output spectra of the second-order polyphase BP-ΣΔM in Figure 2.35 for $\beta = -1.75, -1.25, 0, 1.25, 1.75$.

where k stands for an integer number that satisfies $0 \leq kP + j \leq L$, with L being the overall modulator order [89].

As an illustration, Figure 2.35 shows a second-order polyphase BP-ΣΔM with $P = 2$ and an NTF with zeros at $z = e^{\pm j\alpha}$, given by:

$$\mathrm{NTF}_{\mathrm{PF2}} = 1 + \beta z^{-1} + z^{-2} \tag{2.42}$$

In this case, the polyphase ($P = 2$) decomposition of Equation (2.42) leads to $N_0(z^{-2}) = 1 + z^{-2}$ and $N_1(z^{-2}) = \beta$, where $\beta = -2\cos(\alpha)$ is a programmable coefficient changing the zeros of the NTF from $z = 1$ to $z = -1$ when varied from -2 to 2, as illustrated in Figure 2.36. It can be shown that the spur tones due to path mismatches fall out of the signal band at f_s/P, with P being the number of paths used in the polyphase decomposition [89].

2.6 Continuous-time ΣΔ Modulators: Architecture and Basic Concepts

The majority of ΣΔMs reported in recent years were implemented using DT circuit techniques, mostly based on SC circuits. However, the increasing demand for ever-faster ADCs in broadband communication systems has increased interest in CT-ΣΔMs recently, since they are able to operate at higher sampling rates with lower power consumption than their DT counterparts [23, 65, 82].

Figure 2.37 illustrates the conceptual block diagram of a CT-ΣΔM for the case of low-pass input signals. To facilitate a direct comparison with DT implementations, the same partitioning as in Figure 1.21 has been adopted.

Looking at both figures, several differences between DT- and CT-ΣΔMs are clearly visible. The most significant one is related to the location of the sampling operation, which moves from the modulator input in DT-ΣΔMs to the point before the quantizer in CT-ΣΔMs. The loop filter can thus be realized using CT circuit techniques, but, given that the modulator output is a DT signal and the modulator input is a CT signal, a discrete-to-continuous time (DT–CT) transformation is required in CT-ΣΔMs to create the feedback signal. This can be better visualized by comparing the signal processing involved in the block diagrams (Figures 2.38a and 2.38b).

The reconstruction of the modulator output signal is critical in CT-ΣΔMs, and has a significant impact on the overall modulator behavior [90]. There are a number of DAC waveforms that can be used in CT-ΣΔMs. Figure 2.39 shows a summary of representative possibilities, using the nomenclature for the feedback waveforms developed by Ortmanns and Gerfers [23]. The most commonly used DACs incorporate rectangular feedback pulses of three principal types:

- nonreturn-to-zero (NRZ; Figure 2.39a)
- return-to-zero (RZ; Figure 2.39b)
- half-delay return-to-zero (HRZ; Figure 2.39c).

The DAC impulse response of these rectangular pulses can thus be globally expressed as,

$$r_{\rm DAC}(t) = r_{(\alpha,\beta)}(t) = \begin{cases} 1, & \alpha T_s \leq t < \beta T_s \\ 0, & \text{otherwise} \end{cases} \tag{2.43}$$

where (α, β) equals (0, 1), (0, 1/2) and (1/2, 1) for NRZ, RZ, and HRZ DACs, respectively. Their Laplace S-transforms can be also written generally as:

$$R_{\rm DAC}(s) = \frac{e^{-s\alpha T_s} - e^{-s\beta T_s}}{s} \tag{2.44}$$

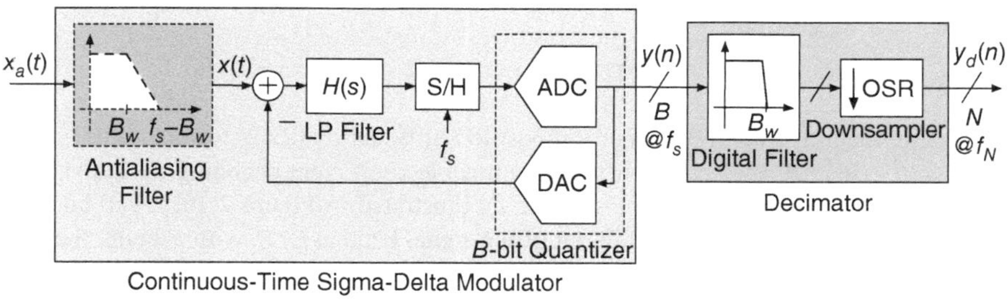

Figure 2.37 Block diagram of a continuous-time ΣΔ ADC. A low-pass ΣΔM is assumed.

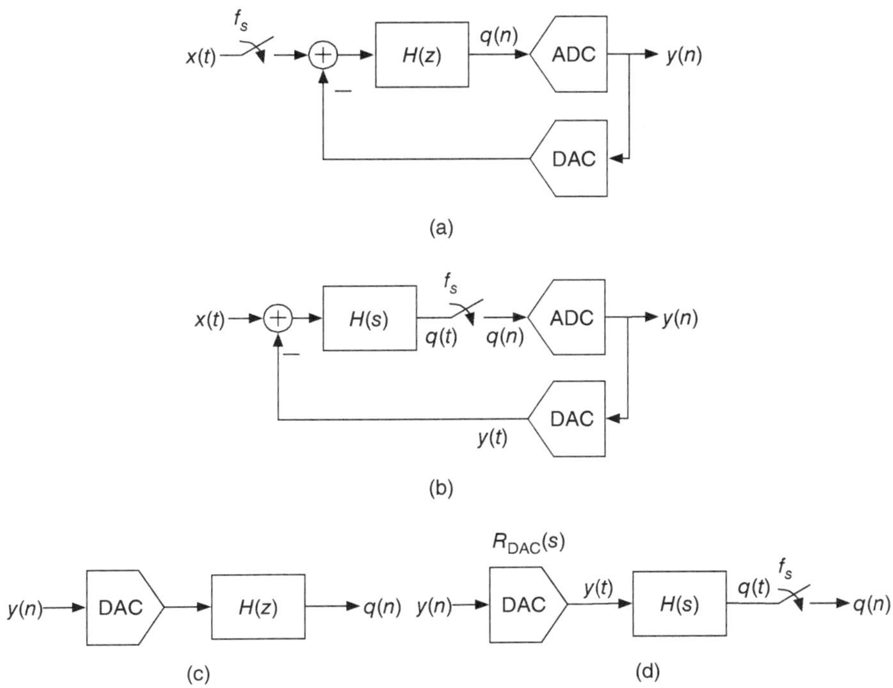

Figure 2.38 Conceptual block diagrams of ΣΔMs: (a) DT modulator; (b) CT modulator; (c) open-loop representation of a DT-ΣΔM; (d) open-loop representation of a CT-ΣΔM.

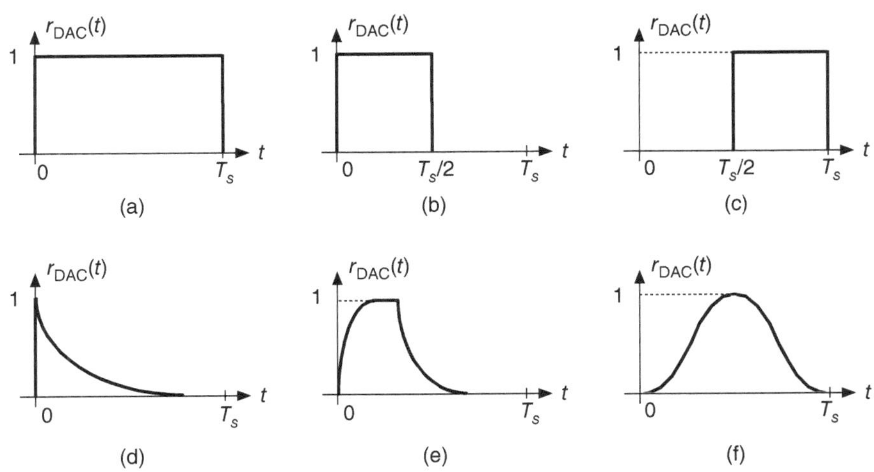

Figure 2.39 Most common DAC impulse responses: (a) nonreturn-to-zero (NRZ); (b) return-to-zero (RZ); (c) half-delay return-to-zero (HRZ); (d) switched-capacitor (SC); (e) exponential slope; (f) cosine.

Although they combine DT and CT signals, modulators fitting into the architecture in Figure 2.38b are generically considered to be CT-ΣΔMs. Due to the combination of both types of dynamics, together with the intrinsic nonlinearity associated with quantization, the mathematical analysis of CT-ΣΔMs becomes more difficult than that for DT-ΣΔMs [23, 65, 81, 82, 90]. However, they have several advantages over DT-ΣΔMs, such as:[12]

- An explicit antialiasing filter can be avoided (Figure 2.37). Since the sampling operation takes place before the quantizer, the resulting STF is affected by a sinc($\pi f/f_s$) function [5]. This sinc characteristic attenuates the signal spectrum exactly at multiples of the sampling frequency and thus leads to an *implicit antialiasing filter* in CT-ΣΔMs.
- Errors associated with the sampling process have less impact on the modulator performance. Since the sampling operation takes place before the quantizer, the resulting errors attenuate in a similar way as the quantization error does.
- There is no *settling error* associated with the loop filter circuitry. As will be shown in Chapter 3, signals in DT circuits must settle to their steady-state values within a given accuracy, since complete settling would require infinite time.
- The operation speed is faster. This is inherent to the operation of CT circuits, in which the circuit dynamics are not parasitic, as they are in DT circuits, but a design primitive.
- They are not affected by kT/C noise as SC-ΣΔMs are.

2.6.1 An Intuitive Analysis of CT-ΣΔMs

Before analysing the effect of the DAC waveform and describing the synthesis method usually employed to design CT-ΣΔMs, it is interesting to get an intuitive insight into these kinds of ΣΔMs. To this end, let us consider the conceptual block diagram of a CT-ΣΔM, as shown in Figure 2.40. For

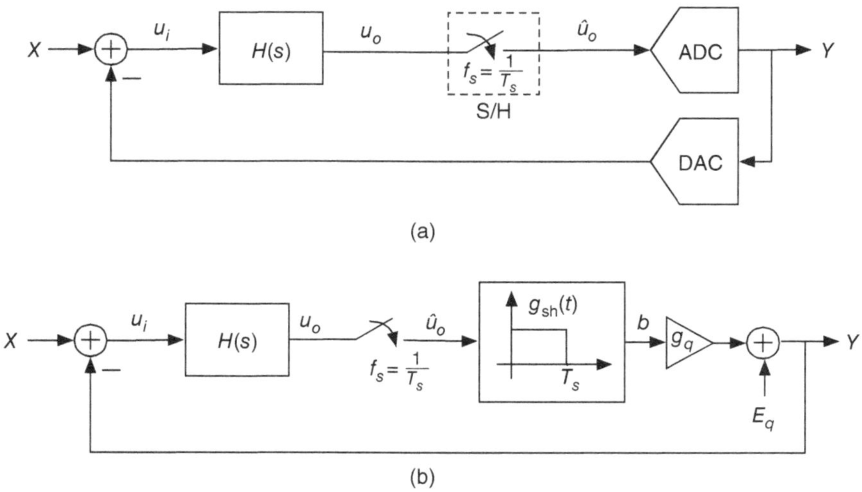

Figure 2.40 Conceptual diagram of a CT-ΣΔM for an intuitive analysis. (a) nonlinear diagram; (b) equivalent linear model.

[12]CT-ΣΔMs also present several drawbacks compared to DT-ΣΔMs, such as larger errors in the practical realization of the modulator coefficients, larger sensitivity to jitter and to loop delay, and so on. These issues will be discussed in Chapter 4.

the analysis that follows, it will be assumed that the feedback DAC is ideal and that it has a gain of approximately unity in the signal band. Under this assumption and considering a linear model for the quantizer, the modulator in Figure 2.40a can be transformed into the one depicted in Figure 2.40b.

A simple analysis of Figure 2.40b shows that the frequency transform of the the loop-filter input and the modulator output are respectively given by:

$$\begin{aligned} U_i(f) &= g_1 X(f) - g_1{}' Y(f) \\ Y(f) &= E_q(f) + g_q B(f) \end{aligned} \tag{2.45}$$

where $B(f)$ is the Fourier transform of the quantizer input, $b(t)$.

The sampling operation can be expressed as:

$$b(t) = \hat{u}_o(t) * g_{\text{sh}}(t) \tag{2.46}$$

where the symbol $*$ denotes the convolution operator and $\hat{u}_o(t)$ is the sampled version of the integrator output, $u_o(t)$, given by:

$$\hat{u}_o(t) = \sum_n u_o(nT_s)\delta(t - nT_s) \tag{2.47}$$

and $g_{\text{sh}}(t)$ is the impulsive response of the S/H block, given by:

$$g_{\text{sh}}(t) = \begin{cases} 1, & 0 \le t \le T_s \\ 0, & \text{otherwise} \end{cases} \tag{2.48}$$

Applying the Fourier transform to Equation 2.46 yields:

$$B(f) = \text{sinc}(fT_s) \sum_n U_o(f - nf_s) \tag{2.49}$$

Assuming that OSR $\gg 1$, then $\text{sinc}(fT_s) \simeq fT_s$ and, neglecting the spectral components above f_s, the Fourier transform of the modulator output can be expressed as:

$$Y(f) \simeq \text{STF}(f)X(f) + \text{NTF}(f)E_q(f) \tag{2.50}$$

where

$$\text{STF}(f) = \frac{g_1 g_q H(f)}{1 + g_1{}' g_q H(f)} \text{ and } \text{NTF}(f) = \frac{1}{1 + g_1{}' g_q H(f)} \tag{2.51}$$

which are similar to the expressions obtained for DT-ΣΔMs: STF $\to 1$ and NTF$(f) \to 0$ if $H(f) \to \infty$ within the band of interest. Note that the simplest filter transfer function $H(f)$ that satisfies this condition is a CT integrator transfer function, given by:

$$H(f) = \frac{1}{2\pi j f \tau} \tag{2.52}$$

where τ stands for the time constant of the circuit implementing $H(f)$. This will be considered in detail in Chapter 4.

Let us consider the modulator in Figure 2.41a, which implements a second-order CT-ΣΔM, where $g_q = 1/(g_2 g_1{}')$ is assumed and $H(f)$ is given by Equation 2.52. Figure 2.42 plots the SNR

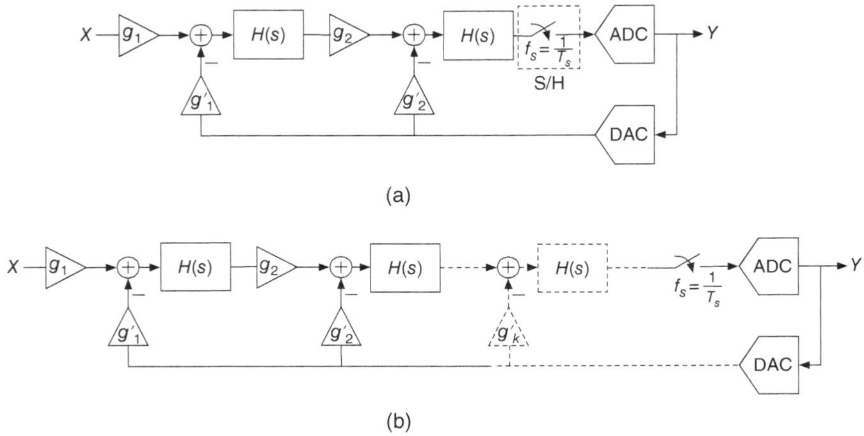

Figure 2.41 Block diagram of single-loop CT-ΣΔMs: (a) second-order; (b) L-th order.

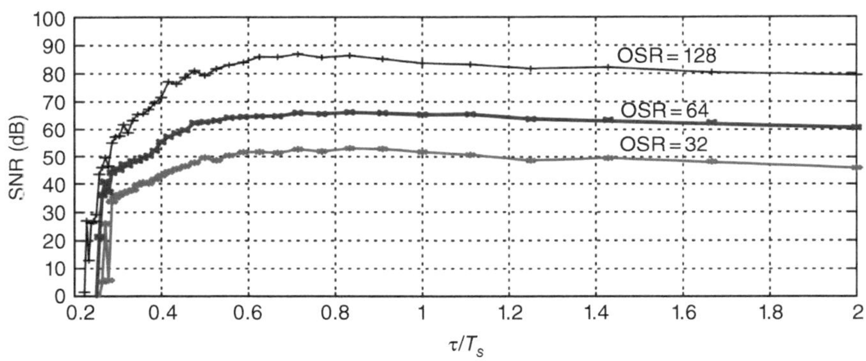

Figure 2.42 SNR vs. τ/T_s for a second-order CT-ΣΔM and different values of OSR.

versus τ/T_s for different values of OSR. Note that the maximum value of SNR is achieved for $\tau \simeq T_s$. Taking this into account, and assuming a linear model for the quantizer, it can be demonstrated that the output of the modulator in Figure 2.41a has a Fourier transform approximately given by:

$$Y(f) \simeq \frac{g_1}{g_1{}'}X(f) + (2\pi jf\tau)^2 E_q(f) \tag{2.53}$$

In the more general case of a L-th order single-loop CT-ΣΔM like that conceptually shown in Figure 2.41b, the output of the modulator can be ideally approximated by:

$$Y(f) \simeq \frac{g_1}{g_1{}'}X(f) + (2\pi jf\tau)^L E_q(f) \tag{2.54}$$

which corresponds to the same ideal noise-shaping performance obtained by a second-order and a L-th order DT-ΣΔM, respectively.

2.6.2 Some Words about Alias Rejection in CT-ΣΔMs

As stated above, one of the benefits of CT-ΣΔMs is that they provide an implicit AAF. In order to analyze this feature, let us consider the conceptual diagram of a CT-ΣΔM, as shown in Figure 2.38. This block diagram can be transformed into the one shown in Figure 2.43a, where the loop filter and the implicit S/H block are reallocated across the input adder and placed in front of the CT-ΣΔM and the feedback loop filter. Note that the resulting forward filter, $FF(s)$, is not necessarily identical to the feedback loop filter, $LF(s)$. This is a direct consequence of the fact that in a CT-ΣΔM, the input signal is CT and the output signal is DT [5, 23, 90].

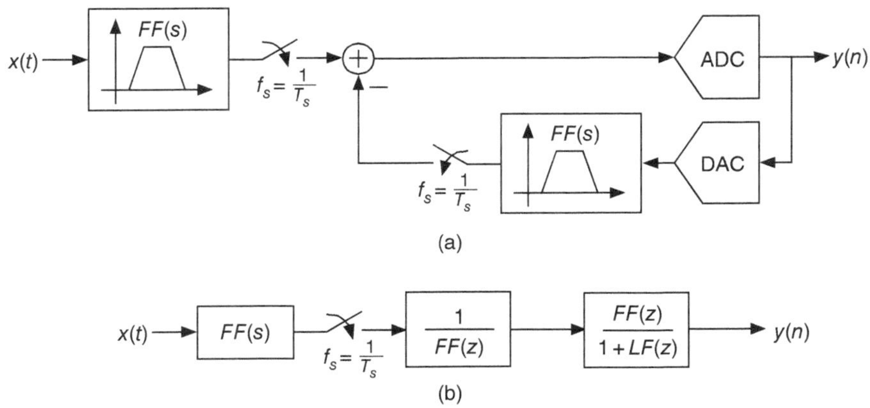

Figure 2.43 Implicit AAF in CT-ΣΔMs: (a) equivalent system representation of a CT-ΣΔM; (b) AAF block diagram.

The block diagram shown in Figure 2.43a reveals that, as was discussed in previous sections, a CT-ΣΔM behaves as a DT-ΣΔM due to the sampling operation in the feedback path, but with an additional CT forward filter, $FF(s)$. So, the corresponding STF frequency response can be obtained from the block diagram in Figure 2.43b. From this figure, the AAF can be approximately written as [90]:

$$F_{\text{AAF}}(\omega) = \frac{FF(j\omega)}{FF(e^{j\omega})} \tag{2.55}$$

where $FF(j\omega)$ is the FF loop filter of the CT-ΣΔM and $FF(e^{j\omega})$ is its DT equivalent. This equation shows that the AAF-behaviour can be defined for different loop-filter characteristics: low-pass, band-pass, and so on [23]. Thus, based on this analysis, it can be shown that the AAF transfer function of an L-th order LP CT-ΣΔM is given by [23]:

$$F_{\text{AAF}}(\omega) = \left| \frac{f_s(1 - e^{j\omega})}{j\omega e^{-j\omega}} \right|^L \simeq \left[\text{sinc}\left(\frac{\pi f}{f_s} \right) \right]^L \tag{2.56}$$

As an illustration, Figure 2.44 shows the effect of the AAF in a second-order CT-ΣΔM, comparing the output spectrum with the corresponding one obtained by a DT-ΣΔM. In both cases two input sinewave signals, one in-band and the other out-of-band, are applied. Note that thanks to the implicit AAF in the CT-ΣΔM, the out-of-band (aliased) tone is markedly attenuated (around 60 dB) compared to that for the DT-ΣΔM, where both the input and the aliased signals have the same magnitude.

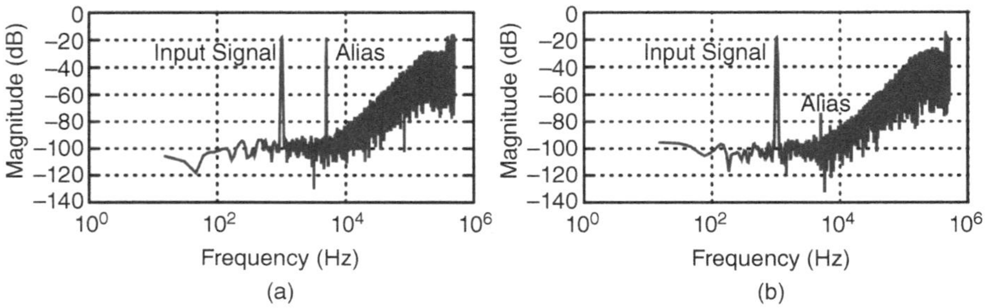

Figure 2.44 Antialiasing filtering in second-order CT-ΣΔMs for an input signal at 1 kHz and an aliased signal at f_s-5 kHz with $f_s = 1$ MHz: (a) DT-ΣΔM output; (b) CT-ΣΔM output.

According to Equation 2.56, the effect of AAF increases with the modulator order, so that the aliased signal can be reduced to below the noise floor for $L > 3$ [23]. However, in practical cases, the AAF feature depends also on the DAC waveform; for SC DACs, the alias rejection can be completely destroyed unless circuit compensation techniques are used [91].

2.7 DT–CT Transformation of ΣΔMs

The analysis of CT-ΣΔMs carried out in Section 2.6.1 assumed an ideal DAC impulse response. Although this approach is intuitive and allows us to gain some insights about the signal processing in CT-ΣΔMs, it is not an accurate methodology for synthesizing CT-ΣΔMs since the output signal must be transformed from DT to CT in the feedback loop. This signal reconstruction is critical, and has a significant impact on the overall behaviour of the modulator [90]. As stated above, a lot of work has been carried out in the past on DT-ΣΔM implementations, and great collective effort has been applied to developing innovative architectures, precise models, and design and simulation CAD tools for DT-ΣΔMs. Therefore, a straightforward procedure for designing CT-ΣΔMs comes from considering an equivalent DT loop filter as an starting point, designing the DT-ΣΔM to meet the required performance, and then applying a DT–CT transformation.

2.7.1 The Impulse-invariant Transformation

Let us consider again the block diagram of the DT- and CT-ΣΔMs in Figures 2.38a and 2.38b respectively, and the signal processing involved. The underlying principle of the DT–CT transformation is to achieve equivalence of both modulators by ensuring that the inputs of the quantizers are the same at the sampling instants:

$$q(n) = q(t)|_{t=nT_s} \tag{2.57}$$

If this condition is met, the output bitstreams of both modulators and the noise performance will be identical. As illustrated in Figures 2.38c and 2.38d, the condition in Equation (2.57) can be translated into an equivalence of the input-output signal processing that is performed in both ΣΔMs in open-loop configuration. This yields,

$$\mathcal{Z}^{-1}\{H(z)\} = \mathcal{L}^{-1}\{R_{\text{DAC}}(s)H(s)\}|_{t=nT_s} \tag{2.58}$$

where $\mathcal{Z}^{-1}\{\cdot\}$ and $\mathcal{L}^{-1}\{\cdot\}$ stand for the inverse Z- and L-transform operators, respectively, and R_{DAC} stands for the CT transfer function of the DAC – see Equation (2.44) for the case of rectangular waveforms. In the time domain this leads to the condition,

$$h(n) = [r_{\text{DAC}}(t) * h(t)]|_{t=nT_s} = \int_{-\infty}^{+\infty} r_{\text{DAC}}(\tau)h(t-\tau)\mathrm{d}\tau|_{t=nT_s} \tag{2.59}$$

where $r_{\text{DAC}}(t)$ stands for the impulse response of the specific DAC – see Equation (2.43) for rectangular waveforms. This transformation between the DT and CT domains is known as the *invariant-impulse transformation* (IIT), because it makes the open-loop impulse responses of both ΣΔMs equal at the sampling instants.

Based on the IIT in Equation (2.58), the most common procedure to design CT-ΣΔMs consists of:

- matching the equivalent filter $H(z)$ with a reference DT loop filter chosen to fulfill the specifications;
- solving for the coefficients of the CT loop filter $H(s)$ with the specific DAC response considered;
- implementing $H(s)$ with CT techniques, usually based on Gm-C or active-RC techniques.[13]

In theory, any arbitrary ΣΔM topology, either single-loop or cascade, can be synthesized this way.

The information compiled in Tables 2.3 and 2.4 for the equivalence of low-pass loop filter poles at DC ($z = 1$, $s = 0$) in the DT and CT domains is useful in such a methodology [23]. For general poles, similar information is available in the literature [92]. This can also be useful for the design of band-pass CT-ΣΔMs. Finally, the use of these tables of equivalent poles is easy and can be automated and, moreover, extended to DAC impulse responses other than rectangular waveforms [23].

Table 2.3 CT equivalents of first-order to fourth-order DT low-pass loop filter poles for the rectangular feedback DAC pulses defined in Equation (2.43) [23].

***Z*-domain**	***S*-domain equivalents with $f_s(\text{Hz}) = 1/T_s$**
$\frac{1}{z-1}$	$\frac{\omega_0}{s}, \quad \omega_0 = \frac{f_s}{\beta-\alpha}$
$\frac{1}{(z-1)^2}$	$\frac{\omega_1 s+\omega_0}{s^2}, \quad \omega_0 = \frac{f_s^2}{\beta-\alpha}, \quad \omega_1 = \frac{f_s(\alpha+\beta-2)}{2(\beta-\alpha)}$
$\frac{1}{(z-1)^3}$	$\frac{\omega_2 s^2+\omega_1 s+\omega_0}{s^3}, \quad \omega_0 = \frac{f_s^3}{\beta-\alpha}, \quad \omega_1 = \frac{f_s^2(\alpha+\beta-3)}{2(\beta-\alpha)},$
	$\omega_2 = \frac{f_s[\beta(\beta-9)+\alpha(\alpha-9)+4\alpha\beta+12]}{12(\beta-\alpha)}$
$\frac{1}{(z-1)^4}$	$\frac{\omega_3 s^3+\omega_2 s^2+\omega_1 s+\omega_0}{s^4}, \quad \omega_0 = \frac{f_s^4}{\beta-\alpha}, \quad \omega_1 = \frac{f_s^3(\alpha+\beta-4)}{2(\beta-\alpha)},$
	$\omega_2 = \frac{f_s^2[(\beta-\alpha)^2+2\alpha\beta-12(\alpha+\beta)+22]}{12(\beta-\alpha)},$
	$\omega_3 = \frac{f_s[\beta^2(\alpha-2)+\alpha^2(\beta-2)-8\alpha\beta+11(\alpha+\beta)-12]}{12(\beta-\alpha)}$

[13]Careful choice of the CT filter structure is required to ensure sufficient degrees of freedom to implement the reference DT loop filter [81].

Table 2.4 DT equivalents of first-order to fourth-order CT low-pass loop filter poles for the rectangular feedback DAC pulses defined in Equation (2.43) [23].

S-domain	**Z**-domain equivalents with f_s(Hz) $= 1/T_s$
$\frac{f_s}{s}$	$\frac{\omega_0}{z-1}, \quad \omega_0 = \beta - \alpha$
$\frac{f_s^2}{s^2}$	$\frac{\omega_1 z + \omega_0}{(z-1)^2}, \quad \omega_0 = \frac{(\beta^2-\alpha^2)}{2}, \quad \omega_1 = \frac{[\beta(1-\beta)-\alpha(1-\alpha)]}{2}$
$\frac{f_s^3}{s^3}$	$\frac{\omega_2 z^2 + \omega_1 z + \omega_0}{(z-1)^3}, \quad \omega_0 = \frac{(\beta^3-\alpha^3)}{6},$ $\omega_1 = -\frac{(\beta^3-\alpha^3)}{3} + \frac{(\beta^2-\alpha^2)}{2} + \frac{(\beta-\alpha)}{2},$ $\omega_2 = -\frac{(\beta^3-\alpha^3)}{6} - \frac{(\beta^2-\alpha^2)}{2} + \frac{(\beta-\alpha)}{2}$
$\frac{f_s^4}{s^4}$	$\frac{\omega_3 z^3 + \omega_2 z^2 + \omega_1 z + \omega_0}{(z-1)^4}, \quad \omega_0 = \frac{(\beta^4-\alpha^4)}{24},$ $\omega_1 = -\frac{(\beta^4-\alpha^4)}{8} + \frac{(\beta^3-\alpha^3)}{6} + \frac{(\beta^2-\alpha^2)}{4} + \frac{(\beta-\alpha)}{6},$ $\omega_2 = \frac{(\beta^4-\alpha^4)}{8} - \frac{(\beta^3-\alpha^3)}{3} + \frac{2(\beta-\alpha)}{3},$ $\omega_3 = -\frac{(\beta^4-\alpha^4)}{24} + \frac{(\beta^3-\alpha^3)}{6} - \frac{(\beta^2-\alpha^2)}{4} + \frac{(\beta-\alpha)}{6}$

2.7.2 DT–CT Transformation of a Second-order ΣΔM

For the sake of illustration, Figure 2.45 shows a second-order CT-ΣΔM obtained using such a method from its DT equivalent in Figure 2.1a. Note from the earlier figure that the DT integrators – with transfer function given by Equation (1.19) – are transformed into CT integrators with *unscaled* transfer functions given by

$$\text{ITF}(s) = \frac{1}{sT_s} = \frac{f_s}{s} \tag{2.60}$$

and that the scaling coefficients k_i of the CT-ΣΔM are associated with the feedback paths [23]. This notation has been adopted because it can easily account for nonidealities in the loop filter, as will be shown in Chapter 4. Note from Figure 2.45 that the discrimination of the signal scaling coefficient

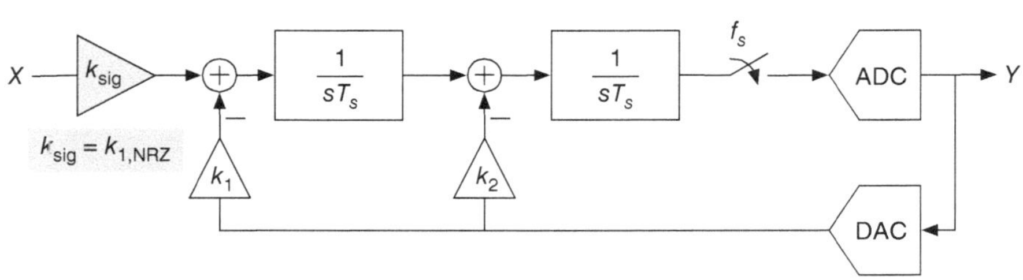

Figure 2.45 Block diagram of a second-order CT-ΣΔM using the notation in [23].

$k_{\rm sig}$ and the first feedback scaling coefficient k_1 accounts for changing feedback coefficients when adopting DAC feedback pulses different from the rectangular NRZ [23]. Therefore, the input scaling coefficient is always equal to $k_{1,\rm NRZ}$ regardless of the particular feedback waveform that is used.

The CT equivalent of the DT-ΣΔM in Figure 2.1a can be easily calculated by, on the one hand, applying the IIT in Equation (2.58) to the DT loop filter,

$$H(z) = -\frac{a_1 a_2}{(z-1)^2} - \frac{a_2}{z-1} \xrightarrow[\text{NRZ}]{\text{DT–CT}} H(s) = -a_1 a_2 \frac{f_s^2}{s^2} - \left(a_2 - \frac{a_1 a_2}{2}\right)\frac{f_s}{s} \tag{2.61}$$

where an NRZ rectangular waveform has been assumed, as an example, in the feedback DAC of the CT equivalent. To that end, rows 1 and 2 of Table 2.3 have been adopted, with $(\alpha, \beta) = (0, 1)$. On the other hand, the CT loop filter in Figure 2.45 is obtained as:

$$H(s) = -k_1 \frac{f_s^2}{s^2} - k_2 \frac{f_s}{s} \tag{2.62}$$

Equating the CT coefficients in Equations (2.61) and (2.62) results in the following relations between the coefficients of the second-order DT-ΣΔM and the CT coefficients for the case of an NRZ DAC:

$$\begin{aligned} k_{1,\rm NRZ} &= a_1 a_2 \\ k_{2,\rm NRZ} &= a_2 - \frac{a_1 a_2}{2} \end{aligned} \tag{2.63}$$

The DT scaling coefficients $(a_1, a_2) = (0.5, 0.5)$ typically used – see Table 2.1 and optimal coefficients reported in [1] – thus result in $k_{1,\rm NRZ} = 0.25$ and $k_{2,\rm NRZ} = 0.375$.

For the sake of illustration, Figure 2.46 shows a possible implementation of the resulting CT-ΣΔM using active-RC integrators. Note that the corner frequency ω_I of an active-RC integrator is determined by the value of $1/(RC)$, so that the scaling coefficients of the CT-ΣΔM yield [23]:

$$\text{ITF}_i(s) = \frac{\omega_{Ii}}{s} = k_i \frac{f_s}{s} = \frac{1}{sRC} \quad \Rightarrow \quad k_i f_s = \omega_{Ii} = \frac{1}{RC} \tag{2.64}$$

For the second-order CT-ΣΔM in Figures 2.45 and 2.46, the relations to be fulfilled for the correct implementation of the modulator coefficients are thus,

$$\begin{aligned} R_{\rm in1} C_{I1} &= \frac{1}{k_{\rm sig} f_s}, & R_{\rm DAC1} C_{I1} &= \frac{1}{k_1 f_s} \\ R_{\rm in2} C_{I2} &= \frac{1}{f_s}, & R_{\rm DAC2} C_{I2} &= \frac{1}{k_2 f_s} \end{aligned} \tag{2.65}$$

which, for the particular case of NRZ DAC considered, yield:

$$\begin{aligned} R_{\rm in1} &= R_{\rm DAC1}, & R_{\rm DAC1} C_{I1} &= \frac{1}{0.25 f_s} \\ R_{\rm in2} &= 0.375 R_{\rm DAC2}, & R_{\rm DAC2} C_{I2} &= \frac{1}{0.375 f_s} \end{aligned} \tag{2.66}$$

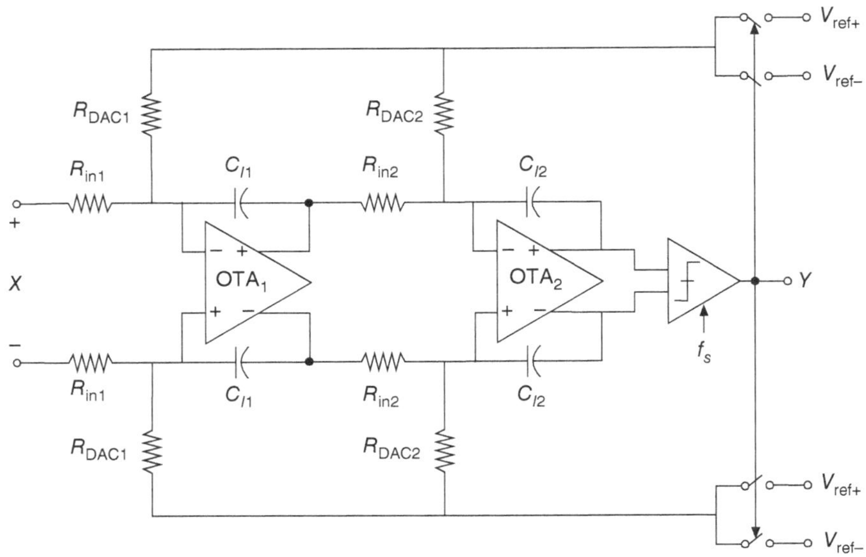

Figure 2.46 Active-RC implementation of a second-order CT-ΣΔM with a 1-bit NRZ DAC.

2.8 Direct Synthesis of CT-ΣΔMs

An alternative method for designing the CT loop filter directly uses the desired NTF(f) as a starting point, in the same way as described in Section 2.2 for synthesizing optimized NTFs in the DT case. This synthesis method is often referred to as *direct CT synthesis method*. Usually, an inverse Chebyshev distribution of the NTF(f) zeros is considered because it has advantages in terms of SNR and stability. Once the desired NTF(f) has been chosen, the loop filter can be derived from the linearized model [82]. As an illustration, Figure 2.47a shows a third-order single-loop CT-ΣΔM example [23], synthesized using a direct synthesis method. The scaling coefficients, k_i, have been obtained considering a Chebyshev type II distribution of NTF zeroes, giving $\{k1, k2, k3, \gamma\} = \{0.51, 0.97, 1.95, 0.04/(k_2k_3)\}$. The corresponding output spectrum is shown in Figure 2.47b, using a sampling frequency of $f_s = 100$ MHz, an NRZ DAC waveformn and a 3-bit quantizer.

One of the main drawbacks of using the direct synthesis method is that the know-how about DT-ΣΔMs is not reused, which explains why the majority of CT-ΣΔM designers use the DT–CT transformation method. However, that method may sometimes yield an increase in complexity in the analog circuitry, with a resultant penalty in sensitivity to variations on the technological process parameters. This is particularly critical in the case of CT cascades, where, to get a functional modulator while keeping the DCL of the original DT-ΣΔM, every integrator and DAC output must be connected to the integrator input of later stages [23].

Alternatively, cascade CT-ΣΔMs can be directly synthesized in the CT domain in order to get architectures that are more efficient, in terms of analog circuit complexity and robustness to mismatch [93]. In order to explain this methodology, let us consider the block diagram of an m-stage cascade CT-ΣΔM, as shown in Figure 2.48. The overall modulator output can be written as:

$$y_o(z) = \sum_{k=1}^{m} y_k(z)\mathrm{CL}_k(z) \tag{2.67}$$

where $y_k(z)$ and $\mathrm{CL}_k(z)$ denote, respectively, the output and the DCL transfer function of the kth stage.

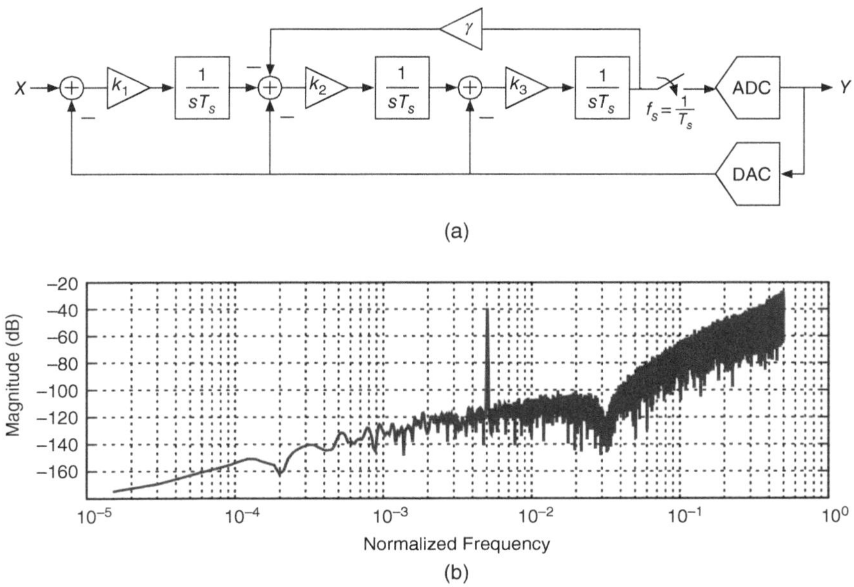

Figure 2.47 Third-order CT-ΣΔM with Chebyshev loop filter approximation [23]: (a) block diagram; (b) output spectrum.

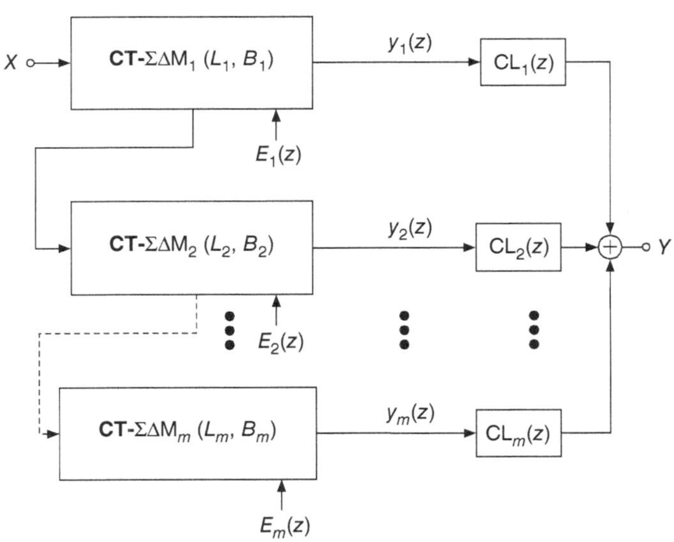

Figure 2.48 Block diagram of a cascade m-stage CT-ΣΔM.

It can be shown that the quantization errors of the first $m-1$ stages are cancelled out if $\mathrm{CL}_k(z)$ are given by:

$$\mathrm{CL}_k(z) = \frac{-\mathcal{Z}\{\mathcal{L}^{-1}\{R_{\mathrm{DAC}}(s)F_{km}(s)\}|_{t=nT_s}\}}{1-\mathcal{Z}\{\mathcal{L}^{-1}\{R_{\mathrm{DAC}}(s)F_{mm}(s)\}|_{t=nT_s}\}} \cdot \mathrm{CL}_m(z) \tag{2.68}$$

where $F_{km}(s)$ represents the transfer function from $y_k(s)$ to the input of the final, mth stage quantizer. The DCL transfer function of the final stage, $\mathrm{CL}_m(z)$, can be chosen to be the simplest form that preserves the required noise shaping.

Note that the synthesis equations (2.67) and (2.68) not only take into account the single-stage loop filter transfer functions, but also the inter-stage loop filter transfer functions. The latter must be included in the synthesis methodology to obtain a functional modulator with a minimum number of interstage paths. In this way, a cascade CT-ΣΔM can be directly synthesized in the CT domain by optimizing the NTF of its stages and then deriving the DCL transfer functions from Equation (2.68) [93].

2.9 Summary

This chapter has presented an overview of ΣΔ architectures from a system-level point of view. The existing methods for increasing the effective resolution of a ΣΔ modulator have been discussed, presenting practical ways to implement stable high-order noise-shaping topologies, considering both single-loop and cascade topologies, and multi-bit quantization. The discrete- and continuous-time implementation of ΣΔ modulators have been addressed, as well as their application to convert low- or band-pass signals. The implications of these different alternatives at the architectural- and circuit-level were presented in an incremental way, from the initial case of discrete-time low-pass single-bit modulators, to continuous-time topologies.

These basic concepts were mainly discussed assuming quantization noise as the only source of error limiting the resolution of a ΣΔ modulator. The effect of nonidealities associated wtih the practical circuit implementation of the modulator blocks will be analyzed in Chapters 3 and 4, for SC and CT-ΣΔMs, respectively.

References

[1] A. Marques, V. Peluso, M. S. Steyaert, and W. M. Sansen, "Optimal Parameters for ΔΣ Modulator Topologies," *IEEE Trans. on Circuits and Systems II: Analog and Digital Signal Processing*, vol. 45, pp. 1232–1241, September 1998.

[2] Y. Geerts, M. Steyaert, and W. Sansen, *Design of Multi-bit Delta-Sigma A/D Converters.* Kluwer Academic Publishers, 2002.

[3] F. Medeiro, B. Pérez-Verdú, and A. Rodríguez-Vázquez, *Top-Down Design of High-Performance Sigma-Delta Modulators.* Kluwer Academic Publishers, 1999.

[4] R. del Río, F. Medeiro, B. Pérez-Verdú, J. M. de la Rosa, and A. Rodríguez-Vázquez, *CMOS Cascade ΣΔ Modulators for Sensors and Telecom: Error Analysis and Practical Design.* Springer, 2006.

[5] J. Candy, "A Use of Double Integration in Sigma-Delta Modulation," *IEEE Transactions on Communications*, vol. 33, pp. 249–258, March 1985.

[6] B. E. Boser and B. A. Wooley, "The Design of Sigma-Delta Modulation Analog-to-Digital Converters," *IEEE J. of Solid-State Circuits*, vol. 23, pp. 1298–1308, December 1988.

[7] G. Yin and W. Sansen, "A High-Frequency and High-Resolution Fourth-Order ΣΔ A/D Converter in BiCMOS Technology," *IEEE J. of Solid-State Circuits*, vol. 29, pp. 857–865, August 1994.

[8] F. Medeiro, B. Pérez-Verdú, J. M. de la Rosa, and A. Rodríguez-Vázquez, "Multi-Bit Cascade ΣΔ Modulator for High-Speed A/D Conversion with Reduced Sensitivity to DAC Errors," *IET Electronics Letters*, vol. 34, pp. 422–424, March 1998.

[9] A. M. Marques, V. Peluso, M. S. J. Steyaert, and W. M. Sansen, "A 15-b Resolution 2 MHz Nyquist Rate ΔΣ ADC in a 1-μm CMOS Technology," *IEEE J. of Solid-State Circuits*, vol. 33, pp. 1065–1075, July 1998.

[10] P. Benabes, A. Gauthier, and D. Billet, "New Wideband Sigma-Delta Convertor," *IET Electronics Letters*, vol. 27, pp. 1575–1577, August 1993.

[11] J. Silva, U. Moon, J. Steensgaard, and G. C. Temes, "Wideband Low-Distortion Delta-Sigma ADC Topology," *IET Electronics Letters*, vol. 37, pp. 737–738, June 2001.

[12] R. Gaggl, M. Inversi, and A. Wiesbauer, "A Power Optimized 14-Bit SC ΔΣ Modulator for ADSL CO Applications," *IEEE ISSCC Digest of Technical Papers*, pp. 82–83, February 2004.

[13] K. Y. Nam, S. M. Lee, D. K. Su, and B. A. Wooley, "A Low-Voltage Low-Power Sigma-Delta Modulator for Broadband Analog-to-Digital Conversion," *IEEE J. of Solid-State Circuits*, vol. 40, pp. 1855–1864, September 2005.

[14] T. Christen, T. Burger, and Q. Huang, "A 0.13 μm CMOS EDGE/UMTS/WLAN Tri-Mode ΣΔ ADC with −92 dB THD," *IEEE ISSCC Digest of Technical Papers*, pp. 240–241, February 2007.

[15] T. Christen and Q. Huang, "A 0.13 μm CMOS 0.1-20 MHz Bandwidth 86–70 dB DR Multi-Mode DT ΔΣ ADC for IMT-Advanced," *Proc. of the IEEE European Solid-State Circuits Conf.*, pp. 414–417, September 2010.

[16] A. Morgado, R. del Río, J. M. de la Rosa, L. Bos, J. Ryckaert, and G. van der Plas, "A 100 kHz-10 MHz BW, 78-to-52 dB DR, 4.6-to-11 mW Flexible SC ΣΔ Modulator in 1.2-V 90-nm CMOS," *Proc. of the IEEE European Solid-State Circuits Conf.*, pp. 418–421, September 2010.

[17] R. W. Adams, P. F. Ferguson, A. Ganesan, S. Vincelette, A. Volpe, and R. Libert, "Theory and Practical Implementation of a Fifth-Order Sigma-Delta A/D Converter," *J. Audio Eng. Soc.*, vol. 39, pp. 515–528, July 1991.

[18] R. W. Adams and R. Schreier, "Stability Theory in ΣΔ Modulators," in *Delta-Sigma Data Converters: Theory, Design and Simulation* (S.R. Norsworthy, R. Schreier, and G.C. Temes, editors), IEEE Press, 1997.

[19] W. L. Lee and C. G. Sodini, "A Topology for Higher Order Interpolative Coders," *Proc. of the IEEE Intl. Symp. on Circuits and Systems*, pp. 459–462, May 1987.

[20] F. O. Eynde, *High-Performance Analog Interfaces for Digital Signal Processors*. Ph.D. Thesis, Katholieke Universiteit Leuven, 1990.

[21] S. H. Ardalan and J. J. Paulos, "An Analysis of Nonlinear Behavior in Delta-Sigma Modulators," *IEEE Trans. on Circuits and Systems*, vol. 34, pp. 593–603, June 1987.

[22] L. A. Williams and B. A. Wooley, "Third-Order Cascaded Sigma-Delta Modulators," *IEEE Trans. on Circuits and Systems*, vol. 38, pp. 489–498, May 1991.

[23] M. Ortmanns and F. Gerfers, *Continuous-Time Sigma-Delta A/D Conversion: Fundamentals, Performance Limits and Robust Implementations*. Springer, 2006.

[24] R. W. Adams, "The Design of High-Order Single-Bit ΣΔ ADCs," in *Delta-Sigma Data Converters: Theory, Design and Simulation* (S.R. Norsworthy, R. Schreier, and G.C. Temes, editors), IEEE Press, 1997.

[25] D. R. Welland, B. P. del Signore, E. J. Swanson, T. Tanaka, K. Hamashita, S. Hara, and K. Takasuka, "A Stereo 16-Bit Delta-Sigma A/D Converter for Digital Audio," *J. Audio Eng. Soc.*, vol. 37, pp. 476–486, July 1989.

[26] P. F. Ferguson, A. Ganesan, and R. W. Adams, "One Bit Higher Order Sigma-Delta A/D Converters," *Proc. of the IEEE Intl. Symp. on Circuits and Systems*, vol. 2, pp. 890–893, May 1990.

[27] R. Schreier, "An Empirical Study of Higher Order Single Bit Sigma Delta Modulators," *IEEE Trans. on Circuits and Systems – II: Analog and Digital Signal Processing*, vol. 40, pp. 461–466, August 1993.

[28] Y. Matsuya, K. Uchimura, A. Iwata, T. Kobayashi, M. Ishikawa, and T. Yoshitome, "A 16-bit Oversampling A-to-D Conversion Technology Using Triple-Integration Noise Shaping," *IEEE J. of Solid-State Circuits*, vol. 22, pp. 921–929, December 1987.

[29] L. Longo and M. Copeland, "A 13 bit ISDN-Band Oversampled ADC Using Two-Stage Third-Order Noise Shaping," *Proc. of the IEEE Custom Integrated Circuit Conf.*, pp. 21.2.1–4, 1988.

[30] W. Chou, P. Wong, and R. Gray, "Multi-Stage Sigma-Delta Modulation," *IEEE Trans. on Information Theory*, vol. 35, pp. 784–796, July 1989.

[31] M. Rebeschini, N. R. van Bavel, P. Rakers, R. Greene, J. Caldwell, and J. R. Haug, "A 16-b 160 kHz CMOS A/D Converter Using Sigma-Delta Modulation," *IEEE J. of Solid-State Circuits*, vol. 25, pp. 431–440, April 1990.

[32] T. Karema, T. Ritoniemi, and H. Tenhunen, "An Oversampled Sigma-Delta A/D Converter Circuit Using Two-Stage Fourth Order Modulator," *Proc. of the IEEE Intl. Symp. on Circuits and Systems*, vol. 4, pp. 3279–3282, 1990.

[33] H. Baher and E. Afifi, "Novel Fourth-Order Sigma-Delta Convertor," *IET Electronics Letters*, vol. 28, pp. 1437–1438, July 1992.

[34] K. Vleugels, S. Rabii, and B. Wooley, "A 2.5-V Sigma-Delta Modulator for Broadband Communications Applications," *IEEE J. of Solid-State Circuits*, vol. 36, pp. 1887–1899, December 2001.

[35] R. del Río, F. Medeiro, B. Pérez-Verdú, and A. Rodríguez-Vázquez, "High-Order Cascade Multibit ΣΔ Modulators for xDSL Applications," *Proc. of the IEEE Intl. Symp. on Circuits and Systems*, vol. 2, pp. 37–40, 2000.

[36] I. Dedic, "A Sixth-Order Triple-Loop ΣΔ CMOS ADC with 90 dB SNR and 100 kHz Bandwidth," *IEEE ISSCC Digest of Technical Papers*, pp. 188–189, 1994.

[37] K. Cornelissens and M. Steyaert, "A 1-V 84 dB DR 1 MHz Bandwidth Cascade 3–1 Delta-Sigma ADC in 65-nm CMOS," *Proc. of the IEEE European Solid-State Circuits Conf.*, pp. 332–335, 2009.

[38] R. Tortosa, A. Aceituno, J. M. de la Rosa, A. Rodríguez-Vázquez, and F. V. Fernández, "A 12-bit, 40MS/s Gm-C Cascade 3-2 Continuous-Time Sigma-Delta Modulator," *Proc. of the IEEE Intl. Symp. on Circuits and Systems*, pp. 1–4, 2007.

[39] R. del Río, F. Medeiro, J. M. de la Rosa, B. Pérez-Verdú, and A. Rodríguez-Vázquez, "Reliable Analysis of Settling Errors in SC Integrators: Application to ΣΔ Modulators," *IET Electronics Letters*, vol. 36, pp. 503–504, March 2000.

[40] P. Benabes, A. Gauthier, and R. Kielbasa, "New High-Order Universal ΣΔ Modulator," *IET Electronics Letters*, vol. 31, pp. 1575–1577, January 1995.

[41] N. Maghari *et al.*, "Sturdy MASH ΔΣ Modulator," *IET Electronics Letters.*, vol. 42, pp. 1269–1270, October 2006.

[42] N. Maghari *et al.*, "Multi-Loop Efficient Sturdy MASH Delta-Sigma Modulators," *Proc. of the IEEE Intl. Symp. on Circuits and Systems*, pp. 1216–1219, May 2008.

[43] A. Morgado, R. del Río, and J. M. de la Rosa, "A New Cascade ΣΔ Modulator for Low-voltage Wideband Applications," *IET Electronics Letters*, vol. 43, pp. 910–911, August 2007.

[44] T. Brooks, "Architecture Considerations for Multi-Bit ΣΔ ADCs," in *Analog Circuit Design – Structured Mixed-Mode Design, Multi-Bit Sigma-Delta Converters, Short Range RF Circuits* (M. Steyaert, A. H. M. van Roermund, and J. H. Huijsing, editors), Kluwer Academic Publishers, 2002.

[45] L. R. Carley, R. Schreier, and G. C. Temes, "Delta-Sigma ADCs with Multibit Internal Converters," in *Delta-Sigma Data Converters: Theory, Design and Simulation* (S. R. Norsworthy, R. Schreier, and G. C. Temes, editors), IEEE Press, 1997.

[46] A. A. Hamoui and K. W. Martin, "High-Order Multibit Modulators and Pseudo Data-Weighted-Averaging in Low-Oversampling ΔΣ ADCs for Broad-Band Applications," *IEEE Trans. on Circuits and Systems – I: Regular Papers*, pp. 72–85, January 2004.

[47] L. R. Carley and J. Kenney, "A 16-bit 4th Order Noise-Shaping D/A Converter," *Proc. of the IEEE Custom Integrated Circuits Conf.*, pp. 21.7.1–4, 1988.

[48] K. B. Klaasen, "Digitally Controlled Absolute Voltage Division," *IEEE Trans. on Instrumentation and Measurement*, vol. 24, pp. 106–112, June 1975.

[49] B. Leung and S. Sutarja, "Multibit Σ–Δ A/D Converter Incorporating a Novel Class of Dynamic Element Matching Techniques," *IEEE Trans. on Circuits and Systems – II: Analog and Digital Signal Processing*, vol. 39, pp. 35–51, January 1992.

[50] R. T. Baird and T. Fiez, "Linearity Enhancement of Multibit ΔΣ A/D and D/A Converters using Data Weighted Averaging," *IEEE Trans. on Circuits and Systems – II: Analog and Digital Signal Processing*, vol. 42, pp. 753–762, December 1995.

[51] R. Schreier and B. Zhang, "Noise-shaped multibit D/A converter employing unit elements," *IET Electronics Letters*, vol. 31, pp. 1712–1713, September 1995.

[52] R. E. Radke, A. Eshraghi, and T. S. Fiez, "A 14-bit Current-Mode ΣΔ DAC Based Upon Rotated Data Weighted Averaging," *IEEE J. of Solid-State Circuits*, vol. 35, pp. 1074–1084, August 2000.

[53] M. Vadipour, "Techniques for Preventing Tonal Behavior of Data Weighted Averaging Algorithm in Σ–Δ Modulators," *IEEE Trans. on Circuits and Systems – II: Analog and Digital Signal Processing*, vol. 47, pp. 1137–1144, November 2000.

[54] I. Fujimori, L. Longo, A. Hairapetian, K. Seiyama, S. Kosic, J. Cao, and S.-L. Chan, "A 90 dB SNR 2.5 MHz Output-Rate ADC Using Cascaded Multibit Delta-Sigma Modulation at 8× Oversampling Ratio," *IEEE J. Solid-State Circuits*, vol. 35, pp. 1820–1828, December 2000.

[55] A. A. Hamoui and K. Martin, "Linearity Enhancement of Multibit ΔΣ Modulators Using Pseudo Data-Weighted Averaging," *Proc. of the IEEE Intl. Symp. on Circuits and Systems*, pp. III.285–288, 2002.

[56] A. Hairapetian, G. C. Temes, and Z. X. Zhang, "Multibit sigma-delta modulator with reduced sensitivity to DAC nonlinearity," *IET Electronics Letters*, vol. 27, pp. 990–991, May 1991.

[57] B. P. Brandt and B. A. Wooley, "A 50 MHz Multibit Sigma-Delta Modulator for 12-b 2 MHz A/D Conversion," *IEEE J. of Solid-State Circuits*, vol. 26, pp. 1746–1756, December 1991.

[58] S. R. Norsworthy, R. Schreier, and G. C. Temes, *Delta-Sigma Data Converters: Theory, Design and Simulation.* IEEE Press, 1997.

[59] A. Rodríguez-Vázquez, F. Medeiro, J. M. de la Rosa, R. del Río, R. Tortosa, and B. Pérez-Verdú, "Sigma-Delta CMOS ADCs: An Overview of the State-of-the-Art," in *CMOS Telecom Data Converters* (A. Rodríguez-Vázquez, F. Medeiro, and E. Janssens, editors), Kluwer Academic Publishers, 2003.

[60] R. Schreier and G. C. Temes, *Understanding Delta-Sigma Data Converters.* IEEE Press, 2005.

[61] R. Reutemann, P. Balmelli, and Q. Huang, "A 33 mW 14b 2.5 MSample/s ΣΔ A/D Converter in 0.25 μm Digital CMOS," *IEEE ISSCC Digest of Technical Papers*, vol. 1, p. 316, 2002.

[62] R. Schreier and M. Snelgrove, "Bandpass Sigma-Delta Modulation," *IET Electronics Letters*, vol. 25, pp. 1560–1561, November 1989.

[63] P. H. Gailus, "Method and Arrangement for a Sigma Delta Converter for Bandpass Signals," *U.S. Patent 4,857,828, Aug. 1988, filed Jan. 28 1988*, 1989.

[64] J. M. de la Rosa, B. Pérez-Verdú, R. del Río, F. Medeiro, and A. Rodríguez-Vázquez, "Bandpass Sigma-Delta A/D Converters: Fundamentals, Architectures and Circuits," in *CMOS Telecom Data Converters* (A. Rodríguez-Vázquez, F. Medeiro, and E. Janssens, editors), Kluwer Academic Publishers, 2003.

[65] J. Cherry, W. Snelgrove, and W. Gao, "On the Design of a Fourth-Order Continuous-Time LC Delta-Sigma Modulator for UHF A/D Conversion," *IEEE Trans. on Circuits and Systems – II: Analog and Digital Signal Processing*, vol. 47, pp. 518–530, June 2000.

[66] B. Thandri and J. Silva-Martinez, "A 63 dB 75-mW Bandpass RF ADC at 950 MHz Using 3.8-GHz Clock in 0.25-μm SiGe BiCMOS Technology," *IEEE J. of Solid-State Circuits*, vol. 42, pp. 269–279, February 2007.

[67] J. Ryckaert, J. Borremans, B. Verbruggen, L. Bos, C. Armiento, J. Craninckx, and G. van der Plas, "A 2.4 GHz Low-Power Sixth-Order RF Bandpass ΔΣ Converter in CMOS," *IEEE J. of Solid-State Circuits*, vol. 44, pp. 2873–2880, November 2009.

[68] N. Beilleau, H. Aboushady, F. Montaudon, and A. Cathelin, "A 1.3V 26 mW 3.2 GS/s Undersampled LC Bandpass ΣΔ ADC for a SDR ISM-band Receiver in 130 nm CMOS," *Proc. of the IEEE Radio Frequency Integrated Circuits Symp.*, 2009.

[69] J. Ryckaert *et al.*, "A 6.1 GS/s 52.8 mW 43 dB DR 80 MHz Bandwidth 2.4 GHz RF Bandpass ΔΣ ADC in 40 nm CMOS," *Proc. of the IEEE Radio Frequency Integrated Circuits Symp.*, pp. 443–446, June 2010.

[70] K. Koli *et al.*, "A 900 MHz Direct Delta-Sigma Receiver in 65-nm CMOS," *IEEE J. of Solid-State Circuits*, pp. 2807–2818, Dec. 2010.

[71] E. Martens *et al.*, "RF-to-Baseband Digitization in 40 nm CMOS With RF Bandpass ΔΣ Modulator and Polyphase Decimation Filter," *IEEE J. of Solid-State Circuits*, vol. 47, pp. 990–1002, April 2012.

[72] H. Shibata *et al.*, "A DC-to-1 GHz Tunable RF ΔΣ ADC Achieving DR=74 dB and BW=150 MHz at f_0 =450 MHz Using 550 mW," *IEEE ISSCC Digest of Technical Papers,* pp. 150–151, February 2012.

[73] S. Gupta *et al.*, "A 0.8-2 GHz Fully-Integrated QPLL-Timed Direct-RF-Sampling Bandpass ΣΔ ADC in 0.13 μm CMOS," *IEEE J. of Solid-State Circuits*, vol. 47, pp. 1141–1153, May 2012.

[74] A. Ashry and H. Aboushady, "A 4th Order 3.6 GS/s RF ΣΔ ADC with a FoM of 1 pJ/bit," *IEEE Trans. on Circuits and Systems – I: Regular Papers*, vol. 60, pp. 2606–2617, October 2013.

[75] M. Englund *et al.*, "A Programmable 0.7-2.7 GHz Direct ΔΣ Receiver in 40 nm CMOS," *IEEE J. of Solid-State Circuits*, pp. 644–655, March 2015.

[76] S. A. Jantzi *et al.*, "Quadrature Bandpass ΔΣ Modulation for Digital Radio," *IEEE J. of Solid-State Circuits*, vol. 32, pp. 1935–1950, December 1997.

[77] T. Paulus, S. S. Somayajula, T. A. Miller, B. Trotter, C. Kyong, and D. A. Kerth, "A CMOS IF Transceiver with Reduced Analog Complexity," *IEEE J. of Solid-State Circuits*, vol. 33, pp. 2154–2159, December 1998.

[78] L. Louis, J. Abcarius, and G. W. Roberts, "An Eighth-Order Bandpass ΔΣ Modulator for A/D Conversion in Digital Radio," *IEEE J. of Solid-State Circuits*, vol. 34, pp. 423–431, April 1999.

[79] S. A. Jantzi and W. M. Snelgrove, "Bandpass Sigma-Delta Analog-to-Digital Conversion," *IEEE Trans. on Circuits and Systems*, vol. 38, pp. 1406–1409, November 1991.

[80] S. A. Jantzi, W. M. Snelgrove, and P. F. Ferguson, "A Fourth-Order Bandpass Sigma-Delta Modulator," *IEEE J. of Solid-State Circuits*, vol. 28, pp. 282–291, March 1993.

[81] J. V. Engelen and R. van de Plassche, *BandPass Sigma-Delta Modulators: Stability Analysis, Performance and Design Aspects.* Kluwer Academic Publishers, 1999.

[82] L. Breems and J. H. Huijsing, *Continuous-Time Sigma-Delta Modulation for A/D Conversion in Radio Receivers.* Kluwer Academic Publishers, 2001.

[83] P. G. R. Silva and J. H. Huijsing, *High Resolution IF-to-Baseband ΣΔ ADC for Car Radios*. Springer, 2008.

[84] R. H. van Veldhoven and A. H. M. van Roermund, *Robust Sigma Delta Converters*. Springer, 2011.

[85] R. F. Cormier, T. L. Sculley, and R. H. Bamberger, "A Fourth Order Bandpass Delta-Sigma Modulator with Digitally Programmable Pass-band Frequency," *Analog Integrat. Circuits Signal Process.*, vol. 12, pp. 217–229, 1997.

[86] K. Yamamoto, A. C. Carusone, and F. P. Dawson, "A Delta-Sigma Modulator With a Widely Programmable Center Frequency and 82 dB Peak SNDR," *IEEE J. of Solid-State Circuits*, vol. 43, pp. 1772–1782, August 2008.

[87] A. K. Ong and B. A. Wooley, "A Two-Path Bandpass ΣΔ Modulator for Digital IF Extraction at 20 MHz," *IEEE J. of Solid-State Circuits*, vol. 32, pp. 1920–1934, December 1997.

[88] V. Ferragina *et al.*, "Gain and Offset Mismatch Calibration in Time-Interleaved Multipath A/D Sigma-Delta Modulators," *IEEE Trans. on Circuits and Systems – I: Regular Papers*, vol. 51, pp. 2365–2373, December 2004.

[89] D. Feng *et al.*, "Polyphase Decomposition for Tunable Band-Pass Sigma-Delta A/D Converters," *IEEE J. on Emerging and Selected Topics in Circuits and Systems*, vol. 5, pp. 537–547, December 2015.

[90] O. Shoaei, *Continuous-Time Delta-Sigma A/D Converters for High Speed Applications.* PhD. dissertation, Carleton University, 1995.

[91] S. Pavan, "The Inconvenient Truth about Alias Rejection in Continuous-Time ΔΣ Converters," *Proc. of the IEEE Intl. Symp. on Circuits and Systems (ISCAS)*, pp. 526–529, May 2011.

[92] J. Cherry and W. Snelgrove, "Excess Loop Delay in Continuous-Time Delta–Sigma Modulators," *IEEE Trans. on Circuits and Systems – II: Analog and Digital Signal Processing*, vol. 46, pp. 376–389, April 1999.

[93] R. Tortosa, J. M. de la Rosa, F. V. Fernández, and A. Rodríguez-Vázquez, "A New High-Level Synthesis Methodology of Cascaded Continuous-Time ΣΔ Modulators," *IEEE Trans. on Circuits and Systems – II: Express Briefs*, vol. 53, pp. 739–743, August 2006.

CHAPTER 3

CIRCUIT ERRORS IN SWITCHED-CAPACITOR $\Sigma\Delta$ MODULATORS

As discussed in previous chapters, ADCs based on $\Sigma\Delta$ modulation offer key advantages for practical implementation in present-day CMOS processes compared to other data-conversion techniques. Unlike Nyquist-rate converters, which require high precision in their building blocks to achieve overall high accuracy, oversampling and quantization noise shaping allow speed to be traded for accuracy. In this way, an operation that is relatively insensitive to imperfections in the analog circuit can be obtained at the cost of increased complexity and speed in the associated digital circuitry [1].

The principles of $\Sigma\Delta$ modulation were presented in Chapter 1 and alternative $\Sigma\Delta$M topologies and implementation techniques (DT and CT) were presented in Chapter 2. However, it was only the quantization error that was taken into account when considering the achievable performance of the different alternatives. Besides this error – which is inherent to any analog-to-digital conversion technique – only the effect of DAC errors was considered when comparing the performance of single-bit and multi-bit $\Sigma\Delta$Ms at the architectural level.

This chapter analyzes the main nonideal mechanisms affecting the performance of SC $\Sigma\Delta$Ms. Although it is commonly accepted that $\Sigma\Delta$ ADCs are less sensitive to nonidealities in the analog circuitry than other conversion techniques, their impact will be larger the more demanding the ADC specifications. Therefore, the influence of these errors on the modulator performance must be carefully considered during early design phases. However, this chapter is not intended to be an exhaustive description of all nonideal effects, but rather a practical description of the main ones. The aim is to provide sufficient insight into the problem and to present analytical procedures that can be applied to other error mechanisms.

Circuit errors with a large influence on the behavior of SC-$\Sigma\Delta$Ms – such as integrator leakage, capacitor mismatch, settling errors, and kT/C noise – are analyzed, and the main sources of distortion

Sigma-Delta Converters: Practical Design Guide, Second Edition. José M. de la Rosa.

in SC-ΣΔMs are also discussed. System-level considerations, behavioral models, and closed-form expressions are obtained for the influence of each nonideality. From these, useful guidelines for the design of ΣΔMs can be extracted. These are put into practice in a case study at the end of the chapter.

3.1 Overview of Nonidealities in Switched-capacitor ΣΔ Modulators

There are a number of circuit nonidealities and nonlinearities that degrade the performance of analog ΣΔ modulator blocks. The way in which these nonidealities affect the performance of ΣΔMs depends on many different factors: the nature of the error itself, the influence of the specific circuit, its effect on the modulator noise transfer function, and so on.

In the case of SC implementations, the main nonideal effects, as illustrated in Figure 3.1, can be grouped according to the ΣΔM circuit they affect:

- amplifiers: output swing, finite gain, dynamic limitations, and circuit noise;
- switches: on-resistance, thermal noise, charge injection, and clock feedthrough;
- capacitors: mismatch and nonlinearity;
- multi-bit ADCs and DACs: offset, gain error, and nonlinearity;
- clock: jitter.

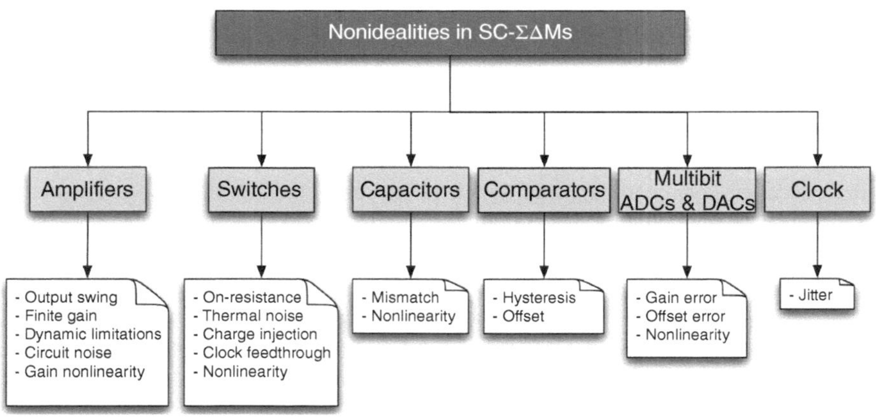

Figure 3.1 The main nonidealities affecting the performance of switched-capacitor ΣΔMs.

The influence of nonidealities on the performance of ΣΔMs strongly depends on the location of the corresponding noise source in the modulator. According to this criterion, the above errors can be classified into two main families:

- **Errors that modify the modulator NTF**, such as the finite amplifier gain and gain-bandwidth product and capacitor mismatch. Their effect strongly depends on the modulator topology. For instance, cascade ΣΔMs are more sensitive to capacitor mismatch and finite amplifier gain than single-loop architectures. The same applies for low-pass ΣΔMs with optimized zeros and band-pass ΣΔMs with local feedback.

- **Errors that can be modeled as additive noise sources at the modulator input** and so are not in-band attenuated by the noise shaping. Their effect is thus independent of the modulator topology. Some errors belonging to this family are clock jitter, circuit noise, and distortion caused by circuit nonlinearities.

For those nonideal effects affecting the modulator NTF, the procedure that is commonly used to analyze the impact on the modulator performance is [2, 3]:

- Obtain an equivalent integrator circuit taking into account the nonideal effect under study.
- Analyze the impact of the nonideality on the integrator transfer function ITF(z), such that ITF(z) → $\text{ITF}_\epsilon(z)$, with ϵ being the error vector including all nonideal parameters involved in the equivalent integrator circuit obtained in the previous step.
- To compute the effect of ϵ on a ΣΔM, the integrator transfer functions are replaced with $\text{ITF}_\epsilon(z)$ and a linear quantizer model is considered to obtain the nonideal $\text{NTF}_\epsilon(z)$.
- The nonideal NTF is integrated within the signal band to obtain the degraded in-band noise power IBN(ϵ). Usually, some approximations are required to obtain closed-form expressions for IBN(ϵ) as a function of ϵ.

This procedure, which can be formulated as:

$$\text{ITF}(z) \rightarrow \text{ITF}_\epsilon(z) \rightarrow \text{NTF}_\epsilon(z) \rightarrow \text{IBN}(\epsilon) \tag{3.1}$$

is applied to low-pass SC-ΣΔMs throughout this section, although it can be easily generalized to:

- band-pass SC-ΣΔMs, – working on the resonator transfer functions RTF(z);
- CT-ΣΔMs – working on S-domain, as will be shown in the next chapter.

As case studies, single-loop ΣΔMs with distributed feedback (see Section 2.2) and ΣΔ cascades (see Section 2.3) are considered. For these modulator topologies, optimal coefficients for second-, third-, and fourth-order ΣΔMs can be found in the literature for both the single-bit case [4] and the multi-bit case [5].

Wherever behavioral simulation results corresponding to single-bit single-loop implementations are presented throughout this section, the following modulator coefficients have been used for the distributed feedback topologies (see Figure 2.7):

- $(a_1, a_2) = (0.5, 0.5)$ for the second-order ΣΔM (SL2 for short);
- $(a_1, a_2, a_3) = (0.2, 0.5, 0.5)$ for the third-order ΣΔM (SL3 for short);
- $(a_1, a_2, a_3, a_4) = (0.2, 0.2, 0.5, 0.5)$ for the fourth-order ΣΔM (SL4 for short).

The performance of single-bit cascade ΣΔMs is demonstrated throughout this section with behavioral simulation results on a 2-1-1 ΣΔM (see Figure 2.14), with the following coefficients:

- $(a_1, a_2, a_3, a_4, b_1, c_1, b_2, c_2) = (0.5, 0.5, 0.5, 0.5, 2, 0.5, 1, 1)$

3.2 Finite Amplifier Gain in SC-ΣΔMs

In the previous chapters, the ideal performance of the different low-pass SC-ΣΔMs presented was derived when considering the ideal transfer function of an ideal SC FE integrator:

$$\mathrm{ITF}(z) = \frac{z^{-1}}{1 - z^{-1}} \tag{3.2}$$

If the finite amplifier gain A_v and the parasitic capacitor C_P at the amplifier summing node are accounted for in the charge transfer of an SC FE integrator, as shown in Figure 3.2, the difference equation can be written as [3]:

$$\begin{aligned} v_o(nT_s) = {} & \frac{1 + \left(1 + \frac{C_P}{C_I}\right)\frac{1}{A_v}}{1 + \left(1 + \frac{C_P}{C_I} + \sum_{i=1}^{N_i} \frac{C_{Si}}{C_I}\right)\frac{1}{A_v}} v_o[(n-1)T_s] \\ & + \frac{\sum_{i=1}^{N_i} \frac{C_{Si}}{C_I} v_i[(n-1)T_s]}{1 + \left(1 + \frac{C_P}{C_I} + \sum_{i=1}^{N_i} \frac{C_{Si}}{C_I}\right)\frac{1}{A_v}} \end{aligned} \tag{3.3}$$

Transforming Equation (3.3) to the Z-domain and identifying terms as in the following expression,

$$v_o(z) = \mathrm{ITF}_{A_v}(z) \left[\sum_{i=1}^{N_i} \frac{C_{Si}}{C_I} v_i(z)\right] \tag{3.4}$$

the transfer function of the integrator when affected by finite amplifier gain – that is, the transfer function of a *leaky* integrator – yields:

$$\begin{aligned} \mathrm{ITF}_{A_v}(z) = {} & \frac{1}{1 + \left(1 + \frac{C_P}{C_I} + \sum_{i=1}^{N_i} \frac{C_{Si}}{C_I}\right)\frac{1}{A_v}} \\ & \times \frac{z^{-1}}{1 - z^{-1}\left[\frac{1+\left(1+\frac{C_P}{C_I}\right)\frac{1}{A_v}}{1+\left(1+\frac{C_P}{C_I}+\sum_{i=1}^{N_i}\frac{C_{Si}}{C_I}\right)\frac{1}{A_v}}\right]} \end{aligned} \tag{3.5}$$

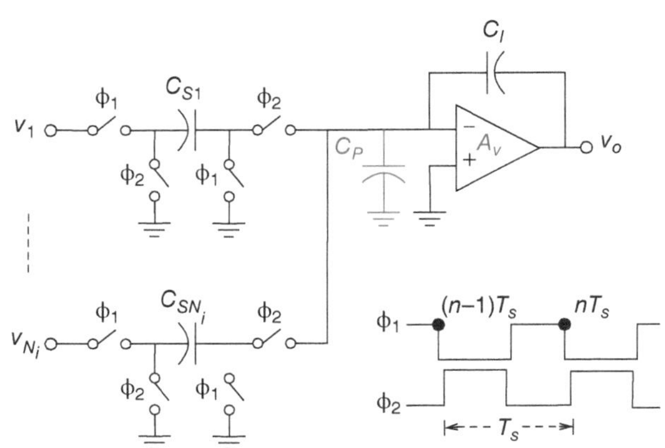

Figure 3.2 SC FE integrator with N_i input paths and finite amplifier gain A_v.

Therefore, when compared to the ideal case in Equation (3.2), amplifier finite gain introduces a gain error in the ITF and a shift of its pole from its ideal position at DC ($z = 1$). Neglecting the gain error and noting that $\sum_{i=1}^{N_i} \frac{C_{Si}}{C_I} = \sum_{i=1}^{N_i} g_i$, Equation (3.5) can be approximated to a more compact expression:

$$\mathrm{ITF}_{A_v}(z) \approx \frac{z^{-1}}{1 - z^{-1}\left(1 - \frac{\sum_{i=1}^{N_i} g_i}{A_v}\right)} \tag{3.6}$$

As an example, let us consider the effect of the modified ITF on the second-order DT-ΣΔM in Figure 2.3b. Assuming that $\sum_{i=1}^{N_i} g_i \sim 1$ for the two integrators in the modulator, the NTF affected by finite amplifier gain can be easily derived [3]. Given that the poles of the ITFs become the zeros of the NTF, both zeros of the NTF move away from DC as the amplifier gain decreases. Figure 3.3 depicts the PSD of the quantization error that is obtained for several values of A_v and illustrates the degradation of the modulator noise-shaping. The following approximate expression can be derived for the in-band noise when integrator leakage is accounted for [3]:

$$\mathrm{IBN}_2(A_v) \approx \frac{\Delta^2}{12} \frac{1}{(k_q a_1 a_2)^2} \left(\frac{1}{{A_v}^4 \mathrm{OSR}} + \frac{2\pi^2}{3{A_v}^2 \mathrm{OSR}^3} + \frac{\pi^4}{5\mathrm{OSR}^5} \right) \tag{3.7}$$

Note that the DC gain of the amplifiers should be in the range of the oversampling ratio ($A_v \approx \mathrm{OSR}$) in order to keep every term in Equation (3.7) proportional to OSR^{-5} and retain the ideal noise shaping. For usual values of the oversampling ratio and the amplifier DC gain, this equation can be further simplified to:

$$\mathrm{IBN}_2(A_v) \approx \frac{\Delta^2}{12} \frac{1}{(k_q a_1 a_2)^2} \left(\frac{2\pi^2}{3{A_v}^2 \mathrm{OSR}^3} + \frac{\pi^4}{5\mathrm{OSR}^5} \right) \tag{3.8}$$

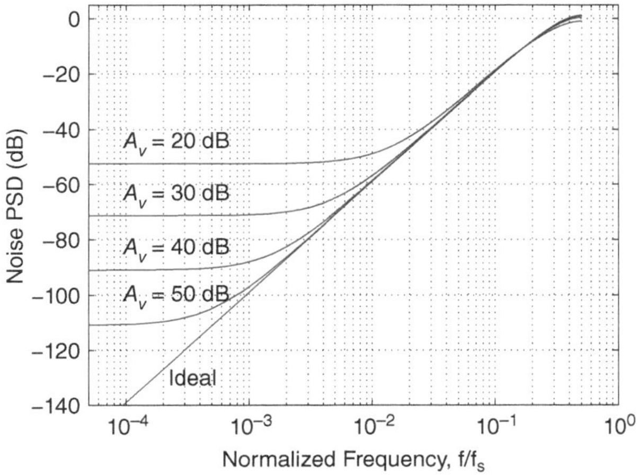

Figure 3.3 Degradation of the noise-shaping of a second-order SC-ΣΔM with finite amplifier gain.

A similar procedure can be followed to derive the modified NTF of Lth-order loops affected by integrator leakage and thus calculate the increased in-band noise. For an Lth-order SC-ΣΔM with distributed feedback, the in-band noise can be obtained as follows [3]:

$$\mathrm{IBN}_L(A_v) \approx \frac{\Delta^2}{12}\frac{1}{(k_q\prod_{i=1}^{L}a_i)^2} \times \left[\frac{1}{A_v{}^{2L}\mathrm{OSR}} + \sum_{i=1}^{L-1}\frac{L(L-1)\cdots(L-i+1)\pi^{2i}}{i!(2i+1)A_v{}^{2(L-i)}\mathrm{OSR}^{(2i+1)}} + \frac{\pi^{2L}}{(2L+1)\mathrm{OSR}^{2L+1}}\right] \tag{3.9}$$

where again the amplifier DC gains must be in the range of the oversampling ($A_v \approx$ OSR) to keep every term in Equation (3.9) proportional to $\mathrm{OSR}^{-(2L+1)}$ and retain the ideal Lth-order noise shaping. The expression above can usually be further simplified to [3]:

$$\mathrm{IBN}_L(A_v) \approx \frac{\Delta^2}{12}\frac{1}{(k_q\prod_{i=1}^{L}a_i)^2} \times \left[\frac{L\pi^{2(L-1)}}{(2L-1)A_v{}^2\mathrm{OSR}^{(2L-1)}} + \frac{\pi^{2L}}{(2L+1)\mathrm{OSR}^{2L+1}}\right] \tag{3.10}$$

Integrator leakage can be foreseen to have a stronger impact on cascade ΣΔMs, since the degradation of the ITF filtering leads to a modification of the cascaded loop filters (on the analog side) that is not compensated for by the cancelation logic (DCL on the digital side) – see Figure 2.12. This imbalance will cause the quantization errors of all the cascaded stages to appear at the modulator output.

For the particular case of a 2-1-1 DT cascade, like the one illustrated in Figure 2.14, the IBN when integrator leakage is accounted for can be obtained as [3]:

$$\mathrm{IBN}_{211}(A_v) \approx \frac{\Delta_1^2}{12}\frac{1}{(k_q a_1 a_2)^2}\frac{4\pi^2}{3A_v^2\mathrm{OSR}^3} + \frac{\Delta_2^2}{12}\frac{1}{c_1^2}\frac{\pi^4}{5A_v^2\mathrm{OSR}^5} + \frac{\Delta_3^2}{12}\frac{1}{(c_1c_2)^2}\left(\frac{\pi^6}{7A_v^2\mathrm{OSR}^7} + \frac{\pi^8}{9\mathrm{OSR}^9}\right) \tag{3.11}$$

Note that the amplifier DC gain required to retain the ideal noise shaping increases while moving from the back-end to the front-end stages of the cascade. Therefore, for the third-stage amplifier $A_v \approx$ OSR is sufficient, but for the second-stage amplifier, $A_v \approx \mathrm{OSR}^2$ and for the first-stage amplifiers $A_v \approx \mathrm{OSR}^3$. Note also that, if multi-bit quantization of B bits is employed in the last stage of the cascade, these requirements further increase by a factor $(2^B - 1)^2$.

A detailed analysis of the effect of integrator leakage on the IBN of generic cascade SC-ΣΔMs, as well as of particular cascade configurations, can be found in the literature [3]. For an Lth-order N-stage cascade DT-ΣΔM like the one illustrated in Figure 2.12, the IBN when considering integrator leakage yields [3]:

$$\mathrm{IBN}_{\mathrm{casc}}(A_v) \approx \frac{\Delta_1^2}{12}\frac{1}{(k_q\prod_{i=1}^{L_1}a_i)^2}\frac{\alpha_1\pi^{2(L_1-1)}}{(2L_1-1)A_v^2\mathrm{OSR}^{2L_1-1}}$$

$$+ \frac{\Delta_2^2}{12}\frac{1}{c_1^2}\frac{\alpha_2\pi^{2(L_1+L_2-1)}}{[2(L_1+L_2)-1]A_v^2\text{OSR}^{2(L_1+L_2)-1}}$$
$$+\cdots+\frac{\Delta_N^2}{12}\frac{1}{\prod_{i=1}^{N-1}c_i^2}$$
$$\times\left[\frac{\alpha_N\pi^{2(L-1)}}{(2L-1)A_v^2\text{OSR}^{2L-1}}+\frac{\pi^{2L}}{(2L+1)\text{OSR}^{2L+1}}\right] \quad (3.12)$$

where L_i corresponds to the order of the ith stage in the cascade and the values of coefficients α_i depend on L_i.

Figure 3.4 shows the effect of the finite DC gain of amplifiers on single-loop and cascade SC-ΣΔMs, illustrated for second-, third- and fourth-order single-loops and for a 2-1-1 cascade. In

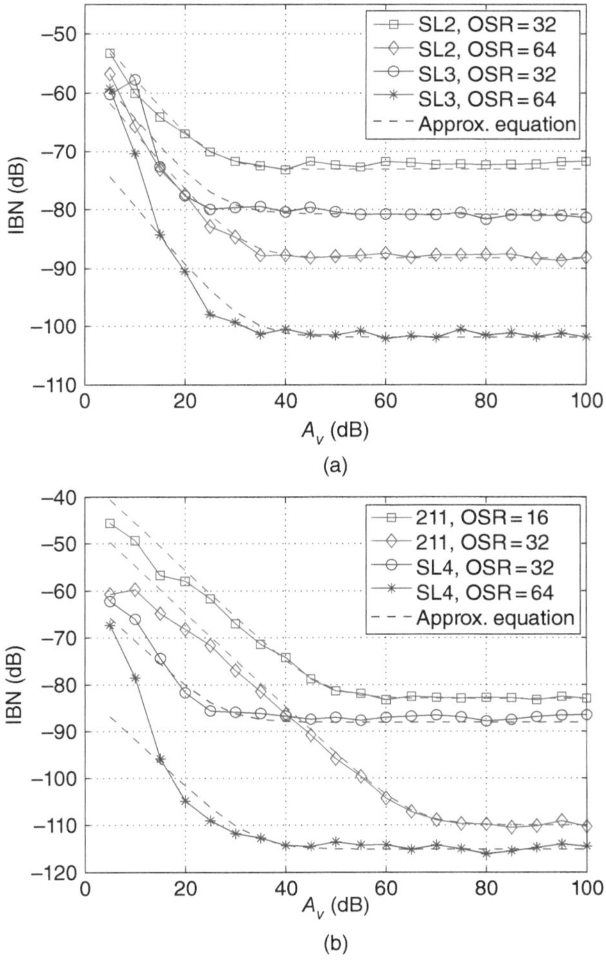

Figure 3.4 Influence of finite amplifier gain on the in-band noise of SC-ΣΔMs: (a) second- and third-order loops; (b) 2-1-1 cascade and fourth-order loop. Approximate results obtained from Equations (3.10) and (3.11).

all cases, the in-band noise is computed from the modulator NTF using the nonapproximated ITF in Equation (3.5), as well as from the approximated closed-form expressions in Equations (3.10) and (3.11). Note that both results are in good accordance. Note also that the larger sensitivity of cascade ΣΔMs to integrator leakages is evident from Figure 3.4b.

3.3 Capacitor Mismatch in SC-ΣΔMs

As illustrated in Figure 2.4, in SC-ΣΔMs, integrator gain coefficients g_i are implemented as capacitor ratios C_{Si}/C_I and the implemented values will thus deviate from the nominal ones due to variations in technological process parameters. For the case of a particular gain coefficient g that is implemented as the ratio of m to n unit capacitors C_u, the actual implemented coefficient g_ϵ will exhibit an error ϵ_g that can be estimated as [3]:

$$\left.\begin{aligned} g_\epsilon &= g(1 \pm \epsilon_g) \\ g &= \frac{C_S}{C_I} = \frac{mC_u}{nC_u} \end{aligned}\right\} \Rightarrow \epsilon_g = 3\frac{\sigma_g}{g} = 3\sqrt{\frac{1}{m} + \frac{1}{n}}\frac{\sigma_{C_u}}{C_u} \leq 3\sqrt{2}\sigma_C \tag{3.13}$$

where the integrator gain error is estimated, in the worst case, as three times its relative standard deviation. This is related in turn to the relative standard deviation of the unit capacitor used, σ_C . Note that the estimation for ϵ_g in Equation (3.13) should be divided by $\sqrt{2}$ in fully-differential SC integrators.

Nowadays, SC-ΣΔMs are mostly implemented in mixed-mode technological processes that include precise capacitor primitives, such as MiM or MoM capacitors, with a mismatch typically lower than 0.1%. This means that integrator gain errors in SC-ΣΔMs will normally be lower than 0.3%, or even less if a large number of unit capacitors and common-centroid layout techniques are used for implementing the coefficients.

These small deviations of the integrator coefficients due to capacitor mismatch can be foreseen to have little impact on the in-band noise of single-loop ΣΔMs, since the filtering provided by the integrators remains unchanged. Indeed, if integrator gain errors are accounted for, the IBN of an Lth-order SC-ΣΔM with distributed feedback can be estimated as,

$$\mathrm{IBN}_L(\epsilon_g) \approx \frac{\Delta^2}{12}\frac{1}{(k_q \prod_{i=1}^{L} a_i)^2}\frac{\pi^{2L}}{(2L+1)\mathrm{OSR}^{2L+1}}\prod_{i=1}^{L}(1+\epsilon_{gi})^2 \tag{3.14}$$

where ϵ_{gi} refers to the gain error of the ith integrator, which can be estimated in the worst case as $3\sigma_C$. Note from Equation (3.14) that, for the IBN of a second-order ΣΔM to increase by 3 dB, integrator gain errors should be as large as 20%: too large indeed to be considered an actual mismatch!

Conversely, capacitor mismatch has a strong impact on cascade ΣΔMs, since the deviation of the integrator gains is not compensated for by the digital coefficients of the DCL. Therefore, quantization errors of the cascaded stages will leak to the modulator output with low-order shapings, considerably increasing the modulator IBN.

For the general case of an Lth-order N-stage cascade SC-ΣΔM, the IBN if integrator gain errors are taken into account can be approximated to [3],

$$\mathrm{IBN}_{\mathrm{casc}}(\epsilon_g) \approx \frac{\Delta_1^2}{12}\frac{1}{(k_q \prod_{i=1}^{L_1} a_i)^2}\frac{\pi^{2L_1}}{(2L_1+1)\mathrm{OSR}^{2L_1+1}}\left(\sum_{i=1}^{L_1}\epsilon_{gi}\right)^2$$

$$+\frac{\Delta_2^2}{12}\frac{1}{c_1^2}\frac{\pi^{2(L_1+L_2)}}{[2(L_1+L_2)+1]\mathrm{OSR}^{2(L_1+L_2)+1}}\left(\sum_{i=L_1+1}^{L_1+L_2}\epsilon_{gi}\right)^2$$

$$+\cdots+\frac{\Delta_N^2}{12}\frac{1}{\prod_{i=1}^{N-1}c_i^2}\frac{\pi^{2L}}{(2L+1)\mathrm{OSR}^{2L+1}}\left(1+\sum_{i=L-L_N}^{L}\epsilon_{gi}\right)^2 \tag{3.15}$$

whereas, for the particular case of a 2-1-1 cascade, the former equation yields [3]:

$$\mathrm{IBN}_{211}(\epsilon_g)\approx\frac{\Delta_1^2}{12}\frac{1}{(k_q a_1 a_2)^2}\frac{\pi^4}{5\mathrm{OSR}^5}(\epsilon_{g1}+\epsilon_{g2})^2+\frac{\Delta_2^2}{12}\frac{1}{c_1^2}\frac{\pi^6}{7\mathrm{OSR}^7}\epsilon_{g3}^2$$

$$+\frac{\Delta_3^2}{12}\frac{1}{(c_1c_2)^2}\frac{\pi^8}{9\mathrm{OSR}^9}(1+\epsilon_{g4})^2 \tag{3.16}$$

Note that – similar to what happened in Section 3.2 for the case of amplifier DC gain – the requirements on the integrator gain error for retaining the ideal noise shaping get more stringent as we move from the back-end to the front-end stages. Thus, for the integrator in the second stage of the cascade $\epsilon_{g3} \approx \mathrm{OSR}^{-1}$ is sufficient, but for the first-stage integrators $\epsilon_{g1}, \epsilon_{g2} \approx \mathrm{OSR}^{-2}$. Note also that, if multi-bit quantization of B bits is employed in the last stage of the cascade, these requirements again increase by a factor $(2^B - 1)^2$.

Figure 3.5 illustrates this increasing impact of capacitor mismatch with OSR and with the stage of the cascade being considered. Note that the results displayed correspond to worst-case estimates of the IBN based on Equation (3.16) with $\epsilon_{gi} = 3\sigma_C$. More accurate estimates would require Monte Carlo simulation of the modulator behavioral model considering the particular implementation of each integrator gain coefficient in terms of unit capacitors.

3.4 Integrator Settling Error in SC-ΣΔMs

Speed limitations in SC integrators due to the limited dynamic response of amplifiers cause errors in the charge transfer. The impact of the resulting error in the integrator output voltage settling error on the modulator performance will be higher, the higher the sampling frequency. As the clock frequency increases in SC-ΣΔMs to cope with larger conversion bandwidths, integrator settling error becomes one of the bottlenecks in practical implementation. On the one hand, the time slot for the integrator operation gets reduced; on the other, the amplifier dynamic requirements must be minimized to optimize the modulator power consumption. Therefore, an adequate understanding of the mechanisms degrading the settling of SC integrators and an accurate quantification of the generated errors become mandatory in obtaining efficient ΣΔ designs.

3.4.1 Behavioral Model for the Integrator Settling

The behavioral model of the transient response for SC FE integrators included in this section is based on the analysis presented in an earlier book [6]. The model includes the effect of the amplifier dynamic limitations, such as finite gain–bandwidth product (GB) and slew rate (SR), on the charge transfer during both the integration and sampling phases. Also, parasitic capacitors associated with both amplifiers and switches, as well as the capacitor load at the integrator output – which changes from integration to sampling – are taken into account. To accurately describe the dynamic performance and

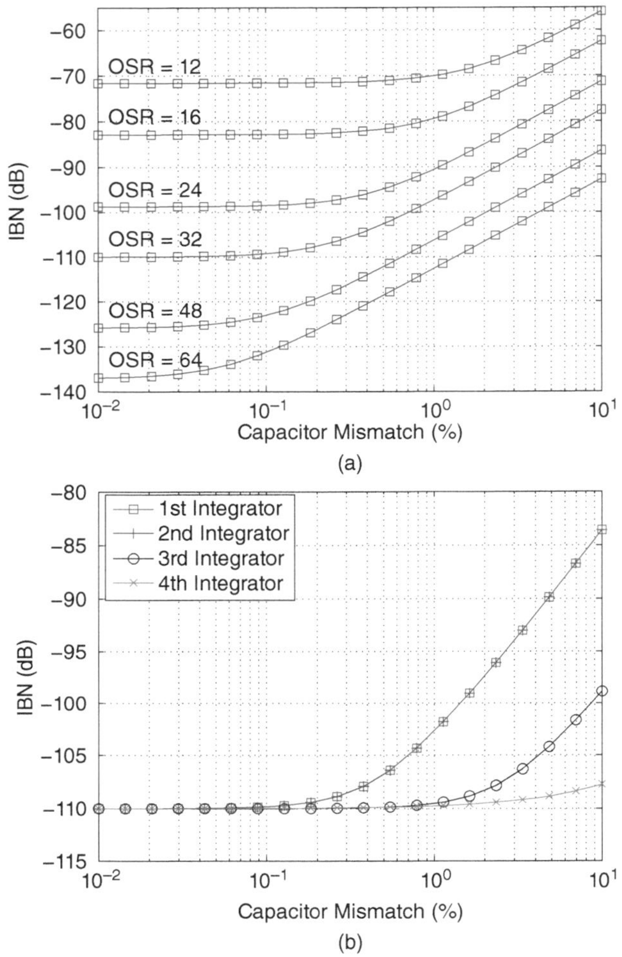

Figure 3.5 Influence of capacitor mismatch on the in-band noise of a 2-1-1 SC-ΣΔM: (a) considering the same mismatch error in all integrators; (b) individual impact of the mismatch error in each integrator for OSR = 32. Worst-case estimations of IBN considering $\epsilon_{gi} = 3\sigma_C$ in Equation (3.16).

determine the integrator output voltage, the equivalent circuit shown in Figure 3.6 is solved in the behavioral model. In the circuit scheme, the SC integrator under study is considered to have N_i input branches and another SC integrator acting as a load; that is, with the N_o input branches connected to its output. On the other hand, the amplifier is modeled as shown in Figure 3.7, with a single-pole dynamic (to account for the finite bandwidth) and a nonlinear characteristic with maximum output current I_o (to account for the limited SR).[1]

The analysis of this model for the incomplete settling error begins with the computation of the equivalent capacitive load at the amplifier output node during both the sampling (ϕ_1) and integration

[1]Note that C_o in Figure 3.7 is merged with C_L in Figure 3.6. Note also that r_o is included in the amplifier model in Figure 3.7 for completeness, but its effect is actually neglected in the analysis of the integrator dynamics in Figure 3.6 under the assumption $g_o \ll g_m$.

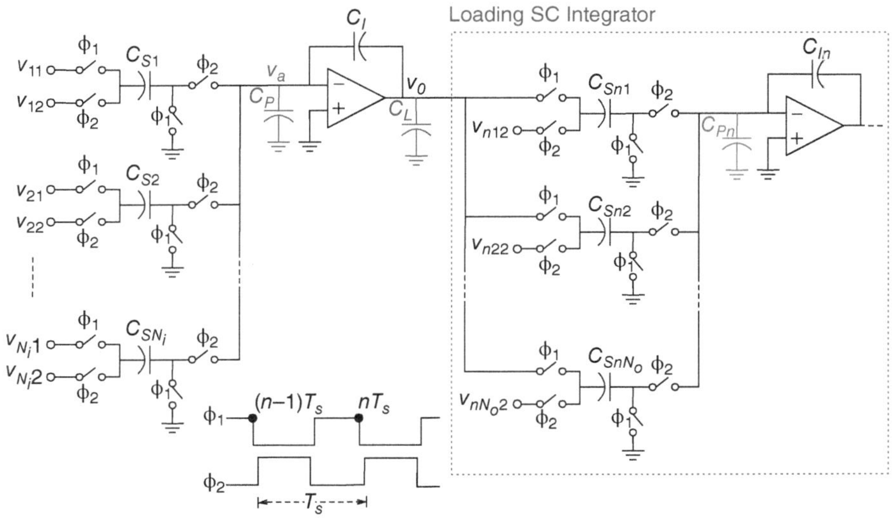

Figure 3.6 SC FE integrator under consideration followed by a loading SC integrator.

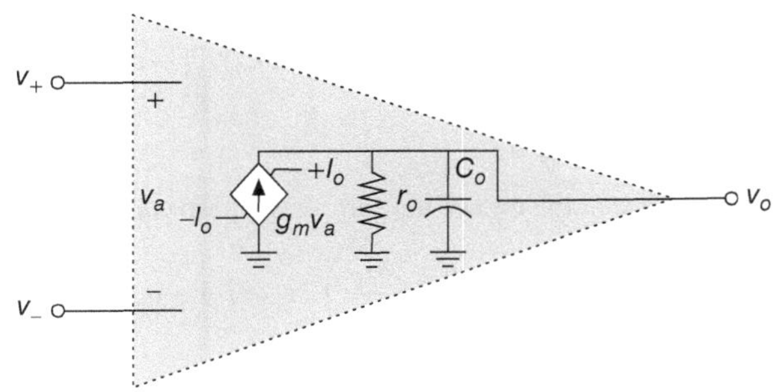

Figure 3.7 Amplifier single-pole model with limited output current.

(ϕ_2) phases. These are respectively given by:

$$C_{\text{eq},\phi_1} = C_P + \left(C_L + \sum_{i=1}^{N_o} C_{Sni}\right)\left(1 + \frac{C_P}{C_I}\right)$$

$$C_{\text{eq},\phi_2} = C_P + \sum_{i=1}^{N_i} C_{Si} + C_L\left(1 + \frac{C_P + \sum_{\mathrm{i}=1}^{N_i} C_{Si}}{C_I}\right) \tag{3.17}$$

where C_{Si} stands for the sampling capacitor of the ith SC input branch of the integrator under consideration, C_{Sni} is the sampling capacitor of the ith SC input branch of the load integrator, C_P is the parasitic capacitor at the summation node of the input SC branches, and C_L is the amplifier load capacitor.

The settling model is analyzed during a complete clock cycle (during both clock phases) considering the different possibilities for the amplifier dynamic operation – i.e., linearly or in slew – and keeping track of the voltage at both the integrator output v_o and the amplifier summation node v_a. Therefore, the error in the integrator output voltage at the end of one sampling–integration process can be accurately obtained.

Let $v_a[(n-1/2)T_s]$ and $v_o[(n-1/2)T_s]$ be the respective amplifier input and output voltages at the end of a preceding integration phase. These will serve as initial conditions to derive the integrator evolution during a complete clock cycle. The voltage at the amplifier summation node at the end of the next sampling phase – that is, at $t = nT_s$ – can be accurately obtained as [6]:

$$v_a(nT_s) = \begin{cases} v_{a0,\phi_1} e^{-\frac{g_m}{C_{\mathrm{eq},\phi_1}}\frac{T_s}{2}} & |v_{a0,\phi_1}| \le \frac{I_o}{g_m} \\ \frac{I_o}{g_m}\mathrm{sgn}(v_{a0,\phi_1}) e^{-\frac{g_m}{C_{\mathrm{eq},\phi_1}}\left(\frac{T_s}{2}-t_{o,\phi_1}\right)} & |v_{a0,\phi_1}| > \frac{I_o}{g_m}, t_{o,\phi_1} \le \frac{T_s}{2} \\ v_{a0,\phi_1} - \frac{I_o}{C_{\mathrm{eq},\phi_1}}\mathrm{sgn}(v_{a0,\phi_1})\frac{T_s}{2} & |v_{a0,\phi_1}| > \frac{I_o}{g_m}, t_{o,\phi_1} > \frac{T_s}{2} \end{cases} \tag{3.18}$$

where t_{o,ϕ_1} is the duration of the slew-rate-limited integrator settling (relative to $T_s/2$), given by:

$$t_{o,\phi_1} = \frac{C_{\mathrm{eq},\phi_1}}{g_m}\left(\frac{g_m |v_{a0,\phi_1}|}{I_o} - 1\right) \tag{3.19}$$

where sgn() is the sign function, and v_{a0,ϕ_1} stands for the value of v_a at the beginning of the sampling phase. It can be computed as,

$$v_{a0,\phi_1} = v_a[(n-1/2)T_s] - \sum_{i=1}^{N_o} \frac{C_{Sni}}{C_{\mathrm{eq},\phi_1}}\{v_o[(n-1/2)T_s] - v_{C_{Sni}}[(n-1/2)T_s]\} \tag{3.20}$$

where $v_{C_{Sni}}$ is the voltage across capacitor C_{Sni}.[2]

The integrator output voltage at the end of the sampling phase can be obtained as,

$$v_o(nT_s) = v_o[(n-1/2)T_s] + \left(1 + \frac{C_P}{C_I}\right)\{v_a(nT_s) - v_a[(n-1/2)T_s]\} \tag{3.21}$$

as opposed to the ideal situation in which $v_o(nT_s) = v_o[(n-1/2)T_s]$.

Note from Equations (3.18) and (3.21) that, for the integrator model in Figure 3.6, the amplifier GB product and output SR during sampling are obtained as:

$$\begin{aligned} \mathrm{GB}_{\phi_1}(\mathrm{rad/s}) &= \frac{g_m}{C_{\mathrm{eq},\phi_1}} \\ \mathrm{SR}_{\phi_1}(\mathrm{V/s}) &= \left(1 + \frac{C_P}{C_I}\right)\frac{I_o}{C_{\mathrm{eq},\phi_1}} \end{aligned} \tag{3.22}$$

During the integration phase, the incomplete settling model is evaluated by proceeding in a similar way as for the sampling phase. Thus, at the end of the subsequent integration phase – that is, at $t = (n+1/2)T_s$ – the value of v_a is given by [3]:

$$v_a[(n+1/2)T_s] = \begin{cases} v_{a0,\phi_2} e^{-\frac{g_m}{C_{\mathrm{eq},\phi_2}}\frac{T_s}{2}} & |v_{a0,\phi_2}| \le \frac{I_o}{g_m} \\ \frac{I_o}{g_m}\mathrm{sgn}(v_{a0,\phi_2}) e^{-\frac{g_m}{C_{\mathrm{eq},\phi_2}}\left(\frac{T_s}{2}-t_{o,\phi_2}\right)} & |v_{a0,\phi_2}| > \frac{I_o}{g_m}, t_{o,\phi_2} \le \frac{T_s}{2} \\ v_{a0,\phi_2} - \frac{I_o}{C_{\mathrm{eq},\phi_2}}\mathrm{sgn}(v_{a0,\phi_2})\frac{T_s}{2} & |v_{a0,\phi_2}| > \frac{I_o}{g_m}, t_{o,\phi_2} > \frac{T_s}{2} \end{cases} \tag{3.23}$$

[2]Note that the behavioral model also makes it necessary to keep track of the summation node voltage of the loading integrator v_{an}, since $v_{C_{Sni}}[(n-1/2)T_s] = v_{ni2}[(n-1/2)T_s] - v_{an}[(n-1/2)T_s]$.

where t_{o,ϕ_2} is the duration of the slew-rate-limited integrator settling (relative to $T_s/2$), given by,

$$t_{o,\phi_2} = \frac{C_{\mathrm{eq},\phi_2}}{g_m}\left(\frac{g_m|v_{a0,\phi_2}|}{I_o} - 1\right) \tag{3.24}$$

and v_{a0,ϕ_2} stands for the value of v_a at the beginning of the integration phase. The latter can be computed as

$$\begin{aligned} v_{a0,\phi_2} = {} & \frac{1}{C_{\mathrm{eq},\phi_2}}\left(1 + \frac{C_L}{C_I}\right)\sum_{i=1}^{N_i} C_{Si}\{v_{i2}(nT_s) - v_{i1}[(n-1/2)T_s]\} \\ & + \frac{C'}{C_{\mathrm{eq},\phi_2}} v_a(nT_s) \end{aligned} \tag{3.25}$$

where v_{i1}, v_{i2} are the voltages connected to the input of the ith SC branch during ϕ_1 and ϕ_2, respectively, and C' stands for:

$$C' = C_P + C_L\left(1 + \frac{C_P}{C_I}\right) \tag{3.26}$$

The integrator output voltage at the end of the integration phase can be obtained as,

$$\begin{aligned} v_o[(n+1/2)T_s] = {} & v_o(nT_s) + \sum_{i=1}^{N_i} \frac{C_{Si}}{C_I}\{v_{i1}[(n-1/2)T_s] - v_{i2}(nT_s)\} \\ & - \left(1 + \frac{C_P}{C_I}\right) v_a(nT_s) \\ & + \left(1 + \frac{C_P + \sum_{i=1}^{N_i} C_{Si}}{C_I}\right) v_a[(n+1/2)T_s] \end{aligned} \tag{3.27}$$

as opposed to the ideal integration process with no dynamic limitations, in which the last two terms in Equation (3.27) are null.

The amplifier GB product and output SR during this phase can be obtained in a similarly way as for the sampling phase:

$$\begin{aligned} \mathrm{GB}_{\phi_2}(\mathrm{rad/s}) &= \frac{g_m}{C_{\mathrm{eq},\phi_2}} \\ \mathrm{SR}_{\phi_2}(\mathrm{V/s}) &= \left(1 + \frac{C_P + \sum_{i=1}^{N_i} C_{Si}}{C_I}\right)\frac{I_o}{C_{\mathrm{eq},\phi_2}} \end{aligned} \tag{3.28}$$

Figure 3.8 illustrates how the equations above can be concatenated to accurately keep track of the summation and output voltages of an SC integrator over the clock periods. They can be easily incorporated into CAD tools for the behavioral simulation of SC-ΣΔMs or SC circuits in general. Moreover, the previous model can be easily extrapolated to other operating conditions: integration and sampling phases with different durations, different switching loading conditions at the integrator output, inclusion of the parasitic capacitance of the switches, and so on.

3.4.2 Linear Effect of Finite Amplifier Gain–Bandwidth Product

The model for the transient response of SC integrators described above can be easily incorporated into behavioral simulators for SC-ΣΔMs. This enables the influence of settling errors on the modulator performance – in terms of both the increased in-band noise and the generated distortion – to be

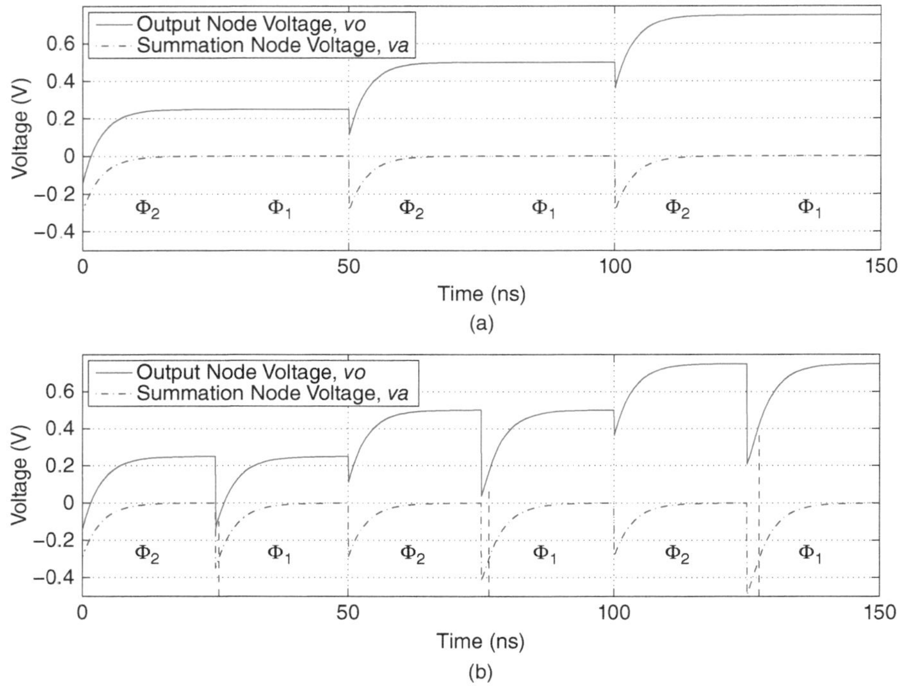

Figure 3.8 Illustration of the influence of switching load conditions on the transient response of an SC integrator: (a) loading SC branches are not considered; (b) one loading SC branch with a 0.5-pF capacitor is considered. (Vertical dashed lines indicate time positions $t_{o,\phi 1}$ where the integrator ends a slew-rate-limited response and starts evolving linearly). Parameters used are (see Figures 3.6 and 3.7): $v_{11} = 0$, $v_{12} = -1$ V, $C_{S1} = 0.25$ pF, $C_I = 1$ pF, $C_P = 0.1$ pF, $C_L = 1$ pF, $g_m = 0.5$ mA/V, and $I_o = 0.15$ mA for the SC integrator under consideration, $v_{n12} = -1$ V and $C_{Sn1} = 0.5$ pF for the loading SC integrator, and $T_s = 50$ ns.

accurately quantified. In the early design stages (high-level design) it is often useful to work with closed-form expressions which, although being coarse approximations of the behavioral model, can help to give an insight into the influence of settling parameters on different modulator topologies. To this end, a linear transient response will be assumed for SC integrators in this section, as if settling error was determined only by the finite amplifier GB with no limitation on the SR.

With these considerations in mind, the finite difference equation of an SC FE integrator can be obtained from Equations (3.18), (3.21), (3.23), and (3.27) to be,

$$v_o[(n+1/2)T_s] \approx v_o[(n-1)T_s] + \frac{C_S}{C_I}(1-\epsilon_{st})\{v_1[(n-1)T_s] - v_2[(n-1/2)T_s]\} \tag{3.29}$$

where only one input branch is considered for simplicity. The settling error associated with the linearly-limited transient response is represented by ϵ_{st}, which thus contains terms in $e^{-\mathrm{GB}_{\phi 1} T_s/2}$ and in $e^{-\mathrm{GB}_{\phi 2} T_s/2}$, with GB in rad/s. If settling errors associated with integration dominate the overall defective settling rather than those originating during sampling, the linear settling error can be simply reduced to:

$$\epsilon_{st} \approx e^{-\mathrm{GB}_{\phi 2} \frac{T_s}{2}} = e^{-\pi \frac{\mathrm{GB}_{\phi 2}(\mathrm{Hz})}{f_s}} \tag{3.30}$$

Transforming Equation (3.29) to the Z-domain, the integrator output results in,

$$v_o(z) \approx \frac{C_S}{C_I}(1-\epsilon_{st})\frac{z^{-1}v_1(z) - z^{-1/2}v_2(z)}{1-z^{-1}} \tag{3.31}$$

so that, under the assumptions above, settling error translates into a gain error in the ideal ITF, whose effect on the IBN of SC-ΣΔMs can be computed in a similar way as was done for capacitor mismatch in Section 3.3. Therefore, Equations (3.14)–(3.15) still hold for quantifying the effect of linear defective settling to first order, just by replacing ϵ_g with ϵ_{st}.

Figure 3.9 illustrates the effect of the finite amplifier GB product on single-loop and cascade SC-ΣΔMs. The in-band noise of second- and third-order single-loops and of a 2-1-1 cascade is computed using both the behavioral model for the integrator settling in Section 3.4.1 and the approximate closed-form expressions for the generated gain error. Large amplifier output currents

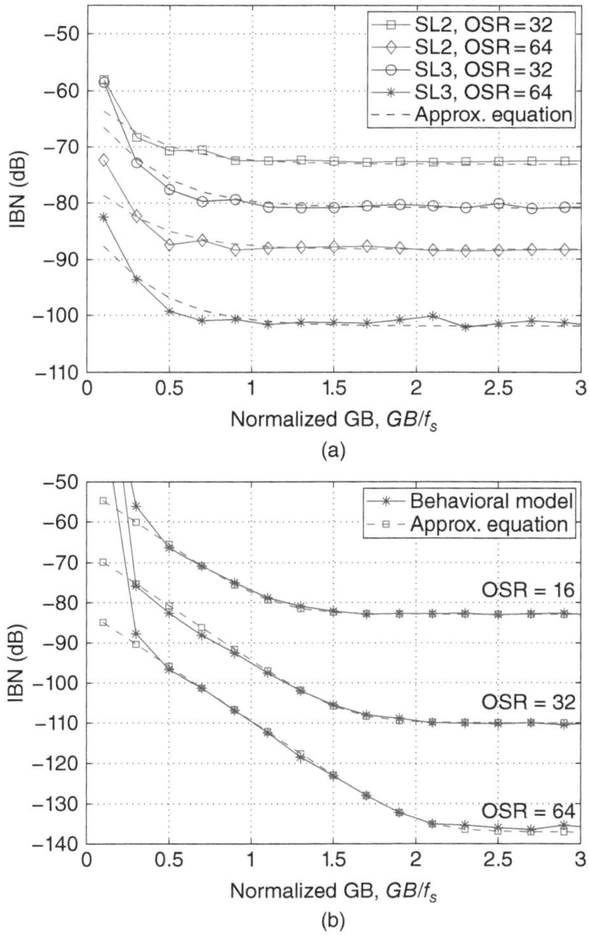

Figure 3.9 Simulation results for the influence of amplifier GB on the in-band noise of SC-ΣΔMs: (a) second- and third-order loops; (b) 2-1-1 cascade. Approximate results have been obtained from Equations (3.14) and (3.16) with $\epsilon_{st} = e^{-\pi \mathrm{GB(Hz)}/f_s}$.

have been used in the behavioral simulations to make the influence of SR limitation negligible, so it is necessary only to consider linear errors only. Note that, as expected, cascade ΣΔMs are more sensitive to GB limitations than single-loops. Usually, an amplifier GB of 1–$2f_s$ is sufficient for single-loop modulators to achieve full performance, whereas the requirement increases to 3–$10f_s$ for cascade ΣΔMs as the oversampling ratio increases.

3.4.3 Nonlinear Effect of Finite Amplifier Slew Rate

In contrast to errors arising from finite amplifier GB, finite amplifier SR due to limited output current capability has a purely nonlinear effect on the performance of ΣΔMs, generating distortion and an increase in the noise floor.

For the case of single-loop SC-ΣΔMs, slew-rate limited integrator dynamics essentially translate into distortion. Figure 3.10 illustrates the impact of amplifier SR on a single-bit third-order ΣΔM

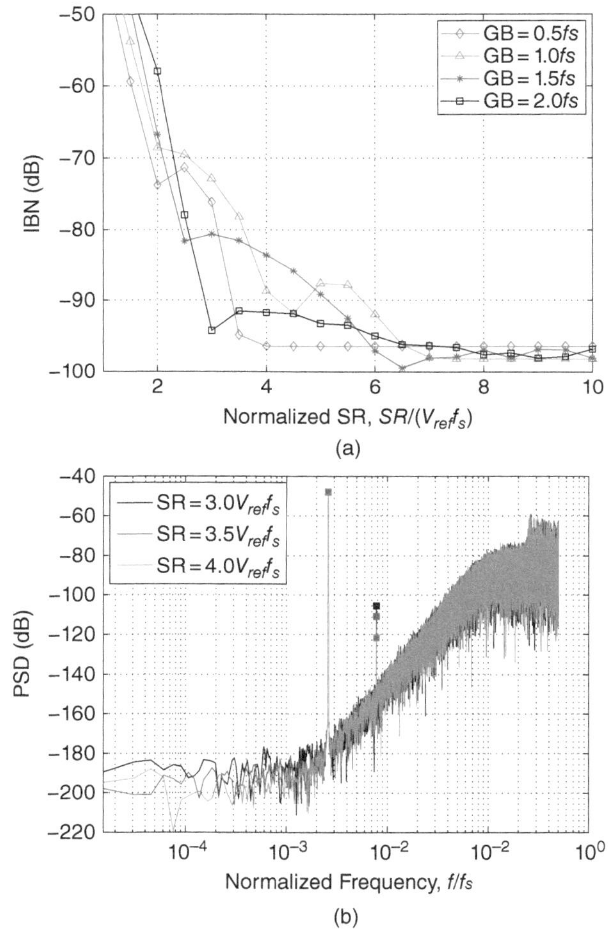

Figure 3.10 Simulation results for a third-order SC-ΣΔM with OSR = 64 under the influence of finite amplifier slew rate: (a) effect on the in-band noise; (b) effect on the output spectrum for GB $= f_s$. Input signal with $P_{\text{in}} = -6$ dBFS and $f_{\text{in}} = B_w/3$. Generated distortion is included in the IBN computation.

operating with an oversampling ratio of 64. An input tone with −6 dBFS ($0.5V_{\text{ref}}$ amplitude) and frequency equal to $B_w/3$ is applied to the modulator in the behavioral simulations. Note from the results that, depending on the amplifier GB, an SR of 4–8$V_{\text{ref}}f_s$ is enough to reduce the power of the generated distortion to a level that does not affect the IBN.

For the case of cascade SC-ΣΔMs, finite amplifier SR generates distortion as well as an increase in the noise floor due to noise leakages, as shown in Figure 3.11. For that reason, SR requirements are larger than for single-loops and usually range from 4 to $10V_{\text{ref}}f_s$ depending on the amplifier GB: the larger the GB, the lower the required SR.

Finally, note that the slew-rate limited integrator dynamic is a nonlinear signal-dependent phenomenon, which occurs during modulator operation at a frequency directly determined by the signal level at the integrators' inputs. Therefore, the ultimate way to reduce SR requirements on an SC-ΣΔM is to resort to multi-bit internal quantization.

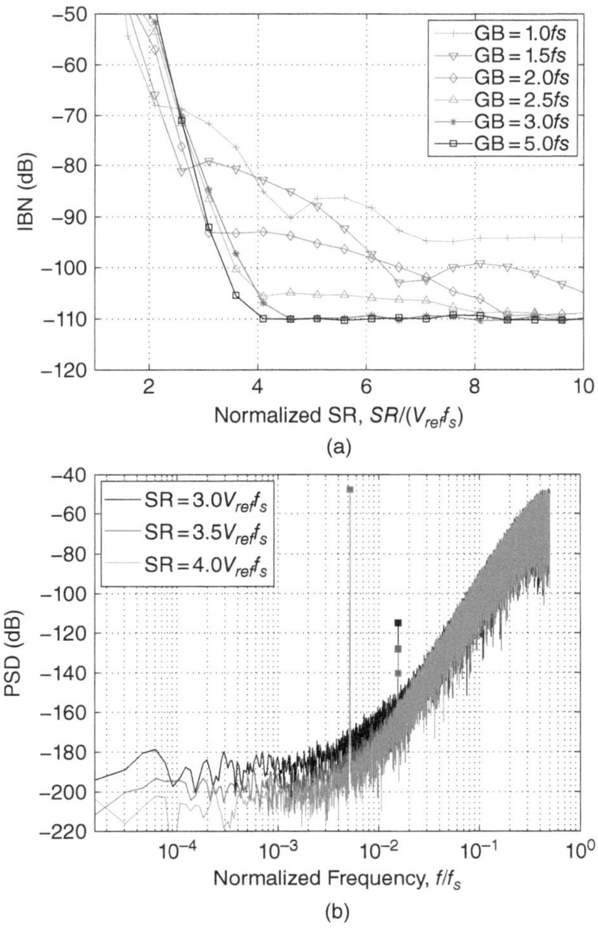

Figure 3.11 Simulation results for a 2-1-1 SC-ΣΔM with OSR = 32 under the influence of finite amplifier slew rate: (a) effect on the in-band noise; (b) effect on the output spectrum for GB = $3f_s$. Input signal with $P_{\text{in}} = -6\text{dBFS}$ and $f_{\text{in}} = B_w/3$. Generated distortion is included in the IBN computation.

3.4.4 Effect of Finite Switch On-resistance

Switches in the SC branches of ΣΔMs are implemented with MOSFETs, using either single nMOS or pMOS transistors, or CMOS transmission gates. These operate in the triode region when on and thus exhibit, in practice, a nonzero on-resistance.

If the on-resistance of the switches is the only nonideality accounted for in the operation of an SC integrator, it clearly leads to an incomplete charge transfer due to the *RC* time constant that is created in the SC branch. Considering, for instance, the scheme in Figure 3.12, the integrator output voltage can be obtained as [3]:

$$v_o(z) = \frac{C_S}{C_I}(1 - \epsilon_{R_{\text{on},\phi_2}}) \left[\frac{(1 - \epsilon_{R_{\text{on},\phi_1}})z^{-1}v_1(z) - z^{-1/2}v_2(z)}{1 - z^{-1}} \right] \quad (3.32)$$

where $\epsilon_{R_{\text{on},\phi_1}}$ stands for the charging error in C_S during ϕ_1 related to the on-resistance of switches S_1 and S_2, and $\epsilon_{R_{\text{on},\phi_2}}$ represents the error in the charge transfer from C_S to C_I during ϕ_2 related to the on-resistance of switches S_3 and S_4. If R_{on} is the on-resistance of a single switch, assuming that all switches are of the same size and that both clock phases have the same duration leads to:

$$\epsilon_{R_{\text{on},\phi_1}} = \epsilon_{R_{\text{on},\phi_2}} = e^{-\frac{1}{2R_{\text{on}}C_S}\frac{T_s}{2}} = e^{-\pi\frac{f_{R_{\text{on}}}(\text{Hz})}{f_s}} \quad (3.33)$$

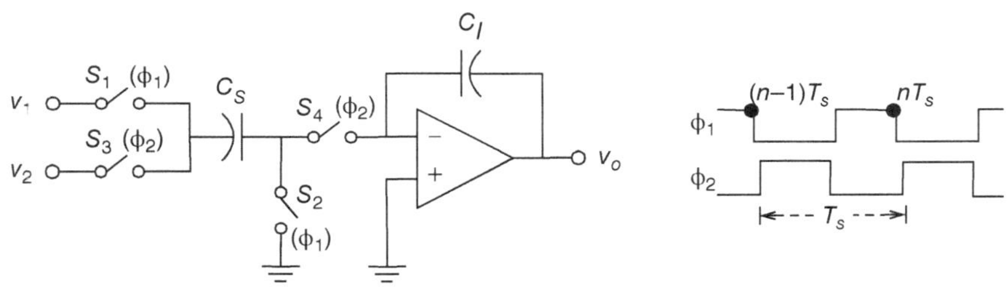

Figure 3.12 SC FE integrator with a single input branch.

Therefore, charge transfer error due to the on-resistance of the switches translates into a gain error in the ideal ITF, whose effect on the IBN of SC-ΣΔMs can be computed in a similar way as in Sections 3.3 or 3.4.2. Therefore, finite switch on-resistance will have a considerably lower impact on ΣΔ single-loops than on cascades.

Besides this consideration of how the switch on-resistance as a standalone nonideality affects an SC integrator, the effect can also be considered in combination with limited amplifier dynamics, and this gives a better account of the situation in practice. Figure 3.13 shows electrical simulation results to illustrate the influence of R_{on} on the transient response of the same SC integrator that was considered in Figure 3.8. Only one clock cycle is shown here, to give visibility of the R_{on} effect. Note that the linear amplifier response is slowed down as the on-resistance increases, affecting the integrator settling during both the sampling and the integration phases. To incorporate this effect into behavioral simulations, the effective amplifier GB during both clock phases can be approximated to [3]:

$$\begin{aligned} \text{GB}_{R_{\text{on}},\phi_1}(\text{Hz}) &\approx \frac{\text{GB}_{\phi_1}(\text{Hz})}{1 + \text{GB}_{\phi_1}(\text{Hz})/f_{R_{\text{on}},\phi_1}}, \quad \text{with } f_{R_{\text{on}},\phi_1} = \frac{1}{2\pi \cdot 2R_{\text{on}}C_{Sn}} \\ \text{GB}_{R_{\text{on}},\phi_2}(\text{Hz}) &\approx \frac{\text{GB}_{\phi_2}(\text{Hz})}{1 + \text{GB}_{\phi_2}(\text{Hz})/f_{R_{\text{on}},\phi_2}}, \quad \text{with } f_{R_{\text{on}},\phi_2} = \frac{1}{2\pi \cdot 2R_{\text{on}}C_S} \end{aligned} \quad (3.34)$$

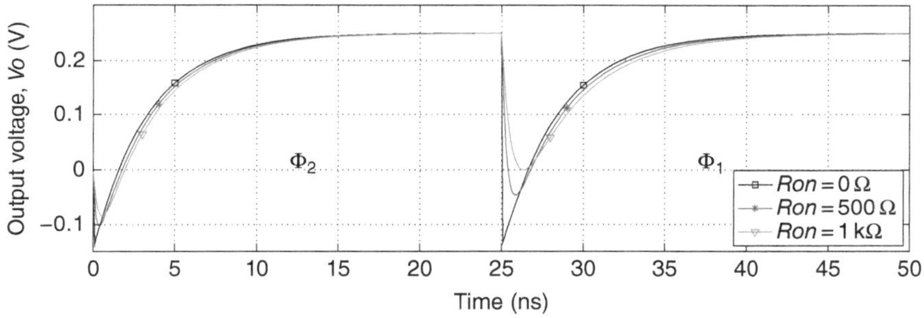

Figure 3.13 Illustration of the influence of the switch on-resistance on the transient response of an SC integrator with a loading SC branch. Simulation parameters used are the same as those for Figure 3.8.

where $f_{R_{\mathrm{on}},\phi_1}$ and $f_{R_{\mathrm{on}},\phi_2}$ stand for the RC poles introduced by the loading SC branch and by the input SC branch, respectively, and GB_{ϕ_1} and GB_{ϕ_2} are given by Equations (3.22) and (3.28) respectively.

In addition, the results presented in Section 3.4.2 for the linear effect of finite amplifier GB on the IBN of SC-ΣΔMs can be easily refined to include the slow-down effect of the switches R_{on}. To that end, the following gain error

$$\epsilon_{st,R_{\mathrm{on}}} = e^{-\pi \frac{\mathrm{GB(Hz)}}{1+\mathrm{GB(Hz)}/f_{R_{\mathrm{on}}}}} \tag{3.35}$$

can be considered in Equations (3.14) and (3.15). Figure 3.14 illustrates the combined linear effect of the finite amplifier GB and finite switch on-resistance on a third-order single-loop and on a 2-1-1 cascade SC-ΣΔM. The lower sensitivity of single-loops to these errors is clear and a switch on-resistance such that $f_{R_{\mathrm{on}}}$ is 4–5f_s is sufficient, in combination with the limited amplifier GB, to achieve full performance. This requirement for the R_{on} usually increases to 10–20f_s for cascade ΣΔMs as the oversampling increases.

3.5 Circuit Noise in SC-ΣΔMs

Electronic noise generated by transistors and resistors is present in any circuit implementation and imposes an ultimate limit to the resolution of ADCs. However, its impact is more severe in DT-ΣΔMs that employ SC-techniques due to the white spectrum of the main circuit-noise sources. In a SC-ΣΔM, these broadband noise components are sampled together with the input signal at the clock frequency, so that they fold over the modulator passband and may cause a considerable increase in the modulator in-band noise due to aliasing.

As stated in Section 3.1, the influence of nonidealities on the IBN of ΣΔMs is mainly determined by the location of the corresponding noise source in the modulator. With respect to circuit noise, all SC integrators in a ΣΔM add noise in the modulator passband, but the role of the front-end integrator is indeed dominant. When referred to the modulator input, noise power contributed by the remaining integrators is divided by the gain of preceding integrators in the modulator passband, so their influence strongly diminishes while moving from front-end to back-end integrators. Conversely, no shaping takes place at the modulator input and the first integrator has thus to fulfill the noise and linearity requirements of the complete ΣΔM.

Let us consider the SC integrator in Figure 3.15a to be the front-end integrator of a SC-ΣΔM. Two input SC branches are considered: the one including capacitor C_{S1} is assumed to sample the

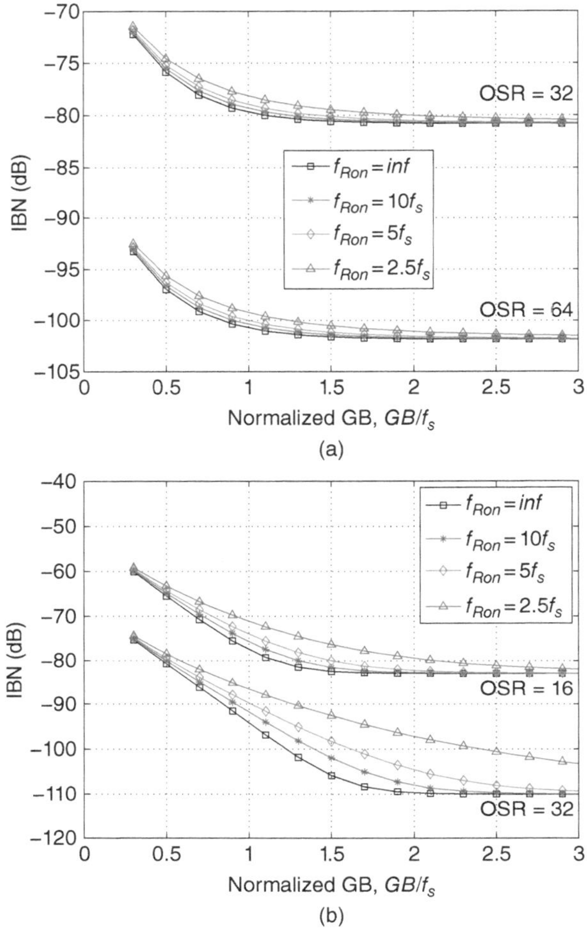

Figure 3.14 Influence of switch on-resistance on the in-band noise of SC-ΣΔMs: (a) third-order loop; (b) 2-1-1 cascade. Approximate results have been obtained from Equations (3.14) and (3.16) with $\epsilon_{st,R_{\rm on}} = e^{-\pi {\rm GB(Hz)}/[1+{\rm GB(Hz)}/f_{R_{\rm on}}]}$.

modulator input signal ($v_1 = v_{\rm in}$), whereas the one including capacitor C_{S2} samples the DAC feedback signal ($v_2 = v_{\rm fb}$). The main sources of circuit noise in SC integrators – thermal noise generated in the switches and noise generated in the amplifier – have been incorporated in the equivalent models in Figures 3.15b and 3.15c during each of the clock phases. Both thermal and flicker components will be considered.[3]

Figure 3.15b shows the model for the thermal noise introduced by switches controlled by clock phase ϕ_1. For both SC branches, the two active switches are assumed to have the same on-resistance ($R_{\rm on}$) and they are in series with a noise voltage source $v_{\rm sw}$. The PSD of each of these noise sources in

[3]The noise associated with the DAC reference voltage is not considered here for simplicity, but it can be incorporated in a similar way to the amplifier noise [3].

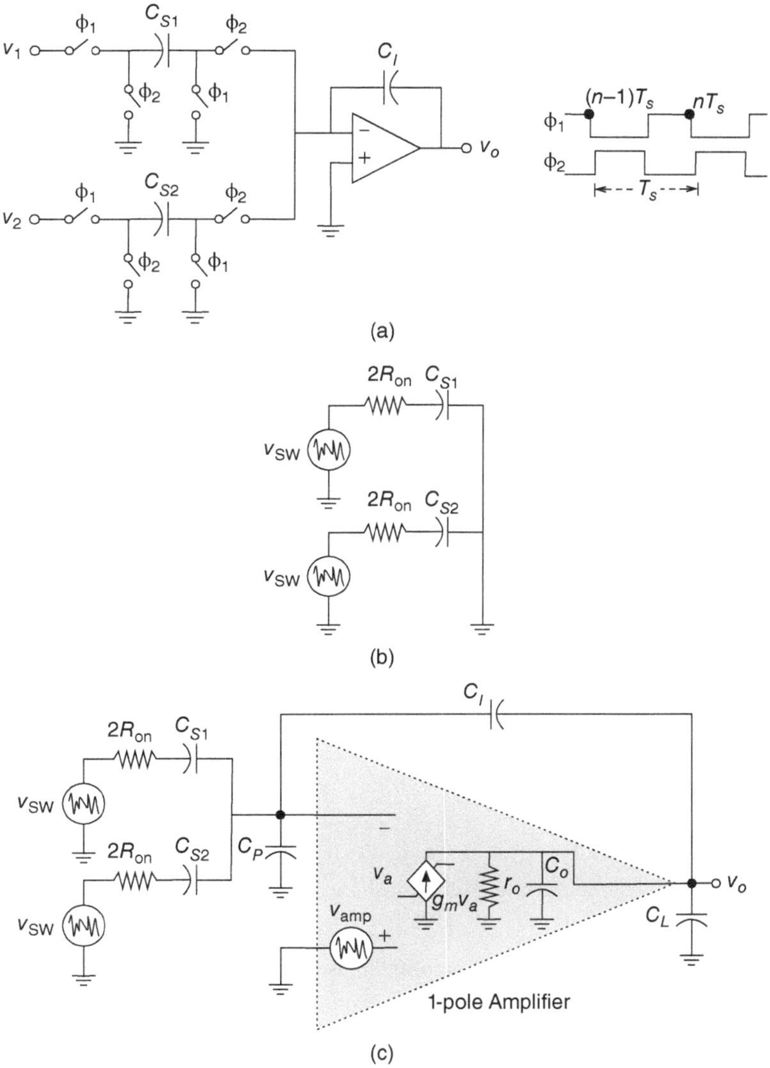

Figure 3.15 Circuit noise analysis in an SC integrator: (a) SC FE integrator with two input paths (single-ended version); (b) equivalent circuit model for sampling, with noise sources due to switches active during ϕ_1; (c) equivalent circuit model for integration, with noise sources due to switches active during ϕ_2 and due to the amplifier.

a single-sided frequency representation is thus $S_{\rm sw} = 4kT(2R_{\rm on})$, where k is Boltzmann's constant and T is the absolute temperature. Each of the noise sources generates a sample-and-held noise component in the corresponding capacitor voltage, given by the well-known kT/C expression due to the foldover effect [7–9]:

$$S_{{\rm sw},C_{Si}}(f) \approx \frac{2kT}{C_{Si}f_s}\,{\rm sinc}^2(\pi f/f_s) \tag{3.36}$$

Figure 3.15c shows the model for the thermal noise introduced by switches controlled by clock phase ϕ_2 and for the noise in the amplifier. These switches bring an additional noise component in the capacitor voltage of the corresponding SC branch, similar to Equation (3.36). On the other hand, a single-pole model is assumed for the amplifier; its equivalent input noise is modeled by a voltage source v_{amp} at the positive input terminal. As illustrated in Figure 3.16, the amplifier noise is essentially determined by a broadband thermal component and a narrowband flicker component, so that

$$S_{\rm amp}(f) \approx S_{\rm amp,th}(f) + S_{{\rm amp},1/f}(f) \approx S_{\rm amp,th}\left(1 + \frac{f_{\rm cr}}{f}\right) \tag{3.37}$$

where $S_{\rm amp,th}$ stands for the amplifier thermal noise PSD referred to its input and $f_{\rm cr}$ stands for the amplifier corner frequency: the frequency at which $1/f$ noise is equal to the thermal noise.[4] The amplifier noise generates correlated sample-and-held noise components in the integrator sampling capacitors, which can be obtained as

$$S_{{\rm amp},C_{Si}}(f) \approx S_{\rm amp,th}\left(\frac{2B_{w,{\rm noise}}}{f_s} + \frac{f_{\rm cr}}{f}\right)\,{\rm sinc}^2(\pi f/f_s) \tag{3.38}$$

where the amplifier noise bandwidth $B_{w,{\rm noise}}$ is required to account for the foldover effect of thermal noise. It can be estimated as $B_{w,{\rm noise}}({\rm Hz}) \approx {\rm GB}_{\phi_2}({\rm rad/s})/4$ [3], with ${\rm GB}_{\phi_2}$ given by Equation (3.28).

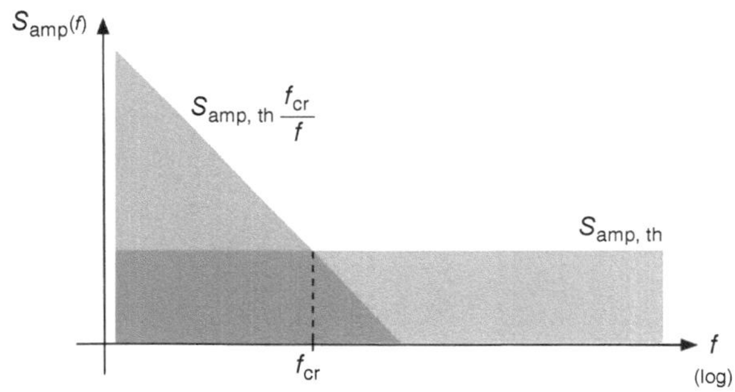

Figure 3.16 Illustration of the PSD of the amplifier noise showing the contributions of $1/f$ and thermal noise.

Adding up these circuit noise components in the SC integrator, the total input-referred[5] noise PSD yields [3]:

$$S_{\rm noise,in}(f) \approx 2\left[2S_{{\rm sw},C_{S1}}(f) + 2S_{{\rm sw},C_{S2}}(f)\frac{C_{S2}^2}{C_{S1}^2}\right] + S_{{\rm amp},C_{Si}}(f)\left(1 + \frac{C_{S2}}{C_{S1}}\right)^2 \tag{3.39}$$

where the factor 2 multiplying $S_{{\rm sw},C_{Si}}(f)$ accounts for the contributions of switches controlled by ϕ_1 and ϕ_2, whereas the factor 2 before the brackets accounts for the actual fully-differential implementation of the SC integrator, in which the number of SC branches and thus of switches doubles in comparison

[4]For a single-stage amplifier, the thermal noise component can be approximated to $S_{\rm amp,th} \approx \frac{8kT}{3g_m}2(1 + n_{\rm th})$, where g_m is the transconductance of the amplifier input transistor, and $n_{\rm th}$ stands for the noise factor contributed by the remaining transistors in the amplifier, and a factor 2 accounts for the fully-differential amplifier implementation. The amplifier corner frequency strongly depends on the size of the input transistors, their type, and the technological process, and it is thus not easy to accurately estimate it with a closed-form expression. However, both noise parameters can be easily characterized by means of electrical simulation.

[5]That is, referred to the voltage across capacitor C_{S1} that samples the modulator input signal.

to the single-ended scheme in Figure 3.15. Replacing Equations (3.36) and (3.38) in (3.39), the total input-referred noise PSD of the front-end integrator of a SC-ΣΔM can thus be approximated as:

$$S_{\text{noise,in}}(f) \approx \frac{8kT}{C_{S1}f_s}\left(1+\frac{C_{S2}}{C_{S1}}\right) + S_{\text{amp,th}}\left(\frac{\text{GB}_{\phi_2}}{2f_s}+\frac{f_{\text{cr}}}{f}\right)\left(1+\frac{C_{S2}}{C_{S1}}\right)^2 \tag{3.40}$$

in which the approximation $\text{sinc}^2(\pi f/f_s) \approx 1$ for $f \ll f_s$ has been used for simplicity.

The input-referred IBN of a SC-ΣΔM due to circuit noise can be easily obtained by integrating the former expression over the input signal bandwidth, so that

$$\begin{aligned}\text{IBN}_{\text{noise}} &= \int_0^{B_w} S_{\text{noise,in}}(f)\text{d}f \approx \frac{4kT}{C_{S1}\text{OSR}}\left(1+\frac{C_{S2}}{C_{S1}}\right)\\ &+ S_{\text{amp,th}}\left[\frac{\text{GB}_{\phi_2}}{4\text{OSR}}+f_{\text{cr}}\ln\left(\frac{B_w}{f_0}\right)\right]\left(1+\frac{C_{S2}}{C_{S1}}\right)^2\end{aligned} \tag{3.41}$$

Here the $1/f$ noise component has been integrated from a frequency $f_0 > 0$ to exclude DC due to its logarithmic nature.

For a SC-ΣΔM to achieve a given noise performance, the sum of all three components in Equation (3.41) has to meet the required noise floor. Note that:

- for a given OSR, to reduce the contribution of the switches' thermal noise, the size of the sampling capacitors at the modulator input must be increased, which results in larger speed requirements for the amplifier and thus in higher power consumption;
- for a given OSR, to reduce the contribution of the amplifier thermal noise, its GB must be reduced as much as the integrator settling requirements allow;
- to reduce the flicker contribution the amplifier corner frequency must be kept low – in low-bandwidth applications, cancelation techniques such as correlated double sampling (CDS) or chopper are often required for further reduction of the $1/f$ component [10].

3.6 Clock Jitter in SC-ΣΔMs

SC-ΣΔMs are affected in practice by timing uncertainties[6] in the clock phases that control the SC operation. However, they exhibit larger tolerance to clock jitter than Nyquist converters, since jitter sensitivity is reduced by the modulator OSR [12].

The effect of clock jitter in SC-ΣΔMs is mainly limited to a sampling uncertainty of the modulator input signal. Timing uncertainties during the integration phase only cause an extra error to be added to the integrator settling error and their influence can be neglected in practice, whereas the contributions of other integrators than the front-end one will be reduced by the noise shaping. Therefore, different SC-ΣΔMs exhibit similar sensitivity to clock jitter [13].

Sampling time uncertainty causes a nonuniform sampling of the modulator input signal, which results in an increase in the in-band error power. The magnitude of this increase is usually estimated for SC-ΣΔMs by assuming random statistical properties for the clock jitter [12]. For a modulator

[6]This effect is inherent to every clock generation circuitry – crystal oscillators, PLL-based oscillators, and so on – and is mostly caused by thermal noise, phase noise, and spurious tones that degrade the spectral purity of clock signal [11].

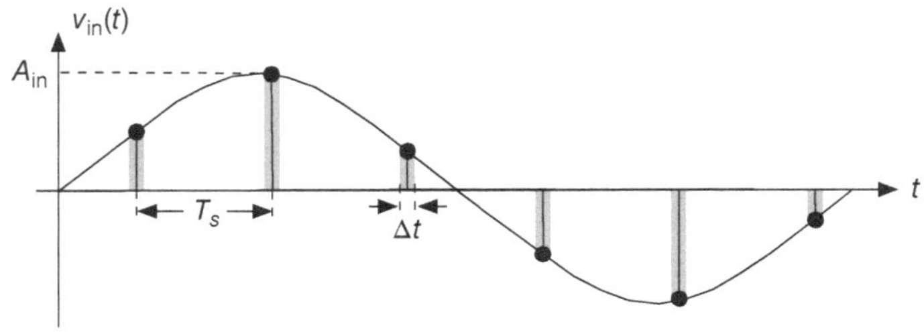

Figure 3.17 Illustration of nonuniform sampling of a signal due to clock jitter. The shaded areas represent the timing uncertainties.

input sine wave $v_{in}(t) = A_{in} \sin(2\pi f_{in} t)$, as shown in Figure 3.17, an uncertainty of Δt in the sampling instant causes an error in the sampled signal, given by:

$$\epsilon_j = v_{in}(nT_s + \Delta t) - v_{in}(nT_s) \approx \frac{dv_{in}(t)}{dt}\Big|_{nT_s} \Delta t = (2\pi f_{in}) A_{in} \Delta t \cos(2\pi f_{in} n T_s) \tag{3.42}$$

Under the assumption of white jitter, the power of this modulated error distributes uniformly, so that only a fraction of it is located within the ΣΔM passband. The in-band noise due to clock jitter can thus be easily obtained as

$$\text{IBN}_j = \int_{-B_w}^{+B_w} \frac{A_{in}^2}{2} \frac{(2\pi f_{in} \sigma_j)^2}{f_s} df = \frac{A_{in}^2}{2} \frac{(2\pi f_{in} \sigma_j)^2}{\text{OSR}} \tag{3.43}$$

where σ_j stands for the standard deviation of the timing uncertainty. Taking into account that $A_{in} \leq \Delta/2$ and $f_{in} \leq f_s/(2\text{OSR})$, an upper bound can be calculated for Equation (3.43) as:

$$\text{IBN}_j \leq \frac{\Delta^2}{8} \frac{(\pi f_s \sigma_j)^2}{\text{OSR}^3} \tag{3.44}$$

showing that the sensitivity of SC-ΣΔMs to clock jitter is reduced as OSR^{-3}. Nevertheless, clock jitter can be a strong limiting factor in applications such as digital wireless receivers, where BP-ΣΔMs are used to digitize RF signals with high sampling frequencies: typically $f_s = 4 \cdot f_{in}$, with f_{in} being the input signal frequency. This is illustrated in Figure 3.18 for a fourth-order BP-ΣΔM, where it can be

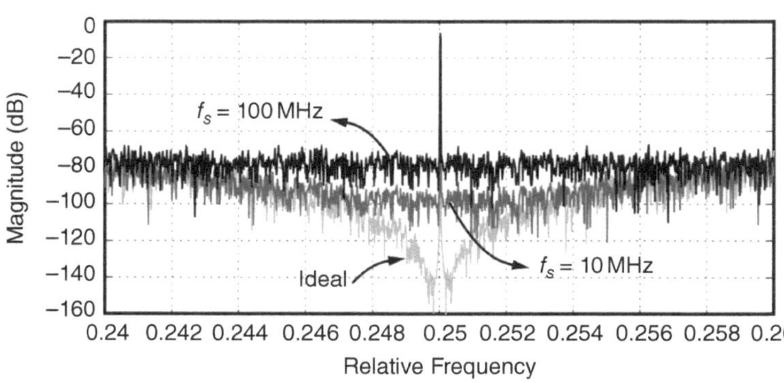

Figure 3.18 Illustrating the effect of clock jitter ($\sigma_j = 0.1$) in a fourth-order BP-ΣΔM.

noted how, by increasing f_s, the performance of the modulator is degraded due to the IBN increase caused by clock jitter.

3.7 Sources of Distortion in SC-ΣΔMs

Analog devices used for the implementation of ΣΔMs exhibit in practice a certain nonlinearity. These nonlinearities generate distortion and thus limit the peak SNDR attainable for high input amplitudes. Nevertheless, deriving closed-form expressions for the distortion generated in a ΣΔM is in general much more awkward than analyzing the effect of linear errors. Therefore, several simplifications are often made in order to handle nonlinearities. First, only the sources of distortion associated with the front-end integrator in the ΣΔM are considered, since they directly add to the modulator signal with no attenuation and thus dominate the overall modulator nonlinearity. Distortion generated in subsequent integrators is suppressed by the increasing noise shaping when referred to the modulator input, so that their contributions can be considered negligible in practice. Second, each source of nonlinearity is conceived as a small deviation from the ideal linear behavior – that is, as a weak nonlinearity – that affects the modulator performance in an additive way.

Figure 3.19 illustrates the main sources of distortion in an SC integrator in which a fully-differential topology is assumed for the suppression of even-order harmonics. In SC-ΣΔMs, linearity is essentially limited by the voltage dependency of capacitors, of the switches on-resistance, and of the amplifier gain, as well as by the SR-limited integrator dynamics (as discussed in Section 3.4.3). Distortion arising from charge injection in the switches can be neglected if clock phases with delayed falling edges are employed [14]. Moreover, given the highly linear capacitors that modern mixed-mode technological processes offer – for example, MiM and MoM capacitors – their effect will not further considered here.[7] The influence of the nonlinearity of the switches and of the amplifier gain will be discussed below.

3.7.1 Nonlinear Amplifier Gain

In practice, the DC gain of an amplifier is voltage-dependent, since the output resistance of the amplifier's output transistors decreases as the amplifier output voltage deviates from the quiescent

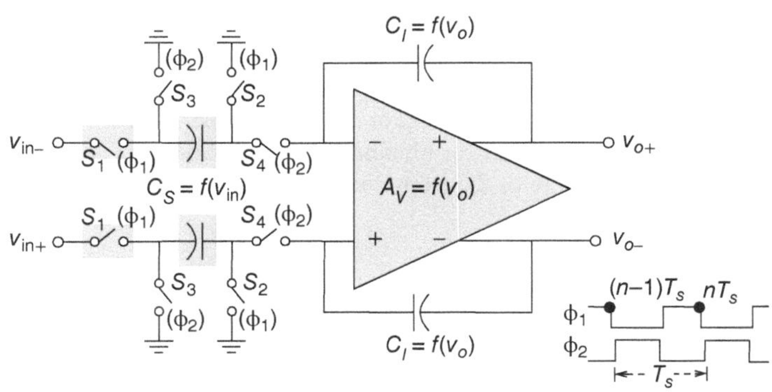

Figure 3.19 Main sources of distortion in a fully-differential SC integrator.

[7]If different capacitor primitives are to be considered for the SC implementation, the interested reader can find details on the generated distortion in [3, 13].

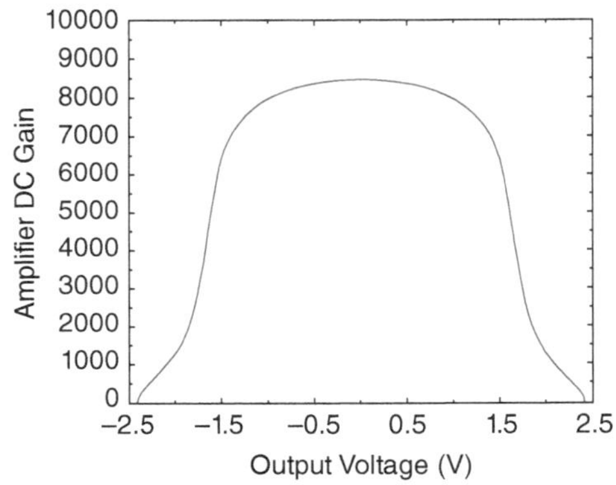

Figure 3.20 Illustration of the dependency of the amplifier gain on the output voltage level.

point. Figure 3.20 illustrates this effect with the electrical simulation results of a folded-cascode amplifier. Note that the amplifier DC gain is about 8500 (78.5 dB) at the common-mode output voltage, but it decreases for increasing output levels and drops abruptly near the amplifier saturation region.

The influence of the amplifier gain nonlinearity can be easily incorporated into the leaky integrator model in Section 3.2, so that the Equation (3.3) can be re-written as [3]:

$$v_o(nT_s) = \frac{1+\left(1+\frac{C_P}{C_I}\right)\frac{1}{A_v(v_{o,n-1})}}{1+\left(1+\frac{C_P}{C_I}+\sum_{i=1}^{N_i}\frac{C_{Si}}{C_I}\right)\frac{1}{A_v(v_{o,n})}} v_o[(n-1)T_s] + \frac{\sum_{i=1}^{N_i}\frac{C_{Si}}{C_I}v_i[(n-1)T_s]}{1+\left(1+\frac{C_P}{C_I}+\sum_{i=1}^{N_i}\frac{C_{Si}}{C_I}\right)\frac{1}{A_v(v_{o,n})}} \quad (3.45)$$

where $A_v(v_{o,n-1})$ represents the effective amplifier gain at the output voltage corresponding to clock cycle $(n-1)T_s$ and $A_v(v_{o,n})$ corresponds to that of clock cycle nT_s. As will be shown in Section 5.3, solving this difference equation in an iterative way, together with a table look-up for the amplifier gain, enables accurate accounting for the voltage dependency of the amplifier gain in transient behavioral simulations, whether the nonlinearity is weak or strong!

For weak nonlinearities, a polynomial approximation can be used for modeling the voltage dependency near the quiescent point and for obtaining rough estimates of the generated distortion. Let us assume that the amplifier gain of the front-end integrator in an SC-ΣΔM, such as that shown in Figure 3.19, is expressed as

$$A_v(v_o) = A_v(1 + c_{\text{nl1}}v_o + c_{\text{nl2}}v_o^2 + \ldots) \quad (3.46)$$

where $c_{\text{nl}i}$ stands for the ith-order voltage-gain coefficient of the amplifier DC gain. If a sine wave with amplitude A_{in} is applied at the modulator input, the input-referred distortion of the third-order harmonic can be estimated as [3, 15]:

$$\text{HD}_3 \approx \frac{c_{\text{nl2}}}{4}\frac{\left(1+\frac{C_P}{C_I}+\frac{C_S}{C_I}\right)}{A_v}\left(\frac{C_S}{C_I}\right)^2 A_{\text{in}}^2 \quad (3.47)$$

where C_P is the parasitic capacitor at the amplifier input nodes. Note that decreasing the integrator gain coefficient clearly helps to reduce distortion. However, the most direct way to reduce the effect of the amplifier gain nonlinearity is by increasing the value of the amplifier gain itself [3, 13].

3.7.2 Nonlinear Switch On-Resistance

Switches in SC-ΣΔMs are usually implemented as CMOS transmission gates, so that either the nMOS or the pMOS transistors are on for a given voltage level to be transmitted. Figure 3.21a sketches the on-conductance of nMOS and pMOS switches, assuming that they exhibit a resistance in the triode region that can be approximated as [3, 16]:

$$
\begin{aligned}
R_{\mathrm{on}N} &\approx \frac{1}{K_N \frac{W_N}{L_N}\left(V_{GS} - V_{TN} - \frac{V_{DS}}{2}\right)} \approx \frac{1}{K_N \frac{W_N}{L_N}\left(V_{DD} - V_{TN} - V_{\mathrm{in}}\right)} \\
R_{\mathrm{on}P} &\approx \frac{1}{K_P \frac{W_P}{L_P}\left(V_{SG} - |V_{TP}| - \frac{V_{SD}}{2}\right)} \approx \frac{1}{K_P \frac{W_P}{L_P}\left(V_{\mathrm{in}} - V_{SS} - |V_{TP}|\right)}
\end{aligned}
\tag{3.48}
$$

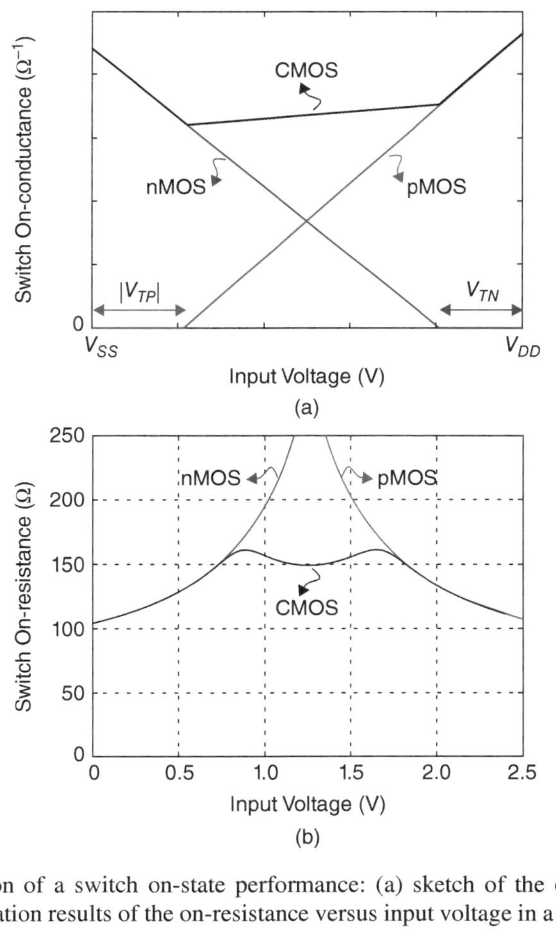

Figure 3.21 Illustration of a switch on-state performance: (a) sketch of the on-conductance versus input voltage; (b) simulation results of the on-resistance versus input voltage in a 2.5-V 0.25-μm CMOS process.

where $V_{in} = (V_D + V_S)/2$ represents the switch input voltage – that is, the common-mode voltage of the drain and source terminals. The on-resistance of the CMOS transmission gate is thus obtained as $R_{on} = R_{onN}||R_{onP}$, warranting a rail-to-rail operation of the switch as long as $V_{DD} - V_{SS} > V_{TN} + |V_{TP}|$.[8] Figure 3.21b shows the electrical simulation results of a CMOS transmission gate, in which the voltage dependency of the switch on-resistance is clearly visible.

To analyze the relative influence of the different switches in the front-end integrator of an SC-ΣΔM on the generated distortion, let us consider the schematic in Figure 3.19. The modulator input signal is sampled on capacitors C_S through switches S_1 and S_2 during ϕ_1. Since switches S_1 are connected to the modulator input, their on-resistances directly depend on the modulator input level and are the dominant source of distortion. However, switches S_2 have one of their terminals connected to the common-mode voltage – that is, to a voltage that remains approximately constant over time – so that the voltage level of these switches is not expected to change much over the clock periods [5]. As a result, the distortion introduced by switches S_2 will be considerably lower than that of switches S_1. The same reasoning can be applied to switches S_3 and S_4 during ϕ_2: switches S_3 have one terminal connected to a fixed voltage – the common-mode voltage, as depicted in Figure 3.19, or the DAC feedback voltage – and switches S_4 are connected to the virtual ground of the amplifier. Their influence on the generated distortion can thus be neglected in practice.

The distortion generated by nonlinear sampling in an SC-ΣΔM can be accurately evaluated through transistor-level electrical simulations of the equivalent circuit in Figure 3.22. A tone with large amplitude can be applied at the differential input and the differential voltage stored in capacitors C_S can be collected at the clock rate to compute the FFT and measure the total harmonic distortion (THD).

Finally, as will be demonstrated in Chapter 8, the generated distortion can be reduced, not only by keeping the switch as linear as possible, but also by reducing the value of the on-resistance itself. Figure 3.23 illustrates the switch on-resistance for different alternative relative sizings of the switch transistors. If the sizes compensate for the difference in the transconductance parameter of the nMOS and pMOS transistors – that is, $K_N W_N = K_P W_P$, as used for instance in [3] – the nonlinearity of the on-resistance is low, but its average value is larger than for $W_N = W_P$, as used in [5]. In the latter case, the switch area and its parasitic capacitors increase, but the slow-down effect of the switch R_{on} on the

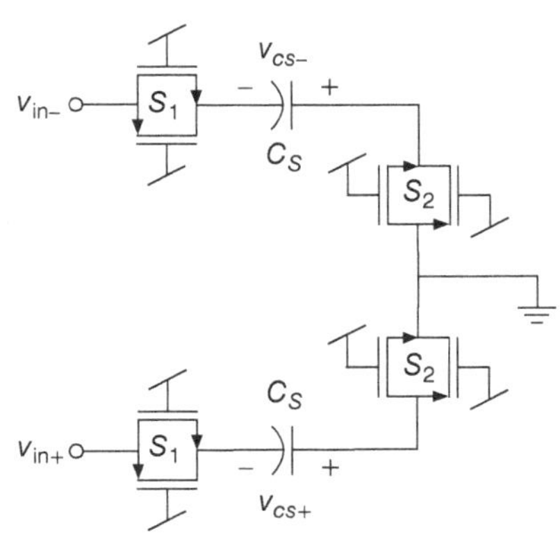

Figure 3.22 Equivalent circuit for evaluating distortion during sampling due to switch nonlinearity.

[8] Note that, if $V_{DD} - V_{SS} < V_{TN} + |V_{TP}|$, a gap will appear at the mid-range in Figure 3.21a, since neither the nMOS switch nor the pMOS will conduct. In a low-voltage environment, this problem is often overcome using clock-boosting techniques [17, 18] or low-V_T transistors.

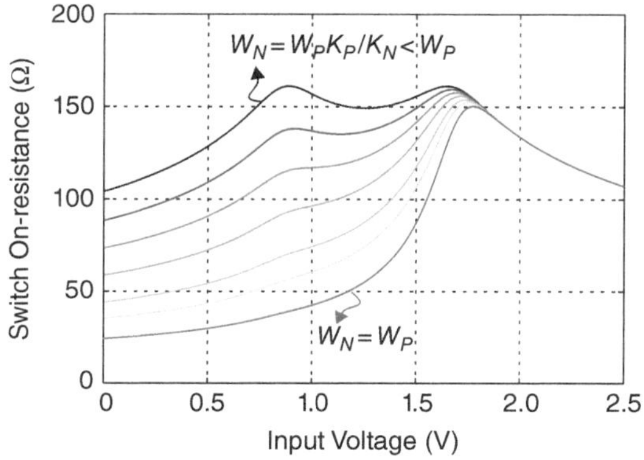

Figure 3.23 Illustration of the switch on-resistance nonlinearity for different transistor sizings in a 2.5-V 0.25-μm CMOS process.

integrator settling will decrease, as discussed in Section 3.4.4. Note that the design trade-offs above can be solved in opposite directions depending on the particular linearity and speed requirements of an SC-ΣΔM, as well as the modulator input range relative to the supply voltage.

3.8 Case Study: High-level Sizing of a ΣΔM

For the sake of illustration, this section provides guidelines on how to approach the design of a ΣΔM to fulfill a given set of specifications by means of the closed-form expressions and the behavioral models derived in this chapter. The impact of the different circuit nonidealities on the ΣΔM performance will be considered in a cumulative way, trying to map the modulator specifications onto the electrical requirements of its main building blocks in a *top-down design methodology* – often referred to as the *high-level sizing* of a ΣΔM.

This approach will be exemplified in a 2-1-1 cascade SC-ΣΔM operating at 100-MHz clock frequency, and intended to achieve an effective resolution of around 12 bits in the A/D conversion of low-pass signals with 4-MHz bandwidth. To that end, let us consider the Z-domain block diagram of a 2-1-1 DT cascade illustrated in Figure 2.15, together with the modulator scaling coefficients in the first column of Table 2.2 [6]. Let us also assume a modulator full-scale range of 2 V.

3.8.1 Ideal Modulator Performance

The ideal in-band quantization error of the 2-1-1 cascade ΣΔM can be particularized from Equation (2.23) to:

$$\mathrm{IBN}_Q = \frac{\Delta_3^2}{12}\frac{1}{(c_1c_2)^2}\frac{\pi^8}{9\mathrm{OSR}^9} \tag{3.49}$$

If single-bit quantization is used in the three cascaded stages ($B_1 = B_2 = B_3 = 1$), Equation (3.49) yields an IBN_Q of −67.2 dB for $\Delta_3 = 2$ (since $B_3 = 1$) and OSR = 12.5 (since f_s = 100 MHz and B_w = 4 MHz). Therefore, an oversampling ratio of 12.5 is clearly insufficient only in combination with a fourth-order shaping for the cascade to achieve the required IBN; for example −83 dB for an effective number of bits (ENOB) of 13, as obtained from Equations (1.17) and (1.18).

Taking advantage of the ease of use of dual-quantization schemes in cascade ΣΔMs (see Section 2.4.3), IBN_Q can be easily reduced by resorting to a multi-bit quantizer in the last stage. For $B_3 = 3$, the last-stage quantization step reduces to $\Delta_3 = 2/(2^3 - 1)$ and IBN_Q thus yields −84.1 dB. Figure 3.24 illustrates the spectra of the ΣΔ cascade for an input sinusoid of −6 dBFS at 1.33 MHz ($f_{in} = B_w/3$). The figure compares the spectrum of the overall output of the fourth-order multi-bit cascade (labeled as 211mb) with those of its partial second- and third-order single-bit outputs (labelled as SL2 and 21, respectively). The increase in the shaping order is evident from the slopes of the spectra, whereas the reduction of the quantization error power for the multi-bit case is noticeable in the high-frequency region.

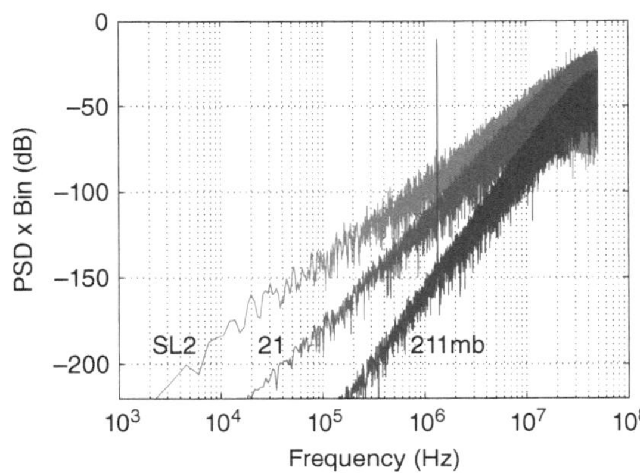

Figure 3.24 Illustration of: (SL2) second-order; (21) third-order; (211mb) fourth-order shaping in the 2-1-1 ΣΔM. Quantization noise in the three ΣΔ stages is considered as the only source of error. Input signal with $P_{in} = -6$ dBFS and $f_{in} = B_w/3$.

Figure 3.25 illustrates a possible implementation of the 2-1-1 multi-bit ΣΔM using fully-differential SC techniques. Note that only the mixed-signal section of the cascade is shown; the digital cancellation logic is omitted (see Figure 2.15). Note also that the stages would operate with a differential reference voltage $V_{ref} = V_{ref+} - V_{ref-} = 1$ V to obtain the 2-V differential full-scale range. The values of the sampling and integrating capacitors in Figure 3.25 are expressed as multiples of a unit capacitor C_u, whose ratios implement the modulator scaling coefficients in the first column of Table 2.2 [6]. To adapt these to the multi-bit quantization in the last stage, the corresponding coefficients are doubled ($g_4 = 2$, $g'_4 = g''_4 = 1$ in the multi-bit case).

3.8.2 Noise Leakages

As discussed in Section 2.4.1, errors in the multi-bit DAC will be mostly determined by the mismatch between its unit elements, which would be resistors for the SC-ΣΔM in Figure 3.25. From Equation (2.30), the worst-case errors in the 3-bit DAC output can be estimated as,

$$\epsilon_{DAC} \approx \frac{1}{2\sqrt{2^{B_3}}}(3\sigma_R) \tag{3.50}$$

where σ_R stands for the relative error in the value of the unit resistor and three sigmas are considered. Thanks to the dual-quantization scheme, these errors are injected at the last-stage input and will be

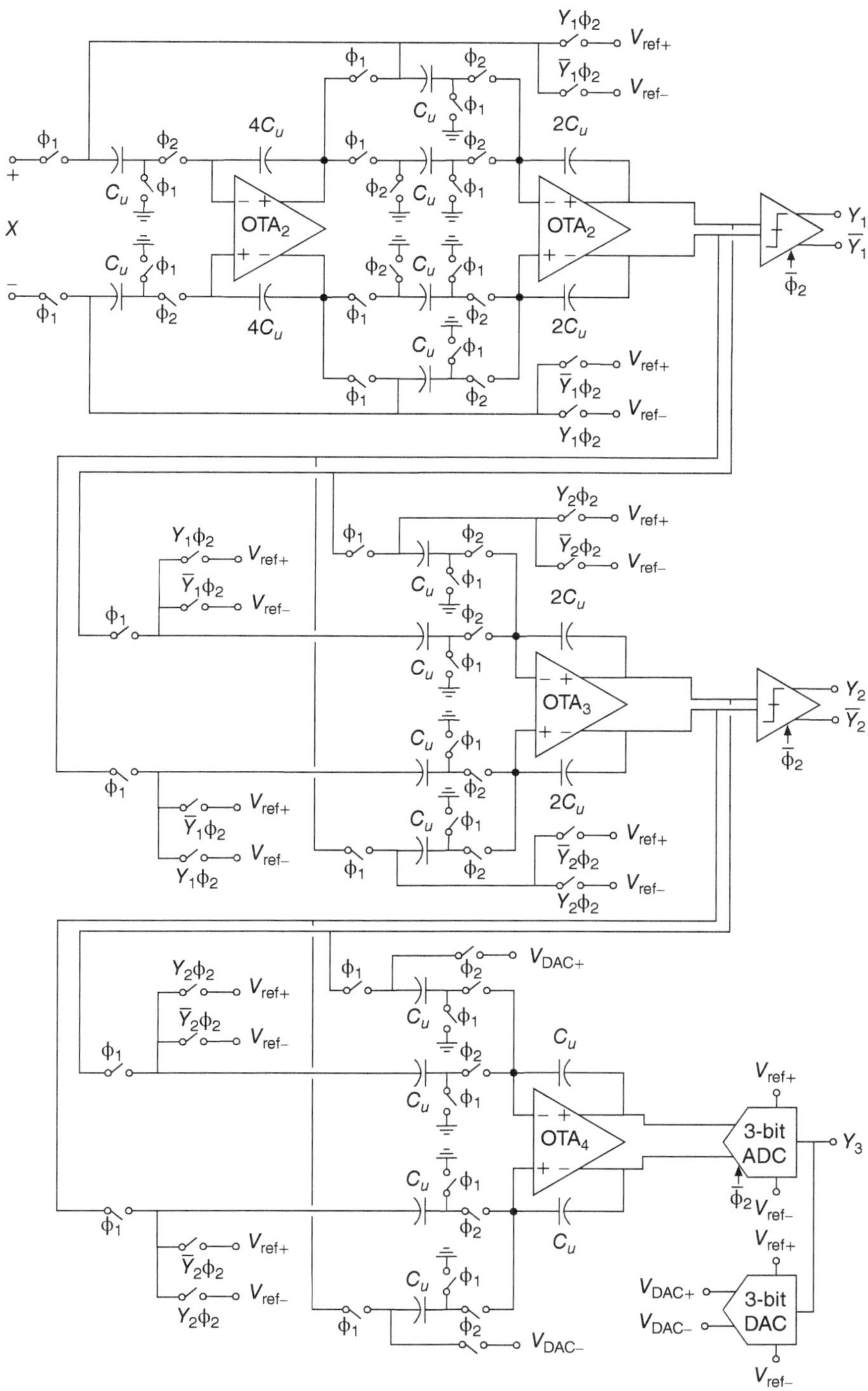

Figure 3.25 Fully-differential SC implementation of a 2-1-1 multi-bit ΣΔ modulator.

third-order shaped at the modulator output, so that their contribution to the total IBN can be estimated from Equation (2.32) as:

$$\mathrm{IBN_{DAC}} \approx \epsilon_{\mathrm{DAC}}^2 \frac{1}{(c_1c_2)^2} \frac{\pi^6}{7\mathrm{OSR}^7} \tag{3.51}$$

Assuming a standard deviation of 0.1% in the unit resistors – which is a reasonable value in present-day mixed-signal CMOS processes – $\mathrm{IBN_{DAC}}$ equals −113.6 dB; in other words, it is negligible compared to $\mathrm{IBN}_Q = -84.1$ dB.

As discussed throughout this chapter, cascade ΣΔMs are especially sensitive to leakages of low-order quantization errors due to circuit-level nonidealities. For the case of SC ΣΔ cascades, the most-critical error mechanisms are the finite gain of the amplifiers and the mismatching between the unit capacitors used for implementing the modulator coefficients. Figure 3.26 illustrates the impact of noise leakages on the output spectrum of the 2-1-1 multi-bit ΣΔM for a DC gain of 65 dB in the amplifiers and a capacitor mismatch of 0.1%, in addition to the multi-bit DAC errors. Results are compared to a purely ideal behavioral simulation only considering quantization errors (as done in Figure 3.24). The integrated error powers versus frequency are also included for comparison purposes. Note that, in spite of the degradation of the shaping performance due to noise leakages, the resulting error power integrated in the 4-MHz bandwidth (−82.6 dB) is only slightly larger than the simulated IBN_Q (−84.1 dB). The influence of both error mechanisms can be separately estimated from Equations (1.17) and (1.18) as:

$$\mathrm{IBN}_{A_v} \approx \frac{\Delta_1^2}{12} \frac{1}{(k_q a_1 a_2)^2} \frac{4\pi^2}{3A_v^2\mathrm{OSR}^3} + \frac{\Delta_2^2}{12} \frac{1}{c_1^2} \frac{\pi^4}{5A_v^2\mathrm{OSR}^5} + \frac{\Delta_3^2}{12} \frac{1}{(c_1c_2)^2} \frac{\pi^6}{7A_v^2\mathrm{OSR}^7} \tag{3.52}$$

$$\mathrm{IBN}_{\sigma_C} \approx \frac{\Delta_1^2}{12} \frac{1}{(k_q a_1 a_2)^2} \frac{\pi^4}{5\mathrm{OSR}^5} (2 \cdot 3\sigma_C)^2 + \frac{\Delta_2^2}{12} \frac{1}{c_1^2} \frac{\pi^6}{7\mathrm{OSR}^7} (3\sigma_C)^2 \tag{3.53}$$

The increase in the modulator IBN due to integrator leakages can thus be estimated as −91.3 dB for $A_v = 65$ dB, whereas that due to capacitor mismatch equals −91.0 dB for $\sigma_C = 0.1\%$.[9]

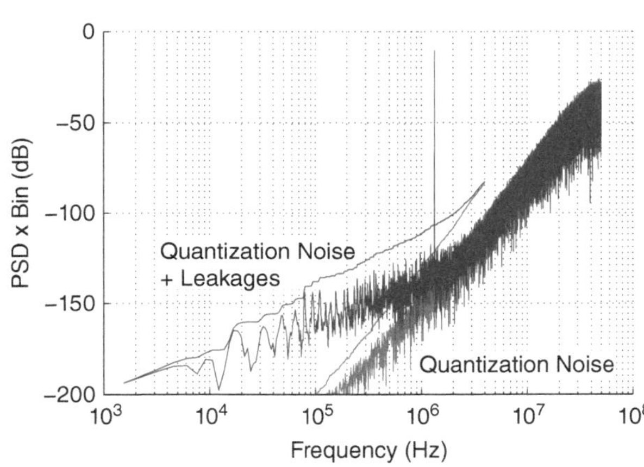

Figure 3.26 Effect of noise leakages on the modulator output spectrum ($A_v = 65$ dB, $\sigma_C = 0.1\%$, $\sigma_R = 0.1\%$). Input signal with $P_{\mathrm{in}} = -6$ dBFS and $f_{\mathrm{in}} = B_w/3$.

[9]Note that the closed-form expressions in Equations (3.51), (3.52), and (3.56) provide an accurate estimation of the increase in the modulator IBN, since the result obtained from $\mathrm{IBN}_Q + \mathrm{IBN_{DAC}} + \mathrm{IBN}_{A_v} + \mathrm{IBN}_{\sigma_C}$ (−84.1 dB − 113.6 dB − 91.3 dB − 91.0 dB = −82.7 dB) is in close accordance with behavioral simulation results in Figure 3.26 (−82.6 dB).

The former equations can also be employed to evaluate the influence of noise leakages on the performance of the 2-1-1 ΣΔM at different clock rates and with different multi-bit quantizations. This is illustrated in Figure 3.27, which shows the effective resolution of the ΣΔ cascade against B_3 for varying OSR. Note that the ENOB curves tend to saturate in the presence of nonidealities, leading to a practical useful limit of multi-bit quantization. However, for the values assumed for the nonideal parameters A_v, σ_C, and σ_R, the initial selection of $B_3 = 3$ for OSR = 12.5 provides a good trade-off for achieving the targeted modulator performance.[10]

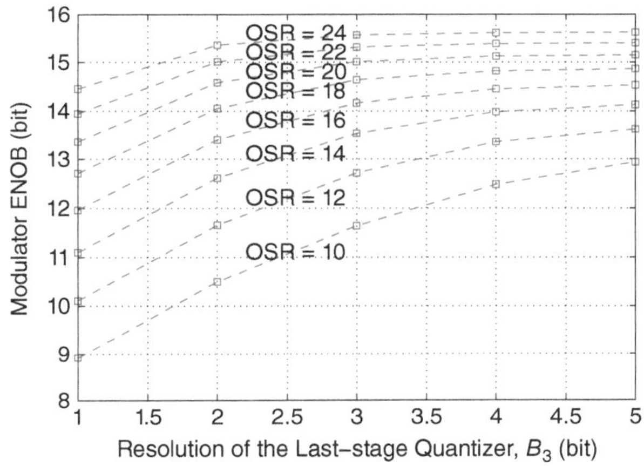

Figure 3.27 Modulator effective resolution versus the number of bits in the last-stage quantizer for varying OSR. All noise leakages are considered ($A_v = 65$ dB, $\sigma_C = 0.1\%$, $\sigma_R = 0.1\%$).

3.8.3 Circuit Noise

As discussed in Section 3.5, the contribution of circuit noise to the overall IBN is dominated by the front-end integrator. Only kT/C noise and amplifier thermal noise will be considered in this high-level sizing of the 2-1-1 SC-ΣΔM.

On the one hand, the kT/C noise budget determines the minimum value of the sampling capacitor (C_S) to be used. A small value will be selected to reduce the capacitive load of the integrators and, hence, their settling requirements. This noise contribution can be estimated from Equation (3.41) as:

$$\text{IBN}_{kT/C} \approx \frac{4kT}{C_S \text{OSR}} \tag{3.54}$$

For $C_S = C_u = 0.4$ pF in the front-end integrator (see Figure 3.25), $\text{IBN}_{kT/C}$ equals −84.8 dB; in other words, it yields an in-band contribution similar to that of the ideal quantization noise.[11]

On the other hand, the contribution of the amplifier thermal noise can be estimated from Equation (3.41) as,

$$\text{IBN}_{\text{amp,th}} \approx S_{\text{amp,th}} \frac{\text{GB}_{\phi_2}}{4\text{OSR}} \tag{3.55}$$

[10]Note that a selection of $B_3 = 2$ for OSR = 14 ($f_s = 112$ MHz) leads to similar modulator ENOB under the influence of the same nonidealities.

[11]A similar budget is often allocated for IBN_Q and $\text{IBN}_{kT/C}$ to optimize the power consumption of SC-ΣΔMs.

which requires knowing a priori the amplifier GB product to quantify the foldover effect. The required GB_{ϕ_2} can be estimated by considering the integrator settling error to be a linear gain error, as discussed in Section 3.4.2. Therefore, the in-band noise due to limited linear dynamics[12] can be obtained as,

$$\mathrm{IBN}_{st} \approx \frac{\Delta_1^2}{12}\frac{1}{(k_q a_1 a_2)^2}\frac{\pi^4}{5\mathrm{OSR}^5}(2\epsilon_{st})^2 + \frac{\Delta_2^2}{12}\frac{1}{c_1^2}\frac{\pi^6}{7\mathrm{OSR}^7}(\epsilon_{st})^2 \tag{3.56}$$

where ϵ_{st} stands for the gain error, given by:

$$\epsilon_{st} \approx e^{-\pi\frac{\mathrm{GB}_{\phi_2}(\mathrm{Hz})}{f_s}} \tag{3.57}$$

For $\mathrm{GB}_{\phi_2} \approx 2.1f_s$ (around 200 MHz), IBN_{st} equals −96.8 dB. This value is considerably smaller than IBN_Q and $\mathrm{IBN}_{kT/C}$ so as to have some margin in the noise budget for allocating more accurate settling error estimates in the next steps.

Based on this value of GB_{ϕ_2} for settling considerations, the input-referred thermal noise of the amplifier can be estimated for a given noise budget. For $S_{\mathrm{amp,th}} = (6\mathrm{nV}/\sqrt{\mathrm{Hz}})^2$, $\mathrm{IBN}_{\mathrm{amp,th}}$ equals −90.3 dB. The total in-band white noise will thus be dominated by kT/C noise and can be found to equal −83.7 dB.

3.8.4 Settling Error

Assuming that the parasitics at the amplifier input and output nodes are $C_P = 0.1$ pF and $C_L = 0.5$ pF, respectively, the equivalent load at the amplifier output during integration can be found from Equation (3.17)

$$C_{\mathrm{eq},\phi_2} = C_P + C_S + C_L\left(1 + \frac{C_P + C_S}{C_I}\right) \tag{3.58}$$

to be around 1.2 pF. Previously, the required amplifier GB was estimated for only a linear settling error to be $\mathrm{GB}_{\phi_2} = 2.1f_s$. According to Equation (3.28), the amplifier transconductance for a dominant-pole model yields around 1.5 mA/V.

Figure 3.28 shows simulation results for the considered 2-1-1 ΣΔM using the behavioral model for the transient response of SC integrators presented in Section 3.4.1, which accurately accounts for both linear and nonlinear integrator dynamics. The simulated modulator IBN against the amplifier transconductance is shown in Figure 3.28a for different values of the amplifier output current. Note that, for $g_m = 1.5$ mA/V, the modulator performance is not limited by nonlinear integrator dynamics as long as $I_o \geq 250$ μA. According to Equation (3.28), the required amplifier SR can be estimated as

$$\mathrm{SR}_{\phi_2}(\mathrm{V/s}) = \left(1 + \frac{C_P + C_i}{C_I}\right)\frac{I_o}{C_{\mathrm{eq},\phi_2}} \tag{3.59}$$

which is around 275 V/μs.

Figure 3.28b illustrates the effect of the amplifier limited output current on the modulator output spectra for a −6 dBFS@1.33 MHz input tone. Note that, for $I_o \leq 250\mu$A, the limited amplifier SR clearly degrades the shaping performance of the modulator.

[12] Assuming thus nonlimited nonlinear settling – in other words, infinite SR.

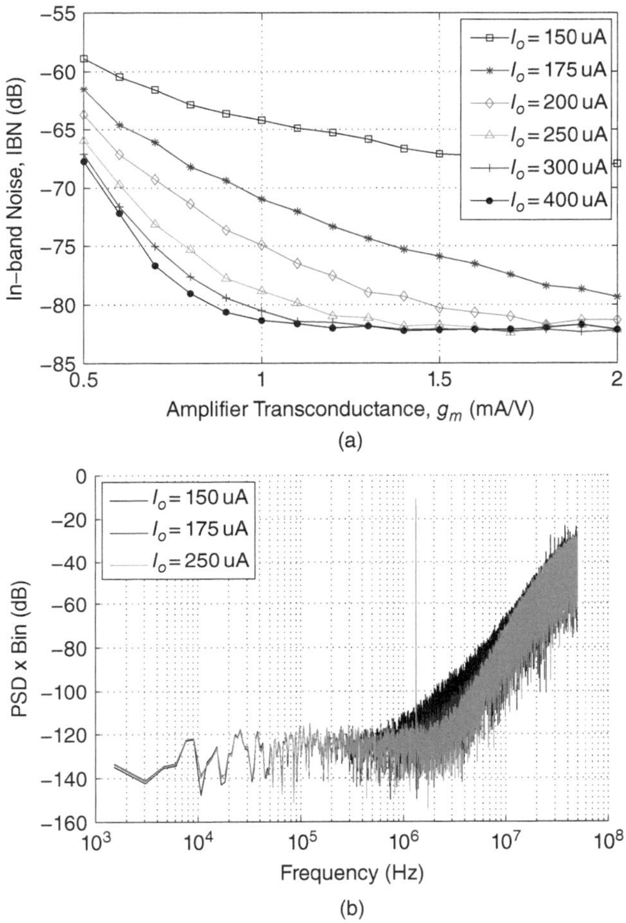

Figure 3.28 Influence of the amplifier dynamics on the modulator performance: (a) IBN against the amplifier transconductance for different values of the amplifier output current; (b) output spectrum for $g_m = 1.5$ mA/V and different amplifier output currents. Input signal with $P_{in} = -6$ dBFS and $f_{in} = B_w/3$. Noise leakages and thermal noise are also accounted for in the behavioral simulations.

3.8.5 Overall High-Level Sizing and Noise Budget

Table 3.1 summarizes the electrical specifications of the main modulator blocks derived in the presented high-level sizing of the 2-1-1 SC-ΣΔM. Accordingly, Table 3.2 shows the selected noise budget for achieving the targeted resolution of 12bit at 4-MHz conversion bandwidth. According to the closed-form expressions used, the total in-band noise is −80.1 dB, leading to a DR of 77.1dB (12.5bit).

Note from Table 3.1 that the derived requirements apply for the front-end integrator, whereas those of the remaining integrators can usually be scaled down to some extent. Extensive behavioral simulations are often performed for this purpose.

Table 3.1 High-level sizing of the 2-1-1 SC-ΣΔM.

Modulator	Topology	2-1-1
	Dual quantization, B_1, B_2, B_3	1 bit, 1 bit, 3 bit
	Signal bandwidth, B_w	4MHz
	Clock frequency, f_s	100MHz
	Oversampling ratio, OSR	12.5
	Reference voltage, V_{ref}	1 V
Front-end integrator	Unit capacitor, C_u	0.4 pF
	Capacitor standard deviation, σ_C	0.1%
	Sampling capacitor, C_S	0.4 pF
	Integration capacitor, C_I	1.6 pF
Front-end amplifier	DC gain, A_v	65 dB
	Input capacitor, C_P	0.1 pF
	Output capacitor, C_L	0.5 pF
	Equivalent output load, C_{eq,ϕ_2}	1.2 pF
	Transconductance, g_m	1.5 mA/V
	Gain-bandwidth product, GB_{ϕ_2}	206.5 MHz$(= 2.1f_s)$
	Slew rate, SR_{ϕ_2}	275 V/μs$(= 2.7V_{\text{ref}}f_s)$
	Input-referred thermal noise, $S_{\text{amp,th}}$	$(6\text{nV}/\sqrt{\text{Hz}})^2$
	Output swing	$\pm V_{\text{ref}}$
Last-stage quantizer	Resistor standard deviation, σ_R	0.1%

Table 3.2 Noise budget for the high-level sizing of the 2-1-1 SC-ΣΔM proposed in Table 3.1.

		dB	dB
Quantization noise			−84.1
	Multi-bit DAC error	−113.6	
	Integrator leakage	−91.3	
	Capacitor mismatch	−91.0	
Total noise leakages			−88.1
Settling error (linear)			−96.8
	kT/C noise	−84.8	
	Amplifier thermal noise	−90.3	
Total thermal noise			−83.7
Total in-band noise			−80.1

Note also that the generated distortion has not been strictly evaluated, so that the derived values for the gain and SR of the front-end amplifier may increase, depending on the involved nonlinearities, so as not to degrade the modulator SNDR.

However, the requirements obtained for the modulator blocks can be used as the starting point for their electrical design. A *bottom-up* approach is then often followed to ensure the targeted modulator specifications are met.

3.9 Summary

In this chapter the main error mechanisms degrading the performance of SC-ΣΔMs have been analyzed. These errors are caused by nonidealities that affect the circuit implementation of the modulator analog blocks and produce extra error components. These add to the in-band quantization error noise and can severely limit the modulator performance. Therefore, nonidealities have been studied in detail; behavioral models and, if possible, practical closed-form expressions have been presented to estimate their influence on ΣΔM behavior.

Errors modifying the noise transfer functions of SC-ΣΔMs have been analyzed, including finite amplifier gain, capacitor mismatch, and integrator settling. Single-loop modulator topologies are less sensitive to these errors than cascades, which suffer from noise leakage. Nonidealities whose effect can be modeled as an additive noise source at the modulator input have been studied, such as circuit noise and clock jitter, and the main sources of distortion have also been addressed. The next chapter will focus on the effect of the main circuit errors on CT-ΣΔMs.

References

[1] S. R. Norsworthy, R. Schreier, and G. C. Temes, *Delta-Sigma Data Converters: Theory, Design and Simulation.* IEEE Press, 1997.

[2] M. Ortmanns and F. Gerfers, *Continuous-Time Sigma-Delta A/D Conversion: Fundamentals, Performance Limits and Robust Implementations.* Springer, 2006.

[3] R. del Río, F. Medeiro, B. Pérez-Verdú, J. M. de la Rosa, and A. Rodríguez-Vázquez, *CMOS Cascade ΣΔ Modulators for Sensors and Telecom: Error Analysis and Practical Design.* Springer, 2006.

[4] A. Marques, V. Peluso, M. S. Steyaert, and W. M. Sansen, "Optimal Parameters for ΔΣ Modulator Topologies," *IEEE Trans. on Circuits and Systems II: Analog and Digital Signal Processing*, vol. 45, pp. 1232–1241, September 1998.

[5] Y. Geerts, M. Steyaert, and W. Sansen, *Design of Multi-bit Delta-Sigma A/D Converters.* Kluwer Academic Publishers, 2002.

[6] R. del Río, F. Medeiro, J. M. de la Rosa, B. Pérez-Verdú, and A. Rodríguez-Vázquez, "Reliable Analysis of Settling Errors in SC Integrators: Application to ΣΔ Modulators," *IET Electronics Letters*, vol. 36, pp. 503–504, March 2000.

[7] J. Fischer, "Noise Sources and Calculation Techniques for Switched Capacitor Filters," *IEEE J. of Solid-State Circuits*, vol. 17, pp. 742–752, August 1982.

[8] C.-A. Gobet and A. Knob, "Noise Analysis of Switched Capacitor Networks," *IEEE Trans. on Circuits and Systems*, vol. CAS-30, pp. 37–43, January 1983.

[9] R. Gregorian and G. C. Temes, *Analog MOS Integrated Circuits for Signal Processing.* John Wiley & Sons, 1986.

[10] C. C. Enz and G. C. Temes, "Circuit Techniques for Reducing the Effects of Op-Amp Imperfections: Autozeroing, Correlated Double Sampling, and Chopper Stabilization," *Proceedings of the IEEE*, vol. 84, pp. 1584–1614, November 1996.

[11] T. H. Lee, *The Design of CMOS Radio-Frequency Integrated Circuits.* Cambridge University Press, 2004.

[12] B. E. Boser and B. A. Wooley, "The Design of Sigma-Delta Modulation Analog-to-Digital Converters," *IEEE J. of Solid-State Circuits*, vol. 23, pp. 1298–1308, December 1988.

[13] B. Brandt, P. Ferguson, and M. Rebeschini, "Analog Circuit Design of ΣΔ ADCs," in *Delta-Sigma Data Converters: Theory, Design and Simulation* (S.R. Norsworthy, R. Schreier, and G.C. Temes, editors), IEEE Press, 1997.

[14] D. Haigh and B. Singh, "A Switching Scheme for Switched Capacitor Filters which Reduces the Effect of Parasitic Capacitances Associated with Switch Control Terminals," *Proc. of the IEEE Intl. Symp. on Circuits and Systems*, pp. 586–589, 1983.

[15] G. Yin and W. Sansen, "A High-Frequency and High-Resolution Fourth-Order ΣΔ A/D Converter in BiCMOS Technology," *IEEE J. of Solid-State Circuits*, vol. 29, pp. 857–865, August 1994.

[16] K. R. Laker and W. M. C. Sansen, *Design of Analog Integrated Circuits and Systems*. McGraw-Hill, 1994.

[17] J.-T. Wu and K.-L. Chang, "MOS Charge Pumps for Low-Voltage Operation," *IEEE J. of Solid-State Circuits*, vol. 33, pp. 592–597, April 1998.

[18] K. Bult, "Analog Design in Deep Sub-Micron CMOS," *Proc. of the IEEE European Solid-State Circuits Conf.*, pp. 11–17, 2000.

[19] J. Cherry and W. Snelgrove, "Clock Jitter and Quantizer Metastability in Continuous-Time Delta–Sigma Modulators," *IEEE Trans. on Circuits and Systems – II: Analog and Digital Signal Processing*, vol. 46, pp. 661–676, June 1999.

[20] S. Luschas and H. S. Lee, "High-Speed ΣΔ Modulators with Reduced Timing Jitter Sensitivity," *IEEE Trans. on Circuits and Systems – II: Analog and Digital Signal Processing*, vol. 49, pp. 712–720, November 2002.

[21] O. Oliaei, "Design of Continuous-Time Sigma-Delta Modulators with Arbitrary Feedback Waveform," *IEEE Trans. on Circuits and Systems – II: Analog and Digital Signal Processing*, vol. 50, pp. 437–444, August 2003.

[22] O. Oliaei, "Sigma-Delta Modulator with Spectrally Shaped Feedback," *IEEE Trans. on Circuits and Systems II: Analog and Digital Signal Processing*, vol. 50, pp. 518–530, September 2003.

[23] L. Hernández, A. Wiesbauer, S. Patón, and A. DiGiandomenico, "Modelling and Optimization of Low Pass Continuous-Time Sigma-Delta Modulators for Clock Jitter Noise Reduction," *Proc. of the IEEE Intl. Symp. on Circuits and Systems*, pp. 1072–1075, 2004.

[24] L. Hernández, P. Rombouts, E. Prefasi, S. Patón, and C. L. M. García, "A Jitter Insensitive Continuous-Time ΣΔ Modulator Using Transmission Lines," *Proc. of the IEEE Intl. Conf. on Electronics, Circuits and Systems*, pp. 109–112, 2004.

[25] R. H. van Veldhoven and A. H. M. van Roermund, *Robust Sigma Delta Converters*. Springer, 2011.

[26] P. G. R. Silva and J. H. Huijsing, *High Resolution IF-to-Baseband ΣΔ ADC for Car Radios*. Springer, 2008.

[27] J. Cherry and W. Snelgrove, "Excess Loop Delay in Continuous-Time Delta–Sigma Modulators," *IEEE Trans. on Circuits and Systems – II: Analog and Digital Signal Processing*, vol. 46, pp. 376–389, April 1999.

[28] J. F. Jensen, G. Raghavan, A. E. Cosand, and R. H. Walden, "A 3.2-GHz Second-Order Delta-Sigma Modulator Implemented in InP HBT Technology," *IEEE J. of Solid-State Circuits*, vol. 30, pp. 1119–1127, October 1995.

[29] J. van Engelen, R. J. van de Plassche, E. Stikvoort, and A. G. Venes, "A Sixth-Order Continuous-Time Bandpass Sigma-Delta Modulator for Digital IF Radio," *IEEE J. of Solid-State Circuits*, vol. 34, pp. 1753–1764, December 1999.

[30] S. Yan and E. Sánchez-Sinencio, "A Continuous-Time ΣΔ Modulator With 88-dB Dynamic Range and 1.1-MHz Signal Bandwidth," *IEEE J. of Solid-State Circuits*, vol. 39, pp. 75–86, January 2004.

[31] P. Benabes, M. Keramat, and R. Kielbasa, "A Methodology for Designing Continuous-Time Sigma-Delta Modulators," *Proc. of the IEEE European Design and Test Conf.*, pp. 46–50, 1997.

[32] P. Fontaine, A. N. Mohieldin, and A. Bellaouar, "A Low-Noise Low-Voltage CT ΔΣ Modulator with Digital Compensation of Excess Loop Delay," *IEEE ISSCC Digest of Technical Papers*, pp. 498–499, 2005.

[33] G. Mitteregger, C. Ebner, S. Mechnig, T. Blon, C. Holuigue, and E. Romani, "A 20-mW 640-MHz CMOS Continuous-Time ΣΔ ADC With 20-MHz Signal Bandwidth, 80-dB Dynamic Range and 12-bit ENOB," *IEEE J. of Solid-State Circuits*, vol. 41, pp. 2641–2649, December 2006.

[34] S. Pavan, N. Krishnapura, R. Pandarinathan, and P. Sankar, "A Power Optimized Continuous-Time ΔΣ ADC for Audio Applications," *IEEE J. of Solid-State Circuits*, vol. 43, pp. 351–360, February 2008.

[35] M. Keller, A. Buhmann, J. Sauerbrey, M. Ortmanns, and Y. Manoli, "A Comparative Study on Excess-Loop-Delay Compensation Techniques for Continuous-Time Sigma–Delta Modulators," *IEEE Trans. on Circuits and Systems – I: Regular Papers*, vol. 55, pp. 3480–3487, December 2008.

[36] F. Gerfers, M. Ortmanns, and Y. Manoli, "A 1.5-V 12-bit Power-Efficient Continuous-Time Third-Order ΣΔ Modulator," *IEEE J. of Solid-State Circuits*, vol. 38, pp. 1343–1352, August 2003.

[37] S. Patón, A. di Giandomenico, L. Hernández, A. Wiesbauer, T. Poetscher, and M. Clara, "A 70-mW 300-MHz CMOS Continuous-Time ΣΔ ADC With 15-MHz Bandwidth and 11 Bits of Resolution," *IEEE J. of Solid-State Circuits*, vol. 39, pp. 1056–1063, July 2004.

[38] L. Breems, R. Rutten, R. van Veldhoven, and G. van der Weide, "A 56 mW Continuous-Time Quadrature Cascaded ΣΔ Modulator With 77 dB DR in a Near Zero-IF 20 MHz Band," *IEEE J. of Solid-State Circuits*, vol. 42, pp. 2696–2705, December 2007.

[39] P. Kiss, J. Silva, A. Wiesbauer, T. Sun, U.-K. Moon, J. T. Stonick, and G. C. Temes, "Adaptive Digital Correction of Analog Errors in MASH ADCs – Part II: Correction Using Test-Signal Injection," *IEEE Trans. on Circuits and Systems – II: Analog and Digital Signal Processing*, vol. 47, pp. 629–638, July 2000.

[40] L. Breems, R. Rutten, and G. Wetzker, "A Cascaded Continuous-Time ΣΔ Modulator with 67-dB Dynamic Range in 10-MHz Bandwidth," *IEEE J. of Solid-State Circuits*, vol. 39, pp. 2152–2160, December 2004.

[41] Y.-S. Shu, J. Kamiishi, K. Tomioka, K. Hamashita, and B.-S. Song, "LMS-Based Noise Leakage Calibration of Cascaded Continuous-Time ΣΔ Modulators," *IEEE J. of Solid-State Circuits*, vol. 45, pp. 368–379, February 2010.

[42] M. Ortmanns, F. Gerfers, and Y. Manoli, "Compensation of Finite Gain-Bandwidth Induced Errors in Continuous-Time Sigma–Delta Modulators," *IEEE Trans. on Circuits and Systems – I: Regular Papers*, vol. 51, pp. 1088–1099, June 2004.

[43] L. Breems and J. H. Huijsing, *Continuous-Time Sigma-Delta Modulation for A/D Conversion in Radio Receivers*. Kluwer Academic Publishers, 2001.

[44] J. Cherry and W. Snelgrove, *Continuous-Time Delta-Sigma Modulators for High-Speed A/D Conversion*. Kluwer Academic Publishers, 1999.

CHAPTER 4

CIRCUIT ERRORS AND COMPENSATION TECHNIQUES IN CONTINUOUS-TIME ΣΔ MODULATORS

Following the analysis of nonidealities in SC-ΣΔMs in Chapter 3, this chapter focuses on the main circuit and system effects on the operation of CT-ΣΔMs. The first part of the chapter is devoted to the nonideal performance caused by CT-ΣΔM building blocks, in particular integrators and resonators. These are briefly described in Section 4.2. This group of errors includes finite DC gain (Section 4.3), time-constant variation (Section 4.4), transient response (Section 4.5), nonlinear harmonic distortion (Section 4.6), and circuit noise (Section 4.7). The second part of the chapter covers the dominant sources of so-called "architectural timing errors" in CT-ΣΔMs, namely clock jitter (Section 4.8), excess loop delay (Section 4.9), and quantizer metastability (Section 4.10).

4.1 Overview of Nonidealities in Continuous-time ΣΔ Modulators

As stated in Section 2.6, CT-ΣΔMs are potentially faster than SC ones, leading to much more relaxed designs (in terms of power consumption) when high-speed operation is required. In addition, they do not suffer from kT/C noise. However, SC implementations have intrinsically lower parameter variations, since most circuit parameters are defined by capacitor ratios, instead of by absolute parameter values, as is the case for CT-ΣΔMs.

Sigma-Delta Converters: Practical Design Guide, Second Edition. José M. de la Rosa.

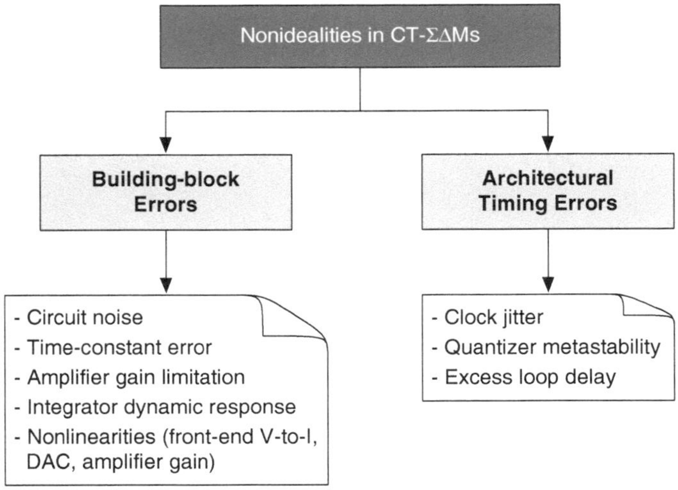

Figure 4.1 Main nonidealities affecting the performance of continuous-time ΣΔMs.

Circuit errors in CT-ΣΔMs can be divided into two main categories, as illustrated in Figure 4.1:

- building-block errors, which are the nonideal effects derived from the modulator loop filter implementation – similar to the SC case – such as finite amplifier gain (for active-RC implementations), integrator time-constant error, circuit noise, nonlinearities, and so on;
- architectural timing errors, namely clock jitter, excess loop delay, and quantizer metastability.

The former group of errors have similar effects on the performance of CT-ΣΔMs as on their SC counterparts. Therefore, they can also be classified according to how they degrade the modulator performance: either causing a deviation in NTF or an additive noise component at the modulator input. The procedure to analyze the impact of these errors is similar to that followed in Chapter 3. Thus, Equation 3.1 is transformed for CT-ΣΔMs into:

$$\mathrm{ITF}(s) \rightarrow \mathrm{ITF}_{\epsilon}(s) \rightarrow \mathrm{NTF}_{\epsilon}(z) \rightarrow \mathrm{IBN}(\epsilon) \tag{4.1}$$

Based on that procedure, the main nonideal effects of CT-ΣΔMs are described in the following sections, putting emphasis on the most critical design issues. Let us first start by reviewing some basic concepts of CT integrators and resonators, which are the main building blocks of CT-ΣΔM loop filters.

4.2 CT Integrators and Resonators

CT integrators are the basic building blocks of the modulator loop filter of LP CT-ΣΔMs, and they can be combined to build resonators in order to implement the loop filter of BP CT-ΣΔMs. There are three main practical CT integrator topologies – conceptually depicted in Figure 4.2. All these circuit

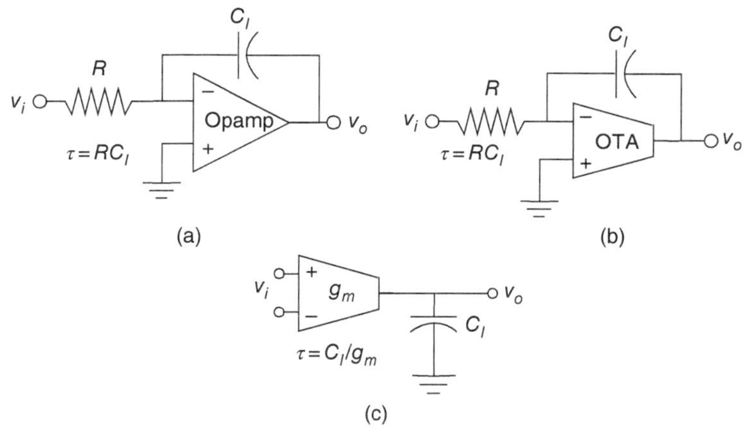

Figure 4.2 Conceptual schematic of most common circuit techniques for realizing CT integrators: (a) active-RC; (b) OTA-RC; (c) Gm-C.

implementations have the same ideal transfer function, given by:

$$\text{ITF}_i(s) = \frac{1}{s\tau} \tag{4.2}$$

where τ is the integrator time constant, which is the product of the resistive and capacitive elements of the CT integrators shown in Figure 4.2.

In practice, the operation of CT integrators involves a number of design trade-offs as will be discussed later in this book. Thus active-RC circuits have better linearity and a larger output swing, whereas Gm-C integrators are – a priori – potentially faster and can be easily tuned, thus being better suited to reconfigurable applications, as will be discussed in Chapter 9.

Resonators can be easily built by connecting Gm-C integrators in a loop, as illustrated in Figure 4.3. As discussed in Chapter 2, these circuits are the basic building blocks required to build BP CT-ΣΔMs. Analyzing the circuit in Figure 4.3a, it can be shown that the resonant frequency, which in turns corresponds to the notch frequency, f_n, is given by:

$$f_n = \frac{1}{2\pi}\sqrt{\frac{g_{m2}g_{mR}}{C_1C_2}} \tag{4.3}$$

which can be programmed by tuning the feedback transconductance, g_{mR} [1]. Resonators can be also implemented using active-RC circuits with only one opamp, as illustrated in Figure 4.3b. The circuit uses a passive network to implement a positive feedback, thus achieving a higher quality factor [2].

The main limitation of Gm-C circuits is caused by their stronger nonlinearity, which is associated with the voltage-to-current conversion. Circuit linearization strategies such as source degeneration can be used [3], although they are normally penalized by an increase in the power consumption. Alternatively, the CT integrator placed at the front-end of the loop filter can be replaced with an active-RC integrator; the remaining integrators in the loop are Gm-C integrators. In this way, the effect of nonlinearity is mostly attenuated by the virtual ground implemented in active-RC integrators.

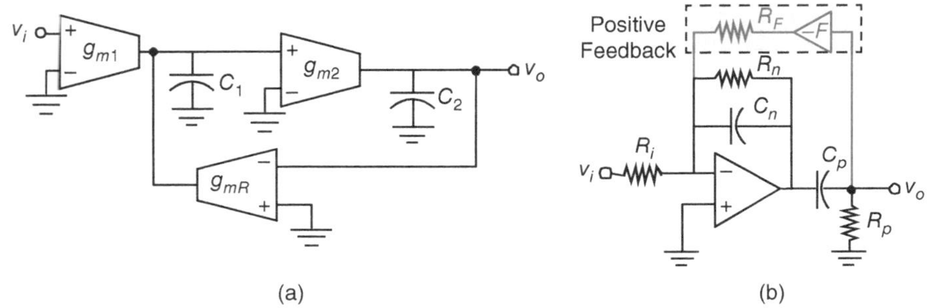

Figure 4.3 Examples of CT resonators using: (a) Gm-C integrators [1]; (b) single-opamp active-RC integrators [2].

Active-RC integrators are in general more robust than Gm-C integrators, thanks in part to their virtual ground. These circuits are usually the best design choice to solve the trade-offs between linarity, speed, power consumption, and thermal noise. Moreover, the use of circuit techniques such as assisted opamps [4] or multi-stage feedforward compensated opamps [5], usually leads to more power-efficient solutions. For this reason, this book mostly focuses on active-RC circuit implementations, although Gm-C circuits will be also considered in order to demonstrate some nonideal phenomena.

4.3 Finite Amplifier Gain in CT-ΣΔMs

Finite DC gain of amplifiers affects CT-ΣΔMs exactly in the same way as shown in Section 3.2 for SC-ΣΔMs. Let us consider an ideal active-RC integrator in which the transfer function of the ith input path to the output is:

$$\mathrm{ITF}_i(s) = \frac{-1}{sR_iC_I} = \frac{-k_if_s}{s} \tag{4.4}$$

If the finite amplifier gain A_v is accounted for in the CT integrator, as shown in Figure 4.4, the transfer function of the ith input path to the output can be derived as [6]:

$$\mathrm{ITF}_{i,A_v}(s) = \frac{-k_if_s}{s\left(1+\frac{1}{A_v}\right)+\frac{1}{A_v}\sum_{i=1}^{N_i}k_if_s} \approx \frac{-k_if_s}{s+\frac{1}{A_v}\sum_{i=1}^{N_i}k_if_s} \tag{4.5}$$

i.e., as a single-pole transfer function with a DC gain of $A_vk_i/\sum_{i=1}^{N_i}k_i$ and a pole at $f_s\sum_{i=1}^{N_i}k_i/A_v$, which is displaced away from its ideal DC position.

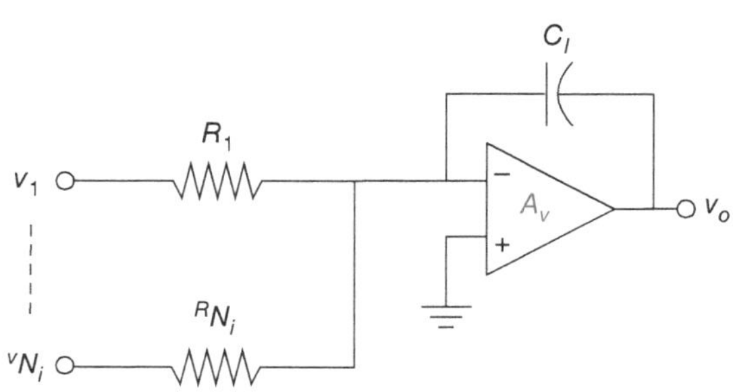

Figure 4.4 Active-RC integrator with N_i input paths and finite amplifier gain.

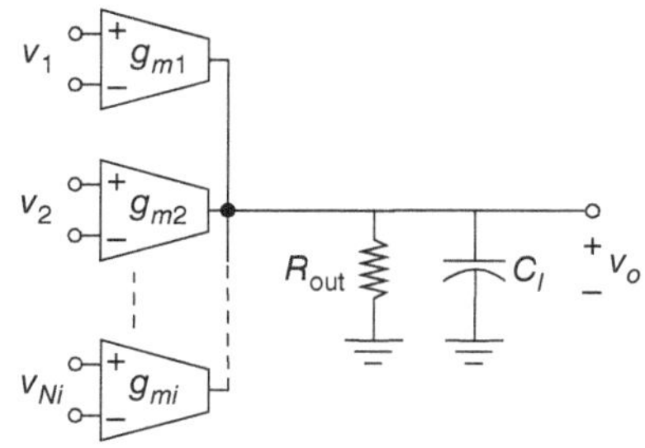

Figure 4.5 Gm-C integrator with N_i input paths and finite amplifier gain.

Note that similar expressions can be derived for a Gm-C integrator with multiple input branches like that shown in Figure 4.5, giving:

$$\mathrm{ITF}_{i,A_{(v,g)}}(s) = \frac{k_{i,g} f_s}{s + \frac{1}{A_{(v,g)}} k_{i,g} f_s} \tag{4.6}$$

where $k_{i,g} \equiv g_{m_i}/C_I$, $A_{(v,g)} \equiv g_{m,i} R_{\mathrm{out}}$ and $R_{\mathrm{out}} = 1/(\sum_{i=1}^{N_i} g_{\mathrm{out,i}})$ is the overall output resistance of the transconductor, with $g_{\mathrm{out,i}}$ being the output conductance of the ith branch (see Figure 4.5).

In the following analysis, active-RC integrators will be considered without loss of generality. Thus, assuming that $\sum_{i=1}^{N_i} k_i \sim 1$ for all integrators in the CT-ΣΔM, the NTF degraded by finite amplifier gain can be derived following a procedure similar to that described in Section 3.2 for SC-ΣΔMs. The in-band noise of a second-order CT-ΣΔM, like that shown in Figure 2.45, can thus be obtained as

$$\mathrm{IBN}_2(A_v) \approx \frac{\Delta^2}{12} \frac{1}{(k_q k_{1,\mathrm{NRZ}})^2} \left(\frac{2\pi^2}{3{A_v}^2 \mathrm{OSR}^3} + \frac{\pi^4}{5\mathrm{OSR}^5} \right) \tag{4.7}$$

which matches Equation (3.8) for SC implementations while taking into account that the DT–CT equivalence yields $k_{1,\mathrm{NRZ}} = a_1 a_2$ – see Equation (2.63).[1]

Similarly, the in-band noise of an Lth-order CT-ΣΔM with distributed feedback under the influence of leaky integrators yields,

$$\mathrm{IBN}_L(A_v) \approx \frac{\Delta^2}{12} \frac{1}{(k_q k_{1,\mathrm{NRZ}})^2} \left[\frac{L\pi^{2(L-1)}}{(2L-1){A_v}^2 \mathrm{OSR}^{(2L-1)}} + \frac{\pi^{2L}}{(2L+1)\mathrm{OSR}^{2L+1}} \right] \tag{4.8}$$

which matches Equation (3.10) if $k_{1,\mathrm{NRZ}} = \prod_{i=1}^{L} a_i$ is accounted for in the DT–CT equivalence.

The influence of finite amplifier gain on cascade CT-ΣΔMs is similar to that for their SC counterparts. For instance, if a 2-1-1 cascade CT-ΣΔM is considered, the in-band noise under the influence of A_v yields,

$$\begin{aligned}\mathrm{IBN}_{211}(A_v) \approx\ & \frac{\Delta_1^2}{12} \frac{1}{(k_q k_{1,\mathrm{NRZ}})^2} \frac{4\pi^2}{3A_v^2 \mathrm{OSR}^3} + \frac{\Delta_2^2}{12} \frac{1}{c_1^2} \frac{\pi^4}{5A_v^2 \mathrm{OSR}^5} \\ & + \frac{\Delta_3^2}{12} \frac{1}{(c_1 c_2)^2} \left(\frac{\pi^6}{7A_v^2 \mathrm{OSR}^7} + \frac{\pi^8}{9\mathrm{OSR}^9} \right)\end{aligned} \tag{4.9}$$

which again matches Equation (3.11) of the equivalent SC-ΣΔM.

[1] Note that the notation given in [6] and adopted in Chapter 2 (see Figure 2.45) in terms of distributed feedback coefficients k_i facilitates the equivalence of expressions for the effect of loop-filter nonidealities in both DT and CT domains.

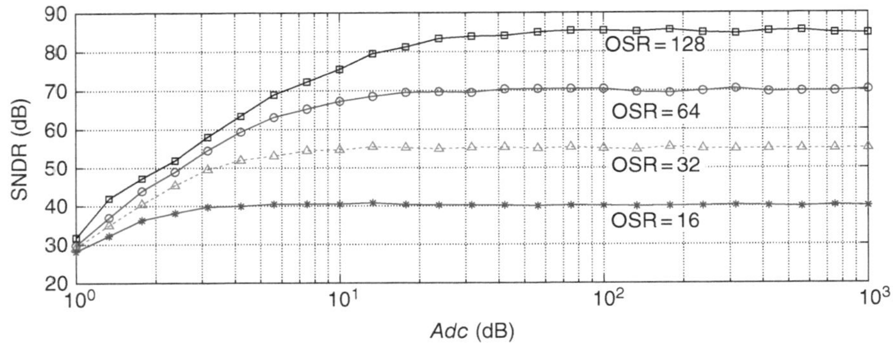

Figure 4.6 Effect of A_v on the SNR of a second-order single-bit CT-ΣΔM.

As an illustration, Figure 4.6 depicts the influence of A_v on the IBN of single-loop CT-ΣΔMs for different values of L and OSR. Note that the conclusions derived in Section 3.2 for the influence of finite amplifier gain on the performance of SC-ΣΔMs with either single-loop or cascade topologies are thus directly applicable to the CT case [6].

4.4 Time-constant Error in CT-ΣΔMs

As discussed in Section 3.3, deviations of the integrator gain coefficients from their nominal values due to variations in technological process parameters affect the performance of ΣΔMs in different ways depending on the modulator architecture. Single-loop topologies deliver quiet, robust performance against integrator gain variations, whereas cascades suffer from the leakage of low-order-shaped quantization noise. However, the impact of this nonideality in SC-ΣΔMs is not critical, since integrator gains are implemented as capacitor *ratios* and, in modern CMOS processes, make use of components with a *matching* better than 0.1%.

Conversely, as shown in Section 4.2, integrator gains in CT-ΣΔMs are mapped onto either $1/(RC)$ or g_m/C values, thus involving *absolute* component values. In modern CMOS technologies, *tolerances* of 10–20% under process and temperature variations are not unusual, thus increasing the possible variation of integrator gains to more than 30%. Although the nature of the nonideality is essentially the same as for SC-ΣΔMs – that is, an integrator gain error – the magnitude of the impact is clearly considerably larger for CT-ΣΔMs.

Let us denote the tolerance error of a CT-integrator gain as ϵ_τ. Taking into account this error, the transfer function of a CT integrator deviates from the ideal one as:

$$\mathrm{ITF}_{i,\epsilon_\tau}(s) = \frac{1}{s\tau(1+\epsilon_\tau)} \tag{4.10}$$

Thus, if a tolerance ϵ_{RC} is accounted for in the time constant of an active-RC integrator, the transfer function of the ith input path to the output modifies to

$$\mathrm{ITF}_{i,\epsilon_{RC}}(s) = \frac{1}{sR_iC_I(1+\epsilon_{RC})} = \frac{k_i}{(1+\epsilon_{RC})}\frac{f_s}{s} \tag{4.11}$$

which can be used for re-computing the NTF and the IBN of a particular CT-ΣΔM architecture under the influence of this nonideality – assuming a modulator linear model.

For an Lth-order CT-ΣΔM with distributed feedback, the IBN can thus be estimated as,

$$\mathrm{IBN}_L(\epsilon_{RC}) \approx \frac{\Delta^2}{12}\frac{1}{(k_q k_{1,\mathrm{NRZ}})^2}\frac{\pi^{2L}}{(2L+1)\mathrm{OSR}^{2L+1}}\prod_{i=1}^{L}(1+\epsilon_{RCi})^2 \tag{4.12}$$

where ϵ_{RCi} refers to the time-constant error of the ith integrator. Note that the above equation is similar to Equation (3.14) for the influence of integrator gain errors in SC-ΣΔMs, taking into account that $k_{1,\mathrm{NRZ}} = \prod_{i=1}^{L} a_i$ in the DT–CT equivalence.

For a 2-1-1 CT cascade, the IBN under the influence of time-constant errors yields,

$$\begin{aligned}\mathrm{IBN}_{211}(\epsilon_{RC}) \approx\ & \frac{\Delta_1^2}{12}\frac{1}{(k_q k_{1,\mathrm{NRZ}})^2}\frac{\pi^4}{5\mathrm{OSR}^5}(\epsilon_{RC1}+\epsilon_{RC2})^2 + \\ & + \frac{\Delta_2^2}{12}\frac{1}{c_1^2}\frac{\pi^6}{7\mathrm{OSR}^7}\epsilon_{RC3}^2 \\ & + \frac{\Delta_3^2}{12}\frac{1}{(c_1c_2)^2}\frac{\pi^8}{9\mathrm{OSR}^9}(1+\epsilon_{RC4})^2\end{aligned} \tag{4.13}$$

which is similar to Equation (4.13) for the equivalent SC-ΣΔM [6].

Figure 4.7 shows simulation results for the impact of time-constant errors on the performance of single-loop and cascade CT-ΣΔMs, and compares them with the approximated expressions in Equations (4.12) and (4.13).[2] Note from the simulation results that single-loop modulators are affected by time-constant errors in two different ways:

- For positive tolerances ($\epsilon_{RC} > 0$), the integrator gain coefficients k_i in Equation 4.11 are reduced or, similarly, coefficients a_i in the equivalent DT-ΣΔM. This leads only to less aggressive noise shaping and thus to an increase in the IBN.
- For negative tolerances ($\epsilon_{RC} < 0$), integrator gain coefficients are increased, leading to more aggressive noise shaping. Therefore, after a short decrease of the IBN, negative time-constant errors reduce the modulator overload level and jeopardize its stability, as is clearly visible for the third-order loop in Figure 4.7b.

The larger sensitivity of CT-ΣΔMs to component tolerances is also noticeable from Figure 4.7c. Note that the IBN of the 2-1-1 cascade with OSR = 32 increases by nearly 30 dB for a time-constant error of 20%.

Concluding, the design of high-order single-loop CT-ΣΔMs is in practice limited more by their tendency to instability (for $\epsilon_{RC} < 0$) than by the degradation of the IBN (for $\epsilon_{RC} > 0$). Therefore,

[2]To achieve accurate matching between analytical and simulation results for large error gains, the effective quantizer gain of single-bit ΣΔMs in Equation (2.10) has to be modified to [6]:

$$k_q = \begin{cases}(1+\epsilon_1)/a_1, & \text{single-bit first-order } \Sigma\Delta\text{M} \\ 2(1+\epsilon_L)/a_L, & \text{single-bit } L\text{th-order } \Sigma\Delta\text{M}\end{cases} \tag{4.14}$$

where ϵ_L is the error in the last integrator gain.

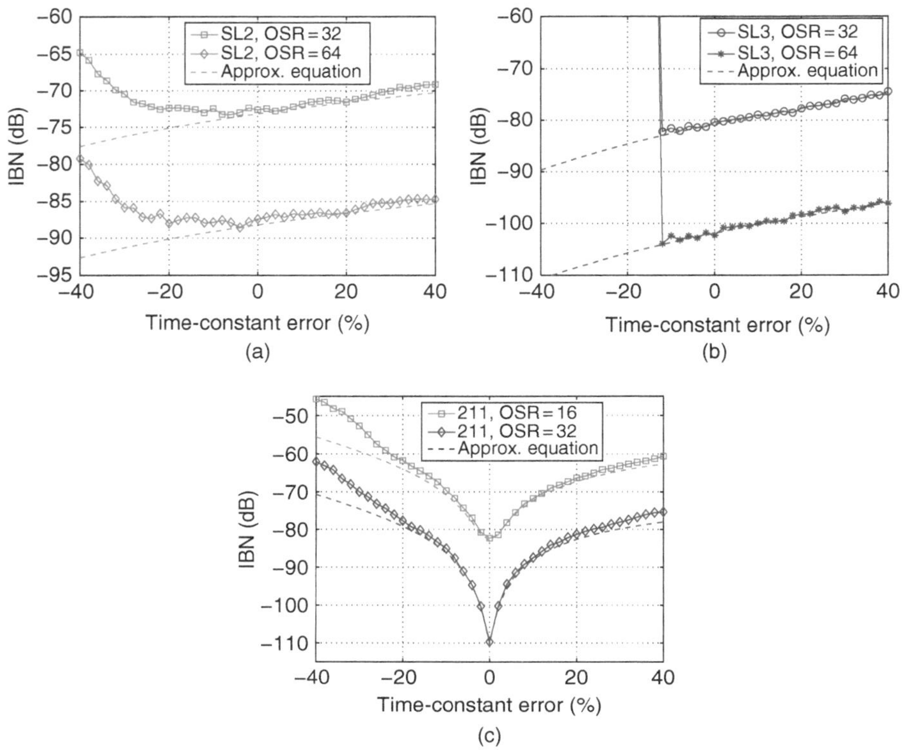

Figure 4.7 Simulation results for the in-band noise of CT-ΣΔMs under the influence of time-constant errors: (a) second-order single-loop. (b) third-order single-loop. (c) 2-1-1 cascade. Simulation results have been obtained for an input signal with $P_{\text{in}} = -20$ dBFS and $f_{\text{in}} = B_w/3$ and for an NRZ rectangular feedback. Approximate results have been obtained from Equations (4.12) and (4.13).

if no countermeasures are taken against the influence of gain errors, less aggressive noise shaping must be implemented using a set of scaling coefficients that are suboptimal in terms of the modulator performance, but for which stability can be guaranteed over all possible ϵ_{RC} variations [7, 8]. Alternatively, time-constant errors can be reduced in CT-ΣΔMs by tuning the absolute value of passive components with programmable banks of capacitors [9] or resistors [10].[3]

In contrast to tuning the analog filter, cascade architectures also offer the possibility of digitally correcting the error cancelation logic to reduce the power of low-order-shaped quantization noise that leaks to the modulator output [11]. Digital correction techniques have been successfully used to reduce the impact of time-constant errors in CT-ΣΔMs [12, 13].

4.5 Finite Integrator Dynamics in CT-ΣΔMs

As CT-ΣΔMs loop-filter circuits operate in the CT domain, their transient response is not strictly limited to be settled within a finite time slot – usually half of the clock signal period – as happens in

[3]Transconductance tuning in gm-C filters and MOSFET tuning in MOSFET-C implementations are also widely used in CT analog filter design. However, they are not so common in CT-ΣΔMs, since the nonlinearity associated with active components considerably limits the modulator performance [9].

SC-ΣΔMs. Therefore, the finite integrator dynamics are in general not critical, although they must be taken into account in order to avoid high-frequency poles of the integrator transient response degrading the performance of the modulator, especially in those applications digitizing high-frequency signals, where there are various design trade-offs involving stability, speed, and power consumption.

4.5.1 Effect of Finite Gain–Bandwidth Product on CT-ΣΔMs

The influence of finite amplifier GB can easily be incorporated into CT modulators by replacing the amplifier in Figure 4.4 by a single-pole model:

$$A(s) = \frac{A_v}{1 + \frac{s}{w_A}}, \quad \text{with} \quad \text{GB} = A_v \omega_A \tag{4.15}$$

where ω_A is the dominant pole of the amplifier inside the integrator. The integrator transfer function of the ith input path to the output is thus modified to [6]:

$$\text{ITF}_{i,\text{GB}}(s) \approx \frac{k_i f_s}{s} \frac{\dfrac{\text{GB}}{\text{GB} + \sum_{i=1}^{N_i} |k_i f_s|}}{1 + \dfrac{s}{\text{GB} + \sum_{i=1}^{N_i} |k_i f_s|}} \tag{4.16}$$

where an integrator gain error arises due to the finite amplifier GB, as well as a second pole. For the sake of clarity, the modified ITF can thus be re-written as,

$$\text{ITF}_{i,\text{GB}}(s) \approx \frac{k_i}{(1 + \epsilon_{\text{GB}})} \frac{f_s}{s} \frac{1}{\left(1 + \frac{s}{\omega_p}\right)}, \quad \text{with} \quad \begin{cases} \epsilon_{\text{GB}} = \sum_{i=1}^{N_i} \omega_{Ii} / \text{GB} \\ \omega_p = \text{GB} + \sum_{i=1}^{N_i} \omega_{Ii} \end{cases} \tag{4.17}$$

where $\omega_{Ii} = |k_i f_s|$.

Regarding Equation (4.17), the influence of the second integrator pole is often neglected to derive closed-form expressions for the influence of finite amplifier GB on CT-ΣΔMs. Here, the expressions for the IBN due to finite GB can be approximated from the corresponding expressions for the gain error introduced by time-constant variations – Equations (4.12) and (4.13) – just by replacing ϵ_{RC} with the expression for ϵ_{GB} in Equation (4.17). The observations already made in Section 4.4 about the different sensitivities of single-loop and cascade architectures and about possible correction techniques also apply here.

The approximation can a priori be considered too coarse, but it provides results that are in accordance to the simulation of NRZ CT-ΣΔMs if the amplifier GB is not very low (GB $\geq f_s$). However, this effect has to be taken into account in order to optimize the design in terms of power dissipation and speed, and becomes more and more important in those applications that operate in the gigahertz range. In these cases, the effect of high-frequency poles should be considered in the design process, although the mathematical analysis becomes more complex and does not always give accurate design equations, thus requiring the use of simulation techniques such as those described later in this book.

An intuitive analysis of this effect has been given by Ortmanns *et al.* [14], who model the effect of GB on CT-ΣΔMs as two different nonideal errors: an integrator gain error and extra loop delay.[4]

[4] These extra loop delays can be in turn modeled as excess loop delays, which can be compensated as described in Section 4.9.

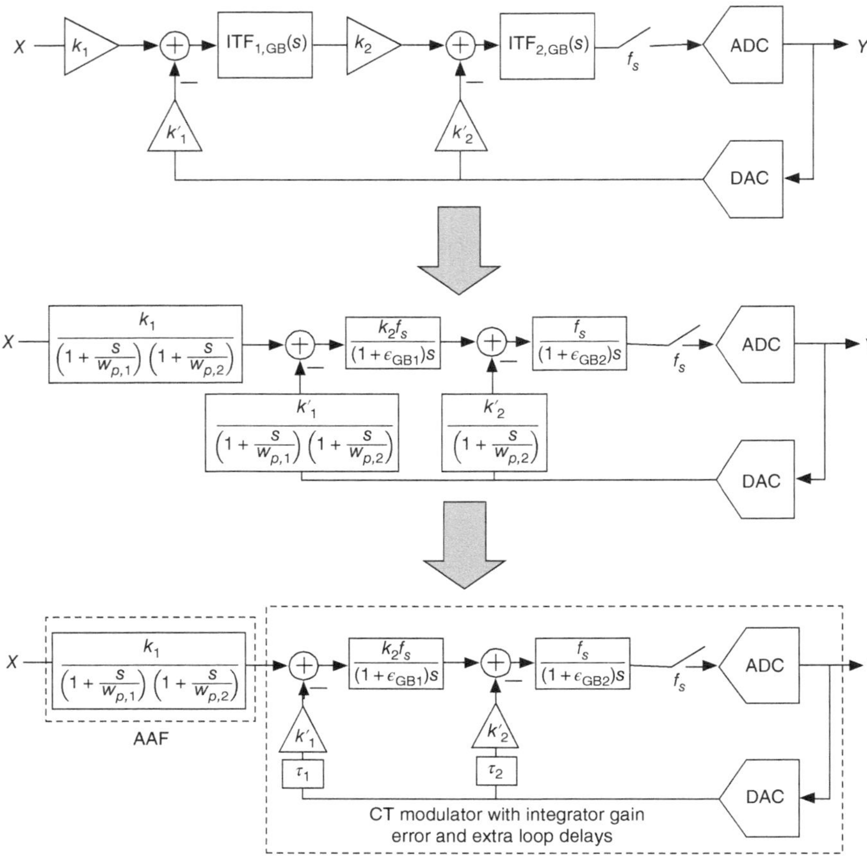

Figure 4.8 Model of GB in CT-ΣΔMs, as proposed by Ortmanns *et al.* [14].

The model is illustrated in Figure 4.8 for a single-loop second-order CT-ΣΔM, where the extra loop delays, τ_i, are given by:

$$\tau_1 = \frac{1 - e^{(-\omega_{p,2}/f_s)}}{\omega_{p,2}}, \tau_2 = \frac{(1 - e^{(-\omega_{p,2}/f_s)})\omega_{p,1}^2 - \omega_{p,2}^2(1 - e^{(-\omega_{p,1}/f_s)})}{\omega_{p,1}\omega_{p,2}(\omega_{p,1} - \omega_{p,2})} \tag{4.18}$$

and $\omega_{p,i}$ stands for the second pole of the ith integrator [14]. One of the benefits of this model is that it allows the effect of finite GB to be reduced with the same techniques used to compensate the excess loop delay, as discussed in Section 4.9.

However, finite amplifier GB is a common nonideality in the dynamics of a ΣΔM, and so by changing the shape and the dynamic behavior of the feedback pulses, its influence on the modulator performance is expected to change [6]. Conversely, there are few published studies that extend the basic single-pole model in Equation (4.15) and/or apply it to different feedback pulse forms [14]. There are even fewer on the influence of the finite amplifier SR on the performance of CT-ΣΔMs and most designs rely greatly on simulations. The reason for this may be found in the fact that, generally speaking, CT-ΣΔMs can work with finite amplifier GB and SR in the integrators that is lower than for SC implementations (see Section 3.4.2 and 3.4.3), due to the absence of the high-current peaks seen in the latter. Indeed, state-of-the-art CT-ΣΔMs are clocked in the gigahertz range, requiring use of power-efficient opamp topologies – for example, multi-stage feedforward-compensated opamps – where it becomes more difficult to set a dominant pole, and usually requiring long settling times. However, this is not a particularly limiting factor in CT-ΣΔMs, where the entire integrator output waveform is processed by the rest of the loop filter.

4.5.2 Effect of Finite Slew Rate on CT-ΣΔMs

Another component of the finite transient response of CT integrators is the maximum output rate that results from the limited (output) current used to charge the integrating capacitor. This is conceptually illustrated in Figure 4.9. As discussed in Chapter 3, a well-known expression used by designers to estimate the maximum SR is given by:

$$\mathrm{SR} \equiv \left|\frac{\mathrm{d}v_o(t)}{\mathrm{d}t}\right|_{\max} = \frac{I_o}{C_{\mathrm{eq}}} \tag{4.19}$$

where I_o and C_{eq} stand respectively for the maximum current and equivalent capacitance at the output of the integrator (see Figure 4.9).

Similar to the case of finite GB, the effect of SR on CT-ΣΔMs is less significant than on SC-ΣΔMs, even if single-bit quantization is used. Indeed, the use of multi-bit quantization reduces the integrators' input/output swings, thus attenuating the effect of SR. However, due to the strong dependence of SR on the exact waveforms of the diverse signals involved in the dynamics of CT-ΣΔMs, the analysis of SR becomes mathematically too complex. Designers therefore usually rely on simulation to obtain the required SR specifications, as will be explained in Chapters 5 and 6.

As an illustration, Figure 4.10 shows the impact of SR on a second-order Gm-C single-bit CT-ΣΔM. Note that in addition to increasing the in-band noise floor, harmonic distortion appears in the output spectrum of the modulator as a consequence of the strong nonlinearity introduced by the SR. There are other sources of nonlinearity causing harmonic distortion in CT-ΣΔMs, as discussed in the next section.

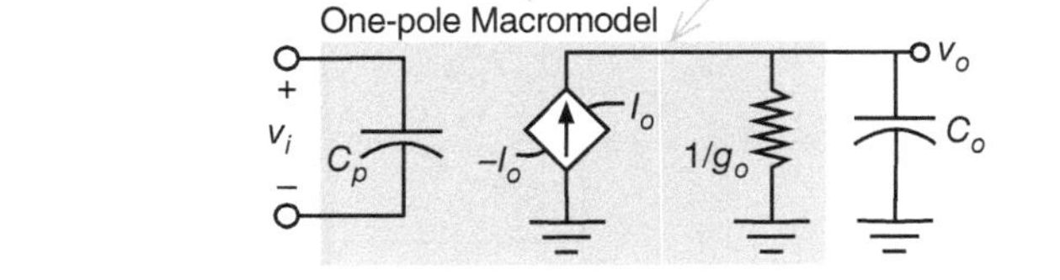

Figure 4.9 Conceptual model of finite slew rate in CT integrators.

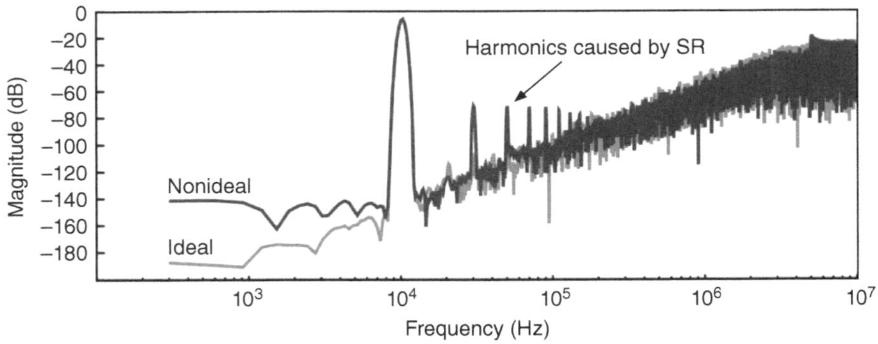

Figure 4.10 Illustrating the effect of SR on a second-order single-bit Gm-C ΣΔM (f_s = 20MHz, SR = 45V/μs).

4.6 Sources of Distortion in CT-ΣΔMs

Apart from the effect of SR, the linearity of a CT-ΣΔM is ultimately limited by that of the input stage and by signal-dependent errors arising from the feedback DAC. Both sources of distortion are briefly discussed next.

4.6.1 Nonlinearities in the Front-end Integrator

As stated in Section 3.7, the linearity of a ΣΔ modulator is ultimately limited by the nonlinearities associated with the front-end integrator. Therefore, for medium- and high-resolution CT-ΣΔMs, active-RC integrators are usually employed at the modulator front end rather than Gm-C integrators [6, 15, 16], since a more linear voltage-to-current conversion can be obtained in practice with resistors than with active devices.

Similar to Figure 3.19 for SC implementations, Figure 4.11 illustrates the main sources of distortion in an active-RC integrator. These errors are a direct consequence of the nonlinear characteristics of CMOS integrated devices: transistors, resistors, and capacitors. However, the amplification block – the amplifier in active-RC integrators and the transconductor in Gm-C integrators – usually play the dominant role, due to the high linearity achieved by passive devices in

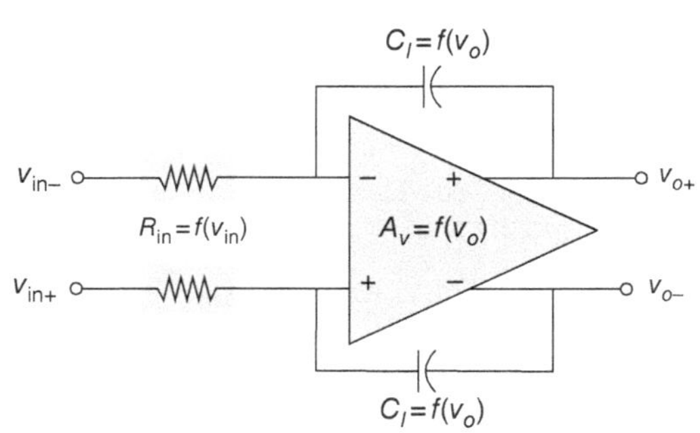

Figure 4.11 The main sources of distortion in active-RC integrators.

modern nanometer technological processes. Moreover, nonlinearities in the differential input pair of the amplifier are accentuated by the finite DC gain.

In contrast, the DC gain of an amplifier exhibits in practice a high dependency on the common-mode output voltage due to degradation of its output impedance, as illustrated in Figure 3.20 and mathematically expressed as:

$$A_v(v_o) = f(v_o) \simeq A_{v_{\text{nom}}}(1 + A_{v1}v_o + A_{v2}v_o^2 + A_{v3}v_o^3 + ...) \tag{4.20}$$

where $A_{v_{\text{nom}}}$ denotes the nominal DC gain and A_{vi} stands for the ith nonlinear coefficient. Breems and Huijsing [15] have shown how a small residual differential voltage at the input pair due to finite gain is modulated, and that the third-order harmonic distortion is derived as

$$\text{HD}_3 \approx \frac{1}{64 g_m R_{\text{in}}^3 I_B^2}\left(1 + \frac{R_{\text{in}}}{R_{\text{DAC}}}\right) A_{\text{in}}^2 \tag{4.21}$$

where R_{in} is the input resistor, R_{DAC} is the feedback resistor, and g_m and I_B are, respectively, the transconductance and the bias current of the input transistors of the amplifier. Note from the above equation that the modulator linearity can be considerably improved by increasing R_{in} by as much as the thermal noise limit allows. Although giving lower efficiency, a larger amplifier input transconductance benefits linearity and at the same time helps to reduce thermal noise, as will be shown in Section 4.7.

In the case of Gm-C integrators, the main source of nonlinearity is associated with the voltage-to-current (V/I) conversion carried out by the nonlinear transconductance, given by [15]:

$$g_m(v_i) = f(v_i) \simeq g_{m_{\text{nom}}} v_i + g_{m2}v_i^2 + g_{m3}v_i^3 + ... \tag{4.22}$$

where $g_{m_{\text{nom}}}$ denotes the nominal transconductance, and g_{mi} stands for the ith nonlinear coefficient. Assuming a fully-differential circuit implementation, it can be shown that the third-order harmonic distortion at the output of CT-ΣΔMs due to nonlinear V/I conversion is given by:

$$\text{HD}_3 \approx \frac{g_{m3}}{g_{m1}} A_{\text{in}}^2 \tag{4.23}$$

This can be used to approximately predict the harmonic distortion, which is ultimately the limiting factor of the linearity of CT-ΣΔMs, although it is difficult to model with the accuracy required for design purposes. However, in order to quantify in a more accurate way the effect of nonlinear errors in CT-ΣΔMs, many authors rely on simulations. To address this problem, an intuitive analytical model for nonlinearities of CT integrators was proposed by Sankar and Pavan [17]. Figure 4.12 illustrates this model for a Gm-C (Figure 4.12a) and an active-RC (Figure 4.12b) integrator.

In essence, the model involves replacing the ideal transfer function of a CT integrator with the following one:

$$\text{ITF}(s)_{\text{NL}} = f(x)\text{ITF}_i(s) \tag{4.24}$$

where $f(x)$ denotes the nonlinear function, which is given by:

$$f(x) = \begin{cases} x - \dfrac{g_3}{4g_m}x^3, & \text{Gm-C} \\ x - \left[\dfrac{2g_3}{g_m}(2 + g_m R^3)\right] x^3, & \text{Active-RC} \end{cases} \tag{4.25}$$

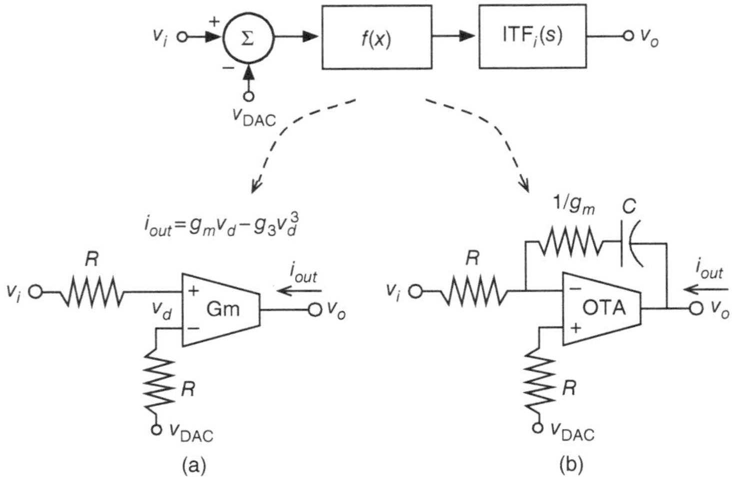

Figure 4.12 Behavioral model of nonlinearity in CT integrators [17]: (a) Gm-C integrator; (b) active-RC integrator.

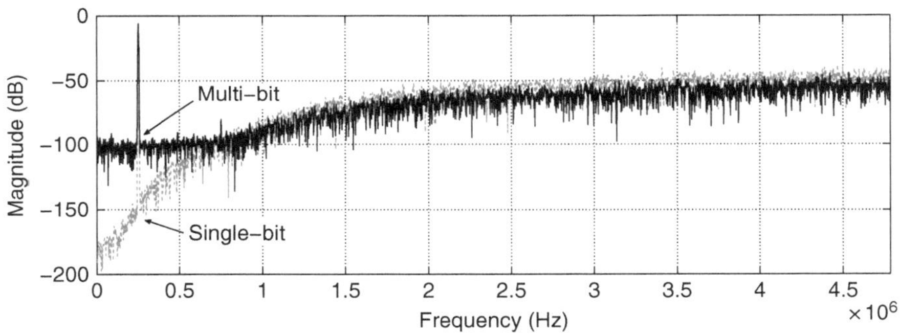

Figure 4.13 Illustrating the effect of nonlinearities on single-bit and multi-bit CT-ΣΔMs as predicted by Sankar and Pavan [17]. A fifth-order single-loop with feedforward topology is considered, using 5-bit quantization and a 10^{-3} third-order nonlinear coefficient.

Using this model, the authors demonstrated that nonlinear CT integrators cause not only harmonic distortion but also an increase of the IBN. Figure 4.13 illustrates this effect on a fifth-order single-loop CT-ΣΔM with feedforward summation considering two cases, namely single-bit and multi-bit quantization. Note that in the latter case, the noise floor increases together with the third-order harmonic distortion caused by the nonlinear integrator.

4.6.2 Intersymbol Interference in the Feedback DAC

In practical realizations of CT-ΣΔMs, besides timing errors due to excess loop delay (discussed in Section 4.9) and nonlinearities arising from mismatches between the output levels of multi-bit DACs (discussed in Section 2.4), the implemented feedback DAC also exhibits SR limitations as well as different rise and fall times. These nonidealities cause additional errors that directly add to the modulator input and tend to degrade the performance, unless they are made small enough.

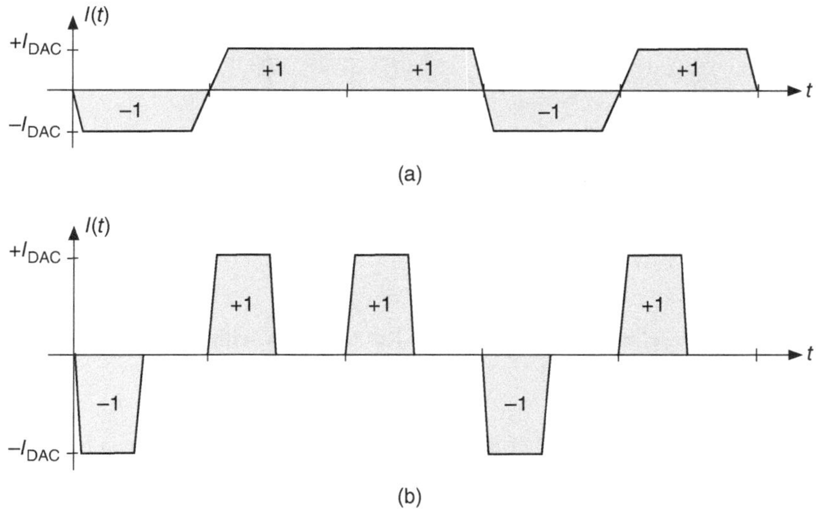

Figure 4.14 Influence of the DAC slew rate limitation on the feedback charge for: (a) NRZ DAC, causing a signal-dependent charge error (ISI); (b) RZ DAC, causing a constant charge error.

For the case of CT–ΣΔMs employing an NRZ-DAC, these nonidealities introduce intersymbol interference (ISI) [18], as illustrated in Figure 4.14a. Note that any limitation on the DAC SR or mismatch between its rise and fall times makes the area of one 1-symbol differ from half of the area of two consecutive 1-symbols. Therefore, the resulting errors in the feedback charge depend on the modulator output sequence and thus create a signal-dependent distortion that limits the modulator performance.

The most common way to prevent ISI is to resort to RZ feedback pulses, so that, in spite of different rise and fall times in the DAC response, the error remains constant in every clock period and independent of the output sequence, as shown in Figure 4.14b. In this way, the resulting error can be compensated by tuning the modulator scaling coefficients [6]. This is similar to the approach to compensation for excess loop delay, which will be discussed in Section 4.9.

4.7 Circuit Noise in CT-ΣΔMs

As discussed in Section 3.5, circuit noise strongly impacts the IBN of SC-ΣΔMs due to the aliasing of broadband noise components that are sampled together with the modulator input signal. Conversely, sampling capacitors are not used in purely CT-ΣΔMs, so that kT/C noise is not present, which is a clear advantage. Broadband noise components are thus filtered before they get to the sampler, so that the fold-over effect is attenuated by the modulator noise shaping [6].

4.7.1 Noise Analysis Considering NRZ Feedback DACs

Let us consider the active-RC integrator in Figure 4.15 to be the front-end integrator[5] of a CT-ΣΔM. Two input branches are considered: one for the modulator input signal (v_{in}) and the other for the

[5]Without loss of generality, the analysis described here is based on the most common situation in practical implementations, in which the front-end integrator of CT-ΣΔMs is realized using active-RC, rather than Gm-C techniques, for improved linearity and low noise. A similar noise study can be carried out by considering a Gm-C integrator instead of an active-RC integrator.

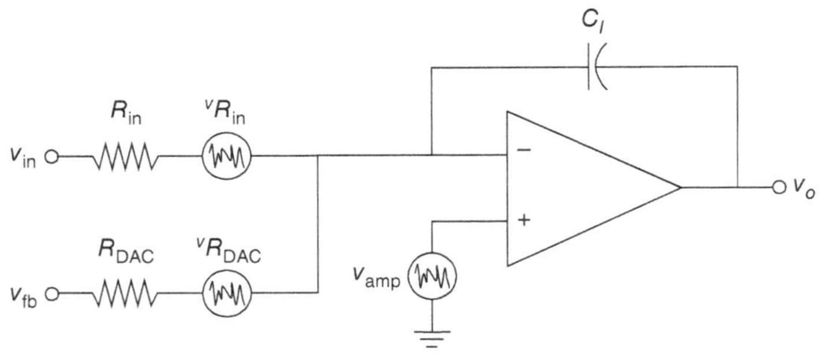

Figure 4.15 Circuit noise sources in a CT active-RC integrator with two input paths. Single-ended version.

DAC feedback signal (v_{fb}). Resistor thermal noise and amplifier noise have been incorporated in the scheme in Figure 4.15.[6] On the one hand, each resistor has a noise voltage source v_{R_i} in series, whose PSD in a single-sided representation is thus

$$S_{R_i} = 4kTR_i \tag{4.26}$$

where k is Boltzmann's constant and T is the absolute temperature. On the other hand, the voltage source v_{amp} represents the amplifier input noise. As illustrated in Figure 3.16, it is essentially determined by a broadband thermal component and a narrowband $1/f$ component,[7] so that $S_{amp}(f)$ is given by Equation (3.37).

Adding up the former circuit noise components in the CT integrator, the total input-referred noise PSD yields [16, 19]

$$S_{\text{noise,in}}(f) \approx 2\left(S_{R_{\text{in}}} + S_{R_{\text{DAC}}}\frac{R_{\text{in}}^2}{R_{\text{DAC}}^2}\right) + S_{\text{amp}}(f)\left[(2\pi f R_{\text{in}} C_I)^2 + \left(1 + \frac{R_{\text{in}}}{R_{\text{DAC}}}\right)^2\right] \tag{4.27}$$

where the factor 2 before the brackets accounts for the actual fully-differential implementation of the CT integrator, in which the number of resistors doubles compared to the single-ended scheme in Figure 4.15. Replacing Equations (3.37) and Equation (4.26) in (4.27), the PSD of the total input-referred noise of the front-end integrator of the CT-ΣΔM can thus be approximated to:

$$S_{\text{noise,in}}(f) \approx 8kT\left(R_{\text{in}} + R_{\text{DAC}}\frac{R_{\text{in}}^2}{R_{\text{DAC}}^2}\right) + S_{\text{amp,th}}\left(1 + \frac{f_{\text{cr}}}{f}\right)\left[(2\pi f R_{\text{in}} C_I)^2 + \left(1 + \frac{R_{\text{in}}}{R_{\text{DAC}}}\right)^2\right] \tag{4.28}$$

[6]For simplicity, the noise associated with the DAC reference voltage is not considered here, but it can be incorporated in a similar way to the amplifier noise [16, 19].

[7]Similar to the case of SC-ΣΔMs, the effect of flicker ($1/f$) noise can be addressed either by design or by using chopper techniques [20]. The latter has been demonstrated to be efficient in low-frequency applications [21], but it must be carefully analyzed in wideband applications. In this case, the effect of out-of-band interfering signals can degrade the performance of the modulator due to the frequency translation caused by chopped integrators [22].

The input-referred IBN of a CT-ΣΔM due to circuit noise can be easily obtained by integrating Equation (4.28) over the input signal bandwidth, yielding

$$\mathrm{IBN}_{\mathrm{noise}} = \int_0^{B_w} S_{\mathrm{noise,in}}(f)\mathrm{d}f \approx 8kTR_{\mathrm{in}}B_w\left(1+\frac{R_{\mathrm{in}}}{R_{\mathrm{DAC}}}\right) + S_{\mathrm{amp,th}}\left[B_w + f_{\mathrm{cr}}\ln\left(\frac{B_w}{f_0}\right)\right]\left(1+\frac{R_{\mathrm{in}}}{R_{\mathrm{DAC}}}\right)^2 \quad (4.29)$$

in which the $1/f$ noise component has been integrated from a frequency $f_0 > 0$ to exclude DC. Taking into account that $R_{\mathrm{in}}C_I = 1/f_s = 1/(2B_w\mathrm{OSR})$, resulting terms that are proportional to OSR^{-2} have been neglected in the final calculation [6].

For a CT-ΣΔM to achieve a given noise performance, the sum of all three components in Equation (4.29) have to meet the required noise floor. Note that, to reduce the contribution of the resistors' thermal noise, the integrator resistance at the modulator input must be decreased, which results in a larger power consumption, given that $f_s = 1/(R_{\mathrm{in}}C_I)$. To reduce the amplifier noise, its transconductance can be increased (for the white noise component) and the size of the input transistors can be increased (for the flicker component).

4.7.2 Noise Analysis Considering SC Feedback DACs

The noise analysis presented above assumes a rectangular-shaped NRZ feedback DAC. This results in a time-invariant loop filter, which can be analyzed straightforwardly. However, the effect of the DAC waveform has a direct influence on the modulator dynamics, and hence on the effect that the different noise sources may have on the performance degradation. This is the case for SC, return-to-open (RTO), and RZ DACs – where their noise sources are injected in the input node of the modulator through switches controlled by non-overlapping clock phases, as conceptually depicted in Figure 4.16 [23].

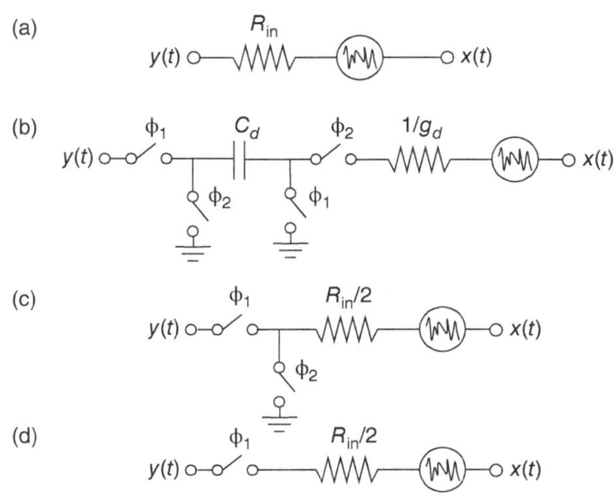

Figure 4.16 Model of noise sources of different feedback DACs on CT-ΣΔMs: (a) NRZ; (b) SC; (c) RZ; (d) RTO [23].

Under these circumstances, the loop filter of CT-ΣΔMs behaves as a time-varying system. Apart from other limitations – such as the impact on AAF [24] – a time-varying loop filter causes aliasing of noise from out-of-band frequencies, thus increasing the IBN due to thermal noise in a similar way as what happens in SC-ΣΔMs, as analyzed in Section 3.5.

This phenomenon has been studied in detail [23] by assuming the time-varying nature of CT-ΣΔM loop filters when SC/RTO feedback DACs are considered. Based on such an assumption, the input-referred PSD of CT-ΣΔMs due to a given noise source can be formulated as [23]:

$$S_{\text{noise},l}(f) = \sum_{m=-\infty}^{\infty} |J_1(2\pi f + 2m\pi f_s)|^2 S_{nl}(2\pi f + 2m\pi) \tag{4.30}$$

where m is an integer number, S_{nl} denotes the PSD of the lth noise source associated with a given circuit element, for instance the input resistance, the DAC, the amplifier, and so on; and $J_1(f)$ is a $1 \times l$ vector whose rth element is the transfer function from the rth source to the modulator output. For instance, considering an active-RC integrator made up of a single-stage operational transconductance amplifier (OTA), it can be demonstrated that the input-referred PSD in the signal band due to thermal noise can be approximately written as:

$$S_{\text{noise, in}}(0)|_{\text{SC-DAC}} \simeq 4kTR_{\text{in}}(1 + |J_1(0)|^2) + \frac{2kTg_d}{g_m(g_m + g_d)} + \frac{12kTg_m}{(g_m + g_d)^2} \tag{4.31}$$

where g_m is the transconductance of the OTA, g_d is the switch-on conductance, and $J_1(0)$ is given by:

$$J_1(0) \simeq \frac{g_m R_{\text{in}}}{1 + g_m R_{\text{in}}} + \exp\left(\frac{-T_s g_m g_d}{2(g_m + g_d)}\right) \tag{4.32}$$

This gives a more accurate estimate of the IBN than if the alias components are not considered in the computation [23].

4.8 Clock Jitter in CT-ΣΔMs

Continuous-time ΣΔMs are more severely affected by timing uncertainties in the clock-signal edges – also known as clock jitter – than their DT counterparts [25]. In contrast to DT-ΣΔMs, clock jitter occurs in CT-ΣΔMs at those points where signals are transformed from the CT to the DT domain or vice versa – i.e at the S/H and reconstruction feedback DAC path – as conceptually depicted in Figure 4.17a. The error introduced by the S/H is attenuated within the signal band by the modulator noise-shaping effect, and can hence be neglected. However, errors resulting from timing uncertainties in the DAC feedback signal – illustrated in Figure 4.17b for a 1-bit NRZ and RZ DAC waveforms – add directly to the modulator input with no suppression, hence being the dominant jitter effect and limiting the overall modulator performance. Due to their higher sensitivity, the analysis of clock jitter in CT-ΣΔMs has attracted significant interest in the technical literature [6, 16, 26–33].

The magnitude of jitter-induced errors depends on the pulse shape of the DAC feedback signal, as well as on the statistical properties of clock jitter. Indeed, the reason why CT-ΣΔMs are more sensitive to clock jitter than SC-ΣΔMs is behind the CT-to-DT waveform transformation carried out in the feedback DAC. As illustrated in Figure 4.17c, in a SC DAC waveform, most of the charge transfer occurs at the start of the clock period, so that the amount of charge lost due to timing error is relatively small. By contrast, in rectangular DACs, charge is transferred at a constant rate over the clock period, and hence charge loss from the same timing error is a larger portion of the total

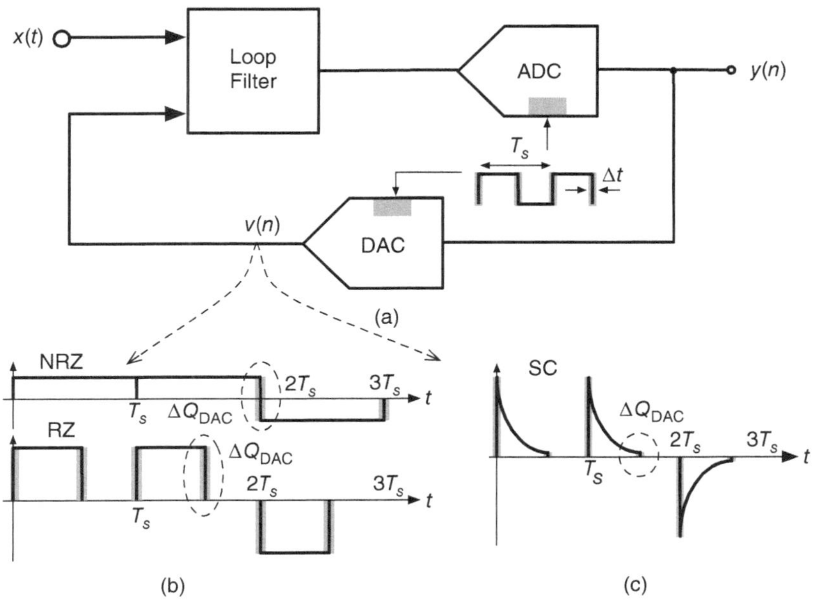

Figure 4.17 Clock jitter in CT-ΣΔMs: (a) main error sources. (b) jittered rectangular (NRZ, RZ) DAC waveforms; (c) jittered SC DAC waverforms.

charge [34]. In Sections 4.8.1–4.8.3, the effects of commonly used DAC pulse shapes are revisited. A random jitter assumption is assumed in order to provide a general overview of its effect and the effect of the loop filter is discussed afterwards.

4.8.1 Jitter in Return-to-zero DACs

Figure 4.18 illustrates the effect of clock jitter on rectangular-shaped DAC feedback signals of a single-bit CT-ΣΔM. Both RZ and NRZ schemes, which are common in SI DACs, are depicted for comparison purposes.

If the timing uncertainty in one side of an RZ pulse is considered, jitter will induce a charge error on the feedback signal $\epsilon_{Q_{\rm DAC}}$ within one clock period. This is given by

$$\epsilon_Q = I_{\rm DAC}\Delta t \tag{4.33}$$

where $I_{\rm DAC}$ stands for the amplitude of the RZ DAC pulses and Δt for the timing uncertainty. If a random jitter is assumed and both sides of the RZ pulse are accounted for (see Figure 4.18b), the variance of the jitter-induced charge error per clock cycle yields

$$\sigma^2_{Q,{\rm RZ}} = 2I^2_{\rm DAC}\sigma^2_j \tag{4.34}$$

where σ_j^2 stands for the variance of the timing uncertainty and the factor 2 results from assuming that the instants of both jittered sides are statistically independent. Relating Equation (4.34) to time, the variance of charge error can be translated into a current error variance:

$$\sigma^2_{I,{\rm RZ}} = \frac{\sigma^2_{Q,{\rm RZ}}}{T_s^2} = 2I^2_{\rm DAC}(f_s\sigma_j)^2 \tag{4.35}$$

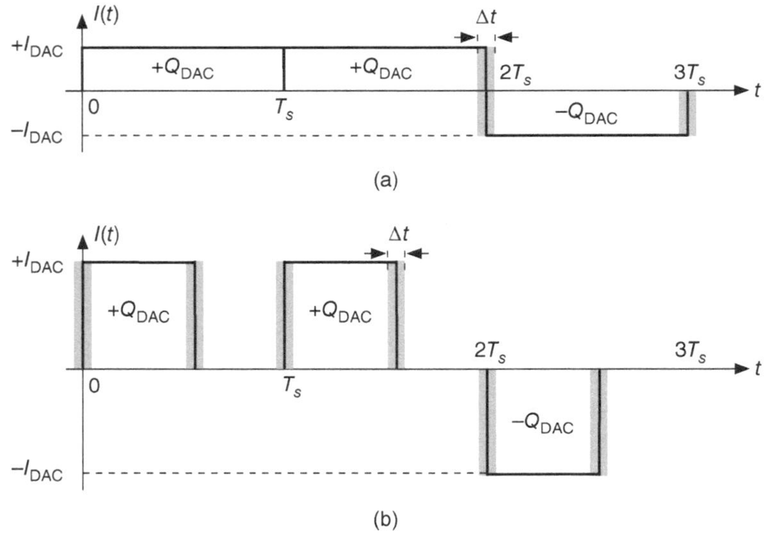

Figure 4.18 Detail of the jitter effect on the feedback signal of a single-bit CT-ΣΔM: (a) NRZ DAC. (b) RZ DAC. The gray shaded areas represent the timing uncertainties. A pulse sequence of (+1, +1, −1) is considered.

The amplitude of the RZ DAC pulse I_{DAC} is determined by the feedback scaling coefficient and the DAC step width Δ. For the first integrator of a single-bit CT-ΣΔM, this is given by [6]:

$$I_{\mathrm{DAC}} = k_{1,\mathrm{RZ}}\, \Delta/2 \tag{4.36}$$

so that:

$$\sigma^2_{I,\mathrm{RZ}} = \left(\frac{\Delta}{2}\right)^2 2k^2_{1,\mathrm{RZ}}(f_s\sigma_j)^2 \tag{4.37}$$

To obtain the resulting IBN, this noise component can be easily related to the modulator input as [6]:

$$\mathrm{IBN}_{j,\mathrm{RZ}} \approx \frac{\sigma^2_{I,\mathrm{RZ}}}{k^2_{1,\mathrm{NRZ}}\mathrm{OSR}} = \left(\frac{\Delta}{2}\right)^2 \frac{k^2_{1,\mathrm{RZ}}}{k^2_{1,\mathrm{NRZ}}} \frac{2(f_s\sigma_j)^2}{\mathrm{OSR}} \tag{4.38}$$

showing that the sensitivity of a CT-ΣΔM with RZ feedback to clock jitter is reduced by only OSR^{-1}. It thus compares unfavorably to DT-ΣΔMs, in which a suppression proportional to OSR^{-3} is achieved; see Equation (3.44).

4.8.2 Jitter in Non-return-to-zero DACs

The effect of clock jitter in NRZ feedback DACs can be obtained in a similar way as was done in Section 4.8.1 for RZ DACs. However, in contrast to RZ feedback pulses, jitter affects the feedback charge only if the feedback signal changes its state, as shown in Figure 4.18a. Therefore, the variance of the jitter-induced charge error per clock cycle for an NRZ DAC is in practice less than half that of its RZ counterpart in Equation (4.34), since a state transition does not necessarily occur in every cycle. For large input signals and single-bit quantization, it can be approximated to [6, 26]:

$$\sigma^2_{Q,\mathrm{NRZ}} \approx 0.7 I^2_{\mathrm{DAC}} \sigma^2_j \tag{4.39}$$

On the other hand, if a transition state takes place in an RZ DAC, the amplitude of the step is Δ instead of $\Delta/2$ in Equation (4.36), so that:

$$I_{\mathrm{DAC}} = k_{1,\mathrm{NRZ}}\ \Delta \tag{4.40}$$

Therefore, the resulting IBN for a single-bit CT-ΣΔM employing NRZ feedback DAC can be estimated as:

$$\mathrm{IBN}_{j,\mathrm{NRZ}} \approx \frac{\sigma^2_{I,\mathrm{NRZ}}}{k^2_{1,\mathrm{NRZ}}\mathrm{OSR}} = \Delta^2 \frac{0.7(f_s\sigma_j)^2}{\mathrm{OSR}} \tag{4.41}$$

Taking into account that $k_{1,\mathrm{RZ}}$ is typically twice the value of $k_{1,\mathrm{NRZ}}$ – i.e., the duration of the RZ pulse is typically $T_s/2$ – the sensitivity to clock jitter of both rectangular-shape feedback DACs can be easily compared from Equations (4.38) and (4.54)

$$\frac{\mathrm{IBN}_{j,\mathrm{NRZ}}}{\mathrm{IBN}_{j,\mathrm{RZ}}} \approx \frac{0.7}{2} \approx -4.5\ \mathrm{dB} \tag{4.42}$$

showing that NRZ feedback leads to lower sensitivity to clock jitter than the RZ counterpart [6, 16].

Resorting to multi-bit quantization is a common approach for reducing the sensitivity to clock jitter of CT-ΣΔMs. Figure 4.19a illustrates a DAC feedback signal employing multi-bit NRZ. Intuitively, the number of state transitions per clock cycle increases from 0.7 (for the single-bit case) to a value close to 1 (for the multi-bit case), but two adjacent NRZ pulses will generally differ by only 1 LSB, thus reducing the jitter-induced charge error per clock cycle compared to Equation (4.39). Therefore, the IBN due to clock jitter of an NRZ CT-ΣΔM is reduced by approximately 6 dB per additional bit [6]. From Figure 4.19b it can also be intuitively seen that the reduction of jitter noise by resorting to multi-bit RZ is only minor compared to multi-bit NRZ, since the feedback signal mostly has to change by more than 1 LSB at each clock cycle.

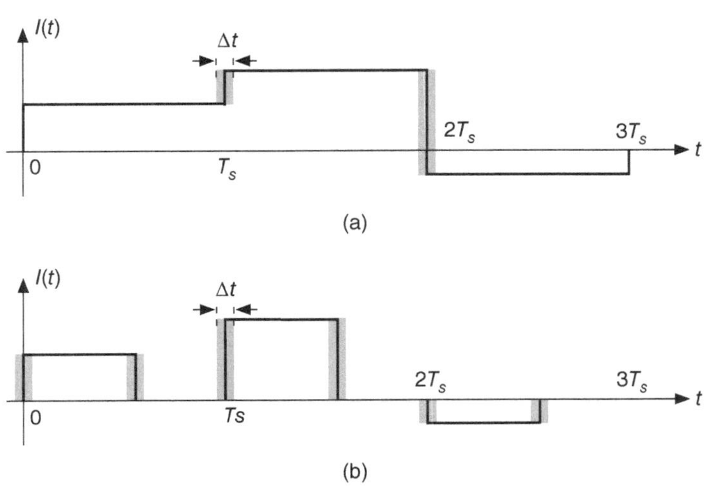

Figure 4.19 Illustration of the jitter effect on the feedback signal of a multi-bit CT-ΣΔM: (a) NRZ DAC; (b) RZ DAC. The gray shaded areas represent the timing uncertainties.

4.8.3 Jitter in Switched-capacitor DACs

The use of shaped pulses in the feedback DAC is an alternative approach commonly employed for reducing the sensitivity of CT-ΣΔMs to clock jitter. Figure 4.20 illustrates a typical feedback signal in an SC DAC, in which the RZ scheme uses an exponentially decaying waveform. If the timing uncertainty on one side of the SC-RZ pulse is considered, the charge error on the feedback signal $\epsilon_{Q_{\rm DAC}}$ induced by jitter within one clock period yields:

$$\epsilon_Q \approx I_{\rm DAC}\, e^{-\frac{T_s}{2\tau_{\rm DAC}}} \Delta t \tag{4.43}$$

where $\tau_{\rm DAC} = R_{\rm DAC} C_{\rm DAC}$ is the time constant[8] of the DAC exponential current. A duration of $T_s/2$ is assumed for the RZ interval in Equation (4.43). If both jittered sides are considered (see Figure 4.20) and their timings are assumed uncorrelated, the variance of the jitter-induced charge error per clock cycle for an SC-RZ DAC can be approximated to:

$$\sigma^2_{Q,\rm SC} = 2\left(I_{\rm DAC}\, e^{-\frac{T_s}{2\tau_{\rm DAC}}}\right)^2 \sigma_j^2 \tag{4.44}$$

Following a similar procedure to that in Section 4.8.1 for an SI-RZ feedback DAC, the resulting IBN can be obtained as [6]:

$$\mathrm{IBN}_{j,\rm SC} \approx \frac{\sigma^2_{I,\rm SC}}{k^2_{1,\rm NRZ}\mathrm{OSR}} = \left(\frac{\Delta}{2}\right)^2 \frac{k^2_{1,\rm SC}}{k^2_{1,\rm NRZ}} \left(e^{-\frac{T_s}{2\tau_{\rm DAC}}}\right)^2 \frac{2(f_s\sigma_j)^2}{\mathrm{OSR}} \tag{4.45}$$

Note that the jitter-induced noise is exponentially reduced compared to SI-RZ feedback – see Equation (4.38).

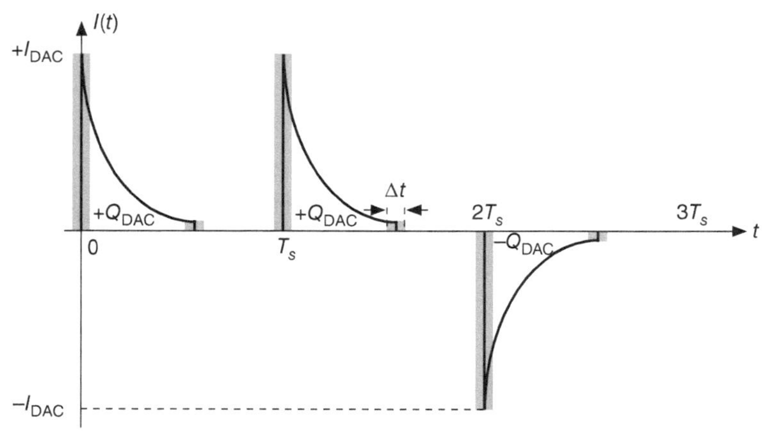

Figure 4.20 Illustration of the jitter effect on the feedback signal of a CT-ΣΔM with an SC DAC. The gray shaded areas represent the timing uncertainties. A pulse sequence of $(+1, +1, -1)$ is considered.

[8] The resistance $R_{\rm DAC}$, via which the DAC capacitor $C_{\rm DAC}$ is discharged, is generally determined by the resistance of the DAC switches and the integrator input impedance in feedback configuration [19].

4.8.4 Lingering Effect of Clock Jitter Error

The analysis presented above assumed that the noise spectrum due to clock jitter error is flat; in other words, that the clock jitter error behaves as a white noise source. Although this assumption can be used to make rough estimates for design purposes, a more accurate analysis reveals that clock jitter has a *lingering* effect on the output spectrum of CT-ΣΔMs, including both white and shaped spectral components [35]. In order to show this effect, let us consider a CT-ΣΔM with a multi-bit NRZ DAC.[9] In this case, the clock jitter error sequence,[10] $\epsilon_Q(n)$, can be related to the modulator output signal, $y(n)$, as [36]:

$$\epsilon_Q(n) = (y(n) - y(n-1))\frac{\Delta t(n)}{T_s} \tag{4.46}$$

where $\Delta t(n)$ stands for the jitter pulse width at time nT_s. Assuming that $\Delta t(n)$ corresponds to a stationary process with zero mean and a variance of σ_j, the total power of the clock jitter error can be expressed as:

$$P_{\epsilon_Q} = \frac{\sigma_j^2}{T_s^2}\mathcal{E}\{[y(n) - y(n-1)]^2\} \tag{4.47}$$

where $\mathcal{E}\{\cdot\}$ represents the *expectation* operator [37].

Assuming that $x(n)$ is the input signal sequence and a unity STF, the modulator output sequence can be written as:

$$y(n) = x(n) + q(n) \tag{4.48}$$

where $q(n)$ is an error signal that contains both the shaped quantization error and the clock jitter error components. Using Equation (4.48), and assuming that $x(n)$ and $q(n)$ are uncorrelated, the expression $\mathcal{E}\{[y(n) - y(n-1)]^2\}$ can be approximated by:

$$\mathcal{E}\{[y(n) - y(n-1)]^2\} \simeq \mathcal{E}\{\Delta x_n{}^2\} + \mathcal{E}\{\Delta q_n{}^2\} \tag{4.49}$$

where $\Delta x_n \equiv x(n) - x(n-1)$ and $\Delta q_n \equiv q(n) - q(n-1)$. Assuming a sinewave input signal of amplitude $A_{\rm in}$ and frequency $f_{\rm in}$, it can be shown that $\mathcal{E}\{\Delta x_n{}^2\}$ and $\mathcal{E}\{\Delta q_n{}^2\}$ can be respectively written as [38]:[11]

$$\mathcal{E}\{\Delta x_n{}^2\} \simeq 2(\pi A_{\rm in} f_{\rm in} T_s)^2 \tag{4.50}$$

$$\mathcal{E}\{\Delta q_n{}^2\} \simeq \frac{X_{\rm FS}^2}{12\pi(2^B-1)^2}\int_0^{\pi}\left|(1-e^{-j2\pi f})NTF(e^{-j2\pi f})\right|^2 df \tag{4.51}$$

From Equations (4.47)–(4.51), the total power of the clock jitter can be derived, yielding:

$$P_{\epsilon_Q} \simeq \frac{\sigma_j^2}{T_s^2}\cdot\Big[2(\pi A_{\rm in} f_{\rm in} T_s)^2 + \frac{X_{\rm FS}^2}{12\pi(2^B-1)^2}\int_0^{\pi}\left|(1-e^{-j2\pi f})NTF(e^{-j2\pi f})\right|^2 df\Big] \tag{4.52}$$

[9]The interested reader can read a detailed analysis of the lingering effect of clock jitter with RZ DACs in the literature [35].
[10]The DAC output waveform with clock jitter can be modeled as the sum of a jitter-free waveform plus a stream of pulses due to clock jitter error, usually called the *jitter error sequence* [25].
[11]A detailed analysis of this derivation can be found in the literature [39].

Note that this expression contains terms that express the dependency of the clock jitter error power on the input signal and the noise shaping implemented by CT-ΣΔMs. Thus, depending on the actual CT-ΣΔM topology, and hence the exact expression of $NTF(e^{-j2\pi f})$, the integration in the second term may become more or less involved.

An alternative approach involves using the state-space formulation to derive $\mathcal{E}\{\Delta q_n^{\,2}\}$. The state-space formulation provides a good mathematical framework to study arbitrary CT-ΣΔMs. Most importantly, it allows the study of jitter-induced noise in the time domain and extraction of the noise autocorrelation function [35]. Based on this approach – which is analyzed in more detail in Appendix A – it can be shown that the IBN due to clock jitter can be expressed as:

$$\mathrm{IBN}_{\epsilon_Q} = \sigma_j^{\,2} B_w \left[\frac{(2\pi A_{\mathrm{in}} f_{\mathrm{in}})^2}{f_s} + \frac{f_s X_{\mathrm{FS}}^2}{3(2^B - 1)^2} \overline{\Psi}(\overline{\lambda}, L) \right] \tag{4.53}$$

where $\overline{\Psi}(\overline{\lambda}, L)$ is a matrix function derived from the state-space representation of the modulator, with $\overline{\lambda}$ being the state matrix [38] (see Appendix A). Note that in Equation (4.53), the primary time uncertainty error term, $\sigma_j^{\,2}$, is multiplied by two different factors which, added together, define what is hereinafter called the *jitter amplification factor*. These two terms are:

- the signal-dependent term (SDT), which depends on the input signal
- the modulator-dependent term (MDT), which depends on the modulator architecture.

Note also in Equation (4.53) that the MDT increases with f_s, while the SDT decreases with f_s. According to this model, an optimum value of f_s can be found that minimizes the jitter amplification factor in a given CT-ΣΔM [38]. This is illustrated in Figure 4.21 for a third-order single-loop CT-ΣΔM with feedforward summation, where both the SDT and MDT are plotted versus f_s for $B_w = 20$ MHz and $B = 5$. Note that in this case, the minimum amplification, and hence the minimum value of $\mathrm{IBN}_{\epsilon_Q}$,

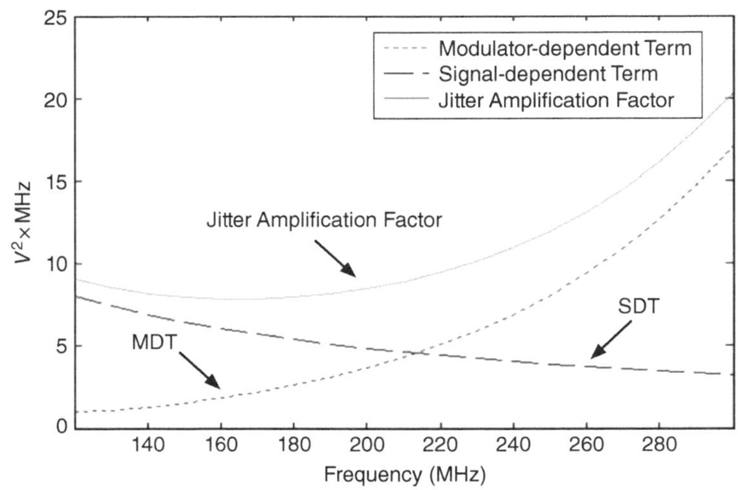

Figure 4.21 Effect of sampling frequency on the various terms affecting the IBN due to clock jitter in a 5-bit third-order CT-ΣΔM [38].

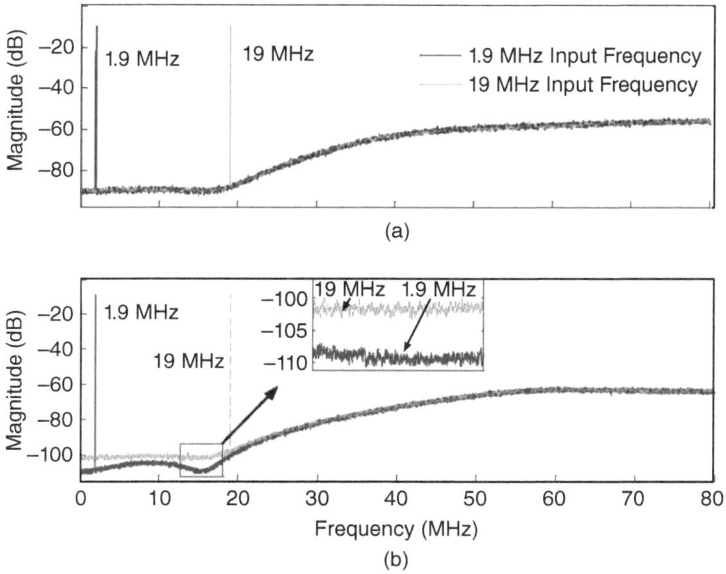

Figure 4.22 Illustrating the effect of signal frequency on the IBN due to clock jitter in a third-order CT-ΣΔM, for $f_s = 160$ MHz and: (a) $B = 2$; (b) $B = 5$ [38].

is obtained for $f_s = 170$ MHz. Of course, the value of f_s is constrained by other design trade-offs involving both architectural and circuital nonidealities, as discussed throughout this book.

The effect of SDT on IBN_{ϵ_Q} becomes dominant as the MDT is reduced by increasing L and/or B. This is illustrated in Figure 4.22, where the output spectrum of a third-order single-loop CT-ΣΔM is shown for different values of f_{in} and B. Note that in Figure 4.22a, corresponding to $B = 2$, the IBN does not change with f_{in} since the clock jitter error is dominated by the MDT. However, as B increases from 2 to 5, the MDT decreases and hence the IBN is dominated by the SDT, as illustrated in Figure 4.22b.

4.8.5 Reducing the Effect of Clock Jitter with FIR and Sine-shaped DACs

As stated in Section 4.8.2, the use of multi-bit quantization and NRZ feedback DAC constitute – a priori – a good combination to mitigate the impact of clock jitter because the transitions of the modulator feedback signal are reduced with the number of bits. However, the use of multi-bit quantization brings a number of inconveniences in terms of circuit complexity, area/power consumption, and the inherent nonlinearity caused by mismatches in the feedback DAC elements, as discussed in Section 2.4.

Another way to address the sensitivity of single-bit CT-ΣΔMs to clock jitter consists of filtering or *shaping* the DAC waveform by using a finite impulsive response (FIR) filter [40–44]. The idea – originally proposed by Oliaei [29] and conceptually illustrated in Figure 4.23 – is to feedback a filtered version of the single-bit quantization output in such a way that, due to the high-frequency attenuation of the FIR filter, the DAC output is a multi-level waveform. The two-level modulator output sequence, $y(n)$, is filtered by an N-tap LP FIR filter, so that the amplitude of the DAC output transitions can be reduced to $X_{\text{FS}}/N = \Delta/N$, which means a reduction of IBN_{ϵ_Q} by $20\log(N)$ dB. In this way, using a single-bit ADC and a FIR DAC allows the low-jitter sensitivity and high linearity to be achieve for a multi-bit ΣΔM, while retaining the simplicity and robustness of single-bit ΣΔMs [45].

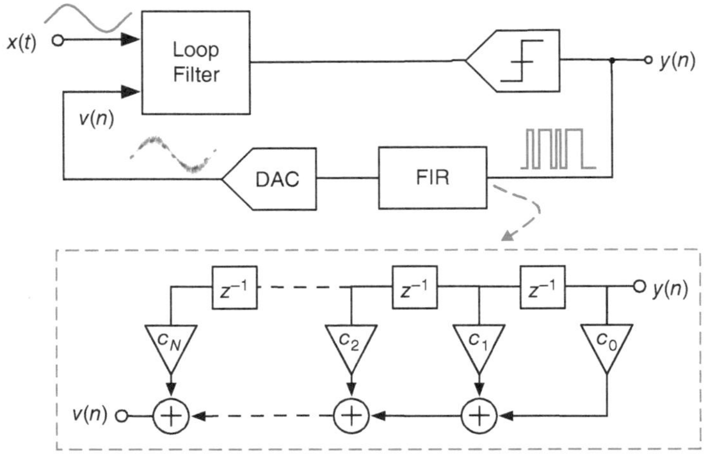

Figure 4.23 FIR-DAC based single-bit CT-ΣΔMs [45].

Note that FIR DACs introduce an extra delay in the modulator loop filter. This delay has to be suitably controlled so as to stabilize the system and guarantee correct performance. This problem can be solved by using the methods of moments, which allows us to perfectly compensate for the FIR DAC's delay [46]. Another issue is related to the number of taps used in the FIR filter. This can be optimized to maximize the modulator performance and efficiency [44].

A similar strategy to reduce the sensitivity of the DAC is conceptually depicted in Figure 4.24. The DAC output waveform is *sine-shaped* [27]. As shown in Figure 4.24, the sine-shaped waveform is obtained by mixing the DAC output with a raised-cosine carrier signal so that the clock jitter is almost removed in the signal fed back to the modulator input [47]. A simple analysis similar to that carried out in previous sections demonstrates that the IBN degraded by clock jitter in CT-ΣΔMs with sine-shaped feedback DACs can be approximated by [48]:

$$\mathrm{IBN}_{j,\mathrm{Sine}} \approx \mathrm{IBN}_{j,\mathrm{NRZ}} \frac{2\pi^2}{9} (f_s \sigma_j)^4 \tag{4.54}$$

which, assuming that $f_s\sigma_j \ll 1$, may result in a significant SNR improvement with respect to rectangular DAC waveforms.

The use of feedback FIR filters and raised-cosine DACs has been shown to be more suitable in BP CT-ΣΔMs using undersampling [50] to digitize RF signals [47, 51], where the NTF is severely

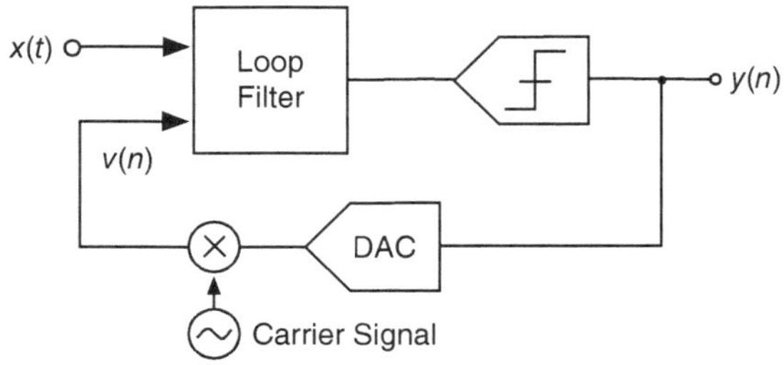

Figure 4.24 Sine-shaped feedback DACs to attenuate the effect of clock jitter error [49].

degraded by the quality factor of the embedded resonators. This problem can be mitigated by raised-cosine FIR DACs thanks to their better frequency response as compared to NRZ DACs [52]. However, in practice, sine-shaped DACs can be severely limited by the phase noise associated with the carrier signal, which must be accounted for in practical applications [49].

In a more general case, the effect of phase noise associated with the sources generating the clock signal in practical CT-ΣΔMs may become a ultimate limiting factor in terms of clock jitter degradation [45]. The mixing process between the clock-phase noise and the modulator input signal gives rises to a *coloured* jitter error, which can severely degrade the SNR of CT-ΣΔMs [53, 54]. This effect does not depend either on the DAC waveform or the modulator topology, since clock jitter IBN is dominated by the SDT, as discussed in Section 4.8.4.

4.9 Excess Loop Delay in CT-ΣΔMs

Excess loop delay (ELD) is a timing error that greatly affects the behavior of CT-ΣΔMs. It can be defined as the *constant* delay between the quantizer sampling edge and the corresponding edge in the feedback DAC pulse and ultimately arises from the nonzero switching time of transistors in the quantizer and DAC circuitry [55].[12] ELD can be expressed as a fraction of the sampling period

$$t_d = \tau_d T_s \tag{4.55}$$

and is often represented as an explicit delay that is inserted prior to the DAC, as shown in Figure 4.25b.

One of the main effects of ELD is to introduce additional poles that increase the order of both the STF and NTF, which may lead to unstable behavior of the CT-ΣΔM. The analysis of ELD particularly depends on the feedback DAC waveform: it might become mathematically complex if a generic *L*th order CT-ΣΔM with an arbitrary DAC waveform is considered. Nevertheless, there have been various attempts to obtain closed-form expressions for the stability condition due to ELD. Some studies are based on analyzing the equivalent DT system, while others apply a direct CT methodology. In this book, the error is first analyzed in an intuitive way by applying a steady-state condition of the modulator operating in the CT domain. Afterwards, a more accurate and conventional study is carried out based on the impulse-invariant DT–CT transformation [55, 58, 59].

4.9.1 Intuitive Analysis of ELD

Consider the generic CT-ΣΔM architecture shown in Figure 2.41a, but with the loop filter having an ELD like that in Figure 4.25. Following the same analytical procedure described in Section 2.6.1, it can be shown that the NTF is degraded by the action of ELD as:

$$\mathrm{NTF_{ELD}}(f) = \frac{1}{1 + g_1{}'g_q H(f) e^{-2\pi f \tau_d}} \tag{4.56}$$

Assuming that $f\tau_d \ll 1$, the expression above simplifies to:

$$\mathrm{NTF_{ELD}}(f) \simeq \frac{1}{1 + g_1{}'g_q H(f)(1 - 2\pi j f \tau_d)} \tag{4.57}$$

where $e^{-2\pi f \tau_d} \simeq 1 - 2\pi f \tau_d$ has been considered.

[12]As will be discussed in Section 4.10, deliberate delays can also be introduced between the quantizer clocking edge and the subsequent latch feeding the DAC to increase the available time slot for the quantizer decision and to mitigate the influence of quantizer metastability [9, 26, 56, 57].

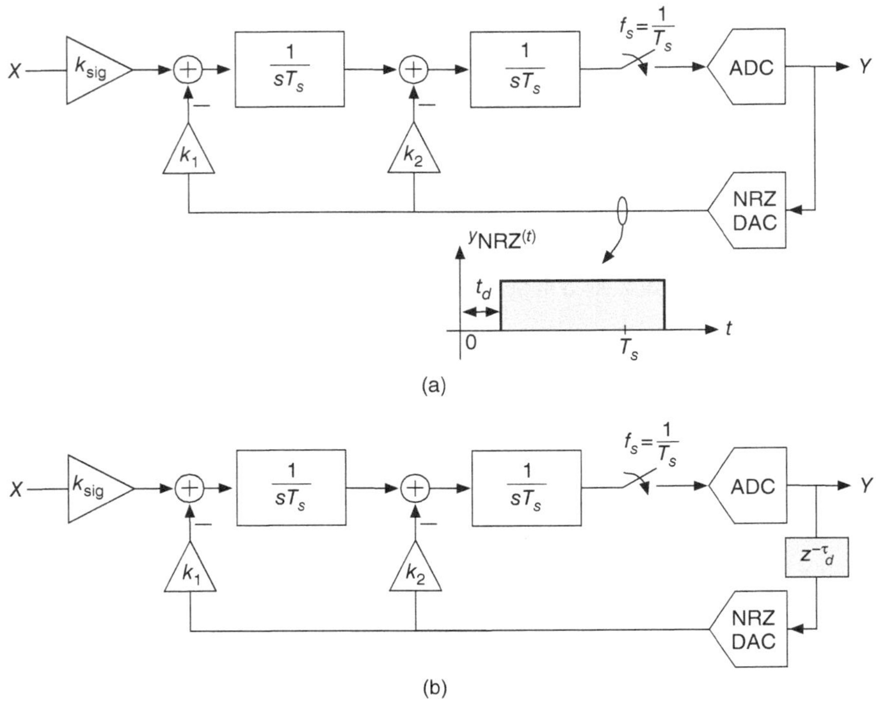

Figure 4.25 Illustration of excess loop delay in a second-order CT-ΣΔM: (a) effect on an NRZ DAC pulse; (b) equivalent diagram with an explicit delay block between the ADC and the DAC.

Thus, in the case of a first-order CT-ΣΔM, Equation 4.57 transforms into:

$$\mathrm{NTF}_{\mathrm{ELD,1st}}(f) \simeq \frac{2\pi j f \tau_d}{g_1' g_q + 2\pi j f (1 - t_d g_1' g_q)} \tag{4.58}$$

According to this simple – but intuitive – analysis, the stability of the modulator can be guaranteed if the quantizer gain is assumed to be $g_q = 1/(t_d g_1')$. However, this is very difficult to implement in practice since there is not robust control of the quantizer gain, specially in single-bit quantization, where the quantizer gain is undetermined. This problem aggravates as the order of the modulator increases. For instance, following the same analytical approach as in the case of a second-order single-loop CT-ΣΔM, the NTF modified by ELD is approximately given by:

$$\mathrm{NTF}_{\mathrm{ELD,2nd}}(f) \simeq \frac{(2\pi f \tau_d)^2}{(2\pi f \tau_d)^2 - g_1' g_2 g_q (1 - 2\pi j f \tau_d) - g_2' g_q 2\pi f \tau_d (1 - 2\pi j f \tau_d)} \tag{4.59}$$

From this expression, assuming OSR ≫ 1 and applying the Ruth–Hurwitz stability criterion, it can be shown that a second-order CT-ΣΔM is stable if the following condition is satisfied:[13]

$$\tau_d \le \frac{g_2'}{2 g_1' g_2} \tag{4.60}$$

[13]Indeed, the condition derived from the analysis based on the linear model of the quantizer yields $\tau_d \le \frac{g_2'}{g_1' g_2}$, although simulation-based analysis considering the nonlinear model of the quantizer results in the condition given in Equation 4.60.

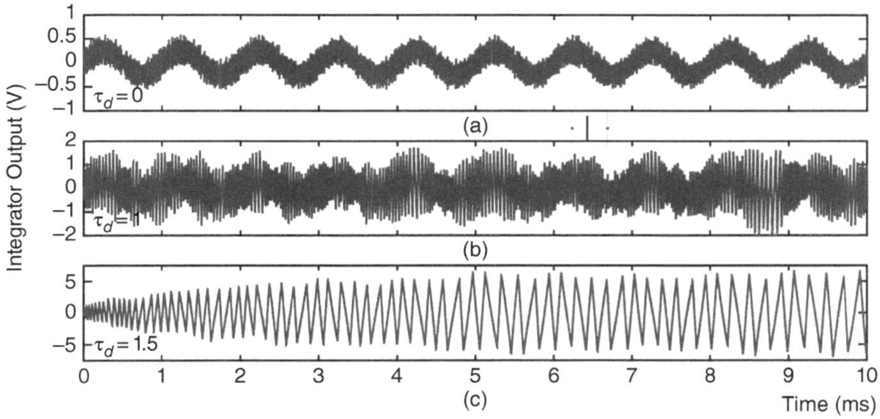

Figure 4.26 Unstable behavior in the front-end integrator of a second-order LP CT-ΣΔM due to ELD for: (a) $\tau_d = 0$ (ideal case); (b) $\tau_d = 1$ (stability condition limit); (c) $\tau_d = 1.5$ (unstable behavior).

The validity of the above expression is illustrated in Figure 4.26, where a single-bit second-order CT-ΣΔM is simulated for different values of τ_d. Note that as τ_d increases up to $\tau_d > 1$ (the condition limit in this example), the output of the front-end integrator increases over time, yielding unstable behavior. The effect of ELD is similar in other CT-ΣΔM topologies with arbitrary DAC waveforms. Figure 4.27 shows a fourth-order single-bit BP CT-ΣΔM with multi-feedback DAC path [18]. Similar to what happens to LP CT-ΣΔMs, the output of the resonators increases with τ_d, thus causing unstable behavior, and completely degrading the output spectrum of the modulator.

In an arbitrary Lth-order CT-ΣΔM topology, it is not possible to obtain an analytical condition for stability. Alternatively, some simulation-based approaches [55] demonstrate that in the more general case of an Lth-order CT-ΣΔM, the *critical* value of ELD, $\tau_d|_{\text{crit}}$, that keeps the modulator stable is given by:

$$\tau_d|_{\text{crit}} \leq \frac{1}{|H(f)|_{\text{out-of-band}}} \tag{4.61}$$

where $|H(f)|_{\text{out-of-band}}$ stands for the out-of-band gain of the loop filter transfer function. However, the use of stability-condition limits is not robust since they do not guarantee stable operation of the modulator in the majority of practical situations. Instead, the effect of ELD can be compensated for, as described in the next section, by modifying the modulator loop filter in order to accommodate this unavoidable error.

4.9.2 Analysis of ELD based on Impulse-invariant DT-CT Transformation

A more accurate analysis of ELD requires consideration of the sampling operation in the loop filter, as well as the feedback DAC waveform, in order to apply the impulse-invariant DT–CT transformation. Note from Figure 4.25a that, if an NRZ pulse waveform is used in the feedback DAC, ELD shifts part of the feedback pulse into the next clock cycle. This actually results in a change of the DAC impulse response from its ideal expression

$$r_{\text{DAC}}(t) = r_{(0,1)}(t) = \begin{cases} 1, & 0 \leq t < T_s \\ 0, & \text{otherwise} \end{cases} \tag{4.62}$$

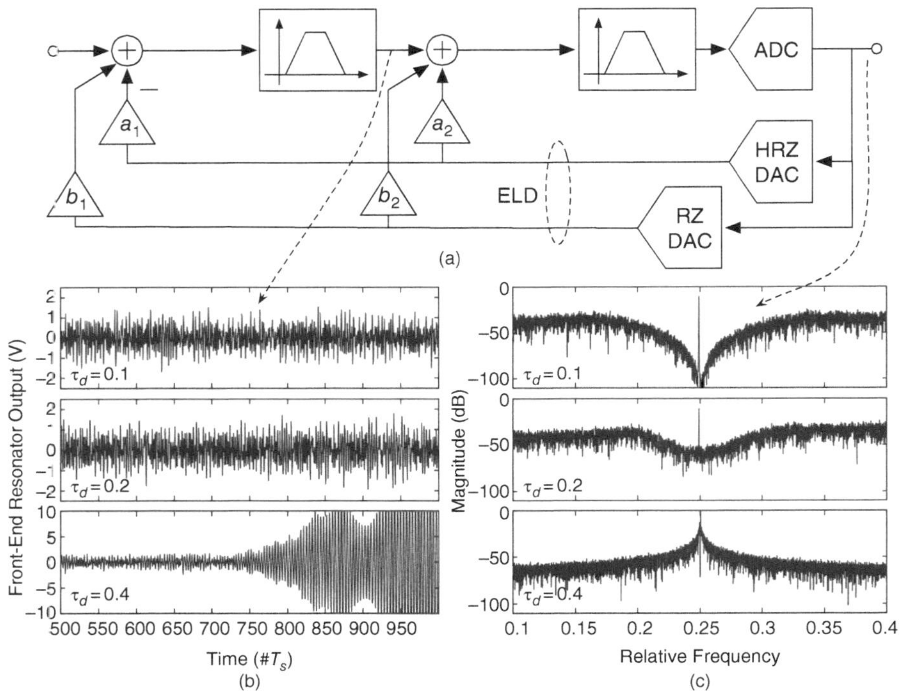

Figure 4.27 Effect of ELD in a fourth-order BP CT-ΣΔM with multi-feedback DAC path [25]: (a) block diagram of the modulator; (b) transient response of the front-end resonator; (b) output spectrum of the modulator.

to,

$$r^*_{\text{DAC}}(t) = r_{(\tau_d,1+\tau_d)}(t) = \begin{cases} 1, & t_d \leq t < T_s + t_d \\ 0, & \text{otherwise} \end{cases} \tag{4.63}$$

This can be expressed as a linear combination of a DAC pulse from τ_d to 1 and a one-sample-delayed DAC pulse from 0 to τ_d [57]:

$$r^*_{\text{DAC}}(t) = r_{(\tau_d,1+\tau_d)}(t) = r_{(\tau_d,1)}(t) + r_{(0,\tau_d)}(t-1) \tag{4.64}$$

Taking into account these two feedback pulses, Equation (2.58) and Table 2.4 can be used to derive the DT-equivalent of the actual CT-ΣΔM affected by ELD,

$$H(z, \tau_d) = \mathcal{Z}\{\mathcal{L}^{-1}[R^*_{\text{DAC}}(s)H(s)]|_{t=nT_s}\} \tag{4.65}$$

This can be used to show that the actual order of the equivalent DT-ΣΔM increases, decreasing the maximum stable input amplitude and degrading the noise shaping performance [57].

As an example, the derivation involved is completed here for the second-order CT-ΣΔM shown in Figure 4.25a [6, 57]. The equivalent DT loop filter of the first and second branches of $H(s)$ yields (with $f_s = 1$),

$$\frac{-k_1}{s^2} \xrightarrow{\text{CT-DT}} \frac{-k_1(1-\tau_d)^2 z - k_1(1-\tau_d^2)}{2(z-1)^2} + z^{-1}\frac{-k_1\tau_d(2-\tau_d)z - k_1\tau_d^2}{2(z-1)^2}$$
$$\frac{-k_2}{s} \xrightarrow{\text{CT-DT}} \frac{-k_2(1-\tau_d)}{z-1} + z^{-1}\frac{-k_2\tau_d}{z-1} \tag{4.66}$$

where the z^{-1} in the second terms indicates the shift by one sample due to ELD. Adding the equivalent DT terms of both modulator branches in Equation (4.66) gives:

$$H(z,\tau_d) = \frac{\alpha_2 z^2 + \alpha_1 z + \alpha_0}{2z(z-1)^2},$$
$$\text{with} \begin{cases} \alpha_2 = -k_1(1-\tau_d)^2 - 2k_2(1-\tau_d) \\ \alpha_1 = -k_1(1+2\tau_d-2\tau_d^2) + 2k_2(1-2\tau_d) \\ \alpha_0 = -k_1\tau_d^2 + 2k_2\tau_d \end{cases} \tag{4.67}$$

Note that for $\tau_d = 0$, Equation (4.67) becomes the ideal DT loop filter $H(z)$ of a second-order modulator, as shown in Figure 2.1a – see Equation 2.63. However, for $\tau_d \neq 0$, the order of the DT transfer function increases by one, so that the equivalence in Equation (2.58) cannot be fulfilled with the original number of scaling coefficients (k_1 and k_2).

The detrimental effect of ELD on the modulator stability and dynamic range can be somewhat overcome by tuning the loop filter coefficients [57]. However, the restoration of the actual equivalence to the ideal DT loop filter under the influence of ELD requires introduction of one more degree of freedom in the CT diagram; that is, one more coefficient.[14] The simplest way to do this is the insertion of an additional feedback path scaled by k_c^* [60]; an example with a second-order CT-ΣΔM is shown in Figure 4.28. Note that the CT loop filter is modified to:

$$H_m(s) = -\frac{k_1^*}{s^2} - \frac{k_2^*}{s} - k_c^* \tag{4.68}$$

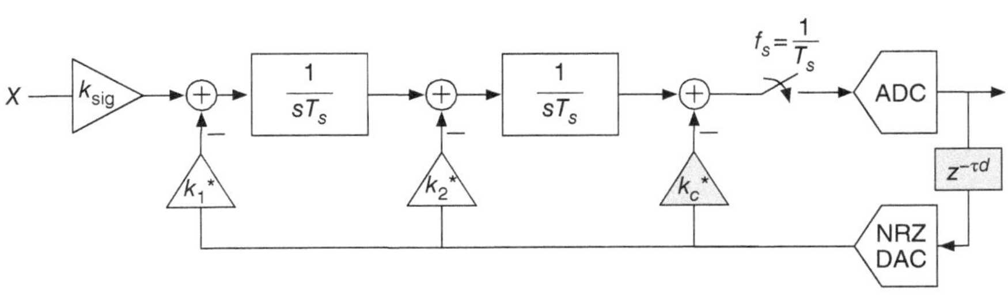

Figure 4.28 Classical approach for compensating ELD in a second-order CT-ΣΔM.

[14]If an RZ-DAC is used in the CT-ΣΔM, ELD shifts part of the feedback pulse into the next clock cycle only if the delay t_d exceeds the time slot between the end of the pulse and the end of the sample; e.g., $t_d > T_s/2$. If that is not the case, the order of equivalent DT loop filter does not increase, but ELD leads to a deviation of the actual modulator coefficients from the desired original values; that is, to a scaling mismatch that increases the IBN. However, in high-speed CT-ΣΔMs, ELD can become so significant that even RZ pulses get shifted to the next clock cycle, thus requiring an extra degree of freedom to compensate the effect [6].

so that the $\mathcal{Z}$-transform of the modified loop filter

$$H_m(z,\tau_d) = \mathcal{Z}\{\mathcal{L}^{-1}[R^*_{\rm DAC}(s)H_m(s)]|_{t=nT_s}\} \tag{4.69}$$

can then be mapped on the $\mathcal{Z}$-transform of the ideal loop filter

$$H(z) = \mathcal{Z}\{\mathcal{L}^{-1}[R_{\rm DAC}(s)H(s)]|_{t=nT_s}\} \tag{4.70}$$

to derive the value of coefficients k_i^*. For the second-order CT-ΣΔM in Figure 4.28, the scaling coefficients for ELD compensation can be found to be [6]:

$$k_1^* = k_1, \quad k_2^* = k_1\tau_d + k_2, \quad k_c^* = \frac{k_1\tau_d^2}{2} + k_2\tau_d \tag{4.71}$$

4.9.3 Alternative ELD Compensation Techniques

Note that the approach to ELD compensation illustrated in Figure 4.28 requires an additional DAC and a summing amplifier prior to the ADC. Other architectural alternatives have been proposed [9, 57, 61–63], although these share the same underlying principle of providing one extra degree of freedom. The interested reader can refer to the literature for a comparison of ELD compensation techniques and their extension to the case of cascaded CT-ΣΔMs [58]. As an illustration, some alternative ELD compensation strategies are shown in Figure 4.29.

Figure 4.29a shows a technique based on adding a feedforward coefficient, k_0, around the modulator loop filter [59]. This simple and elegant approach is similar to that used in the direct synthesis method like discussed in Section 4.9.2, but considering that the loop filter with ELD can be modeled as:

$$H(f)|_{\rm ELD} = H(f)\cdot e^{-2j\pi f\tau_d} \tag{4.72}$$

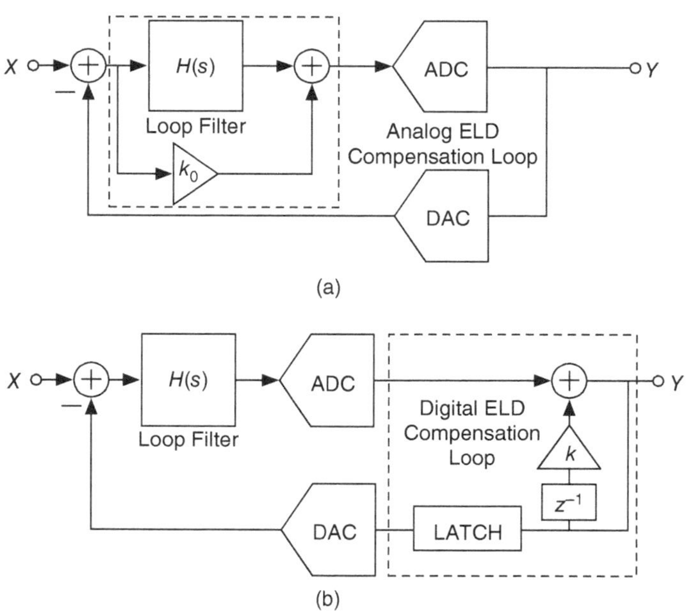

Figure 4.29 ELD compensation techniques proposed by: (a) Pavan [59]; (b) Fontaine *et al.* [61].

Assuming that $H(f)$ can be expressed as a function of the loop-filter coefficients, k_i, as:

$$H(f) = \frac{k_1}{s} + \frac{k_2}{s^2} + \ldots + \frac{k_N}{s^N} \tag{4.73}$$

it can be shown that the ELD in NRZ DACs can be compensated for if k_0 is given by [59]:

$$k_0 = k_1 \tau_d + k_2 \frac{{\tau_d}^2}{2!} + \ldots + k_N \frac{{\tau_d}^N}{N!} \tag{4.74}$$

Note that the strategies discussed above compensate for the ELD in the analog part of the CT-ΣΔM. Alternative solutions have also been proposed for the digital domain. This is the case shown in Figure 4.29b [61], where a digital feedback loop is inserted after the internal quantizer and the analog coefficients are modified to produce an NTF almost equivalent to the ideal *delay-free* loop filter. A half-clock-delay latch is also used to relax the comparator transient response and gives better isolation of the quantizer from the feedback DAC events. The idea is similar to the one presented by Yan and Sánchez-Sinencio [9], but shifting the compensation from the analog to the digital part of the modulator loop filter. The main limitation of this technique is that it requires additional quantization levels in order to fully restore the required SNR [58].

4.10 Quantizer Metastability in CT-ΣΔMs

The analysis of CT-ΣΔMs in previous sections assumed a quantizer that is ideal, except for the inherent quantization error. In practice, quantizers are usually implemented using Nyquist-rate ADCs (usually Flash ADCs),[15] in which non-ideal effects are mainly caused by the comparator errors, namely offset and hysteresis. As is the case for SC-ΣΔMs, the impact of comparator errors is attenuated by the gain of the modulator loop filter and hence they can be considered negligible compared to other circuit errors [64]. Nevertheless, although comparator offset and hysteresis are less critical than other CT-ΣΔM errors, they must be taken into account in high-performance applications, particularly when high resolution is required and multi-bit quantization is used. Thus, there is a trade-off between static resolution (limited by offset and hysteresis) and the transient response of the comparators, which is directly related with the modulator sampling rate [6].

In addition to offset and hysteresis, the quantizer operation is degraded by two timing-induced errors, namely clock jitter and comparator *metastability*. In Section 4.8, the sensitivity of CT-ΣΔMs to random clock jitter was discussed in relation to its impact on the variation in the feedback charge. Even with an ideal sampling clock with no timing uncertainties, quantizer metastability can still introduce a statistical variation in the charge that is fed back. Since real quantizers contain a regenerative stage with a finite regeneration gain, quantizer inputs with a magnitude near zero will take longer to resolve than inputs with a large magnitude [26]. Thus, assuming that the comparator is realized with a regenerative latch,[16] it can be shown that the comparator delay, τ_{dms}, depends on its input voltage, v_q, as [45, 65]:

$$\tau_{dms} = \tau_{\text{latch}} \ln \left(\frac{V_{\text{H}}}{k_c v_q} \right) \tag{4.75}$$

[15] Alternative circuit implementations of the ADCs embedded in ΣΔMs are based on time encoding rather than amplitude encoding, as will be discussed in Section 10.8.

[16] Regenerative latch comparators will be described in Chapter 8.

where V_H stands for voltage level that corresponds to the logic one, k_c is the comparator gain and $\tau = C_{latch}/g_{mlatch}$ denotes the time constant of the regenerative latch, with C_{latch} and g_{mlatch} being respectively the parasitic capacitance and the transconductance of the latch. Based on transient simulation results, the signal-dependent delay derived in Equation (4.75) can be approximated by the following expression [66]:

$$\tau_{dms} = d_0 + \frac{d_1}{|v_q|^\alpha} \tag{4.76}$$

where d_0 stands for the fixed comparison time delay and α is an adjustable parameter. As an illustration, Figure 4.30 represents τ_{dms} versus v_q for $d_0 = 0$ and different values of d_1 and α. Note that this nonideal effect is often referred to as *signal-dependent* delay, as opposed to *constant* excess delay due to signal propagation, as discussed in Section 4.9. Figure 4.31 illustrates these different delay components for a real quantizer as a function of its input signal magnitude v_q.[17]

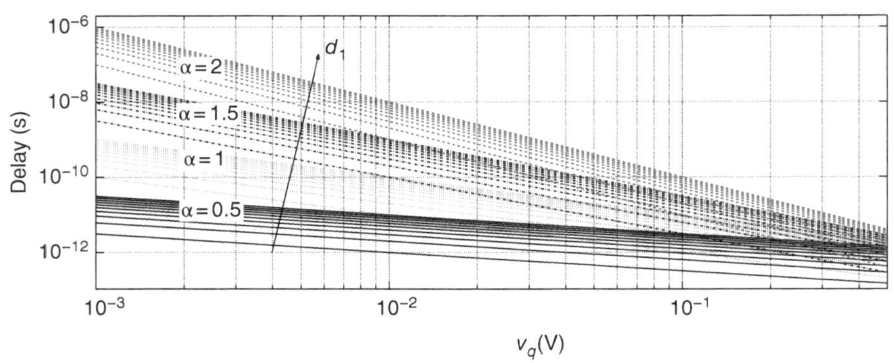

Figure 4.30 Illustrating the signal-dependent delay in a regenerative-latch comparator when considering the model in [66] for $d_0 = 0$ and d_1 varying within (0.1–1) psV^α.

The variation of the comparison time with the signal level at the quantizer input affects the performance of a CT-ΣΔM in a way similar to clock jitter. Since the quantizer input signal in a ΣΔM is ideally decorrelated from the modulator input, the times when the quantizer input is near zero or has a large magnitude appear randomly over time [26]. This results in a random component in the feedback charge from one clock period to another, and thus to an increase of the modulator IBN, which in the case of NRZ feedback DACs can be expressed as[26]:[18]

$$\mathrm{IBN}_{\mathrm{ms,NRZ}} \approx p_{\mathrm{ms}} \Delta^2 \frac{(f_s \sigma_{\mathrm{ms}})^2}{\mathrm{OSR}} \tag{4.77}$$

where p_{ms} is the probability that the comparator output makes a transition and σ_{ms} represents the standard deviation of τ_{dms}. As an illustration, Figure 4.32 compares the effect of the comparator fixed delay and signal-dependent delay (metastability) on the output spectrum of a fifth-order single-loop CT-ΣΔM with feedforward summation [67]. It can be seen that the quantizer metastability can severely degrade the performance of CT-ΣΔMs as compared to the effect of a fixed comparator delay.

[17] The hysteresis component in practice means that the quantizer sometimes does not make a decision to change the output bit when it should. This nonideality is not further considered here since its influence on the performance of ΣΔ modulators is almost negligible [64].

[18] As shown in [26], quantizer metastability can also result in a considerable decrease in the modulator SNDR for small input levels.

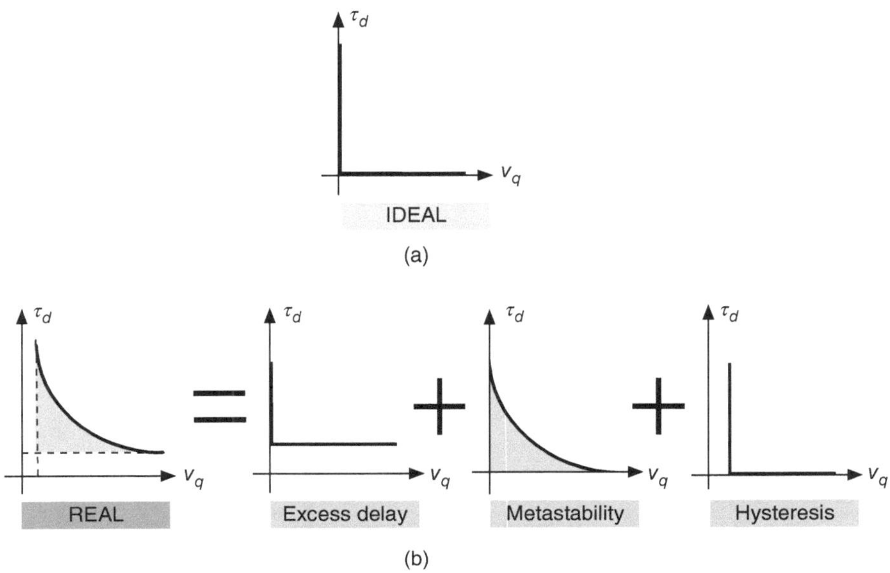

Figure 4.31 Quantizer delay versus input level: (a) ideal characteristic; (b) real characteristic as the addition of components due to constant excess delay, metastability, and hysteresis [26].

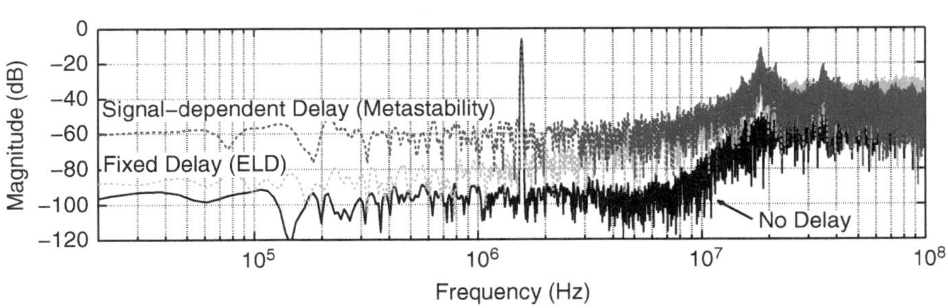

Figure 4.32 Effect of quantizer metastability in a fifth-order single-loop CT-ΣΔM with feedforward summation (f_s = 500 MHz, OSR = 40) [67].

However, although the effect of quantizer metastability on the modulator performance resembles that of random clock jitter, and hence the compensation techniques that address the latter could be also applied to the former, the architectural approaches to circumventing quantizer metastability are in practice closer to the analysis of ELD in Section 4.9. Indeed, the easiest alternative consists of inserting a latch stage between the quantizer and the feedback DAC, clocked differently to the quantizer so as to provide it a constant time to resolve [26, 56]. Figure 4.33b illustrates the approach proposed by Cherry and Snelgrove [26], in which signal-dependent delay is palliated by introducing a constant ELD of $\tau_d = 1/2$. The loop filter coefficients can thus be tuned to reduce the unfavorable effect of this fixed delay on the modulator stability and resolution. Figure 4.33c shows the architectural solution proposed by Yan and Sánchez-Sinencio, [9], in which a full delay is introduced before the DAC of the main loop, whereas a half delay and an additional DAC establish a secondary loop to accommodate ELD: an approach similar to that illustrated in Figure 4.28. Note that the proposed

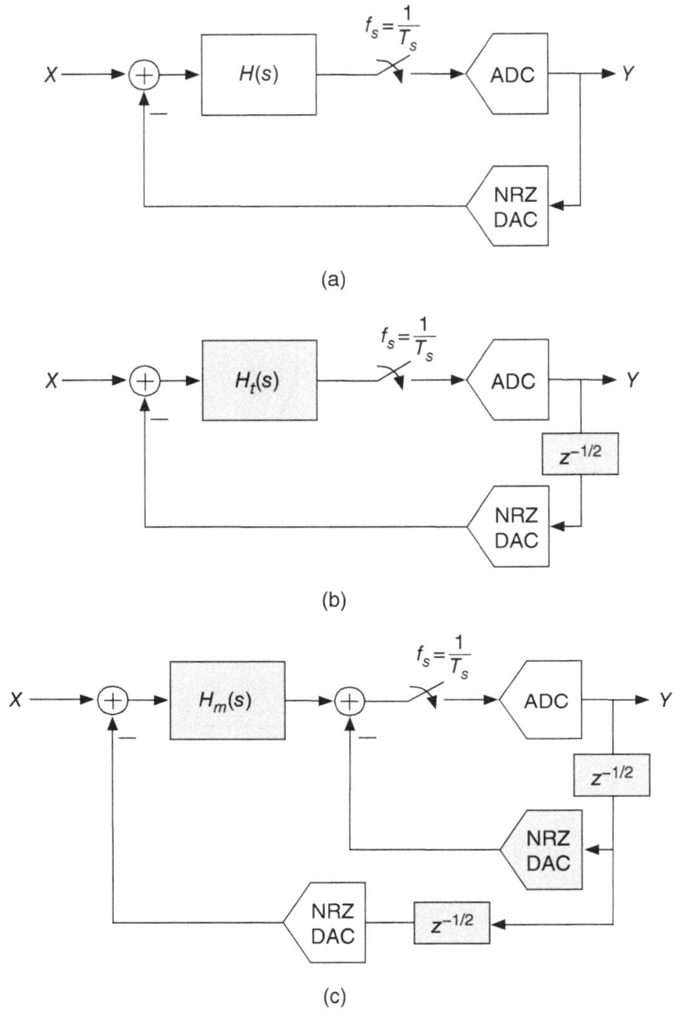

Figure 4.33 Towards the architectural compensation of timing errors in CT-ΣΔMs: (a) conventional architecture suffering from signal-dependent loop delay; (b) alternative architecture with extra half-clock delay and tuning of loop coefficients [26]; (c) alternative architecture for compensating both excess loop delay and quantizer metastability [9].

architecture provides enough degrees of freedom for modifying the CT loop filter and restoring the actual equivalence to the ideal DT version of Figure 4.33a, so that both nonzero excess loop delay and signal-dependent delay are compensated in practice.

Therefore, the effect of metastability can be circumvented by including latches between the quantizer and the feedback DAC. Clocking each latch stage on the opposite clock phase to the previous stage gives the previous stage a good deal of time to settle. However, the price to pay is the additional delay introduced in the feedback loop [25]. Indeed, the effect of metastability is a critical

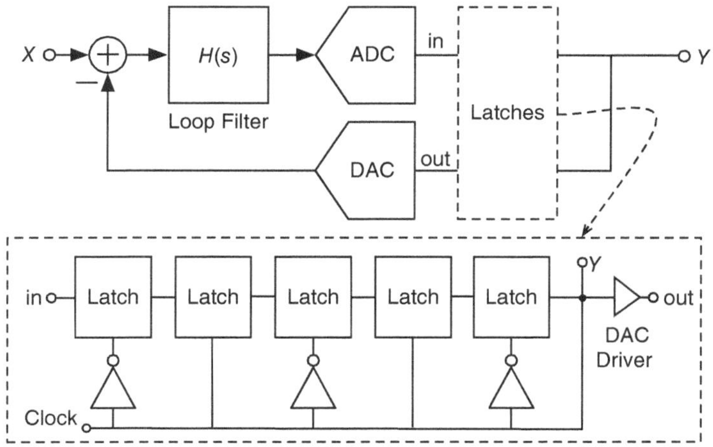

Figure 4.34 Second-order CT-ΣΔM with additional latches in the quantizer to reduce metastability [68].

issue as the clock frequency increases and the timing errors become limiting factors. However, the use of multiple latches has been demonstrated to be very effective, even for sampling frequencies in the gigahertz range. One of the first attempts to successfully implement this idea was a CT-ΣΔM using a 2-GHz clock signal [68]. As illustrated in Figure 4.34, additional latches are used in the quantizer to synchronize the last latch's output transitions with the sampling clock times. These additional latches can reduce comparator's metastability. The number of latches varies depending on a number of practical issues. Thus, for typical process parameters, a three-latch solution is adequate to meet the metastability specification. However, to meet the metastability specification over all process corners, a five-latch comparator has proven necessary [68].

4.11 Summary

The main nonideal effects degrading the performance of CT-ΣΔMs were analyzed in this chapter. These nonidealities can be classified according to their associated error mechanisms into two main groups: building-block (or circuit) errors and architectural errors. The first category includes the finite DC gain of amplifiers, the time-constant variation, incomplete settling responses of integrators, harmonic distortion, and electrical noise. The second category deals with those timing effects inherent in the practical implementation of CT-ΣΔMs, namely clock jitter error, excess loop delay, and quantizer metastability. The noise-shaping degradation of CT-ΣΔMs due to circuit-level errors is similar to that obtained in SC-ΣΔMs. However, timing errors constitute the most important limiting factors affecting the performance of CT-ΣΔMs.

Intuitive analytical procedures were described to give insights into the cause and effect of all the error mechanisms, and some of the most common circuits and systems compensation techniques proposed in literature were reviewed. The knowledge derived from the analyses presented so far will be applied in the next few chapters to provide the tools required to complete a systematic and optimized design of a ΣΔM.

References

[1] G. Raghavan *et al.*, "Architecture, Design, and Test of Continuous-Time Tunable Intermediate-Frequency Bandpass Delta-Sigma Modulators," *IEEE J. of Solid-State Circuits*, vol. 36, pp. 5–13, January 2001.

[2] H. Chae *et al.*, "A 12 mW Low Power Continuous-Time Bandpass ΔΣ Modulator With 58 dB SNDR and 24 MHz Bandwidth at 200 MHz IF," *IEEE J. of Solid-State Circuits*, vol. 49, pp. 405–415, February 2014.

[3] Y. Tsividis, "Integrated Continuous-Time Filter Design – An Overview," *IEEE J. of Solid-State Circuits*, vol. 29, pp. 166–176, March 1994.

[4] S. Pavan and P. Sankar, "Power Reduction in Continuous-Time Delta-Sigma Modulators Using the Assisted Opamp Technique," *IEEE J. of Solid-State Circuits*, vol. 45, pp. 1365–1379, July 2010.

[5] H. Shibata *et al.*, "A DC-to-1GHz Tunable RF ΔΣ ADC Achieving DR = 74dB and BW = 150MHz at f_0 = 450 MHz Using 550 mW," *IEEE J. of Solid-State Circuits*, vol. 47, pp. 2888–2897, December 2012.

[6] M. Ortmanns and F. Gerfers, *Continuous-Time Sigma-Delta A/D Conversion: Fundamentals, Performance Limits and Robust Implementations*. Springer, 2006.

[7] F. Gerfers, M. Ortmanns, and Y. Manoli, "A 1.5-V 12-bit Power-Efficient Continuous-Time Third-Order ΣΔ Modulator," *IEEE J. of Solid-State Circuits*, vol. 38, pp. 1343–1352, August 2003.

[8] S. Patón, A. di Giandomenico, L. Hernández, A. Wiesbauer, T. Poetscher, and M. Clara, "A 70-mW 300-MHz CMOS Continuous-Time ΣΔ ADC With 15-MHz Bandwidth and 11 Bits of Resolution," *IEEE J. of Solid-State Circuits*, vol. 39, pp. 1056–1063, July 2004.

[9] S. Yan and E. Sánchez-Sinencio, "A Continuous-Time ΣΔ Modulator With 88-dB Dynamic Range and 1.1-MHz Signal Bandwidth," *IEEE J. of Solid-State Circuits*, vol. 39, pp. 75–86, January 2004.

[10] L. Breems, R. Rutten, R. van Veldhoven, and G. van der Weide, "A 56 mW Continuous-Time Quadrature Cascaded ΣΔ Modulator With 77 dB DR in a Near Zero-IF 20 MHz Band," *IEEE J. of Solid-State Circuits*, vol. 42, pp. 2696–2705, December 2007.

[11] P. Kiss, J. Silva, A. Wiesbauer, T. Sun, U.-K. Moon, J. T. Stonick, and G. C. Temes, "Adaptive Digital Correction of Analog Errors in MASH ADCÕs – Part II: Correction Using Test-Signal Injection," *IEEE Trans. on Circuits and Systems – II: Analog and Digital Signal Processing*, vol. 47, pp. 629–638, July 2000.

[12] L. Breems, R. Rutten, and G. Wetzker, "A Cascaded Continuous-Time ΣΔ Modulator with 67-dB Dynamic Range in 10-MHz Bandwidth," *IEEE J. of Solid-State Circuits*, vol. 39, pp. 2152–2160, December 2004.

[13] Y.-S. Shu, J. Kamiishi, K. Tomioka, K. Hamashita, and B.-S. Song, "LMS-Based Noise Leakage Calibration of Cascaded Continuous-Time ΣΔ Modulators," *IEEE J. of Solid-State Circuits*, vol. 45, pp. 368–379, February 2010.

[14] M. Ortmanns, F. Gerfers, and Y. Manoli, "Compensation of Finite Gain-Bandwidth Induced Errors in Continuous-Time Sigma–Delta Modulators," *IEEE Trans. on Circuits and Systems – I: Regular Papers*, vol. 51, pp. 1088–1099, June 2004.

[15] L. Breems and J. H. Huijsing, *Continuous-Time Sigma-Delta Modulation for A/D Conversion in Radio Receivers*. Kluwer Academic Publishers, 2001.

[16] R. H. van Veldhoven and A. H. M. van Roermund, *Robust Sigma Delta Converters*. Springer, 2011.

[17] P. Sankar and S. Pavan, "Analysis of Integrator Nonlinearity in a Class of Continuous-Time Delta–Sigma Modulators," *IEEE Transactions on Circuits and Systems II: Express Briefs*, vol. 54, pp. 1125–1129, December 2007.

[18] J. Cherry and W. Snelgrove, *Continuous-Time Delta-Sigma Modulators for High-Speed A/D Conversion*. Kluwer Academic Publishers, 1999.

[19] P. G. R. Silva and J. H. Huijsing, *High Resolution IF-to-Baseband ΣΔ ADC for Car Radios*. Springer, 2008.

[20] C. C. Enz and G. C. Temes, "Circuit Techniques for Reducing the Effects of Op-Amp Imperfections: Autozeroing, Correlated Double Sampling, and Chopper Stabilization," *Proceedings of the IEEE*, vol. 84, pp. 1584–1614, November 1996.

[21] S. Billa, A. Sukumaran, and S. Pavan, "A 280μW 24kHz-BW 98.5dB-SNDR Chopped Single-Bit CT $\Delta\Sigma$M Achieving <10Hz 1/f Noise Corner Without Chopping Artifacts," *IEEE ISSCC Digest of Technical Papers*, pp. 276–277, February 2016.

[22] S. Pavan, "Analysis of Chopped Integrators, and its Application to Continuous-Time Delta-Sigma Modulator Design," *IEEE Transactions on Circuits and Systems I – Regular Papers*, vol. 64, pp. 1953–1965, August 2017.

[23] R. S. Rajan and S. Pavan, "Device Noise in Continuous-Time Oversampling Converters," *IEEE Transactions on Circuits and Systems –I: Regular Papers*, vol. 59, pp. 1829–1839, September 2012.

[24] S. Pavan, "The Inconvenient Truth about Alias Rejection in Continuous-Time $\Delta\Sigma$ Converters," *Proc. of the IEEE Intl. Symp. on Circuits and Systems (ISCAS)*, pp. 526–529, May 2011.

[25] J. Cherry, W. Snelgrove, and W. Gao, "On the Design of a Fourth-Order Continuous-Time LC Delta-Sigma Modulator for UHF A/D Conversion," *IEEE Trans. on Circuits and Systems – II: Analog and Digital Signal Processing*, vol. 47, pp. 518–530, June 2000.

[26] J. Cherry and W. Snelgrove, "Clock Jitter and Quantizer Metastability in Continuous-Time Delta–Sigma Modulators," *IEEE Trans. on Circuits and Systems – II: Analog and Digital Signal Processing*, vol. 46, pp. 661–676, June 1999.

[27] S. Luschas and H. S. Lee, "High-Speed $\Sigma\Delta$ Modulators with Reduced Timing Jitter Sensitivity," *IEEE Trans. on Circuits and Systems – II: Analog and Digital Signal Processing*, vol. 49, pp. 712–720, November 2002.

[28] O. Oliaei, "Design of Continuous-Time Sigma-Delta Modulators with Arbitrary Feedback Waveform," *IEEE Trans. on Circuits and Systems – II: Analog and Digital Signal Processing*, vol. 50, pp. 437–444, August 2003.

[29] O. Oliaei, "Sigma-Delta Modulator With Spectrally Shaped Feedback," *IEEE Trans. on Circuits and Systems II: Analog and Digital Signal Processing*, vol. 50, pp. 518–530, January 2003.

[30] L. Hernández, A. Wiesbauer, S. Patón, and A. DiGiandomenico, "Modelling and Optimization of Low Pass Continuous-Time Sigma-Delta Modulators for Clock Jitter Noise Reduction," *Proc. of the IEEE Intl. Symp. on Circuits and Systems*, pp. 1072–1075, 2004.

[31] L. Hernández, P. Rombouts, E. Prefasi, S. Patón, and C. L. M. García, "A Jitter Insensitive Continuous-Time $\Sigma\Delta$ Modulator Using Transmission Lines," *Proc. of the IEEE Intl. Conf. on Electronics, Circuits and Systems*, pp. 109–112, 2004.

[32] K. Reddy and S. Pavan, "Fundamental Limitations of Continuous-Time Delta-Sigma Modulators Due to Clock Jitter," *IEEE Transactions on Circuits and Systems - I: Regular Papers*, vol. 54, pp. 2184–2194, October 2007.

[33] V. Vasudevan, "Analysis of Clock Jitter in Continuous-Time Sigma-Delta Modulators," *IEEE Transactions on Circuits and Systems - I: Regular Papers*, vol. 56, pp. 519–528, October 2009.

[34] R. van Veldhoven, "A Triple-mode Continuous-time $\Sigma\Delta$ Modulator with Switched-capacitor Feedback DAC for a GSM-EDGE/CDMA2000/UMTS Receiver," *IEEE J. of Solid-State Circuits*, vol. 38, pp. 2069–2076, December 2003.

[35] O. Oliaei, "State-Space Analysis of Clock Jitter in Continuous-Time Oversampling Data Converters," *IEEE Trans. on Circuits and Systems II: Analog and Digital Signal Processing*, vol. 50, pp. 31–37, January 2003.

[36] L. Risbo, $\Sigma\Delta$ *Modulators: Stability and Design Optimization.* PhD dissertation, Technical University of Denmark, 1994.

[37] J. G. Proakis and D. G. Manolakis, *Digital Signal Processing. Principles, Algorithmics and Applications.* Prentice-Hall, 1998.

[38] R. Tortosa *et al.*, "Analysis of Clock Jitter Error in Multibit Continuous-Time $\Sigma\Delta$ Modulators with NRZ Feedback Waveform," *Proc. of the IEEE Intl. Symp. on Circuits and Systems*, pp. 3103–3106, May 2005.

[39] R. Tortosa, J. M. de la Rosa, F. V. Fernández, and A. Rodríguez-Vázquez, "Clock Jitter in Multi-bit Continuous-time $\Sigma\Delta$ Modulators with Non-Return-to-Zero Feedback Waveform," *Elsevier Microelectronics Journal*, vol. 39, pp. 137–151, January 2008.

[40] B. Putter, "ΣΔ ADC with Finite Impulse Response Feedback DAC," *IEEE ISSCC Digest of Technical Papers*, February 2004.

[41] N. Beilleau, H. Aboushady, and M. Loureat, "Using Finite Impulse Response Feedback DACs to design ΣΔ Modulators based on LC Filters," *Proc. of the IEEE Intl. Midwest Symp. on Circuits and Systems (MWSCAS)*, pp. 696–699, August 2005.

[42] P. Shettigar and S. Pavan, "Design Techniques for Wideband Single-Bit Continuous-Time ΔΣ Modulators With FIR Feedback DACs," *IEEE J. of Solid-State Circuits*, vol. 47, pp. 2865–2879, December 2012.

[43] V. Srinivasan *et al.*, "A 20mW 61dB SNDR (60MHz BW) 1b 3rd-Order Continuous-Time Delta-Sigma Modulator Clocked at 6GHz in 45nm CMOS," *IEEE ISSCC Dig. of Tech. Papers*, pp. 158–159, February 2012.

[44] A. Sukumaran and S. Pavan, "Low Power Design Techniques for Single-Bit Audio Continuous-Time Delta Sigma ADCs Using FIR Feedback," *IEEE J. of Solid-State Circuits*, vol. 49, pp. 2515–2525, November 2014.

[45] S. Pavan, R. Schreier, and G. C. Temes, *Understanding Delta-Sigma Data Converters.* Wiley-IEEE Press, 2nd ed., 2017.

[46] S. Pavan, "Continuous-Time Delta-Sigma Modulator Design Using the Method of Moments," *IEEE Transactions on Circuits and Systems I – Regular Papers*, vol. 61, pp. 1629–1637, June 2014.

[47] N. Beilleau, H. Aboushady, F. Montaudon, and A. Cathelin, "A 1.3V 26mW 3.2GS/s Undersampled LC Bandpass ΣΔ ADC for a SDR ISM-band Receiver in 130 nm CMOS," *Proc. of the IEEE Radio Frequency Integrated Circuits Symp.*, 2009.

[48] A. Ashry and H. Aboushady, "Jitter Analysis of Bandpass Continuous-Time ΣΔMs for Different Feedback DAC Shapes," *Proc. of the IEEE Intl. Symp. on Circuits and Systems (ISCAS)*, pp. 3997–4000, May 2010.

[49] A. Ashry, D. Belfort, and H. Aboushady, "Phase Noise Effect on Sine-Shaped Feedback DAC Used in Continuous-Time ΣΔ ADCs," *IEEE Transactions on Circuits and Systems I: Regular Papers*, vol. 62, pp. 717–724, March 2015.

[50] A. I. Hussein and W. Kuhn, "Bandpass ΣΔ Modulator Employing Undersampling of RF Signals for Wireless Communications," *IEEE Trans. on Circuits and Systems II: Analog and Digital Signal Processing*, vol. 47, pp. 614–620, July 2000.

[51] S. Gupta *et al.*, "A 0.8-2GHz Fully-Integrated QPLL-Timed Direct-RF-Sampling Bandpass ΣΔ ADC in 0.13μm CMOS," *IEEE J. of Solid-State Circuits*, vol. 47, pp. 1141–1153, May 2012.

[52] S. Asghar, R. del Rio, and J. M. de la Rosa, "Undersampling RF-to-Digital CT ΣΔ Modulator with Tunable Notch Frequency and Simplified Raised-Cosine FIR Feedback DAC," *Proc. of the IEEE Intl. Symp. on Circuits and Systems (ISCAS)*, pp. 1994–1997, May 2013.

[53] A. Edward and J. Silva-Martinez, "General Analysis of Feedback DACÕs Clock Jitter in Continuous-Time Sigma-Delta Modulators," *IEEE Transactions on Circuits and Systems II: Express Briefs*, vol. 61, pp. 506–510, July 2014.

[54] Jiazuo *et al.*, "Phase Noise vs. Jitter Analysis in Continuous-Time LP and BP ΣΔ Modulators with Interferers," *Proc. of the IEEE Intl. Conf. on Electronics, Circuits and Systems (ICECS)*, December 2016.

[55] J. Cherry and W. Snelgrove, "Excess Loop Delay in Continuous-Time Delta–Sigma Modulators," *IEEE Trans. on Circuits and Systems – II: Analog and Digital Signal Processing*, vol. 46, pp. 376–389, April 1999.

[56] J. F. Jensen, G. Raghavan, A. E. Cosand, and R. H. Walden, "A 3.2-GHz Second-Order Delta-Sigma Modulator Implemented in InP HBT Technology," *IEEE J. of Solid-State Circuits*, vol. 30, pp. 1119–1127, October 1995.

[57] J. van Engelen, R. J. van de Plassche, E. Stikvoort, and A. G. Venes, "A Sixth-Order Continuous-Time Bandpass Sigma-Delta Modulator for Digital IF Radio," *IEEE J. of Solid-State Circuits*, vol. 34, pp. 1753–1764, December 1999.

[58] M. Keller, A. Buhmann, J. Sauerbrey, M. Ortmanns, and Y. Manoli, "A Comparative Study on Excess-Loop-Delay Compensation Techniques for Continuous-Time Sigma–Delta Modulators," *IEEE Trans. on Circuits and Systems – I: Regular Papers*, vol. 55, pp. 3480–3487, December 2008.

[59] S. Pavan, "Excess Loop Delay Compensation in Continuous-Time Delta-Sigma Modulators," *IEEE Transactions on Circuits and Systems II: Express Briefs*, vol. 55, pp. 1119–1123, November 2008.

[60] P. Benabes, M. Keramat, and R. Kielbasa, "A Methodology for Designing Continuous-Time Sigma-Delta Modulators," *Proc. of the IEEE European Design and Test Conf.*, pp. 46–50, 1997.

[61] P. Fontaine, A. N. Mohieldin, and A. Bellaouar, "A Low-Noise Low-Voltage CT ΔΣ Modulator with Digital Compensation of Excess Loop Delay," *IEEE ISSCC Digest of Technical Papers*, pp. 498–499, 2005.

[62] G. Mitteregger, C. Ebner, S. Mechnig, T. Blon, C. Holuigue, and E. Romani, "A 20-mW 640-MHz CMOS Continuous-Time ΣΔ ADC With 20-MHz Signal Bandwidth, 80-dB Dynamic Range and 12-bit ENOB," *IEEE J. of Solid-State Circuits*, vol. 41, pp. 2641–2649, December 2006.

[63] S. Pavan, N. Krishnapura, R. Pandarinathan, and P. Sankar, "A Power Optimized Continuous-Time ΔΣ ADC for Audio Applications," *IEEE J. of Solid-State Circuits*, vol. 43, pp. 351–360, February 2008.

[64] B. E. Boser and B. A. Wooley, "The Design of Sigma-Delta Modulation Analog-to-Digital Converters," *IEEE J. of Solid-State Circuits*, vol. 23, pp. 1298–1308, December 1988.

[65] A. Rodríguez-Vázquez *et al.*, "Comparator Circuits," in *Wiley Encyclopedia of Electrical and Electronics Engineering*, pp. 577–600, John Wiley & Sons, 1999.

[66] J. Cherry, W. Snelgrove, and P. Schvan, "Signal-dependent Timing Jitter in Continuous-time ΣΔ Modulators," *Electronics Letters*, vol. 33, pp. 1118–1119, June 1997.

[67] J. Ruiz-Amaya *et al.*, "High-Level Synthesis of Switched-Capacitor, Switched-Current and Continuous-Time ΣΔ Modulators Using SIMULINK-based Time-Domain Behavioral Models," *IEEE Trans. on Circuits and Systems – I: Regular Papers*, pp. 1795–1810, Sep. 2005.

[68] E. H. Dagher *et al.*, "A 2-GHz Analog-to-Digital Delta-Sigma Modulator for CDMA Receivers With 79-dB Signal-to-Noise Ratio in 1.23-MHz Bandwidth," *IEEE J. of Solid-State Circuits*, vol. 38, pp. 1819–1828, November 2004.

CHAPTER 5

BEHAVIORAL MODELING AND HIGH-LEVEL SIMULATION

The analysis of nonideal effects in Chapters 3 and 4 allows us to derive precise equivalent circuits and models for the different ΣΔM building blocks. This chapter shows how these models can be used to improve the accuracy and computational efficiency of system-level simulations. These kinds of simulations constitute an essential tool for the systematic design of ΣΔMs.

This chapter is organized as follows. Section 5.1 describes the systematic *top-down/bottom-up* synthesis methodology, which is commonly followed in the design of ΣΔMs. Section 5.2 compares the different simulation approaches that can be used for the evaluation of ΣΔMs, putting emphasis on the benefits of behavioral simulation techniques. Section 5.3 focuses on the behavioral modeling approach and Section 5.4 shows an efficient way of implementing ΣΔM block models in SIMULINK. Section 5.5 describes SIMSIDES, a time-domain behavioral simulator implemented in the MATLAB/SIMULINK environment, and which is based on the behavioral modeling techniques described in this chapter. Finally, two case studies are described in Section 5.6. These illustrate the use of SIMSIDES for the high-level sizing and simulation of CT- and SC-ΣΔMs.

5.1 Systematic Design Methodology of ΣΔ Modulators

One of the most common approaches to the systematic design of high-performance ΣΔMs is the *top-down/bottom-up* hierarchical synthesis methodology, which is conceptually illustrated in Figure 5.1 [1]. In this approach, a given system is divided into several *hierarchical levels*, so that at each *abstraction* level of the system hierarchy, a design (or *sizing*) process takes place, thus transmitting (or *mapping*) the system specifications in a hierarchical way, from the top level to the

Sigma-Delta Converters: Practical Design Guide, Second Edition. José M. de la Rosa.

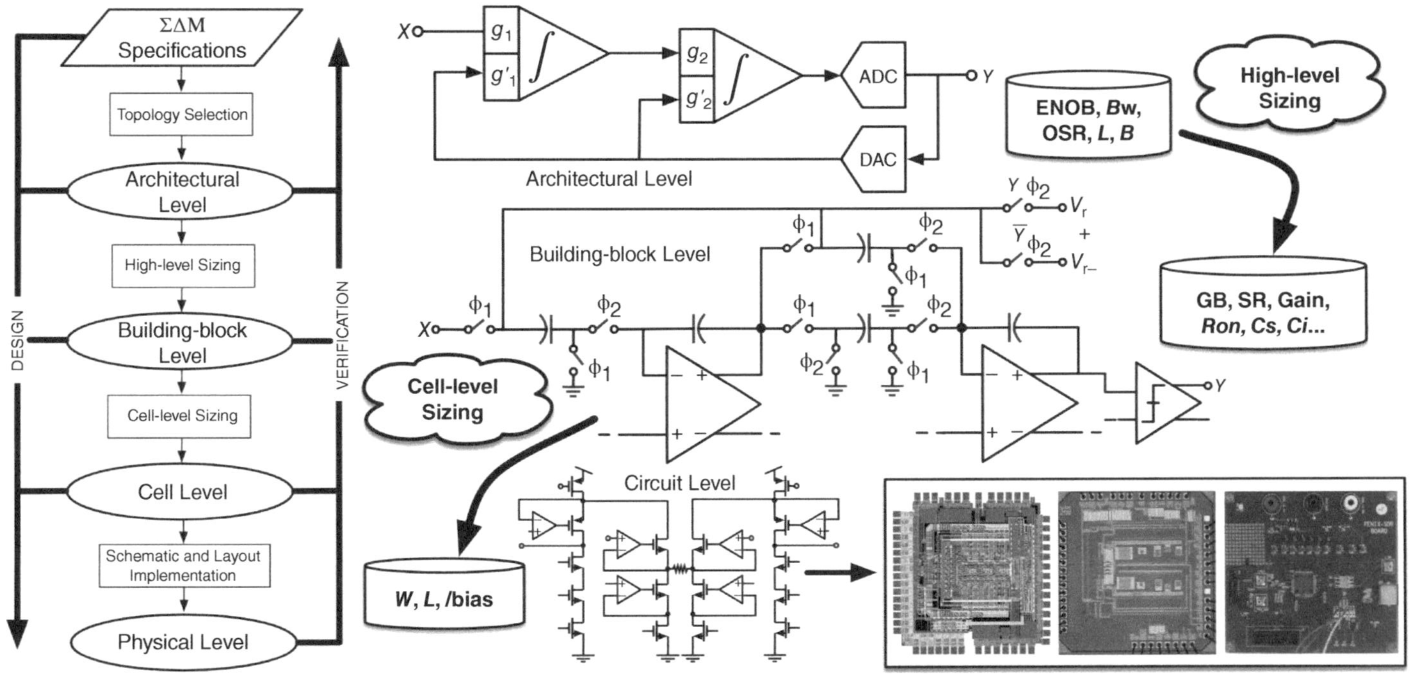

Figure 5.1 Hierarchical synthesis methodology: (a) conceptual block diagram; (b) system partitioning commonly used in **ΣΔ**Ms [1].

bottom level. The reverse path in Figure 5.1 corresponds to the hierarchical bottom-up verification process of the system performance [2, 3].

5.1.1 System Partitioning and Abstraction Levels

As illustrated in Figure 5.1, a ΣΔM is partitioned into the following hierarchical levels [2–6]:

- *architecture or topology level*: single-loop or cascade ΣΔM, single-bit or multi-bit quantization, low-pass or band-pass, DT or CT implementation, and so on;
- *subcircuit or building-block level*: amplifiers, transconductors, comparators, capacitors, resistors, switches, and so on;
- *cell level*: the circuit topology of a given building block, for instance folded-cascode or telescopic cascode OTA, SC or current-steering DAC, nMOS or CMOS switches, and so on;
- *physical level*: from transistor-level schematics to layout and chip implementation.

The design process of a ΣΔM starts from the system-level *specifications*; in other words, the effective resolution (ENOB) and signal bandwidth (B_w). The first goal is to find the modulator topology that best fulfills these specifications with the minimum power dissipation. To this end, the initial ideal design equations for NTF and IBN – which are based on a linear model of the embedded quantizers, as described in Chapter 1 – are used to calculate approximate values for the main ΣΔM parameters; that is, OSR, L, and B. Once these parameters are known, the architecture topology can be synthesized using more accurate nonlinear model equations. To this end, Schreier's MATLAB Delta-Sigma toolbox [7, 8] is widely used in the ΣΔ community. Usually there are several topologies that are a-priori good candidates to meet a given set of modulator specifications.

In order to determine the best ΣΔM architecture, analytical procedures are normally used for estimating the power consumption of the different ΣΔM topologies [9]. These procedures are based on compact expressions for the in-band noise power, like those derived in Chapter 3, that take into account both architectural and technological features, and include the impact of critical circuit errors such as thermal noise, finite OTA DC gain (or equivalent), or incomplete settling (GB and SR). Recent studies have demonstrated that nonlinearities can be efficiently incorporated into Schreier's toolbox in order to fine tune this architectural exploration procedure [10].

Based on the system-level estimation of the power consumption, a ranking of different candidate ΣΔM architectures can be determined in order to decide the most suitable topology. Other important criteria at this step are the sensitivity to circuit element tolerances and mismatch (especially important in CT cascade topologies), the modulator loop filter stability (particularly critical in high-order single-loop topologies), and/or the impact of the feedback DAC nonlinearity in multi-bit ΣΔMs.

5.1.2 Sizing Process

Once the modulator architecture has been selected, the next step involves mapping the modulator-level specifications (ENOB and B_w) onto building-block specifications; that is, amplifier finite DC gain, output swing, GB, SR, and so on. In this system-level design process – commonly referred to as the *high-level sizing* – the design parameters are the electrical specifications of the different ΣΔM subcircuits: amplifiers, transconductors (in CT-ΣΔMs), comparators, switches, and the passive elements, such as capacitors, resistors, and even inductors in the case of some BP-ΣΔMs [11].

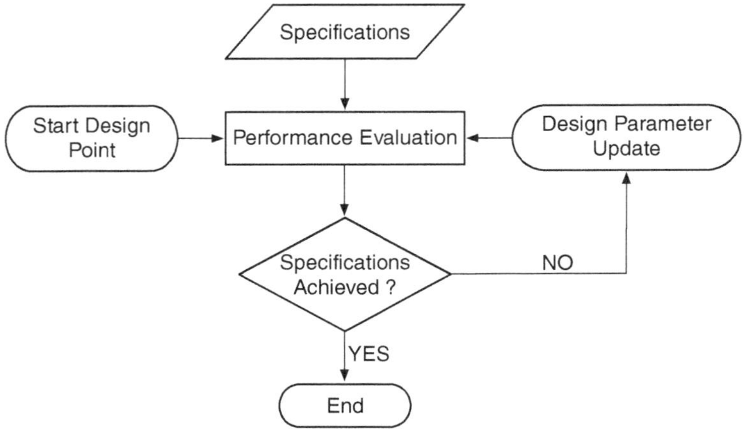

Figure 5.2 Iterative procedure usually followed for high-level and cell-level sizing.

The result of this multidimensional design space exploration is the starting point for the *electrical* or *cell-level sizing* process, in which, after an appropriate topology has been selected for every ΣΔM subcircuit, the corresponding transistor sizes and bias currents are obtained.

Therefore, as shown in Figure 5.1, the design methodology of ΣΔMs can be essentially divided into two main sizing processes: high-level sizing and cell-level sizing. Both tasks require a multitude of *re-design* iterations until the specifications at each level are met [12]. This procedure is conceptually depicted in Figure 5.2 [6]. At each iteration the performance of the circuit is evaluated at a given point of the design space, the design parameters are modified accordingly, and the procedure is repeated again. Note that, although this procedure is the same in both high-level sizing and cell-level sizing, the design problems and their formulations are different in the two cases. High-level sizing is a system-level synthesis process in which the input variables are the *system-level (modulator) specifications* and the output variables are the building-block electrical specifications (*design parameters*). The latter specifications are the input variables for the cell-level sizing process.

The design parameter selection in Figure 5.2 can be carried out "manually"; that is, the entire design space is searched by considering all possible combinations of design parameters. In this case, once the whole design space has been completely explored and checked, the best design can be selected from among those that meet the required specifications with the lowest (estimated) power dissipation and area. However, this "brute force" search approach is quite inefficient – and even infeasible in many cases – in terms of computational resources and CPU time.[1] Therefore, an *optimization engine* is normally used for guiding the exploration of the design space. To this end, approaches such as *simulated annealing* [13] or *genetic* algorithms [5] may be used. Usually, the same optimization methodology – normally based on the formulation and minimization of a design-oriented *cost function* – is used in both sizing tasks; that is, high-level sizing and cell-level sizing [4].

Different performance-evaluation strategies can be seen in Figure 5.2. Essentially, the cost function can be evaluated by means of equations or simulations [3]. Although the former approach is much faster than the latter, the accuracy of the results depends strongly on the topology; that is,

[1] At the end of this chapter a simulation-based high-level synthesis methodology is explained and applied to some case studies. As will be shown, this synthesis method does not use any optimizer and it is based on parametric simulations for checking both isolated and cumulative effects of different building-block errors in order to evaluate the performance of a ΣΔM. The resulting synthesis is reasonably efficient in terms of CPU time and it can be fined tuned by using optimization, as will be described in Chapter 6.

the ΣΔM architecture (for high-level sizing) and the circuit schematic (for cell-level sizing). For that reason, the most common approach followed by the ΣΔM community uses simulation for performance evaluation [4–6, 14–20]. However, in contrast to the optimization process – which is used in both sizing tasks – a different simulation approach is required to evaluate the performance of either the entire (system-level) modulator or just a single building block, like an amplifier or a comparator (cell-level). The latter can be analyzed using an electrical (SPICE-like) simulator with a high degree of accuracy and computational efficiency. In contrast, the evaluation of the system-level (modulator) performance can be carried out using different simulation approaches, as discussed in the next section.

5.2 Simulation Approaches for the High-level Evaluation of ΣΔMs

The high-level synthesis described above requires a large number of simulations to be performed. Depending on the design and the required specifications, hundreds or even thousands of iterations may be needed. Therefore, in order to make this part of the design feasible, simulations should consume small amounts of CPU time. This is critical for ΣΔMs due to their nonlinear *oversampled* data nature. For that reason, in order to compute the IBN power at the output of a ΣΔM with sufficient numerical accuracy, a transient simulation of at least 2^{16} clock cycles is usually required.[2] Therefore, a transient analysis of a ΣΔM using a common transistor-level (SPICE-like) circuit simulation is too slow to give an efficient space exploration, regardless of whether it is performed using manual parameter sweeps or guided by an optimizer. As an example, a 2^{16}-point transient analysis in HSPICE of a cascade 2-1 SC-ΣΔM including the clock-phase generator and other auxiliary subcircuits takes over 85 h of CPU time in a 2.2-GHz core machine with 4-GB RAM. This means that a synthesis process based on this performance evaluation approach would take months or even years! Hence, transistor-level simulation is obviously computationally infeasible for synthesis purposes and it is normally used only for the final design verification [20].

In addition to the long CPU times required to simulate a ΣΔM at the transistor level, another practical issue associated with this simulation approach is related to the convergence problems that typically arise in electrical simulators like HSPICE [22] or Spectre [23]. Moreover, transistor-level simulation is not a suitable simulation method for the high-level sizing process because the design parameters – the device sizes and biasing – are hierarchically far from the system-level specifications. Therefore, it is quite difficult for a designer to get an insight into what is happening at the system level by running simulations based on transistor-level models: the impact of a given transistor width or length has a direct impact on many different performance metrics of ΣΔM building blocks. For instance, the size of the input differential-pair transistor of an OTA affects the values of finite DC gain, nonlinearity, input parasitic capacitance, and GB, among others.

5.2.1 Alternatives to Transistor-level Simulation

These considerations suggest that increasing the abstraction level of the simulation is important, both to speed up the simulations and to work with design parameters that are closer to the system-level specifications. However, the price to pay for increasing the abstraction level (and consequently the simulation speed) is the loss of accuracy. This problem has motivated the exploration of different simulation approaches in order to optimize the trade-off between CPU time and precision. This is conceptually illustrated in Figure 5.3 [5].

[2]The FFT algorithms that are traditionally used to obtain the output spectra of ADCs in general, and of ΣΔMs in particular, are more efficiently computed if the number of points of the data sequence is a power of 2 [21]. Therefore, the number of points is often expressed as 32k-point or 64k-point, if 2^{15} or 2^{16} clock cycles are considered, respectively.

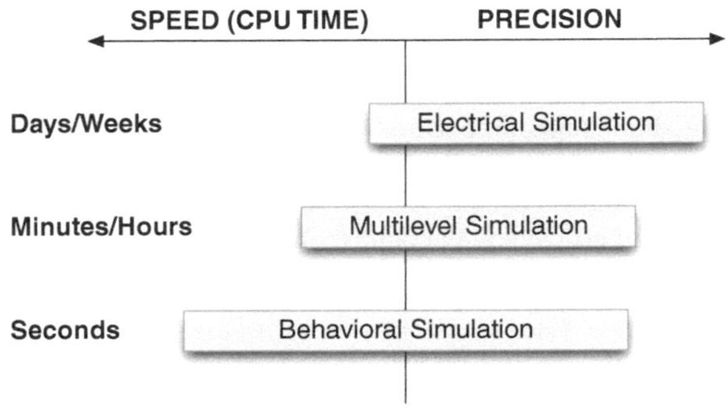

Figure 5.3 Comparison of simulation approaches in terms of CPU time and accuracy.

An alternative to the transistor-level simulation approach that also runs on electrical simulators is to use circuit *macromodels*. These macromodels are usually implemented as equivalent circuits based on ideal voltage- and current-controlled sources, which allow the main nonideal effects to be modeled. This approach has the advantage that it can be combined with transistor-level schematics using the same (electrical) simulator. Thus, critical parts of the system can be modeled at transistor level, whereas the remaining ones can be modeled less accurately with macromodels, leading to what is commonly known as a *multi-level simulation* technique [2]. Obviously, the simulation CPU time will increase with the number of parts being modeled at transistor level. Therefore, in most practical cases, the use of this approach does not really represent a significant improvement over full transistor-level simulation.[3]

As an example, Table 5.1 compares the CPU times required to simulate a fourth-order single-loop $\Sigma\Delta M$ in the Cadence Spectre circuit simulator, considering 2^{15}-point transient analysis with the `moderate` setting [23]. The modulator topology consists of a cascade of resonators with feedforward summation – like that shown in Figure 2.32b. This $\Sigma\Delta M$ architecture requires five amplifiers to implement the four loop-filter integrators and the active adder. Simulations were carried out considering ideal macromodels for the switches and the digital blocks (including the clock-phase generator). It can be seen from Table 5.1 that the CPU time increases by approximately 30–40 min with every additional amplifier simulated when considering a transistor-level implementation. The overall CPU time goes from 45 min required to simulate the entire system with macromodels to 4 h when all amplifiers are implemented at transistor level. These CPU times are more than doubled if `conservative` mode is used for the simulation; over 8 h are required to complete the simulation.

Table 5.1 Example of CPU time required to simulate a fourth-order single-loop $\Sigma\Delta M$ considering different situations in a multi-level approach. Simulations were carried out in a SUN Fire X2200 M2 server with 4-GB RAM and a 2.2-GHz Dual Core AMD Opteron CPU, running a 64-bit Linux operating system.

Simulation approach	CPU time (`moderate` mode)	CPU time (`conservative` mode)
All opamps macromodeled	45 min.	2 hours
One transistor-level opamp	1 hour, 20 min.	4 hours
Two transistor-level opamps	2 hours	5 hours, 30 min.
Five transistor-level opamps	4 hours	8 hours, 20 min.

[3]The use of macromodels may also be a good strategy to implement the required *bridge* between the system level and the electrical level, as will be described in Chapter 7.

As stated above, it is important to note that the data in Table 5.1 do not account for the time consumed to simulate the digital part of the system – the feedback DAC, DEM, clock-phase generator, digital output buffers, and so on – since this was considered ideal in these simulations. If transistor-level implementation is considered for those circuits, the overall CPU time for a conservative simulation mode is increased to one day or even more. Under these conditions, a 2^{16}-point transient simulation would take over three days of CPU time!

5.2.2 Event-driven Behavioral Simulation Technique

As illustrated in Figure 5.3, the best trade-off between accuracy and speed is achieved by the so-called *event-driven behavioral simulation* technique [15]. In this approach – conceptually illustrated in Figure 5.4 – the modulator is partitioned into a set of subcircuits, often called *basic* or *building blocks*, with independent functionality [3, 13]. In ΣΔMs, the most important building blocks are integrators and resonators,[4] and the embedded quantizers, made up of an ADC and a DAC. The behavioral modeling technique thus involves describing each of these building blocks with a model, often referred to as a *behavioral model* or *behavioral law*, that emulates their actual operation and takes into account the effect of the main nonideal circuit error mechanisms described in Chapters 3 and 4.

Note that, strictly speaking, the definition given above for a behavioral model does not necessarily imply that the model is implemented using equations. Indeed, there is an alternative approach that involves using so-called *look-up table* models [24, 25]. The idea behind these models is to extract the

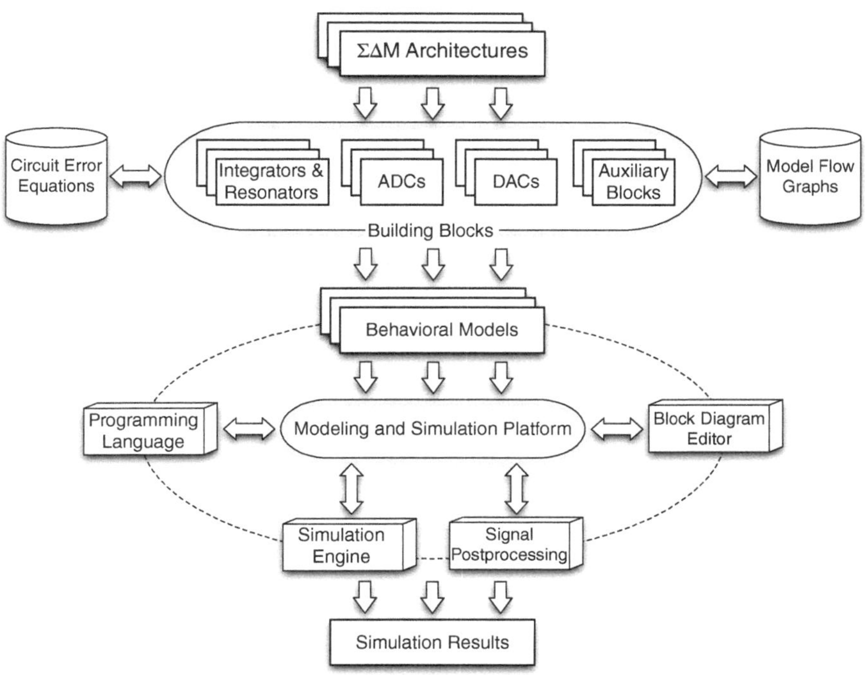

Figure 5.4 Conceptual block diagram of the behavioral modeling and simulation process.

[4]Integrators are the basic elements of the loop filter in LP-ΣΔMs. These circuits can be combined to implement resonators in order to implement BP-ΣΔMs, as described in Chapter 1.

input–output characteristics of a given building block from electrical simulations and then to map the extracted information in tables, which can be used for modeling the functionality of that building block. In this way, these tables are an alternative to the original transistor-level circuits, and they can accelerate the simulations with a high degree of accuracy. However, the price to pay is the loss of generality and of reusability of the models, because they depend on the circuit topology. Indeed, new tables have to be generated whenever the architecture itself or any circuit parameter is modified. For these reasons, look-up table methodologies are more suited to bottom-up verification than high-level synthesis.

The most commonly used behavioral modeling approach is based on finite-difference equations. These equations describe the functionality of building blocks by expressing their output signals in terms of their internal state variables and their input signals. Therefore, the accuracy of the behavioral simulation strongly depends on how precisely those equations describe the actual behavior of the corresponding ΣΔM subcircuit [6, 16, 20].

5.2.3 Programming Languages and Behavioral Modeling Platforms

Behavioral models and the corresponding simulation engine (see Figure 5.4) can be codified and implemented in a number of platforms using different programming/modeling languages, including C. The latter is a general purpose and universal programming language that exhibits high flexibility to describe behavioral models and allows behavioral simulation tools to be implemented in many different operating systems and platforms. Early approaches to behavioral-level simulation of ΣΔMs were completely compiled in C, which was shown to be highly appropriate for fast simulation and high-level synthesis of ΣΔMs [13].

The main disadvantage of C-coded behavioral modeling and simulation is that it is restricted to a limited number of building-block models – those included in the corresponding (previously-programmed) libraries – thus reducing the kinds of ΣΔM architectures that can be simulated. From this point of view, this approach is inflexible, because building-block models cannot easily be modified and the extension to new building blocks and architectures is constrained by the capabilities of the simulation engine, as well as by the designer's skills in C programming. This may explain why the majority of reported C-based behavioral simulators are intended to simulate SC-ΣΔMs [5, 13].

In order to overcome all these problems, several alternative approaches for the implementation of behavioral models have been followed. One approach that has gained popularity with ΣΔM designers is based on the use of standard hardware description languages (HDL), such as VHDL [26] and its analog extensions [27] or Verilog and Verilog AMS [20]. As will be shown in Chapter 7, VHDL-based behavioral model descriptions can be combined with HDL models of other analog, digital, and mixed-signal circuits, and they can be integrated in the design flow of commercial design environments such as Cadence Design FrameWork II [28].

An alternative and widely-used approach for the behavioral modeling and simulation of ΣΔMs uses the MATLAB/SIMULINK platform [29, 30]. This well-known mathematical software – which constitutes a standard CAD platform today in science and engineering – presents a number of advantages in terms of friendliness of the user interface: flexibility for the extension of building block libraries and for the simulation of either DT or CT systems, good trade-off between accuracy and simulation speed, direct access to very powerful tools for signal postprocessing, and a good interface to HDL modeling [6, 16].

The rest of this chapter is devoted to the behavioral modeling and simulation of ΣΔMs in the MATLAB/SIMULINK environment, explaining the different approaches to modeling ΣΔM building blocks and how to use these models for the efficient simulation of ΣΔMs.

5.3 Implementing ΣΔM Behavioral Models

The analysis of error mechanisms in ΣΔMs allows designers to obtain a expressions that shows the degradation caused by circuit-level electrical paramet(of the modulator hierarchy. On the one hand, the analytical procedure described is used for propagating the effect of errors from the building-block (either int(transfer function to the modulator NTF, in order to obtain the ΣΔM performanc(IBN and SNDR. As stated in Section 5.1, simplified versions of these equations parameters (L, OSR, and B) to circuit-level errors are highly appropriate for estimates of the power consumption and preliminary architecture selection. On th equations describing the functionality of building blocks as a function of erro basis for building accurate behavioral models. To this end, those equations r into computational flow charts that can be implemented in programming langu: is illustrated in the next section for two basic building blocks: an SC FE int(integrator. These two circuits are basic building blocks of SC- and CT-ΣΔMs, re

5.3.1 From Circuit Analysis to Computational Algorithms

Figure 5.5 shows the conceptual (single-ended) schematic of an SC FE integrat(Gm-C integrator (Figure 5.5b). Considering ideal circuit elements, the output vol is given by the following finite-difference equation,

$$v_o(nT_s) = \frac{C_s}{C_i} v_i[(n-1)T_s] + v_o[(n-1)T_s]$$

where v_i is the input voltage, C_s and C_i are respectively the sampling and integ nT_s stands for the nth sampling time instant, with T_s being the sampling period.

Note that Equation (5.1) is easy to solve numerically since the value c at a given time instant nT_s can be computed by adding the value of the ou

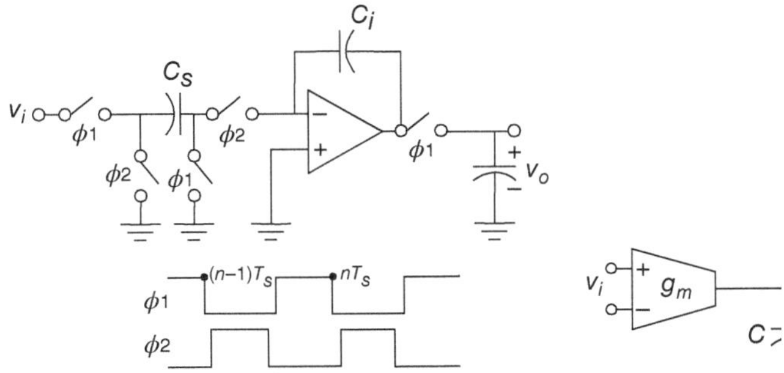

Figure 5.5 Conceptual schematic of: (a) SC FE integrator; (b) Gm-C

[5]The procedure described in this section focuses on a behavioral model considering firstly ideal be the effect of the nonlinear finite amplifier gain in SC integrators and of the finite OTA gain in complete description of the behavioral model, as well as its implementation in the MATLAB/SIMU detailed in Section 5.4.

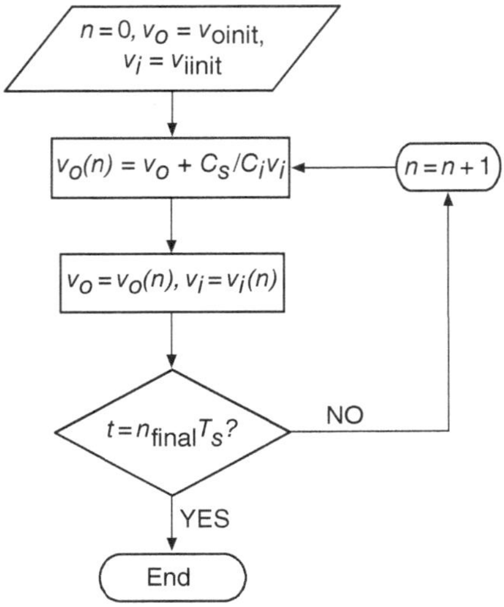

Figure 5.6 Flow chart used for computing behavioral Equation (5.1).

pling period $(n-1)T_s$ and the value of the input voltage at $(n-1)T_s$ multiplied by C_s/C_i. operation – conceptually represented by the flow chart in Figure 5.6 – can be expressed in a putational model as follows:

```
it=0;
1e-12;
1e-12;
;
nal=10;
1:10)=1;
1)=vinit;
le n<=nfinal
 n=n+1;
 vo(n)=vo(n-1)+Cs/Ci*vi(n-1);
;
```

re `vinit` stands for the initial condition, `nfinal` stands for the final sampling instant, and $C_i = 1$pF has been assumed.[6] A `while` statement has been considered, although other iterative ments such as `if` or `for` can be used as well.

The ideal functionality of Figure 5.5b can be mathematically expressed in the following rential equation,

$$g_m v_i(t) = C\frac{dv_o(t)}{dt} \tag{5.2}$$

e g_m is the transconductance and C is the integration capacitance. The above expression can be rated in MATLAB/SIMULINK solvers very efficiently [29].

The ideal models described above for SC and CT integrators in the time domain can be emented very easily in the frequency domain using SIMULINK elementary library blocks [30]

MATLAB code used in this example can be easily translated to other programming languages, such as C, Verilog-A, or -AMS [20].

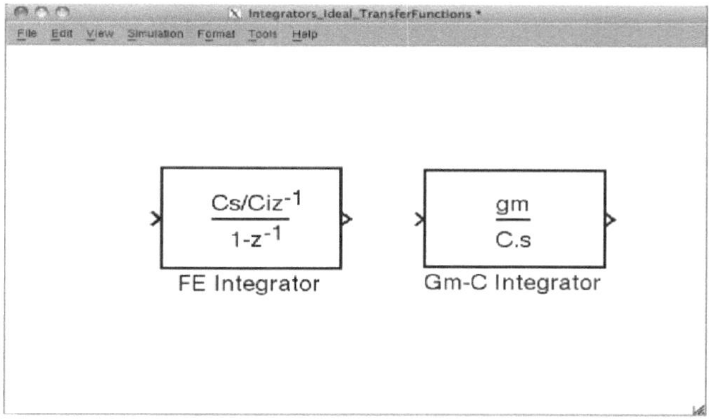

Figure 5.7 Ideal SIMULINK models of the integrators shown in Figure 5.5.

as illustrated in Figure 5.7. The corresponding transfer functions can be obtained by applying a Z-transform and an S-transform to Equations (5.1) and (5.2), respectively, giving:

$$\begin{aligned} v_o(z) &= \frac{C_s/C_i \cdot z^{-1}}{(1-z^{-1})} \cdot v_i(z) \\ v_o(s) &= \frac{g_m}{sC} \cdot v_i(s) \end{aligned} \tag{5.3}$$

where $v_i(z)$ and $v_o(z)$ stand for the Z-transforms of $v_i(nT_s)$ and $v_o(nT_s)$ in Figure 5.5a, and $v_i(s)$ and $v_o(s)$ are the S-transforms of $v_i(t)$ and $v_o(t)$ in Figure 5.5b. Both models can be computed very efficiently using the discrete-time and continuous-time solvers provided in SIMULINK [30].

5.3.2 Time-domain versus Frequency-domain Behavioral Models

The ideal behavioral models described above can be represented in the time domain by Equations (5.1) and (5.2) or in the frequency domain by Equation (5.3). There is an exact correspondence between the two representations, giving rise to identical results. Indeed, provided that the ΣΔM building blocks can be treated as linear time-invariant systems, the most convenient (and simplest) way of modeling their behavior is using frequency-domain transfer functions. However, this is not very useful in most practical situations because, as stated in Chapters 3 and 4, the performance of ΣΔMs is degraded by the effect of circuit-level errors. As will be shown in this chapter, the majority of circuit errors are modeled in a more accurate way if behavioral models are described in the time domain rather than in the frequency domain.

As an example, let us consider the effect of OTA finite DC gain on the integrators in Figure 5.5. This effect can be modeled as shown in Figure 5.8, where A_v stands for the finite DC voltage gain of the opamp in the SC FE integrator and g_o is the finite output conductance of the transconductor in the Gm-C integrator, such that the finite DC gain is given by g_m/g_o. The time-domain equations describing the behavior of the integrators in Figure 5.8 are:

$$\begin{aligned} v_o(nT_s) &= \frac{1+\mu}{1+(1+g)\mu} v_o[(n-1)T_s] + \frac{g}{1+(1+g)\mu} v_i[(n-1)T_s] \\ g_m v_i(t) &= C\frac{\mathrm{d}v_o(t)}{\mathrm{d}t} + g_o v_o(t) \end{aligned} \tag{5.4}$$

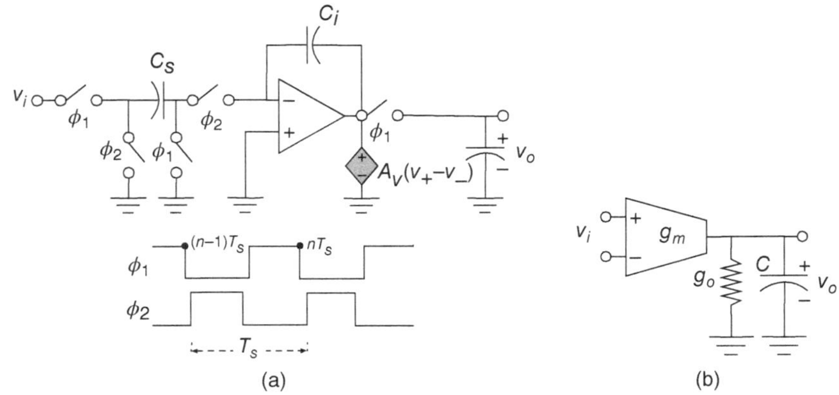

Figure 5.8 Modeling OTA finite DC gain in: (a) SC FE integrators; (b) Gm-C integrators.

where $\mu \equiv 1/A_v$ and $g = C_s/C_i$. Note that the behavioral model of the SC integrator with finite DC gain can be computed using a flow chart similar to the ideal one shown in Figure 5.6 by simply modifying the corresponding multiplication factors of v_i and v_o according to Equation (5.4). For CT integrators, the differential equation can be numerically solved using either C or MATLAB as the programming language.[7]

Operating in a similar way as in the previous section, the (frequency-domain) transfer functions of both circuits in Figure 5.8 can be obtained, yielding:

$$\begin{aligned} v_o(z) &= \frac{gz^{-1}}{1+(1+g)\mu-(1+\mu)z^{-1}} \cdot v_i(z) \\ v_o(s) &= \frac{g_m/g_o}{1+sC/g_o} \cdot v_i(s) \end{aligned} \tag{5.5}$$

The above transfer functions can be easily modeled using elementary blocks of the SIMULINK library, as shown in Figure 5.9. However, as stated above, frequency-domain models are not useful when nonideal and nonlinear effects are considered.[8] Instead, a time-domain model considering the operation during different clock phases is a more accurate way to include the various circuit-level phenomena. In order to illustrate this, let us analyze the circuit in Figure 5.8a in clock phases ϕ_1 and ϕ_2.

At the end of clock phase ϕ_1 (sampling phase) of period $(n-1)T_s$, the sampling capacitor C_s and the integration capacitor C_i are charged at the following voltages, respectively,

$$\begin{aligned} v_{C_s,\phi_1} &= -v_i[(n-1)T_s] \\ v_{C_i,\phi_1} &= -(1+\mu)\cdot v_{o1}[(n-1)T_s] = -(1+\mu)\cdot v_{o1}[(n-3/2)T_s] \end{aligned} \tag{5.6}$$

where $v_{o1}(nT_s)$ denotes the voltage at the output of the opamp in Figure 5.8a.

[7]The interested reader can find detailed documentation related to the implementation of the continuous-time and discrete-time *stateflow* charts available in MATLAB [31].

[8]Since the mathematical treatment of nonlinear errors becomes very complicated when the analysis is carried out in the time domain, some authors have proposed alternative ways to model the effect of *weak nonlinearities* by combining MATLAB functions with linear SIMULINK models [16] in SC-ΣΔMs, or by modifying Schreier's Delta-Sigma Toolbox to simulate CT-ΣΔMs in discrete time [10].

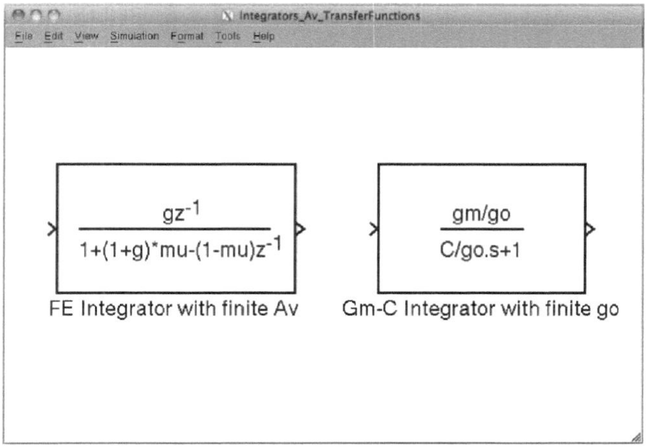

Figure 5.9 Modeling the effect of finite DC gain on the integrators transfer function in SIMULINK.

At the end of clock phase ϕ_2 (integration phase) of period $(n-1/2)T_s$, the voltages across C_s and C_i are respectively given by:

$$\begin{aligned} v_{C_s,\phi_2} &= -\mu \cdot v_{o1}[(n-1/2)T_s] \\ v_{C_i,\phi_2} &= -(1+\mu) \cdot v_{o1}[(n-1/2)T_s] \end{aligned} \tag{5.7}$$

The charge-balance equation at the negative input node of the opamp in Figure 5.8a can be written as:

$$C_s \cdot (v_{C_s,\phi_2} - v_{C_s,\phi_1}) + C_i \cdot (v_{C_i,\phi_2} - v_{C_i,\phi_1}) = 0 \tag{5.8}$$

Substituting Equations (5.6) and (5.7) into Equation (5.8) gives:

$$[1+(1+g)\mu] \cdot v_{o1}[(n-1/2)T_s] = g \cdot v_i[(n-1)T_s] + (1+\mu) \cdot v_{o1}[(n-3/2)T_s] \tag{5.9}$$

Taking into account that $v_o(nT_s) = v_{o1}[(n-1/2)T_s]$, the above expression transforms into the SC part of Equation (5.4), which, after taking the Z-transform, results in Equation (5.5).

Note that the charge-balance analysis described in Equations (5.6)–(5.8) is more complex than using Z-domain transfer functions to model simple phenomena, such as the ideal behavior of simple building blocks and/or the effect of linear errors, such as finite DC gain. However, as the model becomes more accurate – including more nonideal circuit errors and/or nonlinearities – the charge-balance model is the most suitable approach for implementation in a behavioral simulator. As an illustration, let us assume the voltage-dependence of the finite DC gain in SC FE integrators.[9] This nonlinear effect can be modeled by replacing the voltage linear gain A_v parameter in Figure 5.8a with a nonlinear function of the output voltage $A_v(v_o) \simeq A_v \cdot (1 + \text{avnl}_1 \cdot v_o + \text{avnl}_2 \cdot v_o^2 + \ldots)$, where avnl_i

[9]Precise time-domain behavioral models of CT-ΣΔM building blocks including the main circuit effects will be described in Section 5.4.

represents the ith-order nonlinear voltage-gain coefficient [32]. Taking these nonlinear effects into account, the charge-balance equations transform into:

$$\begin{aligned}
v_{C_s,\phi_1} &= -v_i[(n-1)T_s] \\
v_{C_i,\phi_1} &= -[1 + 1/A_{v(n-3/2)}] \cdot v_{o1}[(n-1)T_s] = -(1+\mu) \cdot v_{o1}[(n-3/2)T_s] \\
v_{C_s,\phi_2} &= -[1/A_{v(n-1/2)}] \cdot v_{o1}[(n-1/2)T_s] \\
v_{C_i,\phi_2} &= -[1 + A_{v(n-1/2)}] \cdot v_{o1}[(n-1/2)T_s]
\end{aligned} \tag{5.10}$$

where $A_{v(n-1/2)} = A_v(v_{o1}[(n-1/2)T_s])$ and $A_{v(n-3/2)} = A_v(v_{o1}[(n-3/2)T_s])$.

From Equations (5.9) and (5.10), it can be shown that the nonlinear finite-difference equation describing the behavior of the integrator can be written as:

$$\begin{aligned}
\left(1 + \frac{1+g}{A_{v(n-1/2)}}\right) \cdot v_{o1}[(n-1/2)T_s] &= g \cdot v_i[(n-1)T_s] + \\
&+ \left(1 + \frac{1}{A_{v(n-3/2)}}\right) v_{o1}[(n-3/2)T_s]
\end{aligned} \tag{5.11}$$

Note that an iterative procedure is needed to compute the behavioral model in Equation (5.11), because the output voltage of the opamp $v_{o1}(nT_s)$ depends on the amplifier nonlinear gain, which in turns changes with $v_{o1}(nT_s)$.

5.3.3 Implementing Time-domain Behavioral Models in MATLAB

The behavioral model described by Equations (5.10) and (5.11) can be programmed using a so-called M-file [29] and with the MATLAB code shown in Figure 5.10, where `C1` and `C2` are respectively the sampling capacitance (C_s) and integration capacitance (C_i), and `count` is a parameter that determines the clock phase, with `count=0` corresponding to the sampling phase and `count=1` to the integration phase. In this way, as illustrated in Figure 5.11, the clock-phase scheme can be selected to be either the one shown in Figure 5.8a or the complementary one; that is, with the input switch being clocked at ϕ_2. The input signal is modeled by a matrix made up of two vectors, `u(1)` and `u(2)`, while the output is stored in a variable `y`, with `yold` modeling the output sample stored in the previous clock phase and `ytemp` being a temporal state variable used until convergence is achieved.

Note that the convergence criterion used in the iterative procedure is `AVVAR=abs [(AVOLD -AVNEW)/AVNEW] < thrs`, where `thrs` models the threshold value selected for convergence (usually `thrs=0.01`), `abs(x)` stands for the absolute value of `x`, and `AVOLD` and `AVNEW` are, respectively, the old and new values of the parameter to be solved – A_v in this example. With this criterion, convergence is reached normally in three or four iterations, so CPU time is not excessive [6].

Figure 5.12 shows a SIMULINK model of the SC FE integrator with nonlinear DC gain shown in Figure 5.8a. This is based on the MATLAB code described in Figure 5.10. Note that a MATLAB function block – called the `MATLAB Fcn` block in SIMULINK [30] – is used for this purpose. Figure 5.12b shows the MATLAB Fcn `intfescavnl` used in this example, which applies the M-file shown in Figure 5.10 to the input of the block. The input arguments of the MATLAB functions (e.g., `AV,AVNL1,AVNL2...C1,C2,PHI`) are included in the `Function Block Parameters` dialogue window [30], as illustrated in Figure 5.12c. This can be configured by the user. Note that two additional blocks in Figure 5.12b are used for suitably sampling the output

```
function y = intfeavnl(u,AV,AVNL1,AVNL2,AVNL3,AVNL4,C1,C2,PHI)
persistent VC1 VC2 VC1OLD count yold
if (isempty(VC1))
    VC1=0;
    VC2=0;
    VC1OLD=0;
    count=0;
    yold=0;
end
     if (PHI==1)
            if (count==0)
                VC1 =u(1);
                VC2 =yold;
                y=yold;
                count=count+1;
            else
                VC1OLD = VC1;
                VC1 = u(2);
                AVNEW = AV;
                ITER = 0;
                AVVAR=1;
                while (AVVAR>0.01&ITER<50),
                    ytemp = ((C1/C2)*(VC1OLD-VC1)*(AVNEW/(1+AVNEW+(C1/C2)))) + (VC2*((AVNEW+1)/(1+AVNEW+(C1/C2))));
                    AVOLD = AVNEW;
                    VO2 = ytemp^2;
                    VO3 = ytemp^3;
                    VO4 = ytemp^4;
                    AVNEW = AV*(1+(AVNL1*ytemp)+(AVNL2*VO2)+(AVNL3*VO3)+(AVNL4*VO4));
                    AVVAR = abs((AVOLD-AVNEW)/AVNEW);
                    ITER = ITER + 1;
                end
                y = ytemp;
                yold=y;
                count=0;
            end
    elseif (count==0)
                VC1OLD = VC1;
                VC1 = u(2);
                AVNEW = AV;
                ITER = 0;
                AVVAR=1;
                while (AVVAR>0.01&ITER<50),
                    ytemp = ((C1/C2)*(VC1OLD-VC1)*(AVNEW/(1+AVNEW+(C1/C2)))) + (VC2*((AVNEW+1)/(1+AVNEW+(C1/C2))));
                    AVOLD = AVNEW;
                    VO2 = ytemp^2;
                    VO3 = ytemp^3;
                    VO4 = ytemp^4;
                    AVNEW = AV*(1+(AVNL1*ytemp)+(AVNL2*VO2)+(AVNL3*VO3)+(AVNL4*VO4));
                    AVVAR = abs((AVOLD-AVNEW)/AVNEW);
                    ITER = ITER + 1;
                end
                y = ytemp;
                yold=y;
                count=count+1;
            else
                VC1 = u(1);
                VC2 = yold;
                y=yold;
                count=0;
            end
```

Figure 5.10 MATLAB code for the behavioral model of the SC FE integrator in Figure 5.8a including the effect of the amplifier gain nonlinearity.

signal at the correct sampling instant. One of these blocks is a `Unit Delay` block, which adds an extra delay of $T_s/2$. The other block is a so-called SIMULINK `S-function` block, which will be explained in Section 5.4.

One of the problems of using M-files is that the MATLAB interpreter is called at each time step, thus slowing down the simulation [29]. This problem is aggravated as the model complexity increases. Let us consider, for instance, a cascade 2-1-1 SC-ΣΔM in which all building blocks are ideal except for the nonlinear DC gain – modeled by the M-file shown in Figure 5.10. Figure 5.13 shows the SIMULINK block diagram of the modulator, highlighting its main parts. A simulation of 2^{16} clock periods of this modulator takes 148 s in a 2.4-GHz core with 4-GB RAM.[10]

There is an alternative approach to the simulation of ΣΔMs based on MATLAB functions [16, 33]. The models included in this toolbox [34] are based on the interconnection of SIMULINK standard library blocks, and are very intuitive and useful for system-level evaluation. However,

[10]This simulation was carried out using the normal mode in SIMULINK. The use of the accelerator mode available in SIMULINK speeds up the simulation regardless the modeling strategy used. However, the accelerator mode cannot be used for some types of M-files, for instance when these files use parameters that have more than one dimension – which is common in the modeling of ΣΔM building blocks [30].

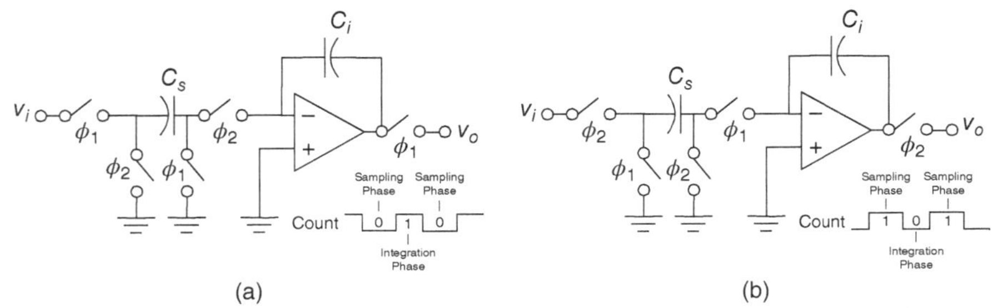

Figure 5.11 Meaning of count in the behavioral model of an SC FE integrator, considering that the input switch clock phase is: (a) ϕ_1. (b) ϕ_2.

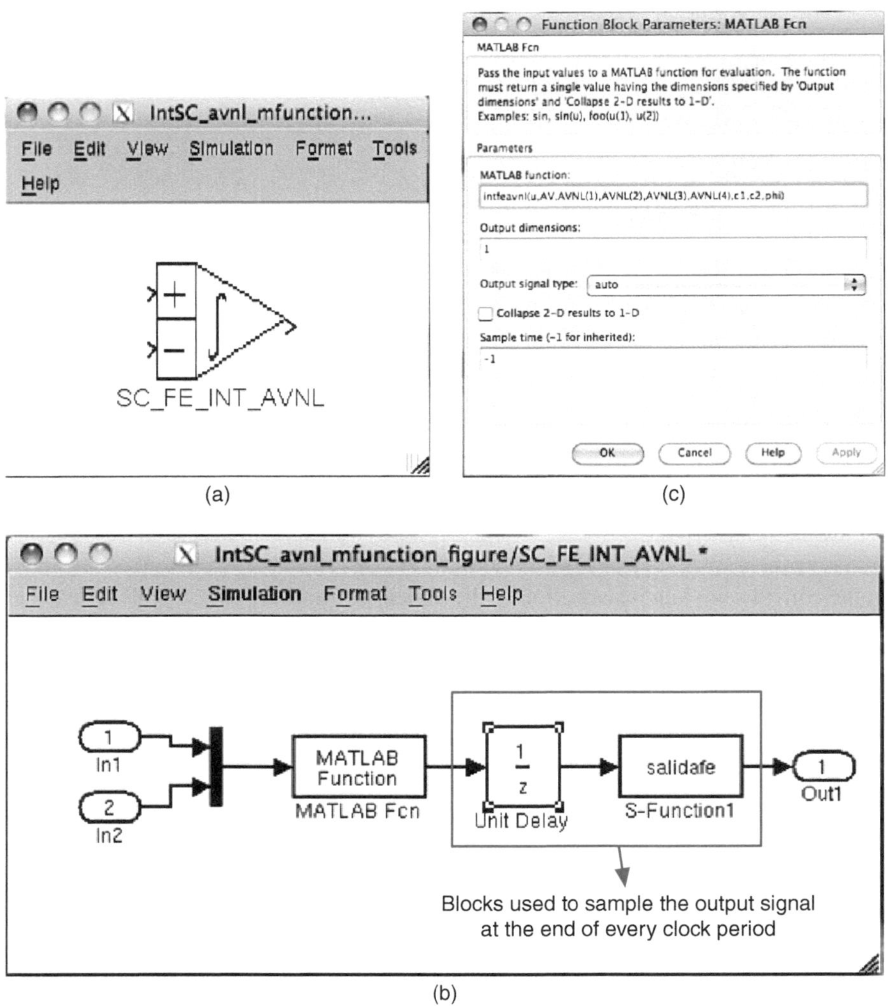

Figure 5.12 SIMULINK model of the SC FE integrator shown in Figure 5.8a: (a) SIMULINK mask; (b) SIMULINK block diagram including a MATLAB function; (c) associated dialogue window.

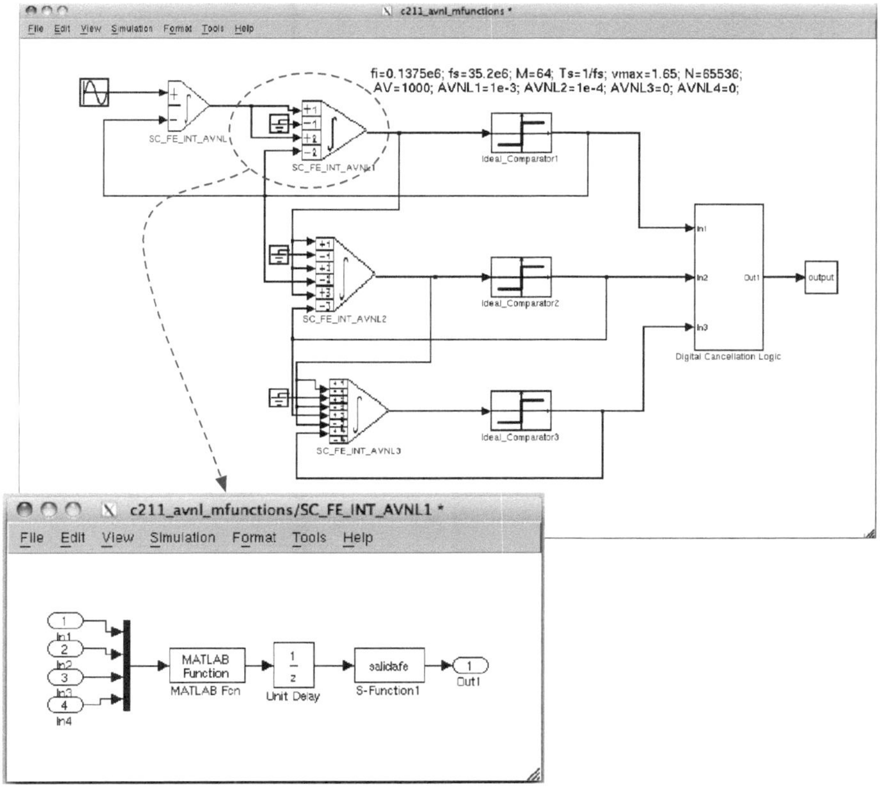

Figure 5.13 SIMULINK model of a cascade 2-1-1 SC-ΣΔM with nonlinear DC gain modeled using the M-file shown in Figure 5.10.

the block library is limited to SC circuits and uses relatively simple models that do not take into account limitations such as the nonlinearities associated with OTA DC gain and capacitors. Although, the models are very intuitive, the implementation of each building block requires several sets of elementary SIMULINK blocks using MATLAB functions, with a subsequent penalty in computation time. This issue is critical for an optimization-based synthesis process as stated in previous sections. Moreover, most parts of the models are implemented in the *Z*-domain, and so the circuit behavior during different clock phases is not taken into account. This may lead to a rather imprecise modeling of errors associated with the integrator transient response, as discussed in Section 3.4.

Figure 5.14 shows an example: a SIMULINK block diagram of a cascade 2-1-1 SC-ΣΔM modeled with the SIMULINK toolbox developed by Brigati *et al.* [34]. The integrator behavioral models already include finite open-loop opamp DC gain, incomplete settling error, slew rate, mismatch capacitor ratio error, and thermal noise. In addition, the main nonlinear effects were added to the original model, namely nonlinear sampling switch on-resistance, nonlinear capacitors, and nonlinear open-loop opamp DC gain, the latter modeled using the M-function given in Figure 5.10. The main parts of this block diagram, as well as those parameters required to simulate the modulator, are shown in Figure 5.14. A 2^{16}-point simulation in a 2.4-GHz core with 4 GB RAM takes 80 s.

This CPU time can be reduced to 8 s if the ΣΔM building-block behavioral modes are implemented in C using the so-called SIMULINK *S-functions*[35]. A *system-function* or *S-function* is a computer

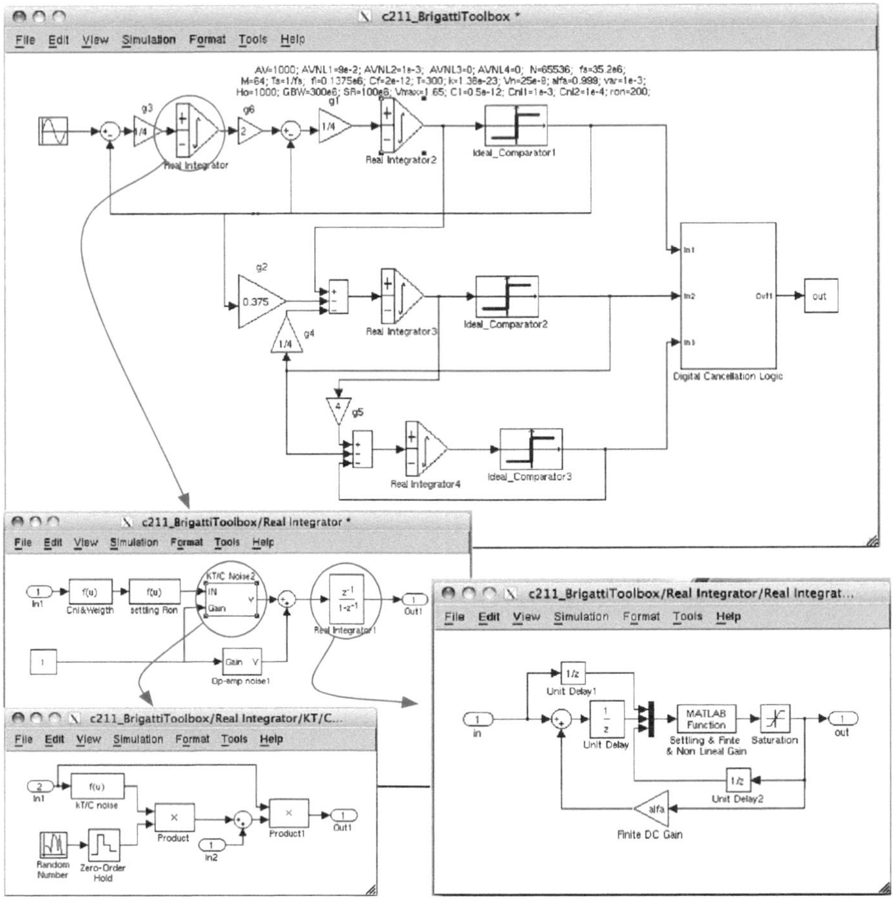

Figure 5.14 SIMULINK model of a cascade 2-1-1 SC-ΣΔM using Brigatti's toolbox [34].

language description of a SIMULINK block that can be written in MATLAB code, Ada, Fortran, or C code [35]. The latter are special-purpose source files which allow computation algorithms written in C to be included in SIMULINK models. This approach speeds up the simulations – up to 50 times faster in some cases – compared to the use of MATLAB functions or M-files to code the behavioral models, even when the accelerator mode is used [6]. In addition to the benefits in terms of CPU time, the use of S-functions for the behavioral simulation of ΣΔMs allows designers to model circuit-level error mechanisms in a more accurate way, as will be described in the rest of this chapter.

5.3.4 Building Time-domain Behavioral Models as SIMULINK C-MEX S-functions

Figure 5.15a illustrates a step-by-step procedure to implement the behavioral model of a ΣΔM building block in the SIMULINK enviroment by using C-coded S-functions. The main steps are the following [6]:[11]

[11]Recent versions of MATLAB allow us to automatically build C-coded S-functions from specifications and code fragments that the user can supply using a GUI [36]. The procedure explained in this book allows C MEX S-functions to be written from scratch, thus giving the designer complete control of the behavioral model to be built.

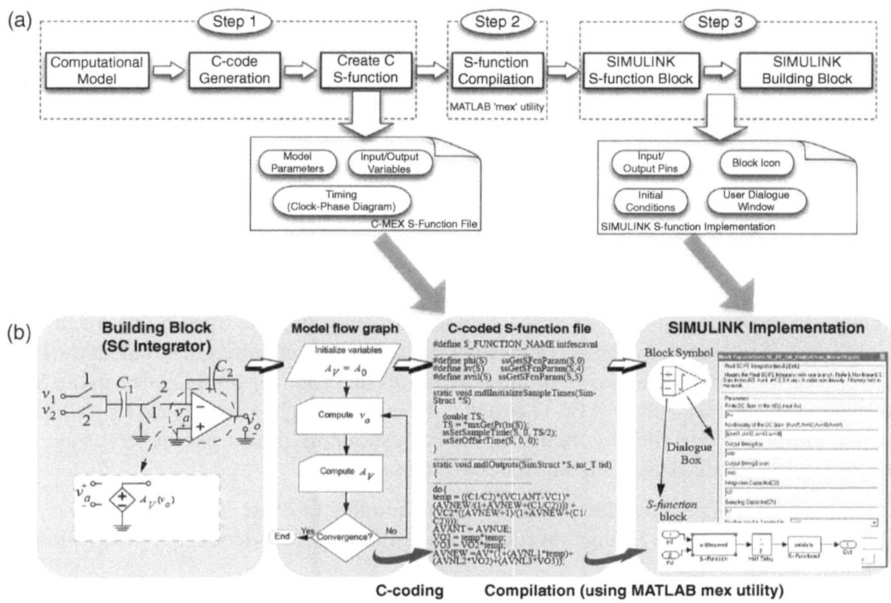

Figure 5.15 Procedure to incorporate a behavioral model into the MATLAB/SIMULINK environment using C-MEX S-functions: (a) Conceptual step-by-step flow. (b) Illustrating the process for an SC FE integrator with finite nonlinear DC gain.

- *Definition of a computational model*: Given a ΣΔM building block that includes a set of nonidealities, a computational model which allows the output to be calculated as a function of the input(s) and the internal states (if any) is defined.
- *Generation of the C code* corresponding to the computational model defined in the previous step.
- *Implementation of the computational model into a C-MEX S-function*: To this end, SIMULINK provides various S-function template files, which can accommodate the C-coded computational model generated in the previous step. Both DT and CT building blocks can be modeled, as will be shown in Section 5.4. S-function template files are composed of several subsections, called *callback methods*, which are routines that perform the different tasks required in each simulation stage. These tasks include variable initialization, computation of output variables, and update of state variables [35].
- *Compilation of the S-function*: This is done by using the `mex` utility provided in MATLAB [35]. The utility can be run from the MATLAB command prompt window, by typing `mex filename.c`, where `filename.c` can be either a single C-coded S-function file or a combination of different source C files.[12] The resulting object files are dynamically compiled and linked in SIMULINK when needed in a given simulation.
- *Incorporation of the model into the SIMULINK environment*. This can be done by using the `S-function` block of the SIMULINK libraries [35]. A block diagram containing the S-function block is created including the input/output pins. A dialogue box is used for specifying the name of the underlying S-function. In addition, model parameters can be changed in this box.

[12]The MATLAB `mex` utility can be used for compiling one or more C/C++ or Fortran source files. This book focuses on C/C++ source files.

As an example, Figure 5.15b illustrates how to apply the main steps listed above to create the S-function of an SC FE integrator with finite and nonlinear DC gain, like the one shown in Figure 5.8. Note that the computational model flow graph in Figure 5.15b shows only the iterative procedure used for computing the nonlinear DC gain. The entire related MATLAB code is listed in Figure 5.10. When more nonidealities are to be considered, a more complex computation model – taking all nonidealities into account in the right sequence – is needed, as will be detailed in Section 5.4. For the sake of simplicity, the example in Figure 5.15b only shows some significant sections of the S-function file associated with the SC integrator model and how this S-function is incorporated into the SIMULINK environment.

Figure 5.16 shows the entire C-coded S-function file of an SC integrator with finite nonlinear DC gain, highlighting its main parts. Note that, in addition to the computational model itself (shown in Figure 5.17), an S-function includes the other important functions required to simulate the behavioral model using SIMULINK. The majority of these functions and routines are included in the S-function template files provided by MATLAB. Therefore, the easiest way to model ΣΔM building blocks is to modify those template files by including the corresponding model parameters, state variables, input/output signals, clock-phase diagram scheme, and so on.[13]

MATLAB provides detailed documentation and a number of examples, which are very useful to this end [35]. For instance, the S-function file shown in Figure 5.16 is made up of the following parts and functions:

- `S-function name` and definitions: This section of the S-function is used for defining the name of the block model and its corresponding parameters. In the example of Figure 5.16, the model parameters are `phi`, `ts`, `c1`, `c2`, `av`, `avnl`, `cp`, `opspos`, and `opsneg`, the latter being the positive and negative limits of the amplifier output swing.
- `mdlInitializeSizes`, where the number of inputs, outputs, and internal states are specified. Note that building blocks used in SIMULINK may have a vector of inputs, a vector of outputs, and a vector of states. The dimensions of these vectors are also specified in this part of the S-function. In this example, two input ports, zero-state variables, and one output are defined, the size of all of them being one. The number of sampling rates (or sampling times) – one in this case – is also specified in the `mdlInitializeSizes` function.
- `mdlInitializeSampleTimes`, where the sampling time and the clock-phase scheme is defined. This function is used for defining the building block at single-rate, at multi-rate, or if it is a CT circuit. The number of clock phases, the on-time period of each, as well as the delays – called *offset* in the S-function – among them are also specified in this routine. For instance, in this example, a nonoverlapping two-phase clock with a 50% duty cycle is considered.
- `mdlStart` performs the initialization of the S-function, setting up the required model parameters and the initial values of the internal variable states.
- `mdlOutputs` is the main part of the S-function because the C code of the behavioral model is introduced in this section. The routine computes the model and stores the corresponding results in an output array.
- `mdlTerminate` is a routine that must be included in the S-function file in order to keep the required template structure, because SIMULINK calls it at the end of the simulation. In the more general case, this function is used for performing any action required at the end of the simulation, such as freeing up memory. In this example, this function is not used and hence it is empty.

[13] Although the procedure described in this section can be easily implemented by using the S-function template files provided by MATLAB, note also that there is a SIMULINK block, called S-function Builder, that builds an S-function from specifications and C code provided by the user. Although the S-function templates may depend on the MATLAB releases, the same modeling approach applies [30].

```
#define S_FUNCTION_NAME intfesc1branchavnl
#define S_FUNCTION_LEVEL 2

#include "tmwtypes.h"
#include "simstruc.h"
#include <math.h>
#include<string.h>

#define phi(S)      ssGetSFcnParam(S,0)
#define ts(S)       ssGetSFcnParam(S,1)
#define c1(S)       ssGetSFcnParam(S,2)
#define c2(S)       ssGetSFcnParam(S,3)
#define av(S)       ssGetSFcnParam(S,4)
#define avnl(S)     ssGetSFcnParam(S,5)
#define opspos(S)   ssGetSFcnParam(S,6)
#define opsneg(S)   ssGetSFcnParam(S,7)
#define cp(S)       ssGetSFcnParam(S,8)

#define NPARAM 9
```

Definition of **S-Function name** and model parameters

```
static void mdlInitializeSizes(SimStruct *S)
{

    ssSetNumSFcnParams(S, NPARAM);
    if (ssGetNumSFcnParams(S)  != NPARAM)
       return;
    ssSetNumContStates(S, 0);
    ssSetNumDiscStates(S, 0);

    if (!ssSetNumInputPorts(S, 2)) return;
    ssSetInputPortWidth (S, 0, 1);
    ssSetInputPortWidth (S, 1, 1);
    ssSetInputPortDirectFeedThrough(S, 0, 1);
    ssSetInputPortDirectFeedThrough(S, 1, 1);

    if (!ssSetNumOutputPorts(S, 1)) return;
    ssSetOutputPortWidth (S, 0, 1);

    ssSetNumSampleTimes(S, 1);
    ssSetNumRWork(S, 5);
    ssSetNumIWork(S, 0);
    ssSetNumPWork(S, 0);
    ssSetNumModes(S, 0);
    ssSetNumNonsampledZCs(S, 0);
    ssSetOptions(S,
SS_OPTION_EXCEPTION_FREE_CODE);
}
```

Function: **'mdlInitializeSizes'**
Determines the number of inputs, outputs, internal states, etc.

```
static void mdlInitializeSampleTimes(SimStruct
*S)
{
    double TS;

    TS = *mxGetPr(ts(S));
    ssSetSampleTime(S, 0, TS/2);
    ssSetOffsetTime(S, 0, 0);
}
```

Function: **'mdlInitializeSampleTimes'**
Defines sampling rate parameters and clock-phase scheme

```
static void mdlStart(real_T *x0,SimStruct *S)
{
    real_T *work = ssGetRWork(S);
    work[0]=0;
    work[1]=0;
    work[2]=0;
    work[3]=0;
    work[4]=1000;
}
```

Function: **'mdlStart'**
Initializes the state vectors of an S-function

```
static void mdlOutputs(SimStruct *S, int_T
tid)
{
/*
 C-code of the behavioral model goes here
*/
}
```

Function: **'mdlOutputs'**
Computes the behavioral model included in the S-function

```
static void mdlTerminate(SimStruct *S)
{
}

#ifdef  MATLAB_MEX_FILE
#include "simulink.c"
#else
#include "cg_sfun.h"
#endif
```

Function: **'mdlTerminate'**
Required to terminate the simulation of the behavioral model

Figure 5.16 MATLAB C-coded S-function file of the integrator in Figure 5.10.

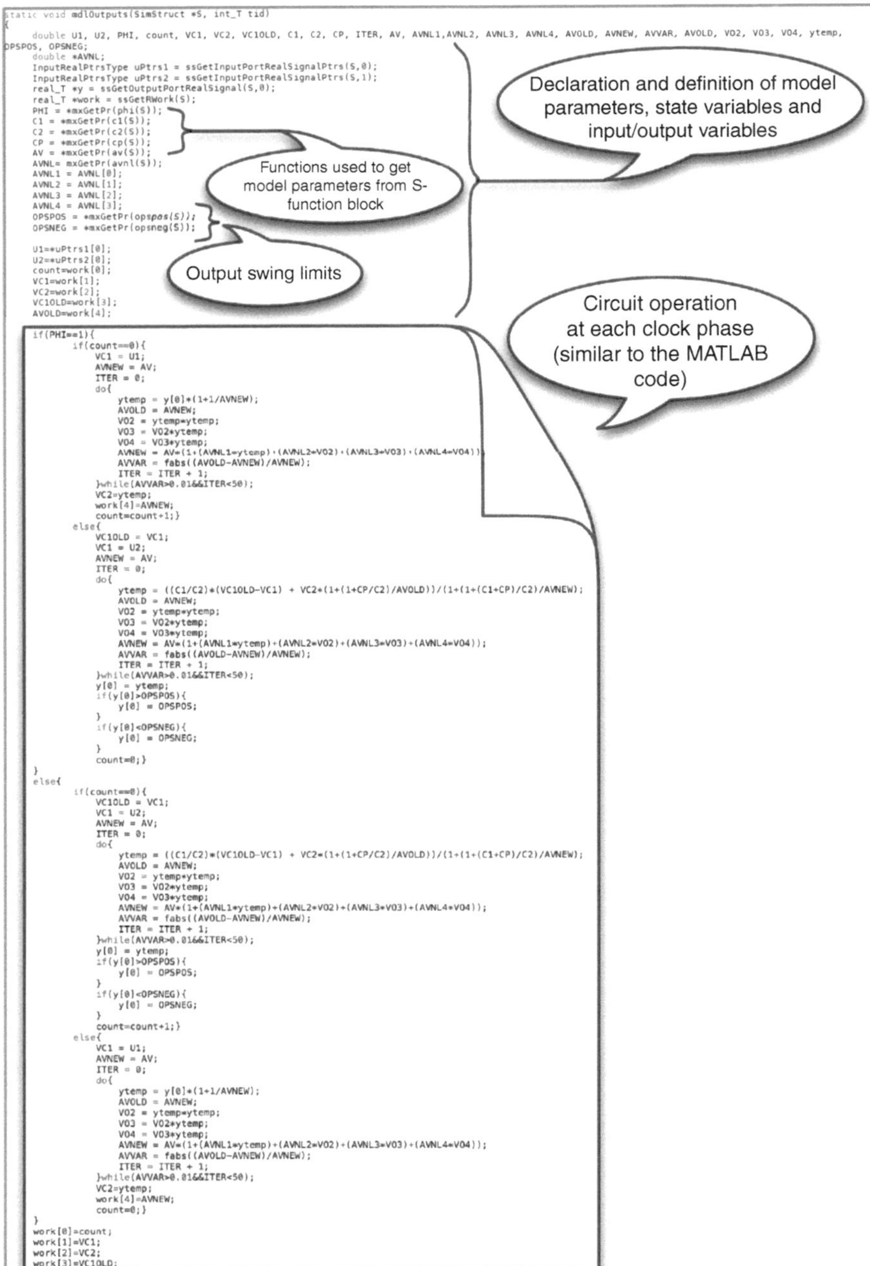


```c
static void mdlOutputs(SimStruct *S, int_T tid)
{
    double U1, U2, PHI, count, VC1, VC2, VC1OLD, C1, C2, CP, ITER, AV, AVNL1,AVNL2, AVNL3, AVNL4, AVOLD, AVNEW, AVVAR, AVOLD, VO2, VO3, VO4, ytemp,
OPSPOS, OPSNEG;
    double *AVNL;
    InputRealPtrsType uPtrs1 = ssGetInputPortRealSignalPtrs(S,0);
    InputRealPtrsType uPtrs2 = ssGetInputPortRealSignalPtrs(S,1);
    real_T *y = ssGetOutputPortRealSignal(S,0);
    real_T *work = ssGetRWork(S);
    PHI = *mxGetPr(phi(S));
    C1 = *mxGetPr(c1(S));
    C2 = *mxGetPr(c2(S));
    CP = *mxGetPr(cp(S));
    AV = *mxGetPr(av(S));
    AVNL= mxGetPr(avnl(S));
    AVNL1 = AVNL[0];
    AVNL2 = AVNL[1];
    AVNL3 = AVNL[2];
    AVNL4 = AVNL[3];
    OPSPOS = *mxGetPr(opspos(S));
    OPSNEG = *mxGetPr(opsneg(S));

    U1=*uPtrs1[0];
    U2=*uPtrs2[0];
    count=work[0];
    VC1=work[1];
    VC2=work[2];
    VC1OLD=work[3];
    AVOLD=work[4];
    if(PHI==1){
            if(count==0){
                VC1 = U1;
                AVNEW = AV;
                ITER = 0;
                do{
                    ytemp = y[0]*(1+1/AVNEW);
                    AVOLD = AVNEW;
                    VO2 = ytemp*ytemp;
                    VO3 = VO2*ytemp;
                    VO4 = VO3*ytemp;
                    AVNEW = AV*(1+(AVNL1*ytemp)+(AVNL2*VO2)+(AVNL3*VO3)+(AVNL4*VO4));
                    AVVAR = fabs((AVOLD-AVNEW)/AVNEW);
                    ITER = ITER + 1;
                }while(AVVAR>0.01&&ITER<50);
                VC2=ytemp;
                work[4]=AVNEW;
                count=count+1;}
            else{
                VC1OLD = VC1;
                VC1 = U2;
                AVNEW = AV;
                ITER = 0;
                do{
                    ytemp = ((C1/C2)*(VC1OLD-VC1) + VC2*(1+(1+CP/C2)/AVOLD))/(1+(1+(C1+CP)/C2)/AVNEW);
                    AVOLD = AVNEW;
                    VO2 = ytemp*ytemp;
                    VO3 = VO2*ytemp;
                    VO4 = VO3*ytemp;
                    AVNEW = AV*(1+(AVNL1*ytemp)+(AVNL2*VO2)+(AVNL3*VO3)+(AVNL4*VO4));
                    AVVAR = fabs((AVOLD-AVNEW)/AVNEW);
                    ITER = ITER + 1;
                }while(AVVAR>0.01&&ITER<50);
                y[0] = ytemp;
                if(y[0]>OPSPOS){
                    y[0] = OPSPOS;
                }
                if(y[0]<OPSNEG){
                    y[0] = OPSNEG;
                }
                count=0;}
    }
    else{
            if(count==0){
                VC1OLD = VC1;
                VC1 = U2;
                AVNEW = AV;
                ITER = 0;
                do{
                    ytemp = ((C1/C2)*(VC1OLD-VC1) + VC2*(1+(1+CP/C2)/AVOLD))/(1+(1+(C1+CP)/C2)/AVNEW);
                    AVOLD = AVNEW;
                    VO2 = ytemp*ytemp;
                    VO3 = VO2*ytemp;
                    VO4 = VO3*ytemp;
                    AVNEW = AV*(1+(AVNL1*ytemp)+(AVNL2*VO2)+(AVNL3*VO3)+(AVNL4*VO4));
                    AVVAR = fabs((AVOLD-AVNEW)/AVNEW);
                    ITER = ITER + 1;
                }while(AVVAR>0.01&&ITER<50);
                y[0] = ytemp;
                if(y[0]>OPSPOS){
                    y[0] = OPSPOS;
                }
                if(y[0]<OPSNEG){
                    y[0] = OPSNEG;
                }
                count=count+1;}
            else{
                VC1 = U1;
                AVNEW = AV;
                ITER = 0;
                do{
                    ytemp = y[0]*(1+1/AVNEW);
                    AVOLD = AVNEW;
                    VO2 = ytemp*ytemp;
                    VO3 = VO2*ytemp;
                    VO4 = VO3*ytemp;
                    AVNEW = AV*(1+(AVNL1*ytemp)+(AVNL2*VO2)+(AVNL3*VO3)+(AVNL4*VO4));
                    AVVAR = fabs((AVOLD-AVNEW)/AVNEW);
                    ITER = ITER + 1;
                }while(AVVAR>0.01&&ITER<50);
                VC2=ytemp;
                work[4]=AVNEW;
                count=0;}
    }
    work[0]=count;
    work[1]=VC1;
    work[2]=VC2;
    work[3]=VC1OLD;
```

Figure 5.17 C code included in the *mdlOutputs* section of the S-function file in Figure 5.16.

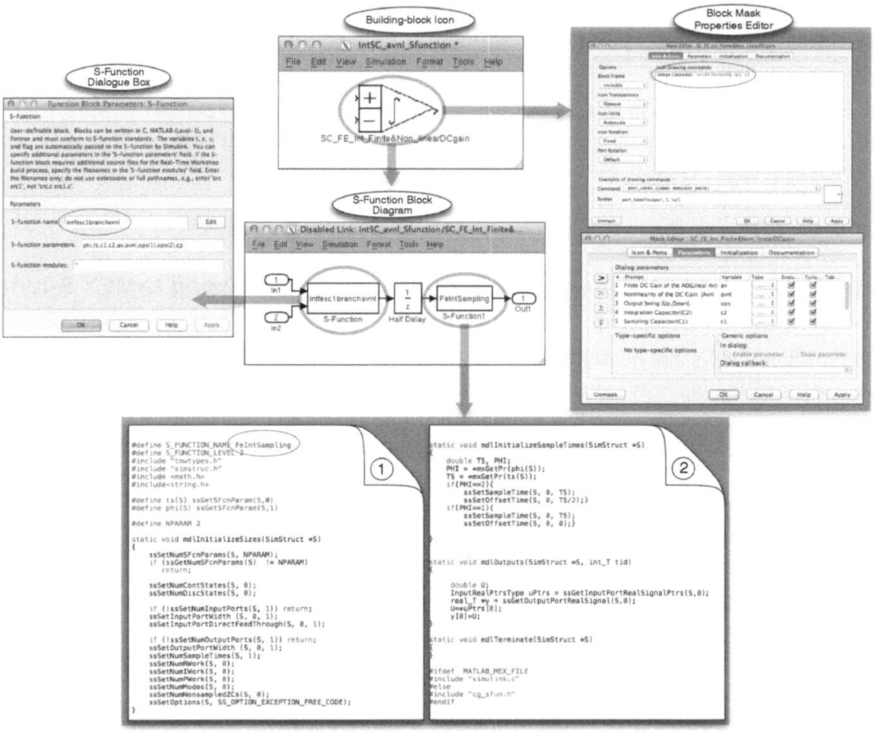

Figure 5.18 Illustrating the implementation of the S-function of Figure 5.16 in SIMULINK.

As stated above, once the C-coded S-function source file (`intfesc1branchavnl.c`) has been generated, it must be compiled to make it executable in MATLAB. In this example, the compilation is done by running `mex intfesc1branchavnl.c` in the MATLAB command window. As a result, a compiled file – called the MEX file – is generated; the term MEX comes from "MATLAB EXecutable" [29].

The compiled MEX S-function is incorporated into a SIMULINK model using the `S-function block` available in the `User-Defined Functions` SIMULINK library. Figure 5.18 illustrates the SIMULINK block associated with the S-function of Figure 5.16, showing the main dialogue windows. The S-function name and the model parameters are entered into the `Block Parameter` dialogue box. It is very important that the block name, as well as the model parameters, are the same as those included in the source file of Figure 5.16. The input/output terminal ports are connected to the S-function block in the SIMULINK block diagram.

Note that an additional S-function block called `FeIntSampling` is included in the block diagram of Figure 5.18. This block – whose C-coded S-function source file is shown in Figure 5.16 – is used for sampling the integrator output at the correct time instant. This additional S-function is combined with a `Half Delay` SIMULINK block to give more precise control of the clock phase at which the integrator output is taken and to collect only one output sample per clock period.

In order to make the use of S-functions easy in more complex systems, a block mask can be created. An specific mask icon can be used instead of the SIMULINK subsystem's standard icon. This is very useful because it enables different building blocks in a given system like a ΣΔM to be identified. The mask has also a dialogue box, allowing the designer to change the model parameters in a simple way. These model parameters are passed to the S-function and dynamically linked into SIMULINK when they are needed during a simulation.

The procedure described in this section must be followed in order to create the behavioral models of the different ΣΔM building blocks. The next section explains in detail some of the most important building-block models, as well as their implementation in the MATLAB/SIMULINK environment using C-coded S-functions.

5.4 Efficient Behavioral Modeling of ΣΔM Building Blocks using C-MEX S-functions

All ΣΔM building blocks can be modeled with C-MEX S-functions using the methodology described in the last section. In this way, any ΣΔM architecture can be simulated at the system level in an efficient way, in terms of accuracy and CPU time. As an illustration, Figure 5.19 shows the SIMULINK block diagram of a second-order ΣΔM implemented using SC integrators (Figure 5.19a) and Gm-C integrators (Figure 5.19b). These block diagrams contain S-function blocks that model integrators, quantizers, and feedback DACs. Other auxiliary subcircuits, such as the output impedances of the Gm-C integrators (made up of R_o and C) and the digital latches, are also included in the models.

Each block in Figure 5.19 can be implemented using different circuit topologies, each of which is affected by a number of circuit error mechanisms. A detailed explanation of all different ΣΔM building-block behavioral models and their corresponding S-functions is beyond the scope of this book. Instead, this section describes the model of the most essential elements of ΣΔMs – integrators, quantizers, and DACs – taking into account the most important nonideal effects and putting special emphasis on those aspects related to their implementation using C-MEX S-functions.

5.4.1 Modeling of SC Integrators using S-functions

Let us consider again the conceptual schematic of an SC FE integrator shown in Figure 5.8a. Note that a single-ended schematic is shown for the sake of simplicity. However, the behavioral model described below takes into account a fully-differential topology.

Its ideal behavior is described by the finite-difference equation given in Equation (5.1). The impact of finite OTA DC gain was considered in Section 5.3, taking into account both linear and nonlinear effects. However, in practice the performance of SC integrators is degraded by a number of error mechanisms, as described in Chapters 3 and 4. An accurate behavioral model must take into account the contribution of the main circuit errors, as well as the clock phases in which they are affecting the performance of the circuit to be modeled – an SC FE integrator in this case. To this end, the effect of all SC circuit nonidealities is computed by following the iterative procedure shown in the flow chart of Figure 5.20 [6].

The model starts by loading the values of the required model parameters, the input signal, and the initial conditions – i.e., the voltages at internal nodes of the integrator (including the values stored on the sampling and integration capacitor) stored in the previous clock period. These are usually set to zero at the beginning of a simulation. Before starting to compute the behavioral model, some initial calculations are made: the equivalent input-referred noise, the actual value of the integrators' weights due to capacitor mismatch, and the parasitic and load capacitances at different nodes of the integrator during sampling and integration clock phases.

The different integrator clock phases, which correspond to the two main branches in Figure 5.20, are selected according to the value of the clock-phase counter `count` and the input switch phase – either

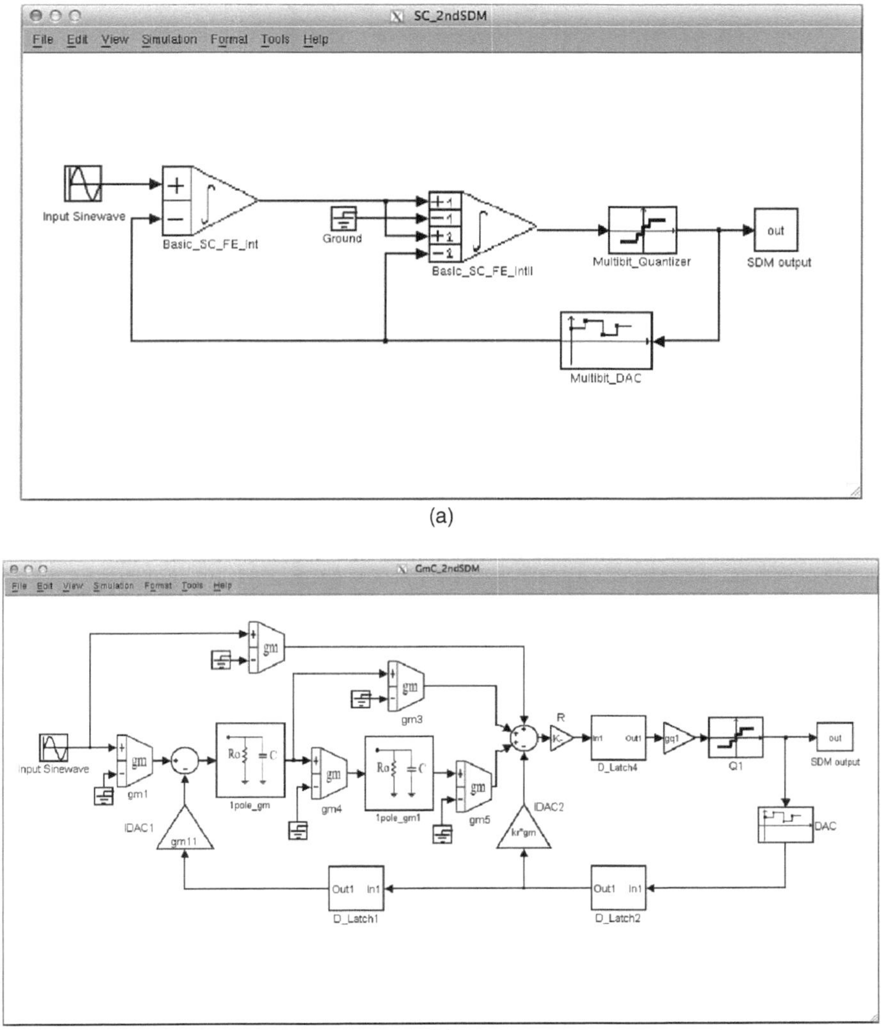

Figure 5.19 Illustrating the behavioral model of a second-order ΣΔM using S-functions: (a) SC implementation; (b) Gm-C implementation.

ϕ_1 or ϕ_2, as illustrated in Figure 5.11. In each clock phase, the effect of main SC circuit nonidealities is taken into account: finite (linear and nonlinear) switch on-resistance, capacitor nonlinearity, thermal noise, incomplete settling, finite (linear and nonlinear) OTA DC gain (modeled as described in Section 5.3), and output swing limitation. The behavioral model of these errors is based on the nonideal equations set out in Chapter 3. These equations can be codified in C and incorporated in an S-function as illustrated in Figures 5.21–5.23.

Note that the S-function file follows the same structure as that shown in Figure 5.16, but incorporating more detailed information about the circuit errors in the behavioral model. For the sake of simplicity, Figures 5.21–5.23 show only the most important parts of the S-function file. Although additional comments have been included in the C code, the following sections explain the main parts

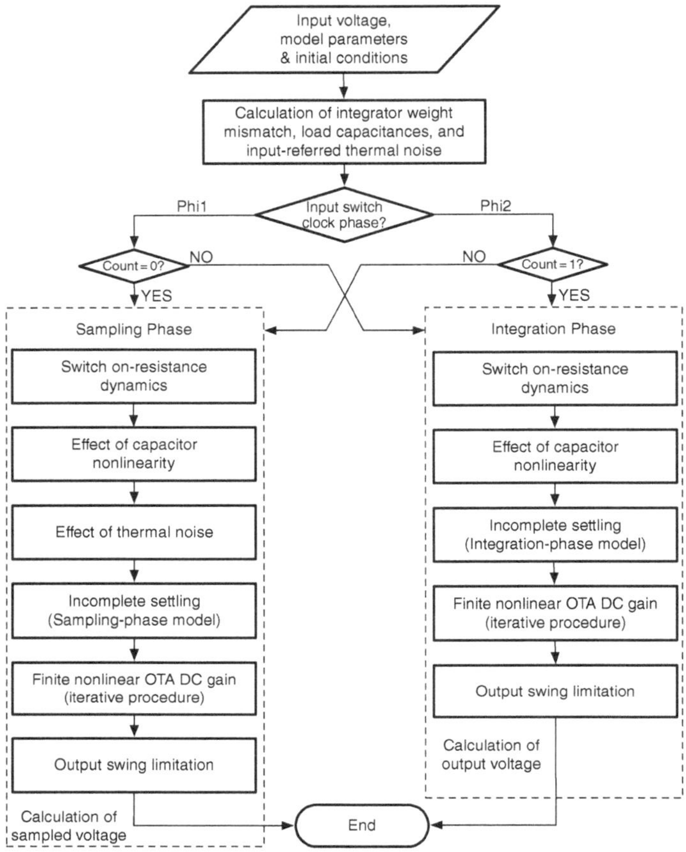

Figure 5.20 Flow chart of the SC FE integrator computational model.

of the S-function by linking the C code with the design equations, described in Chapter 3, on the different SC circuit nonidealities.

5.4.1.1 Capacitor Mismatch and Nonlinearity As discussed in Section 3.3, capacitor mismatch is modeled as a random deviation of the integrators' weights from the nominal value g. This random variation is included in the C code [37] as a normal (or Gaussian) distribution, with a mean value of g (denoted as `MEAN` in Figure 5.21) and a variance (a model parameter `VARIANCE`).

Another nonideal effect associated with the integrator capacitors is the nonlinear dependence of their capacitance on the voltage drop across them (v), which is modeled as a polynomial function:

$$C_{s,i}(v) = C_{s,i_{(\text{nominal})}} \cdot (1 + \text{CNL}_1 \cdot v + \text{CNL}_2 \cdot v^2) \tag{5.12}$$

where $C_{s,i_{(\text{nominal})}}$ are the nominal values of the sampling (C_s) or the integration capacitors (C_i) and $\text{CNL}_{1,2}$ are respectively the first- and second-order nonlinear voltage coefficients represented in the model by `CNL1,2` (see Figure 5.22). In this way, the charge stored in the capacitors is given by [38]:

$$Q_{s,i}(v) = \int_0^v C_{s,i}(v)\text{d}v = C_{s,i_{(\text{nominal})}} \cdot \left(v + \frac{\text{CNL}_1}{2} \cdot v^2 + \frac{\text{CNL}_2}{3} \cdot v^3\right) \tag{5.13}$$

Part 1

```
#define S_FUNCTION_NAME intfesclbranch_allerrors
#define S_FUNCTION_LEVEL 2
...
#define phi(S)          ssGetSFcnParam(S,0)   /* Input switch clock phase */
#define ts(S)           ssGetSFcnParam(S,1)   /* Sampling time */
#define c(S)            ssGetSFcnParam(S,2)   /* Capacitor values and non-linear coefficients */
#define var(S)          ssGetSFcnParam(S,3)   /* Integrator weight's variance, eq. vrms input and temperature */
#define ops(S)          ssGetSFcnParam(S,4)   /* Output swing limits */
#define ao(S)           ssGetSFcnParam(S,5)   /* OTA Finite DC gain, transconductance and maximum output current */
#define setnoise(S)     ssGetSFcnParam(S,6)   /* Flag that activates the effect of thermal noise   */
#define avnl(S)         ssGetSFcnParam(S,7)   /* Non-linear coefficients of the finite DC gain */
#define cp(S)           ssGetSFcnParam(S,8)   /* OTA parasitic capacitances and load capacitances */
#define settlingon(S)   ssGetSFcnParam(S,9)   /* Flag that activates incomplete settling model */
#define setin(S)        ssGetSFcnParam(S,10)  /* Integrator identifier used in the incomplete settling model */
#define idown(S)        ssGetSFcnParam(S,11)  /* Integrator identifier */
#define idnext(S)       ssGetSFcnParam(S,12)  /* Identifier of the Integrator connected at the Integrator output */
#define setloads(S)     ssGetSFcnParam(S,13)  /* Flag that activates additional load capacitances  */
#define CLOAD_S(S)      ssGetSFcnParam(S,14)  /* Additional load capacitance in the sampling phase */
#define CLOAD_I(S)      ssGetSFcnParam(S,15)  /* Additional load capacitance in the integration phase */
#define setphases(S)     ssGetSFcnParam(S,16) /* Flag that activates the reduction of the clock phase switch-on time */
#define phases_reduction(S)    ssGetSFcnParam(S,17) /* Duty cycle, reduction of the integration and sampling time */
#define BW(S)    ssGetSFcnParam(S,18)         /*  SIGNAL BANDWIDTH */
#define numid(S)    ssGetSFcnParam(S,19)      /* Integrator identification number, i.e. int1, int2, int3... */
#define NPARAM 20

static void mdlInitializeSizes(SimStruct *S)
{...
      ssSetInputPortWidth (S, 3, DYNAMICALLY_SIZED); /* Input port where the "From" block is connected */
...}

static void mdlInitializeSampleTimes(SimStruct *S) /* Clock-phase definition (the same as for ideal SC Integ. */
{...}

static void mdlStart(real_T *x0,SimStruct *S)  /*Initial conditions (similar to ideal SC integrator */
{...}

static void mdlOutputs(SimStruct *S, int_T tid)
{    /* Declaration of model variables and parameters */
     double U1, U2, PHI, cont, VC1, VC2, VC1ANT
...
       InputRealPtrsType uPtrs1 = ssGetInputPortRealSignalPtrs(S,0);
...
       InputRealPtrsType uPtrs4 = ssGetInputPortRealSignalPtrs(S,3);
     real_T *y = ssGetOutputPortRealSignal(S,0);
...
      /* DEFINITION OF MODEL PARAMETERS */
     PHI = *mxGetPr(phi(S)); /* Input switch clock phase */
     setloads = *mxGetPr(setloads(S)); /* Additional Load capacitances during the sampling (S) and integration phase (I) */
     if (setloads) { CLOAD_S = *mxGetPr(CLOAD_S(S)); CLOAD_I = *mxGetPr(CLOAD_I(S));}
     else {CLOAD_S = 0; CLOAD_I = 0;}

     setphases = *mxGetPr(setphases(S));  /* Duration of Sampling/Integration clock-phase */
     if (setphases==1){
       phases_reduction = mxGetPr(phases_reduction(S)); duty_cycle=phases_reduction[0];
       int_red=phases_reduction[1];samp_red=phases_reduction[2];}
     else{duty_cycle=50; int_red=0; samp_red=0;}

     p = mxGetPr(c(S)); /* Capacitor values and non-linear coefficients */
     C1 = p[0]; C2 = p[1]; CNL1 = p[2]; CNL2 = p[3];
     p = mxGetPr(ao(S)); /* OTA Finite DC gain (AV), transconductance (GM) and maximum output current (IO) */
     AV = p[0]; GM = p[1]; IO = p[2];
     p = mxGetPr(avnl(S)); /* Non-linear coefficients of the finite DC gain */
     AVNL1 = p[0]; AVNL2 = p[1]; AVNL3 = p[2]; AVNL4 = p[3];
     p = mxGetPr(var(S)); /* Integrator weight's variance, eq. vrms input and temperature */
     VARIANCE = p[0]; INPSD = p[1]; TEMP = p[2]; RON = p[3];
     p = mxGetPr(cp(S)); /* OTA Input parasitic capacitances and load capacitance */
       CP1 = p[0]; CP2 = p[1]; CLOAD = p[2];
     TS = *mxGetPr(ts(S)); /* Sampling time */
     FS = 1/TS; /* Sampling frequency */
     p = mxGetPr(ops(S)); /* Output swing limits */
     OPSPOS = p[0]; OPSNEG = p[1];
     SETTLINGON = *mxGetPr(settlingon(S)); /* Settling flag that activates incomplete settling model */
     SETIN = *mxGetPr(setin(S)); /* Integrator identifier to be used in the incomplete settling error computation */
     U1=*uPtrs1[0]; /* Value of signal connected to input branch #1 */
     U2=*uPtrs2[0]; /* Value of signal connected to input branch #2 */
     UNOISE=*uPtrs3[0]; /* Value of noise source connected at the input */
     N=*uPtrs4[0]; /* Number of input branches */
     cont=work[0]; /* Counter used to determine clock phase: sampling phase / integration phase */

/*      Initial values of VC1(n), VC1(n-1)... */
     VC1=work[1]; VC2=work[2]; VC1ANT=work[3];VAN=work[4]; VAN1MED=work[5];VAINII=work[6]; VAINIS=work[7]; AV_OLD=work[8];

/*   Reduction of clock-phase duration */
     TINT=TS*(duty_cycle/100)-int_red; /* Integration phase duration */
     TSAMP=TS*(1-duty_cycle/100)-samp_red; /*Sampling phase duration */

/*   Mismatch Parameters and calculation of integrator's weight       */
     MEAN = C1/C2; /* Nominal value of the integrator weight */
     SEMILLA = 1; /*Gaussian random number generator */
     a=16807; m=2147483647; seed=SEMILLA; zno=seed; zn=(a*zno)%m;noise_int=zn/m;
     zno=zn; zn=(a*zno)%m; seed=zn;noise_int2=zn/m;
     weigth=MEAN+(VARIANCE*sqrt(-2*log(noise_int))*cos(2*3.1416*noise_int2));
```

Figure 5.21 Excerpt of the MATLAB C-coded S-function file of an SC FE integrator considering all circuit errors: Model parameters, definition of clock-phase timing, and calculation of integrator weight including mismatch (Part 1 of 3).

Note that the resulting expression can be viewed as the charge stored in a nominal linear capacitor $C_{s,i_{(\mathrm{nominal})}}$ with a voltage across it given by $(v + \mathrm{CNL}_1/2 \cdot v^2 + \mathrm{CNL}_2/3 \cdot v^3)$.

5.4.1.2 Input-referred Thermal Noise The circuit noise model takes into account the thermal noise generated in the switches and the opamp [32]. Flicker (or $1/f$) noise is not considered in

```
/*   Calculation of the equivalent load capacitance in the integration (CEQI) and sampling phase (CEQS)*/
     CP=CP2; CSUM=0;
     if (setloads==1)
     {
         CEQI = C1+CP+((CLOAD+CLOAD_I)*(1+((C1+CP)/C2))); CPRIMA = CP + ((CLOAD+CLOAD_I)*(1+(CP/C2)));
         if (SETIN==1)
         {
             for(i=N+1;i<=2*N;i++){CSUM=CSUM + (*uPtrs4[i]);}
             CEQS = CP+(CLOAD+CLOAD_S+CSUM)*(1+CP/C2);
         }
         else {CEQS= CP + ((CLOAD+CLOAD_S)*(1+(CP/C2)));}
     }
     else
     {
         CEQI = C1+CP+(CLOAD*(1+((C1+CP)/C2))); CPRIMA = CP + (CLOAD*(1+(CP/C2)));
         if (SETIN==1)
         {
             for(i=N+1;i<=2*N;i++){ CSUM=CSUM + (*uPtrs4[i]);}
             CEQS = CP+(CLOAD+CSUM)*(1+CP/C2);
         }
         else
         {
             CEQS=CPRIMA;
         }
     }

/*   Thermal Noise Parameters and Calculation of the input referred noise voltage */

   setnoise = *mxGetPr(setnoise(S)); K = 1.380658*pow(10,(-23));
   SIN_PHI1 = K*TEMP/(C1*FS); /*   Switches' contribution (PHI1) */
   z1I=GM/CPRIMA; p1I=GM/CEQI; p2=CEQI/CPRIMA/(2*RON*C1);
      BWn_SPHI2=p1I*p2*(p1I*p2+(z1I*z1I))/(4*z1I*z1I*(p1I+p2));
   SS=2*K*TEMP*2*RON; SIN_PHI2 = 2*BWn_SPHI2*SS/FS; /*   Switches' contribution (PHI2) */
   SOP = INPSD*INPSD; BWn_OP= p1I*p2/(4*(p1I+p2));SIN_OP=2*BWn_OP*SOP/FS; /*   Amplifier contribution */

   if(RON==0){DELTANOISE=sqrt(12*SIN_OP*FS);}
   else {DELTANOISE = sqrt(12*(2*SIN_PHI1+2*SIN_PHI2+SIN_OP)*FS);} /* SIN_PHI1,2 ARE DOUBLED TO TAKE INTO ACCOUNT THE DIFFERENTIAL IMPLEMENTATION */

   if (setnoise!=1) {DELTANOISE=0;}

   VNOISE = UNOISE*DELTANOISE;

   /* Code that accounts for the effect of the RON on the dynamics */
      ...
      GMSEFF=GM*CEQS/(CEQS+(2*RNON*CSUM*GM)); /* RNON = switch-on resistance of the switches connected at the output of the integrator */
      GMIEFF=GM*CEQI/(CEQI+(2*RON*C1*GM));

     /*  Input switch clocked at phi1: PHI=1     */

   if(PHI==1){
         if(cont==0){ /* SAMPLING PHASE */
           if(SETTLINGON==1){ /* Effect of Linear switch-on resistance (Ron), capacitor non-linearity and thermal noise */
             if(RON==0){VC1 = -((U1*(1+(CNL1*U1/2)+(CNL2*U1*U1/3)))+VNOISE);}
             else{VC1 = -((U1*(1+(CNL1*U1/2)+(CNL2*U1*U1/3)))+VNOISE)*(1-exp((-1)*TSAMP/(RON*C1)))}
             if (SETIN==1) /* Incomplete Settling - Sampling Phase */
                 {
                     VAINIS = VAN;
                     for(i=1;i<=N;i++){VAINIS = VAINIS - (*uPtrs4[N+i]/CEQS)*(y[0]- (*uPtrs4[i]));}
                 }
             else {VAINIS = VAN;}
             if(VAINIS>=0){SIGNO=1;}
             if(VAINIS<0){SIGNO=-1;}
                            TOS = (CEQS*fabs(VAINIS)/IO)-(CEQS/GMSEFF); /* TOS and opamp input-node voltage */
                            if(fabs(VAINIS)<=(IO/GMSEFF)){VAN1MED = VAINIS*exp(-GMSEFF*TSAMP/(CEQS));}
             if((fabs(VAINIS)>(IO/GMSEFF))&&(TOS<(TSAMP))){VAN1MED = (IO/GMSEFF)*SIGNO*exp(-GMSEFF*((TSAMP)-TOS)/(CEQS));}
             if((fabs(VAINIS)>(IO/GMSEFF))&&(TOS>=(TSAMP))){VAN1MED = VAINIS-((IO/CEQS)*SIGNO*(TSAMP));}
             AVNEW = AV; /* Non-linear Finite DC Gain - Iterative procedure */
             ITER = 0;
             do{
                 temp = y[0]*(1+1/AVNEW)+((1+(CP/C2))*(VAN1MED-VAN)); /* opamp output-node voltage */
                 temp = temp*(1+(CNL1*temp/2)+(CNL2*temp*temp/3)); /* Capacitor non-linearity */
                 AVANT = AVNEW; VO2 = temp*temp; VO3 = VO2*temp; VO4 = VO3*temp;
                 AVNEW = AV*(1+(AVNL1*temp)+(AVNL2*VO2)+(AVNL3*VO3)+(AVNL4*VO4));
                                    AVVAR = fabs((AVANT-AVNEW)/AVNEW); ITER = ITER + 1;
             }while(AVVAR>0.01&&ITER<50);
             VC2=temp;
             work[8]=AVNEW;
             if(VC2>OPSPOS){ /* Effect of output swing */
                 VC2 = OPSPOS; /* MAXIMUM VALUE OF THE SATURATION VOLTAGE */
             }
             if(VC2<OPSNEG){
                 VC2 = OPSNEG; /* MINIMUM VALUE OF THE SATURATION VOLTAGE */
             }
             cont=cont+1;
           }
           else{
             if(RON==0){VC1 = -((U1*(1+(CNL1*U1/2)+(CNL2*U1*U1/3)))+VNOISE);}
             else{VC1 = -((U1*(1+(CNL1*U1/2)+(CNL2*U1*U1/3)))+VNOISE)*(1-exp((-1)*TSAMP/(RON*C1)));}
             AVNEW = AV;
             ITER = 0;
             do{
                 temp = y[0]*(1+1/AVNEW);
                 temp = temp*(1+(CNL1*temp/2)+(CNL2*temp*temp/3));
                 AVANT = AVNEW; VO2 = temp*temp; VO3 = VO2*temp; VO4 = VO3*temp;
                 AVNEW = AV*(1+(AVNL1*temp)+(AVNL2*VO2)+(AVNL3*VO3)+(AVNL4*VO4));
                 AVVAR = fabs((AVANT-AVNEW)/AVNEW); ITER = ITER + 1;
             }while(AVVAR>0.01&&ITER<50);
             VC2=temp;
             work[8]=AVNEW;
             if(VC2>OPSPOS){ /*Effect of output swing*/
                 VC2 = OPSPOS;
             }
             if(VC2<OPSNEG){
                 VC2 = OPSNEG;
             }
             cont=cont+1;
           }
```

Part 2
Equivalent load calculation
Input-referred thermal noise
Incomplete settling during sampling phase
Nonlinear finite opamp DC gain
Output swing
Nonlinear finite opamp DC gain and output swing computation when settling is not considered

Figure 5.22 Excerpt of MATLAB C-coded S-function file of an SC FE integrator considering all circuit errors: calculation of equivalent load capacitances, input-referred thermal noise, capacitor nonlinearity, incomplete settling during sampling phase, nonlinear finite opamp DC gain, and output swing (Part 2 of 3).

the system-level behavioral model because it is assumed that it will be cancelled out by suitable electrical design at the transistor level. Other noise contributions, like thermal and flicker noise sources associated with the ΣΔM reference voltages, are usually considered during the electrical (transistor-level) design. In this way, based on the analysis described in Section 3.5, the RMS value of the input-referred thermal noise included in the model (`VNOISE` in Figure 5.22) is approximated by,

$$v_{\mathrm{th}} = \mathrm{UNOISE}\sqrt{12 f_s (2S_{in\phi_1} + 2S_{in\phi_2} + S_{\mathrm{op}})} \tag{5.14}$$

```
        /* cont==1 */ /* INTEGRATION PHASE */
        else{
         if(SETTLINGON==1){
           VC1ANT = VC1;/* Effect of Linear switch-on resistance (Ron) and capacitor non-linearity */
           if(RON==0){
               VC1 = -((U2*(1+(CNL1*U2/2)+(CNL2*U2*U2/3))));
           }
           else{
               VC1 = -((U2*(1+(CNL1*U2/2)+(CNL2*U2*U2/3))))*(1-exp((-1)*TINT/(RON*C1)));
           }
              /* Incomplete Settling - Integration Phase */ /* VAINII */
            VAINII = (((1+((CLOAD+CLOAD_I)/C2))/CEQI)*C1*(-VC1+VC1ANT))+((CPRIMA/CEQI)*VAN1MED);
           /**/
           if(VAINII>0){
               SIGNO=1;
           }
           if(VAINII<0){
               SIGNO=-1;
           }
           TOI =(TS/2)+(CEQI*fabs(VAINII)/IO)-(CEQI/GMIEFF); /* TOI and opamp input-node voltage */
           if(fabs(VAINII)<=(IO/GMIEFF)){
               VAN = VAINII*exp(-GMIEFF*TINT/(CEQI));
           }
           if((fabs(VAINII)>(IO/GMIEFF))&&(TOI<(2*TINT))){
               VAN = (IO/GMIEFF)*SIGNO*exp(-GMIEFF*((2*TINT)-TOI)/(CEQI));
           }
           if((fabs(VAINII)>(IO/GMIEFF))&&(TOI>=(2*TINT))){
               VAN = VAINII-((IO/CEQI)*SIGNO*(TINT));
           }
           AVNEW = AV;/* Non-linear Finite DC Gain */
           ITER = 0;
           do{ /* opamp output-node voltage */
               temp = (-(weigth)*(VC1ANT-VC1) + VC2*(1+(1+CP/C2)/AV_OLD))/(1+(1+weigth+CP/C2)/AVNEW) + (1+((CP+C1)/C2))*VAN - (1+(CP/C2))*VAN1MED;
               temp = temp/(1+(CNL1*temp/2)+(CNL2*temp*temp/3)); /* Capacitor non-linearity */
               if(temp>OPSPOS){ /*Effect of output swing*/
                   temp=OPSPOS;
               }
               if(temp<OPSNEG){
                   temp=OPSNEG;
               }
               AVANT = AVNEW;
               VO2 = temp*temp;
               VO3 = VO2*temp;
               VO4 = VO3*temp;
               AVNEW = AV*(1+(AVNL1*temp)+(AVNL2*VO2)+(AVNL3*VO3)+(AVNL4*VO4));
               AVVAR = fabs((AVANT-AVNEW)/AVNEW);
               ITER = ITER + 1;
               }while(AVVAR>0.01&&ITER<50);
           y[0] = temp;
           VC1 = -VC1-VAN;
           y2[0]=VC1;
           cont=0;
         }
         else{
           VC1ANT = VC1;
           if(RON==0){
               VC1 = -((U2*(1+(CNL1*U2/2)+(CNL2*U2*U2/3))));
           }
           else{
               VC1 = -((U2*(1+(CNL1*U2/2)+(CNL2*U2*U2/3))))*(1-exp((-1)*TINT/(RON*C1)));
           }
           AVNEW = AV;
           ITER = 0;
           do{
               temp = (-(weigth)*(VC1ANT-VC1) + VC2*(1+(1+CP/C2)/AV_OLD))/(1+(1+weigth+CP/C2)/AVNEW);
               temp = temp/(1+(CNL1*temp/2)+(CNL2*temp*temp/3));
               if(temp>OPSPOS){
                   temp=OPSPOS;
               }
               if(temp<OPSNEG){
                   temp=OPSNEG;
               }
               AVANT = AVNEW;
               VO2 = temp*temp;
               VO3 = VO2*temp;
               VO4 = VO3*temp;
               AVNEW = AV*(1+(AVNL1*temp)+(AVNL2*VO2)+(AVNL3*VO3)+(AVNL4*VO4));
               AVVAR = fabs((AVANT-AVNEW)/AVNEW);
               ITER = ITER + 1;
               }while(AVVAR>0.01&&ITER<50);
           y[0] = temp;
           y2[0]=-VC1;
           cont=0;
         }
        }
    }

    else /*  Input switch clocked at phi2: PHI=2     */
       {.............. }

    work[0]=cont; work[1]=VC1; work[2]=VC2; work[3]=VC1ANT; work[4]=VAN; work[5]=VAN1MED; work[6]=VAINII; work[7]=VAINIS;
}

static void mdlTerminate(SimStruct *S)
{...}

#ifdef  MATLAB_MEX_FILE
#include "simulink.c"
#else
#include "cg_sfun.h"
#endif
```

Part 3

Incomplete settling during integration phase

Nonlinear finite opamp DC gain and output swing during integration (considering the effect of settling)

Nonlinear finite opamp DC gain and output swing during integration when settling is not considered

Similar code goes here when clock phase of input switch is Phi2

Figure 5.23 Excerpt of MATLAB C-coded S-function file of an SC FE integrator considering all circuit errors: calculation of incomplete settling during integration phase, capacitor nonlinearity, nonlinear finite opamp DC gain, and output swing (Part 3 of 3).

where UNOISE is a random number in the range (−0.5, +0.5) generated by the `Uniform Random Number` building block available in SIMULINK, S_{op} represents the PSD of the thermal noise generated by the opamp, and $S_{in\phi_{1,2}}$ stands for the PSD of the thermal noise generated in the switches during clock phases $\phi_{1,2}$. These PSD functions, respectively denoted as `SIN_OP, SIN_PHI1, SIN_PHI2` in the C code of Figure 5.22, are calculated as a function of different electrical parameters involved in the thermal noise model analyzed in Section 3.5. These parameters include the switch on-resistance R_{on} (denoted as `RON` in the C code of Figure 5.22), the equivalent noise bandwidth of noise sources associated with the switches during clock phase ϕ_2 (denoted as `BWn_SPHI2` in Figure 5.22), and the equivalent bandwidth of the amplifier thermal noise (`BWn_OP` in Figure 5.22). Note that if ideal switches are assumed – that is, $R_{\text{on}} = 0$ – only the amplifier contributes to the thermal noise model.

5.4.1.3 Switch On-resistance Dynamics The effect of finite switch on-resistance R_{on} is considered in both clock phases. For instance, as shown in the C code of Figures 5.21–5.23, the value of the voltage sampled on C_s taking into account the effect of capacitor nonlinearity, thermal noise, and the linear sampling process due to R_{on} is given by,

$$v_{C_s} = -\left[(v_i + \frac{\text{CNL}_1}{2} \cdot v_i^2 + \frac{\text{CNL}_2}{3} \cdot v_i^3) + v_{\text{th}}\right]\left(1 - e^{\frac{T_s}{R_{\text{on}} C_s}}\right) \tag{5.15}$$

where v_i is the value of the input signal at the end of the sampling phase – denoted as `U1` in the C code.

As stated in Section 3.7.2, switches are usually implemented as CMOS transmission gates and the value of R_{on} strongly depends on the value of the voltage drop across the nodes of the switch. This nonlinear phenomenon causes harmonic distortion, which increases with the ratio between the input frequency and the sampling frequency [39], thus being especially important in broadband applications.

In order to model the effect of nonlinear R_{on} in CMOS switches, let us consider again the circuit in Figure 3.22. This circuit – depicted in Figure 5.24 for the sake of clarity – models the sampling operation of a fully-differential SC FE integrator during the sampling phase. Note that only the input CMOS switches connected to C_s on ϕ_1 (sampling phase in Figure 5.24) need to be considered. The behavioral model assumes that the MOS transistors implementing the switches operate in the linear (ohmic) region and that their drain currents $I_{D(N,P)}$ can be expressed as,

$$\begin{aligned} I_{DN} &= \beta_N \left[V_{DS}(V_{GS} - V_{TN}) - \frac{V_{DS}^2}{2}\right] \\ I_{DP} &= \beta_P \left[V_{SD}(V_{SG} - |V_{TP}|) - \frac{V_{SD}^2}{2}\right] \end{aligned} \tag{5.16}$$

where $V_{T(N,P)}$ denotes the threshold voltage of the nMOS and pMOS transistors, V_{DS} and V_{GS} stand for the drain-to-source and gate-to-source voltages of the nMOS transistor, and V_{SD} and V_{SG} are the source-to-drain and source-to-gate voltages of the pMOS transistor.

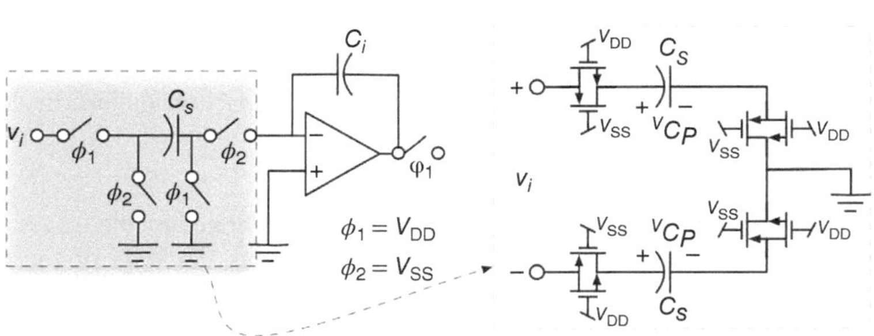

Figure 5.24 Front-end part of a fully-differential SC integrator during the sampling phase.

From Equation (5.16) it can be shown that the currents flowing through the sampling capacitor in the positive and negative branches of Figure 5.24 can be respectively written as,

$$I_{Cp} = C_s \frac{\mathrm{d}v_{Cp}}{\mathrm{d}t} = g_{\text{on}} \left(+\frac{v_i}{2} - v_{Cp} \right) + \Delta\beta \left(\frac{v_{Cp}^2}{2} - \frac{v_i^2}{8} \right)$$

$$I_{Cn} = C_s \frac{\mathrm{d}v_{Cn}}{\mathrm{d}t} = g_{\text{on}} \left(-\frac{v_i}{2} - v_{Cn} \right) + \Delta\beta \left(\frac{v_{Cn}^2}{2} - \frac{v_i^2}{8} \right) \tag{5.17}$$

where $\Delta\beta = (\beta_N - \beta_P)$, $g_{\text{on}} \equiv 1/R_{\text{on}} = [\beta_N(V_{DD} - V_{TN}) + \beta_P(-V_{SS} - |V_{TP}|)]$ is the operating-point switch on-conductance – see Equation (3.48) – and $v_{C_s} = v_{Cp} - v_{Cn}$ represents the differential voltage sampled on C_s.

In order to incorporate the effect of nonlinear sampling caused by CMOS switches in a DT behavioral model, Equations (5.17) can be *discretized* and solved numerically. For instance, if the sampling phase is divided into a number ($j = 1, ..., N_s$) of time intervals δt and a finite-difference approximation is used for integrating Equation (5.17), the sampled voltages of Figure 5.24 can be expressed as,

$$v_{Cp}(t) \simeq \frac{C_s + g_{\text{on}}\delta t - \sqrt{(C_s + g_{\text{on}}\delta t)^2 + \beta\delta t\{\delta t v_i(t)[\beta \frac{v_i(t)}{4} - g_{\text{on}}] - 2C_s v_{Cp}(t - \delta t)\}}}{\beta\delta t}$$

$$v_{Cn}(t) \simeq \frac{C_s + g_{\text{on}}\delta t - \sqrt{(C_s + g_{\text{on}}\delta t)^2 + \beta\delta t\{\delta t v_i(t)[\beta \frac{v_i(t)}{4} + g_{\text{on}}] - 2C_s v_{Cn}(t - \delta t)\}}}{\beta\delta t} \tag{5.18}$$

where $t = j\delta t$, $\delta t = T_s/(2N_s)$, and $v_{C_s}(t - \delta t) \equiv v_{Cp}(t - \delta t) - v_{Cn}(t - \delta t)$ is the differential sampled voltage on C_s at time instant $t - \delta t$.

Figure 5.25 shows C code that can be used in a C-MEX S-function to model the effect of nonlinear R_{on}. It is based on the solution given in Equation (5.18), where `VC1(p,n)` and `VC1OLD(p,n)` denote respectively $v_{C(p,n)}(t)$ and $v_{C(p,n)}(t - \delta t)$. Note that two different cases are considered for the input signal, namely a sinusoidal waveform and a generic waveform. In both cases, the input waveform data can be provided by the user through the SIMULINK S-function. For the sake of completeness, the code in Figure 5.25 also includes a linear version sampling process due to finite R_{on}, in case nonlinearities are not considered.

The C code in Figure 5.25 also shows an alternative approach to modeling the effect of the nonlinear switch on-resistance. This is based on the equivalent circuit shown in Figure 5.26a. In this model, the effect of switches connected to the common voltage node are neglected (see Section 3.7.2) and R_{on} is modeled as a polynomial function of the input signal, given by [6]:

$$R_{\text{on}}(v_i) \simeq \sum_{j=1}^{N_s} p_j \cdot v_i^j(j\delta t) \tag{5.19}$$

Taking into account the above nonlinear characteristic for R_{on}, the voltages sampled at the positive and negative branches of Figure 5.26a can be respectively written as,

$$v_{Cp}(t) \simeq v_{Cp}(t - \delta t) + \left[+\frac{v_i(t - \delta t)}{2} - v_{Cp}(t - \delta t) \right] \left(1 - e^{\frac{T_s}{R_{\text{on}}[+v_i(t)]C_s}} \right)$$

$$v_{Cn}(t) \simeq v_{Cn}(t - \delta t) + \left[-\frac{v_i(t - \delta t)}{2} - v_{Cn}(t - \delta t) \right] \left(1 - e^{\frac{T_s}{R_{\text{on}}[-v_i(t)]C_s}} \right) \tag{5.20}$$

Linear *Ron*

Nonlinear *Ron* (analytic model)

Nonlinear *Ron* lookup table approach

```
switch (model)
    {
    case 1:     /*Finite and Lineal Ron*/
        {
            VC1=(VC1OLD+(*uPtrs1[0]-VC1OLD)*(1-exp(-1*TS/(2*RON*C1))));
            break;
        }
    case 2:     /*Non-linear Ron (analytic model)*/
        {
            g= p[0];
            B= p[1];
            for(i=0;i<k/2;i++) /* Division of the sampling period in "k" dt time intervals */
            {
                if (signal==1){ /* Sinewave input */
                    Vi=A*sin(2*pi*fi*(i*dt+n*TS/2)+phase);}
                else{
                    Vi=sig[k/2*n+i]; /*Generic waveform input (g=gON, B=Beta */
                }
                VC1p=(C1+dt*g-sqrt(pow((C1+g*dt),2)+B*dt*(dt*Vi*(B*Vi/4-g)-2*C1*VC1OLDp)))/(B*dt);
                VC1n=(C1+dt*g-sqrt(pow((C1+g*dt),2)+B*dt*(dt*Vi*(B*Vi/4+g)-2*C1*VC1OLDn)))/(B*dt);
                VC1OLDp=VC1p;
                VC1OLDn=VC1n;
            }
            VC1=VC1p-VC1n;
            break;
        }
    case 3:     /*Non-linear Ron (lookup table model)*/
        {
        for(i=0;i<k/2;i++)
            {
            if (signal==1){
                Vi=A*sin(2*pi*fi*(i*dt+n*TS/2)+phase);}
            else{
                Vi=sig[k/2*n+i];
            }
            Ronp=0;
            Ronn=0;
            for(j=0;j<m;j++)
            {
                Ronp = Ronp + (p[j])*pow(Vi/2,m-j-1);
                Ronn = Ronn + (p[j])*pow(-Vi/2,m-j-1);
            }
            VC1p= (VC1OLDp+(Vi/2-VC1OLDp)*(1-exp(-1*dt/(Ronp*C1))));
            VC1n= (VC1OLDn+(-Vi/2-VC1OLDn)*(1-exp(-1*dt/(Ronn*C1))));
            VC1OLDp=VC1p;
            VC1OLDn=VC1n;
            }
        VC1= VC1p-VC1n;
        break;
        }
}
```

Figure 5.25 Excerpt of the C-coded S-function file including the model of finite (nonlinear) switch on-resistance.

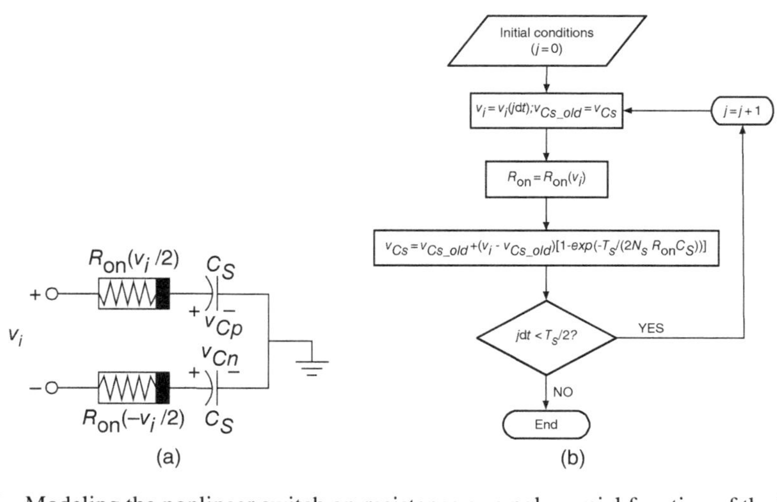

Figure 5.26 Modeling the nonlinear switch on-resistance as a polynomial function of the input signal: (a) equivalent circuit. (b) flow chart.

which can be computed following the flow chart depicted in Figure 5.26b. As shown in Figure 5.25, the corresponding C code has been incorporated in the S-function. Note that the coefficients of the polynomial function in Equation (5.19) can be used either in a synthesis process to evaluate the maximum nonlinearity tolerated for a given set of specifications or to verify a given design using a look-up table approach. In the latter case, the polynomial function coefficients p_j can be obtained from a curve-fitting process applied to the nonlinear R_{on} characteristic obtained in an electrical (transistor-level) simulation of the CMOS switch. The curve-fitting process can be easily performed by using the `polyfit` routine provided by MATLAB [29].

5.4.1.4 Incomplete Settling Error The behavioral model of the transient response for SC FE integrators included in Figures 5.21–5.23 is based on the analysis presented by del Río *et al.* [40], which was discussed in Section 3.4. The model includes the effect of opamp dynamic performance limitations, such as GB and SR, on the charge transfer during both sampling and integration phases. Parasitic capacitors associated with both opamp and CMOS switches, as well as the capacitor load at the integrator output – which changes from sampling to integration phase – are also taken into account. To this end, the equivalent circuit scheme shown in Figure 3.6 is solved in the behavioral model. As shown in Figure 3.6, the model assumes a number N_i of input SC branches – each made up of its corresponding sampling capacitor and switches – and that another SC integrator with N_o input SC branches is connected (as a load) to the output node of the integrator. The model used for the amplifier, shown in Figure 3.7, has a single-pole dynamic and a nonlinear characteristic with maximum output current I_o (`IO` in the C code).[14]

As stated in Section 3.4, the evaluation of the incomplete settling error model begins with the computation of the equivalent capacitive load at the opamp output node during both the integration and sampling phases, given by Equation (3.17). This equation is included in the model of Figure 5.22, where the sampling capacitance of the ith input SC branch of the load integrator is denoted as `uPtrs4[N+i]`, the parasitic capacitor associated with the summation node of the input SC branches is denoted as `CP`, and the load capacitance is `CLOAD`. Note that the model in Figure 5.22 has the possibility of including a different value of C_L during the integration and the sampling phases. Therefore, two different model parameters are used, respectively named `CLOAD_I`, `CLOAD_S`. In this way, the load capacitance during integration and sampling phases are respectively given by `CLOAD+CLOAD_I` and `CLOAD+CLOAD_S`.

After computing the equivalent load capacitances, the settling model is evaluated during both clock phases by considering the different possibilities of the opamp operation; that is, linearly or in slew. In this way, at the end of the sampling phase of period nT_s, the voltage at the input node of the opamp (represented by `VAN` in Figure 5.22) is computed as [40],

$$v_a(nT_s) = \begin{cases} v_{ai,s} e^{-\frac{g_{msef} T_s}{2C_{\mathrm{eq},s}}} & |v_{ai,s}| \le \frac{I_o}{g_{msef}} \\ \frac{I_o}{g_{msef}} \mathrm{sgn}(v_{ai,s}) e^{-\frac{g_{msef}(T_s/2 - t_{o,s})}{C_{\mathrm{eq},s}}} & |v_{ai,s}| > \frac{I_o}{g_{msef}}, t_{o,s} < \frac{T_s}{2} \\ v_{ai,s} - \frac{I_o}{C_{\mathrm{eq},s}} \mathrm{sgn}(v_{ai,s}) \frac{T_s}{2} & |v_{ai,s}| > \frac{I_o}{g_{msef}}, t_{o,s} \ge \frac{T_s}{2} \end{cases} \tag{5.21}$$

where

$$t_{o,s} = \frac{C_{\mathrm{eq},s}}{I_o} |v_{ai,s}| - \frac{C_{\mathrm{eq},s}}{g_{msef}} \tag{5.22}$$

[14] Note that the model described in the S-function of Figures 5.21–5.23 corresponds to an SC FE integrator with one input SC branch. However, the model presented here can be extended to SC integrators with N_i input SC branches.

and $\mathrm{sgn}(v_{ai,s})$ is the sign function of $v_{ai,s}$ (denoted as VAINIS in Figure 5.22), which represents the value of v_a at the beginning of the sampling phase, given by,

$$v_{ai,s} = v_a[(n-1/2)T_s] - \sum_{i=1}^{N_o} \frac{C_{Sni}}{C_{\text{eq},s}} \{v_o[(n-1/2)T_s] - v_{C_{Sni}}[(n-1/2)T_s]\} \tag{5.23}$$

where $v_{C_{Sni}}$ is the voltage across capacitor C_{Sni}, respectively represented by the parameters `uPtrs4[N+i]` and `uPtrs4[i]` in Figure 5.22.

Note that Equations (5.21) and (5.22) are essentially the same as Equations (3.18) and (3.19), but replacing the OTA transconductance g_m by a parameter denoted as g_{msef} (GMSEFF in Figure 5.22). This parameter represents the effective transconductance of the opamp that is used for modeling the GB degradation due to the switch on-resistance during the sampling phase, as described in Section 3.4.4 [32].

Therefore, once $v_a(nT_s)$ has been calculated, the voltage at the output node of the opamp is computed as,

$$\begin{aligned} v_o(nT_s) &= v_o[(n-1/2)T_s] \\ &+ \left\{1 + \frac{1}{A_v[v_o(nT_s)]}\right\} \left(1 + \frac{C_P}{C_I}\right) \{v_a(nT_s) - v_a[(n-1/2)T_s]\} \end{aligned} \tag{5.24}$$

Here, the effect of the nonlinear finite OTA DC gain is also accounted for. This equation is solved by following the iterative procedure described in Section 5.3. Note from Figure 5.22 that the nonlinearity of the capacitor – not shown in Equation (5.24) for the sake of simplicity – is also taken into account in the behavioral model to get a more accurate value of $v_o(nT_s)$.

During the integration phase, the incomplete settling model is evaluated, proceeding in a similar way as during the sampling phase (see Section 3.4). The value of v_a at the end of the integration phase of period $(n+1/2)T_s$ is thus given by [40],

$$v_a[(n+1/2)T_s] = \begin{cases} v_{ai,i} e^{-\frac{g_{mief} T_s}{2C_{\text{eq},i}}} & |v_{ai,i}| \le \frac{I_o}{g_{mief}} \\ \frac{I_o}{g_{mief}} \mathrm{sgn}(v_{ai,i}) e^{-\frac{g_{mief}(T_s - t_{o,i})}{C_{\text{eq},i}}} & |v_{ai,i}| > \frac{I_o}{g_{mief}}, t_{o,i} < T_s \\ v_{ai,i} - \frac{I_o}{C_{\text{eq},i}} \mathrm{sgn}(v_{ai,i}) \frac{T_s}{2} & |v_{ai,i}| > \frac{I_o}{g_{mief}}, t_{o,i} \ge T_s \end{cases} \tag{5.25}$$

where g_{mief} (GMIEFF in Figure 5.23) stands for the effective OTA transconductance during the integration phase and,

$$\begin{aligned} t_{o,i} &= \frac{T_s}{2} + \frac{C_{\text{eq},i}}{I_o} |v_{ai,i}| - \frac{C_{\text{eq},i}}{g_{mief}} \\ v_{ai,i} &= \frac{1}{C_{\text{eq},i}} \left(1 + \frac{C_L}{C_I}\right) \sum_{i=1}^{N_i} C_{S_i} \{v_{i2}(nT_s) - v_{i1}[(n-1/2)T_s]\} + \frac{C'}{C_{\text{eq},i}} v_a(nT_s) \end{aligned} \tag{5.26}$$

with v_{ij} being the voltage at the jth input node of the ith input SC branch, C_{S_i} is the sampling capacitor of the ith input SC branch, and C' (CPRIMA in the C code) is given by:

$$C' = C_P + C_L \left(1 + \frac{C_P}{C_I}\right) \tag{5.27}$$

After computing $v_a[(n + 1/2)T_s]$, the value of the opamp output node is solved in an iteratively way (see Figure 5.23) taking into account the effect of nonlinear OTA DC gain and capacitor nonlinearity. This iterative procedure typically converges in a few iterations and provides very accurate results that are in close agreement with transistor-level simulations [6].

Before concluding this section, it is very important to note a practical issue related to the implementation of the incomplete settling error model in a programming platform in general and in the MATLAB/SIMULINK environment in particular. It should be noted that in order to implement this error, it is not sufficient to provide information about the building block itself (i.e., the SC integrator to be modeled), but also about those building blocks connected to its output. For instance, the values of $v_{C_{Sni}}$ at $(n - 1/2)T_s$ are needed to compute v_a at nT_s – see Equation (5.23). Therefore, an accurate behavioral model must incorporate the required data computed in previous clock cycles that corresponds to those building blocks connected at the output of the integrator. This can be implemented in the SIMULINK environment by using the `From` and `Goto` blocks provided by the `Signal Routing` SIMULINK library [30].

Figure 5.27 illustrates the use of `From` and `Goto` blocks for the behavioral model of SC FE integrators in SIMULINK. In this example, two SC FE integrators are connected to build a second-order ΣΔM. In order to distinguish the two building blocks in their corresponding behavioral models, two different identifiers are used, named `int3` and `int4` in this case. As shown in the corresponding S-function diagram, `From` and `Goto` blocks allow the required information to pass between the two SC integrators without actually connecting them, which greatly simplifies the implementation of arbitrary ΣΔMs without modifying the model code.

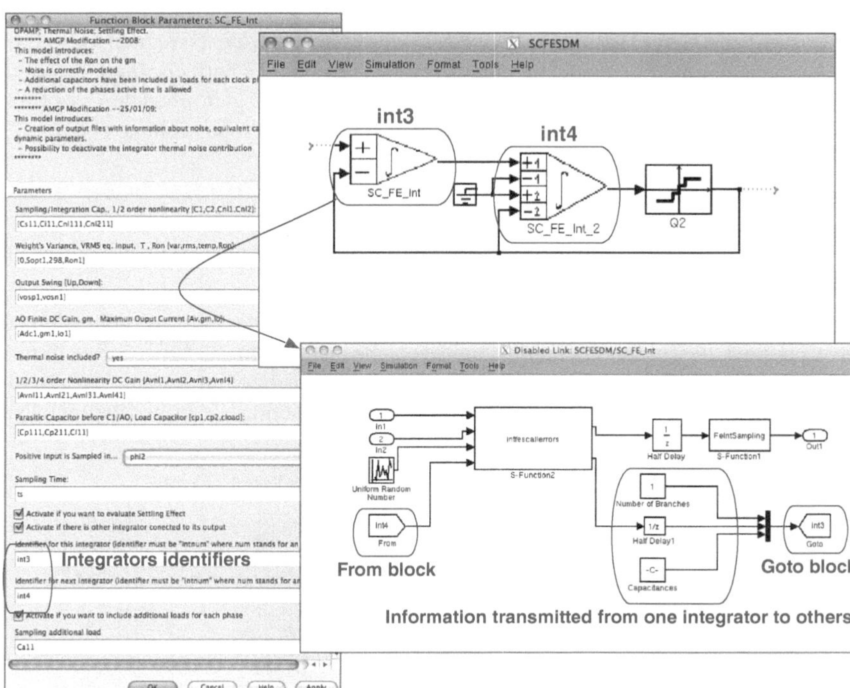

Figure 5.27 Illustrating the use of `From` and `Goto` SIMULINK blocks to transmit information between the models of two SC FE integrators in a second-order ΣΔM.

In the example of Figure 5.27, the information provided by the back-end integrator (identified as `int4`) is included in the model of the front-end integrator (`int3`) by connecting the `From` block to an input port of the front-end integrator. Similarly, the front-end integrator uses the `Goto` block to provide its stored data to another building block. The information stored essentially consists of a data array that contains the values of the sampling capacitors and the sampled voltages of the input SC branches of a given integrator. This data is read by the model using dynamically sized inputs, for instancethe `uPtrs4` array in the C code of Figure 5.21.

5.4.2 Modeling of CT Integrators using S-functions

CT integrators are the critical building blocks of CT-ΣΔMs. As discussed in Chapter 4, these blocks can be implemented using different circuit topologies: active-RC, Gm-C, MOSTFET-RC, Gm-MC, and so on. All of them may have the same ideal characteristics, although involving a number of design trade-offs and error limitations that can be modeled using C-MEX S-functions. This section focuses on the behavioral modeling of S-functions for Gm-C circuits, considering their most important limiting factors. The behavioral model of other CT integrator topologies is considered in Appendices B and C.

Let us consider the Gm-C integrator conceptually depicted in Figure 5.5b, whose ideal behavior is governed by Equation (5.2). In practice, this ideal behavior is degraded by the action of a number of circuit errors: input-referred thermal noise, circuit-element tolerances and mismatch, input and output voltage saturation, transconductance nonlinearity, finite OTA DC gain, transient response (including either a single-pole or a two-pole model for the OTA), and so on [41]. There are several ways of modeling these effects using S-functions, as discussed below.

5.4.2.1 Single-pole Gm-C Model Figure 5.28 shows the equivalent circuit of a Gm-C integrator with a single-pole dynamic. The model includes the following nonideal effects: input-referred thermal noise, input/output voltage saturation, time-constant error, finite DC gain (modeled as a finite output conductance), and a nonlinear transconductance that depends on the input voltage v_i as,

$$g_m \simeq g_{mo} \cdot (1 + g_{m1} \cdot v_i + g_{m2} \cdot v_i^2) \tag{5.28}$$

where g_{mo} is the nominal value of the transconductance and $g_{m(1,2)}$ are the nonlinear transconductance coefficients.

The circuit errors of Figure 5.28 are computed by following the iterative procedure depicted in the flow chart of Figure 5.29. This computational model can be included in a C-MEX S-function, as shown in Figure 5.30, and implemented in an S-function block as illustrated in Figure 5.31. Note that the C-MEX S-function file is made up of the same parts as those S-functions described in previous sections,

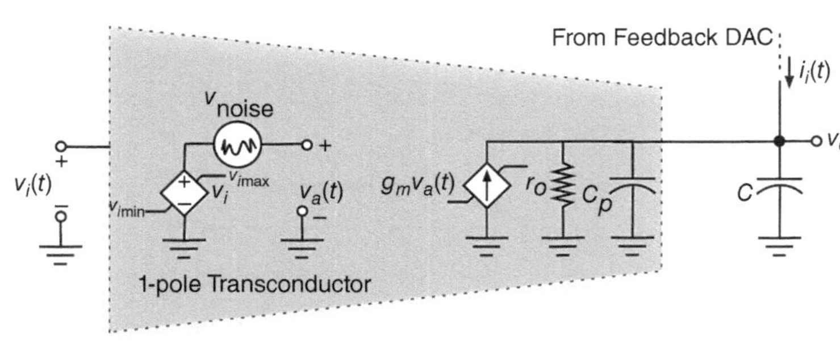

Figure 5.28 Equivalent circuit of a Gm-C integrator considering a single-pole model.

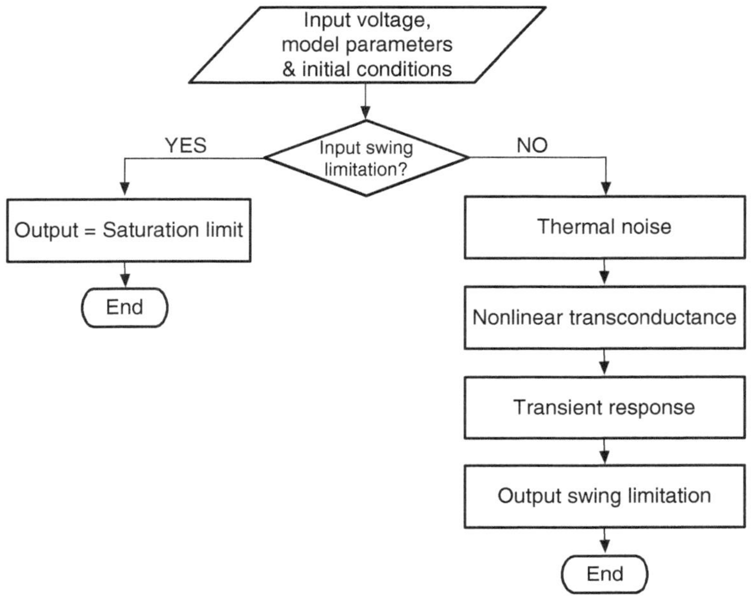

Figure 5.29 Flow chart of the Gm-C integrator computational model.

beginning with model-parameter and state-variable initialization and then computing the different errors according to the flow of Figure 5.29. The main difference is related to those model aspects related to the timing. Thus, the CT nature of the circuit is specified in the `mdlInitializeSampleTimes` structure, where the sample time is defined as `CONTINUOUS_SAMPLE_TIME` to indicate that this S-function corresponds to a CT building block [35].

The output voltage is computed using the `mdlDerivatives` routine [35], which solves the following differential equation,

$$\frac{dv_o(t)}{dt} = \frac{1}{C + C_p} \cdot [g_m v_i(t) - g_o v_o(t) + i_i(t)] \tag{5.29}$$

where g_o is the finite output conductance (denoted as `go` in Figure 5.30), C_p is the parasitic capacitance that models the integration time-constant error (`Cp` in Figure 5.30), and $i_i(t)$ is the current-mode input that can be used for modeling the current provided by a feedback current-steering DAC in a CT-ΣΔM (see Figure 5.31).

5.4.2.2 Two-pole Dynamics Model A two-pole transient response can be included in the model of a Gm-C integrator using the equivalent circuit shown in Figure 5.32. This model can be incorporated in an S-function, as illustrated in Figure 5.33, which shows only an excerpt of the C-MEX file corresponding to the `mdlDerivatives` routine. In this case, the output voltage is computed by solving the following set of differential equations,

$$\begin{aligned} \frac{dv_a(t)}{dt} &= \frac{1}{C_a} \cdot \left[\frac{v_a(t)}{r_a} - g_{ma} v_i(t)\right] \\ \frac{dv_o(t)}{dt} &= \frac{1}{C + C_p} \cdot [g_m v_a(t) - g_o v_o(t) + i_i(t)] \end{aligned} \tag{5.30}$$

```
#define S_FUNCTION_NAME pl_integrator
#define  S_FUNCTION_LEVEL 2
#include "simstruc.h"
#include <math.h>
#define gm(S) ssGetSFcnParam(S,0) /* Nominal transconductance */
#define C(S)  ssGetSFcnParam(S,1) /* Nominal capacitance */
#define et(S) ssGetSFcnParam(S,2) /* Time constant error (%) */
#define Av(S) ssGetSFcnParam(S,3) /* Finite DC gain */
#define ub(S) ssGetSFcnParam(S,4) /* Output swing limitation (max) */
#define lb(S) ssGetSFcnParam(S,5) /* Output swing limitation (min) */
#define iub(S) ssGetSFcnParam(S,6)/* Input swing limitation (max) */
#define ilb(S) ssGetSFcnParam(S,7)/* Input swing limitation (min) */
#define gm1(S) ssGetSFcnParam(S,8)/* First-order transconductance non-linearity */
#define gm2(S) ssGetSFcnParam(S,9)/* Second-order transconductance non-linearity */
#define selector_param(S) ssGetSFcnParam(S,10)
#define N_PARAMS 11

static void mdlInitializeSizes (SimStruct *S)
{...}

/* Continuous-time nature of the circuit is defined here */

static void mdlInitializeSampleTimes(SimStruct *S)
{
        ssSetSampleTime( S, 0, CONTINUOUS_SAMPLE_TIME);
        ssSetOffsetTime ( S, 0, 0.0);
}

#define MDL_INITIALIZE_CONDITIONS
static void mdlInitializeConditions (SimStruct *S)
{
        real_T *x0 = ssGetContStates(S) ;
        x0[0] = 0; /* State variable */
}

static void mdlOutputs (SimStruct *S, int_T tid)
{
        real_T  *y      = ssGetOutputPortRealSignal (S, 0);
        real_T  *x      = ssGetContStates (S);
        real_T ub = *mxGetPr(ub(S));
    real_T lb = *mxGetPr(lb(S));

        /*Checking output saturation*/

    if(x[0]<=lb)
        {x[0]=lb;}
    if(x[0]>=ub)
        {x[0]=ub;}

        y[0] = x[0] ;
}

#define MDL_DERIVATIVES

static void mdlDerivatives(SimStruct *S)
{
    real_T *dx  = ssGetdX(S);
        real_T *x = ssGetContStates (S);
        real_T ub = *mxGetPr(ub(S));
    real_T lb = *mxGetPr(lb(S));
        real_T iub = *mxGetPr(iub(S));
    real_T ilb = *mxGetPr(ilb(S));
        real_T gm1 = *mxGetPr(gm1(S));
    real_T gm2 = *mxGetPr(gm2(S));
        real_T C = *mxGetPr(C(S));
    real_T gm = *mxGetPr(gm(S));
        real_T et = *mxGetPr(et(S));
    real_T Av = *mxGetPr(Av(S));
        real_T  selector = *mxGetPr(selector_param(S));
    InputRealPtrsType uPtrs0 = ssGetInputPortRealSignalPtrs(S,0);
    InputRealPtrsType uPtrs1 = ssGetInputPortRealSignalPtrs(S,1);
    InputRealPtrsType uPtrs2 = ssGetInputPortRealSignalPtrs(S,2);
    real_T vin = *uPtrs0[0]; /* Input voltage */
    real_T iin = *uPtrs1[0]; /* Input current (from the modulator feedback DAC */
    real_T noise = *uPtrs2[0];/* Input noise source */
    real_T input;
    real_T go; /* Output conductance */
    C=C*(1+et); /* Integration capacitance modification due to time-constant error */
    go=gm/Av;   /* Finite DC gain: Av=gm/go */

        if (selector ==1) /* Input voltage without noise contribution */
                {input=vin;}
        else
                {input=vin+noise;} /* Input voltage considering noise contribution */

    input=input+gm1*pow(input,2)+gm2*pow(input,3); /*Non lineal transconductance*/

    if(x[0]<=lb)    /*Output swing limitation */
        {x[0]=lb;}
    if(x[0]>=ub)
        {x[0]=ub;}

    if(input<=ilb)  /*Input swing limitation */
        {input=ilb;}
    if(input>=iub)
        {input=iub;}

    dx[0]=-go/C*x[0]+gm/C*input+1/C*iin; /* Differential equation to evaluate transient response */
                                         /* 1-pole case */
}

static void mdlTerminate(SimStruct *S)
{
}
# ifdef MATLAB_MEX_FILE
#include "simulink.c"
#else
#include "cd_sfun.h"
#endif
```

Model parameters

Timing settings

Initial conditions

Static errors

Input/Output swing

Transient response

Figure 5.30 C-coded S-function file of a Gm-C integrator with a single-pole dynamic.

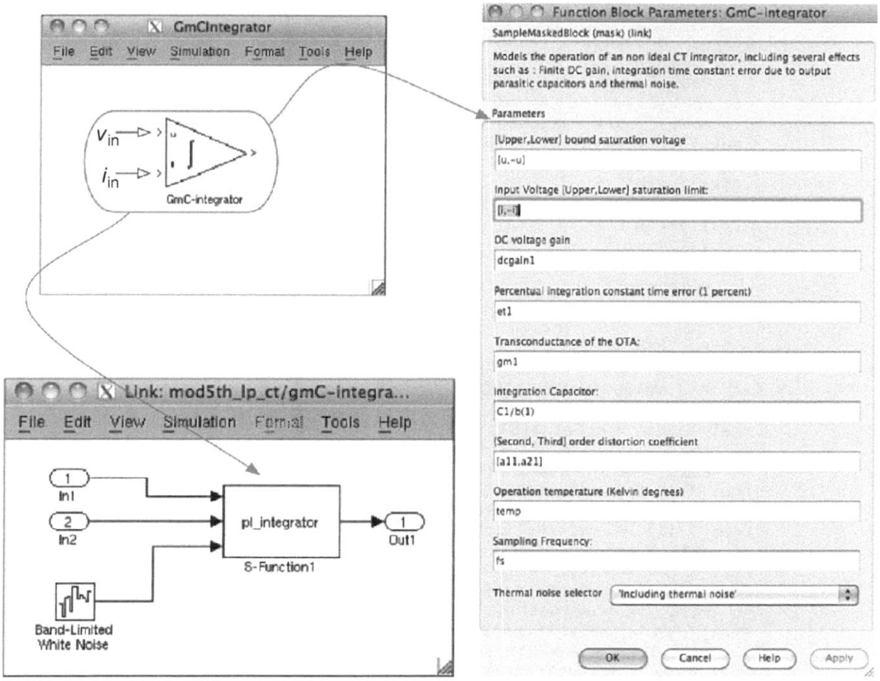

Figure 5.31 Illustrating the incorporation of the Gm-C integrator S-function in SIMULINK.

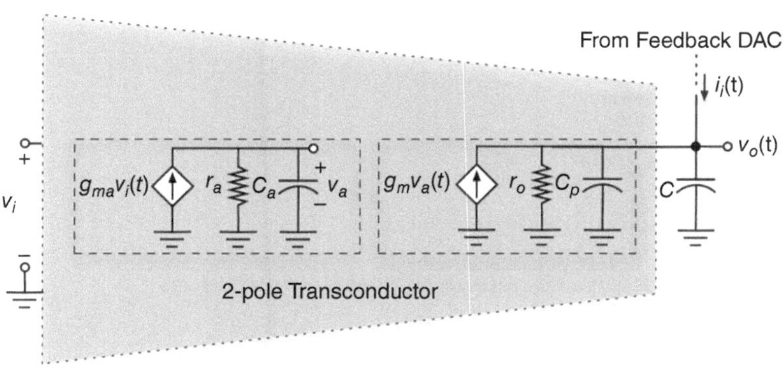

Figure 5.32 Equivalent circuit of a Gm-C integrator considering a two-pole model [6].

where v_a and v_o are represented in Figure 5.33 by two state variables named `x[0]` and `x[1]`, respectively.

5.4.2.3 Modeling Transconductors as S-functions The model of a Gm-C integrator can be also implemented as the cascade of two building blocks, as illustrated in Figure 5.34a: a transconductor and an output impedance circuit made up of the parallel connection of an output resistance ($r_o \equiv 1/g_o$), the integration capacitor C, and its parasitic capacitance C_p. Figure 5.34b shows the implementation

```
static void mdlOutputs (SimStruct *S, int_T tid)
{
        real_T  *y      = ssGetOutputPortRealSignal (S, 0);
        real_T  *x      = ssGetContStates (S);
        real_T ub = *mxGetPr(ub(S));
    real_T lb = *mxGetPr(lb(S));

    if(x[1]<=lb)
        {x[1]=lb;}
    if(x[1]>=ub)
        {x[1]=ub;}

        y[0] = x[1] ;
}

#define MDL_DERIVATIVES

static void mdlDerivatives(SimStruct *S)
{
    real_T *dx  = ssGetdX(S);
        real_T *x = ssGetContStates (S);
        real_T ub = *mxGetPr(ub(S));
    real_T lb = *mxGetPr(lb(S));
        real_T iub = *mxGetPr(iub(S));
    real_T ilb = *mxGetPr(ilb(S));
        real_T gm1 = *mxGetPr(gm1(S));
    real_T gm2 = *mxGetPr(gm2(S));
    real_T fp = *mxGetPr(fp(S));       /* fp= 1/(ra Ca) */
        real_T C = *mxGetPr(C(S));
    real_T gm = *mxGetPr(gm(S));
        real_T et = *mxGetPr(et(S));
    real_T Av = *mxGetPr(Av(S));
        real_T  selector = *mxGetPr(selector_param(S));
    InputRealPtrsType uPtrs0 = ssGetInputPortRealSignalPtrs(S,0);
    InputRealPtrsType uPtrs1 = ssGetInputPortRealSignalPtrs(S,1);
    InputRealPtrsType uPtrs2 = ssGetInputPortRealSignalPtrs(S,2);
    real_T vin = *uPtrs0[0];
    real_T iin = *uPtrs1[0];
    real_T noise = *uPtrs2[0];
    real_T input;
    real_T go;
    C=C*(1+et);
    go=gm/Av;

        if (selector ==1)
                {input=vin;}
        else
                {input=vin+noise;}

    /*Non lineal transconductance*/
    input=input+gm1*pow(entrada,2)+gm2*pow(entrada,3);

    /*Output Saturation*/
    if(x[1]<=lb)
        {x[1]=lb;}
    if(x[1]>=ub)
        {x[1]=ub;}

    /*Input Saturation*/
    if(input<=ilb)
        {input=ilb;}
    if(input>=iub)
        {input=iub;}

    dx[0]=fp*(input-gm*x[0]);
    dx[1]=-go/C*x[1]+gm/C*x[0]+1/C*iin;
}
```

Differential equations ($x[0] = v_a$; $x[1] = v_o$)

Figure 5.33 Excerpt of the S-function of a Gm-C integrator considering a two-pole model.

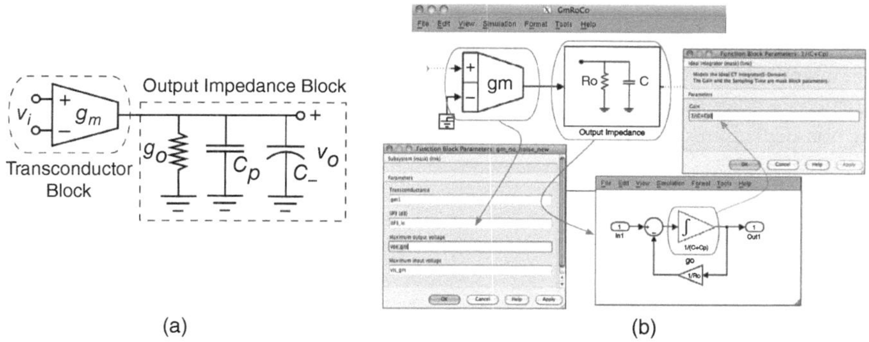

Figure 5.34 Gm-C integrator implemented as a transconductor and an output impedance circuit: (a) conceptual schematic. (b) implementation in SIMULINK.

of Figure 5.34a in SIMULINK. The transconductor S-function block includes the input-referred noise, saturation voltage at both the input and the output nodes, and a transconductance that is a nonlinear function of the input voltage and given by,

$$g_m \simeq g_{mo} \cdot \left(1 - \frac{4 \cdot g_{mo}}{3 \cdot 10^{\frac{\mathrm{IIP3}}{10}}} \cdot v_i^2 \right) \tag{5.31}$$

where IIP3 is the input-referred third-order intercept point, used as a model parameter.

5.4.3 Behavioral Modeling of Quantizers using S-functions

Quantizers are also essential building blocks of ΣΔMs. These blocks are made up of an ADC and a DAC that are embedded in the ΣΔM loop, the former in the feedforward path and the latter in the feedback path.[15] As also happens with integrators, many different ADC and DAC topologies are embedded in ΣΔMs. The choice depends on many factors, such as the circuit nature of the loop filter, the number of bits (or levels) of the quantization process, the circuit topology, and so on. All of them can be modeled with S-functions, following the same philosophy used for integrators. In this section, two commonly used blocks are presented by way of illustration. Other models of embedded ADCs and DACs can be found in Appendix C.

5.4.3.1 Modeling Multi-level ADCs as S-functions As stated in previous chapters, the quantizers embedded in ΣΔMs are usually implemented by multi-bit flash ADCs. These circuits – made up of a bank of comparators (or single-bit quantizers) and a resistive ladder – are subject to a number of circuit errors whose effects on the ΣΔM performance are greatly attenuated by the action of the modulator loop filter. However, they must be taken into account in the behavioral models in order to get efficient and accurate designs. These errors include comparator offset and hysteresis, integral nonlinearity (INL), and gain error, and they can easily be modeled using S-functions following the methodology described in this section.

[15] As stated in Chapter 1, strictly speaking, a quantizer is an analog block with both input and output being analog quantities, except that the output amplitude is discretized into a number of analog levels. For that reason, a quantizer is made up of the cascade of an ADC and a DAC. However, the ADC embedded in a ΣΔM is usually represented by a quantizer symbol, although it is connected in the feedback loop to a DAC to complete the quantization process. This kind of graphical and conceptual representation can be found in many papers and books, although it is not strictly correct.

Figure 5.35a shows the main parts of the S-function block used for modeling a multi-level ADC. This block can be used in either DT- or CT-ΣΔMs for implementing quantizers with an arbitrary number of quantization levels. In this way, if an odd number of levels is set up, a *midtreat* quantization characteristic is implemented. Otherwise, a *midrise* quantizer is computed. The model includes the following nonideal circuit effects: INL, gain error, and offset error. All of these are expressed in terms

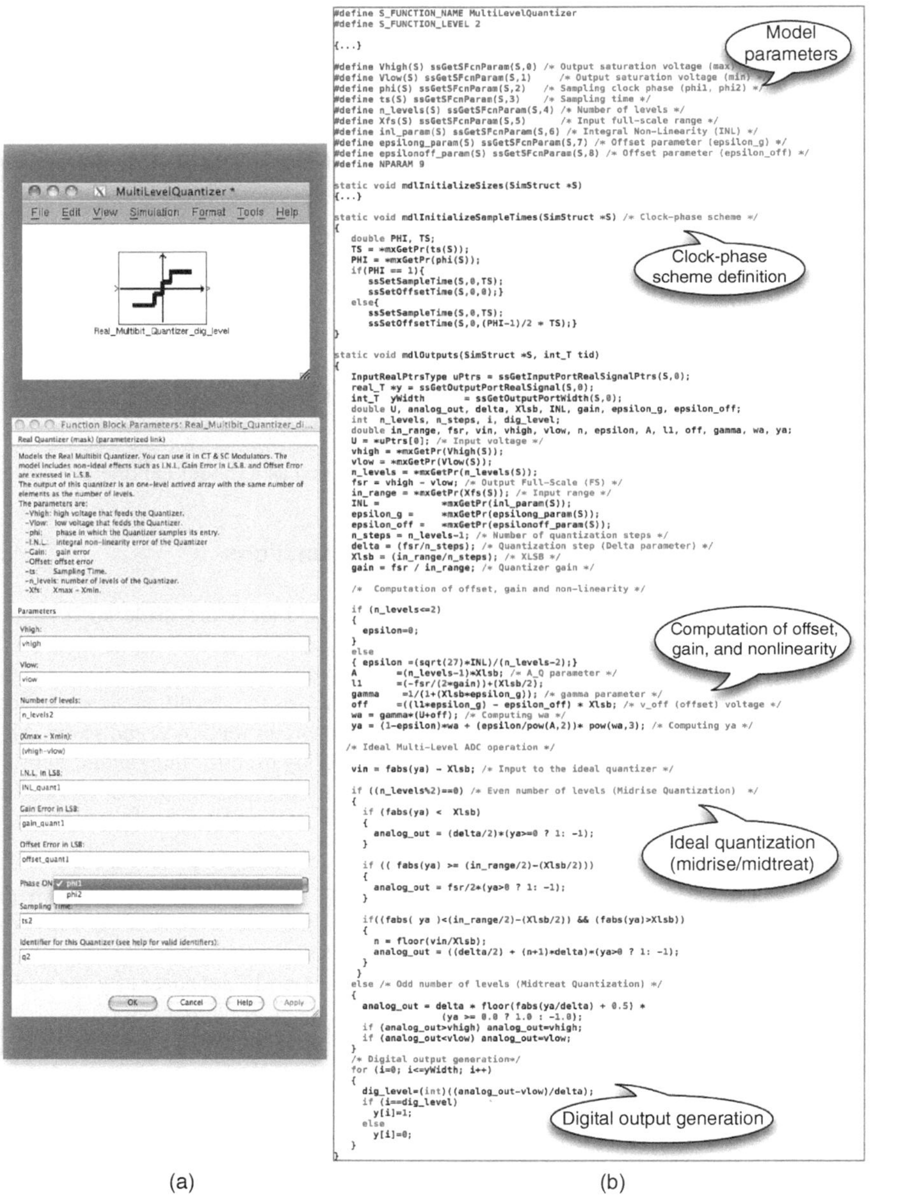

Figure 5.35 Illustrating the behavioral model of multi-level ADCs: (a) S-function block; (b) excerpt of the C code.

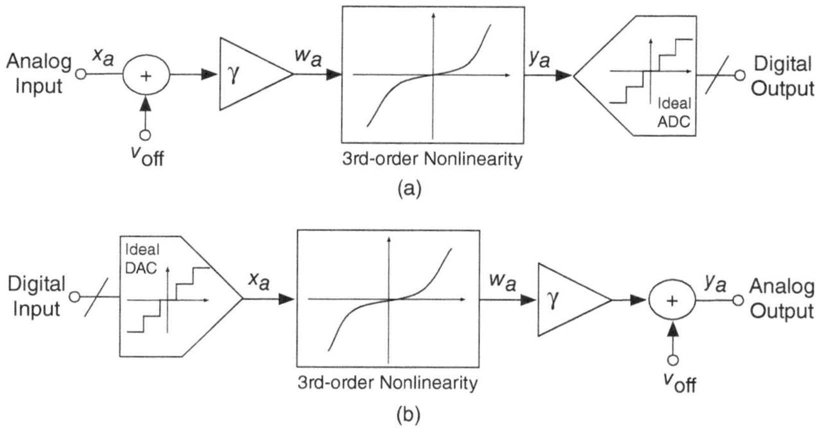

Figure 5.36 Conceptual block diagram of the behavioral model used for the quantizer embedded in ΣΔMs [13]: (a) ADC; (b) DAC.

of the least significant bit (LSB). Other model parameters are the input and output full-scale (FS) range, the clock phase at which the input is sampled, and the sampling time.

Figure 5.35b shows the main parts of the S-function file implementing the model of the multi-level ADC with the conceptual block diagram shown in Figure 5.36a. This model is based on the one presented by Medeiro *et al.* [13], which has been adapted to multi-level quantizers. Essentially, the model consists of the cascade connection of an adder block (which includes the offset error), a linear gain block, a nonlinear transfer function, and an ideal multi-level ADC. The operations of these blocks are sequentially computed in Figure 5.35b, which essentially codifies the following equations,

$$y_a = (1-\epsilon_0)w_a + \frac{\epsilon_0}{A^2}w_a^3 \ , \quad \text{with} \begin{cases} w_a = \gamma(x_a + v_{\text{off}}) \\ \epsilon_0 = \frac{\sqrt{27}}{N_L-2} \cdot \text{INL} \\ A = (N_L - 1)\Delta \end{cases} \tag{5.32}$$

where N_L is the number of levels (named `n_levels` in Figure 5.35b), Δ is the quantization step (`Xlsb` in Figure 5.35b), and v_{off} is the quantizer offset (`off` in Figure 5.35b).

5.4.3.2 Modeling Multi-level DACs as S-functions

Embedded feedback DACs must be accurately modeled at the system level in order to take into account, from the very beginning of the design, circuit errors that can severely degrade the performance of ΣΔMs. This is the case of multi-bit (or multi-level) DACs, where the mismatches among the unit circuit elements (capacitors, resistors, current sources, and so on) used for reconstructing the analog feedback signal give rise to nonlinear input–output characteristics, and consequently to harmonic distortion. This problem is aggravated in CT-ΣΔMs, where the feedback DAC transforms the modulator output signal from the DT domain to the CT domain. This signal reconstruction is critical, and subject to limiting errors, such as clock jitter error and transient response delay, that have a significant impact on the overall behavior of ΣΔMs, as discussed in Chapter 4. Therefore, all these effects must be considered in the behavioral models and can be also implemented in MATLAB S-functions.

As an example, Figure 5.37a shows the S-function of a multi-bit DAC used in CT-ΣΔMs. This model allows implementation of three of the most commonly used DAC waveforms – NRZ, RZ, and

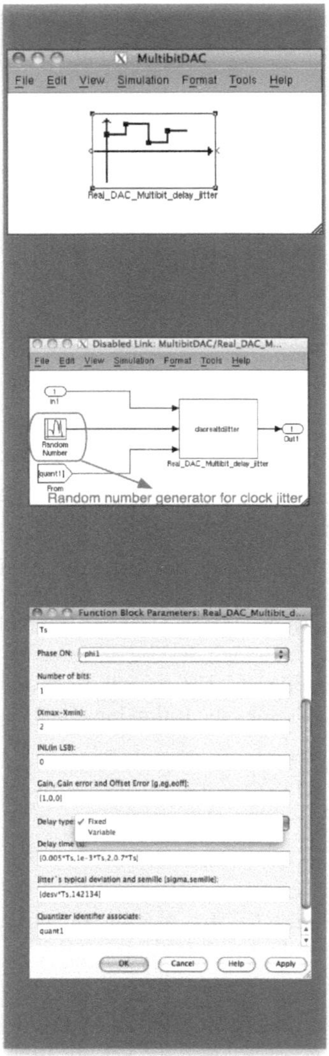

(a)

```c
... static void mdlInitializeSampleTimes(SimStruct *S) /* VARIABLE SAMPLING TIME */
{
   ssSetSampleTime(S, 0, VARIABLE_SAMPLE_TIME);
   ssSetOffsetTime (S, 0, 0.0);
}
#define MDL_GET_TIME_OF_NEXT_VAR_HIT /* DELAY MODEL */
static void mdlGetTimeOfNextVarHit(SimStruct *S)
{
... tdelay = *mxGetPr(tdelay(S));Ts = *mxGetPr(ts(S));phi= *mxGetPr(phi(S));code =
*mxGetPr(code(S));

    if (tdelay==1) /*FIXED DELAY*/
    {
        delay = *mxGetPr(delay(S));
    }
    else    /*VARIABLE DELAY*/
    {
        p = mxGetPr(delay(S)); d0=p[0]; d1=p[1]; x1=p[2];dmax=p[3]; delay=d0+fabs(d1/
(x1*(*uPtrs3[0])));
        if (delay>dmax){
            delay=dmax;}
    }
    if (phi==2){
        delay=delay+Ts/2;}
            nsamp=work[0];action=work[1]; flag=work[4]; /* JITTER MODEL */
    if ( (delay+(*uPtrs1[0]))<=0 || (delay+(*uPtrs2[0]))<=0 ){
        ssSetErrorStatus(S, "Error message");}
    else{
        if (action==0)
        {
            if ( (delay+(*uPtrs1[0]))>=Ts || (delay+(*uPtrs2[0]))>= Ts ){
                ssSetErrorStatus(S, "Error message");}
                action=1;
            if(code == 1)   /*NRZ waveform*/
            {
                ssSetTNext(S, nsamp*Ts + delay + (*uPtrs1[0]));
            }
            else{
                if ( (delay+(*uPtrs1[0]))>=Ts/2 || (delay+(*uPtrs2[0]))>= Ts/2 ){
                    ssSetErrorStatus(S, "Error message");}
                if (flag==0)
                    {
                    ssSetTNext(S, nsamp*Ts/2 + delay +(*uPtrs1[0]));
                    }
                else {ssSetTNext(S, nsamp*Ts/2 + delay +(*uPtrs2[0]));}
                        }
        }
        else
        {
            action=0;
            nsamp=nsamp+1;
            if(code == 1)   /*NRZ waveform*/
            {
                ssSetTNext(S, nsamp*Ts);
            }
            else{
                ssSetTNext(S, nsamp*Ts/2);
                if (flag==0)
                    {
                    flag=1;
                    }
                else {flag=0;}
            }
        }
    }
    work[0]=nsamp; work[1]=action; work[2]=delay; work[4]=flag;
}
...static void mdlOutputs(SimStruct *S, int_T tid)
{
... /********************** Ideal DAC operation *********/
    vin = fabs(U) - Xlsb;
    if(fabs(U) < Xlsb){
        if(U>=0)
                xa = (delta/2);
        else if (U < 0)
                xa = -(delta/2);
    }
    if(U >= ((in_range/2)-(Xlsb/2)) || U <= -(in_range/2)+ (Xlsb/2)){
        if(U>0)
        xa = fsr/2;
    else xa = -fsr/2;
    }
    if(fabs(U) < ((in_range/2)-(Xlsb/2)) && fabs(U)>Xlsb){
        if(U>0){
                n = floor(vin/Xlsb); xa = (delta/2) + (n+1)*delta;
         }else{n = floor(vin/Xlsb); xa = -(delta/2) - (n+1)*delta;}
    }
... /************** DAC waveform selection ***********************/
    if (code==1)     /*NRZ*/
    {
        y[0]=out;
    }
    else if (code==2) /*RZ*/
    {
        if (flag==1)
        {
            y[0]=out;
        }
        else{y[0]=0;}
     }
     else
     {                 /*HRZ*/
        if (flag==1)
        {
            y[0]=0;
        }
        else {y[0]=out;}
     }
  }
```

Callouts: Fixed/signal-dependent delay; Clock-jitter model; DAC ideal characteristic; DAC waveform selection

(b)

Figure 5.37 Illustrating the behavioral model of multi-bit DACs: (a) S-function block; (b) excerpts of the C code, including the model of DAC transient response delay, clock jitter, and waveform selection.

HRZ – and includes the most important errors, such as gain error, offset, INL, transient response delay, and clock jitter.

Figure 5.37b shows the C code of the S-function, highlighting the main parts. In essence, the behavioral model is based on the same concept used for multi-level ADCs, but implemented in a dual way, as depicted in Figure 5.36b; that is, starting from an ideal D/A conversion and then applying the effects of the nonlinearity, gain error, and offset. Note that the DAC model uses the number of bits as a model parameter instead of number of levels, although the latter can be also included in the models, as shown in Appendix C.

Two different cases have been considered for the DAC delay (`delay` in Figure 5.37b): a fixed delay, which is independent of the input voltage, and a signal-dependent delay, which is modeled as described in Chapter 4 [42]:

$$\text{delay}(v_i) = d_0 + \frac{d_1}{x_1|v_i|} \tag{5.33}$$

where d_0 stands for the fixed delay (`d0` in Figure 5.37b) and d_1 and x_1 (`d1,x1` in Figure 5.37b) are curve-fitting parameters that can either be extracted from electrical (transistor-level) simulations or be used for high-level synthesis purposes.

Another important error considered in Figure 5.37 is the clock jitter. This error is modeled as a sampling time instant uncertainty given by,

$$t_n = nT_s + \beta_n \tag{5.34}$$

where $n = 1, 2, ...$; t_n is the time instant and β_n stands for the time uncertainty, implemented in the model as random Gaussian noise source with zero mean and standard deviation provided as a model parameter in Figure 5.37a. This noise source is incorporated as an input port to the S-function block and included in the C code by using `*uPtrs2[0]`, as shown in Figure 5.37b.

The C code modeling the clock jitter, as stated in Chapter 4, strongly depends on the DAC waveform, because a different number of clock-signal edges may occur. For this reason, Equation (5.34) is valid for NRZ DACs, while the time uncertainty in the clock edges in RZ or HRZ DACs is modeled as,

$$\begin{aligned} t_{n1} &= nT_s + \beta_{n1} \\ t_{n2} &= nT_s + \frac{T_s}{2} + \beta_{n2} \end{aligned} \tag{5.35}$$

where t_{n1} and t_{n2} refer to the time instants of the first and second clock edges, respectively, and β_{n1} and β_{n2} stand for the corresponding clock-edge time uncertainties. Note that a variable sampled time is used in Figure 5.37b (modeled by the `VARIABLE_SAMPLE_TIME` parameter) to take into account the effect of clock jitter on the sampling time instant.

5.5 SIMSIDES: A SIMULINK-based Behavioral Simulator for ΣΔMs

All ΣΔM building blocks and their associated error mechanisms can be modeled as C-MEX S-functions following the methodology described in earlier sections. Based on this philosophy, a complete toolbox can be developed in the MATLAB/SIMULINK environment for the time-domain behavioral simulation of ΣΔMs. An example is SIMSIDES, a SIMULINK-based SIgma-DElta simulator, which takes advantage of the benefits provided by MATLAB, namely a user-friendly interface, high flexibility for extension to new models and building blocks, and a powerful set of signal-processing routines [6].

SIMSIDES can be used for simulating any arbitrary ΣΔM topology, considering a circuit implementation with either DT circuits, CT circuits, or a mix of the two; that is, so-called "hybrid" CT/DT-ΣΔMs. In the case of DT-ΣΔMs, behavioral models included in SIMSIDES consider either SC or SI [43] circuit techniques, although the majority of models in SIMSIDES deal with SC circuits, because this is the most commonly used DT circuit technique. As far as CT-ΣΔMs are concerned, the main integrator circuit topologies are included in SIMSIDES' S-functions, namely Gm-C, active-RC, and Gm-MC. Overall, more than 150 S-functions and 250 behavioral models are available in the toolbox. A list of the most important ones is included in Appendix C, where a brief description of the different models, their main functionality, and parameters is given. This section summarizes the most important features of SIMSIDES, overviewing its model libraries, its general structure, and the user interface.

5.5.1 Model Libraries Included in SIMSIDES

The building blocks modeled in SIMSIDES are grouped into a number of SIMULINK libraries and sublibraries. which can be classified according to different criteria, as conceptually depicted in Figure 5.38. The first classification criterion is related to the modulator system hierarchy level in which a given building block is placed. According to this criterion, ΣΔM building blocks are organized in the following libraries: integrators and resonators (basic blocks in ΣΔMs), quantizers, DACs, and auxiliary blocks. The latter include analog and digital mixers, DEM algorithms, latch models, and so on, which are required to simulate specific ΣΔM architectures. A second classification criterion deals with the circuit technique used for implementing ΣΔM building blocks. Thus there are sublibraries including FE and LD integrators, Gm-C and active-RC integrators, and so on.

All SIMSIDES model libraries are in turn classified into two main categories (not shown in Figure 5.38 for the sake of simplicity): ideal libraries and real libraries. The former include only S-functions that contain ideal models of different building blocks – classified attending to the aforementioned criteria. By contrast, real libraries contain behavioral models that include the most important error mechanisms that degrade the performance of ΣΔMs. Table 5.2 summarizes all building blocks modeled in SIMSIDES, as well as the error mechanisms included in their S-function behavioral models.

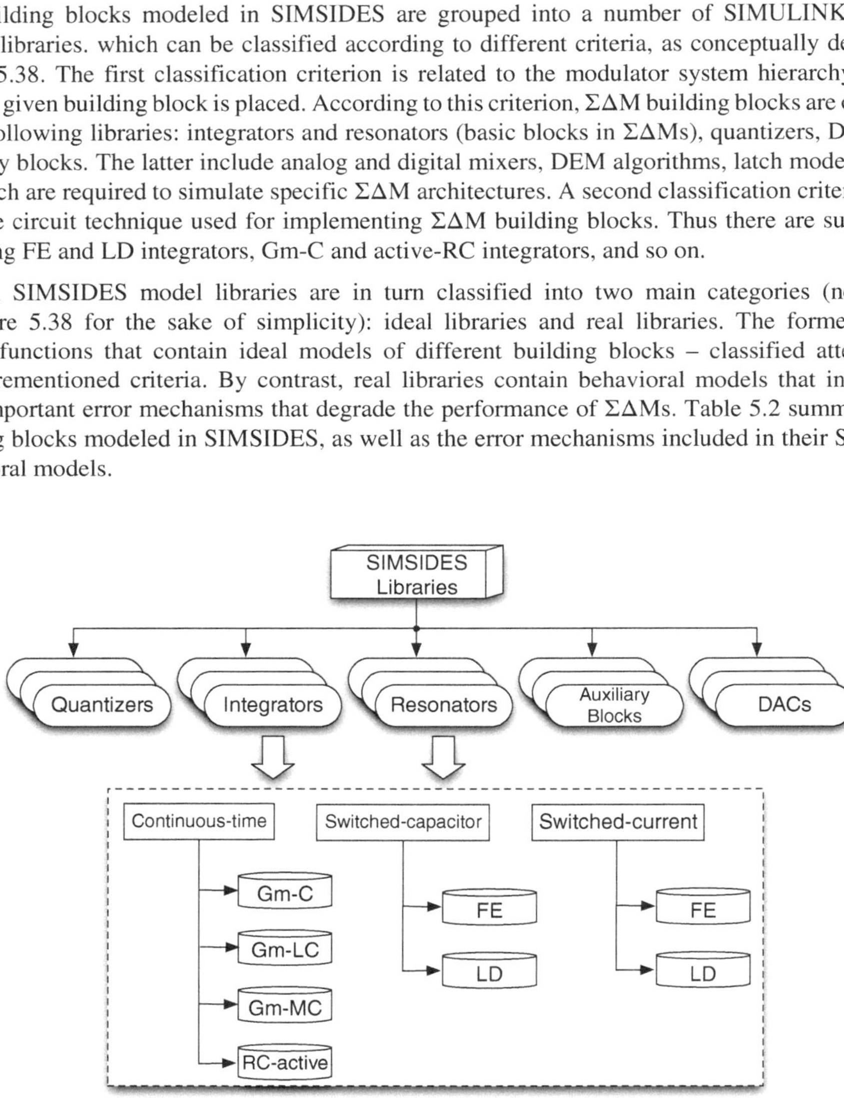

Figure 5.38 Classification of SIMSIDES model libraries.

Table 5.2 Circuits and error mechanisms modeled in SIMSIDES.

Circuit technique	Building block	Error mechanism
Switched-capacitor	Amplifiers	Finite and nonlinear DC gain, incomplete settling error, output swing limitation, thermal noise
	Switches	Thermal noise, finite and nonlinear switch on-resistance
	Capacitors	Mismatch, nonlinearities, parasitic capacitances
Switched-current	Memory cells and Integrators	Linear and nonlinear gain error, thermal noise finite output-input conductance ratio error, charge injection error, incomplete settling error
Continuous-time	Integrators	Finite and nonlinear DC gain, nonlinear transconductance, thermal noise, output swing limitation, transient response
All circuit techniques	Clock generator	Clock jitter
	Comparators	Hysteresis and offset
	Quantizers/DACs	Nonlinearity (INL), gain error, excess loop delay, offset

As an illustration, Figure 5.39 shows some representative libraries and sublibraries, including integrators, resonators, quantizers, and DACs. As illustrated in Figure 5.40, there are a number of different SC integrator S-function blocks, with different numbers of input branches and nonideal effects included in their models, going from ideal models to the most precise ones that include all circuit nonidealities. The benefits of this approach are twofold. On the one hand, it allows the impact of isolated circuit error mechanisms to be evaluated in a very simple way, without dealing with model parameters. This is particularly appropriate for the high-level sizing process, where different error parameters are taken as design variables. On the other hand, the use of different building-block model approaches, which evolve from the simplest ideal approximations to the most accurate and complex ones, may be particularly useful for novice designers who are not familiar with some of the circuit-level parameters used in the most precise behavioral models.

Apart from these libraries containing isolated ΣΔM building blocks, SIMSIDES also includes also some additional libraries with examples of the most commonly used ΣΔM architectures, considering both LP and BP topologies, single-loop and cascade architectures, single-bit and multi-bit embedded quantization, various circuit techniques (DT, CT, and hybrid CT/DT), and so on.

5.5.2 Structure of SIMSIDES and its User Interface

Figure 5.41 shows the general structure of SIMSIDES. First, the modulator architecture is defined by interconnecting the building blocks included in the SIMSIDES libraries in a suitable way, as discussed in the previous section. After the modulator block diagram has been created, the designer can set the model parameters and the simulation options required by the toolbox to do the simulation. Most commonly used analyses can be carried out: time-domain simulations to obtain output spectra and SNR/SNDR-versus-input curves, parametric simulations considering the variation of a given model parameter, Monte Carlo simulations, and so on. The output data generated by the simulator consists of time-domain series, which can be further processed to get typical performance figures.

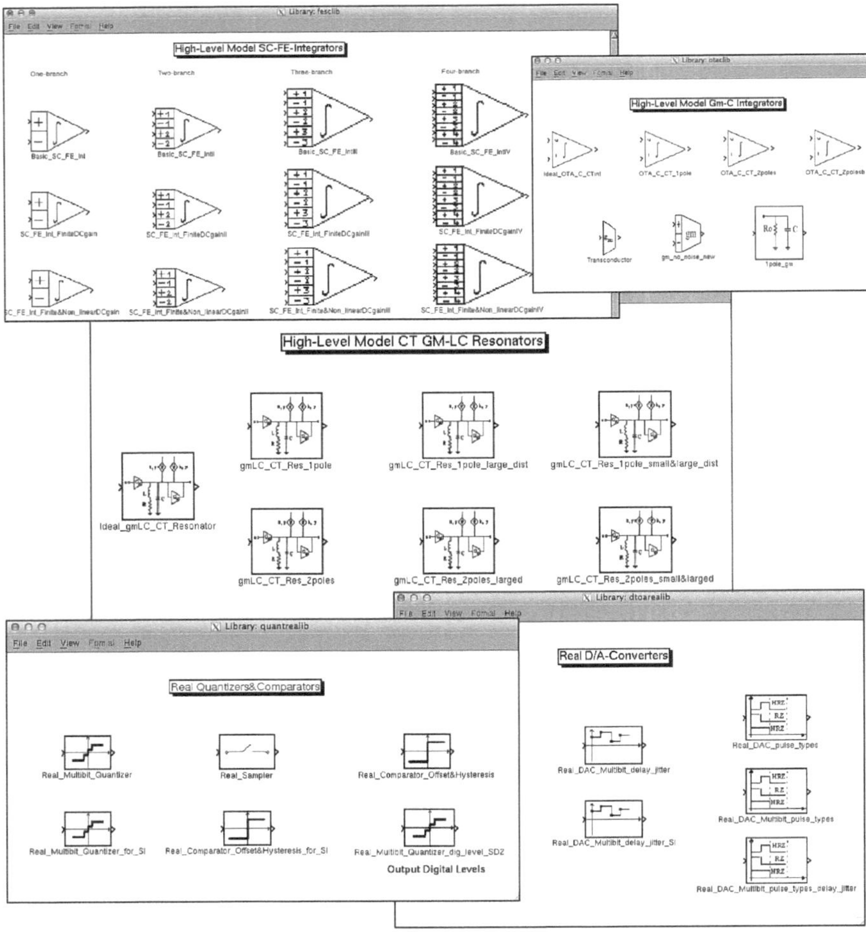

Figure 5.39 Illustrating some SIMSIDES model libraries.

In this way, both integrator input/output histograms and/or output spectra are computed using the routines provided by the signal processing toolbox in MATLAB [29]. Other typical performance metrics can be evaluated, for example the SNR/SNDR or the harmonic and intermodulation distortion. These figures are computed using a collection of internal MATLAB routines specifically developed for SIMSIDES [6].

SIMSIDES includes a GUI that allows designers to browse through all the steps of the simulation and to post-process the simulation results. As an illustration, Figure 5.42 shows some of the most important parts of the SIMSIDES GUI, highlighting some of its menus. Although a more detailed explanation about SIMSIDES GUI is given in Appendix B, the use of this simulator is illustrated in this section through a simple example based on the behavioral model of a cascade 2-1 SC-ΣΔM, on which several types of analysis are carried out.

5.5.2.1 Creating a New ΣΔM Block Diagram

SIMSIDES is launched by typing `simsides` in the MATLAB command window. The main window is displayed. After selecting the pop-up menu

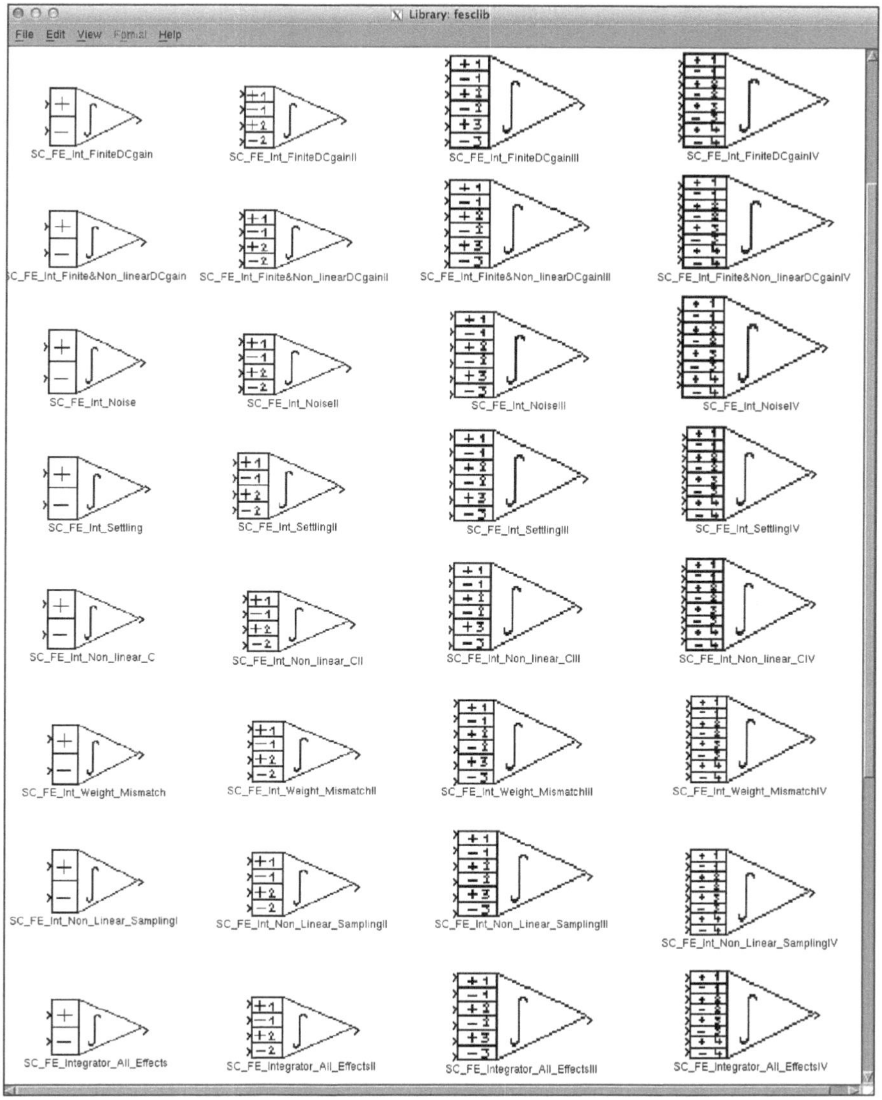

Figure 5.40 SIMSIDES library including different SC FE integrator model approaches.

`File` and then `New Architecture` (see Figure 5.42), a new (empty) SIMULINK model window is displayed. This is illustrated in Figure 5.43 for a cascade 2-1 SC topology. It can be created by adding building blocks from `Edit->Add Block` pop-up menu, and then choosing the circuit technique used to implement the integrator blocks – SC FE integrators in this case. The corresponding model library – `fesclib` in this example (illustrated in Figure 5.40) – is displayed and the appropriate integrator blocks can be included in the new model by dragging and dropping the corresponding blocks. A similar procedure can be followed to incorporate the comparator model and, once all blocks have been properly connected, the block diagram in Figure 5.43 is finished.

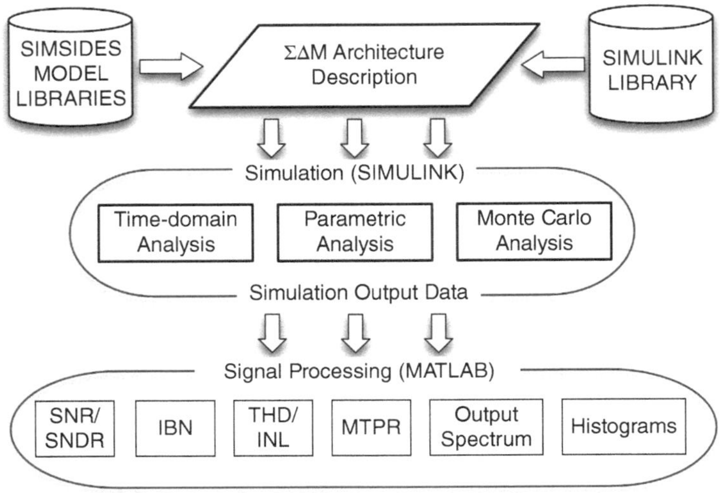

Figure 5.41 General structure of SIMSIDES.

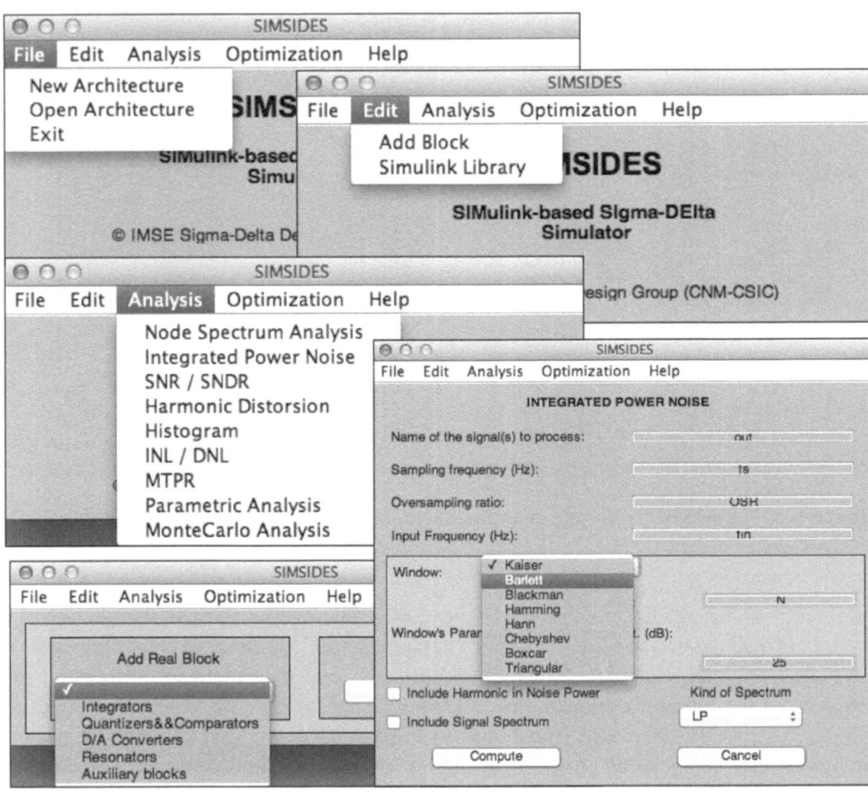

Figure 5.42 Illustrating some parts of the SIMSIDES GUI.

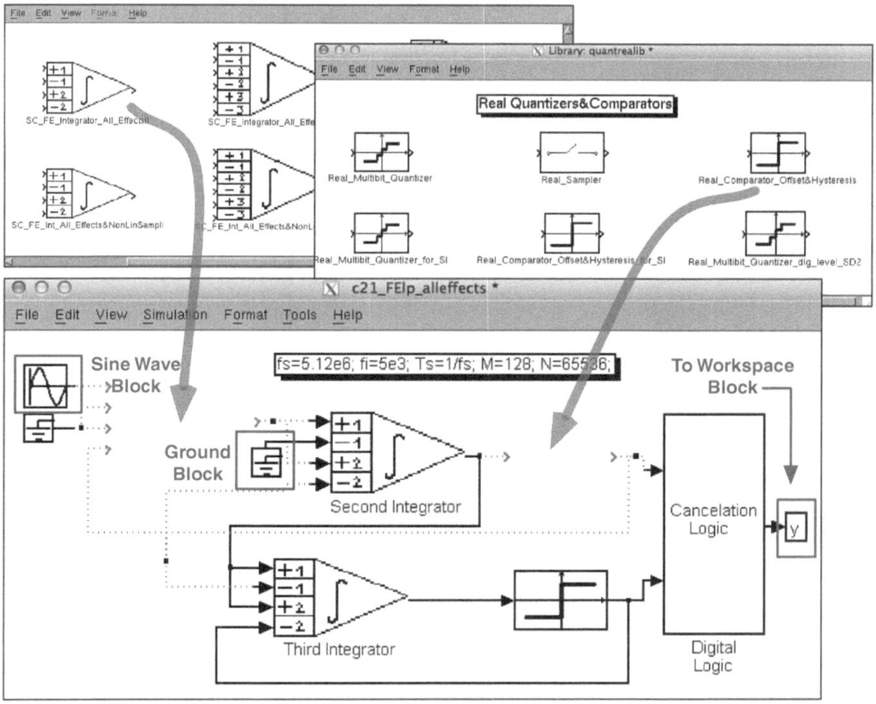

Figure 5.43 Creating the block diagram of a cascade 2-1 SC-ΣΔM in SIMSIDES.

5.5.2.2 Setting Model Parameters Once the block diagram has been defined, the next step involves setting the model parameters. Although this is something that can be done in the MATLAB command window, as the number of model parameters increases it becomes more practical to save their values in an M-file. In addition to the model parameters themselves, some global parameters must be defined in order to run the simulation. These parameters are the input signal frequency (`fi` in Figure 5.43), sampling time and frequency (`Ts,fs`), and the number of simulation clock cycles (`N=65536`). The latter must be included in the `Configuration Parameters` menu of the SIMULINK model window, where the `solver` options are also configured (`Fixed-step` and `discrete` in this example).

5.5.2.3 Simulation Analyses After setting the model parameters, a simulation can be started from the `Simulation ->Start` pop-up menu, as usually done in SIMULINK. This is the usual procedure followed in SIMSIDES to run a one-sample simulation, used for computing an output spectrum for instance. Thus, once the simulation has finished, the output spectrum can be computed and plotted from the `Analysis-> Node Spectrum Analysis` pop-up menu (see Figure 5.42). Selecting this option, a new window is displayed to set the parameters required to compute the output spectrum, such as the sampling frequency, the signal to be processed, as well as parameters related to the window used for FFT computation. In the example shown in Figure 5.44a, a Kaiser window with `N` points and a beta of 30 has been used. As an illustration, Figure 5.44b shows the in-band output spectrum of the modulator in Figure 5.43. This figure also plots the IBN corresponding to OSR = 128. This data has been obtained by choosing the `Analysis->Integrated Power Noise` pop-up menu option. In a similar way, the SNR/SNDR can be computed by selecting `Analysis->SNR/SNDR` in the SIMSIDES main window.

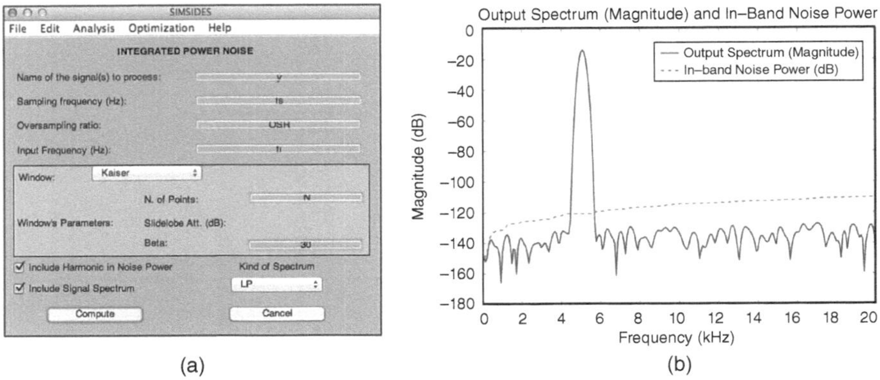

Figure 5.44 Computing IBN in SIMSIDES: (a) user window; (b) in-band output spectrum.

Apart from these simulation analyses, other useful performance metrics can also be evaluated in SIMSIDES, such as linearity – considering either static measures like INL or dynamic ones like THD or MTPR. All these analyses can be combined with a parametric analysis in order to evaluate the impact of a given model parameter. This kind of analysis is specially useful for high-level sizing, as will be illustrated in the next section. Here an example of this analysis is shown, in which a typical SNR-versus-input figure is obtained.

To this end, the `Analysis->Parametric Analysis` pop-up menu option is chosen and a new window is displayed, as shown in Figure 5.45a. There is the possibility to select between one-parameter analysis or two-parameter analysis. In this example, the former is chosen and the input-signal amplitude range is defined. After that, the next step is to define the parameters required to compute the SNDR. Once all parameters have been defined, the parametric simulation is started. A simulation-progress window is graphically displayed. The results are plotted in Figure 5.45b.

5.6 Using SIMSIDES for High-level Sizing and Verification of ΣΔMs

To conclude this chapter, this section illustrates the use of SIMSIDES for the high-level sizing and verification of ΣΔMs. Two different ΣΔM architectures and circuit techniques are considered as case studies:

- an SC second-order single-loop single-bit ΣΔM
- a CT fifth-order cascade 3-2 multi-bit ΣΔM.

5.6.1 SC Second-order Single-Bit ΣΔM

Let us consider the *Z*-domain block diagram of a second-order ΣΔM shown in Figure 5.46a. In this example, an ideal feedback DAC and single-bit quantization with 1-V full-scale range will be assumed. This block diagram can be implemented in SIMSIDES, as shown in Figure 5.46b, where *Z*-domain transfer functions have been replaced by SC FE integrator S-function blocks, as illustrated in Figure 5.46c.

Figure 5.47 represents the ideal modulator output spectrum and the IBN for OSR = 128, considering a sampling frequency of $f_s = 2.56$ MHz and an input tone with half-scale amplitude

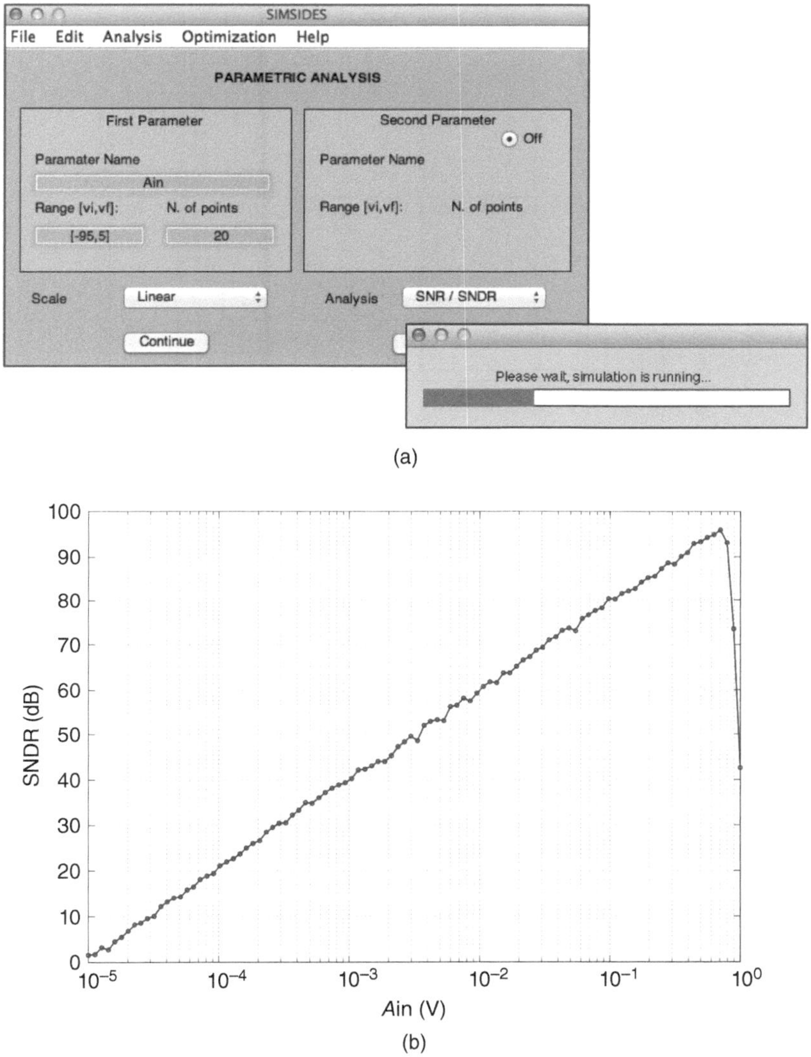

Figure 5.45 Computing SNDR versus input amplitude in SIMSIDES: (a) user window; (b) SNDR versus input amplitude.

(0.5 V). The shaped quantization noise increases at a rate of 15 dB/octave, in accord with theoretical predictions. The ideal SNR computed by SIMSIDES is 87 dB ($\simeq$ 14 bit), 103 dB ($\simeq$ 16.8 bit), and 119 dB ($\simeq$ 19.4 bit), respectively for OSR $= 128, 256, 512$, corresponding to a signal bandwidth of $B_w = 10, 5$, and 2.5 kHz, respectively. Alternatively, the same ideal effective resolutions can be obtained for a signal bandwidth of $B_w = 10$ kHz, if f_s =2.56, 5.12, and 10.24 MHz. In practice, this ideal performance is degraded by the action of circuit errors. In this example, the impact of the following nonidealities will be evaluated: opamp finite DC gain, thermal noise, and incomplete transient response. For each nonideality, the error bound that allows ideal effective resolution to be achieved will be found, in order to map modulator-level specifications onto building-block specifications.

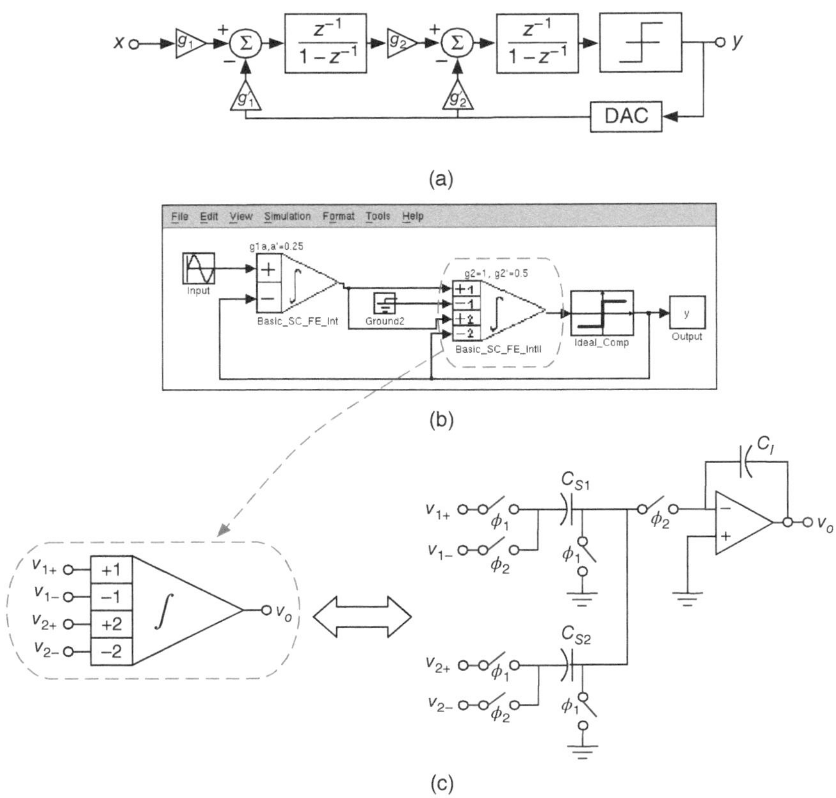

Figure 5.46 Second-order SC-ΣΔM case study: (a) Z-domain block diagram; (b) SIMSIDES implementation; (c) symbol of a two-branch SC integrator in SIMSIDES and its equivalent circuit.

5.6.1.1 Effect of Amplifier Finite DC Gain

There are two possibilities to analyze the effect of a given error in SIMSIDES: either using a behavioral model that contains only that isolated effect, or using a complete behavioral model in which all error parameters are set to be ideal except for those related to the nonideality that is going to be evaluated. In this example, the first approach will be followed. Thus, in order to simulate the effect of finite DC gain A_v, the corresponding S-function models of SC FE integrators available in SIMSIDES are used.

A parametric analysis can be carried out in SIMSIDES in order to obtain the minimum (or critical) value of A_v – represented by $A_{v_{\text{crit}}}$ – required to achieve the ideal resolution. This is illustrated in Figure 5.48, where the SNDR is plotted versus A_v for OSR = 128, 256, and 512, resulting in $A_{v_{\text{crit}}} > 100, 200$, and 400, respectively.[16]

5.6.1.2 Effect of Thermal Noise

Following the same procedure as in the previous section, the effect of circuit (thermal) noise can be simulated. Two main thermal noise sources are considered: the input-referred noise of the opamp (denoted as V_n) and the kT/C noise, evaluated by varying the value

[16]The parametric analysis varying A_v was carried considering bandwidths of 10 kHz and 20 kHz (and their corresponding values of f_s for OSR = 128, 256, and 512). As this nonideality is static, note from Figure 5.48 that the absolute value of B_w does not have any influence on the results obtained.

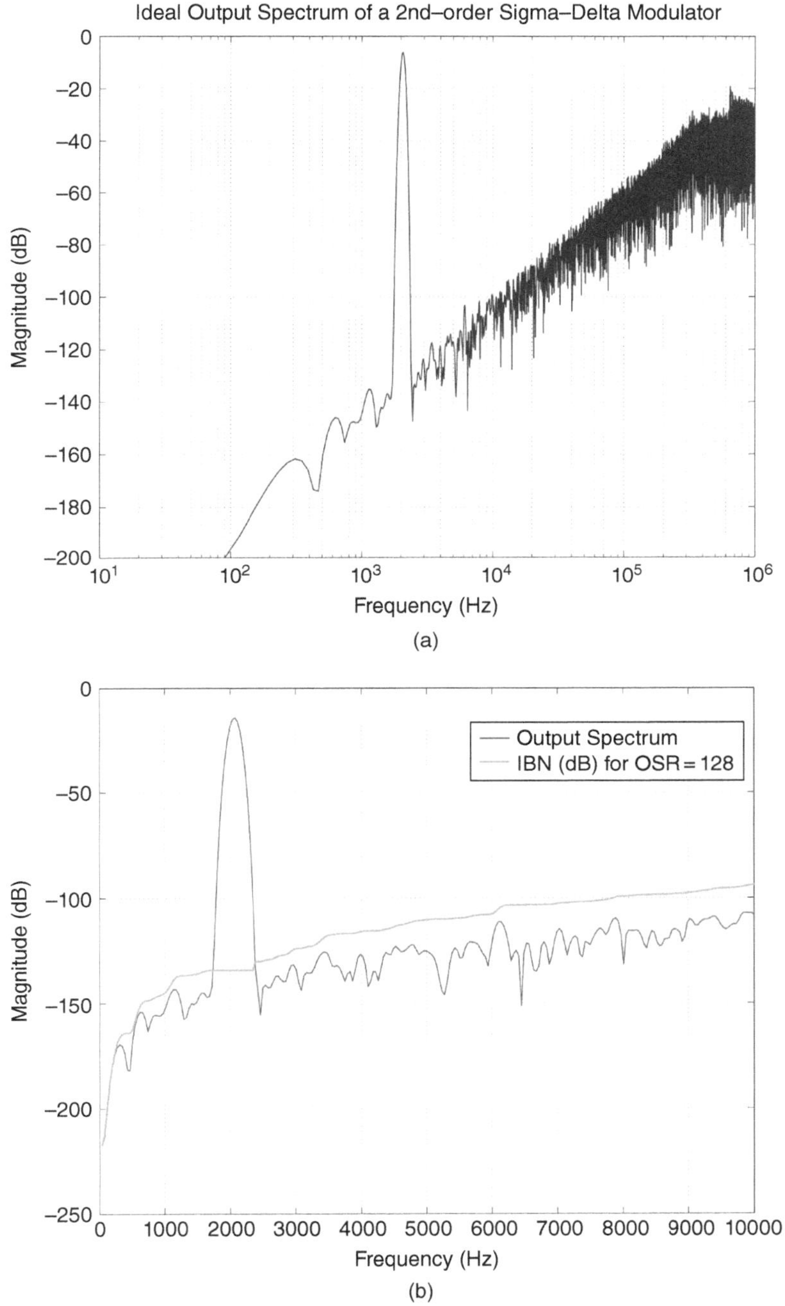

Figure 5.47 (a) Ideal output spectrum of the modulator in Figure 5.46; (b) IBN for OSR = 128.

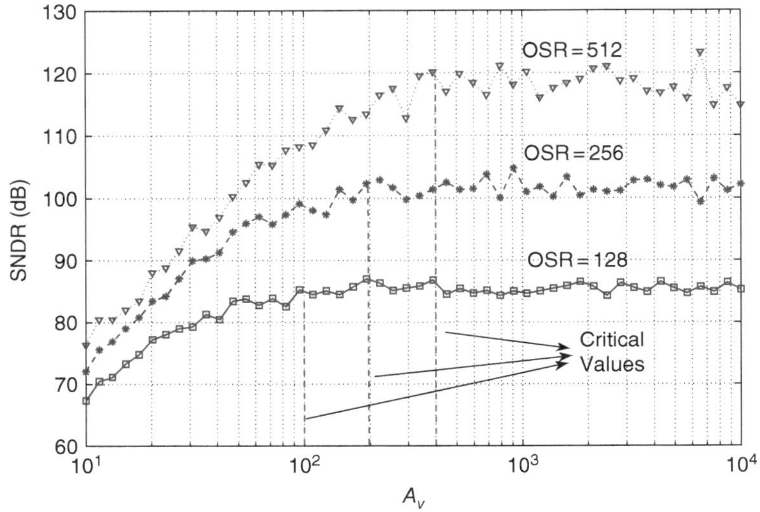

Figure 5.48 SNDR versus A_v of the ΣΔM in Figure 5.46 for different values of OSR.

of the sampling capacitance C_s. In this example, only the contribution of the front-end integrator will be taken into account.[17]

Figure 5.49 depicts SNR versus V_n and C_s for OSR = 128, 256, and 512. According to Equation (5.14), ideal values for other model parameters except V_n and C_s are considered. Note from Figure 5.49 that a critical value of $V_{n_{\text{crit}}} < 1\text{nV}/\sqrt{\text{Hz}}$ can be tolerated in all cases, while $C_{s_{\text{crit}}} > 2, 7$, and 10 pF is approximately required for OSR =128, 256, and 512, respectively. Note that such large values required for C_s are a direct consequence of the high effective resolutions with low modulator filter order and single-bit embedded quantizer. These capacitor values may force a reduction in the digitized signal bandwidth B_w in order to reduce the impact of incomplete settling with feasible power consumption.

5.6.1.3 Effect of the Incomplete Settling Error

In order to evaluate the impact of the incomplete settling error, two model parameters are considered: the OTA transconductance g_m and the maximum OTA output current I_o. The former determines the minimum requirements in terms of GB, whereas the latter limits the maximum SR that can be achieved for a given value of the equivalent load capacitance – automatically computed by the SIMSIDES model as discussed in Section 5.4. As in previous errors, only the impact of the front-end integrator is considered here for illustration purposes.

Figure 5.50 shows the effect of g_m and I_o on the performance of the modulator for different values of OSR. Note that in this case the absolute value of f_{in} and B_w are important, since the incomplete setting affects the dynamic response of the integrators. In this example, an input tone at $f_{\text{in}} = B_w/5$ is considered, for B_w = 10 kHz. Figure 5.50a depicts the SNDR versus g_m, for C_s =10 pF and I_o = 5 mA. Considering these simulation conditions, the effect of SR can be neglected. Alternatively, Figure 5.50b depicts the SNDR versus I_o considering g_m = 5 mA/V, so that only the effect of SR is evaluated and the impact of GB is nulled. The critical values of g_m and I_o for the different cases are highlighted in the figure.

[17]Note that the noise sources of the second integrator are attenuated in the signal band by the gain of the front-end integrator.

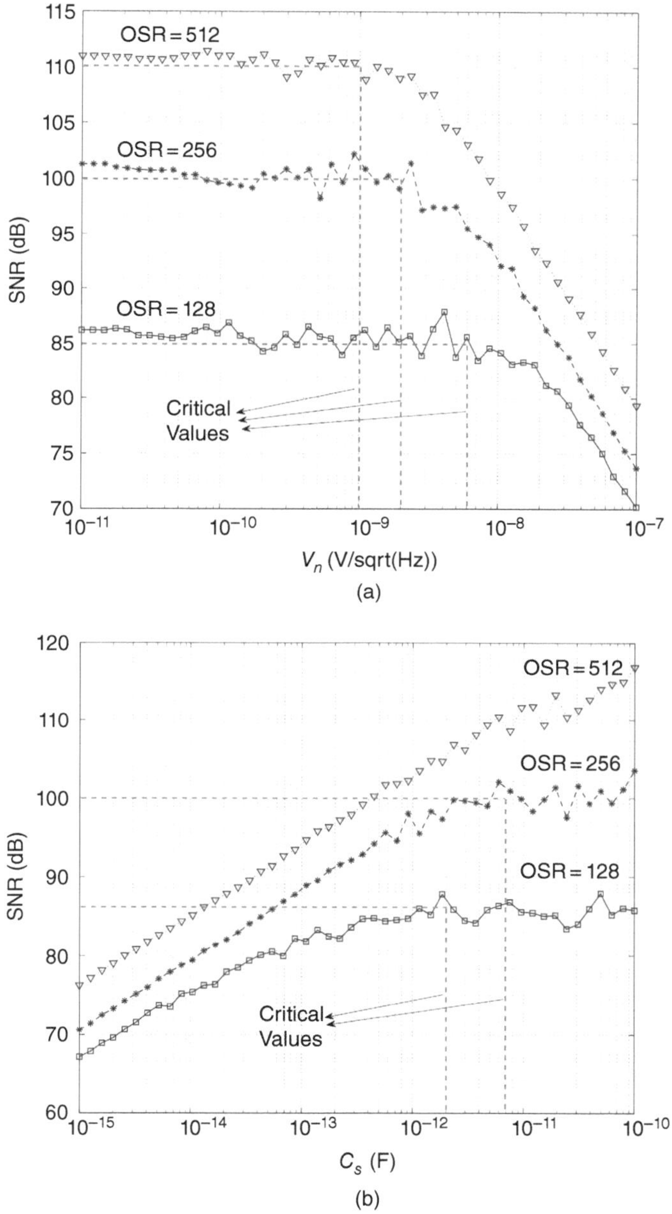

Figure 5.49 Effect of thermal noise sources on the ΣΔM in Figure 5.46: (a) SNR versus V_n; (b) SNR versus C_s.

5.6.1.4 Cumulative Effect of All Errors Simulations carried out in previous sections analyzed the isolated impact of the most important circuit errors limiting the performance of the ΣΔM shown in Figure 5.46. However, that performance can be further degraded as a consequence of the cumulative effect of the different errors acting together. This is of course the case in practical situations. To illustrate this effect, Figure 5.51 depicts the modulator SNDR versus I_o considering both its isolated

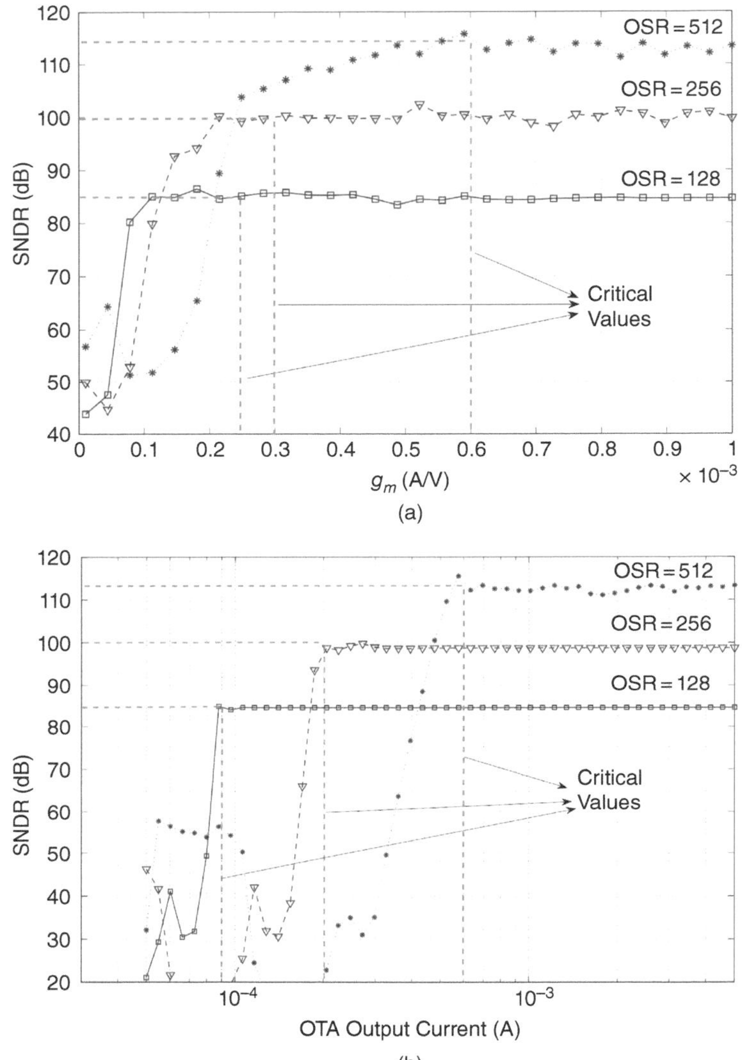

Figure 5.50 Effect of incomplete settling on the ΣΔM in Figure 5.46 for $B_w = 10$ kHz: (a) SNDR versus g_m for $I_o = 5$mA; (b) SNR versus I_o for $g_m = 5$ mA/V.

effect and the cumulative effect of this parameter; that is, considering other nonideal model parameters. Note that in the latter case, a higher value of I_o is required to achieve the specified resolution.

Figure 5.52 plots the output spectrum of the modulator considering all circuit errors, taking into account the critical values of error model parameters obtained for OSR = 512. Figure 5.53 depicts the SNDR versus input signal level for different values of OSR and B_w. Note that the effective resolution achieved is in accord with the specifications.

The results of the high-level sizing process are summarized in Table 5.3, where system-level (modulator) specifications are mapped into building-block (integrator) specifications. For the sake of

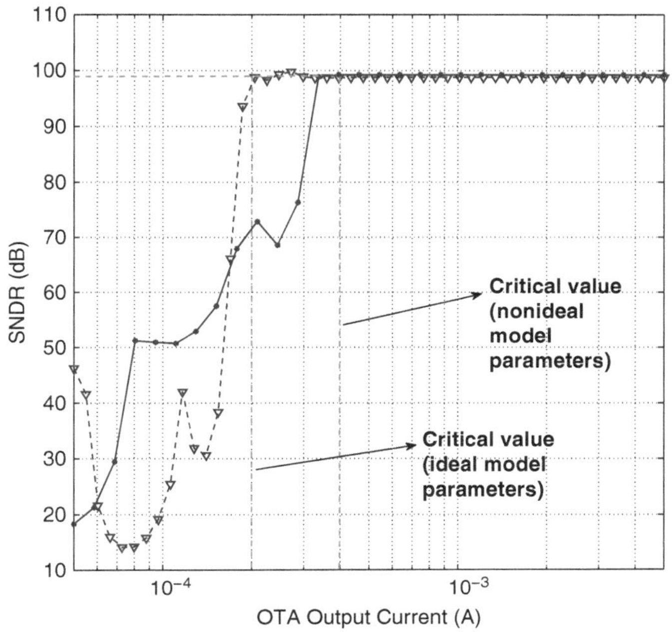

Figure 5.51 Illustrating the cumulative and isolated effect of the amplifier output current I_o.

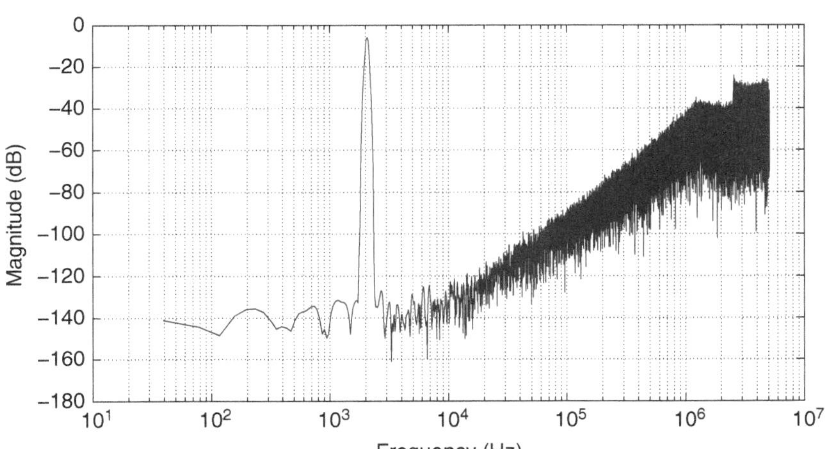

Figure 5.52 Output spectrum of the ΣΔM in Figure 5.46 considering the effect of all circuit errors.

completeness, the required values for GB and SR are also included in the table. Note that this data takes into account the interaction of different circuit errors considered in the example, and hence may be more restrictive than that obtained from the isolated analysis described above. In this case, the effect of switch on-resistance on the OTA transconductance and GB described in Section 5.4.1 has not been taken into account for the sake of simplicity. However, this effect should be considered in practice, resulting in more demanding integrator dynamic specifications. As an illustration, Figure 5.54 depicts the information provided by the corresponding SIMSIDES integrator model if the mentioned degradation

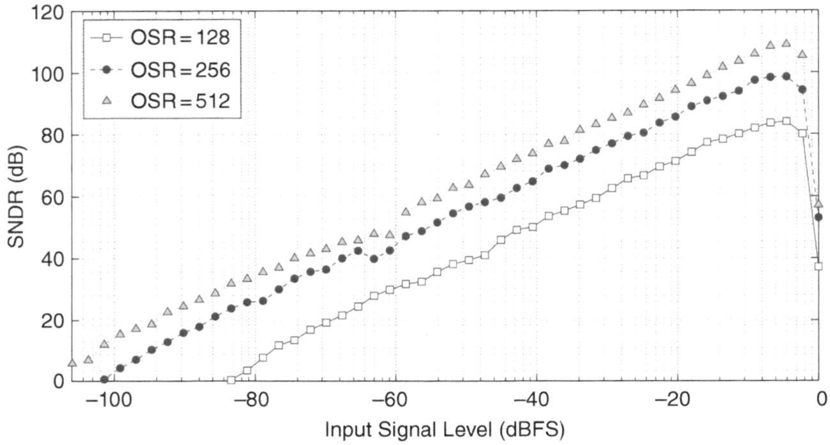

Figure 5.53 SNDR versus input signal level for $B_w = 10$ kHz considering all circuit errors.

Table 5.3 High-level sizing of the second-order SC-ΣΔM shown in Figure 5.46 ($C_{eq} = 2.6$ pF).

Model parameter	Units	OSR value		
		128	256	512
A_v	—	100	400	400
V_n	$nV/\sqrt{Hz}$	6	2	1
C_s	pF	2	7	10
g_m	mA/V	0.25	0.3	0.6
I_o	mA	0.09	0.5	1
GB	MHz	100	120	240
SR	V/μs	46	200	400

is considered. Among other data, the model provides the effective values of the equivalent load capacitance, GB, and SR obtained for each clock phase. Taking into account these model parameters, the half-scale SNDR for OSR = 128 is 78.3 dB; that is, approximately 9 dB less than the ideal one.

5.6.2 CT Fifth-order Cascade 3-2 Multi-bit ΣΔM

The second case study is a CT two-stage cascade ΣΔM consisting of a third-order front-end stage and a second-order back-end stage. Figure 5.55a shows the conceptual block schematic of the modulator; its corresponding implementation in SIMSIDES is depicted in Figure 5.55b [44]. The front-end stage is made up of an integrator and a resonator, while the second-stage loop filter is essentially a resonator. Multi-bit (4-bit) quantization and NRZ feedback DACs are used in both stages and DEM techniques are also included to reduce the impact of DAC mismatch on the modulator linearity. In both stages, an extra feedback branch – made up of a DAC and a D-latch – is connected from the output to the input of the quantizer to compensate for the effect of excess loop delay [45], as described in Section 4.9.

The loop filter of both stages is implemented using Gm-C integrators, whereas current-mode DACs are used in the feedback loop. The modulator has been synthesized in the CT domain following

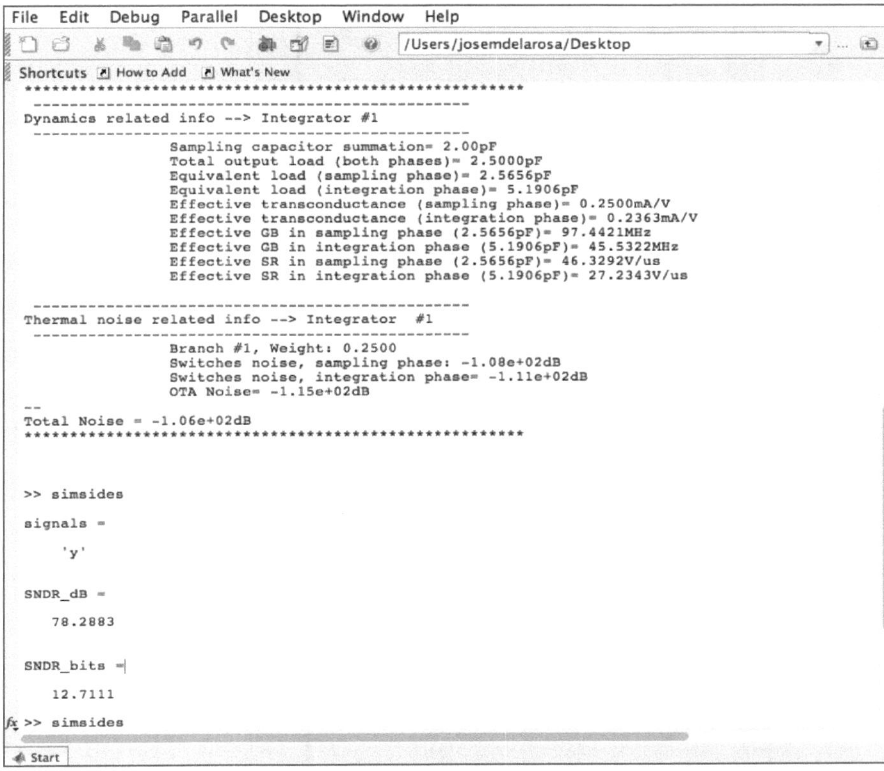

Figure 5.54 Information provided by the SIMSIDES SC FE integrator model when the effect of switch on-resistance is taken into account in the transient response.

the methodology described by Tortosa *et al.* [46] and the resonator poles are placed at an optimum position to minimize the magnitude of NTF within the signal bandwidth. Similar to any other cascade ΣΔM, the DCL functions of each stage (i.e., $\mathrm{DCL}_{1,2}$, not explicitly shown in Figure 5.55 for the sake of simplicity) can be derived from the STF and NTF of both modulator stages, by annulling the first-stage quantization error at the overall modulator output [46].

Note from Figure 5.55b that Gm-C integrators are implemented in SIMSIDES using the model described in Section 5.4.2.3 (see Figure 5.34); that is, as the cascade of a transconductor and an output impedance circuit (output resistance and capacitance in parallel). Alternatively, Gm-C integrators can be also modeled in SIMSIDES as a single building block, as shown in Figure 5.31. This approach is more suited to those ΣΔMs with a reduced number of loop-filter coefficients – commonly implemented as transconductors. In contrast, the modulator in this example has a number of feedforward coefficients, which are also commonly implemented as transconductors.

Table 5.4 sums up the values of loop-filter transconductances g_i, as well as the capacitances C_i used in the modulator. These values have been found through an iterative simulation-based process, which, starting from the nominal values required to place the NTF zeros, optimizes the modulator performance in terms of DR and stability within the full-scale range. Unit transconductors are used in the majority of loop-filter transconductors, and these can be tuned to keep the time constant C/g_m unchanged over C variations.

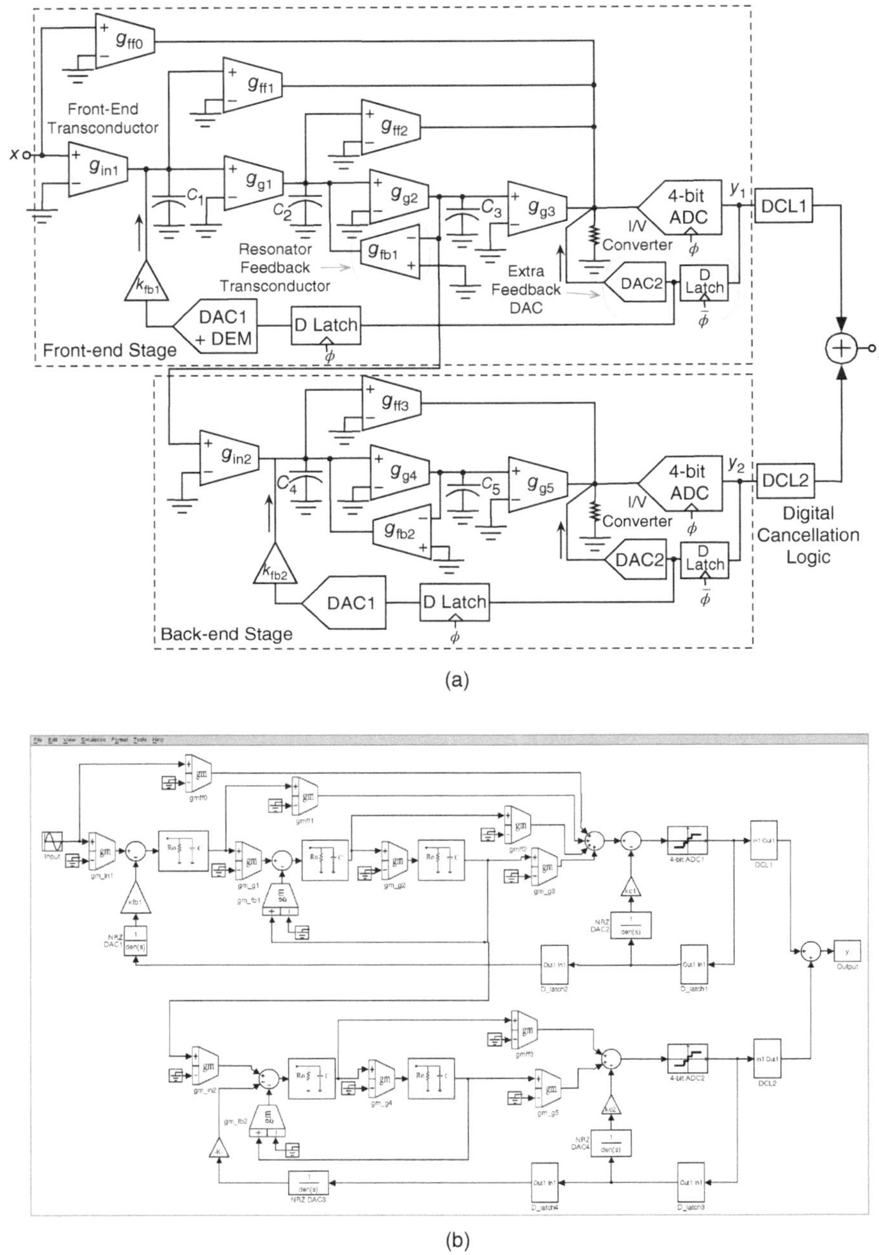

Figure 5.55 Gm-C fifth-order cascade 3-2 ΣΔM with 4-bit quantization in both stages: (a) conceptual schematic; (b) SIMSIDES implementation.

Table 5.4 Loop-filter coefficients of the CT cascade 3-2 ΣΔM in Figure 5.55.

Coefficient	Values
Unit circuit elements	$C_u = 3.65$ pF, $g_u = 190\ \mu$A/V
Capacitors	$C_1 = C_2 = C_3 = C_u$, $C_4 = C_5 = 2C_u$
Feedforward transconductances	$g_{\rm in1} = 852\ \mu$A/V, $g_{\rm ff0} = g_{\rm ff2} = 2g_u$, $g_{\rm ff1} = 4g_u$, $g_{\rm in2} = g_{\rm ff3} = 5g_u$, $g_{\rm g1} = g_{\rm g5} = 3g_u$, $g_{\rm g2} = 5g_u$, $g_{\rm g2} = g_u$, $g_{\rm g4} = 7g_u$
Feedback transconductances	$g_{\rm fb1} = g_{\rm fb2} = g_u$, $k_{\rm fb1} = 730\ \mu$A/V, $k_{\rm fb2} = 6g_u$

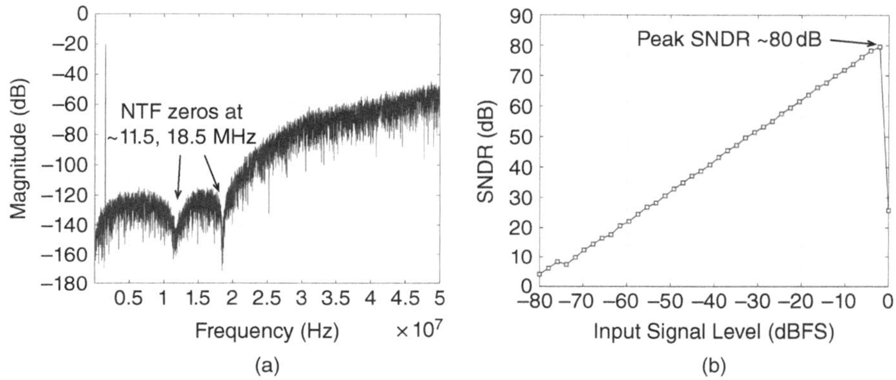

Figure 5.56 Ideal simulation results of the modulator in Figure 5.55: (a) output spectrum; (b) SNDR versus input signal level for OSR = 6 ($B_w = 20$ MHz, $f_s = 240$ MHz), considering a full-scale reference of 0.5 V.

Figure 5.56a shows the ideal output spectrum of the modulator considering an −20 dBFS input tone at 1.49 MHz, when clocked at $f_s = 240$ MHz. Note that shaped quantization noise presents two notches, at 11.5 MHz and 18.5 MHz, which minimize the IBN in the target signal bandwidth (B_w = 20 MHz). The ideal effective resolution achieved by the modulator is illustrated in Figure 5.56b, which depicts the SNDR versus input signal level for OSR = 6, corresponding to $f_s = 240$ MHz and $B_w = 20$ MHz. Under these conditions, the maximum effective resolution that can be achieved is approximately 13 bits. However, this performance is degraded in practice by the action of circuit nonidealities, as analyzed below.

5.6.2.1 Effect of Nonideal Effects The behavioral model used for the transconductors in SIMSIDES takes into account several nonideal circuit effects,[18] including finite DC gain, output saturation voltage, and the input-referred third-order intercept point IIP3. As an illustration, Figure 5.57 shows the effect of the finite DC gain of loop-filter transconductors on the SNDR for OSR = 6, 12, 24, when clocked at $f_s = 240$ MHz. It can be seen that as the OSR increases, the effect of this error is attenuated, as predicted by the theoretical analysis described in Section 3.2. For a signal bandwidth of $B_w = 20$ MHz (OSR = 6), a finite DC gain larger than 50 dB is required.

Figure 5.58 illustrates the effect of IIP3 on the modulator performance, comparing the degradation caused by different transconductors: front-end transconductor ($g_{\rm in1}$ in Figure 5.55), input feedforward

[18] A more detailed explanation of the nonideal circuit effects and electrical parameters included in SIMSIDES block models can be found in Appendix C.

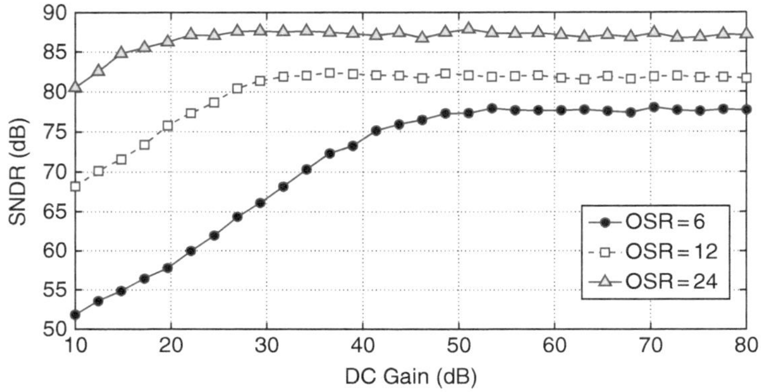

Figure 5.57 Effect of finite DC gain of loop-filter transconductors on the SNDR of Figure 5.55 for different values of OSR.

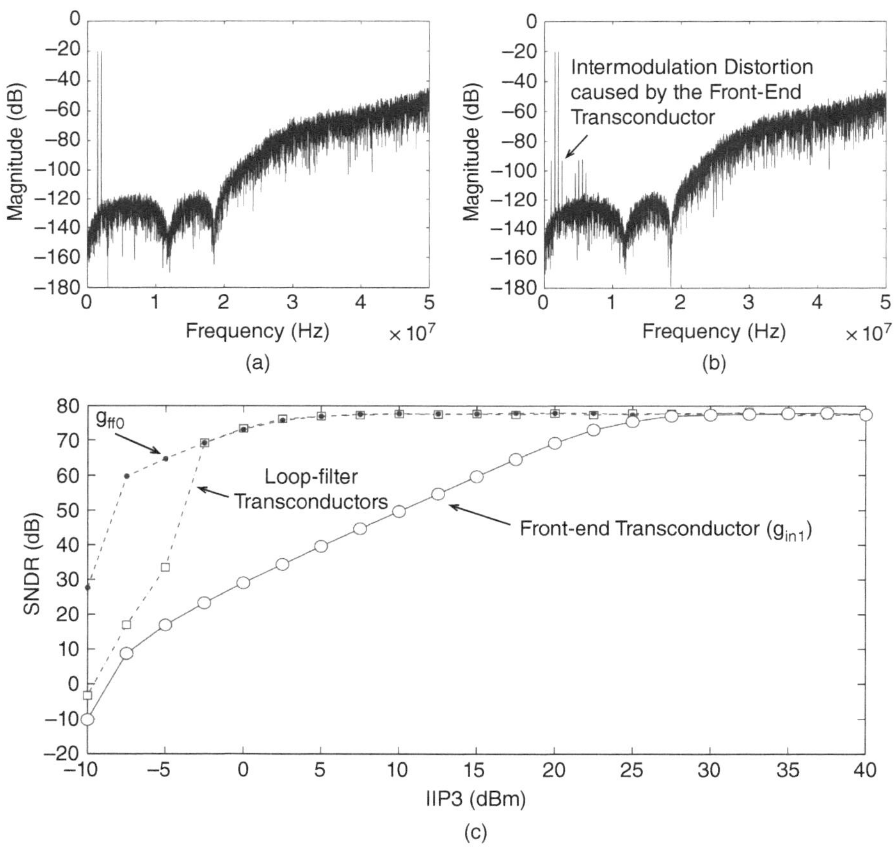

Figure 5.58 Effect of transconductance nonlinearity on the performance of the CT-ΣΔM in Figure 5.55: (a) output spectrum assuming that g_{in1} and g_{ff0} are ideal and that all loop-filter transconductors have an IIP3 = 20 dBm; (b) output spectrum considering that all transconductors are ideal, except for g_{in1} that has an IIP3 = 20 dBm; (c) SNDR versus IIP3 for the different transconductors.

transconductor (g_{ff0}), and the rest of loop-filter transconductors. Figures 5.58a and 5.58b illustrate the output spectrum of the modulator considering two −10 dBFS input tones located at 1.49 MHz and 2.02 MHz. The output spectrum in Figure 5.58a assumes an IIP3 = 20 dBm for all loop-filter transconductors and that g_{in1} and g_{ff0} are ideal. In contrast, Figure 5.58b assumes that all transconductors are ideal, except for the front-end transconductor, which has an IIP3 = 20 dBm. As expected, the front-end transconductor severely degrades the linearity of the modulator, causing a number of intermodulation products to appear in the signal band. The effect of the nonlinearity of the different transcondutors on the modulator resolution is better illustrated in Figure 5.58c, which represents the SNDR versus IIP3 of the different types of transconductors in the modulator. Note that the effective resolution is not degraded if IIP3 > 28 dBm in the front-end transconductor. This specification is relaxed for the rest of the transconductors, in which IIP3 > 5 dBm is enough to achieve the ideal SNDR.

Apart from the impact of the aforementioned nonidealities, the effect of circuit element tolerances and component mismatch are particularly critical in the design of cascade CT-ΣΔMs. The absolute tolerances can be controlled by using time-constant tuning, as is the case for this design example. However, the impact of mismatch error still remains and must be taken into account at the early design stages. Behavioral simulation using SIMSIDES can be used for this purpose. Thus, in order to evaluate the impact of this error on the performance of the modulator in Figure 5.55, maximum values of mismatch were estimated from a 130-nm CMOS process. The results of this analysis are shown in Figure 5.59. The SNR is represented as a function of the standard deviation of transconductances and capacitances (σ_{gm} and σ_C, respectively). Note that, for each point of this surface, a Monte Carlo analysis consisting of 150 simulations was carried out. The value of SNR represented in this figure corresponds to that obtained by more than 90% of the simulations for each value of σ_{gm} and σ_C. Note that even for the worst-case mismatch, an effective resolution higher than 12 bits can be achieved.

5.6.2.2 High-level Synthesis and Verification

As a case study, let us assume that the modulator in Figure 5.55 is designed to fulfill the following specifications: 12-bit effective resolution within 20-MHz signal bandwidth [46]. These specifications can be mapped into the modulator building-block specifications by following a parametric-based analysis, as described in previous sections. Alternatively, an optimization-based procedure can be followed, in which an optimization engine is used for design parameter selection and a behavioral simulator (SIMSIDES in this case) is

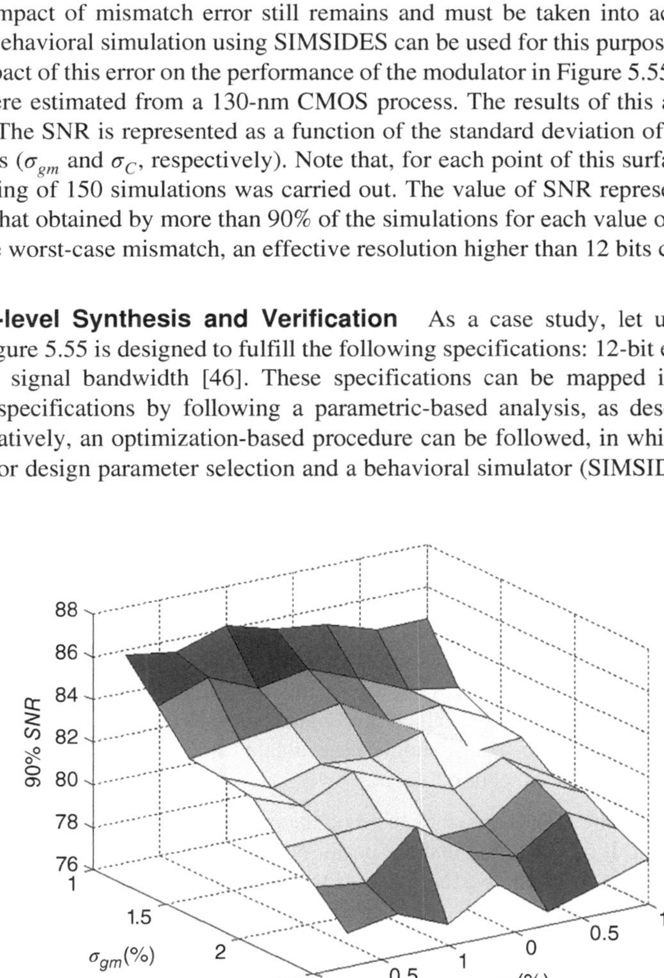

Figure 5.59 Illustrating the effect of mismatch on the SNR of the ΣΔM in Figure 5.55.

Table 5.5 High-level sizing of the CT cascade ΣΔM in Figure 5.55.

Front-end transconductor	
Finite DC gain	≥ 70dB
Linear input swing	≤ 0.3 V
Linear output swing	≤ 0.3 V
Third-order nonlinear coefficient	≤ −86 dB
Loop-filter transconductors	
Finite DC gain	≥ 50 dB
Linear input swing	≤ 0.3 V
Linear output swing	≤ 0.3 V
Third-order nonlinear coefficient	≤ −56 dB
Multi-bit embedded ADCs	
Comparator offset	≤ 20 mV
Comparator hysteresis	≤ 20 mV
Comparator resolution time	≤ 1 ns
Feedback DACs	
Unit current standard deviation	≤ 0.15%LSB
Finite output resistance	≥ 12 MΩ
Settling time	≤ 0.5 ns

the performance evaluator [6]. As an illustration of the latter approach, an statistical optimizer was combined with SIMSIDES for the high-level sizing of the ΣΔM in Figure 5.55.

Table 5.5 sums up the results of the sizing process, showing the critical (maximum/minimum) values of the circuit electrical performance parameters that can be tolerated to meet the required modulator performance. As stated above, the specifications of the front-end transconductor – specifically the finite DC gain and the third-order nonlinearity – are more demanding than for the other transconductors. For this reason, different circuit topologies are in practice used for implementing the front-end transconductor and the loop-filter transconductors in Gm-C ΣΔMs, as will be discussed in Chapter 8.

Apart from the transconductor specifications, the electrical parameters of other building blocks – such as the comparators used in embedded multi-bit flash ADCs or the current-mode feedback DACs – are also given in Table 5.5. A description of the behavioral models of these blocks and their associated parameters can be found in Appendix C. In addition to the requirements listed in Table 5.5, the modulator building blocks must be designed so that their thermal noise contribution does not limit the modulator resolution. This is particularly critical for the front-end transconductor and feedback DAC1 in the first stage, which are connected to the modulator input. Therefore, the impact of noise sources in these blocks should be evaluated in SIMSIDES following a parametric analysis similar to that shown in Section 5.6.1.2.

Figure 5.60 shows the SNDR curve of the modulator in Figure 5.55b when all nonideal effects are taken into account. It can be noted that the peak SNDR is 76.8 dB (12.5 bit), which meets target specifications. Therefore, once the performance of the modulator has been validated by behavioral simulation considering the impact of the main circuit error mechanisms, the resulting building-block electrical parameters in Table 5.5 can be used as initial design specifications for the modulator subcircuits at transistor level. Their implications in the transistor-level design will be discussed in Chapters 7 and 8.

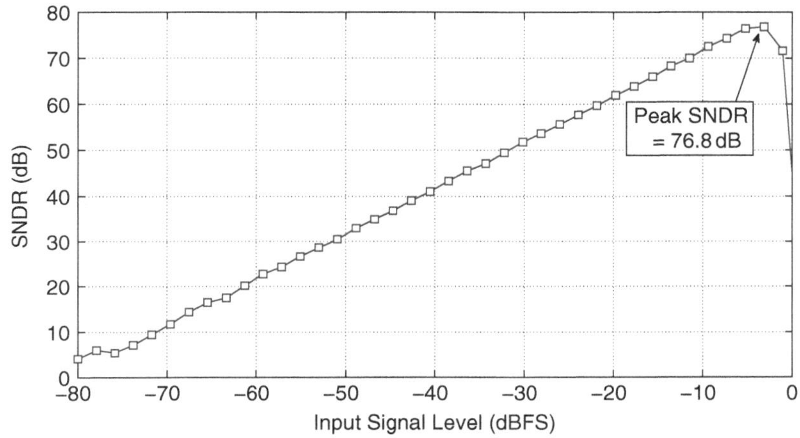

Figure 5.60 SNDR versus input signal level considering all nonideal effects in Figure 5.55.

5.7 Summary

This chapter discussed the use of behavioral modeling and simulation techniques for the high-level analysis and synthesis of ΣΔMs. After examining different approaches and alternatives to the transistor-level electrical simulation of ΣΔMs, it has been demonstrated that using time-domain behavioral models implemented with MATLAB C-MEX S-functions is a very efficient technique, in terms of accuracy, CPU simulation time, and flexibility, for incorporating new circuit effects and building-block models. A step-by-step procedure to implement precise ΣΔM behavioral models using this technique was described in detail and applied to the fundamental ΣΔM building blocks: integrators, quantizers, and embedded DACs. Based on this modeling approach, a SIMULINK-based time-domain behavioral simulator for ΣΔMs – called SIMSIDES – was described and some case studies were presented to illustrate the use of this simulator in the high-level sizing and verification of ΣΔMs. The results obtained from this process constitute the starting point for the electrical (transistor-level) design and verification of ΣΔMs described later in this book.

References

[1] J. M. de la Rosa, "Design Guidelines of ΣΔ Modulators: From System to Chip and Application to Reconfigurable ADCs," *Proc. of the IEEE Intl. Conf. on Electron Devices and Solid-State Circuits (EDSSC)*, pp. 144–148, August 2017.

[2] V. F. Dias *et al.*, "Design Tools for Oversampling Data Converters: Needs and Solutions," *Microelectronics Journal*, vol. 23, pp. 641–650, 1992.

[3] G. Gielen and J. Franca, "CAD Tools for Data Converter Design: An Overview," *IEEE Trans. on Circuits and Systems II: Analog and Digital Signal Processing*, vol. 43, pp. 77–89, February 1996.

[4] F. Medeiro *et al.*, "A Vertically Integrated Tool for Automated Design of ΣΔ Modulators," *IEEE J. of Solid-State Circuits*, vol. 30, pp. 762–772, July 1995.

[5] K. Francken *et al.*, "A high-level simulation and synthesis environment for delta-sigma modulators," *IEEE Trans. on Computer-Aided Design of Integrated Circuits and Systems*, vol. 22, pp. 1049–1061, August 2003.

[6] J. Ruiz-Amaya *et al.*, "High-Level Synthesis of Switched-Capacitor, Switched-Current and Continuous-Time ΣΔ Modulators Using SIMULINK-based Time-Domain Behavioral Models," *IEEE Trans. on Circuits and Systems – I: Regular Papers*, pp. 1795–1810, Sep. 2005.

[7] R. Schreier and G. C. Temes, *Understanding Delta-Sigma Data Converters*. IEEE Press, 2005.

[8] R. Schreier, *The Delta-Sigma Toolbox*. [Online]. Available: http://www.mathworks.com/matlabcentral, 2017.

[9] J. M. de la Rosa *et al.*, "A CMOS 110-dB@40-kS/s Programmable-Gain Chopper-Stabilized Third-Order 2-1 Cascade Sigma-Delta Modulator for Low-Power High-Linearity Automotive Sensor ASICs," *IEEE J. of Solid-State Circuits*, vol. 40, pp. 2246–2264, November 2005.

[10] S. Pavan, "Efficient Simulation of Weak Nonlinearities in Continuous Time Oversampling Converters," *IEEE Transactions on Circuits and Systems I – Regular Papers*, vol. 57, pp. 1925–1934, August 2010.

[11] J. Ryckaert, J. Borremans, B. Verbruggen, L. Bos, C. Armiento, J. Craninckx, and G. van der Plas, "A 2.4 GHz Low-Power Sixth-Order RF Bandpass ΔΣ Converter in CMOS," *IEEE J. of Solid-State Circuits*, vol. 44, pp. 2873–2880, November 2009.

[12] F. V. Fernandez *et al.*, "Design Methodologies for Sigma-Delta Converters," in *CMOS Telecom Data Converters* (A. Rodríguez-Vázquez, F. Medeiro and E. Janssens, editors), Kluwer Academic Publishers, 2003.

[13] F. Medeiro, B. Pérez-Verdú, and A. Rodríguez-Vázquez, *Top-Down Design of High-Performance Sigma-Delta Modulators*. Kluwer Academic Publishers, 1999.

[14] C. H. Wolff and L. Carley, "Simulation of Δ-Σ Modulators using Behavioral Models," *Proc. of the IEEE Intl. Symp. on Circuits and Systems*, pp. 376–379, May 1990.

[15] V. Liberali *et al.*, "TOSCA: A Simulator for Switched-capacitor Noise-shaping A/D Converters," *IEEE Trans.Comput.-Aided Des. Integr. Circuits Syst.*, vol. 12, pp. 1376–1386, September 1993.

[16] P. Malcovati *et al.*, "Behavioral Modeling of Switched-capacitor Sigma-Delta Modulators," *IEEE Trans. on Circuits and Systems – I: Regular Papers*, vol. 50, pp. 352–364, March 2003.

[17] G. Gielen *et al.*, "An Analytical Integration Method for the Simulation of Continuous-Time ΔΣ Modulators," *IEEE Trans. on Computer-Aided Design of Integrated Circuits and Systems*, vol. 23, pp. 389–399, March 2004.

[18] H. Zare-Hoseini, I. Kale, and O. Shoaei, "Modeling of Switched-Capacitor Delta-Sigma Modulators in SIMULINK," *IEEE Transactions on Instrumentation and Measurement*, vol. 54, pp. 1646–1654, August 2006.

[19] M. Keller *et al.*, "A Method for the Discrete-Time Simulation of Continuous-Time Sigma-Delta Modulators," *Proc. of the IEEE Intl. Symposium on Circuits and Systems*, pp. 241–244, May 2007.

[20] G. Suárez, M. Jiménez, and F. O. Fernández, "Behavioral Modeling Methods for Switched-Capacitor ΣΔ Modulators," *IEEE Transactions on Circuits and Systems - I: Regular Papers*, vol. 54, pp. 1236–1244, June 2007.

[21] A. V. Oppenheim and R. W. Schafer, *Discrete-Time Signal Processing*. Prentice Hall, 3rd ed., 2009.

[22] Synopsys, "HSPICE Simulation and Analysis User Guide," *Synopsys Inc.*, 2006.

[23] Cadence, "Spectre Circuit Simulator User Guide," *Cadence Design Systems Inc.*, 2002.

[24] R. Bishop *et al.*, "Table-based Simulation of Delta-Sigma Modulators," *IEEE Trans. on Circuits and Systems – I*, vol. 37, pp. 447–451, March 1990.

[25] G. Brauns *et al.*, "Table-based Modeling of Delta-Sigma Modulators using ZSIM," *IEEE Trans. on Computer-Aided Design of Integrated Circuits and Systems*, vol. 9, pp. 142–150, February 1990.

[26] IEEE-Standards, "IEEE VHDL Language Reference Manual," *IEEE Std 1076-2002*, 2002.

[27] IEEE-Standards, "IEEE VHDL 1076.1 Language Reference Manual," *IEEE Std 1076.1-1999*, 2002.

[28] Cadence, "Cadence Design Framework II," Cadence Design Systems Inc., 2010.

[29] Mathworks, "Using MATLAB Version 6," The Mathworks Inc., 2002.

[30] Mathworks, "Using SIMULINK Version 5," The Mathworks Inc., 2002.

[31] Mathworks, "Stateflow 7.7," The Mathworks Inc., 2002.

[32] R. del Río, F. Medeiro, B. Pérez-Verdú, J. M. de la Rosa, and A. Rodríguez-Vázquez, *CMOS Cascade ΣΔ Modulators for Sensors and Telecom: Error Analysis and Practical Design.* Springer, 2006.

[33] S. Brigati *et al.*, "Modeling Sigma-Delta Modulator Nonidealities in SIMULINK," *Proc. of the IEEE Intl. Symp. on Circuits and Systems*, pp. 2384–2387, May 1999.

[34] S. Brigati, *SD Toolbox*. [Online]. Available: http://www.mathworks.com/matlabcentral, 2002.

[35] Mathworks, "Writing S-Functions Version 5," The Mathworks Inc., 2002.

[36] Mathworks, "Simulink: Developing S-Functions," The Mathworks Inc., 2016.

[37] W. Press *et al.*, *Numerical Recipes in C. The Art of Scientific Computing*. Cambridge University Press, 2nd ed., 1992.

[38] B. Razavi, *Principles of Data Conversion System Design*. IEEE Press, 1995.

[39] W. Yu, S. Sen, and B. Leung, "Distortion Analysis of MOS Track-and-hold Sampling Mixers using Time-varying Volterra Series," *IEEE Trans. on Circuits and Systems – II: Analog and Digital Signal Processing*, vol. 46, pp. 101–113, February 1999.

[40] R. del Río, F. Medeiro, J. M. de la Rosa, B. Pérez-Verdú, and A. Rodríguez-Vázquez, "Reliable Analysis of Settling Errors in SC Integrators: Application to ΣΔ Modulators," *IET Electronics Letters*, vol. 36, pp. 503–504, March 2000.

[41] M. Ortmanns and F. Gerfers, *Continuous-Time Sigma-Delta A/D Conversion: Fundamentals, Performance Limits and Robust Implementations.* Springer, 2006.

[42] J. Cherry and W. Snelgrove, "Excess Loop Delay in Continuous-Time Delta–Sigma Modulators," *IEEE Trans. on Circuits and Systems – II: Analog and Digital Signal Processing*, vol. 46, pp. 376–389, April 1999.

[43] J. M. de la Rosa, B. Pérez-Verdú, and A. Rodríguez-Vázquez, *Systematic Design of CMOS Switched-Current Bandpass Sigma-Delta Modulators for Digital Communication Chips.* Kluwer Academic Publishers, 2002.

[44] R. Tortosa, A. Aceituno, J. M. de la Rosa, A. Rodríguez-Vázquez, and F. V. Fernández, "A 12-bit, 40MS/s Gm-C Cascade 3-2 Continuous-Time Sigma-Delta Modulator," *Proc. of the IEEE Intl. Symp. on Circuits and Systems*, pp. 1–4, 2007.

[45] S. Yan and E. Sánchez-Sinencio, "A Continuous-Time ΣΔ Modulator With 88-dB Dynamic Range and 1.1-MHz Signal Bandwidth," *IEEE J. of Solid-State Circuits*, vol. 39, pp. 75–86, January 2004.

[46] R. Tortosa, J. M. de la Rosa, F. V. Fernández, and A. Rodríguez-Vázquez, "A New High-Level Synthesis Methodology of Cascaded Continuous-Time ΣΔ Modulators," *IEEE Trans. on Circuits and Systems – II: Express Briefs*, vol. 53, pp. 739–743, August 2006.

CHAPTER 6

AUTOMATED DESIGN AND OPTIMIZATION OF ΣΔMS

As stated in Chapter 5, the design of ΣΔMs involves a number of tasks at several abstraction levels, usually addressed by following a top-down/bottom-up systematic methodology – from specifications to die implementation. In such an iterative process, it is a common practice for designers to put a lot of effort into *optimizing* the design of their circuits at transistor level, trying to reduce the power dissipation as much as possible, while keeping the electrical characteristics required by the target specifications. However, in the majority of practical situations, the actual bottleneck is not at the circuit or device level but at the system level, where correct selection of the modulator architecture, loop-filter topology, coefficients, and so on can lead to a more efficient and robust design, while relaxing the design margins in the most critical metrics at electrical level.

Bearing this in mind, this chapter deals with some of the relevant design methods and CAD tools intended for the automated high-level synthesis and optimization of ΣΔMs, considering how they can be combined to help designers find the best ΣΔM design for a given set of requirements. Describing in detail all the reported design methodologies goes beyond the scope of this book. Instead, this chapter focuses on those CAD tools that capture the knowledge derived from the state-of-the-art synthesis and optimization techniques, and considers how they can be applied to the most critical phases of the high-level synthesis.

Following this introduction, Section 6.1 starts with the problem of architecture selection and how it can be addressed and assisted using the well-known Schreier's Delta-Sigma toolbox. Section 6.2 describes how to combine behavioral simulation and optimization to automate the high-level sizing

Sigma-Delta Converters: Practical Design Guide, Second Edition. José M. de la Rosa.

procedure, considering different optimization engines and SIMSIDES as performance evaluator. The use of the so-called "lifting" method and hardware emulation to speed up the simulation and optimization of CT-ΣΔMs is described in Section 6.3. Finally, Section 6.4 shows how to apply multi-objective evolutionary algorithms to automate the high-level design of ΣΔMs.

6.1 Architecture Exploration and Selection: Schreier's Toolbox

The first problem faced by ΣΔM designers is to find the architecture that best satisfies the specifications required for a given application – the ENOB and B_w – with the minimum energy consumption. In general, there are multiple candidate ΣΔM topologies that can – a priori – fullfil the requirements, with diverse combinations of L, OSR and B.

In order to undertake this architecture exploration, Schreier's MATLAB Delta-Sigma toolbox is widely used by ΣΔM designer community [1]. This tool uses nonlinear models to automatically synthesize a ΣΔM loop filter that provides the desired quantization NTF while retaining system stability. Regardless of whether the ΣΔM is implemented using SC or CT circuits, the loop filter and NTF obtained with Schreier's toolbox serves as a starting point for the design. However, additional iterations based on heuristic search and expert knowledge are needed at this abstraction level in order to fine-tune the loop-filter implementations and maximize the peformance of the ΣΔM in terms of circuit realization, DR and robustness against process variations. The latter is critical in CT-ΣΔMs, whose performance relies on the absolute values of circuit elements – such as resistors or capacitors – rather than capacitor ratios, as in SC-ΣΔMs [2, 3].

A complete description of all functions included in the Schreier's toolbox goes beyond the scope of this book and the interested reader can find more detailed information in the user guide, and the demos and tutorials that are available for it [4]. Instead, some of the most useful scripts and procedures in the toolbox are considered here.

6.1.1 Basic Functions of Schreier's Delta-Sigma Toolbox

Schreier's Delta-Sigma toolbox can be downloaded from www.mathworks.com/matlabcentral/fileexchange/19-delta-sigma-toolbox. The toolbox is compressed in a file called `delsig.zip`, which has to be uncompressed in the hard disk and added to the MATLAB path. There are some C-coded files – for example `SimulateDSM.c` and `SimulateMS.c` – which should be compiled using the `mex` utility, as described in Chapter 5, in order to speed-up simulations. As stated above, there is plenty of tutorial documentation provided with the toolbox, as well as examples that readers may find useful in getting to know the different facilities provided. Instead of going into the ins and outs of all of the procedures and scripts, this section will focus on the most representative functions and facilities, illustrating their use with some practical examples.

One of the key functions of the toolbox is the so-called `synthesizeNTF` function,[1] which allows a given NTF to be synthesized for a given order, L, OSR, and out-of-band gain (OBG). The syntax of this function is as follows:

```
ntf = synthesizeNTF(order,OSR,opt,H_inf,f0)
```

where `order` stands for the order of the ΣΔM loop filter (the order of NTF), `opt` is a flag used to set the placement of the zeros of NTF, `H_inf` is the maximum OBG, and `f0` is the center frequency of the ΣΔM, being `f0=0` for LP-ΣΔMs.

[1]There are some alternative functions to synthesize the NTF, such as *synthesizeChebyshevNTF*, which uses a Chebyshev approximation, or *clans*, which is based on the closed-loop analysis of noise-shaper methodology proposed in [5].

Another important function is `realizeNTF`, which transforms the synthesized NTF into a set of coefficients for a particular modulator topology. This function is used as follows:

```
[a,g,b,c] = realizeNTF(ntf,form,stf)
```

where `stf` and `ntf` are respectively the STF and NTF of the ΣΔM, given in zero-pole form, and `form` specifies the modulator topology: cascade-of-resonators in feedback form (CRFB), cascade-of-resonators in feedforward form (CRFF), cascade-of-integrators in feedback form (CIFB), or cascade-of-integrators in feedforward form (CIFF) [1].

Applying the `realizeNTF` function, the coefficient matrix `[a,g,b,c]` is obtained, where `a` is the $[1 \times n]$ vector corresponding to the feedback/feedforward coefficients from/to the quantizer (with n being the modulator order), `g` is the $[1 \times n/2]$ vector of resonator coefficients, `b` is $[1 \times (n+1)]$ vector of feed-in coefficients from the modulator input to the each integrator input, and `g` is the $[1 \times n]$ vector of integrator weights, being all ones in unscaled ΣΔMs [4]. As an illustration, Figure 6.1 shows the block diagram of a CRFF ΣΔM including the coefficient nomenclature used in the toolbox. This modulator architecture corresponds to the high-order single-loop ΣΔM with feedforward summation and local feedback loops shown in Figure 2.10.

Another powerful script included in the toolbox is `stuffABCD`, which gives a state-space representation of an arbitrary ΣΔM loop filter – conceptually shown in Figure 6.2 – using the following syntax:

```
ABCD = stuffABCD(a,g,b,c,form)
```

where `ABCD` contains a state-space representation of the ΣΔM loop filter used in the toolbox. The state-space equations used to update the loop-filter state, $x(n)$, and to compute the output of the loop filter, $y(n)$, are given by:

$$x(n+1) = A \cdot x(n) + B \cdot \begin{bmatrix} u(n) \\ v(n) \end{bmatrix}$$
$$y(n) = C \cdot x(n) + D \cdot \begin{bmatrix} u(n) \\ v(n) \end{bmatrix} \tag{6.1}$$

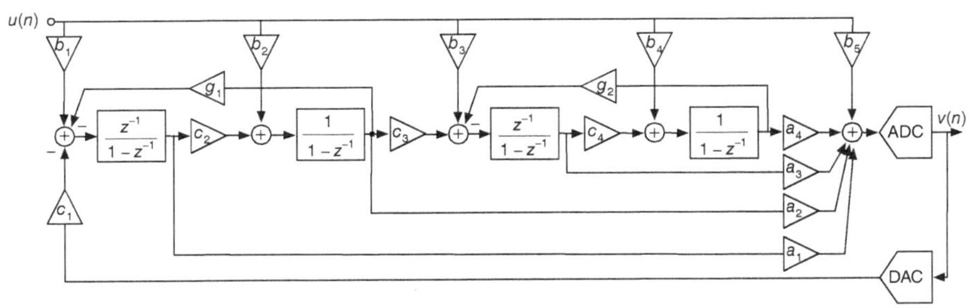

Figure 6.1 Block diagram of CRFF ΣΔM with the loop-filter coefficients used in Schreier's MATLAB toolbox [1].

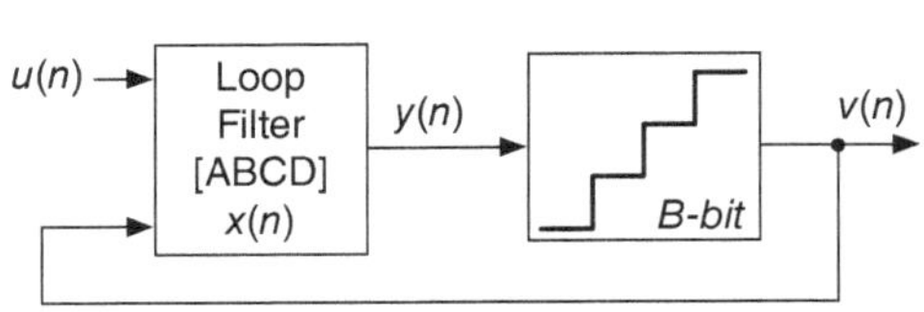

Figure 6.2 State-space representation of a ΣΔM [1].

where $u(n)$ and $v(n)$ are respectively the input and output of the ΣΔM (see Figure 6.2), and A, B, C, and D are respectively the sub-matrices obtained by `stuffABCD` function, given in the form:

$$[ABCD] = \begin{bmatrix} A & B \\ C & D \end{bmatrix} \tag{6.2}$$

The opposite transformation can be carried out by using `mapABCD` function as follows:

```
[a,g,b,c] = mapABCD(ABCD,form)
```

which gives the matrix of loop-filter coefficients `[a,g,b,c]` for a given ΣΔM topology corresponding to a state-space representation defined in the matrix `ABCD`.

Once the NTF has been synthesized and implemented in a given ΣΔM architecture, the loop-filter coefficients can be scaled so that their internal nodes – i.e. the states – have an amplitude lower than a specified limit. This guarantees stability and optimizes the dynamic range and power consumption. It can be done with `scaleABCD` using the following syntax:

```
[ABCDs,umax] = scaleABCD(ABCD,nlev,f,xlim,ymax,umax,N)
```

where `nlev` is the number of quantizer levels, `f` is the input frequency (normalized to f_s), `xlim` is the limit set on the states and `ymax` is a threshold used to keep the ΣΔM stable so that if the quantizer input exceeds this value, stability is not guaranteed [4].

Other useful functions included in the toolbox are `calculateTF` and `simulateDSM`, respectively used to calculate NTF and STF and simulate the synthesized ΣΔM. The use of these functions and procedures is illustrated in the following sections.

6.1.2 Synthesis of a Fourth-order CRFF LP/BP SC-ΣΔM with Tunable Notch

As an illustration, let us consider a fourth-order ΣΔM with a 5-level quantizer and CRFF loop-filter topology, as shown in Figure 6.1. In this example, the ΣΔM is intended to be applied as an ADC for wireless RF receivers, with a constant f_s and a variable f_n. The modulator will be implemented using SC circuits and it will be synthesized so that the notch frequency, f_n, can be programmed from DC to $f_s/4$, to make NTF reconfigurable either as a LP or a BP ΣΔ-ADC. To this end, the matrix of loop-filter coefficients `[a,g,b,c]` has to be programmed to make f_n variable in the range DC-to-25-MHz, considering frequency steps of 1 MHz [6].

Figure 6.3 shows the MATLAB script used to synthesize the loop filter of the reconfigurable LP/BP ΣΔM using the Schreier's toolbox. The first section of the script defines the main parameters (OSR, L, OoBG, etc.) needed to apply `synthesizeNTF`. In this case, as the notch frequency is programmable, the loop-filter, and consequently STF and NTF, needs to be reconfigured. This is implemented using a `for` loop that generates a matrix (`H(i)`) with a set of loop-filter solutions for the different values of f_n (a total of 48 filters are generated in this example).

Once the set of loop filters has been synthesized, the next section of the script uses `realizeNTF` to gives the matrix `[a,g,b,c]` corresponding to the CRFF topology and a state-space representation (`ABCD`) of this topology is generated using `stuffABCD`. At this point of the synthesis procedure – still at an ideal, higher-abstraction level – it is clearer to represent STF and NTF with the selected modulator topology (CRFF in this example). This can be done using the functions `calculateTF` and `evalTF`, as illustrated in Figure 6.4, where the synthesized NTF is depicted for different notch frequencies within the range of interest.

The next step in the script of Figure 6.3 is to scale down the derived coefficients `[a,g,b,c]` in order to maximize the integrators' output swings so that the output swing specifications for the

```
%%%%%%%%%%%%%%%%%%%%%%%%%%%%%%%%%%%%%%%%%%%%%%%
%% Synthesis example: A 4th-order CRFF LP/BPSDM   %
%% Based on [S. Asghar et al., IEEE VLSI-SoC 2012]%
%%%%%%%%%%%%%%%%%%%%%%%%%%%%%%%%%%%%%%%%%%%%%%%
%% Synthesis of NTF (synthesizeNTF)
close all; clear all;
OSR=50; % Oversampling ratio
L = 4;  % Loop-filter order
H_inf=2.048; % OBG (can be different for each notch)
opt =2;  % Optimum placement of NTF zeros
Nlev =5; % No. of levels in embedded quantizer
tun=0;   % Notch frequency parameter used in synthesizeNTF
notch=12; % Notch frequency
for (i=1:48) % Synthesis of NTF for 100 notch frequencies
H(i) = synthesizeNTF(L,OSR,opt,H_inf,0.01*tun);
stf(i)=H(i); % Set STF=1 with the same poles as NTF
stf(i).z=[];
tun=i;
end

%% Loop-filter realization in CRFF form (realizeNTF & stuffABCD)
p=notch+1;
H=H(p);
form = 'CRFF';
[a,g,b,c] = realizeNTF(H,form)
ABCD = stuffABCD(a,g,b,c,form);

%% Compute and plot NTF, STF (calculateTF, evalTF)
[ntf, stf] = calculateTF(ABCD);
f = linspace(0,0.48,1000);
z = exp(2i*pi*f);
magNTF = dbv(evalTF(ntf,z));
magSTF = dbv(evalTF(stf,z));
figure;
plot(f,magNTF,'b',f,magSTF,'m','LineWidth',1)
xlabel('Normalized Frequency');
ylabel('Magnitude');
legend('Noise Transfer Function','Signal TransferFunction')

%% Scale loop-filter coefficients (scaleABCD, mapABCD)
% Limits on the states extracted from simulations
x = [0.85 0.75 1.4 1.45] % Limit on the states (depends on notch)
Np=1e4;
[ABCDs,umax] = scaleABCD(ABCD,Nlev,0.01*(p-1),x,Nlev+5,[],Np)
[as,gs,bs,cs] = mapABCD(ABCDs,form)

%% Simulation (simulateDSM)
fs=100e6;fi=0.01*(p-1)*fs;A=1;ts=1/fs;Ts=1/fs;N=65536 ;i=1;sigma=0.005
u=0.1*sin(2*pi*fi/(fs)*[0:N-1]);
v=simulateDSM(u,ABCDs,Nlev);
spec=fft(v.*ds_hann(N))/(N/4);
figure; plot(linspace(0,0.5,N/2+1),dbv(spec(1:N/2+1)));

%% Map loop-filter coefficient onto feasible SC implementation
R = 100; % Rounding Factor
a2 = round([as].*R)./R
g2 = round([gs].*R)./R
b2 = round([bs].*R)./R
c2 = round([cs].*R)./R

asn_f(p,:)=a2; asn=a2
gsn_f(p,:)= g2; gsn= g2
bsn_f(p,:)=b2; bsn=b2
csn_f(p,:)=c2; csn=c2
```

Synthesis

CRFF realization & NTF, STF evaluation

Loop-filter scaling & SC implementation

Figure 6.3 MATLAB script with Schreier's toolbox functions used for the loop-filter synthesis of a CRFF SC-ΣΔM with reconfigurable notch frequency [6].

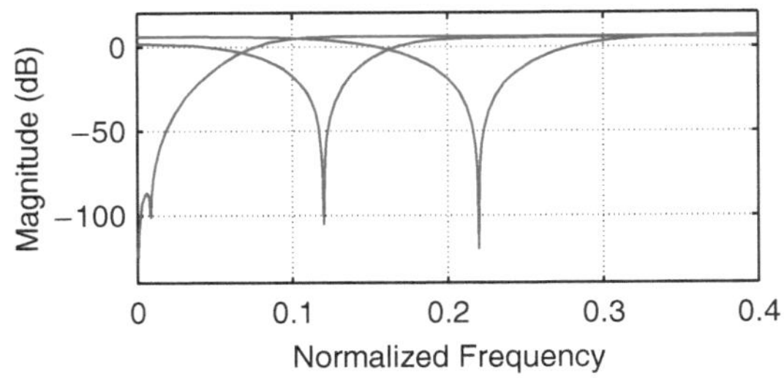

Figure 6.4 Ideal NTF of the CRFF reconfigurable LP/BP ΣΔM synthesized with the script shown in Figure 6.3 for $f_n = 0, f_n = 0.12f_s$, and $f_n = 0.22f_s$.

operational amplifiers can be reduced and the electrical design can be more efficient in terms of power consumption and dynamic range. In this case, the output swing is maximized (below 20% of the FS reference voltage). The limits on the states are determined for each notch frequency based on simulation results using `simulateDSM`.

The resulted scaled coefficients – represented by matrix `[as,gs,bs,cs]` – are rounded using the script `round` in order to obtain coefficient values that can be realized with feasible capacitor ratios. The process gives rise to the final set of coefficients `[asn,gsn,bsn,csn]` in the script of Figure 6.3. As an illustration, the NTF realization and the loop-filter matrices obtained for DC are shown in Figure 6.5. This set of coefficients leads to a straightforward binary-weighted SC implementation of the modulator, whose conceptual circuit implementation is shown in Figure 6.6. The corresponding simulated output spectra are depicted in Figure 6.7, illustrating the correct noise-shaping for different notches.[2] The synthesized modulator constitutes the starting point of the high-level synthesis, in which the modulator specifications are mapped onto circuit-level specifications, as will be discussed later in this chapter.

6.1.3 Synthesis of a Fourth-order BP CT-ΣΔM with Tunable Notch

Schreier's toolbox has also useful functions for synthesizing CT-ΣΔMs, following a similar procedure to that carried out for DT-ΣΔMs. One of these functions is `realizeNTF_ct`, which allows an NTF with a CT loop filter to be realized by providing the corresponding state-space description for a given modulator topology – in either a feedback or a feedforward form – and selected timing for the feedback DAC [4].

As discussed earlier in this book, an alternative synthesis methodology consists of synthesizing the DT filter and applying a CT-to-DT transformation in order to get the desired CT-ΣΔM modulator. To this end, the functions provided by Schreier's toolbox can be used to perform the initial architecture exploration, considering DT systems that can be transformed later to CT systems. This process is illustrated in this section using the example of a BP CT-ΣΔM with widely tunable notch frequency, intended for RF-to-digital conversion in wireless receivers [7].

Figure 6.8 shows the modulator architecture under study. It consists of a fourth-order BP CT-ΣΔM with 4-bit quantizer and a loop filter made up of two resonators with a transfer function, $R(s) = \omega s/(s^2 + \omega^2)$, where normalized values of s and ω are considered with respect to f_s, so that

[2]The interested reader can run the script of Figure 6.3 in MATLAB to generate the different sets of loop-filter coefficients and try different alternative implementations.

```
>> ntf

          (z-1)^2 (z^2 - 1.997z + 1)
  ---------------------------------------------
  (z^2 - 1.186z + 0.3664) (z^2 - 1.409z + 0.6498)

>> ABCD

    1.0000         0         0         0    1.0000   -1.0000
    1.0000    1.0000         0         0         0         0
    1.0000    1.0000    0.9972   -0.0028         0         0
         0         0    1.0000    1.0000         0         0
    1.4022    0.6403    0.2552    0.0417    1.0000         0

>> ABCDs

    1.0000         0         0         0    0.2648   -0.2648
    0.7177    1.0000         0         0         0         0
    0.3210    0.4472    0.9972   -0.0100         0         0
         0         0    0.2831    1.0000         0         0
    5.2954    3.3690    3.0027    1.7309    1.0000         0

>> [as gs bs cs]

    2.8774    3.3690    2.5126    1.7309
         0    0.0100
    0.2648         0         0         0    1.0000
    0.2648    0.7177    0.4472    0.2831

>> [asn gsn bsn csn]

    2.8800    3.3700    2.5100    1.7300
         0    0.0100
    0.2600         0         0         0    1.0000
    0.2600    0.7200    0.4500    0.2800
```

Figure 6.5 NTF and loop-filter coefficients of the CRFF LP/BP SC-ΣΔM ($f_n = 0$).

$s = 2\pi f/f_s$ and $\omega \equiv 2\pi f_n/f_s$. A signal gain coefficient, $k \equiv 1/(2\omega)^2$, can be used to equalize the STF over the notch frequency tuning range [7]. The feedback path consists of a NRZ DAC and a two-tap half-delayed FIR filter, with scaling coefficients, c_i and c_{id}, that provide the necessary degrees of freedom in the synthesis of the NTF when applying a CT-to-DT equivalence. Additional feedback paths with scaling gains, c_{0j}, are included to compensate for the ELD error [8].

The modulator in Figure 6.8 has been synthesized by applying a CT-to-DT transformation to a BP DT-ΣΔM with the required NTF, considering input signals placed at a programable notch frequency. The high-level synthesis of this modulator uses Schreier's toolbox, as depicted in the MATLAB script shown in Figure 6.9. The first step consists of using `synthesizeNTF` to obtain the DT version of the NTF that satisfies the requested specifications (OSR, OBG, f_n, etc.).

The CT version of the loop-filter transfer function, $H(s)$, of the desired BP CT-ΣΔM can be derived from the impulse-invariant transformation as follows:

$$H(z) \equiv Z\{L^{-1}[H(s) \cdot H_{\text{FIR-DAC}}(s)]\} \tag{6.3}$$

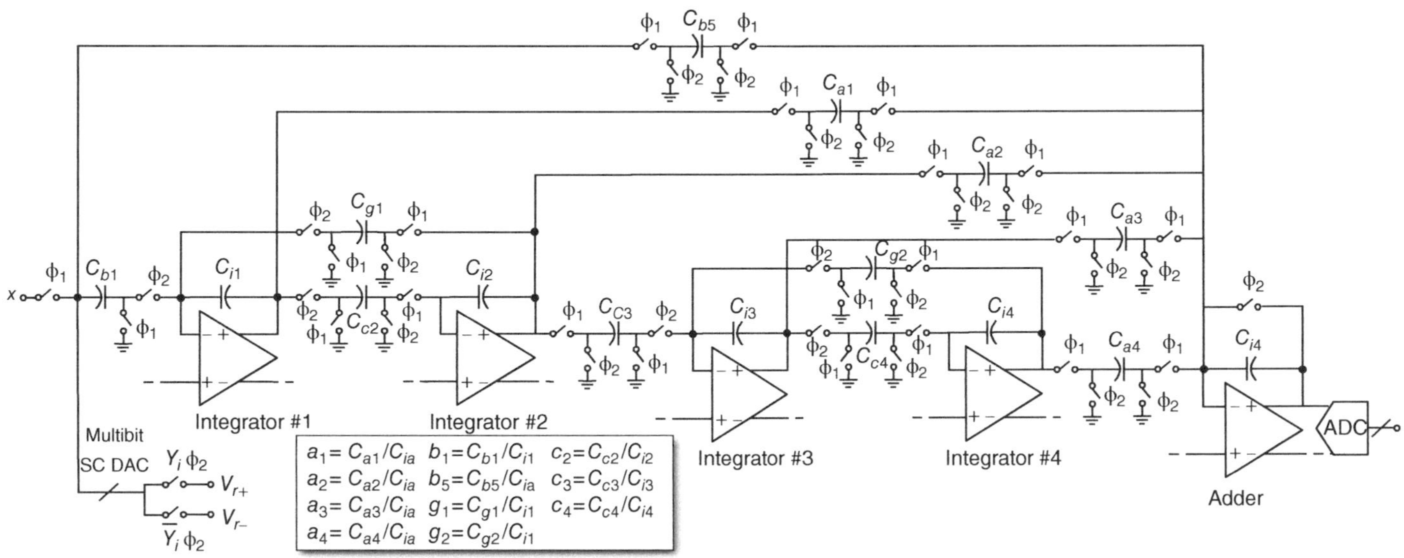

Figure 6.6 Conceptual schematic of the synthesized LP/BP CRFF SC-ΣΔM.

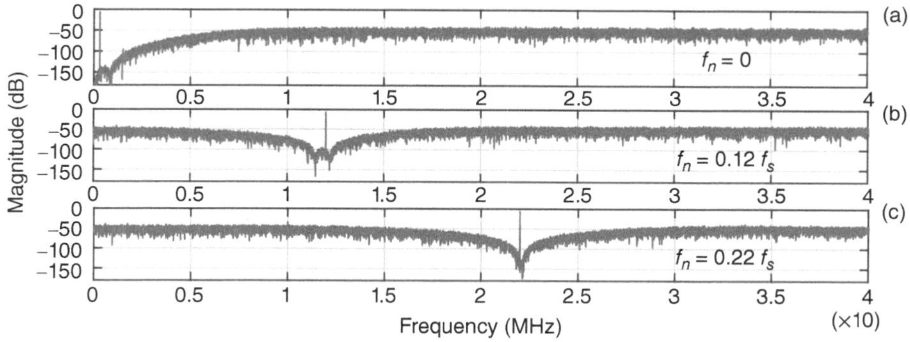

Figure 6.7 Ideal output spectra of the synthesized CRFF LP/BP SC-ΣΔM for: (a) LP case ($f_n = 0$); (b) $f_n = 0.12 f_s$; (c) $f_n = 0.22 f_s$.

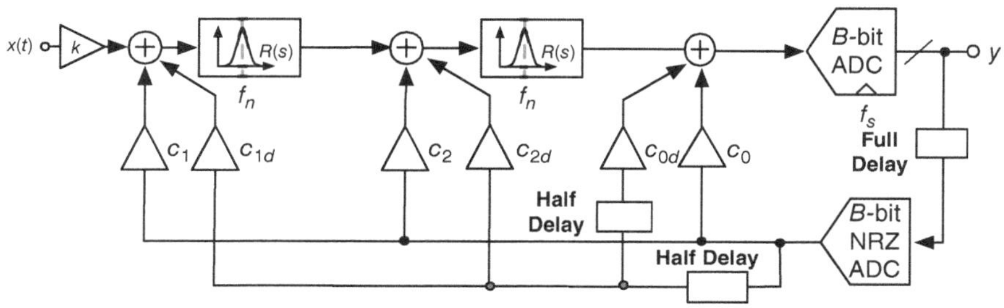

Figure 6.8 Block diagram of a fourth-order BP CT-ΣΔM with tunable notch [7].

where $Z(\cdot)$ and $L(\cdot)$ denote the Z-transform and L-transform symbols, respectively, and $H_{\text{FIR-DAC}}(s)$ is the transfer function of the FIR-based DAC.

Note that, in this case, as an FIR-based NRZ DAC is considered, the modified Z-transform is more suitable for computing Equation (6.3) since there are delays that are not integer multiples of the sampling period, T_s [9] (see Figure 6.8). The so-called *notch-aware synthesis* approach [7], is used to compute the modified Z-transform of each modulator loop-filter path considering that f_n is a design (variable) parameter.

The whole synthesis procedure can be automated in a procedure that combines functions from MATLAB and Schreier's toolbox. Thus, after obtaining the DT version of NTF using `synthesizeNTF`, the next step consists of obtaining the transfer function from the modulator output to the input of the quantizer, computed for the different feedback branches with gain c_i as shown in Figure 6.8, as follows:

$$H_{c_i}(s, v) = c_i \cdot e^{-sp} \cdot \left[\frac{(\frac{\pi}{v})s}{s^2 + (\frac{\pi}{v})^2} \right]^{\lfloor \frac{i}{2} \rfloor} \cdot H_{\text{NRZ-DAC}}(s)$$

$$(p = 2 \text{for } i = 1 \text{and } p = 1 \text{for } i = 0, 2, 3, 4, 5) \tag{6.4}$$

where $\lfloor \cdot \rfloor$ denotes the floor operator, and $H_{\text{NRZ-DAC}}(s)$ stands for the transfer function of the NRZ DAC, respectively given by:

$$H_{\text{NRZ-DAC}}(s) = T_s \cdot \frac{1 - e^{s/2}}{s} \tag{6.5}$$

with $\upsilon \equiv f_s/(2f_n) = \pi/\omega$ being a design parameter used in the synthesis script to account for the variation of f_n with respect to f_s.

As shown in the script of Figure 6.9, as NRZ DACs are used in this example, the expressions of $H_{c_i}(z, \upsilon)$ resulting from a CT-to-DT transformation of Equation (6.5) can be computed using the `c2d` function provided by the Control Toolbox in MATLAB [10]. This function can convert any delayed analog versions into digital ones, provided that the shape of the feedback is of NRZ type. Therefore,

```
%%%%%%%%%%%%%%%%%%%%%%%%%%%%%%%%%%%%%%%%%%%%%%%%%%%%%%%%%%%%%%%%%%%%%%%%%%%
% Calculation of the coefficients of a LC-based 4th-order BP CT-SDM   %
% with variable notch frequency, [Molina et al., IEEE TCAS-I, 2014]   %
%%%%%%%%%%%%%%%%%%%%%%%%%%%%%%%%%%%%%%%%%%%%%%%%%%%%%%%%%%%%%%%%%%%%%%%%%%%
clear;
fs=4e9;                                  %Modulator's Sampling frequency.
fi=1e9;                                  %Input/carrier frequency.
Ts=1/fs;                                 %Sampling period.
notch= fi/fs;                            %Relative notch frequency.
OSR=128;                                 %Oversampling Ratio.
H = synthesizeNTF(4,OSR,0,1.5,notch);    %Synthesis of DT version of NTF
y=tf(1-1/H);
[num,den] = tfdata(y,'v');               %DT Loop Filter in Vector Form.
v=1/(2*notch);                           %For calculation of Hci(z).
w=pi/v;                                  %Normalized freq. of the LC-filter.
k=0.25/w^2;                              %Equalization factor.

%% Computation of H_ci(s,v) and H_ci(z,v) by using 'c2d' function

Hc2_s = tf([w 0], [1 0 w^2]);
Hc3_s = tf([w 0], [1 0 w^2], 'inputdelay', 0.5);
Hc4_s = tf([w^2 0 0], [1 0 2*w^2 0 w^4]);
Hc5_s = tf([w^2 0 0], [1 0 2*w^2 0 w^4], 'inputdelay', 0.5);

Hc2_z = c2d(Hc2_s, 1); %Discretisation of Hc2_s
Hc3_z = c2d(Hc3_s, 1); %Discretisation of Hc3_s
Hc4_z = c2d(Hc4_s, 1); %Discretisation of Hc4_s
Hc5_z = c2d(Hc5_s, 1); %Discretisation of Hc5_s

[numc2,denc2] = tfdata(Hc2_z,'v'); %Vector form of Hc2_z
[numc3,denc3] = tfdata(Hc3_z,'v'); %Vector form of Hc3_z
[numc4,denc4] = tfdata(Hc4_z,'v'); %Vector form of Hc4_z
[numc5,denc5] = tfdata(Hc5_z,'v'); %Vector form of Hc5_z

H4=[numc4(2) numc4(3) numc4(4) numc4(5)];
H5=[numc5(1) numc5(2) numc5(3) numc5(4) numc5(5)];
H2=conv(numc2,denc2);
H3=conv(numc3,denc3);

%% Partial fraction expansion of H_ci(z,v) (Residues method)

[r2,p2,k2] = residue(H2,[den 0 ]);
[a2,b2] = residue(r2(1:4),p2(1:4), k2);
[r3,p3,k3] = residue(H3,[den 0 0 ]);
[a3,b3] = residue(r3(1:4),p3(1:4), k3);
[r4,p4,k4] = residue(H4,[den 0]);
[a4,b4] = residue(r4(1:4),p4(1:4), k4);
[r5,p5,k5] = residue(H5,[den 0 0 ]);
[a5,b5] = residue(r5(1:4),p5(1:4), k5);

%% Solving the CT-to-DT equivalence and loop-filter coefficients

matrix_coeff=real([a2(1) a3(1) a4(1) a5(1);
    a2(2) a3(2) a4(2) a5(2);
    a2(3) a3(3) a4(3) a5(3) ;
    a2(4) a3(4) a4(4) a5(4)]); % ELD and notch Compensation
dt=[num(2);num(3);num(4);num(5)];
ci=matrix_coeff\dt;   %Coefficients Determination
c2=ci(1); % c2
c3=ci(2); % c2d
c4=ci(3); % c1
c5=ci(4); % c1d
c0=-(r2(5)*c2 + r3(5)*c3 + r4(5)*c4 + r5(5)*c5); % c0
c1=-(r3(6)*c3 + r5(6)*c5); % c0d
```

Synthesis of DT NTF

Synthesis of H_ci (s,v)

Residues method

Loop-filter coefficients

Figure 6.9 MATLAB script with Schreier's toolbox functions used for the loop-filter synthesis of the BP CT-ΣΔM [7].

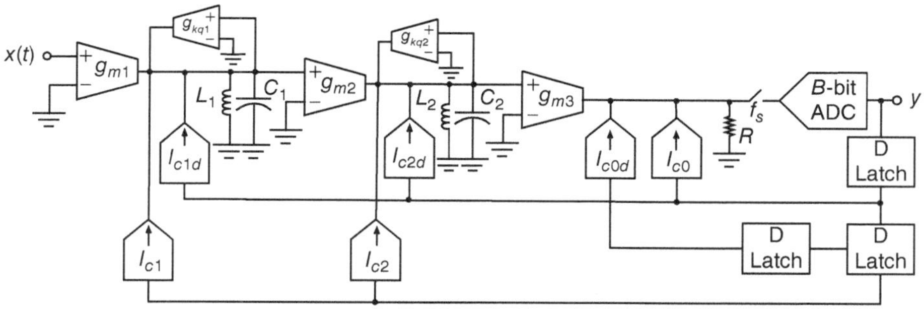

Figure 6.10 Conceptual Gm-LC realization of the synthesized BP CT-ΣΔM [8].

all the required transformations to derive $H_{c_i}(z, \upsilon)$ and the expressions of the loop-filter coefficients, c_i, can be obtained by applying the residues theorem and c2d. Therefore, the whole high-level synthesis procedure can be automated by running the script in Figure 6.9.

The resulting modulator loop filter can be implemented using any arbitrary CT circuit technique: either active-RC or Gm-LC. As an illustration, a Gm-LC implementation is depicted in Figure 6.10. This circuit implementation increases the programmability of the loop-filter coefficients, which are realized as switchable multiples of a unitary transconductance element, g_{mu}, as will be detailed later in this book. It can be shown that the block diagrams in Figures 6.8 and 6.10 are equivalent if the following relationships are satisfied:

$$\begin{aligned} g_{m1,2} &= k_{1,2} \cdot C \cdot \omega \cdot f_s, \quad g_{m3} = k_3/R \\ I_{c0,c0d} &= (c_0, c_{0d}) \cdot V_{\text{FS}} \\ I_{c1,c1d} &= (c_1, c_{1d}) \cdot g_{m1} \cdot V_{\text{FS}} \\ I_{c2,c2d} &= (c_2, c_{2d}) \cdot g_{m2} \cdot V_{\text{FS}}/s_{\text{r}1} \end{aligned} \tag{6.6}$$

where V_{FS} stands for the full-scale reference voltage, $k_{1,2} = k/s_{\text{r}1,2}$ and $k_3 = 1/(s_{\text{r}1} \cdot s_{\text{r}2})$ are scaling forward-path coefficients, and $s_{\text{r}1,2}$ are the weight coefficients that scale the resonator gain, $R_{\text{gain}} = \omega \cdot s$. As will be discussed in Chapter 9, extra feedback transconductances, $g_{kq1,2}$, are needed to enhance the quality factor, Q, of the Gm-LC resonators. These coefficients can be calibrated in order to maximize the performance of the modulator. The modulator loop is completed with a 4-bit quantizer made up of a flash ADC in the forward path and a NRZ current-steering FIR-DAC in the feedback loop [8].

Note that the ΣΔM architectures synthesized so far consider only ideal building blocks. However, as already discussed in this book, the noise-shaping performance of ΣΔMs is degraded in practice by circuit nonideal effects. Therefore, the next step in the synthesis procedure of any ΣΔM is to optimize its performance by considering its main circuit limitations. This process can be automated to some extent by properly combining optimization and behavioral simulation, as discussed in the next sections.

6.2 Optimization-based High-level Synthesis of ΣΔ Modulators

As stated in Chapter 5, the next step after selecting the ΣΔM architecture consists of transmitting or *mapping* the system-level specifications onto circuit-level requirements in order to determine the electrical characteristics that ΣΔM building blocks must satisfy to maximize the modulator performance. This process, often referred to as high-level synthesis or high-level sizing, can be carried out using a behavioral simulator – like for instance SIMSIDES – to explore the design space, as discussed in Chapter 5.

However, the high-level sizing procedure explained in Section 5.6 was carried out manually, i.e. by means of parametric analyses, in order to determine the electrical requirements of ΣΔM subcircuits, such as the opamp DC gain, incomplete settling, thermal noise, etc. Such a procedure can be automated and enhanced by combining an optimization engine that guides the simulator through the design space. This section explains how to do this process. Although the methodology presented here is focused on the use of SIMSIDES, it can be extended to any other simulator that can be used as performance evaluator, according to the optimization flow in Figure 6.11.

6.2.1 Combining Behavioral Simulation and Optimization

Figure 6.11 shows the flow diagram of the optimization-based high-level synthesis of ΣΔMs based on the combination of behavioral simulation (for instance SIMSIDES) as performance evaluation, and an optimizer to explore the design space and find the optimum set of building-block electrical parameters to get the maximum modulator performance with minimum cost in terms of energy and silicon area. The starting point is the modulator topology, which can be synthesized using Schreier's Delta-Sigma toolbox, as discussed in the last section. Here, the design parameters are the building-block specifications; that is, the nonideal parameters used to model the main error mechanisms – as analyzed in Chapters 3 and 4 – that affect the ΣΔM's performance and define the electrical specifications of its subcircuits: integrators, comparators, DAC elements, and so on.

Considering arbitrary initial conditions in Figure 6.11, a set of perturbations of the design parameters is generated by the optimizer. With the new design parameters, appropriate simulations are carried out to evaluate the modulator performance and the process is repeated in an iterative way until the cost function – based on the performance metrics – is optimized.

The type and value of the perturbations, as well as the iteration acceptance or rejection criteria, depend on the type of optimization method. Thus, depending on the nature of the algorithm – either deterministic or statistical – the result of the optimization can be more or less dependent on the initial conditions: the initial values given to the electrical parameters of the ΣΔM subcircuits. Therefore

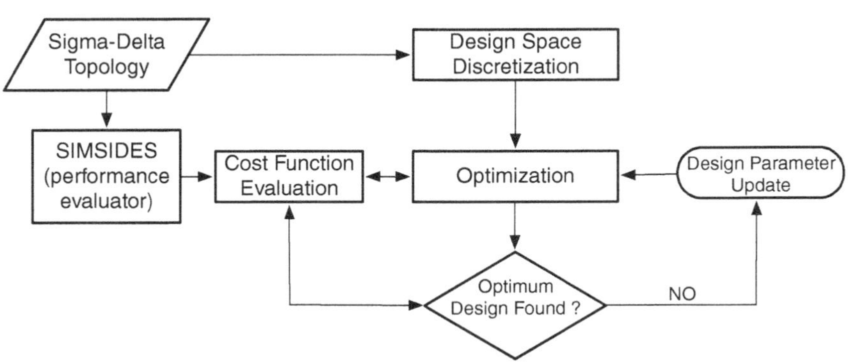

Figure 6.11 Flow diagram of the optimization-based synthesis of ΣΔMs [11].

deterministic algorithms usually require information on the cost function,[3] and on its derivatives, so that only changes of design parameters that improve the cost function are allowed. As a result, the optimization process may be quickly trapped in a local minimum of the cost function, and is thus more useful for fine tuning suboptimal designs, where a good initial design point is known. By contrast, so-called "statistical" techniques do not need information on the derivatives of a cost function, and design parameters are randomly changed, thus avoiding traping in local minima of the cost function and hence leading to global minima. In this way, the best alternative is to use optimization algorithms that consider statistical techniques for an initial, broad exploration of the design space, and deterministic techniques for later fine-tuning. Simulated-annealing and evolutionary algorithms are, a priori, good techniques to be used in the process of high-level synthesis of ΣΔMs, and will be considered here [11–13].

SIMSIDES can be combined with any of the optimization methods provided with MATLAB [14]. Alternatively, other optimization engines can be also coupled to it in order to implement the high-level synthesis methodology shown in Figure 6.11. In the next sections, the different possibilities are presented using some design examples.

6.2.2 Using Simulated Annealing as Optimization Engine

The first step in automating the synthesis process is to formulate the optimization problem in SIMSIDES for a given architecture. As an illustration, let us consider the cascade 2-1 SC-ΣΔM with single-bit quantizer shown in Figure 6.12 as a case study. In this example, only the effect of finite DC gain and incomplete settling error will be considered, and the same opamp design will be used for the second and third integrators. Based on this assumption and taking into account the behavioral models described in Chapter 5, the design parameters are `a01,gm1,io1`, and `a02,gm2,io2`, which stand

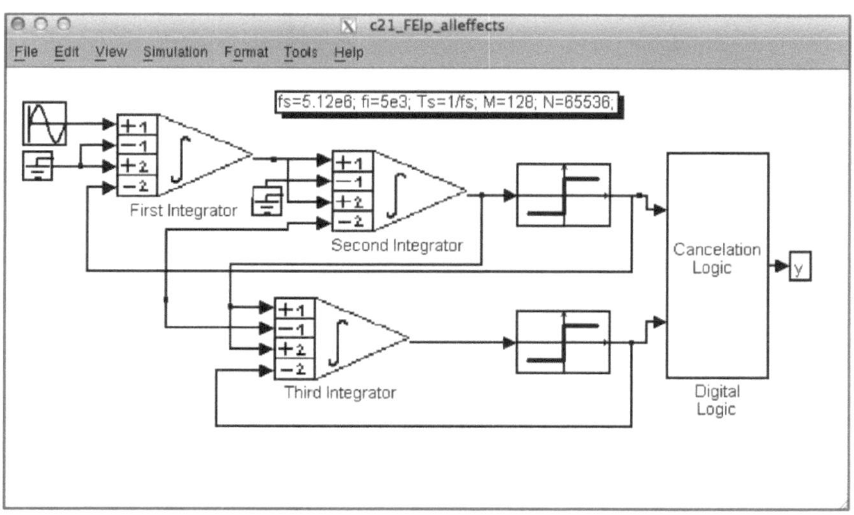

Figure 6.12 SIMSIDES diagram of a cascade 2-1 SC-ΣΔM used as an example for an optimization-based synthesis.

[3]Cost functions are sometimes used in optimization procedures to quantify the degree of compliance with the design objectives, as well as the restrictions imposed on the design variables. These functions are normally defined so that they must be minimized to get an optimum design [12].

respectively for the finite DC gain, finite transconductance, and maximum output currents of the first integrator and the back-end (second and third) integrators. In this example, the modulator parameters are OSR =128 and f_s = 5.12 MHz, and the objective is to obtain an SNR > 100 dB. The high-level sizing problem consists of finding the optimum set of design parameters that maximize the SNR with the minimum values of the design variables, which in turn will result in minimum power consumption.

Before starting the optimization, all parameters required to simulate the modulator in SIMSIDES are defined in a MATLAB script, like the one shown in Figure 6.13. This script sets the values of all model parameters, except those used as design variables in the optimization. Ideal values are given

```
%%%%%%%%%%%%%%%%%%%%%%%%%%%%%%%%%%%%%%%%%%%%%%%%%%%%
% Model parameters for the simulation of 2-1 SC-SDM
%%%%%%%%%%%%%%%%%%%%%%%%%%%%%%%%%%%%%%%%%%%%%%%%%%%%
fs=5.12e6; fi=5e3; Ts=1/fs; OSR=128; N=65536;
kt=0.026*1.6e-19;

%% Model parameters of the first integrator
Cint1=24e-12; % integration capacitor For gain=1
Cs11=6e-12; % sampling capacitor (branch 1)
Cs21=6e-12; % sampling capacitor (branch 2)
innoise1=0; % rms input noise value
ao1=2.63e3; % open-loop DC gain
gm1=4.5e-3; % transconductance
io1=0.977e-3; % maximum output current
ron1=60; % sampling switch-on resistance

%% Model parameters of the second and third integrators
Cint2=3e-12;
Cs12=1.5e-12;
Cs22=1.5e-12;
innoise2=0;
ao2=1.38e3;
gm2=0.87e-3;
io2=0.25e-3;
ron2=650;

%% Common parameters
temp=175; % temperature
osp=2.7; % output swing
cnl1=0; % capacitor first-order non-linear coef.
cnl2=25e-6; % capacitor second-order non-linear coef.
avnl1=0; % DC gain first-order non-linear coef.
avnl2=15e-2; % DC gain second-order non-linear coef.
avnl3=0; % DC gain third-order non-linear coef.
avnl4=0; % DC gain fourth-order non-linear coef.
cpar1=0.6e-12; % parasitic (opamp) input capacitance
cpar2=0.6e-12;
cload=2.28e-12; % opamp (intrinsic) load capacitance

%% Model parameters of comparators
vref=2; % DAC reference voltage
hys=30e-3; % comparator hysteresis
```

Figure 6.13 MATLAB script used to define model parameters required for the simulation of Figure 6.12.

for those design parameters not considered in the optimization, such as the input referred thermal noise: `inoise` in Figure 6.13. The next step consists of defining the ranges of the design variables, the targets, and the solver numerical methods to be considered in the optimization. In this case, an updated version of an optimization core inspired by simulated annealing, called FRIDGE [12] is used.

The information required to run the optimization is provided via the GUI of SIMSIDES, as illustrated in Figure 6.14. The name of the model and the simulation parameters are defined in the first window menu (Figure 6.14a,b), and the design variables and ranges are given in another window menu (Figure 6.14c). The ranges of the design variables can be specified in different ways, as illustrated in this example:

`gm1=[6m,50%]` [nominal value, percentage variation around the nominal value]

`gm2=[6m,100u,6m]` [initial value, minimum, maximum]

`io1=[1m,50%]`

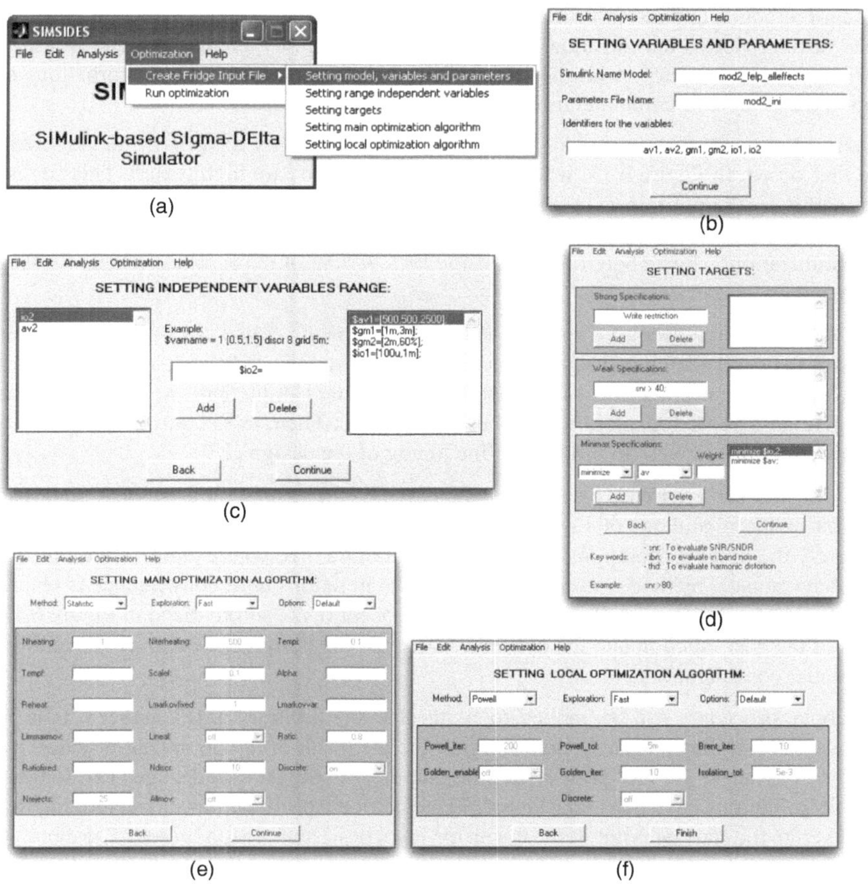

Figure 6.14 Illustrating the use of SIMSIDES to set the optimization of ΣΔMs: (a) launching optimization menu; (b) setting variables and model parameters; (c) setting design variables; (d) optimization targets and objectives; (e) setting main (global) optimization algorithm; (f) setting local optimization algorithm.

```
io2=[1m,100u,5m]
ao1=[2500,50%]
ao2=[2500,100u,2500]
```

Once the design variables and ranges have been defined, the next step is to set the design objectives and targets. Here there are different possibilities depending on the performance metrics to be optimized, for instance SNR, IBN, THD, and so on. These metrics are usually defined as `strong` targets. In addition to the design objectives and targets, a set of design constraints can be defined (normally as `weak` targets) and they can be optionally *weighted* according to the designer's criteria (Figure 6.14d). These weights are used to give priority to the fulfillment of the associated weak restrictions so that the higher the weight, the higher the priority of a given requirement.[4]

The optimization methods – either statistical or deterministic – to be used in the different phases of the optimization process are selected (Figure 6.14e,f). In the case of the FRIDGE optimizer, the optimization procedure is divided into two phases. The first phase – global optimization – explores the design space by dividing it into a multidimensional grid, resulting in a mesh of so-called *hypercubes*. A statistical method is applied in order to escape from local minima. Once the optimum hypercube is reached, a finer (local) optimization is performed inside that hypercube. A deterministic method is usually applied in this step, with the calculation of the design-parameter perturbations requiring information on the cost function and its derivatives [12].

Finally, all the information provided by the user is generated by SIMSIDES in the form of a text file, like that shown in Figure 6.15, where the different sections are highlighted. This file is used as the input netlist by the optimizer. In this case all design objectives and targets are defined as `weak` targets, although a different weight is given to them. A statistical algorithm is selected for the main (global) optimization, while a deterministic method – a variant of the Power method – is used for the final optimization.

The optimization procedure is controlled by a parameter called *temperature* – based on an analogy with the solid-state physics phenomenon that inspired the simulated annealing algorithm. In a similar way to the annealing of solids, when the temperature is high, large movements in the design variables give an extensive exploration of the design space. Then, as the temperature decreases, the variations are reduced as well, in order give fine tuning of the design [12].

As an illustration, Figure 6.16 shows the evolution of the temperature and cost functions for the optimization of the modulator of Figure 6.12. It can be seen that the statistical method is used in approximately the first 300 iterations (see the point in Figure 6.16 where temperature function stops) while the deterministic method is used in the rest of the iterations until convergence is reached, after about 500 iterations. The results of the optimization procedure are summarized in Figure 6.17, giving the values of the cost function and the optimum values of the design objective (SNR) and the design variables under consideration.

The above procedure can be applied to synthesize any arbitrary ΣΔM, either CT or SC, in an optimum way. As an illustration, Table 6.1 shows the the result of the optimization-based high-level synthesis of a cascade 2-1-1 SC-ΣΔM with 3-bit quantizer in the back-end stage [11]. In this case the main circuit limitations have been considered. The electrical specifications characterizing those limitations constitute the starting point of the transistor-level design of the modulator. The circuit example synthesized according to Table 6.1 was implemented on a chip, showing a good agreement between the experimental results and the behavioral simulated performance, as illustrated in Figure 6.18.

[4]In FRIDGE optimizer, strong restrictions are specifications whose fulfillment is considered essential by the designer. Otherwise, weak restrictions are those that have a certain degree of nonfulfillment that can be tolerated by the design. The interested reader can found a more detailed description of these and other details of the FRIDGE optimizer in the literature [12].

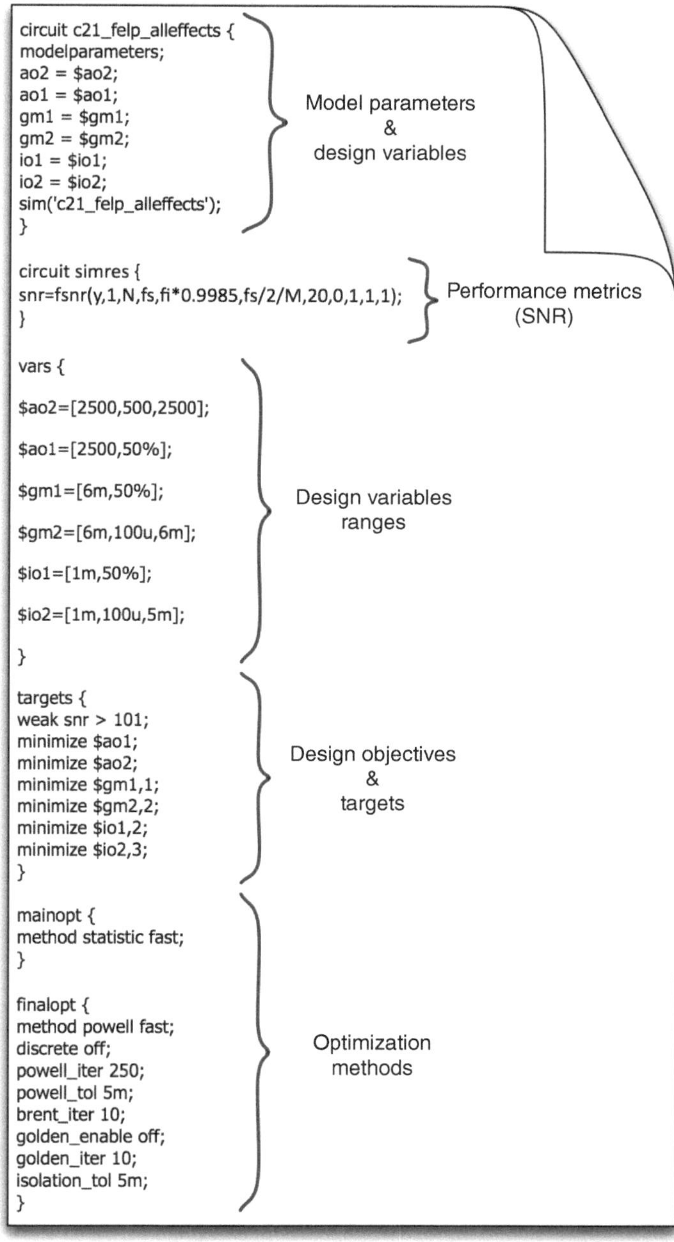

Figure 6.15 Example of FRIDGE input netlist.

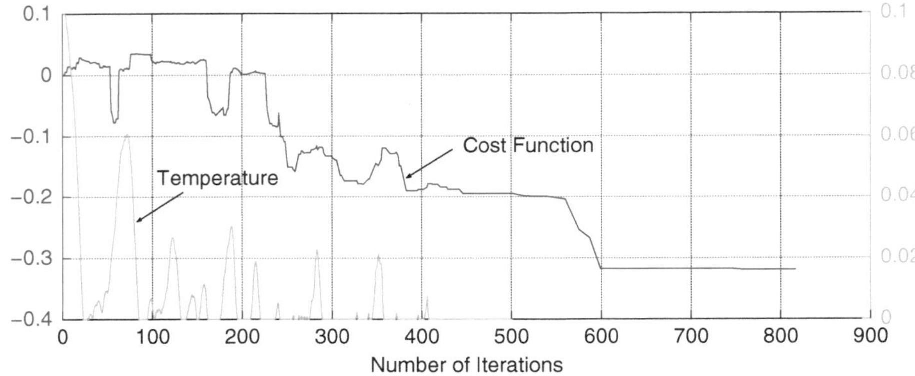

Figure 6.16 Evolution of temperature and cost function of the synthesis procedure of a cascade 2-1 SC-ΣΔM using simulated annealing optimization [11].

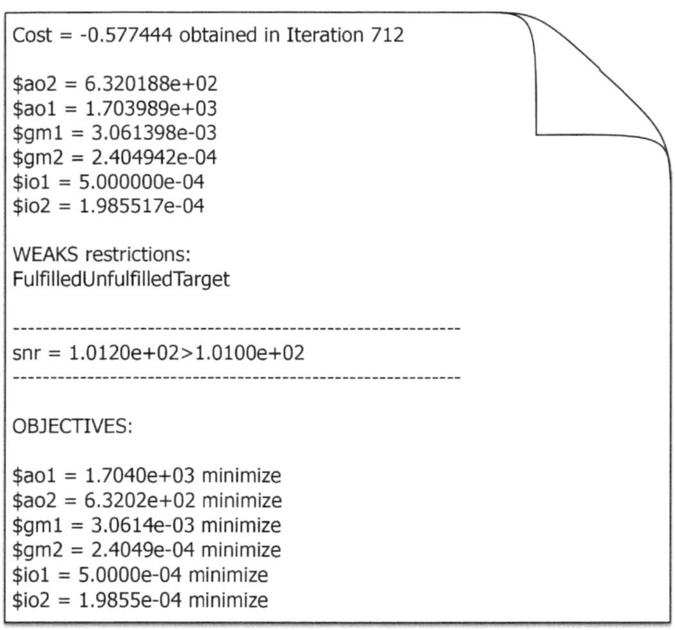

Cost = -0.577444 obtained in Iteration 712

$ao2 = 6.320188e+02
$ao1 = 1.703989e+03
$gm1 = 3.061398e-03
$gm2 = 2.404942e-04
$io1 = 5.000000e-04
$io2 = 1.985517e-04

WEAKS restrictions:
FulfilledUnfulfilledTarget

----
snr = 1.0120e+02>1.0100e+02

OBJECTIVES:

$ao1 = 1.7040e+03 minimize
$ao2 = 6.3202e+02 minimize
$gm1 = 3.0614e-03 minimize
$gm2 = 2.4049e-04 minimize
$io1 = 5.0000e-04 minimize
$io2 = 1.9855e-04 minimize

Figure 6.17 Result of the high-level synthesis of a cascade 2-1 SC-ΣΔM combining simulated annealing and SIMSIDES.

Table 6.1 High-level synthesis results of a cascade 2-1-1 SC-ΣΔM, 3-bit in the back-end stage (f_s = 70.4 MHz, OSR =16, V_{FS} = 1.5 V).

Model parameter	Integrator 1	Integrator 2	Integrators 3,4
Output swing (V)	±1.8	±1.8	±1.8
A_v (dB)	≥ 81	≥ 65	≥ 54
I_o (mA)	≥ 1.5	≥ 2.2	≥ 1.6
g_m (mA/V)	≥ 6.4	≥ 7	≥ 3.4

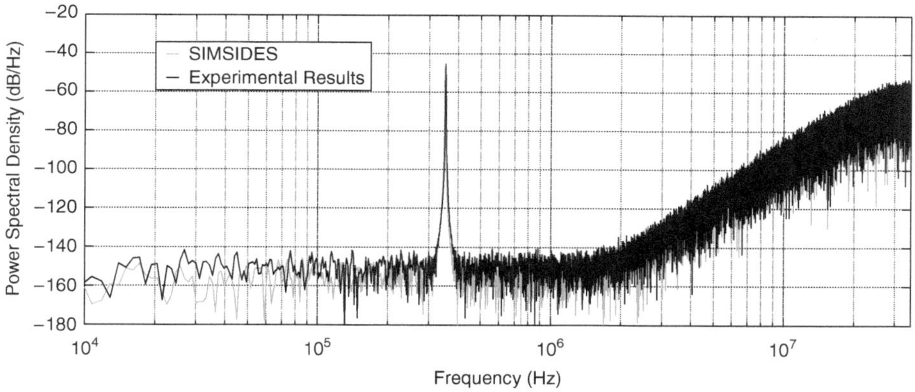

Figure 6.18 Output spectrum of the cascade 2-1-1 SC-ΣΔM synthesized in Table 6.1. Experimental measurement matches behavioral simulations obtained from the optimization-based synthesis process [11].

6.2.3 Combining SIMSIDES with MATLAB Optimizers

In order to take advantage of the diverse optimization algorithms available in MATLAB, an updated version of SIMSIDES includes an optimization menu that guides designers through the main steps to set up and run an optimization, as illustrated in Figure 6.19. Starting from the main optimization menu (Figure 6.19a), designers can choose different ways to automate the high-level synthesis of a given ΣΔM, the behavioral model of which has been built using SIMSIDES. Thus, the most direct option (Figure 6.19b,c) consists of directly launching the `SIMULINK Design Optimization` tool, from which designers can define the optimization problem (design variables, ranges, objectives, constraints, etc.), select the optimization algorithm, run the optimization, and analyze the results.

An alternative way of setting up and running an optimization-based synthesis in SIMSIDES is shown in Figure 6.20. This optimization menu allows designers to customize their optimization problem by defining all the required pieces of information: the name of the SIMSIDES model, the MATLAB script and the main simulation parameters, the number of design variables, the initial values and ranges of variables, and the optimization method. All the main search methods available in MATLAB can be used: `Gradient Descent`, `Simplex Search`, `Pattern`, and so on, as well as various optimization algorithms: `Neider-Mead`, `Genetic`, and so on.

The interested reader can find a detailed description of all these design search methods and optimization algorithms in the MATLAB documentation [15]. As an illustration, let us consider here the same optimization case study considered in Section 6.2.2, but using a cascade 2-1 SC-ΣΔM as a demonstration vehicle and with the genetic-algorithm optimization engine available in MATLAB, which is combined with SIMSIDES for high-level sizing. Two different design targets were considered. Figure 6.21 shows the results of this optimization example. The first (Figure 6.21a) aims to achieve the maximum SNR while the second (Figure 6.21b) targets a design objective of SNR > 100 dB. In both cases, the optimization-based high-level sizing process aims to minimize the building-block electrical requirements, although in general less demanding circuit requirements are obtained in Figure 6.21b.

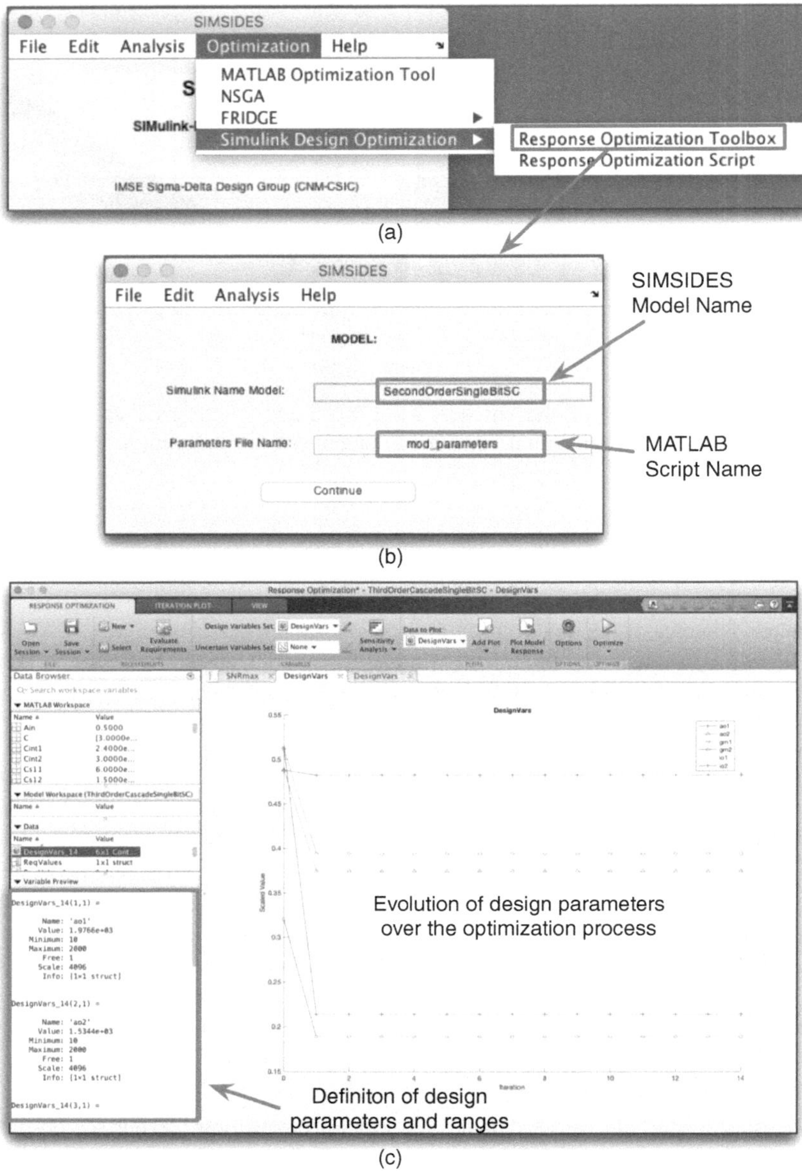

Figure 6.19 Combining SIMSIDES and SIMULINK optimization toolbox: (a) SIMSIDES optimization main menu; (b) launching SIMULINK optimization toolbox from a SIMSIDES model; (c) optimization toolbox main window.

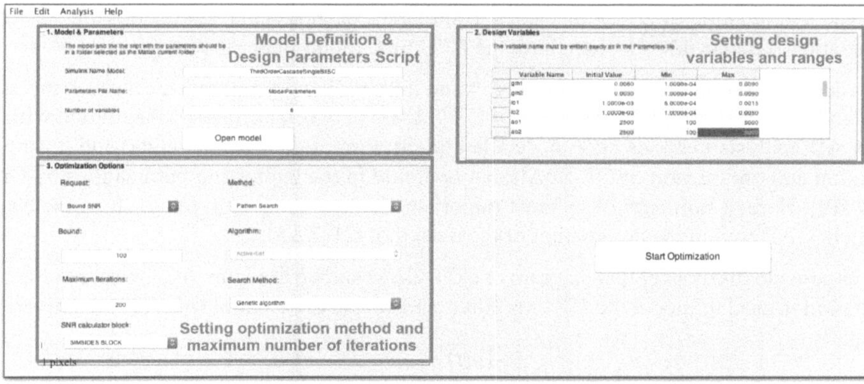

Figure 6.20 SIMSIDES optimization interface.

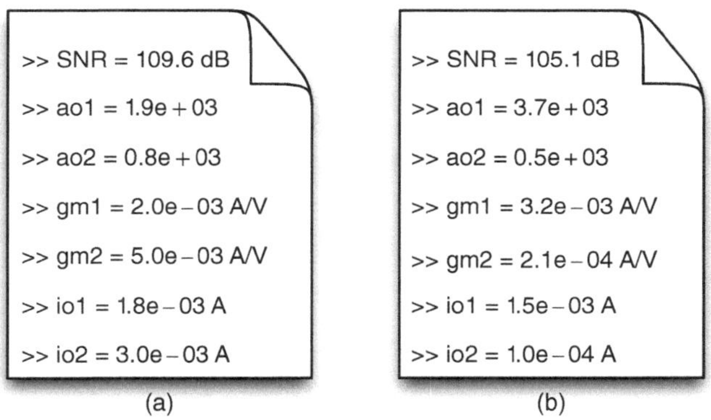

Figure 6.21 Result of the high-level synthesis of a cascade 2-1 SC-ΣΔM by combining SIMSIDES and genetic algorithms available in MATLAB. Two main design objectives are considered: (a) maximize SNR; (b) SNR > 100 dB.

6.3 Lifting Method and Hardware Acceleration to Optimize CT-ΣΔMs

The system-level simulation techniques of CT-ΣΔMs discussed in this book are based on two main approaches. One consists of analysing the DT version of the CT-ΣΔM and applying a DT-to-CT transformation onto the loop filter. This method – mostly used in Schreier's Delta-Sigma toolbox – is widely employed by designers, although it is sometimes necessary to adjust the synthesized CT filter in order to meet the required specifications or to include some nonidealities [16].

Another approach to simulate CT-ΣΔMs is based on analog simulation: this is the approach used in SIMSIDES. In this approach, because the simulation is carried out in the CT domain, the problems associated with the DT–CT transformation can be avoided. However, in order to obtain accurate results, some simulation parameters, such as the solver type, the time-step size, and the tolerance level, need to be adjusted in order to obtain the required accuracy level. Hence, getting a precise simulation involves increasing the CPU time. For instance, using SIMSIDES, a high-order (third-order, fourth-order) CT-ΣΔM considering all circuit errors, typically requires a few seconds to run a transient simulation of 2^{16} clock cycles. If hundreds or thousands simulations are carried out in an optimization-based synthesis procedure, the CPU time might increase up to several hours.

In order to overcome these limitations, some authors have proposed the use of the so-called *lifting* method to accelerate the simulation of CT-ΣΔMs by several orders of magnitude with respect to other CAD analysis methods [17]. A detailed description of the lifting method and its application to simulation and optimizatin of CT-ΣΔMs can be found in the interesting publications by Ortmanns *et al.* [17–19]. Here, a summary of its most important characteristics is given to show the benefits of this efficient CAD tool in the automation of the design of CT-ΣΔMs.

Let us consider the conceptual diagram of a CT-ΣΔM shown in Figure 6.22, where the state-space representation is used to model the CT loop filter, and whose differential equations are given by:

$$\begin{aligned}\frac{\mathrm{d}x}{\mathrm{d}t} &= A \cdot x(t) + B \cdot \begin{bmatrix} u(t) \\ v(t) \end{bmatrix} \\ y(t) &= C \cdot x(t) + D \cdot \begin{bmatrix} u(t) \\ v(t) \end{bmatrix}\end{aligned} \tag{6.7}$$

These differential equations are transformed into finite difference equations in order to simulate the equivalent DT system, as done in Equation (6.1). The majority of commercial CAD tools, like MATLAB, approximate the solution of the above equations by applying a zero-order hold (ZOH) on the input signals. This means that $u(t)$ in Figure 6.22 will be constant over one clock period, which is a good approximation for high values of OSR, but will start to deviate from reality as OSR decreases, for instance in wideband applications.

It can be shown that the correction term, $\Delta x(n+1)$ associated with each discretized state variable $x(n)$ can be formulated as [17]:

$$\Delta x(n+1) = \int_0^1 e^{(1-\zeta)} \cdot B \cdot [u(n+\zeta) - u(n)] d\zeta \tag{6.8}$$

where a normalized sampling frequency, $f_s = 1$, has been asssumed without loss of generality.

Based on this approach, the simulation of any CT-ΣΔM can be carried out by considering the original state-space representation with DT (ZOH) solvers to calculate the values of a ZOH approximation of the CT filter and then adding the required correction terms. This is conceptually depicted in Figure 6.23. Based on this model, the DT version of the state-space representation of a CT-ΣΔM can be written as:

$$\begin{aligned}x(n+1) &= A_{\mathrm{DT}} \cdot x(n) + B_{\mathrm{DT}} \cdot \begin{bmatrix} u(n) \\ v(n) \end{bmatrix} + \Delta x(n) \\ y(n) &= C_{\mathrm{DT}} \cdot x(n) + D_{\mathrm{DT}} \cdot \begin{bmatrix} u(n) \\ v(n) \end{bmatrix}\end{aligned} \tag{6.9}$$

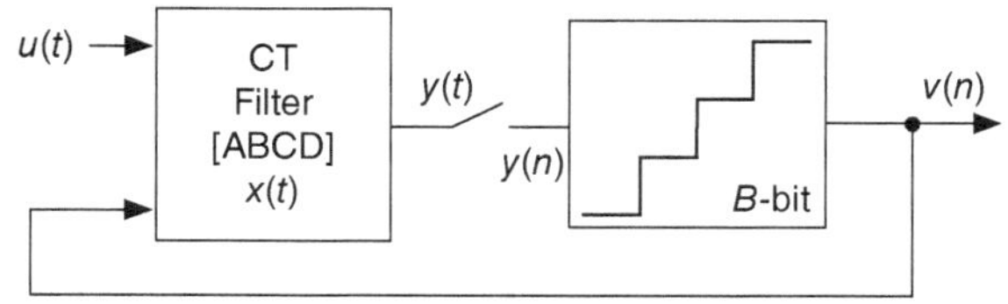

Figure 6.22 CT state-space representation of a CT-ΣΔM.

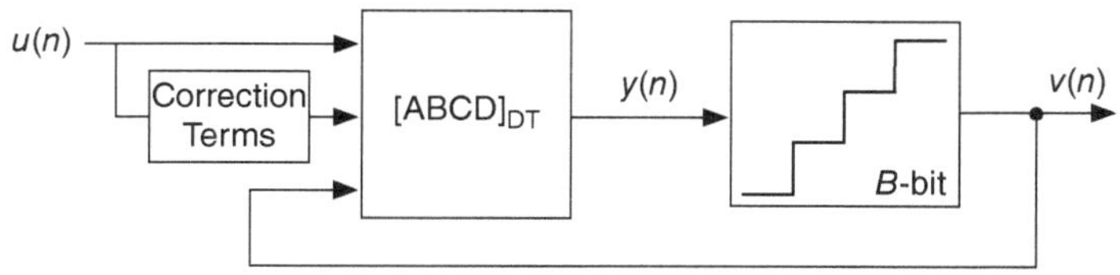

Figure 6.23 DT state-space representation of a CT-ΣΔM with lifting method [17] .

where $[A_{DT}B_{DT}C_{DT}D_{DT}]$ stands for the state-space matrix associated with the DT version of the CT-ΣΔM loop filter. In this way, the lifting method is as precise as the calculation of $\Delta x(n)$, so that the derived DT system approaches the original CT-ΣΔM.

The influence of nonideal circuit effects, such as loop-filter error and ELD, can be accounted for in the above formulation [17]. However, the computation of $\Delta x(n)$ becomes a challenge when handling nonlinear phenomena. Nevertheless, the use of the lifting method allows the simulation of CT-ΣΔMs to be made faster, benefiting from the speed of DT simulation while retaining the accuracy of analog simulation approaches. These characteristics have been applied by some authors to the optimization-based high-level synthesis process by means of emulation – rather than simulation– of CT-ΣΔMs in hardware, as discussed next.

6.3.1 Hardware Emulation of CT-ΣΔMs on an FPGA

The DT representation of CT-ΣΔMs based on the lifting approach allows their building blocks to be emulated in hardware. For instance, DT integrators can be implemented with a digital register in a feedback loop – that is, by its digital representation – which can be in turn be implemented on an field programmable gate array (FGPA) [18], as conceptually depicted in Figure 6.24.

FGPAs usually include the possibility of emulating some other useful ΣΔM building blocks, such as multi-bit quantizers, so that, in principle, an arbitrary CT-ΣΔM topology can be emulated in

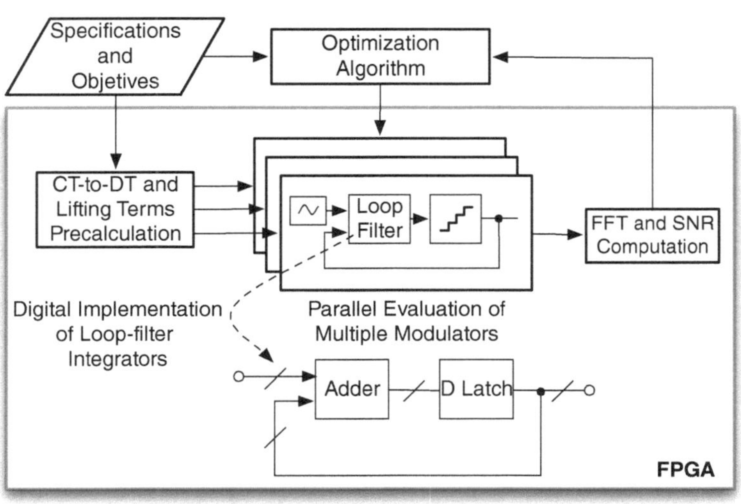

Figure 6.24 Emulation of CT-ΣΔMs on FPGAs for synthesis purposes [18].

hardware. Bruckner *et al.* considered ideal CT-ΣΔMs except for the ELD error [18], although other circuit limitations can be taken into account following the lifting approach [17]. A drastic acceleration factor – of the order of 10^5 – is achieved using this method with respect to conventional simulation of CT-ΣΔMs in MATLAB. Moreover, the emulation speed can be increased by using parallel emulation of different submodules – different CT-ΣΔM designs – as illustrated in Figure 6.24. As a result, millions of CT-ΣΔM designs can be evaluated in only a few seconds, which is computationally very efficient when applied to an optimization-based synthesis methodology, where a huge number of designs are examined in order to find out the best solution that satisfy a given set of specifications.

The possibility of using of FPGAs to emulate CT-ΣΔMs and to use them as performance evaluators in high-level synthesis depends mainly on the memory limitation of the FPGA, as well as the speed of interfaces to external devices. Moreover, additional hardware – an FPGA – is required at the very beginning of the design phase, and these are not always accessible to many designers. In addition, relevant expertise is required to truly benefit from the approach.

6.3.2 GPU-accelerated Computing of CT-ΣΔMs

Following the same philosophy, Bruckner *et al.* proposed another powerful approach to the high-level synthesis of CT-ΣΔMs, based on fast simulation on a graphics processing unit (GPU), instead of on a CPU as is conventionally done [19]. This method allows tens or even hundreds of thousands of simulations to be evaluated per second, thus exploring a whole design space of millions of designs in less than a minute.

The simulation speed-up achieved by GPUs can be enhanced still further because of their parallel computing capabilities, which allows multiple simulations to be evaluated in parallel. This is specially advantageous in optimization-based synthesis procedures guided by genetic algorithms, where many *individuals* – that is, ΣΔM architectures with different sets of loop-filter coefficients – need to be evaluated concurrently. A very powerful and user-friendly synthesis tool for CT-ΣΔMs has been developed by Prof. Ortmanns' group, based on the combination of the lifting method and GPU computing [20]. The tool, called `Uni Ulm Sigma-Delta Synthesis Tool`, is a web-based application that is freely available at `www.sigma-delta.de`.

Figure 6.25 shows a screenshot of the front-end web page of the tool. Starting from initial specifications and basic modulator parameters, designers define the design objectives, variables, constraints, and so on that allows the tool to optimize the high-level synthesis of high-order (up to fourth-order) single-loop CT-ΣΔMs with different loop-filter topologies (CIFB, CIFF, etc.), and single- or multi-bit quantization. Several feedback DAC waveforms (NRZ, RZ, raised cosine, etc.) are supported and ELD can also be included. Both LP and BP CT-ΣΔMs can be computed and loop-filter coefficients can be either enabled or disabled as necessary. The circuit nature of the filter coefficients – either resistive or capacitive – can also be chosen. The tool also allows the loop filter transfer function to be synthesized, as well as STF engineering feature in an automated way, by defining its main characteristics: minimum in-band gain, maximum overall gain, out-of-band attenuation, and so on. Active-RC circuit realization is considered, with different levels of modeling for the loop-filter integrators, spanning everything from ideal models to 1-pole models with PI-type compensation. The main circuit nonideal effects – output swing, finite DC gain and GBW – are also taken into account.

An exhaustive explanation of the synthesis tool goes beyond of this book and the interested reader can find a detailed description in the literature [19], as well as in the documentation available on the tool's website at `www.sigma-delta.de`. Instead, let us consider here a simple example as an illustration of the capabilities of this CAD tool. A fourth-order CT-ΣΔM with CRFF structure, additional feedback DAC for ELD compensation, $B = 4$ bit, $f_s = 1$ GHz and OSR = 20 is considered.

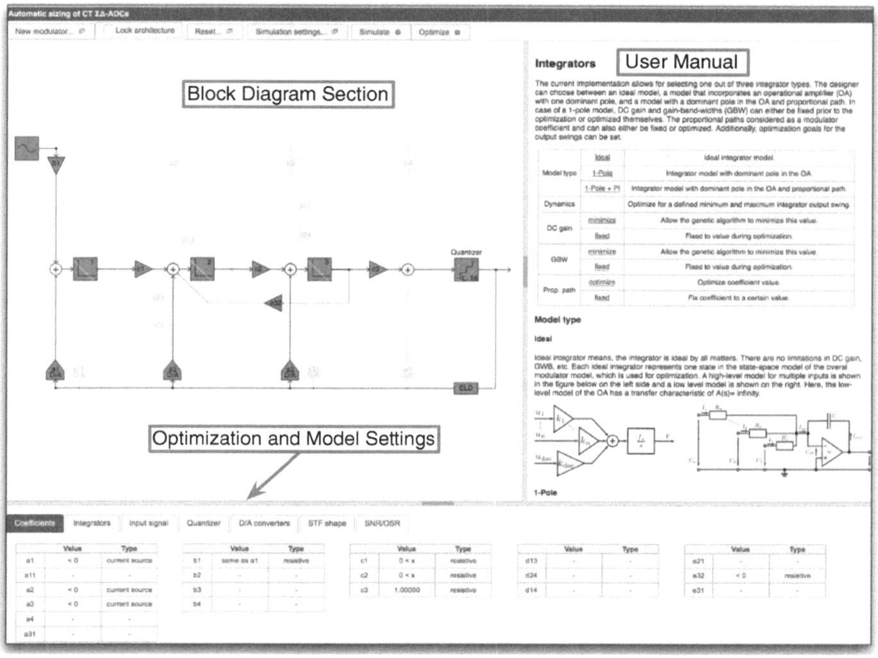

Figure 6.25 Web-based user interface of the synthesis tool for CT-ΣΔMs developed by Ortmanns' group (www.sigma-delta.de) [20].

The objective is to maximize the SNR assuming an ELD of $0.1T_s$ and the following restrictions for the loop-filter integrators:

- one-pole dynamic model
- output swing, OS < 1FS
- finite DC gain, $A_v \in (100, 1000)$
- finite gain-bandwidth product, GBW $\in (0.5, 5)f_s$.

Figure 6.26 shows the outcome of the optimization process, featuring the synthesized set of loop-filter coefficients – which can be implemented using active-RC circuits – as well as the output spectrum and SNR-vs-input-level curve. The optimization tool also gives information about the evolution of the fitness function governing the optimization process. In this case, two million simulations were performed in less than one minute!

6.4 Using Multi-objective Evolutionary Algorithms to Optimize ΣΔMs

In the majority of cases, the result of an optimization-based synthesis procedure like those discussed in previous sections results in a (single) optimum solution in the design space: the best design in terms of the required resolution, signal bandwidth and power consumption. However, such an optimum solution

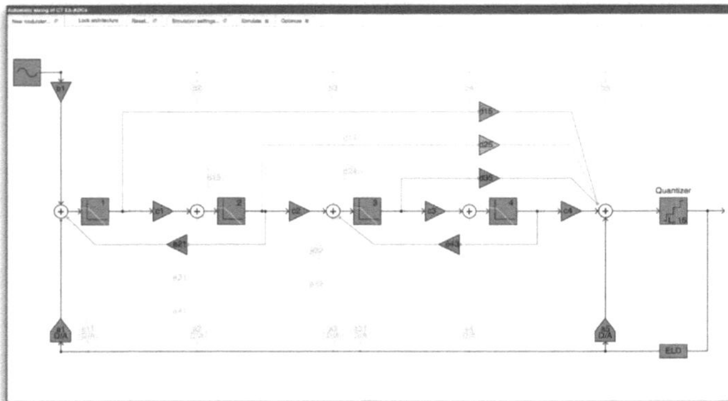

Coefficient	Value
a1	-0.42321
a5	-0.12293
Coefficient	**Value**
b1	0.42321
Coefficient	**Value**
c1	0.84637
c2	0.48488
c3	0.37546
c4	1.00000
Coefficient	**Value**
d15	2.91173
d25	2.14598
d35	1.66167
Coefficient	**Value**
e21	-0.00509
e43	-0.05929

(a)

Integrator	DC gain	GBW	Prop. path value
1	100.63909912109375	0.5009453296661377	-
2	100.25150299072266	0.7618148922920227	-
3	101.5791244506836	0.8792803287506104	-
4	100.0401840209961	0.6083009839057922	-

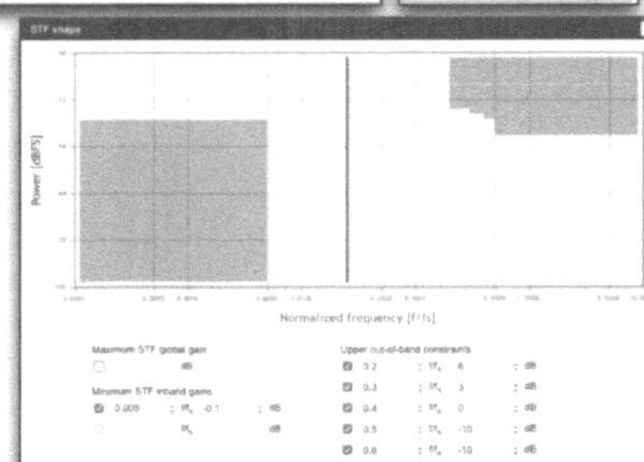

(b)

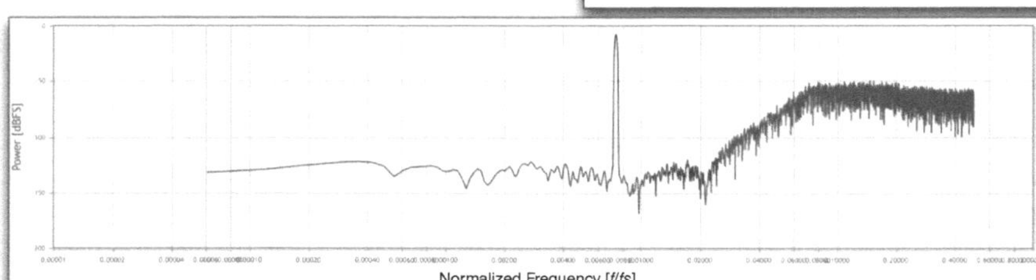

(c)

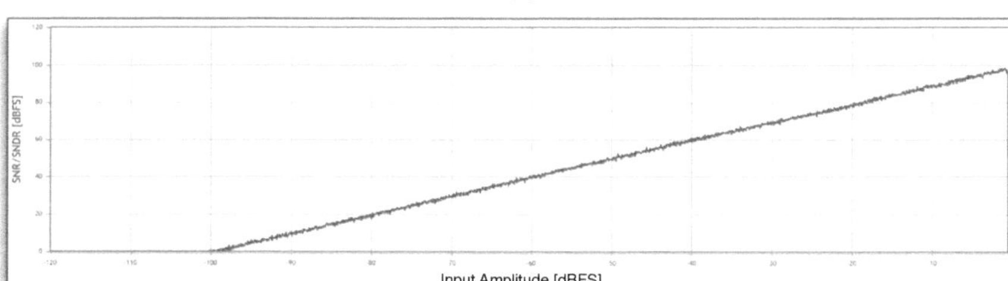

(d)

Figure 6.26 Outcome of the optimization process of a fourth-order CRFF CT-ΣΔM using the web-based synthesis tool for CT-ΣΔMs (www.sigma-delta.de) [20]: (a) block diagram and synthesized loop-filter coefficients; (b) optimized values of A_v and GBW and STF specifications; (c) output spectrum; (d) SNR versus input signal.

is not always the best and definitive result, since there are a number of design constraints that cannot be taken into account at system level, thus requiring multiple iterations between architecture and circuit level in order to obtain the final design – sometimes far from the initial optimum design obtained at system level. This problem is aggravated in some kinds of ΣΔMs, such as high-order or high-speed CT implementations, which are more sensitive to the variability of loop-filter coefficients [3].

In order to address the mentioned limitation, some authors have proposed the use of using so-called multi-objective evolutionary algorithms (MOEAs) to find a set of optimal solutions instead of a single solution, as is the case with conventional optimization approaches [13]. To this end, the synthesis process can be formulated as a multi-objective optimization problem (MOOP), where a MOEA is combined with accurate behavioral simulation to explore the design space, resulting in what is known as a Pareto-optimal front (PoF) of designs [21]. In this way, designers gain more flexibility to select the optimum design that satisfies the required specifications, while guaranteeing robustness and power efficiency.

More generally, the optimization process of any circuit or system can be formulated as a MOOP. In a MOOP a set of conflicting functions (design objectives) must be optimized subject to some constraints, as follows:

$$\begin{aligned}&\text{maximize/minimize } \vec{y} = f_i(\vec{x}), i = 1, 2, \ldots, b\\&\text{subject to } g_j(\vec{x}) \geq 0, j = 1, 2, \ldots, k\end{aligned} \tag{6.10}$$

where $\vec{x}$ is the vector of the design variables (such as transistors sizes), b is the number of design objectives (circuit performances), and k is the number of constraints.

Due to the conflicting nature of the design objectives, solutions of MOOPs are not unique but form a PoF, which can be defined as the set of all possible solutions of a MOOP that represent the best trade-offs between the design objectives. Solutions of a PoF fulfill the Pareto-dominance criterion, which establishes that for two solutions, a and b, a is said to dominate b ($a \prec b$) if all design objectives of a are not worse than those of b and there is at least one design objective of a that is strictly better than the same design objective of b. In the same way, a is a non-dominated solution if there is no other solution that dominates it.

The methodology presented here is based on the use of a MOEA, which is an efficient way to solve MOOPs [21]. MOEAs are based on the evolution of a population of N individuals (that is, different designs) through a number of generations. As in natural selection processes, new solutions are generated using mutation and crossover operators over the generations. Only the best solutions, which are found using tournament and selection mechanisms, prevail. As an example of this approach, an improved version of the so-called nondominated sorting genetic algorithm (NSGA-II)[5] is used [22]. Using this MOEA, the design objectives of the parent population are improved after each generation. An example is depicted in Figure 6.27, where two design objectives are targeted. It can be seen how at each generation the individuals reach better regions of the performance space, resulting in a PoF after 100 generations.

6.4.1 Combining MOEA with SIMSIDES

The methodology described here is based on the use of an evolutionary algorithm (EA), which acts as an optimizer and is coupled to SIMSIDES. Instead of performing the optimization in several

[5]NSGA-II is a multi-objective genetic optimization algorithm which is coded under BSD license. The MATLAB code can be downloaded from the MATLAB file exchange website at www.mathworks.com/matlabcentral/fileexchange.

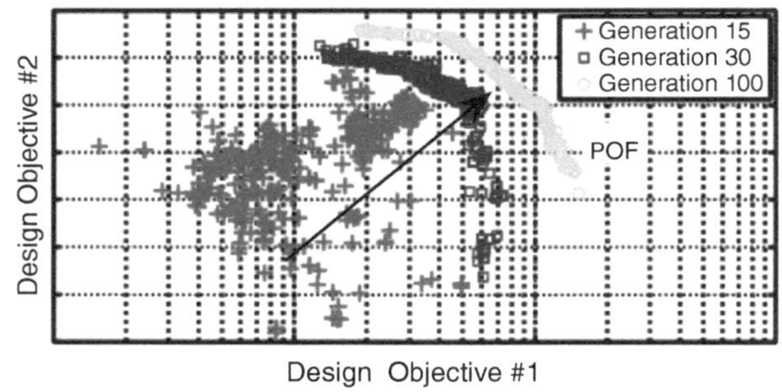

Figure 6.27 An illustration of the evolution towards the Pareto-optimal front [13].

steps, this approach explores different solutions (different combinations of all design parameters) simultaneously. Thus the final result is not a single solution but a set of different optimal solutions that together represent the best achievable trade-off between the SNDR and power consumption. The design-space exploration capabilities of EAs allows discovery of better solutions that, in principle, may be missed by the designer due to time constraints or a bad choice of the sweep range and granularity. The optimization of the design parameters of the modulator can be treated as a MOOP, as stated above, and, therefore, can be solved using the MOEA, NSGA-II. A conceptual representation of this optimization methodology is shown in Figure 6.28, where N is the size of the population and M is the number of generations of the evolutionary process.

Figure 6.29 illustrates how the MOEA optimization is coupled to SIMSIDES. The optimization is launched from the optimization menu (Figure 6.29a), where the user can open a template MATLAB script to set up the optimization process (Figure 6.29b). Once the script has been completed with the information required (Figure 6.29c), the optimization is started and a progress bar indicates the status of the optimization (Figure 6.29e). Alternatively, intermediate information about the results of the optimization can be displayed in the MATLAB command window (Figure 6.29d). The final results can be plotted from SIMSIDES menu, as depicted in Figure 6.29f.

6.4.2 Applying MOEA and SIMSIDES to the Synthesis of CT-ΣΔMs

Without loss of generality, let us consider as a case study, the high-level optimization of a LC-based fourth-order BP CT-ΣΔM like the one shown in Figure 6.10. The design target is to digitize 40-MHz signals with the maximum SNDR and minimum power consumption, P, considering three different

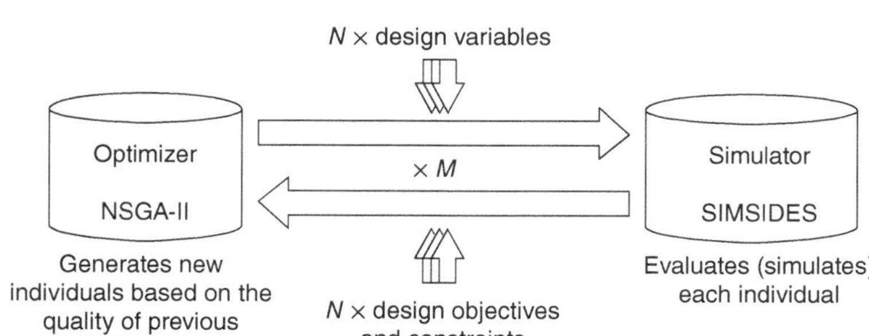

Figure 6.28 Conceptual diagram of the MOAE-based optimization of ΣΔMs.

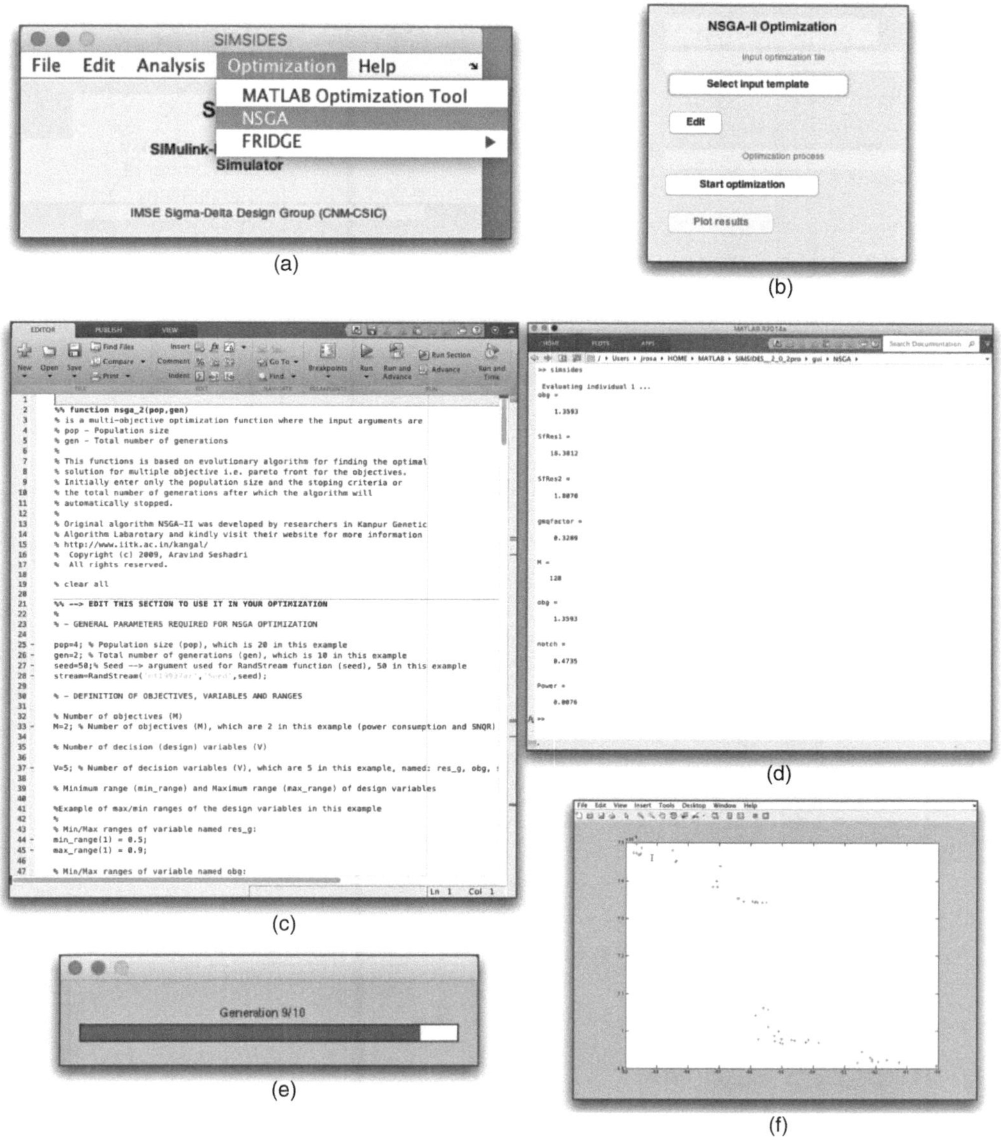

Figure 6.29 Illustrating the use of SIMSIDES with NGSA-II optimization engine: (a) launching optimization menu; (b) selecting the optimization script; (c) example of optimization script; (d) intermediate results of the optimization process displayed at the MATLAB workspace;(e) progress bar showing the status of the optimization; (f) final results plotted as a PoF.

operation modes, corresponding to a signal with B_w = 40 MHz placed at 450 MHz (mode 1), 700 MHz (mode 2), and 950 MHz (mode 3).

Figure 6.30 and Figure 6.31 show an excerpt of the MATLAB script used for the optimization in SIMSIDES with the NSGA-II algorithm, highlighting its main sections. As stated earlier in this chapter, the first step of the synthesis procedure is to define the design variables, as well as their

Part 1

```
%%%%%%%%%%%%%%%%%%%%%%%%%%%%%%%%%%%%%%%%%%%%%%%%%%%%%%
%% function nsga_2(pop,gen)
% is a multi-objective optimization function where the input arguments are
% pop - Population size
% gen - Total number of generations
% ...
%% function nsga_2(pop,gen)
% is a multi-objective optimization function where the input arguments are
% pop - Population size
% gen - Total number of generations
% ... explanation of the optimization algorithm not shown for simplicity

%% EDIT THIS SECTION TO USE IT IN YOUR OPTIMIZATION
% GENERAL PARAMETERS REQUIRED FOR NSGA OPTIMIZATION
pop=20; % Population size (pop), which is 20 in this example
gen=4;  % Total number of generations (gen), which is 10 in this example
seed=50;% Seed --> argument used for RandStream function (seed),
stream=RandStream('mt19937ar','Seed',seed);

%% - DEFINITION OF OBJECTIVES, VARIABLES AND RANGES

M=2; % Number of objectives (M), (power consumption and SNQR in this ex.)
V=5; % Number of design/decision variables (V)

% Min/Max ranges of design variable named res_g:s
min_range(1) = 0.5;
max_range(1) = 0.9;

% Min/Max ranges of design variable named obg:
min_range(2) = 1.25;
max_range(2) = 2.0;
%Min/Max ranges of design variable named SfRes1:
min_range(3) = 2.0;
max_range(3) = 30.0;
%Min/Max ranges of design variable named SfRes2:
min_range(4) = 1.0;
max_range(4) = 12.0;
%Min/Max ranges of design variable named gmqfactor:
min_range(5) = 0.0;
max_range(5) = 0.4;

% Wait/progressing bar:
h = waitbar(0,'Please wait, starting optimization...');

%% Load SIMSIDES model

model='Model_Scaled_Qenh_OoBGSvsResGain.mdl';

load_system(model);

initial_time=cputime;

%% Initialize the population

chromosome = initialize_variables(pop, M, V, min_range, max_range);

waitbar(0,h,sprintf('Evaluating initial population...'))

for i=1:pop
%% EVALUATION SECTION
% Insert here all model parameters required to simulate the model
% and evaluate the objectives(SNQR and Power in this example)
% ...missing code goes here (not shown for simplicity)

    chromosome(i,6)=-SQNRb;
    chromosome(i,7)=Powerb;
% ...
end
```

Population size and number of generations

Design variables, objectives and SIMSIDES model

Initialize optimisation and "chromosomes"

Figure 6.30 Excerpt of MATLAB script used for the optimization of SIMSIDES with NSGA-II (Part 1). The interested reader can see also more details about NSGA-II code at www.mathworks.com/matlabcentral/fileexchange.

Part 2

```
% Sort the population using non-domination-sort.
%...
chromosome = non_domination_sort_mod(chromosome, M, V);
save generations.txt chromosome -ASCII
% Start the evolution process...
waitbar(0,h,sprintf('First generation completed'))
waitbar(0,h,sprintf('Beginning evolutionary optimization...'))
for i = 1 : gen
    pool = round(pop/2);
    tour = 2;
    parent_chromosome = tournament_selection(chromosome, pool, tour);
    mu = 20;
    mum = 20;
    offspring_chromosome = ...
        genetic_operator(parent_chromosome, ...
        M, V, mu, mum, min_range, max_range);
    [offpop,~]=size(offspring_chromosome);

%% EVALUATE OFFSPRING // Model code goes here (not shown for simplicity) ...
    for j=1:offpop
        res_gain=offspring_chromosome(j,1); % variable j -->chromosome(i,j)
        obg=offspring_chromosome(j,2);
        SfRes1=offspring_chromosome(j,3);
        SfRes2=offspring_chromosome(j,4);
        gmqfactor=offspring_chromosome(j,5);
 %%  Missing (model) code goes here (not shown for simplicity) ...
        options = simset('SrcWorkspace','current');
        sim (model,[],options)
        if(length(out)<N-2)
            SQNR=NaN;
        else
            SQNR=fsnr(out,1,N,fs,fin,BW,20,20,5,2,2); %SQNR calculus
        end
        SQNRb=SQNR;
        if SQNRb ~= SQNRb
            SQNRb=-10e3;
            Powerb=10e3;
        end
        offspring_chromosome(j,6)=-SQNRb;
        offspring_chromosome(j,7)=Powerb;
    end
% Intermediate population
    [main_pop,temp] = size(chromosome);
    [offspring_pop,temp] = size(offspring_chromosome);
    clear temp
    intermediate_chromosome(1:main_pop,:) = chromosome;
    intermediate_chromosome(main_pop + 1 : main_pop + offspring_pop,1 : M+V) = ...
        offspring_chromosome;
    intermediate_chromosome = ...
        non_domination_sort_mod(intermediate_chromosome, M, V);

    chromosome = replace_chromosome(intermediate_chromosome, M, V, pop);
    save generations.txt chromosome -ASCII -append

    nodom=0;
    for m=1:pop
        if chromosome(m,8)==1
            nodom=nodom+1;
        end
    end
    if ~mod(i,100)
        clc
    end
    waitbar(i/gen,h,sprintf('Generation %d/%d',i,gen))
end
delete(h)
design_time=cputime-initial_time
save final_pop.txt chromosome -ASCII
```

Sort the population and start evolution process

Perform crossover and mutation

Final selection and results

Figure 6.31 Excerpt of MATLAB script used for the optimization of SIMSIDES with NSGA-II (Part 2). The interested reader can see also more details about NSGA-II code at www.mathworks.com/matlabcentral/fileexchange.

ranges, the design objectives, and the constraints. Note that in this case, no sweep step has to be defined since the NSGA-II can efficiently search over the whole range. Some additional parameters used by genetic algorithms, such as the *population size* and the *number of generations* to be considered, are also set at the beginning of the process. Following the biological analogy, the so-called *chromosomes* are defined as a function of variables, objectives, and ranges, and these chromosomes are crossed over so that the dominant *individuals* are selected over the evolution of different populations.

The design variables of the optimization will be the set of parameters that are adjusted in the design process. In the case study considered here, the design parameters are OBG, R_{gain}, $SfR_{1,2}$, and $g_{\text{kq1,2}}$. The design objectives are SNDR and P, which must be respectively maximized and minimized. The evaluation of the individuals is done as follows. The value of the SNDR can be computed from SIMSIDES, and the power consumption, P, is approximately calculated as the sum of the power consumed by resonators, P_{RES}, by the adder, P_{ADD}, and by the embedded quantizer (this block accounts for both the ADC and DAC power consumption). The latter can be estimated from previous versions of the design from transistor-level simulations, whereas P_{RES} and P_{ADD} can be respectively estimated as:

$$
\begin{aligned}
P_{\text{RES}} &= V_{\text{DD}} \cdot \left[\sum_{i=1}^{3} (I_{ci} + I_{cid}) + \frac{1}{g_{mID}} (g_{m1} + g_{m2}) \right] \\
P_{\text{ADD}} &= V_{\text{DD}} \cdot \left[\frac{g_{m3}}{g_{mID}} + \frac{c_0 + c_{0d}}{R} \right]
\end{aligned}
\tag{6.11}
$$

where V_{DD} is the supply voltage, V_{FS} is the full-scale reference voltage, and g_{mID} stands for the transconductance-versus-current efficiency. In this case study, the same load capacitances, inductances, and Q-enhancement gains are assumed for both resonators in the modulator loop filter (see Figure 6.10). Another important issue in this kind of optimization procedure is the correct choice of the size of the population and the number of generations. In this example, a population size of 50 individuals and 50 generations is assumed.

The modulator operation modes described above impose different frequency constraints in the operation of the modulator. Figure 6.32 shows the outcome of the multi-objective optimization (as Pareto-optimal fronts) of the ΣΔM in each of the three modes. The graph shows the resulting SNDR-P trade-off. Each point in these plots corresponds to a single high-level design of the modulator, following the methodology explained in previous section, which results in a total of 7500 simulations in this case. These fronts have been generated on a 64-bit Linux server with a 2.2-GHz AMD CPU and 16 GB RAM, resulting in an average CPU time of 8 h.

It is interesting to compare the solutions obtained with the MOEA-based optimization methodology with that obtained using a conventional synthesis procedure, like that described in Chapter 5,

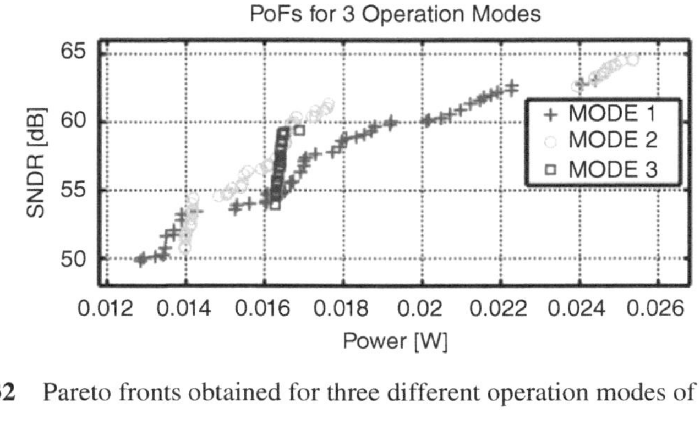

Figure 6.32 Pareto fronts obtained for three different operation modes of Figure 6.10.

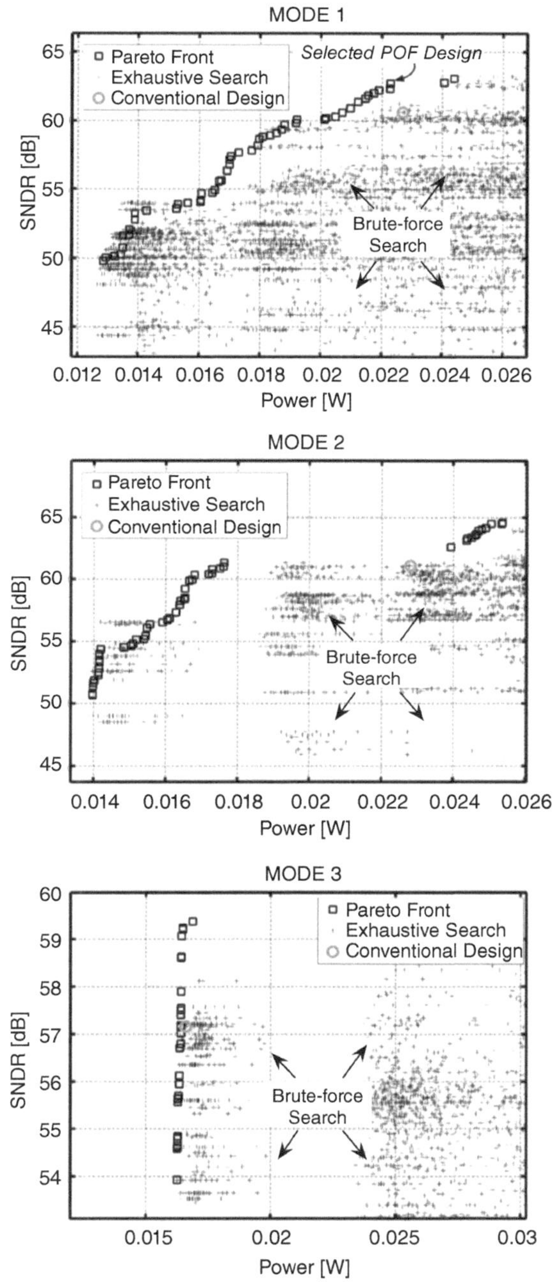

Figure 6.33 Comparison of the MOEA-based optimization methodology with conventional design procedure and exhaustive search, for the three operation modes [13].

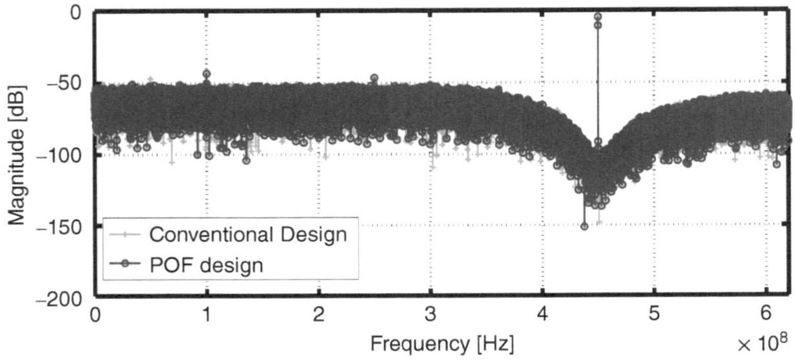

Figure 6.34 Output spectra of the conventional and selected designs for the Mode-1 POF plot of Figure 6.10 [13].

based on a brute-force approach of multiple parametric analysis within the design space. For each mode, both the front of solutions and the conventional design solution are shown in Figure 6.33. Any of the solutions in the fronts is better (in both power and SNDR) than the solutions obtained via the conventional design process. This is due to the more efficient exploration performed by the evolutionary algorithm when compared to the conventional design process.

In terms of CPU time, the conventional process requires less time since it does not implement such an exhaustive exploration of the design space. Moreover, conventional (manual) synthesis finds only a single solution and finds solutions that are worse than any of those obtained using the optimized (MOEA-based) methodology.

It is interesting to know how long would it take to arrive at the PoF of ΣΔM designs without actually using the MOEA-based optimization methodology. This is possible by performing an exhaustive evaluation of all possible ΣΔM designs, sweeping each design variable according to its specified variation grid. The results of the exhaustive exploration for each operation mode are shown in Figure 6.33. Note how, even though all possible combinations of the design variables are evaluated, some regions of the ΣΔM design space are not evaluated due to the variation grids that have to be defined. In other words, it is not possible to reach the whole front using an exhaustive search. In contrast, the MOEA-based methodology does not use a pre-defined grid: the exploration of the design space is carried out according to the quality of the evaluated designs, adapting the search grid accordingly and thus taking the exploration into the regions with the best ΣΔM performance.

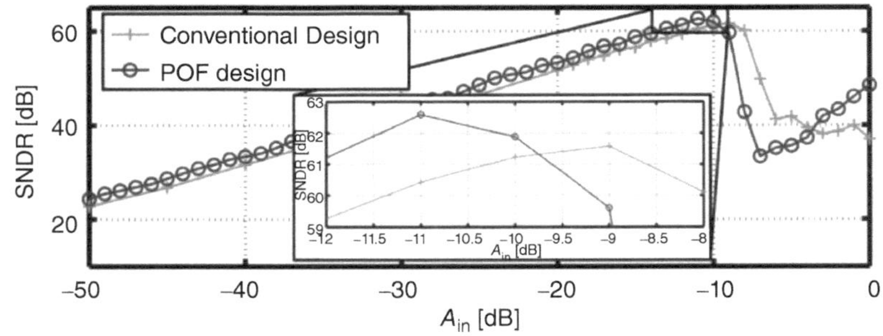

Figure 6.35 SNDR versus input signal amplitude for the conventional and the selected design for the Mode-1 POF plot of Figure 6.10.

Moreover, exhaustive search takes up to 40 h of CPU time while the MOEA-based methodology only takes 20% of this time.

As an illustration, Figure 6.34 compares the half-scale output spectra of modulator designs obtained using the conventional synthesis procedure and a selected design from the front obtained with the MOEA-based design methodology, as indicated in the Mode-1 plot in Figure 6.34. In both cases, correct noise shaping is obtained, although the design solutions found by the MOEA-based methodology are in general better in terms of SNDR and power consumption, as depicted in Figure 6.33. The benefits of using MOAE-based synthesis are better illustrated in Figure 6.35, where SNDR versus input signal is compared against a design obtained by conventional optimization.

6.5 Summary

This chapter has described some representative CAD tools and methodologies for the design automation of ΣΔMs, putting emphasis on system-level optimization. These tools allow designers to explore and select the best ΣΔM topology for a given set of specifications, as well as achieving the most efficient loop-filter circuit realization in terms of SNR and relaxed circuit requirements for their ΣΔM building blocks. In the majority of cases, this constitutes the actual design bottleneck, as demonstrated by state-of-the-art ΣΔ converters. Once the modulator has been synthesized and optimized at the architectural level, designers are in a good position to start the circuit design and implementation. This will be discussed in the next chapters.

References

[1] R. Schreier, *The Delta-Sigma Toolbox.* [Online]. Available: http://www.mathworks.com/matlabcentral, 2017.

[2] K. Francken *et al.*, "A high-level simulation and synthesis environment for delta-sigma modulators," *IEEE Trans. Computer-Aided Design of Integrated Circuits and Systems*, vol. 22, pp. 1049–1061, August 2003.

[3] S. Pavan, "Systematic Design Centering of Continuous Time Oversampling Converters," *IEEE Trans. on Circuits and Systems II: Express Briefs*, vol. 57, pp. 158–162, March 2010.

[4] S. Pavan, R. Schreier, and G. C. Temes, *Understanding Delta-Sigma Data Converters.* Wiley-IEEE Press, 2nd ed., 2017.

[5] J. G. Kenney and L. R. Carley, "Design of Multibit Noise-shaping Data Converters," *Analog Integrated Circuits Signal Processing J.*, vol. 3, pp. 259–272, 1993.

[6] S. Asghar, R. del Rio, and J. M. de la Rosa, "A 0.2-to-2MHz BW, 50-to-86dB SNDR, 16-to-22mW Flexible 4th-Order ΣΔ Modulator with DC-to-44MHz Tunable Center Frequency in 1.2-V 90-nm CMOS," *Proc. of the IEEE Intl. Conference on VLSI and System-on-Chip (VLSI-SoC)*, pp. 47–52, October 2012.

[7] G. Molina *et al.*, "LC-Based Bandpass Continuous-Time Sigma-Delta Modulators with Widely Tunable Notch Frequency," *IEEE Trans. on Circuits and Systems – I: Regular Papers*, pp. 1442–1455, May 2014.

[8] A. Morgado, R. del Río, and J. M. de la Rosa, "Design of a Power-efficient Widely-programmable Gm-LC Band-pass Sigma-Delta Modulator for SDR," *Proc. of the IEEE Intl. Symp. on Circuits and Systems (ISCAS)*, May 2016.

[9] N. Beilleau, H. Aboushady, and M. Loureat, "Using Finite Impulse Response Feedback DACs to design ΣΔ Modulators based on LC Filters," *Proc. of the IEEE Intl. Midwest Symp. on Circuits and Systems (MWSCAS)*, pp. 696–699, August 2005.

[10] Mathworks, "MATLAB R2012b," The Mathworks Inc., 2012.

[11] J. Ruiz-Amaya *et al.*, "High-Level Synthesis of Switched-Capacitor, Switched-Current and Continuous-Time ΣΔ Modulators Using SIMULINK-based Time-Domain Behavioral Models," *IEEE Trans. on Circuits and Systems – I: Regular Papers*, pp. 1795–1810, Sep. 2005.

[12] F. Medeiro, B. Pérez-Verdú, and A. Rodríguez-Vázquez, *Top-Down Design of High-Performance Sigma-Delta Modulators*. Kluwer Academic Publishers, 1999.

[13] M. Velasco, R. Castro-Lopez, and J. M. de la Rosa, "High-Level Optimization of ΣΔ Modulators Using Multi-Objetive Evolutionary Algorithms," *Proc. of the IEEE Intl. Symp. on Circuits and Systems (ISCAS)*, pp. 1494–1497, May 2016.

[14] Mathworks, "Optimization Toolbox User's Guide," The Mathworks Inc., 2016.

[15] Mathworks, "Simulink: Developing S-Functions," The Mathworks Inc., 2016.

[16] S. Pavan, "Efficient Simulation of Weak Nonlinearities in Continuous Time Oversampling Converters," *IEEE Trans. on Circuits and Systems I – Regular Papers*, vol. 57, pp. 1925–1934, August 2010.

[17] M. Keller *et al.*, "A Method for the Discrete-Time Simulation of Continuous-Time Sigma-Delta Modulators," *Proc. of the IEEE Intl. Symposium on Circuits and Systems*, pp. 241–244, May 2007.

[18] T. Bruckner *et al.*, "Hardware-Accelerated Simulation Environment for CT Sigma-Delta Modulators Using an FPGA," *IEEE Trans. on Circuits and Systems – II: Express Briefs*, vol. 59, pp. 471–475, August 2012.

[19] T. Bruckner *et al.*, "A GPU-Accelerated Web-based Synthesis Tool for CT Sigma-Delta Modulators," *IEEE Trans. on Circuits and Systems – I: Regular Papers*, vol. 61, pp. 1429–1441, May 2014.

[20] T. Bruckner *et al.*, *Uni Ulm Sigma-Delta Synthesis Tool*. [Online]. Available: http://www.sigma-delta.de, 2017.

[21] K. Deb, *Multi-Objective Optimization Using Evolutionary Algorithms*. Wiley, 2001.

[22] K. Deb *et al.*, "A Fast and Elitist Multiobjective Genetic Algorithm: NSGA-II," *IEEE Trans. on Evolutionary Computation*, vol. 6, pp. 182–197, April 2002.

CHAPTER 7

ELECTRICAL DESIGN OF ΣΔMs: FROM SYSTEMS TO CIRCUITS

The modeling, simulation and optimization techniques described in Chapters 5 and 6 are needed for the high-level synthesis and verification of ΣΔMs, so that the modulator-level performance metrics are efficiently mapped onto building-block (circuit-level) specifications. Thus, at this stage of the design cycle, the modulator is still realized at system-level, but the electrical performance parameters of all ΣΔM circuit elements (switches, capacitors, amplifiers, transconductors, comparators, and so on) have already been determined in the high-level sizing process. These parameters are in turn the circuit-level specifications, which constitute the start point for the electrical (transistor-level) and physical design of the modulator. This process – conceptually illustrated in Figure 7.1 – comprises a number of successive steps in which the initial (system-level) behavioral-model version of the modulator is transformed into a circuit schematic – initially implemented with macromodels and finally with transistors – afterwards into a layout, and finally into a chip implementation for experimental measurement in a laboratory.

This chapter represents the bridge between the system-level representation and the circuit-level realization of ΣΔMs according to the design flow conceptually illustrated in Figure 7.1. Several practical considerations related to the initial steps of the electrical design are discussed, starting in Sections 7.1 and 7.2 with the macromodel implementation of ΣΔMs as an essential design stage that relates behavioral models with the circuit-level description. This first part of the chapter gives a detailed description of how to represent the main ΣΔM building blocks using SPICE-like macromodels and/or hardware-description languages such as Verilog-A. Some examples of both SC and CT realizations are given to illustrate the use of the presented macromodels.

The second part of the chapter deals with two important practical considerations that must be taken into account from the very beginning of the electrical design phase. Section 7.3 describes

Sigma-Delta Converters: Practical Design Guide, Second Edition. José M. de la Rosa.

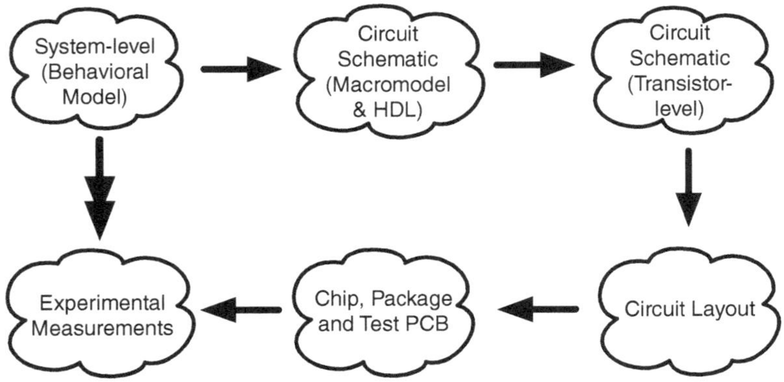

Figure 7.1 Conceptual step-by-step design flow of ΣΔMs.

how to include circuit noise in electrical-level simulations of ΣΔMs, and Section 7.4 shows how to process the modulator output data extracted from electrical simulations in SPICE-like simulators, in order to properly characterize the simulated performance of ΣΔMs in the Cadence® IC Design environment.

7.1 Macromodeling ΣΔMs

The transformation from a behavioral-level description into a circuit schematic as conceptually depicted in Figure 7.1 is carried out in several steps. Thus, at the early stages of the design cycle, HDLs, such as Verilog-A [1], and macromodels are used to represent different modulator building blocks. These models include the main circuit error limitations derived from system-level behavioral models, for instance implemented in SIMSIDES. These models are progressively replaced by transistor-level implementations as the different ΣΔM blocks are designed. In this way, the performance of the modulator is analyzed and checked at different stages of the design cycle, by combining the impact of those subcircuits that have been designed at transistor level with those that have not yet been sized.

This section explains how to use macromodels to implement ΣΔMs at circuit level in electrical simulators. The most important building-block equivalent circuits are derived and some examples are shown to illustrate their use.

7.1.1 SC Integrator Macromodel

Figure 7.2 shows a single-ended equivalent circuit that is frequently used to simulate FE SC integrators with macromodels. The corresponding macromodel for a fully-differential implementation is shown in Figure 7.3 [2]. Both equivalent circuits use ideal capacitors, whereas the switches and the OTA include nonideal circuit effects, as discussed below.

7.1.1.1 Switch Macromodel As illustrated in Figure 7.2, switches are usually modeled as a linear switch on-resistance R_{on} in series with an ideal switch, which is controlled by the corresponding clock phase, either ϕ_1 or ϕ_2. Note that the switch model in SPICE-like simulators consists of a highly

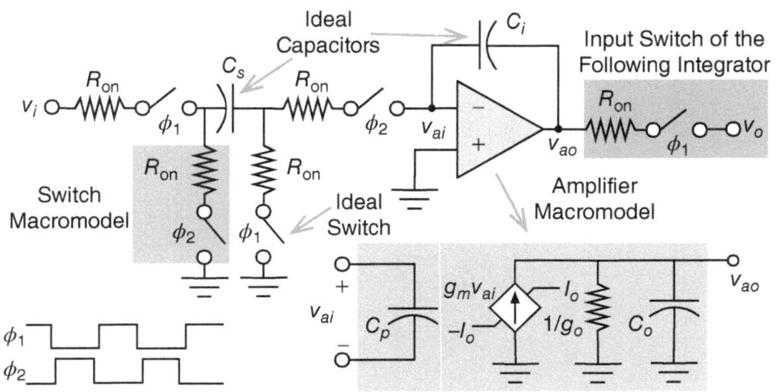

Figure 7.2 Macromodel of a single-ended FE SC integrator.

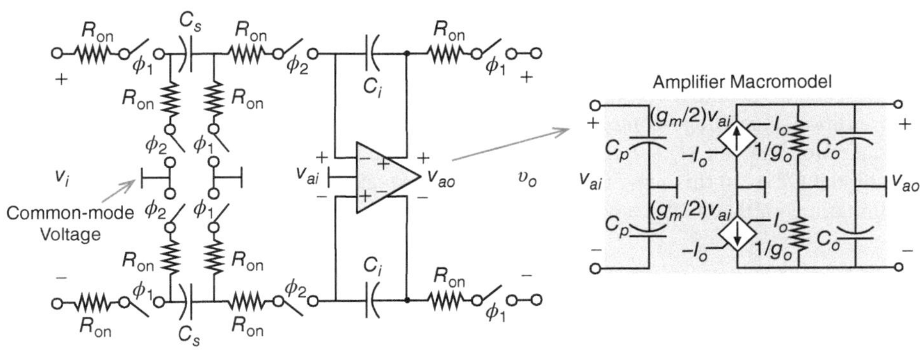

Figure 7.3 Macromodel of a fully-differential FE SC integrator.

nonlinear resistor whose resistance depends on the control clock-phase voltage v_ϕ as follows,

$$R_{\text{switch}} = \begin{cases} R_{\text{on}} & v_\phi \geq v_{\text{TS}} \\ R_{\text{off}} & v_\phi < v_{\text{TS}} \end{cases} \tag{7.1}$$

where R_{off} stands for the switch off-resistance, ideally $R_{\text{off}} \to \infty$, and v_{TS} stands for a threshold voltage that determines if the switch is either closed (on) or open (off). In practice, an almost ideal switch can be modeled in electrical simulators by setting R_{off} and R_{on} just high/low enough to be negligible with respect to other circuit elements [3]. For instance, typical values of R_{on} are chosen to be of the order of micro-ohms, while a very high value of R_{off} is typically of the order of giga-ohms.[1]

Instead of using this simple model, a macromodel which is closer to the transistor-level topology implementation can be used. For instance, let us consider a CMOS switch like those shown in the sampling circuit of Figure 3.22. This switch is made up of a pMOS switch and an nMOS switch connected in parallel. Figure 7.4 shows an equivalent circuit that takes into account the on-resistance of both MOST switches as well as their associated parasitic capacitances, denoted as C_{ps}. Note that this macromodel keeps the symmetry of the original MOST-based circuit.

[1]Note that quasi ideal values of R_{off} and R_{on} can cause tolerance and convergence problems in electrical simulations. This can be controlled by suitable settings of the corresponding numerical tolerance parameters in the simulation options [4].

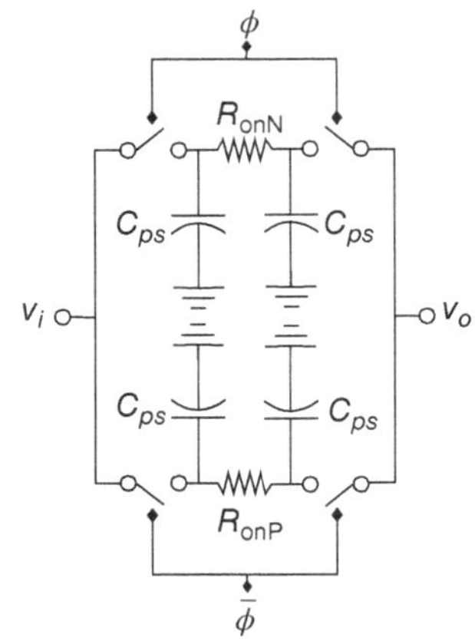

Figure 7.4 Macromodel of a CMOS switch.

7.1.1.2 OTA Macromodel

The OTA circuit is modeled by a well-known single-pole amplifier, where g_m, g_o, and C_o denote respectively the transconductance, output conductance, and output capacitance. In addition, the voltage-controlled current source used for modeling the OTA transconductance has two saturation limits, $(-I_o, +I_o)$. These limits model the minimum and maximum output currents provided by the OTA. In this way, this equivalent circuit takes into account the finite DC gain, GB, and SR limitations of the OTA, respectively given by:

$$A_v \equiv \frac{g_m}{g_o} \quad , \quad \text{GB} \equiv \frac{g_m}{C_o} \quad , \quad \text{SR} \equiv \frac{I_o}{C_o} \tag{7.2}$$

Note that real values of GB and SR deviate in practice from the above expressions due to the effect of the bottom-plate parasitic capacitances, denoted as C_p in Figure 7.3, as well as the capacitive load due to the SC network connected at the output of the integrator.

7.1.2 CT Integrator Macromodel

The equivalent circuit normally used for the macromodeling of CT integrators is based on a one-pole (or two-pole) OTA model together with the additional circuit elements that are required to implement the integrator topology: Gm-C, active-RC, and so on [5].

7.1.2.1 Active-RC Integrators

Figure 7.5 shows a simple macromodel circuit commonly used for active-RC implementations. In this example, a one-pole OTA model similar to that shown in Figures 7.2 and 7.3 is used, although higher dynamics can be implemented if necessary. In a similar way to SC integrators, parasitic and load capacitances of the OTA can also be considered.

7.1.2.2 Gm-C Integrators

Based on the same OTA macromodel circuit, Gm-C integrators can be modeled by the equivalent circuit shown in Figure 7.6a, which includes a second pole modeled by the time constant R_pC_p. In this case, the S-domain transfer function of the integrator is given by:

$$\frac{v_o(s)}{v_i(s)} = \frac{-g_m/g_o}{(1+sC/g_o)(1+sR_pC_p)} \tag{7.3}$$

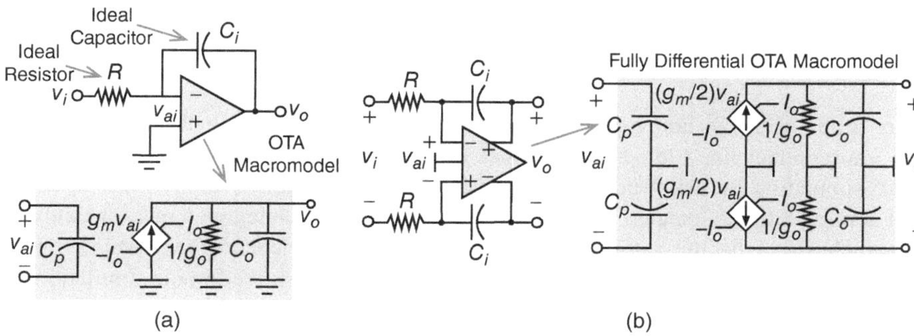

Figure 7.5 Active-RC integrator macromodel: (a) single-ended schematic; (b) fully-differential schematic.

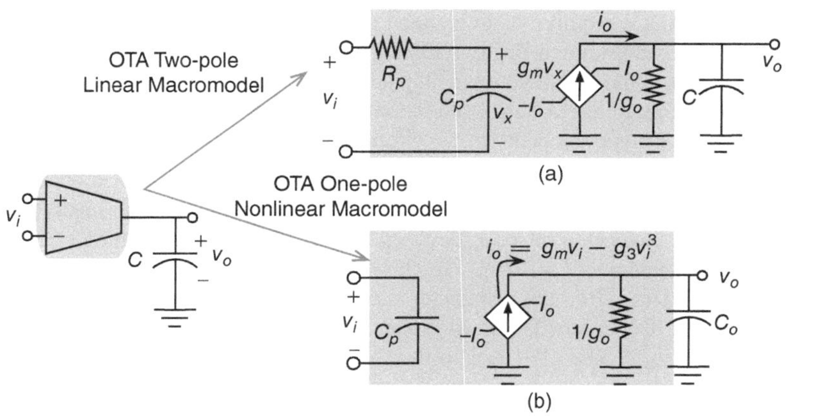

Figure 7.6 Gm-C integrator macromodel: (a) two-pole linear model; (b) one-pole weakly nonlinear model.

7.1.3 Nonlinear OTA Transconductor

The OTA macromodel in the previous sections assumed a linear model for the transconductance g_m considering an output current saturation I_o. In practice, the transconductor of the OTA is assumed to be weakly nonlinear, such that its static output current i_o is related to the OTA input voltage v_i as follows [6]:

$$i_o = \begin{cases} g_m v_i - g_3 v_i^3 & |v_i| \leq \dfrac{\text{IIP3}}{2} \\ \text{sgn}(v_i) I_o & |v_i| > \dfrac{\text{IIP3}}{2} \end{cases} \tag{7.4}$$

where sgn(x) denotes the sign function of x, g_3 is the third-order nonlinear coefficient of the transconductor, IIP3 stands for the input-referred third-order intercept point, and I_o is the maximum output current provided by the transconductor. It can be shown from Equation (7.4) that IIP3 and I_o are related to g_m and g_3 as follows:

$$\text{IIP3} = 2\sqrt{\frac{g_m}{3g_3}} \tag{7.5}$$

$$I_o = \frac{2g_m}{3}\sqrt{\frac{g_m}{3g_3}} \tag{7.6}$$

Figure 7.6b shows an equivalent circuit for the OTA macromodel that includes the weakly nonlinear transconductance element. Note that this equivalent circuit can be easily modeled in SPICE-like simulators by using nonlinear voltage-controlled current sources [3]. Moreover, other sources of nonlinearities associated with the voltage dependency of integrated resistors and capacitors can be also included in the integrator macromodels by using signal-dependent voltage/current sources. However, these kinds of nonlinear effects are more accurately considered in the simulation when the circuit is modeled at device level: taking into account the device models provided by the foundry, such as unsalicided polysilicon resistors or MiM capacitors.

7.1.4 Embedded Flash ADC Macromodel

Multi-bit/multi-level ADCs required to implement the quantizers embedded in ΣΔMs are typically implemented using the well-known flash architecture. This kind of ADC is made up of a bank of comparators that compare the quantizer input signal (i.e., the output of the integrator connected to the quantizer) with a set of reference voltages generated in a resistor ladder [7]. These reference voltages correspond to the transition points between the different adjacent quantization intervals [8].

As an illustration, Figure 7.7a shows the fully-differential schematic of a trilevel flash ADC, which is made up of two comparators and a resistive divider. The latter is made up of four unit resistors R_u, which are connected in series between the negative reference voltage $V_{\text{ref}-} = -v_{\text{ref}}$ and the positive reference voltage $V_{\text{ref}+} = v_{\text{ref}}$. In this example, there are two transition points, $-v_{\text{ref}}$ and $+v_{\text{ref}}$, in the quantizer differential characteristic illustrated in Figure 7.7b, which correspond to the differential tabs of the resistor ladder in Figure 7.7a. Thus the quantizer input $v_i \equiv v_{i+} - v_{i-}$ is compared with $V_{R+} - V_{R-}$ and $V_{R-} - V_{R+}$, in the first and the second comparator respectively. The thermometer code generated at the comparator outputs is converted into a 1-of-3 code (d_{0-2}) using three AND gates.

The circuit shown in Figure 7.7a can be used for simulating ΣΔMs with macromodels. To this end, resistors are considered ideal circuit elements,[2] while comparators and AND gates are modeled

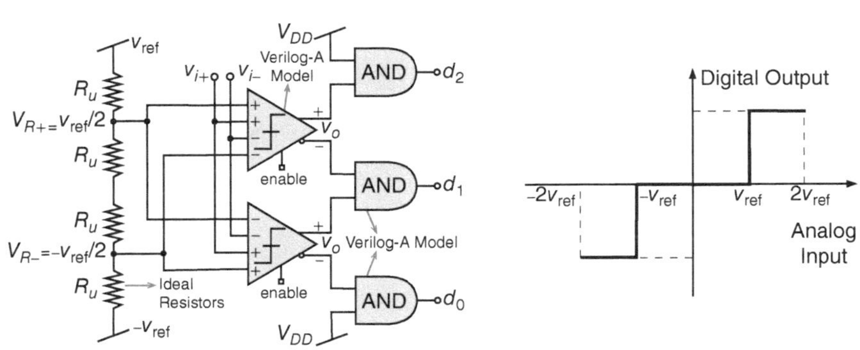

Figure 7.7 Macromodeling multi-level flash ADCs: (a) conceptual schematic of a trilevel flash ADC; (b) static differential transfer characteristic.

[2]These resistors are commonly implemented in practice using unsalicided $p+$ polylayers [9].

```
`include "discipline.h"
`include "constants.h"

//
// Model based on the OVI Verilog-A Language Reference Manual, version 1.0 1996
// and the "comparator" model available in ahdlLib library integrated in Cadence
// Design Framework II
//
//--------------------
// comparator
//
// -  comparator
//
// sigin:   (val,flow)
// sigref:  reference to which 'sigin' is compared (val,flow)
// sigout:  comparator output (val,flow)
//
// INSTANCE parameters
//    sigout_high  = maximum output of the comparator  (val)
//    sigout_low   = minimum output of the comparator  (val)
//    sigin_offset = subtracted from 'sigin' before comparason to sigref (val)
//    comp_slope   = determines the sensitivity of the comparator []
//
// MODEL parameters
//    {none}
//
// Compares ('sigin'-'sigin_offset') to 'sigref' - the output is related to
// their difference by a tanh relationship.
//
// If the difference >>> 'sigref', 'sigout' is 'sigout_high'.
// If the difference = 'sigref', 'sigout' is ('sigout_high' + 'sigout_low')/2.
// If the difference <<< 'sigref', 'sigout' is 'sigout_low'.
// Intermediate points are fitting to a tanh scaled by 'comp_slope'.
//

module Comparator(in, _in, Ref, _Ref, out, _out, enable, phi1d, phi2, phi2d, VDDAA, VSSAA,VDDAD, VSSAD, VDDDD,
VSSDD, bias,cm);
input in, _in, Ref, _Ref, enable, VSSAD, cm;
output out, _out;
electrical in, _in, Ref, _Ref, out, _out, enable, VSSAD, cm;
real vout,vcm;
parameter real sigout_high = 1.2;
parameter real sigout_low = 0;
parameter real sigin_offset = 0;
parameter real comp_slope = 1e10;
parameter real time_rise_1V = 1e-9;
parameter real time_fall_1V = 1e-9;


   analog begin

      @ ( initial_step ) begin
       if (sigout_high <= sigout_low) begin
          $display("Range specification error.  sigout_high = (%E) less than sigout_low = (%E).\n", sigout_high,
sigout_low );
          $finish;
       end
      enh

      vcm= (sigout_high + sigout_low)/2;
      @(above(V(enable,VSSAD)-vcm,1))
      begin
          vout = (sigout_high - sigout_low) * tanh(comp_slope*(V(in,_in)-V(Ref,_Ref)- sigin_offset));
      end

      V(out,VSSAD)<+ vcm+slew(vout/2,(1/time_rise_1V),(-1/time_fall_1V));
      V(_out,VSSAD)<+ vcm-slew(vout/2,(1/time_rise_1V),(-1/time_fall_1V));
   end
endmodule
```


Figure 7.8 Verilog-A code for simulating comparators in quantizer macromodels [10, 11].

using Verilog-A. As an illustration, Figure 7.8 shows the Verilog-A code used for modeling the comparators in Figure 7.7. The model – based on the Open Verilog International (OVI) language reference manual [10, 11] – is quite simple and includes the input offset. The comparison only takes place when the clock signal named `enable` is in the high state. The static input–output transfer characteristic is computed using a hyperbolic tangent function (`tanh`), which is scaled by a parameter called `comp_slope` . The latter determines the static resolution of the comparator by modifying the voltage gain around the input offset voltage, modeled by the `sigin_offset` parameter.

7.1.5 Feedback DAC Macromodel

There are essentially two types of feedback DAC circuits used in ΣΔMs: SC DACs and current-mode DACs, the latter also known as *current-steering* (CS) DACs [12]. The former are mainly used in

SC-ΣΔMs, although they are used also in some CT-ΣΔMs, specifically those implemented with active-RC integrators and intended for low-frequency applications [13]. In contrast, *switched-current*[3] or current-steering feedback DACs are commonly used in wideband CT-ΣΔMs, particularly – but not only – those based on Gm-C loop-filter implementations.

As an example of SC DACs, let us consider again the trilevel ADC shown in Figure 7.7a. The 1-of-3 coded digital output (d_0, d_1, d_2) is fed back to the ΣΔM loop-filter SC integrators using three AND gates, as illustrated in Figure 7.9. In this case, the macromodel of the DAC is simply based on the Verilog-A models of the AND gates as well as the macromodel circuits used for the switches described in Section 7.1.1.1.

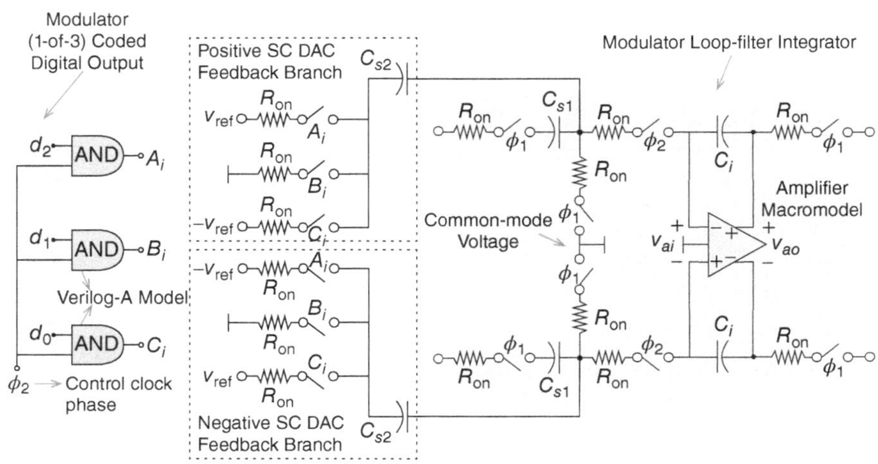

Figure 7.9 Macromodel of a trilevel SC DAC connected to an SC FE integrator within a ΣΔM loop filter. Note that different SC branches, using two different sampling capacitors C_{s1} and C_{s2}, are used in this model. However, there are many situations in practice where a single SC branch is shared by both the input signal and the feedback DAC, as will be shown later in this chapter.

Following the same philosophy, a macromodel circuit that can be used for current-steering DACs is based on simple macromodels of current sources and switches.[4] As an illustration, Figure 7.10 shows the macromodel of a fully-differential trilevel current-steering NRZ DAC. It consists of a set of current sources controlled, through switches, by 1-of-3 coded digital data (d_i). As will be discussed later in this chapter, the ideal operation of the current sources, and consequently the current-steering DAC, is degraded in practice by mismatch errors, finite output impedance (of the current sources), thermal gradients, and so on. The majority of these errors, like mismatch and technology-related errors, require a large number of simulations to be performed in order to evaluate their impact on the performance of the modulator. For that reason, these errors are usually considered in system-level behavioral models, for instance implemented in MATLAB/SIMULINK, as described in Chapter 5. Other nonidealities, like the output impedance of current sources, that also affect the dynamic operation of the modulator

[3]Current-mode feedback DACs are also used in switched-current (SI) ΣΔMs, which are a particular case of DT-ΣΔMs where the loop filter processes current-mode signals and is implemented with SI integrators instead of SC integrators [14]. The difference between switched-current cells and current-steering DACs will be discussed in Section 8.5 .

[4]Alternatively, a more ideal macromodel can be simply based on voltage-controlled current sources as will be illustrated later in this chapter.

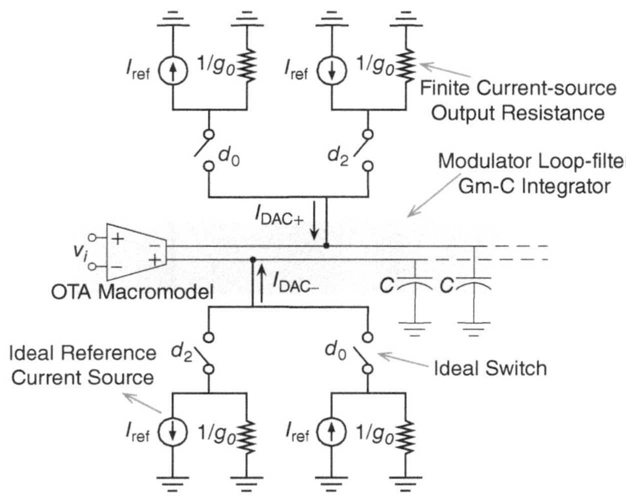

Figure 7.10 Illustrating the macromodel of a trilevel current-steering DAC connected to a Gm-C integrator within a ΣΔM loop filter.

loop filter, can easily be macromodeled by simply including an output resistance in parallel with each unit (or reference) current source, as illustrated in Figure 7.10.

7.2 Examples of ΣΔM Macromodels

A couple of examples are described in this section to illustrate the use of macromodels for the implementation of ΣΔMs. The first one is based on SC circuits while the second is implemented with active-RC circuits. In both cases, the well-known single-loop second-order ΣΔM is used as a demonstration vehicle, considering an embedded trilevel quantizer.

7.2.1 SC Second-order Example

Figure 7.11 shows the conceptual schematic of the second-order SC-ΣΔM considered, which includes a trilevel embedded quantizer. The values of the capacitors used are highlighted in the figure. The circuit elements forming this modulator – switches, amplifiers, a flash (trilevel) ADC, and an SC (trilevel) DAC – can be implemented by using the macromodels described in previous section.

Figure 7.12 shows the schematic of the modulator in Figure 7.11 implemented in Cadence IC Design, using Cadence Virtuoso Schematic. In this example, ideal clock phases (implemented with ideal voltage sources) are used for the sake of simplicity. This is common practice at the very beginning of the design phase since the use of a clock-phase generator circuit – discussed later in this chapter – slows down the simulations unnecessarily.

Note that each modulator building block in Figure 7.12 uses a suitable schematic symbol. This is very useful in practice to clearly identify the different parts of the modulator and to establish an appropriate hierarchic partitioning. This allows the design to be systematized in a

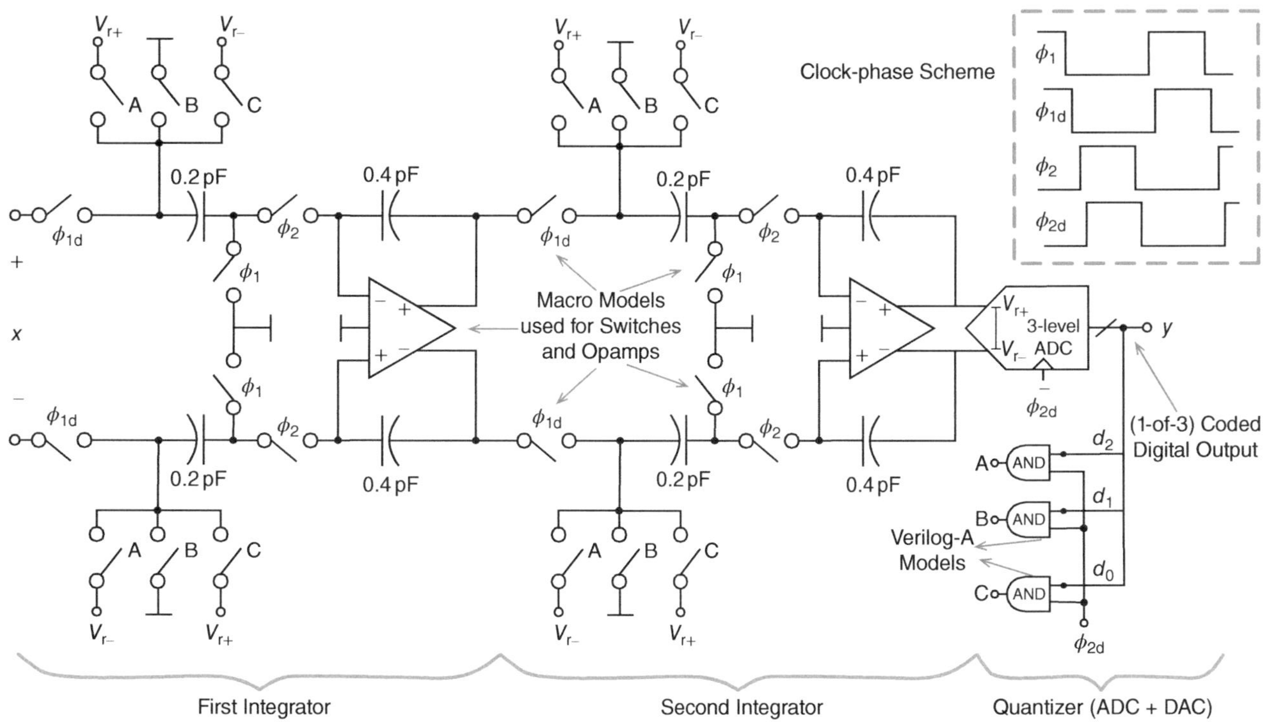

Figure 7.11 Conceptual schematic of a second-order SC-ΣΔM with a trilevel embedded quantizer.

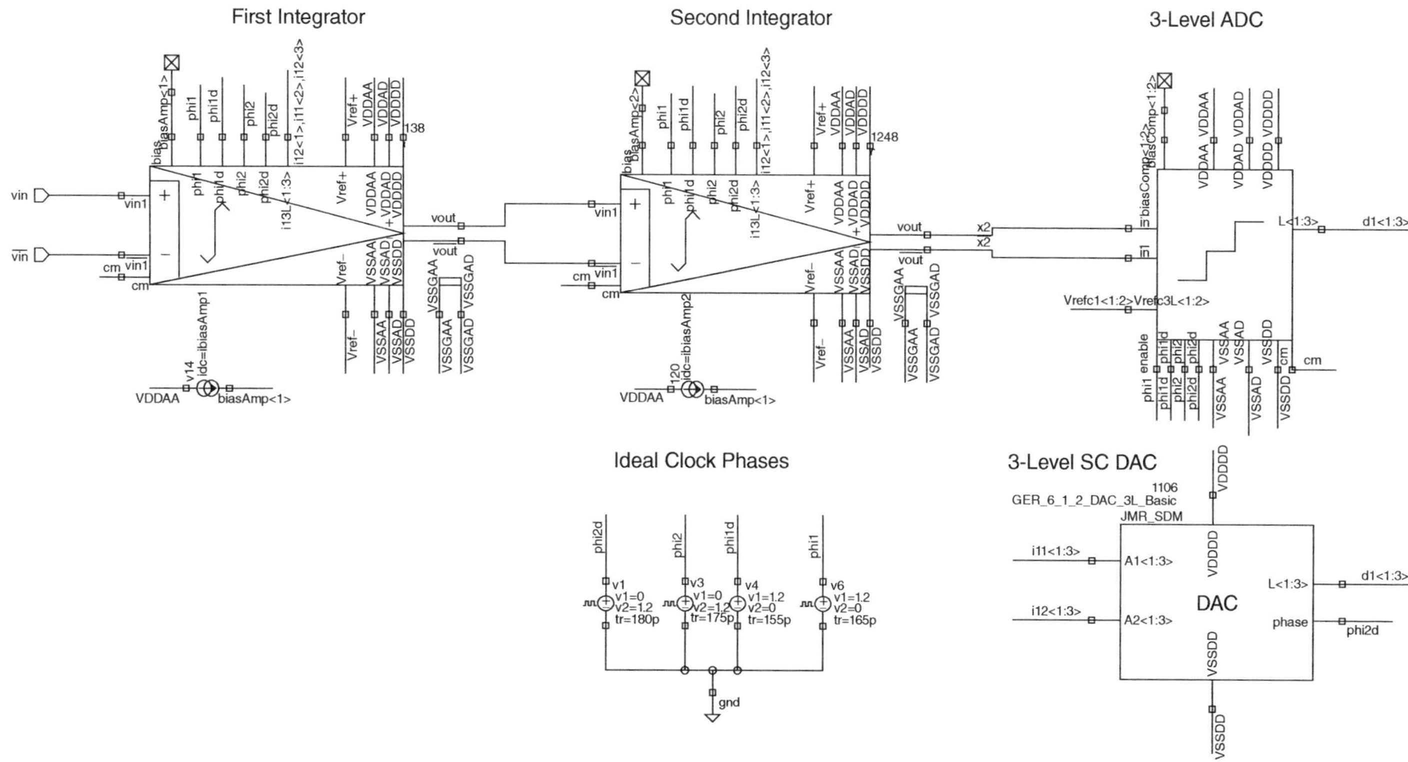

Figure 7.12 Implementation of the modulator in Figure 7.11 using Cadence Virtuoso schematics.

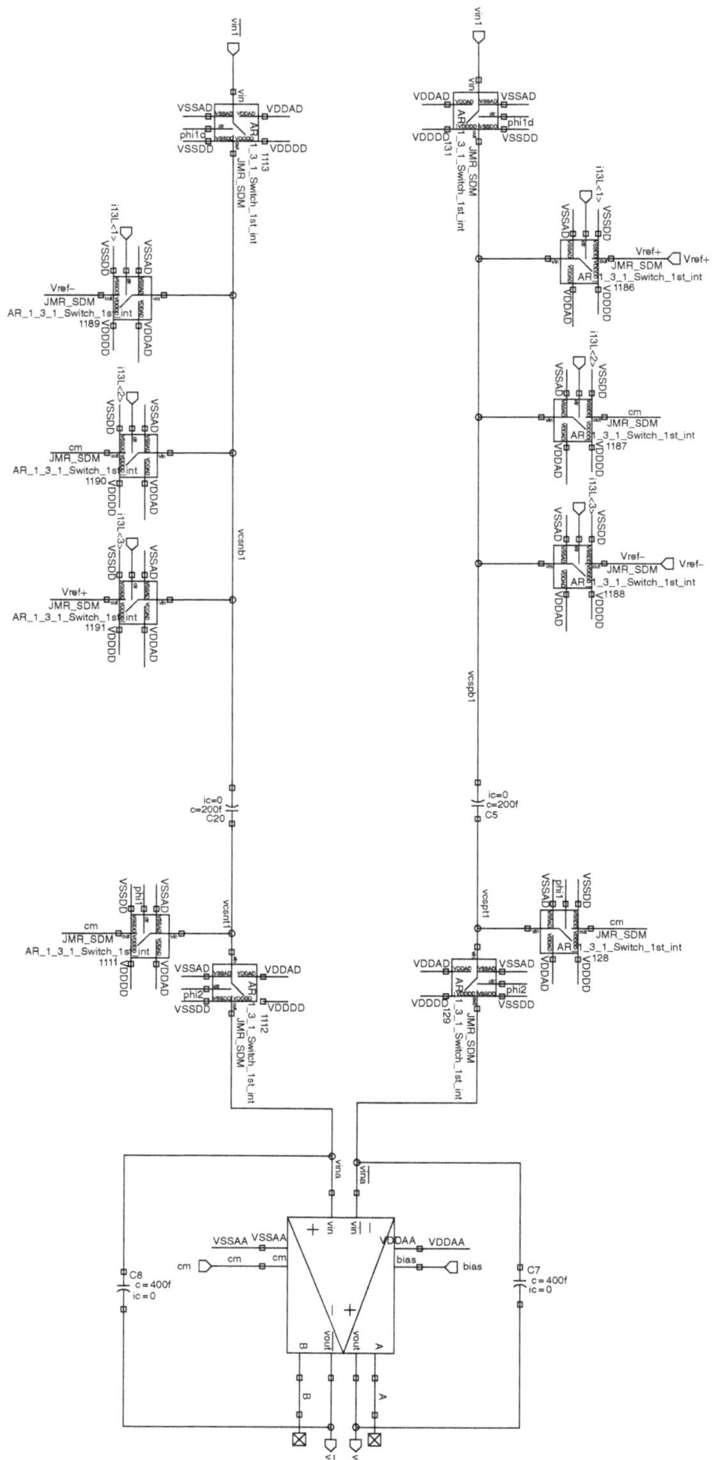

Figure 7.13 Schematic of the SC first integrator of Figure 7.12.

top-down/bottom-up approach. In this way, designers can move through the modulator hierarchy, using the most convenient schematic representation according to the part of the circuit being analyzed. As an illustration, Figure 7.13 shows the macromodel of the first integrator in Figure 7.12; the symbols for the switches, capacitors, and opamps are clearly identified.

As discussed earlier in this chapter, the use of even ideal macromodels allows designers to clearly define the electrical representation of a ΣΔM circuit, including all their nodes and branches. These macromodels can be progressively replaced by their transistor-level implementations as the different building blocks are being sized. This is something that can be done very easily in some circuit design environments, for example the Cadence Design Framework. Figure 7.14 shows how to change the type of implementation, often referred to as the *view*. Note that four different cell view names are available in this example: symbol (which is the highest abstraction level), macromodel, schematic (transistor-level), and layout.

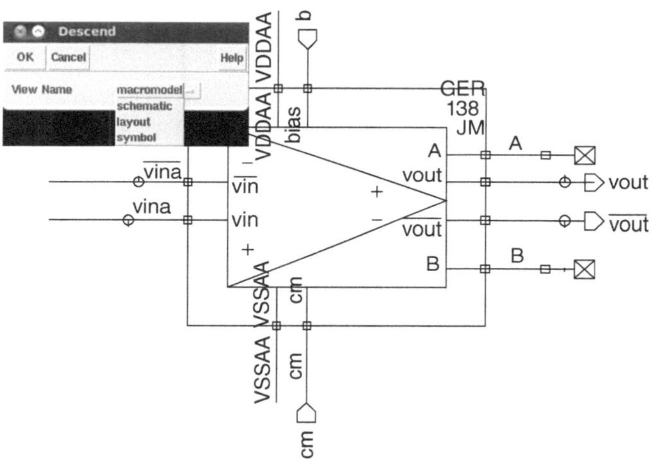

Figure 7.14 Selection of a circuit view name in the Cadence IC Design environment for the operational amplifier shown in Figure 7.13.

As an illustration, Figure 7.15 shows the macromodel of different parts of the SC integrator in Figure 7.13. Figure 7.15a shows the macromodel of the fully-differential amplifier and Figure 7.15b depicts the macromodel of a CMOS switch. In both cases, the different model parameters can be set according to the building-block specifications derived from behavioral simulations, for instance using SIMSIDES, as described in Chapter 5 .

7.2.2 Second-order Active-RC ΣΔM

Figure 7.16 shows the conceptual schematic of a second-order active-RC ΣΔM with trilevel embedded quantizer. For the sake of simplicity, the modulator does not include any excess-loop delay cancellation technique. The values and expressions of the resistances, capacitances, and feedback DAC currents are shown in the figure, and V_{ref} denotes the FS reference voltage of the modulator.

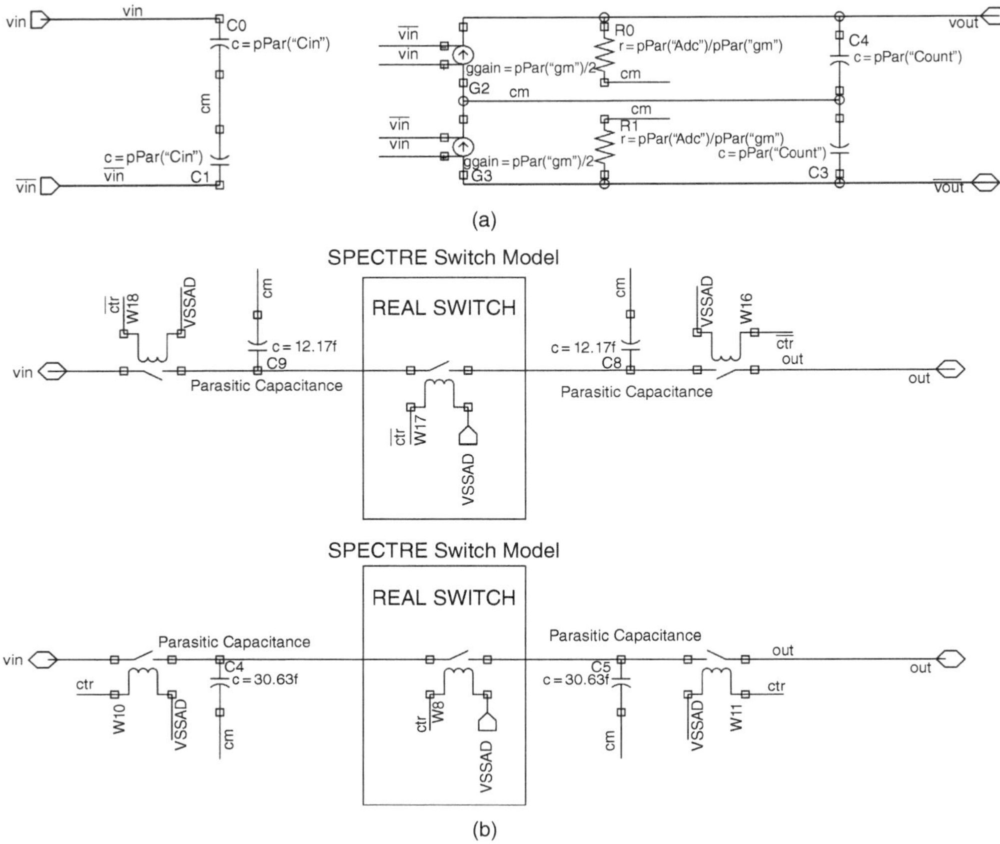

Figure 7.15 Illustrating the macromodel implementation of different circuit elements of SC integrators in the Cadence Design Framework: (a) fully-differential amplifier; (b) switch.

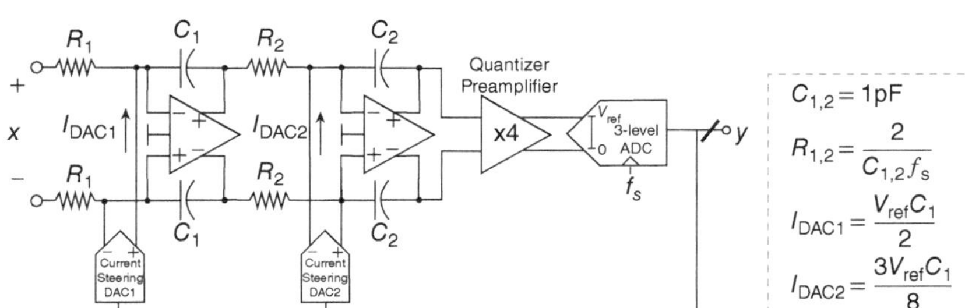

Figure 7.16 Conceptual schematic of a second-order active-RC ΣΔM with a trilevel embedded quantizer and CS feedback DACs.

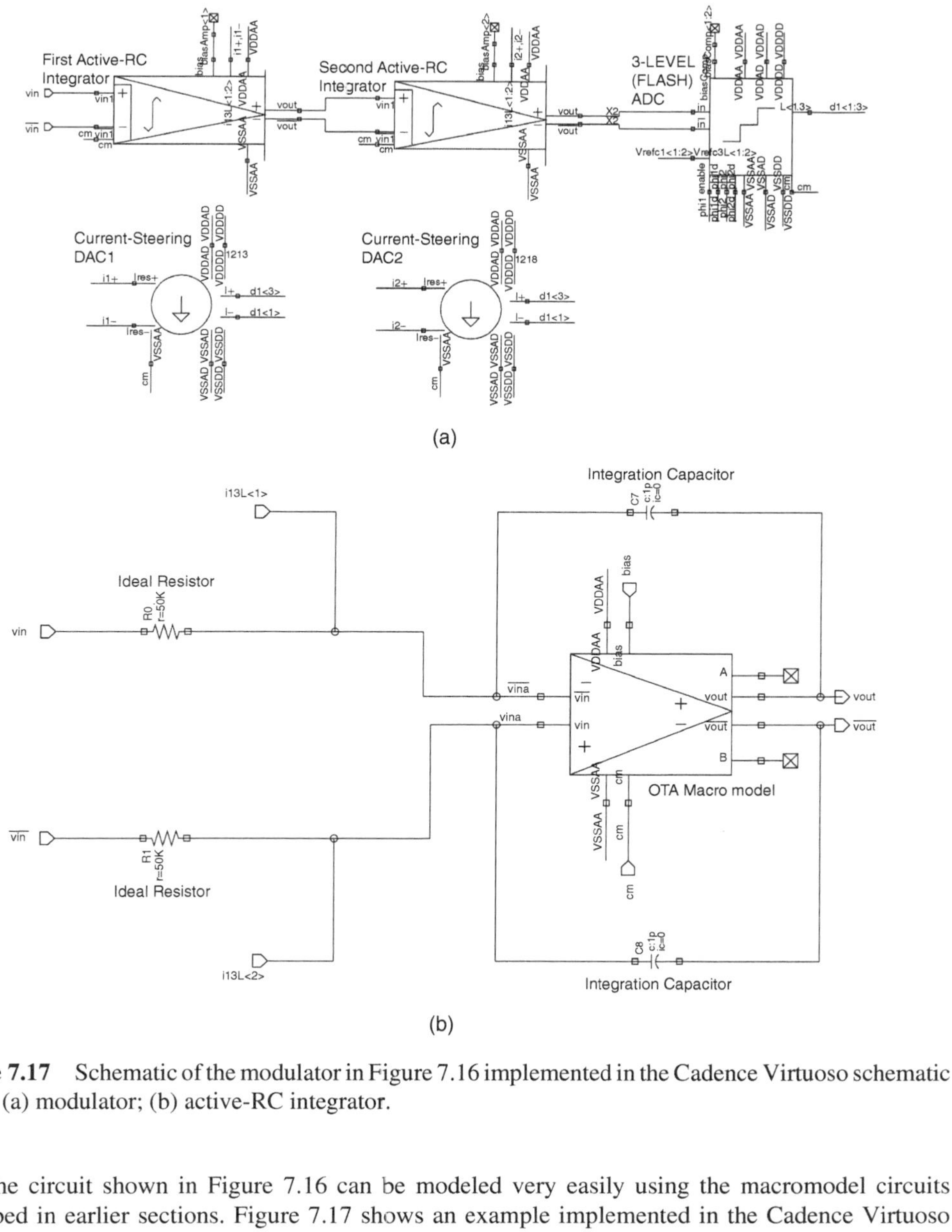

Figure 7.17 Schematic of the modulator in Figure 7.16 implemented in the Cadence Virtuoso schematic editor: (a) modulator; (b) active-RC integrator.

The circuit shown in Figure 7.16 can be modeled very easily using the macromodel circuits described in earlier sections. Figure 7.17 shows an example implemented in the Cadence Virtuoso schematic editor, highlighting the main parts of the circuit. The opamp included in the active-RC model (Figure 7.17b) uses the macromodel described in Figure 7.5b. In this example, the trilevel quantization is implemented as illustrated in Figure 7.18. The ADC, depicted in Figure 7.18a, is modeled as shown in Figure 7.7b. However, for the sake of simplicity, the current-steering DAC, shown in Figure 7.18b, is modeled by two ideal voltage-controlled current sources, which emulate the ideal switches connected in series with the reference currents (see Figure 7.10).

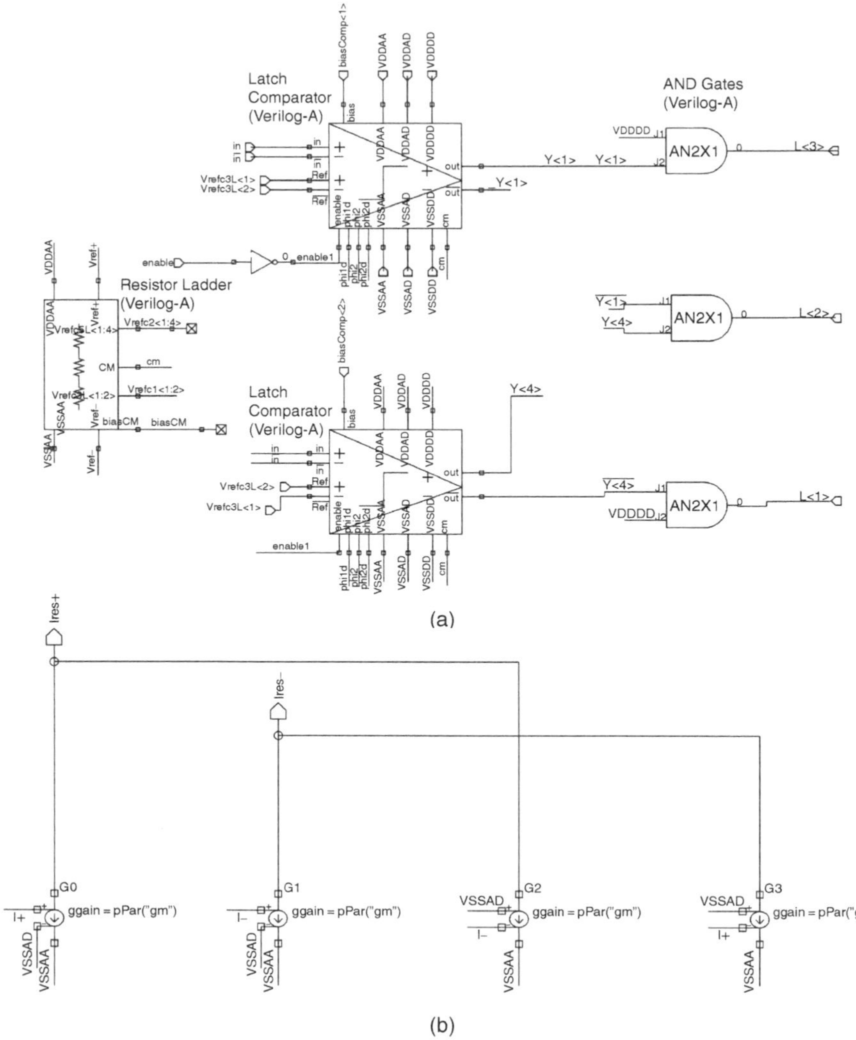

Figure 7.18 Illustrating the macromodel implementation of a trilevel quantizer in the Cadence Virtuoso schematic editor: (a) flash ADC; (b) current-steering DAC.

7.3 Including Noise in Transient Electrical Simulations of ΣΔMs

As stated in Chapters 3 and 4 , circuit noise is an ultimate limiting factor degrading the performance of ΣΔMs. Therefore, it is essential to take into account this effect in all steps of the design flow. At system level, the accurate models described in Chapter 3 can be incorporated in behavioral simulations, for instance using SIMSIDES, as detailed in Chapter 5. At the electrical level, however, the majority of SPICE-like simulators do not include noise sources in the transient analysis,[5] which makes their analysis at transistor-level more complicated.

This section describes a methodology that allows designers to create electrical (transistor-level) simulations of ΣΔMs including noise sources. The method – based on generating a noisy data sequence in MATLAB and then injecting this data sequence into the electrical simulation – can be used in most SPICE-like simulators, for example PSPICE and HSPICE [4, 16].

7.3.1 Generating and Injecting Noise Data Sequences in HSPICE

Let us consider the simple SC circuit shown in Figure 7.19a, in which a noise voltage source v_{ni} is sampled by an ideal switch S_1 and stored on capacitor C_S. This circuit can be simulated using the HSPICE netlist shown in Figure 7.19b. In this example $C_S = 1$pF, the switch off/on-resistances used in the ideal switch model are respectively $R_{\rm off} = 10^{15}\ \Omega$ and $R_{\rm on} = 100\ \Omega$ and the clock frequency is $f_s = 100$ MHz. Note that the noise sequence data is *injected* into the HSPICE transient simulation through the use of the `.DATA` statement [16]. This command allows inclusion of data that has been externally generated. In this example, a two-column (time, voltage) format file, called `noisedata`, is loaded, and the transient analysis uses the time data provided in column 1 of file `noisedata` as the sweep input parameter.

In order to compute the noise data used in the electrical simulations, it should be taken into account that $v_{ni}(t)$ is a random signal, and so its instantaneous value is not known. Instead, it can be described as a random process of zero mean and an amplitude uniformly distributed in the range $(-V_{ni}/2, +V_{ni}/2)$, where V_{ni} denotes the rms value of $v_{ni}(t)$. In this case, the mean-square value of v_{ni}

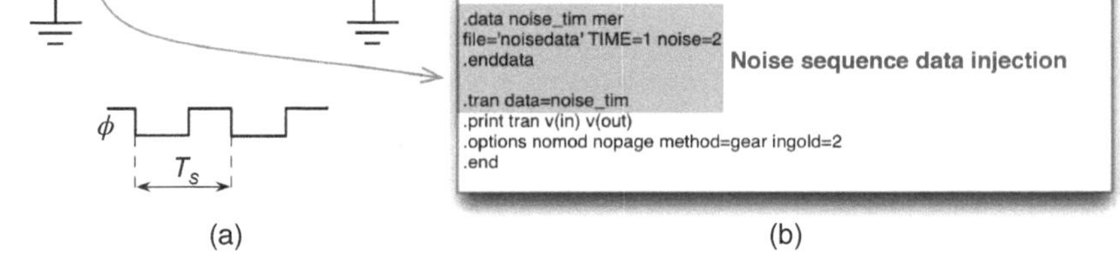

Figure 7.19 Injecting noise data sequences in transient HSPICE simulations: (a) sampling circuit example; (b) SPICE netlist.

[5]Some examples included in this book used Synopsis® HSPICE®. However, recent versions of some SPICE-like simulators, like Cadence Spectre [15] incorporate the possibility to include noise sources in transient analysis. However, the method described in this section allows the effect of a given noise source to be isolated instead of considering all noise sources acting together, as done in a regular noise analysis in Cadence Spectre. Moreover, the methodology explained in this book is well known and can be applied to the majority of SPICE-like simulators, including Spectre.

is given by [17]:

$$\overline{v_{ni}^2} = \frac{1}{V_{ni}} \int_{-V_{ni}/2}^{+V_{ni}/2} v_{ni}^2 \mathrm{d}v_{ni} = \frac{V_{ni}^2}{12} \tag{7.7}$$

Assuming a band-limited noise source, the PSD of v_{ni} can be computed as,

$$S_{v_{ni}} \simeq \frac{\overline{v_{ni}^2}}{B_{wni}} = \frac{V_{ni}^2}{12B_{wni}} \tag{7.8}$$

where B_{wni} stands for the equivalent noise bandwidth of v_{ni}. Hence, if $v_{ni}(t)$ is sampled at f_{sn}, the value of v_{ni} at instant nT_{sn} can be computed as,

$$v_{ni}(nT_s) = \mathrm{rnd}\left(\sqrt{12 f_{sn} S_{v_{ni}}}\right) \tag{7.9}$$

where rnd(x) represents a random number in the range $(-x/2, +x/2)$ and $T_{sn} \equiv 1/f_{sn}$[18].

Figure 7.20 shows the MATLAB code used for generating a N-point data sequence derived from Equation (7.9). Note that the data sequence is saved as a two-column file, in which the first column represents the time series (i.e., $0, T_s, 2T_s, ...$) and the second column is the noise data sequence generated using Equation (7.9).

```
%% GENERATION OF NOISE SEQUENCE OF GIVEN Vrms AMPLITUDE IN A GIVEN BANDWIDTH
%
% noise_Vrms:    Intended amplitude (in Vrms) of the noise integrated
%                in the BW_noise bandwidth
% BW_noise:      Bandwidth in which the noise has been integrated to extract
%                its amplitude
% N:             Numer of samples of the noise time sequence
% fs:            Sampling frequency of the noise time sequence
%
%% SETTING NOISE-SOURCE PARAMETERS
%
noise_Vrms=1e-5;
BW_noise=1e9;
N=65536;
fs=2*BW_noise;
%
%% GENERATION OF NOISE SEQUENCE
%
noise_psd=noise_Vrms^2;
x=sqrt(12*fs/(2*BW_noise)*noise_psd);
y=-x/2+x*rand(N,1);
tstep=1/BW_noise;
%
%% GENERATION OF TIME VECTOR AND SAVING DATA TO THE FILE NAMED noisedata.dat
%
t=0:tstep:(N-1)*tstep;
noisedata=[t' y];
save noisedata.dat noisedata -ascii
%
```

Noise source parameters

Noise data generation

Save data to a file

Figure 7.20 MATLAB code used for generating a N-point data sequence derived from Equation (7.9).

As an illustration, Figure 7.21 shows the PSD of the noise sampled and stored in the capacitor of Figure 7.19a, considering different values of f_s. In this example, a noise source with $V_{ni} = 10\mu$Vrms, $B_{wni} = 1$ GHz, and $f_{sn} = 2B_{wni}$ was considered in order to *emulate* an *unlimited-band* noise source. In this way, the time interval between two consecutive samples is low enough (1 ns in this example), thus enabling the noise source to be modeled as a CT source [19], which is filtered by the circuit made up of $R_{\rm on}$ and C_S ,with $R_{\rm on}$ being the switch on-resistance. This results in an equivalent bandwidth of the sampled noise given by $1/(4R_{\rm on}C_S)$. As shown in Figure 7.21, aliasing occurs in this case because

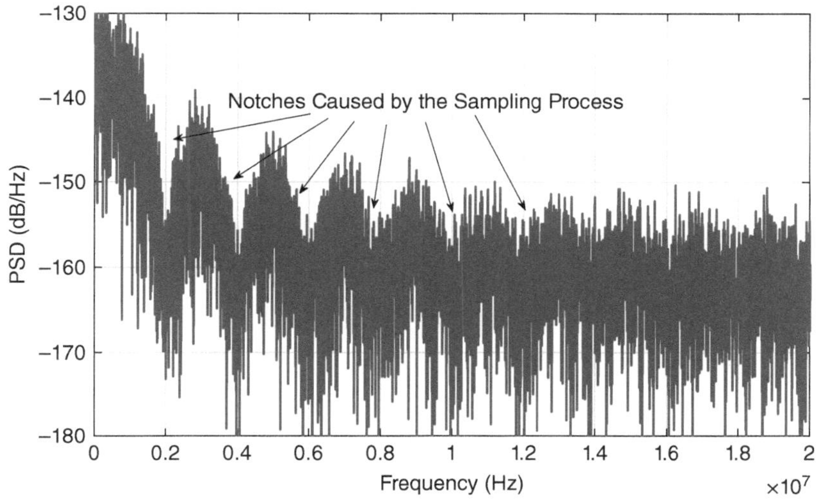

Figure 7.21 Illustrating the effect of sampling noise in HSPICE transient simulations. Output spectrum of the voltage stored in C_S in Figure 7.19a, considering the following simulation data: $R_{\text{on}} = 50\ \Omega$, $C = 0.5$ pF, $f_s = 1$ MHz.

$f_s < 1/(4R_{\text{on}}C_S)$. As a consequence the noise power increases within the Nyquist band; that is, from DC to f_s.

7.3.2 Analyzing the Impact of the Main Noise Sources in SC Integrators

The simulation technique described above can be used for verifying the impact of most critical noise sources on the performance of the main ΣΔ building blocks through transient SPICE simulations. This is particularly important in SC circuits because of the data sampling involved, and the effect this has on their noise sources.

As an illustration, let us consider an SC FE integrator like that shown in Figure 5.5 a. As stated in Chapter 3 , the main sources of circuit noise are generated in the switches and in the amplifier. Figure 7.22 shows the circuit schematics that can be used for evaluating the influence of these noise sources on SC integrators and Figure 7.23 shows the corresponding HSPICE netlist. Note that all circuits are built through simple macromodels for the switches and the amplifier in order to isolate the effect of the different error sources to be considered. Figure 7.22a shows the schematic used for simulating the thermal noise introduced by switches controlled by clock phase ϕ_1 and Figure 7.22b shows the corresponding test-bench schematic for the evaluation of thermal noise generated in the switches controlled by ϕ_2. The test-bench schematic used for the noise sources generated in the amplifier is shown in Figure 7.22c. In the latter case, both thermal and flicker noise components need to be generated. This can be done as detailed in next section.

7.3.3 Generating and Injecting Flicker Noise Sources in Electrical Simulations

The procedure described in Section 7.3.1 assumed white noise sources. However, some sources of noise in ΣΔMs also include flicker ($1/f$) noise components, which might be critical in low-bandwidth applications such as sensors, instrumentation, and biomedical applications.

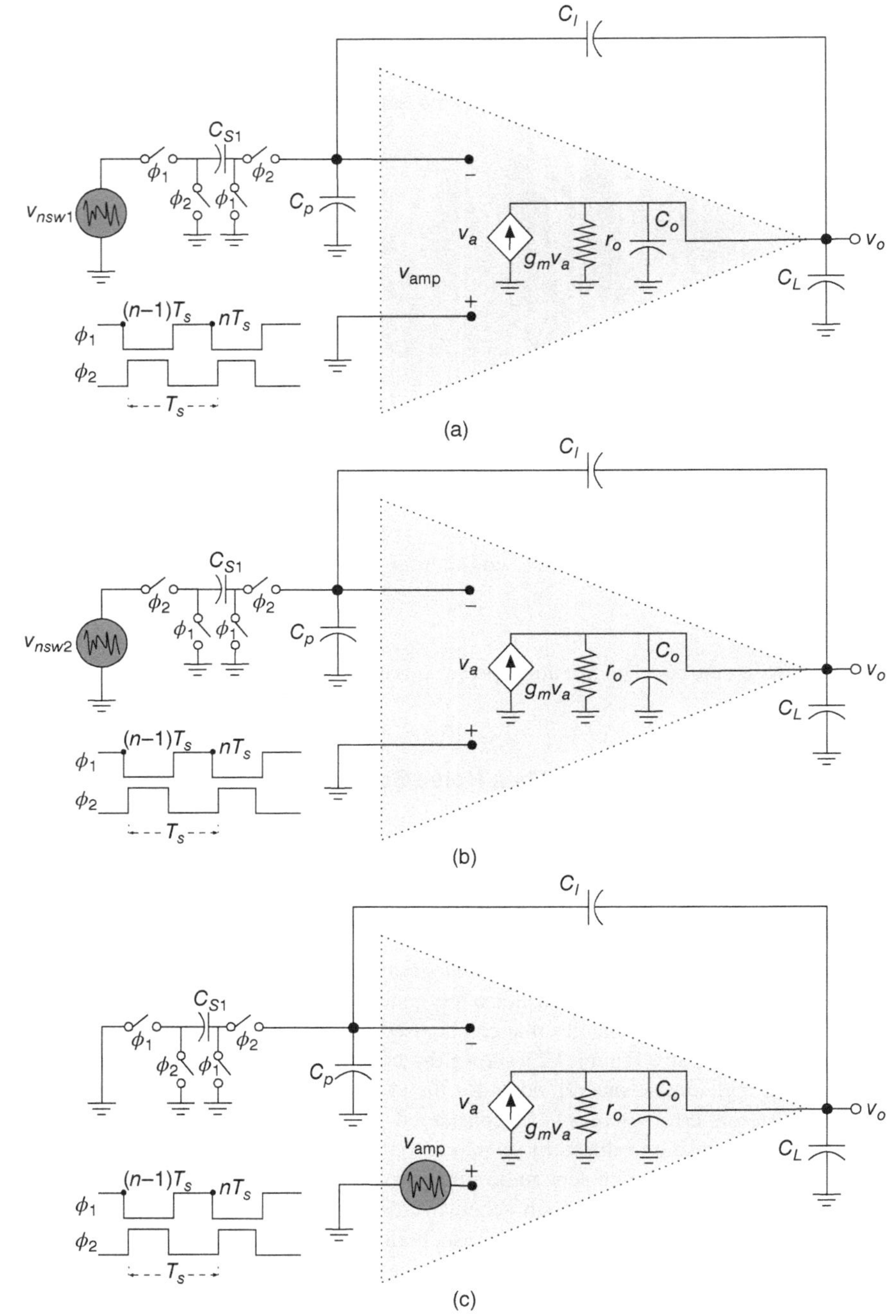

Figure 7.22 Equivalent single-ended circuits used for simulating the effect of the main noise sources in SC integrators: (a) noise contribution of switches controlled by ϕ_1; (b) noise contribution of switches controlled by ϕ_2; (c) noise contribution of the amplifier.

```
** Transient simulation of SC FE Integrators with noise sources

* Noise sources introduced by switches controlled by phi1
* Comment these two lines for the analysis of vnsw2/vamp noise

V11 11 0 pwl (time,noise)
V12 12 0 0

* Noise sources introduced by switches controlled by phi2
* Comment these two lines for the analysis of vnsw1/vamp noise

V11 11 0 0
V12 12 0 pwl (time,noise)

* Comment these five lines for the analysis of vamp noise

Gswitch11 11    c1in   vcr pwl(1) ck1 0 2.4999,1e15 2.5001,150
Gswitch21 c1in  12     vcr pwl(1) ck2 0 2.4999,1e15 2.5001,150
C1        c1in  c1out 'C1'
Gswitch31 c1out in-    vcr pwl(1) ck2  0 2.4999,1e15 2.5001,150
Gswitch41 c1out 0      vcr pwl(1) ck1  0 2.4999,1e15 2.5001,150
```

Noise contribution of switches

```
* Noise sources introduced by the opamp
* Comment these eight lines for the analysis of vnsw1/vnsw2 noise
V11 11 0 0
V12 12 0 0
Gswitch11 11    c1in   vcr pwl(1) ck1 0 2.4999,1e15 2.5001,150
Gswitch21 c1in  12     vcr pwl(1) ck2 0 2.4999,1e15 2.5001,150
C1        c1in  c1out 'C1'
vnoise in- in2- 0 pwl (time,noise)
Gswitch31 c1out in2-   vcr pwl(1) ck2  0 2.4999,1e15 2.5001,150
Gswitch41 c1out 0      vcr pwl(1) ck1  0 2.4999,1e15 2.5001,150
```

Noise contribution of opamp

```
Cp  in- 0   'Cp'
Co  in- out 'Co'
Cl1 out 0   'Cl1'

Rin in- 0 1000e10
Gm out 0 in- 0 'Gm'
Ro out 0  1e15

v2  ck2   0  pulse(0 5 'dc*(1/fclock)'  'to' 'to' '(1-dc)*(1/fclock)-2*to' '1/fclock')
v1  ck1   0  pulse(0 5  0               'to' 'to' 'dc*(1/fclock)-2*to'     '1/fclock')

ec1 c1 0 c1in c1out 1

.param C1=0.5p
.param Cp=0.1p
.param Co=2p
.param Cl1=0.1p
.param Gm=1e-3
.param fclock=70e6
.param to='1/(1000*fclock)'
.param dc=0.90
```

Common section

```
.data noise_tim mer
file='noise20480.dat' TIME=1 noise=2
.enddata
```

Noise sequence injection

```
.tran data=noise_tim
.print tran v(out)
.options nomod nopage method=gear post ingold=2
.end
```

Figure 7.23 HSPICE netlist of the circuits shown in Figure 7.22.

Flicker noise can be generated as a colored noise sequence in MATLAB by filtering a white noise source through a filter with the following transfer function [20],

$$H_f(z) = \frac{1}{(1 - z^{-1})^{\alpha}} \tag{7.10}$$

where α is a real number between 0 and 2. In this way, the corresponding noise sequence data can be injected in HSPICE transient simulations using the same method as described in earlier sections.

Alternatively, a colored noise data sequence, including both $1/f$ and white noise components, can be generated by extracting the PSD data of the corresponding noise source using a `.NOISE` analysis in SPICE and then generating a noise time sequence in MATLAB equivalent to the captured PSD, which can be injected in transient simulations using the `.DATA` statement.

Figure 7.24 shows MATLAB code used for generating a time data sequence according to the electrical data captured from a `.NOISE` simulation in HSPICE. In this example, simulation output data resulting from the noise analysis in HSPICE is stored in a file called

```
%% GENERATION OF COLORED NOISE SEQUENCE EXTRACTING DATA FROM HSPICE
%
% The PSD curve must contain a 1/f and a white component
% The PSD curve must have been obtained from 0 to fs/2
%
clear; clf;
rand('state',0) % Return RAND to its default initial state
%                 Helps to add repeatability to the results
%N=2^18;         % Length of noise time sequences to be generated
N=2^25;          % N must be >> 1 to plot noise PSD at low frequencies
%
%% Loads and plots PSD curve obtained by electrical SPICE simulation
%
data_in=load('psd_eq_ref_1f_white_RRF.dat'); % Input data is assumed to be expressed in V^2 (PSD)
f_p=data_in(:,1);
noise_psd_p=data_in(:,2);

%
%% Estimation of 1/f and white components

% Flicker (1/f) noise
Sf=noise_psd_p(1)*f_p(1);
noise_psd_1f_p=Sf./f_p;
% Thermal (white) noise
Sw=noise_psd_p(length(f_p));
noise_psd_white_p=Sw*ones(length(f_p),1);

%
%% Generation of white noise sequence
fs=2*max(f_p);
noisew_power=Sw*fs/2;
xw=sqrt(12*noisew_power);
yw=-xw/2+xw*rand(N,1);

%% Generation of 1/f noise sequence
% [1] Generation of an initial white noise sequence
fs=2*max(f_p);
noisewi_power=fs/2;
xwi=sqrt(12*noisewi_power);
ywi=-xwi/2+xwi*rand(N,1);
% [2] Computation of its FFT
ywi_freq=fft(ywi); % MATLAB FFT (double-sided spectrum)
% [3] Filtering of the white noise power with the desired H(f)
% Equivalently, the noise % amplitude is filtered with sqrt(H(f)
freq=(fs/N:fs/N:fs)'; % Double-sided frequency vector
H2_freq=freq./freq; % To validate with H(f)=1
H2_freq=Sf./freq;
yf_freq=sqrt(H2_freq).*(ywi_freq);
yf=real(ifft(yf_freq)); % MATLAB IFFT is used for simplicity

%% Addition of white and 1/f noise sequences
y=yw+yf; % Generated noise sequence including both white and 1/f components
```

Estimation of noise components from electrical data

White-noise sequence generation

1/f noise sequence generation

Figure 7.24 MATLAB code used for generating colored noise data sequence extracted from HSPICE `.NOISE` simulation.

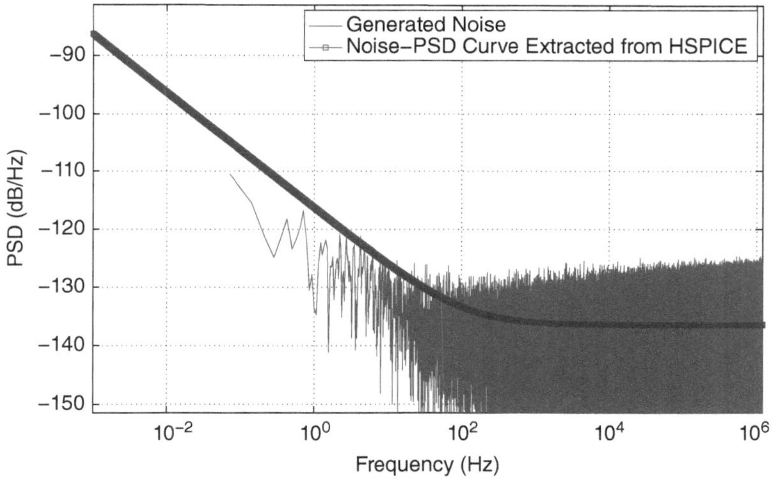

Figure 7.25 MATLAB code used for generating colored noise data sequence extracted from HSPICE `.NOISE` simulation.

`psd_eq_preamp_1f_white_RRF.dat`, which is assumed to be expressed in V^2/Hz; in other words, corresponding to a PSD curve. Both the PSD and rms value of the $1/f$ and white noise components are identified and computed, and the corresponding time data series are generated using Equation (7.9).

As an illustration, Figure 7.25 shows the output spectra generated by the MATLAB routine in Figure 7.24. Note that a huge number of points ($N = 2^{25}$ in this example) are required to generate FFTs at low frequencies and see the flicker noise corner frequency.

7.3.4 Test Bench to Include Noise in the Simulation of ΣΔMs

At the end of the design phase, transistor-level simulations of the whole ΣΔM are mandatory in order to check that the electrical performance agrees with system-level behavioral simulations, and consequently with the target specifications. In this situation, injecting noise sources in SPICE transient simulations is required. To this end, the total input-referred noise source, including the effect of sampling, can be generated and injected at the input node of the modulator following the methodology described in earlier sections.

As an illustration, Figure 7.26a shows how to inject noise in transient simulations carried out using the Cadence Virtuoso Schematic Editor [21] and Cadence Spectre simulator [15]. The input-referred noise source is injected in the modulator (a fourth-order cascade 2-2 SC architecture in this example) by including a piece-wise linear (PWL) voltage source at the input node. The noise data sequence is loaded from a file, as shown in Figure 7.26b. In this case, the noise data sequence is generated using the MATLAB code shown in Figure 7.27, where the standard deviation of the noise source is computed as,

$$v_{ni} = \mathrm{randn}\left(\sqrt{f_s \frac{P_{ni}}{2B_w}}\right) \tag{7.11}$$

where f_s is the sampling frequency of the modulator, B_w is the signal bandwidth, and P_{ni} (`power_IBN` in Figure 7.27) is the IBN due to the noise source – derived from the behavioral simulation data provided

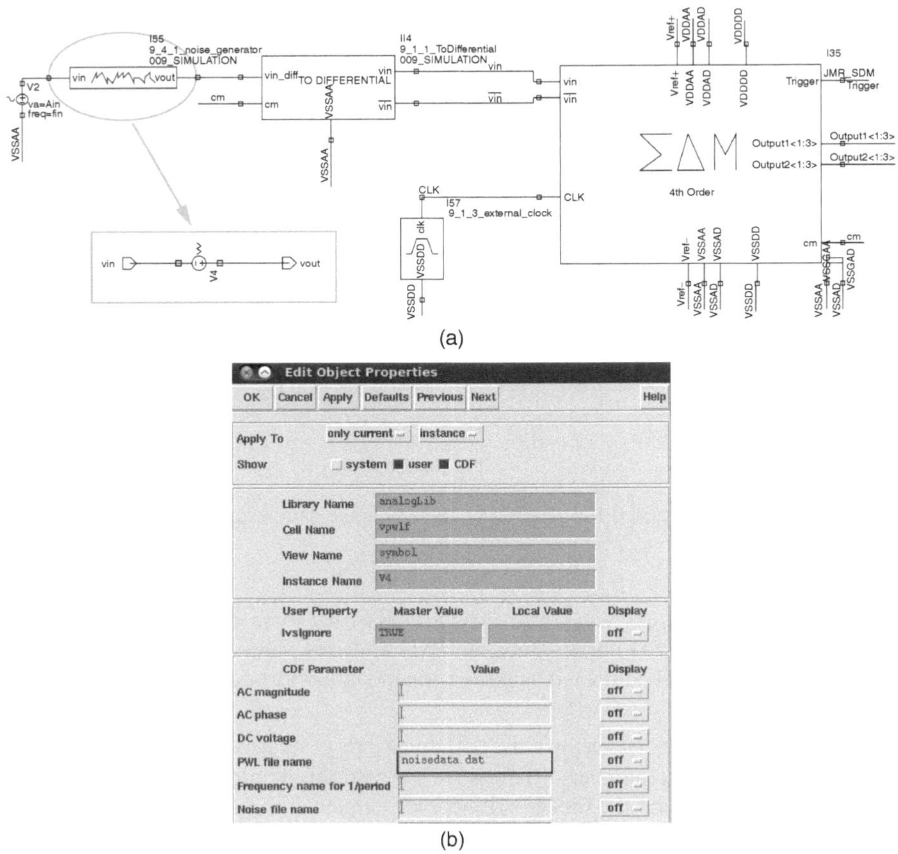

Figure 7.26 Test-bench example to inject noise in transient simulations of ΣΔMs using the Cadence Spectre simulator: (a) schematic in the Virtuoso editor environment; (b) object properties windows highlighting how to load the noise data sequence file.

by SIMSIDES. Note that in this example, Gaussian noise is generated using the `randn` function provided by MATLAB. Several cases of IBN are considered, corresponding to a reconfigurable ΣΔM for multistandard applications. As an illustration, Figure 7.28 shows the output spectrum of a fourth-order cascade 2-2 SC-ΣΔM, in which the effect of injecting thermal noise is highlighted. In this example, macromodels have been considered in all ΣΔ building blocks in order to speed up the simulation.

7.4 Processing ΣΔM Output Results of Electrical Simulations

The ΣΔM output bitstreams obtained from electrical simulations need to be properly processed in order to characterize the main figures of merit: the output spectrum, IBN, SNR/SNDR, THD, and so on. To this end, the following step-by-step procedure can be followed:[6]

[6]The procedure described in this section can be applied to the experimental output results measured in the lab, as discussed later in this chapter.

```
%% GENERATION OF A GAUSSIAN NOISE SEQUENCE WITH GIVEN IBN POWER
clear; close all;
%% Case 1
% power_IBN_dB=-84.2;
% BW=1e5;
% fs=40e6;
% num_cycles=65786;
%% Case 2
% power_IBN_dB=-77.1;
% BW=5e5;
% fs=80e6;
% num_cycles=33269;
%% Case 3
% power_IBN_dB=-73.9;
% BW=1e6;
% fs=120e6;
% num_cycles=16885;
%% Case 4
power_IBN_dB=-65.2;
BW=10e6;
fs=320e6;
num_cycles=16640;

%%%%%%%%%%%%%%%%%%%%%%%%%%%%%%%%%%%%%%%%
power_IBN=10^(power_IBN_dB/10);
std_deviation=sqrt(fs/2*1/BW*power_IBN);

mean=0;
noise=mean+std_deviation*randn(num_cycles,1);
t=0:1/fs:(num_cycles-1)/fs;

data=[t' noise];
save noisedata.dat data -ascii
```

Noise source parameters (four different cases are considered)

Noise data generation

Save data to a file

Figure 7.27 MATLAB code used for generating an *N*-point data sequence derived from Equation (7.11).

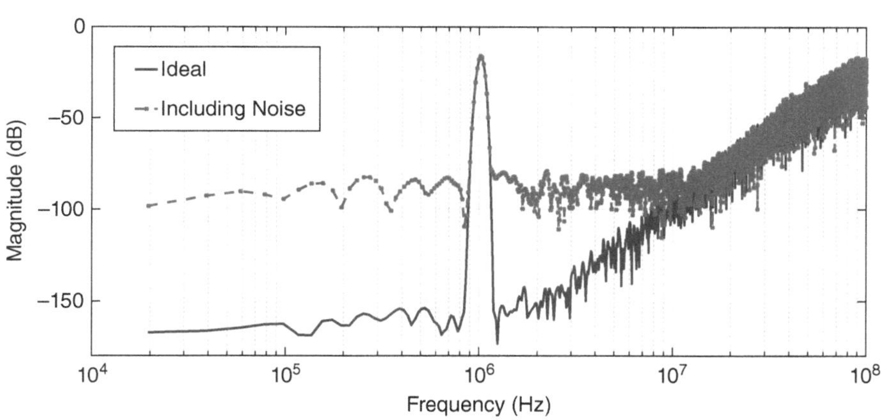

Figure 7.28 Output spectrum of a fourth-order cascade 2-2 SC-ΣΔM. The simulation was carried out in Cadence Spectre, considering macromodels for all building blocks and injecting the input-referred noise source, as illustrated in Figure 7.26.

1) Collect ΣΔM output bitstreams by taking only one sample per clock period during the clock phase in which the comparator outputs are settled.

2) Save the output data in an suitable file format, so that it can be properly loaded and post-processed using signal-processing software such as MATLAB. The most convenient and usual file format is a multiple column text file in which each column corresponds to an output bitstream of the ΣΔM. In cascade topologies, the bitstreams of all the modulator stages must be saved.

3) Load the data in MATLAB and compose the digital outputs from the corresponding bitstreams. For instance, in a ΣΔM with a 3-bit quantizer, the digital output (made up of three binary outputs) is transformed into its equivalent 0-to-7 code and the associated analog level is scaled by the full-scale reference voltage.

4) If the ΣΔM is a cascade topology, the different stage outputs need to be processed by the DCL in order to compose the overall modulator output. Thus, if the DCL is not implemented by hardware, this process can be easily implemented by properly modeling the DCL block diagram in SIMULINK.

5) Compute the different figures of merit (FFT, IBN, SNR/SNDR, THD, and so on) using the corresponding routines, for instance using the MATLAB signal processing toolbox or the SIMSIDES post-processing facilities.

Figure 7.29 shows a conceptual diagram of the step-by-step procedure described above, with the electrical simulations carried out with Cadence Spectre, and the results processed using SIMSIDES. As an example, let us consider a fourth-order cascade 2-2 SC-ΣΔM with trilevel quantization in both stages. Figure 7.30a shows the test-bench schematic used in Cadence to simulate the modulator. In this example, the output of the front-end and the back-end stages of the modulator are respectively named `Output1<1:3>` and `Output2<1:3>`, where `Outputj<i>` corresponds to the *i*th output bit of the *j*th stage quantizer, considering a 1-of-3 digital code as illustrated in Figure 7.7. Overall, six bitstreams are collected and stored in a text file. This task is implemented by the blocks named `WRITE OUTPUT`, which implement the Verilog-A code shown in Figure 7.30b. Note that the output data – consisting of the three bitstreams named `Out1, Out2, Out3` – are collected and stored in the text file at a rate of one sample per clock cycle, as illustrated in Figure 7.30c. The event at

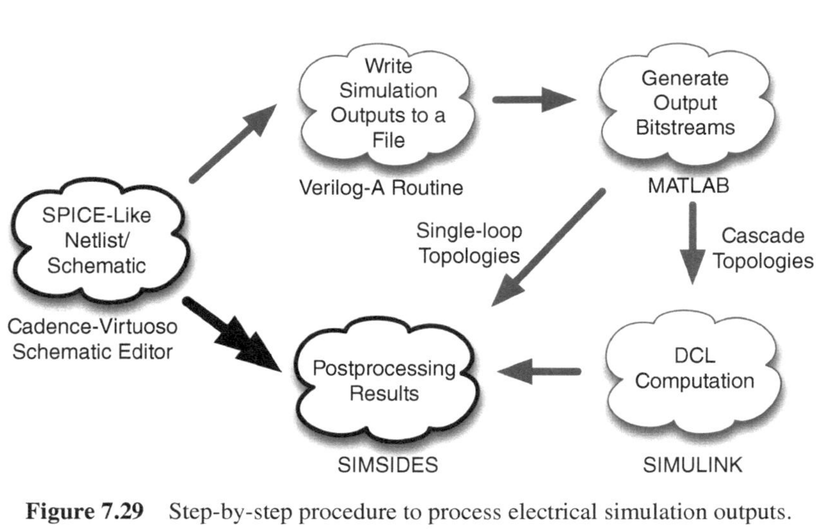

Figure 7.29 Step-by-step procedure to process electrical simulation outputs.

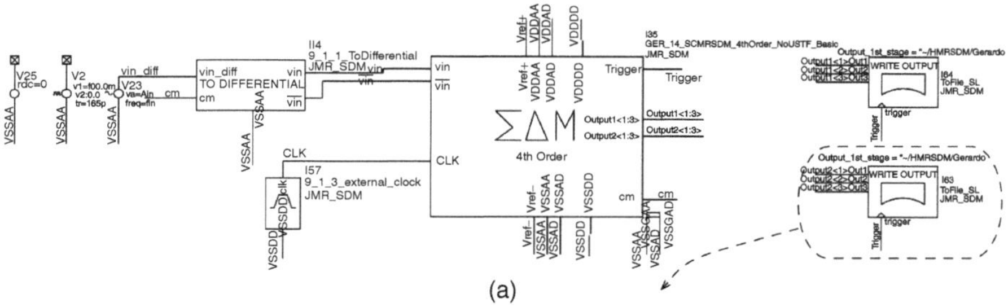

(a)

```
// VerilogA code for WRITE OUTPUT block

`include "constants.vams"
`include "disciplines.vams"
module ToFile_SL(trigger,Out1, Out2, Out3);
input Out1, Out2, Out3;
input trigger;
electrical trigger, Out1, Out2, Out3;
integer wfile1;
real vth;
parameter output_1st_stage = "~/DFWII/out1.txt";
// string constant, must be on one line

      analog begin
             vth=0.6;
          @(initial_step("dc","ac","tran","xf"))begin
                 wfile1=$fopen(output_1st_stage);
          end

      @( cross(V(trigger)-vth,-1)  ) begin
          $fstrobe(wfile1,"%1.1f %1.1f
%1.1f",V(Out1),V(Out2),V(Out3));
      end

      $fclose(wfile1);

      end
endmodule
```

(b)

```
0.0 1.2 0.0
0.0 1.2 0.0
0.0 1.2 0.0
0.0 1.2 0.0
0.0 1.2 0.0
0.0 1.2 0.0
0.0 1.2 0.0
0.0 1.2 0.0
0.0 1.2 0.0
0.0 0.0 1.2
1.2 0.0 0.0
1.2 0.0 0.0
0.0 0.0 1.2
0.0 0.0 1.2
1.2 0.0 0.0
0.0 0.0 1.2
1.2 0.0 0.0
0.0 0.0 1.2
1.2 0.0 0.0
0.0 0.0 1.2
1.2 0.0 0.0
0.0 0.0 1.2
1.2 0.0 0.0

......

0.0 0.0 1.2
1.2 0.0 0.0
```

(c)

Figure 7.30 Collecting and storing the output bitstreams of a ΣΔM in an electrical simulation: (a) test-bench schematic in Cadence Design FrameWork; (b) Verilog-A code to capture simulation output results; (c) excerpt of the generated output file (text format).

```
clear all;

%% Loading and Storing Data from txt (SIMULATION-OUTPUT) files
eval(['load ' 'out1_TB_4thOrder_SR_v01' '.txt']);
eval(['load ' 'out2_TB_4thOrder_SR_v01' '.txt']);
%
Y_1st=eval('out1_TB_4thOrder_SR_v01');
Y_2nd=eval('out2_TB_4thOrder_SR_v01');
%
%% Defining General Simulation Parameters
%
N=2^14; % Number of clock cycles
fs=320e6; % Sampling frequency
ts=1/fs; % Sampling time
t=(1:N)*ts; % Time vector
fin=1e6; % Input signal frequency
nf=length(Y_1st-1); % Length of output streams
ni=nf-N+1; % Removing initial transient data

% Scaling bitstreams from +1.2,0,-1.2 to +1,0,-1
Y1=Y_1st(ni:nf,1)/-1.2;
Y2=Y_1st(ni:nf,3)/1.2;

X1=Y_2nd(ni:nf,1)/-1.2;
X2=Y_2nd(ni:nf,3)/1.2;

% Transforming 1-of-3 coded digital outputs into
a single bitstream

out_1st=Y1+Y2;
out_2nd=X1+X2;

% Generating time-based data series

out1=[t' out_1st];
out2=[t' out_2nd];

% Simulating DCL block diagram

sim('CancellationLogic_casc22_SR.mdl');
```

Loading data — Setting general parameters — Generating 1st- and 2nd-stage output bitstreams — Simulating DCL

(a)

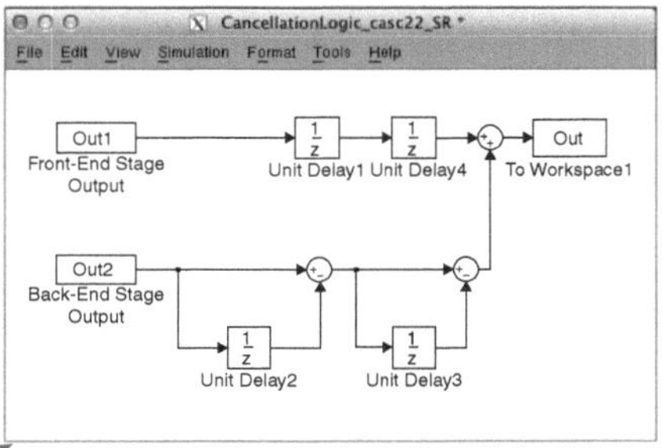

(b)

Figure 7.31 MATLAB routine used for processing ΣΔM outputs from electrical simulations: (a) MATLAB code; (b) DCL block diagram implemented in SIMULINK.

which the data are collected is governed by a trigger signal, named `trigger` in this example. This trigger signal corresponds to the clock phase in which the comparator outputs are settled, which in this example is clock phase ϕ_2.

Once the simulation output data have been stored in a multicolumn text file, it can be loaded and processed using the MATLAB routine shown in Figure 7.31a. Note that the bitstreams of both stages are scaled from the values of the supply voltage (1.2 V) to 1 V, and the 1-of-3 codification is transformed into a single-bit output format. Both output bitstreams are processed by a DCL implemented in SIMULINK, as shown in Figure 7.31b. Once this DCL diagram is simulated, the overall modulator output is saved into the MATLAB workspace and can be processed using the SIMSIDES post-processing facilities. As an illustration, the output spectrum shown in Figure 7.28 was obtained using the procedure and routines described in this section.

7.5 Summary

This chapter presented a design guide that includes a collection of practical recipes to be considered when starting the electrical design of ΣΔMs. The first step in the design flow when going from system to circuit realization consists of using macromodels to represent the main parts of the modulator. Bearing this in mind, a complete description of the macromodels frequently used for the main building blocks was presented, giving some examples of how to use them in both SC and CT-ΣΔMs. Apart from the electrical implementation itself, a number of practical issues related to the electrical characterization of ΣΔMs, such as the injection of noise sources in transient electrical simulations and the post-processing of simulation results, were also addressed.

The next step in the design flow consists of designing the different ΣΔM building blocks at transistor level. This is one of the most important design tasks, which requires a systematic design methodology to be followed, as described in the next chapter.

References

[1] K. Kundert and O. Zinke, *The Designer's Guide to Verilog-AMS*. Kluwer Academic Publishers, 2004.

[2] G. Suárez, M. Jiménez, and F. O. Fernández, "Behavioral Modeling Methods for Switched-Capacitor ΣΔ Modulators," *IEEE Transactions on Circuits and Systems – I: Regular Papers*, vol. 54, pp. 1236–1244, June 2007.

[3] J. M. Rabaey, *SPICE 3 User Guide*. [Online]. Available: http://bwrc.eecs.berkeley.edu/classes/icbook/spice/.

[4] A. Vladimirescu, *The SPICE Book*. John Wiley ç Sons, 1994.

[5] Y. Tsividis, "Integrated Continuous-Time Filter Design – An Overview," *IEEE J. of Solid-State Circuits*, vol. 29, pp. 166–176, March 1994.

[6] S. Pavan, "Efficient Simulation of Weak Nonlinearities in Continuous Time Oversampling Converters," *IEEE Transactions on Circuits and Systems I – Regular Papers*, vol. 57, pp. 1925–1934, August 2010.

[7] B. Razavi, *Principles of Data Conversion System Design*. IEEE Press, 1995.

[8] F. Maloberti, *Data Converters*. Springer, 2007.

[9] A. Morgado, R. del Río, and J. M. de la Rosa, *Nanometer CMOS Sigma-Delta Modulators for Software Defined Radio*. Springer, 2011.

[10] *Verilog-A Language Reference Manual: Analog Extensions to Verilog HDL*. Open Verilog International, 1996.

[11] *Cadence Verilog-A Language Reference*. Cadence Design Systems Inc., 2006.

[12] J. Gonzálex and E. Alarcón, "Current-Steering High-Speed D/A Converters for Communications," in *CMOS Telecom Data Converters* (A. Rodríguez-Vázquez, F Medeiro and E Janssens, editors), Kluwer Academic Publishers, 2003.

[13] M. Ortmanns and F. Gerfers, *Continuous-Time Sigma-Delta A/D Conversion: Fundamentals, Performance Limits and Robust Implementations.* Springer, 2006.

[14] J. M. de la Rosa, B. Pérez-Verdú, and A. Rodríguez-Vázquez, *Systematic Design of CMOS Switched-Current Bandpass Sigma-Delta Modulators for Digital Communication Chips.* Kluwer Academic Publishers, 2002.

[15] *Virtuoso® Spectre® Circuit Simulator User Guide*. Cadence Design Systems Inc., 2010.

[16] *The HSPICE Documentation Set*. Synopsys Inc., 2005.

[17] W. Bennett, "Spectra of Quantized Signals," *Bell System Technical J.*, vol. 27, pp. 446–472, July 1948.

[18] F. Medeiro, B. Pérez-Verdú, and A. Rodríguez-Vázquez, *Top-Down Design of High-Performance Sigma-Delta Modulators.* Kluwer Academic Publishers, 1999.

[19] Y. Dong and A. Opal, "Time-Domain Thermal Noise Simulation of Switched Capacitor Circuits and Delta-Sigma Modulators," *IEEE Transactions on Computer Aided Design of Integrated Circuits and Systems*, vol. 19, pp. 473–481, April 2000.

[20] J. Kasdin, "Discrete Simulation of Colored Noise and Stochastic Processes and $1/f^{\alpha}$: Power Law Noise Generation," *Proceedings of the IEEE*, vol. 83, pp. 802–827, February 1995.

[21] *Virtuoso® Schematic Editor L User Guide*. Cadence Design Systems Inc., 2008.

CHAPTER 8

DESIGN CONSIDERATIONS OF ΣΔM SUBCIRCUITS

Once the modulator has been verified using macromodels and the performance has been evaluated while considering the main nonideal circuit and physical effects (including circuit noise), the next step is the electrical transistor-level design of the ΣΔM building blocks and circuit elements. This chapter moves down to the transistor-level implementation, and focuses on the essential circuit elements that constitute the main building blocks of ΣΔMs. Design considerations concerning these circuits are described, together with the practical simulation test benches that are frequently used for characterizing their main electrical performance metrics, as derived from system-level behavioral simulations.

In this book, we will distinguish between two different categories of ΣΔM building blocks or subcircuits. The first category, referred to as *basic* building blocks, includes the loop filter (essentially based on integrators and resonators) and the embedded quantizer, made up of an ADC (usually a flash ADC made of a bank of comparators) and a DAC. The second category includes a number of so-called *auxiliary* building blocks that are also needed to implement a ΣΔM IC. The most important auxiliary blocks are the clock-phase generator, the master bias generator, the reference voltage generator, and the digital circuits required for buffering and signal processing. This chapter focuses on the first category of ΣΔM subcircuits that constitute their essential building blocks: CMOS switches (Section 8.1), opamps (Section 8.2), transconductors (Section 8.3), comparators (Section 8.4) and DACs (Section 8.5). The remaining (auxiliary) circuits needed for the realization of ΣΔM chips will be discussed in Chapter 9.

Sigma-Delta Converters: Practical Design Guide, Second Edition. José M. de la Rosa.

8.1 Design Considerations of CMOS Switches

Almost all switches used in SC-ΣΔMs are of CMOS type; that is, based on a pMOS and an nMOS transistor connected in parallel, as illustrated in Figure 8.1.[1] As stated in Chapters 3 and 5, the most important design specification of CMOS switches is the switch on-resistance $R_{\rm on}$. The value of $R_{\rm on}$ is mainly constrained by dynamic considerations that affect the integrator transient response, and consequently the effective resolution of the modulator [1].

Let us consider that the CMOS switch of Figure 8.1 is switched on; that is, $\phi = V_{DD}$ and $\overline{\phi} = V_{SS}$. Assuming that the nMOS and pMOS transistors operate in the ohmic region, their on-resistances can be approximated by Equation 3.48, and the overall CMOS switch on-resistance is the result of the parallel connection of resistors $R_{{\rm on}N}$ and $R_{{\rm on}P}$.

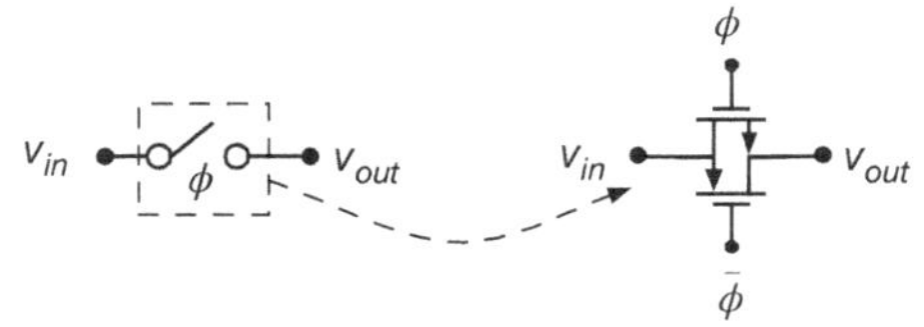

Figure 8.1 Switch symbol and its equivalent CMOS circuit.

8.1.1 Trade-Off Between $R_{\rm on}$ and the CMOS Switch Drain/Source Parasitic Capacitances

As discussed in Chapter 3, the value of $R_{\rm on}$ can be reduced by increasing the aspect ratio (W/L) of both transistors in the CMOS switch. However, this increases the transistors' area, and consequently their associated drain/source parasitic capacitances, with a subsequent penalty for the transient response and integrators' dynamics degradation. Therefore, there is a trade-off between the maximum value of $R_{\rm on}$ that can be tolerated – which can be determined by behavioral simulation as shown in Chapter 5 – and the drain/source parasitic capacitances associated with the CMOS switch, which are in turn conditioned by the value of capacitors used in the SC branches. In this way, switch transistor sizes can be scaled down across the modulator chain, using higher sizes in the front-end switches – where larger capacitors are chosen according to thermal noise considerations – while lower sizes are tolerated in back-end integrators, where smaller capacitors are normally used and hence the influence of switch parasitic capacitances is diminished.

8.1.2 Characterizing the Nonlinear Behavior of $R_{\rm on}$

According to Equation (3.48), the values of $R_{{\rm on}N}$ and $R_{{\rm on}P}$ depend on the switch common-mode voltage, $v_{CM} \equiv (v_{\rm in} + v_{\rm out})/2$, and hence on the drain and source voltages (denoted as $v_{\rm in}$ and $v_{\rm out}$ in Figure 8.1) of the nMOS and pMOS transistors. As a consequence, the value of $R_{\rm on}$ becomes a nonlinear function of the voltage being transmitted, thus generating harmonic distortion, as discussed in Chapters 3 and 5.

In order to evaluate the nonlinear characteristic of $R_{\rm on}$ in CMOS switches, the circuit shown in Figure 8.2a can be used. The gate of the nMOS transistor is connected to V_{DD} and the gate of

[1] Some ΣΔM subcircuits, like some types of SC common-mode feedback (CMFB) circuits or the latches used in some comparators, may use either nMOS or pMOS switches. However, the vast majority of switches in SC-ΣΔMs are implemented as CMOS transmission gates.

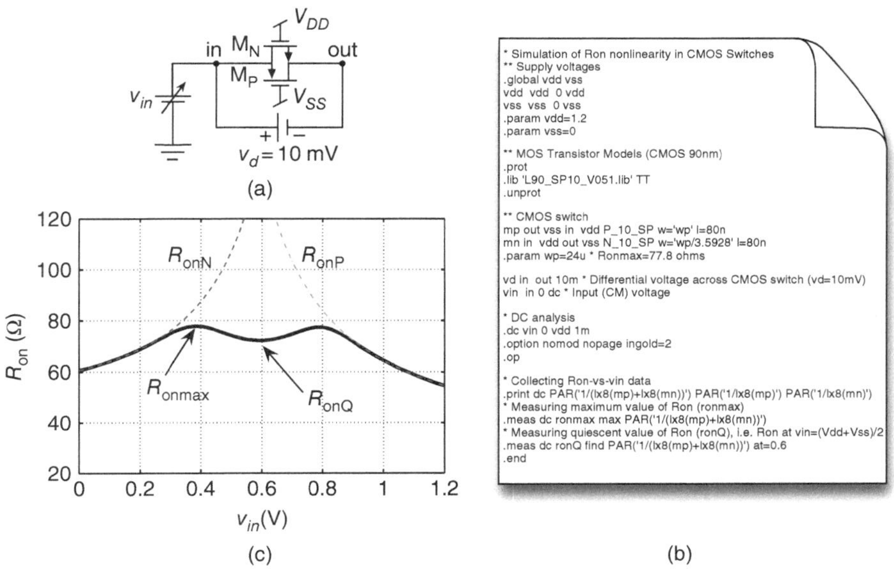

Figure 8.2 Characterizing nonlinear switch on-resistance: (a) test-bench circuit; (b) HSPICE netlist; (c) R_{on} versus v_{in} considering a 90-nm CMOS technology with 1.2-V supply voltage.

the pMOS transistor is connected to V_{SS}, so that both transistors are switched on. A small voltage imbalance, typically of the order of 10–20 mV, is applied across the CMOS switch, to guarantee that both nMOS and pMOS are properly biased and operate in the linear region. A DC common-mode voltage source is connected to the input node of the switch. This voltage is swept in a DC analysis in order to evaluate its impact on the variation of R_{on}.

Figure 8.2b shows the SPICE netlist used for simulating the circuit in Figure 8.2a. In this example, a 10-mV voltage, `vd`, is applied across the switch and the common-mode voltage (`vin` in Figure 8.2a) is swept using a `.DC` analysis in SPICE. In HSPICE, the value of R_{on} can be extracted from each operating point in the `.DC` analysis, by means of a parameter defined as `PAR(1/(lx8(mp)+lx8(mn))`, where `lx8(mp,n)` is an alias parameter used in HSPICE to represent the DC drain-source conductance of MOS transistors (M_N and M_P in Figure 8.2a). Thus the values of R_{onN} and R_{onP} are extracted from `1/lx8(mn)` and `1/lx8(mp)`, respectively [2]. A similar circuit test-bench to that in Figure 8.2a can be used in Cadence Spectre, where the drain-source conductance of a MOS transistor can be extracted from DC simulations as `1/(getData("M.m1:gds"?result"dc")`.

Figure 8.2c represents R_{on} as a function of v_{in}, giving rise to a function similar to that shown in Figure 3.21b. The curves corresponding to both R_{onN} and R_{onP} are also depicted, illustrating the separate contribution of each transistor to the overall switch on-resistance. The maximum value of R_{on}, denoted as `ronmax`, and the quiescent value of R_{on}, denoted as `ronQ`, can be also extracted from HSPICE simulations by using the `.meas` command [2], as detailed in the netlist shown in Figure 8.2b.

The R_{on}-versus-v_{in} characteristic shown in Figure 8.2c has been obtained for $W_P K_P = W_N K_N$, which according to Equation (3.48), gives an almost symmetrical function. An alternative approach consists of increasing W_N to equal W_P, as illustrated in Figure 3.23. For the sake of completeness, a similar figure is depicted in Figure 8.3, with a 90-nm CMOS technology and a 1.2-V supply

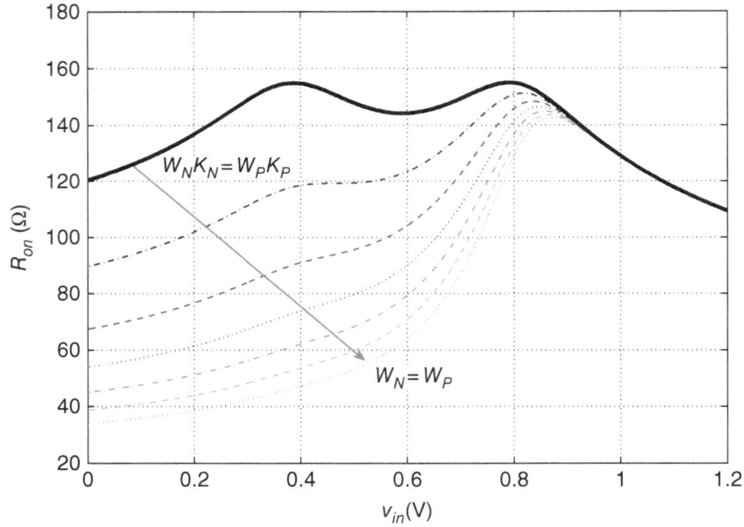

Figure 8.3 R_{on} versus v_{in} for minimum channel length and different values of $W_{N,P}$, considering a 90-nm CMOS technology with 1.2-V supply voltage.

voltage. It can be noted how, as W_N increases, the nonlinearity of the switch on-resistance increases, although its average value decreases. Hence, as discussed in Section 3.7.2, if the sizing of the nMOS and pMOS transistors in a CMOS switch compensates for the difference in their transconductance parameters, the switch on-resistance nonlinearity is reduced, but the average on-resistance is larger than using the same sizes; that is, $W_N = W_P$. In the latter case, the area, and consequently the values of the parasitic drain/source capacitances, increases as well, although the overall effect of the finite switch on-resistance on the settling performance decreases [3]. Therefore, there is a design trade-off involving the switch on-resistance nonlinearity and its average value, which at the modulator level translates into the well-known analog design trade-off between speed (limited by the incomplete settling) and linearity (limited by nonlinear switch on-resistance) [1]. Nevertheless, in the majority of state-of-the-art SC-ΣΔMs, CMOS switches are designed to keep a low enough average value of R_{on}, while keeping a symmetrical R_{on}-versus-v_{in} characteristic, as shown in Figure 8.2c.

8.1.3 Influence of Technology Downscaling on the Design of Switches

According to Equation (3.48), the reduction of the supply voltage caused by CMOS technology downscaling causes an increase of R_{on}. However, this effect can be compensated by the lower channel lengths ($L_{N,P}$) used in smaller technologies. This is illustrated in Figure 8.4, where R_{on} is represented versus W_P for $W_P K_P = W_N K_N$, considering different CMOS processes from 250 nm to 90 nm. Note that, generally speaking, the design of switches benefits from technology downscaling since lower values of R_{on} can be obtained for the same (or even smaller) switch sizing, with the subsequent advantages in terms of silicon area and robustness against parasitic capacitances. It can be noticed how migrating from 180 nm to 130 nm has no effect on R_{on} because the reduction in MOS transistor sizes is compensated by the voltage supply downscaling. However, comparing 130 nm with 90 nm, both using a 1.2-V supply voltage, the downscaling process becomes beneficial.

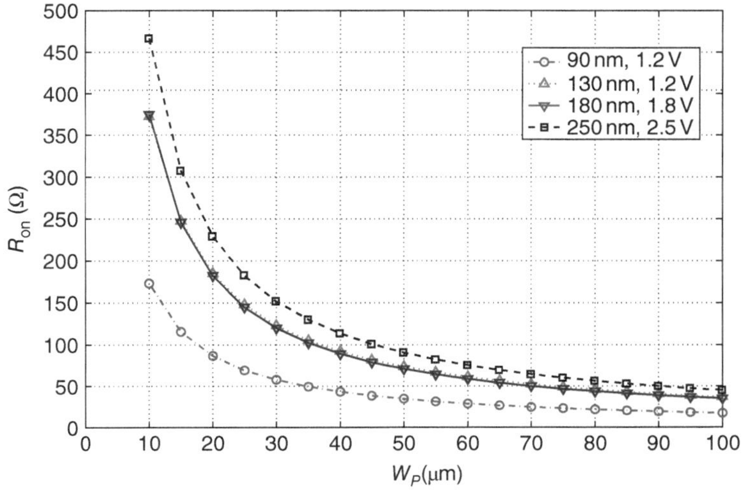

Figure 8.4 Illustrating the effect of technology downscaling on R_{on}.

8.1.4 Evaluating Harmonic Distortion due to CMOS Switches

Figure 8.5a shows the equivalent circuit that can be used for the evaluation of the harmonic distortion caused by the nonlinear sampling process due to the finite switch on-resistance. This test-bench circuit – which contains essentially the equivalent circuit shown in Figure 3.23 – corresponds to the fully-differential implementation of the input SC branch in a typical front-end integrator used in SC-ΣΔMs – conceptually illustrated in Figure 8.5a. As discussed in Section 3.7.2, since switch $S_{1p,n}$ is directly connected to the input node, its nonlinear on-resistance may vary a lot during the sampling period, thus generating considerable harmonic distortion. In contrast, switch $S_{2p,n}$ has one of its terminals connected to a fixed voltage – the analog common-mode voltage – so that the voltage across it remains approximately constant over clock periods. As illustrated in Figure 8.5a, the voltage variations at the input node of S_2 are considerably lower than in S_1. As a consequence, the effect of S_2 can be neglected in practice. Indeed, the test-bench circuit shown in Figure 8.5b will give essentially the same results as the circuit in Figure 8.5a in the majority of practical situations.

The corresponding SPICE netlist is shown in Figure 8.5c. A transient analysis is carried out considering different situations for the input signal: a single tone signal, a two-tone signal, and so on. To this end, a single-ended source is transformed into a differential input signal using voltage-controlled sources, while some subcircuits are used for representing the switches and sampling branches included in the test bench. A `.TRAN` analysis is carried out with a printing time step defined as the sampling period – 10 ns in this example. The stop time is given by `N/fs`, where `N` is the number of simulation clock cycles (4096 in this example) and `fs` is the sampling frequency (100 MHz in this example).

As stated above, harmonic distortion is mainly caused by the switches that are directly connected to the input ($S_{1p,n}$), given that the voltages at the input/output nodes of the remaining ones remain approximately constant over clock periods, and consequently R_{on} is approximately constant during the sampling phase time. The same reasoning applies to those switches that are in any of the following situations:

- Switches that are connected to nodes where the voltage remain constant over the sampling period;

- Switches that are connected to the output of an SC circuit. For instance all switches of the back-end SC integrators; i.e., all integrators except the front-end one. In all these switches, the voltages in their terminals remain constant over clock periods.
- Sampling switches that are connected to the input node of SC-ΣΔMs, whenever the ratio between the input signal frequency and the sampling frequency is small, typically less than one to ten.

The latter situation is illustrated in Figure 8.6, where a sinewave input signal with different values of the input frequency f_{in} is represented within the sampling time period, from 0 to $T_s/2$, with $T_s = 1/f_s$ and $f_s = 100$ MHz in this example. Note that as the ratio between f_{in} and f_s increases, there are larger variations of v_{in} over the sampling period, which translates into a higher variation of R_{on}, thus increasing the harmonic distortion.

The analysis of the harmonic distortion caused by the nonlinear sampling process using the Volterra series method [4, 5] demonstrates that the third-order harmonic distortion caused by this

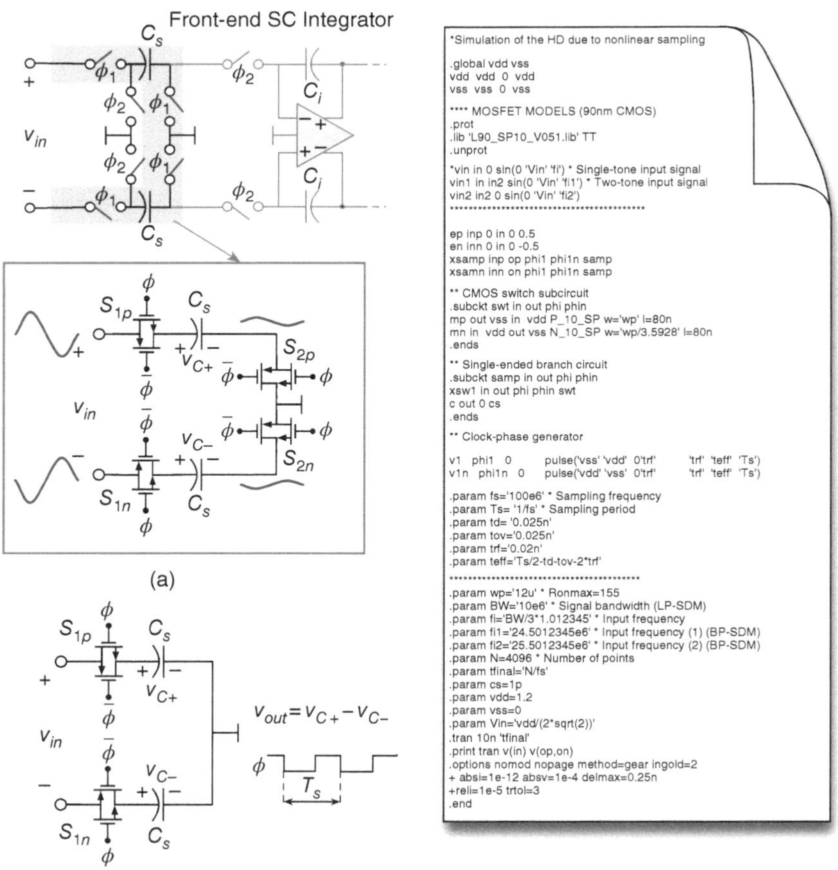

Figure 8.5 Characterizing harmonic distortion caused by nonlinear sampling: (a) test-bench equivalent circuit; (b) practical (simplified) version of the test-bench circuit; (c) HSPICE netlist. A test-bench circuit with $\phi = V_{DD}$ and $\overline{\phi} = V_{SS}$ can be used as well.

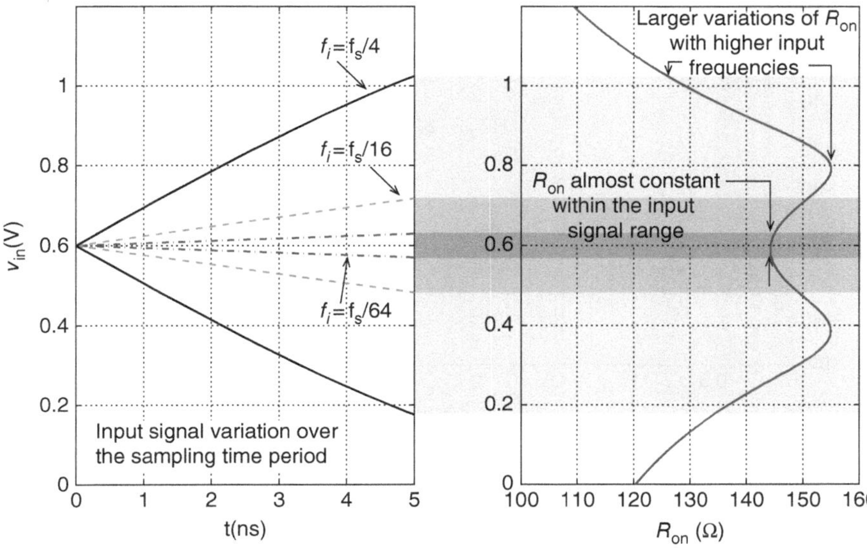

Figure 8.6 Impact of increasing the input signal frequency on the variation of R_{on} over the sampling period. A fully-differential sinewave input signal of frequency f_{in} is considered with $f_s = 100$ MHz. This figure plots both positive and negative single-ended inputs.

dynamic nonlinearity is approximately given by [6]:

$$\mathrm{HD}_3 \simeq \frac{\pi f_{in} C_S R_{on}}{2(V_{ON} - V_T)^2} A_{in}^2 \tag{8.1}$$

where A_{in} is the input signal amplitude, V_{ON} denotes the switch-on voltage (either V_{DD} or $|V_{SS}|$), and V_T is the maximum (worst-case) value of V_{TN} and $|V_{TP}|$.

The expression given in Equation (8.1) is consistent with the results highlighted in Figure 8.6, showing a direct dependency of HD_3 with f_{in}. This is illustrated in Figure 8.7, where several output spectra of the circuit in Figure 8.5b are depicted for different values of f_{in}, with $A_{in} = V_{DD}/2$. These output spectra have been processed with SIMSIDES, based on the use of a Kaiser FFT window.

In practical situations f_{in} is chosen to be $f_{in} = B_w/3$, with B_w being the signal bandwidth. In this way, the third-order harmonic components fall into the signal bandwidth. This is not the case for BP-ΣΔMs, where the notch frequency (i.e., the center signal frequency) is typically placed at $f_s/4$. In this case, other distortion metrics, such us the third-order intermodulation distortion IM_3, are used. Thus, in this case a two-tone input signal is used in the test-bench circuit of Figure 8.5b. As an illustration, Figure 8.8 shows the output spectrum of Figure 8.5b considering two tones with amplitude $V_{DD}/(2\sqrt{2})$ located at $f_{in1} = 24.5$ MHz and $f_{in2} = 25.5$ MHz with $f_s = 100$ MHz.[2]

[2]The input sinewave frequency f_{in} should be located precisely in an FFT bin in order to prevent signal power spreading to adjacent bins [7]. To this end, the number of periods N_p of the input signal within the simulated time (N/f_s, where N stands for the number of points in the FFT and f_s for the sampling frequency) must be an integer. To meet this constraint, the value of the input frequency can be adjusted according to:

$$f_{in} \leftarrow N_p \frac{f_s}{N} \quad , \text{ where } N_p = N \left\lfloor \frac{f_{in}}{f_s} \right\rfloor \tag{8.2}$$

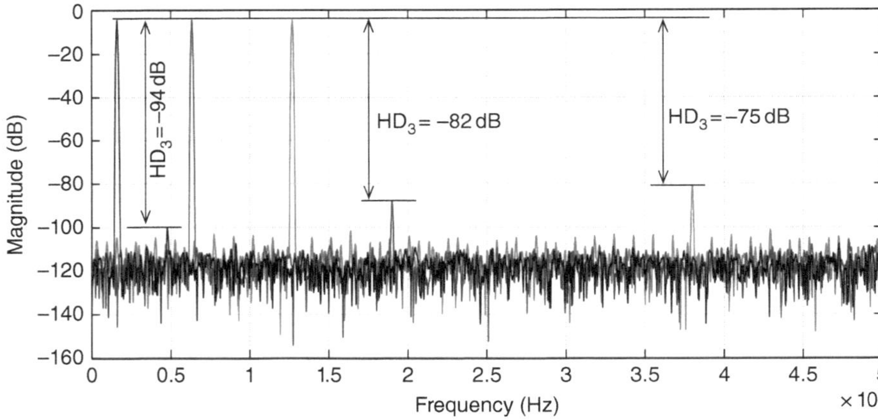

Figure 8.7 Illustrating the impact of increasing the input signal frequency on the harmonic distortion caused by nonlinear sampling process. Three cases of f_{in} are considered: $f_{\text{in}} = f_s/64$, $f_{\text{in}} = f_s/16$, and $f_{\text{in}} = f_s/8$, where $f_s = 100$ MHz.

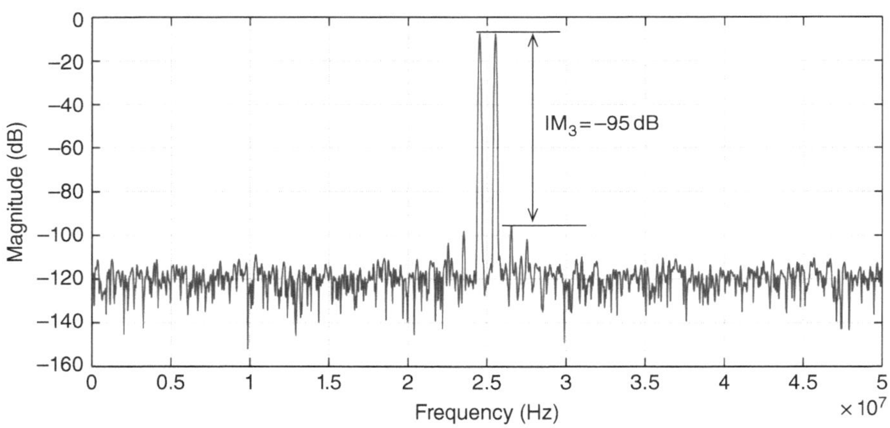

Figure 8.8 Intermodulation distortion caused by the nonlinear sampling operation. Data used in the simulation: 90-nm CMOS technology, $V_{DD} = 1.2$ V, $W_P = 12$ μm, $W_N = 3.34$ μm, $L_P = L_N = 80$ nm, $R_{\text{onQ}} = 155\ \Omega$.

8.2 Design Considerations of Operational Amplifiers

Voltage amplifiers are basic circuits of SC-ΣΔMs, used for building SC integrators and resonators. They are also used for implementing active-RC integrators in CT-ΣΔMs. As a discussed in Chapters 3–5, the main electrical requirements of amplifiers can be determined from closed-form expressions and behavioral simulations, and usually comprise specifications for the DC gain, output swing, dynamic behavior, and input-referred noise.

8.2.1 Typical Amplifier Topologies

Many different topologies can be considered in order to fulfill the derived amplifier specifications at the transistor level. Figure 8.9 depicts some representative amplifier topologies used in ΣΔM design:

- **Telescopic amplifier** (see Figure 8.9a): This single-stage topology is capable of providing moderate DC gain and excellent dynamic behavior while being very power efficient, since it employs a single current branch. However, the topology requires five stacked transistors, which results in a reduced output swing and complicates its design (or even prevents its use) in low-voltage ΣΔ implementations. Nevertheless, the telescopic amplifier should be considered the best option if high DC gain and high output swing are not required.
- **Folded cascode amplifier** (see Figure 8.9b): This single-stage topology exhibits an output swing larger than that of the telescopic amplifier – only four transistors are stacked – but doubles the power consumption because of the two current branches required. It provides very good settling behavior, although its first nondominant pole – and thus its phase margin – is somewhat lower than with a telescopic amplifier. Folded cascode amplifiers are often used if moderate DC gain is required in ΣΔMs with medium- and low-voltage supply.
- **Folded cascode amplifier with gain boosting** (see Figure 8.9c): This topology provides larger DC gain than the conventional folded cascode amplifier in Figure 8.9b by means of increasing its output resistance through the regulation of the cascode transistors. The auxiliary amplifiers are usually designed to be as simple as possible so that their additional power consumption does not penalize that of the overall amplifier. Gain-boosting techniques are often employed in single-stage amplifiers, although especial attention must be paid so that the inner feedback loop does not degrade the amplifier frequency response – or even make it unstable in closed-loop form.
- **Two-stage amplifier with Miller compensation** (see Figure 8.9d): two-stage amplifiers are capable of providing high DC gain as well as a large output swing, since the voltage gain is obtained with two amplification stages rather than cascoding. However, their settling behavior becomes more complex than that of single-stage amplifiers – because of the additional poles and zeros that the internal compensation introduces – and they usually result in higher power consumption.

Apart from these opamp topologies, multi-stage multi-path feedforward amplifiers are widely used in wideband CT-ΣΔMs because these structures can achieve high gain at high frequencies while retaining stability and power efficiency [8–11]. The operating principle behind these amplifiers – conceptually depicted in Figure 8.10 for a third-order structure – is to benefit from the combination of the high-bandwidth of single-stage amplifiers and the high-gain of multi-stage amplifiers by summing their outputs [9]. In this way, the high-order paths provide the required gain, while the lower paths provide the mid-frequency gain and determine the required unity-gain frequency and phase margin. Inter-stage capacitors are used to control the mid-frequency gains and closed-loop stability [12]. Depending on the overall gain requirements, different circuit realizations can be used for each gain stage, going from telescopic amplifiers to the inverter-based transconductors discussed later on in this chapter.

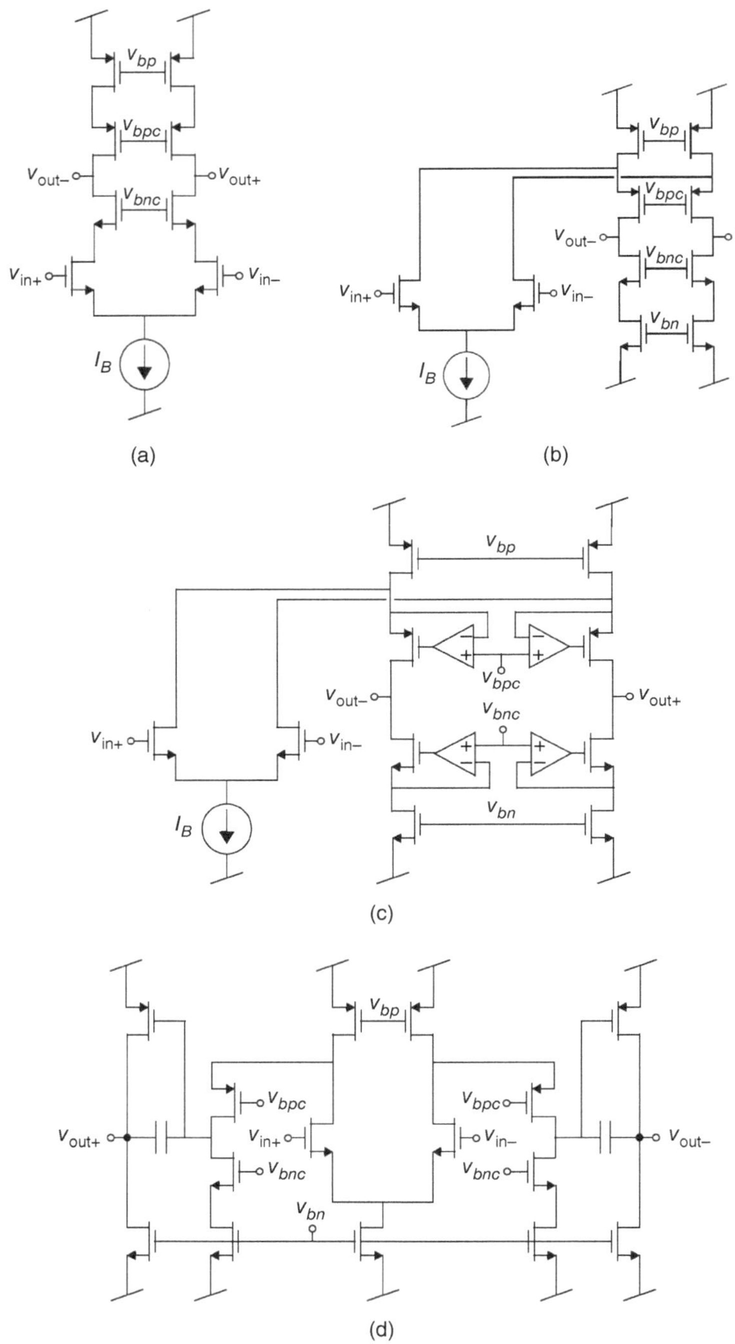

Figure 8.9 Amplifier topologies commonly employed in ΣΔ modulators: (a) telescopic amplifier; (b) folded cascode amplifier; (c) folded cascode amplifier with gain boosting; (d) two-stage amplifier with Miller compensation.

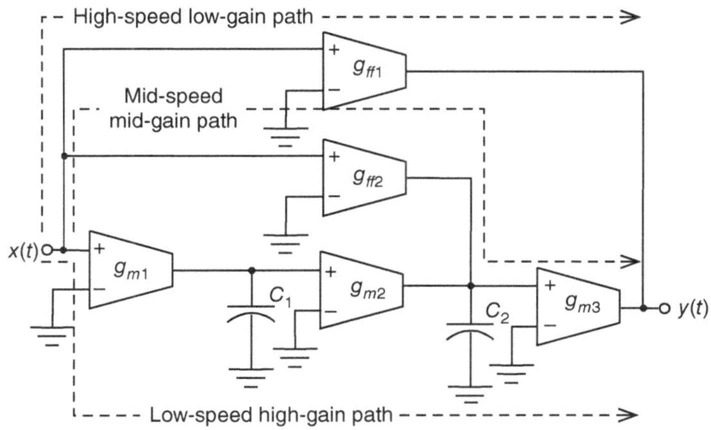

Figure 8.10 Conceptual schematic of a third-stage feedforward amplifier used in wideband CT-ΣΔMs [9].

8.2.2 Common-mode Feedback Networks

Single-ended implementations of ΣΔMs are only rarely seen and the vast majority of reported ΣΔMs – just like in most analog circuits – employ fully-differential circuits because of their reduced sensitivity to even-order harmonics and their better power supply rejection ratio. In the case of SC-ΣΔMs, the differential implementation also helps to reduce the power consumption, since, for the same influence of the kT/C noise, the values of the sampling capacitors can be halved compared to the single-ended case thanks to the doubled input signal range. Also, the clock feedthrough and charge injection of the switches cancel to a common-mode signal.

These reasons lead to the use of fully-differential amplifier topologies – like those previously depicted in Figure 8.9 – which thus require an additional circuit to set the common-mode component of the output voltage to the appropriate level; that is, a CMFB network. These nets operate by means of *sensing* the common-mode level of the output voltages, *comparing* it with the desired common-mode level V_{CM}, and *adapting* the bias conditions of the amplifier through negative feedback so that $(v_{out+} + v_{out-})/2 \approx V_{CM}$. These tasks can be either done continuously or in a discrete-time fashion, leading to CT or SC CMFB nets, respectively.

As an illustration, Figure 8.11 shows an alternative version of the folded cascode amplifier in Figure 8.9b using a pMOS differential input pair, together with its bias circuit. Figures 8.12a and 8.12b respectively depict the DT or CT alternatives for implementing the corresponding CMFB net. Note from the SC CMFB net in Figure 8.12a that the common-mode voltage of amplifier nodes v_{out+} and v_{out-} is sensed and compared with V_{CM} through capacitors that switch according to the nonoverlapping clock phases ϕ_1 and ϕ_2 of the SC-ΣΔM. In the case of CT CMFB nets, as shown in Figure 8.12b, the output common-mode voltage is sensed through resistors and compared with V_{CM} using a differential pair.

SC CMFB nets are often preferred in SC-ΣΔMs to their CT counterparts, since their design is straightforward, they lead to a very small area overhead, and have no static power consumption. Conversely, although designing a CT CMFB net is usually not difficult, it leads to additional power consumption and involves a static inner feedback loop with sufficient gain to affect the dynamic response of the overall amplifier.

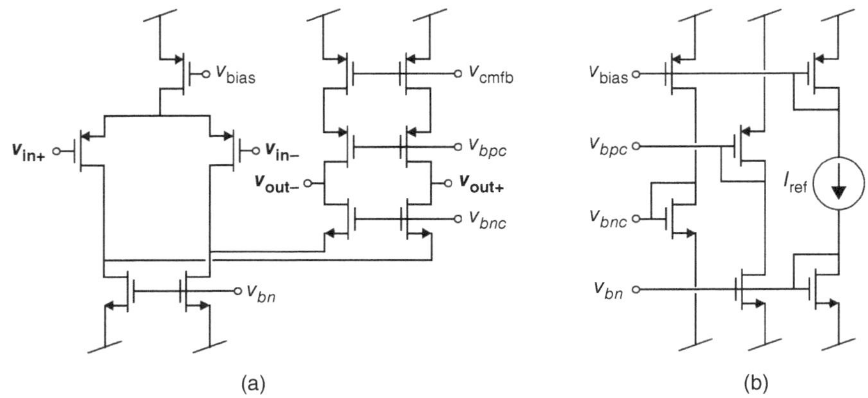

Figure 8.11 Folded cascode amplifier: (a) core circuit; (b) bias circuit.

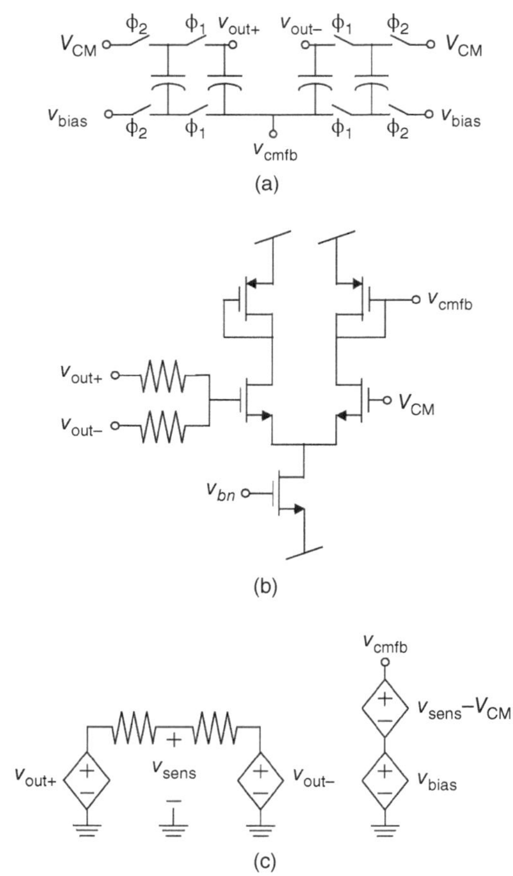

Figure 8.12 Alternative implementations of the common-mode feedback: (a) SC circuit; (b) CT circuit; (c) ideal circuit for simulation purposes.

For the sake of completeness, Figure 8.12c shows an ideal CMFB net that can be employed, together with the amplifier in Figure 8.11, for simulation purposes. Note that operating principle of the CMFB net is only macromodeled using voltage-controlled sources.

8.2.3 Characterization of the Amplifier in AC

Figure 8.13 shows the HSPICE netlist of the folded cascode amplifier in Figure 8.11. The transistor-level sizing corresponds to an amplifier designed for an SC-ΣΔM implemented in a 130-nm CMOS process. The netlist includes the simulation test bench depicted in Figure 8.14a for the electrical characterization of the amplifier performance in an open-loop AC analysis:

- the amplifier frequency response, from which the following features can be directly determined:
  - DC gain (denoted as `Av` in the measurements section at the end of the netlist)
  - 3-dB loss frequency (denoted as `f3dB`)
  - gain-bandwidth product (denoted as `GB`)
  - unity-gain bandwidth (denoted as `UGBW`)
  - phase margin (denoted as `PM`)
- the PSD of the input-referred amplifier noise, from which the following features can be extracted:
  - $1/f$ noise component
  - thermal noise component
  - noise corner frequency.

For that purpose, an AC differential input signal is applied to the amplifier and its differential output voltage ($v_{\text{out}+} - v_{\text{out}-}$) is computed. The test bench includes an ideal CMFB net like that shown in Figure 8.12c, as well as loading capacitors C_L at the amplifier output nodes. The value of C_L (1.4 pF in the netlist in Figure 8.13) corresponds to an estimate of the equivalent amplifier load during the integration phase: C_{eq,ϕ_2} in Equation (3.17).

Figure 8.14b shows the simulated frequency response of the amplifier gain, in which A_v, $f_{3\text{dB}}$, and UGBW are identified. Their measured values are $A_v = 2100 = 66.4$ dB, $f_{3\text{dB}} = 81.5$ kHz, and UGBW = 170 MHz. The amplifier GB can be obtained from the $A_v f_{3\text{dB}}$ product and equals 171 MHz; in other words, GB ≈ UGBW, which is characteristic of a single-stage amplifier with a clearly dominant pole and results in a large phase margin (PM = 82.6° in this case).

Figure 8.14c shows the simulated PSD of the input-referred amplifier noise, in which the flicker and white components can be clearly identified, as well as the corner frequency (around 600 kHz in this case). Note that, as discussed in Section 7.3, the PSD curve obtained can also be used for generating a colored noise data sequence that captures the amplifier noise frequency response and can be injected in transient electrical simulations of the ΣΔM.

8.2.4 Characterization of the Amplifier in DC

Figure 8.15a illustrates a simulation test bench that allows the amplifier I–V transfer characteristic to be obtained, and the amplifier transconductance g_m and maximum output current I_o to be extracted.

```
** NETLIST EXAMPLE FOR THE CHARACTERIZATION OF AN AMPLIFIER

.prot
.LIB  './umc_13/fdk/VB04_PB/Models/Hspice/L130E_HG_RVT33_V131.lib' TT
.unprot

* AMPLIFIER DESCRIPTION
.subckt folded_cascode vin+ vin- vout+ vout- vdd vss
* Amplifier core
M1   vc1   vin+   vc  vc   P_HG_33_L130E w=17.73u  l=0.65u  m=6
M2   vc2   vin-   vc  vc   P_HG_33_L130E w=17.73u  l=0.65u  m=6
M3   vc    vbias  vdd vdd  P_HG_33_L130E w=49.42u  l=1.14u  m=10
M4   vc1   vbn    vss vss  N_HG_33_L130E w=15.00u  l=2.33u  m=11
M5   vc2   vbn    vss vss  N_HG_33_L130E w=15.00u  l=2.33u  m=11
M6   vout- vbnc   vc1 vss  N_HG_33_L130E w= 4.75u  l=0.34u  m=3
M7   vout+ vbnc   vc2 vss  N_HG_33_L130E w= 4.75u  l=0.34u  m=3
M8   vout- vbpc   vc3 vdd  P_HG_33_L130E w= 3.21u  l=0.44u  m=6
M9   vout+ vbpc   vc4 vdd  P_HG_33_L130E w= 3.21u  l=0.44u  m=6
M10  vc3   vcmfb  vdd vdd  P_HG_33_L130E w=49.42u  l=1.14u  m=6
M11  vc4   vcmfb  vdd vdd  P_HG_33_L130E w=49.42u  l=1.14u  m=6
* Amplifier biasing
M3B  vbias vbias  vdd vdd  P_HG_33_L130E w=49.42u  l=1.14u  m=1
M4B  vbn   vbn    vss vss  N_HG_33_L130E w=15.00u  l=2.33u  m=1
M6bp vbnc  vbias  vdd vdd  P_HG_33_L130E w=49.42u  l=1.14u  m=1
M6b  vbnc  vbnc   vss vss  N_HG_33_L130E w= 2.56u  l=3.83u  m=1
M8bn vbpc  vbn    vss vss  N_HG_33_L130E w=15.00u  l=2.33u  m=1
M8b  vbpc  vbpc   vdd vdd  P_HG_33_L130E w= 3.12u  l=1.35u  m=1
Iref vbias vbn 48u
* Ideal common-mode feedback net
eop eout+ 0 vout+ 0 1
eon eout- 0 vout- 0 1
rp  eout+ vsens 1
rn  eout- vsens 1
vcm vcm 0 1.65
e1 vcmfb aux vsens vcm 1
e2 aux   0   vbias 0   1
.ends

* OPEN-LOOP CONFIGURATION
vdd  vdd 0 3.3
vss  vss 0 0
vcm  vcm 0 1.65
evin+ vin+ vcm d 0 0.5
evin- vin- vcm d 0 -0.5
vd  d  0 ac
xfolded  vin+ vin- vout+ vout- vdd vss folded_cascode
clp vout+ 0 1.4p
cln vout- 0 1.4p
eout out 0 vout+ vout- 1
Rout out 0 100x

.options post nomod nopage ingold=2 method=gear
* PERFORMED ANALYSIS
.ac dec 50 10 1g
.noise v(out) vd
.print ac v(out) inoise onoise
.tf v(out) vd
.pz v(out) vd
* PERFORMED MEASUREMENTS
.meas ac  Av     find v(out) at=10
.meas ac  f3dB   when v(out)='Av*0.707' cross=1
.meas ac  GB     param='Av*f3dB'
.meas ac  UGBW   when vdb(out)=0 cross=1
.meas ac  phase  find vp(out) when vdb(out)=0
.meas ac  PM     param='180+phase'

.end
```

Figure 8.13 HSPICE netlist for the AC characterization of the folded cascode amplifier in Figure 8.11.

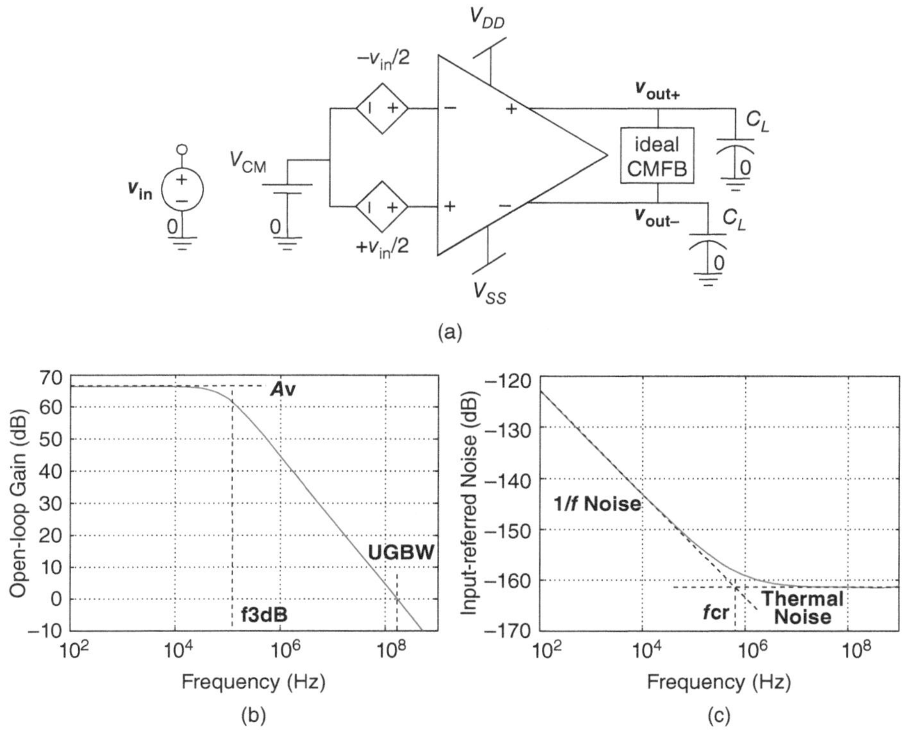

Figure 8.14 Amplifier simulation results obtained from an AC analysis: (a) test-bench circuit; (b) open-loop amplifier frequency response; (c) input-referred noise PSD.

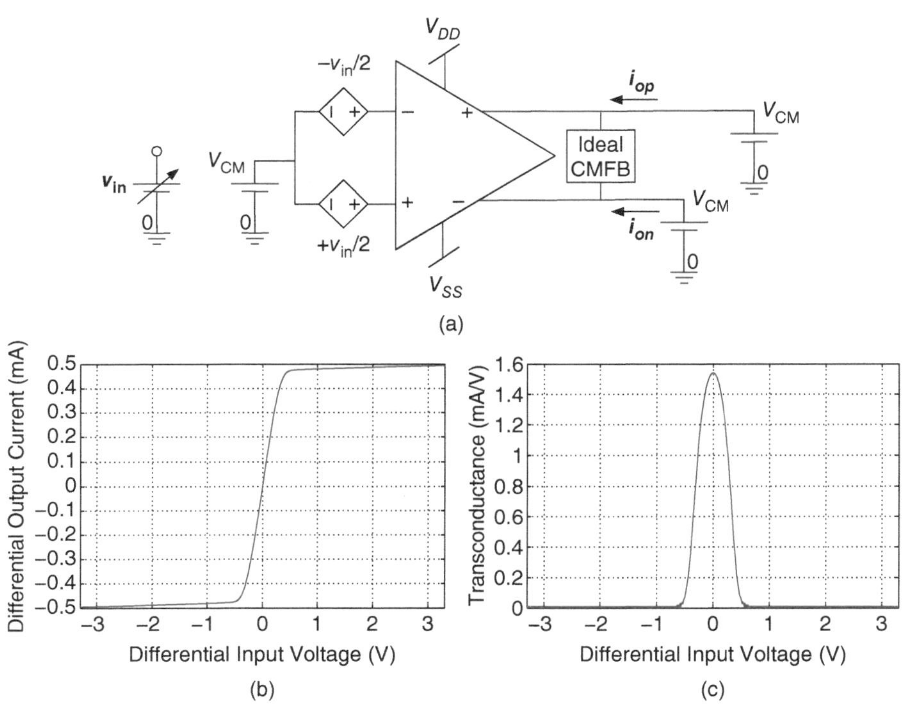

Figure 8.15 Amplifier simulation results obtained from a DC analysis: (a) test-bench circuit. (b) I–V transfer characteristic; (c) transconductance versus input voltage.

To that end, a DC differential signal ranging from $-2(V_{DD} - V_{SS})$ to $+2(V_{DD} - V_{SS})$ is applied at the amplifier input, while its output nodes are fixed to the common-mode level through voltage sources whose current is measured.

Figure 8.15b shows the simulated differential output current ($i_{out} = i_{op} - i_{on}$) of the folded cascode amplifier in Figure 8.13 versus the differential input voltage v_{in}. From this, the maximum amplifier output current can be measured ($\pm I_o = \pm 0.49$ mA in this case).

The amplifier transconductance DC characteristic can be easily obtained from the slope of the I–V curve in the Figure 8.15b by simply computing di_{out}/dv_{in}. The resulting curve is depicted in Figure 8.15c, from which the amplifier transconductance at the quiescent point g_m can be derived ($g_m = 1.54$ mA/V in this example).[3]

8.2.5 Characterization of the Amplifier Gain Nonlinearity

Figure 8.16a illustrates a different DC simulation test bench, in this case for extracting the amplifier voltage transfer characteristic. Using it, the amplifier output swing OS and amplifier DC gain nonlinearity can be easily derived. To do this, a DC differential signal ranging from $-2(V_{DD} - V_{SS})$

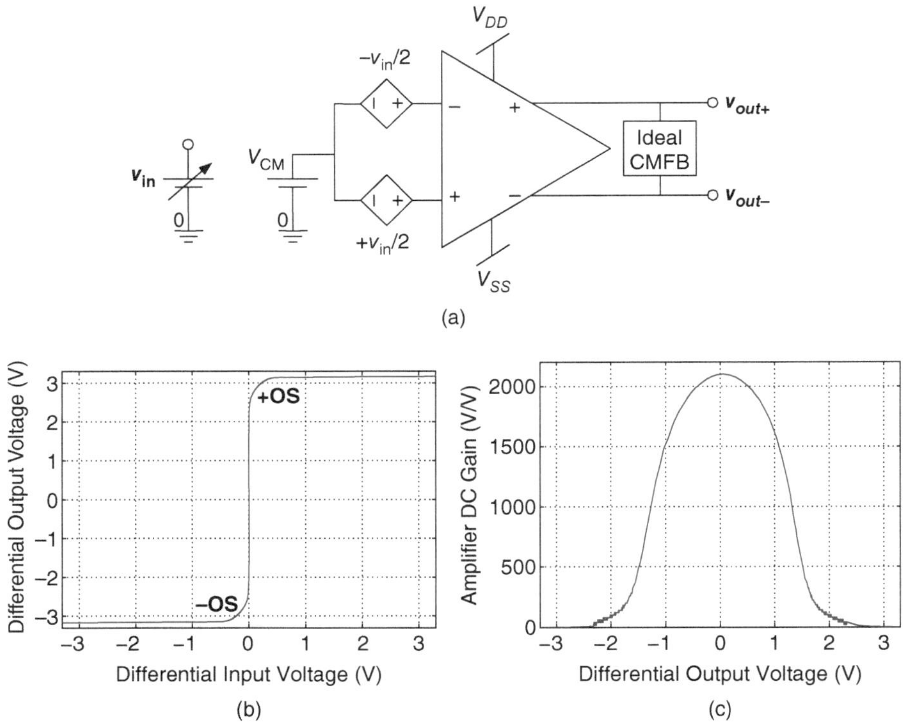

Figure 8.16 Amplifier simulation results obtained from a DC analysis: (a) test-bench circuit; (b) open-loop voltage transfer characteristic; (c) amplifier gain versus output voltage.

[3]Note that the measured values for g_m and GB, and the loading capacitor C_L, can be combined to provide an estimate of the amplifier output parasitic capacitor C_o. Given that $\text{GB} = g_m/(C_L + C_o)$ in a single-stage amplifier, $C_o = 0.3$ pF for the folded cascode amplifier under consideration.

to $+2(V_{DD} - V_{SS})$ is again applied at the amplifier input, but with no loading conditions at its output nodes.

Figure 8.16b shows the simulated differential output voltage ($v_{out} = v_{out+} - v_{out-}$) versus the differential input voltage v_{in} for the folded cascode amplifier in Figure 8.13.

The dependency of the amplifier gain on the output voltage level can be easily obtained from the data contained in the Figure 8.16b, just by computing dv_{out}/dv_{in} from the curve and depicting the results against v_{out} (instead of v_{in}). Figure 8.16c shows the resulting gain curve for the folded cascode amplifier in the example, from which the DC gain at the quiescent point A_v can also be derived ($A_v = 2100$, in accordance with the low-frequency AC results in Figure 8.14b).

Finally, note that the amplifier voltage characteristic in Figure 8.16b can be obtained from a relatively coarse DC sweep of the input voltage, whereas a very fine sweep is required to accurately obtain the amplifier gain curve in Figure 8.16c. For the sake of illustration, the DC input voltage was varied from −3.3 V to +3.3 V in 10-mV steps to obtain the graphical representation in Figure 8.16b. The DC simulation was run again with steps of only 50 μV (but only from −0.3 V to +0.3 V) to give adequate granularity in the v_{out} axis in the results shown in Figure 8.16c.

8.3 Design Considerations of Transconductors

Transconductors are essential building blocks of CT-ΣΔMs. They are used for building Gm-C integrators and also to implement ΣΔM loop-filter coefficients. As an illustration, Figure 8.17 shows two different common examples of CT-ΣΔM implementations. The first one, shown in Figure 8.17a, uses a front-end active-RC integrator whereas the other loop-filter integrators are implemented using Gm-C topologies. This approach is normally chosen due to the better linearity performance achieved by active-RC integrators as compared to Gm-C integrators. If the linearity specification is not very restrictive, purely Gm-C implementations like that shown in Figure 8.17b are usually preferred because of their potentially faster operation. Note that in both examples depicted in Figure 8.17, loop-filter coefficients are implemented using transconductors.

The main electrical characteristics affecting the performance of transconductors used in ΣΔMs are

- input/output swing
- transconductance value and tuning range
- finite DC gain
- finite GB
- mismatch error
- nonlinearity (usually characterized by the IIP3 parameter).

The limit values of these design specifications are derived from system-level behavioral simulations, which are further fine-tuned using macromodel electrical simulations, as described in Section 7.1. Once these specifications have been clearly identified, a suitable circuit topology is chosen and designed at the transistor level in order to meet them.

Obviously, there are plenty of different transconductor circuits that can be used for implementing the loop filter of CT-ΣΔMs. The detailed explanation of all of them is beyond the scope of this

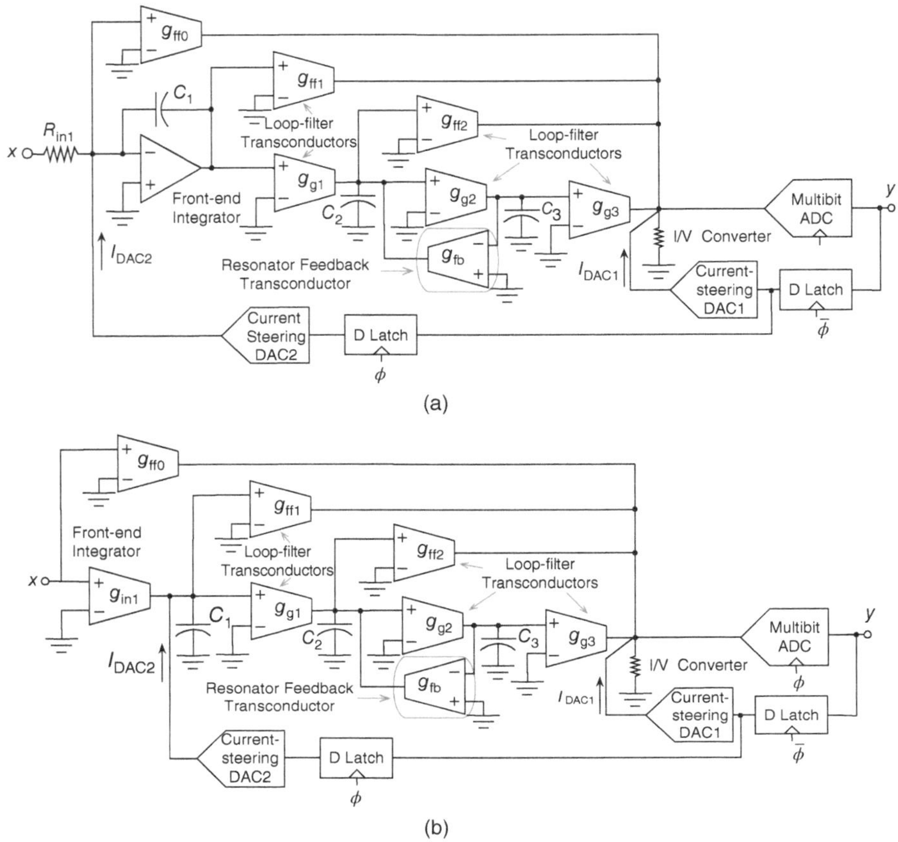

Figure 8.17 Illustrating the use of transconductors in a third-order single-loop CT-ΣΔM, considering: (a) active-RC front-end integrator; (b) Gm-C front-end integrator.

book. Instead, the following sections focus on some circuit examples. The first is suited for the implementation of front-end Gm-C integrators, while the others are more appropriate to build Gm-C integrators embedded in the loop filter.

8.3.1 Highly Linear Front-end Transconductor

One of the main limiting factors in open-loop Gm-C integrators is their poor linearity. Indeed, this issue is critical at the input node of ΣΔMs because harmonic distortion caused by the front-end V/I transconductor is directly translated to the digital domain without any attenuation. Thus, special emphasis should be put on design of a front-end transconductor with sufficiently high linearity, because without this, the performance of the modulator can be severely degraded.

Figure 8.18 shows an example of a front-end transconductor [13]. The circuit combines gain-boosting techniques with resistive source degeneration in order to increase the linearity of the transconductance. Note that in this example nMOS transistors are used in the input differential pair because a triple-well option is available in the technology where the transconductor is going to be integrated. In this way, the MOS body effect can be avoided by simply connecting the source terminal

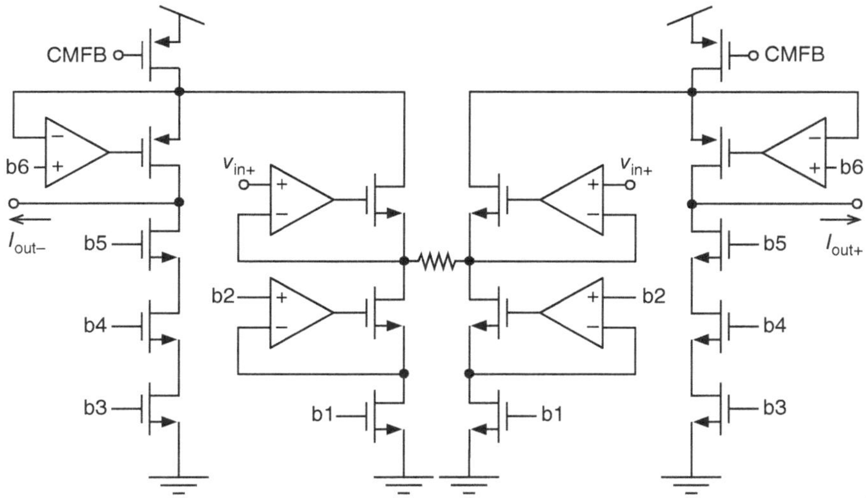

Figure 8.18 Example of ΣΔM front-end transconductor with enhanced linearity [13].

to the substrate terminal. Otherwise, if a triple-well option is not available, pMOS transistors should be used for the differential input pair.

As an example, let us consider that the circuit is placed at the front end of a cascade 3-2 CT-ΣΔM with a first stage like that shown in Figure 8.17b, and that this modulator is intended to digitize 20-MHz signals with 12-bit effective resolution. According to behavioral simulations carried out in SIMSIDES, the following specifications are obtained for the front-end transconductor:

- finite DC gain: >70 dB
- differential input/output swing: 0.3 V
- third-order nonlinearity: > −86 dB.

In addition to the above specifications, input-referred thermal noise must be kept sufficiently low to ensure that the the effective resolution (12-bit) required by the modulator is not degraded. According to the synthesis of the ΣΔM loop-filter coefficients in this example, the nominal input V/I transconductance was chosen to be 850 μA and the integration capacitance is 3.65 pF.

In order to design the circuit in Figure 8.18 to cope with the above specifications, basic OTA design equations for the nominal transconductance, finite DC gain, SR, and GB are usually derived by hand [14] and used as initial point for the electrical design. Transistors are biased and sized in the saturation region by using these "manual" equations, and considering the voltage limits specified by the input/output swing. This initial design is fine-tuned by electrical simulation in order to satisfy the required specifications with minimum power consumption. In addition, as in other essential ΣΔM building blocks, the circuit must be simulated considering the effect of technology corners and mismatch deviations. The circuit in Figure 8.18 was designed in a 130-nm CMOS technology to cope with the aforementioned specifications, achieving a DC gain of 73.8 dB, a maximum differential input/output swing of 0.3 V and $HD_3 = -89$ dB, considering a single 1.2-V supply voltage [13].

8.3.2 Loop-filter Transconductors

Figure 8.19 shows an example of transconductor (with its corresponding CMFB circuit) that can be used for building the loop-filter coefficients of CT-ΣΔMs [13]. High-speed operation is achieved by using only feedforward paths. These paths introduce a high-frequency zero that extends the frequency range. The circuit uses a quadratic-term cancellation technique in order to improve the linearity of the transconductance. The basic idea behind this technique is to include an additional tail current source having its current quadratically dependent on the input signal, as conceptually illustrated in Figure 8.19. It can be shown that the quadratic term of the drain current of the differential-pair transistors can be cancelled if $k_2 = k/k_g^2$, where k_g is a gain factor that takes into account any signal scaling that may be applied to the input signal and k and k_2 stand for the large-signal transconductances of the differential-input pair transistor (M_1) and the extra MOS transistor added (M_2), respectively.

A critical consideration in the design of CT-ΣΔM loop-filter circuit elements is their tunability: therefore, transconductors must be designed so that they can be tuned in order to keep the loop-filter time constants (C/g_m) unchanged as technology parameters vary. To this end, a circuit-tuning strategy must be incorporated in the loop-filter transconductors in order to make the design of CT-ΣΔMs feasible and robust against circuit element tolerances. In the example shown in Figure 8.19, the transconductance can be tuned through the tail bias current I_{tune}. Note that, for voltage headroom reasons, this current source is connected between the positive supply voltage and the common source node. Thus a variation of I_{tune} – through bias voltages named $P_{tune(1,2)}$ – changes the value of the nominal transconductance, without significantly affecting the linearity of the transconductor.

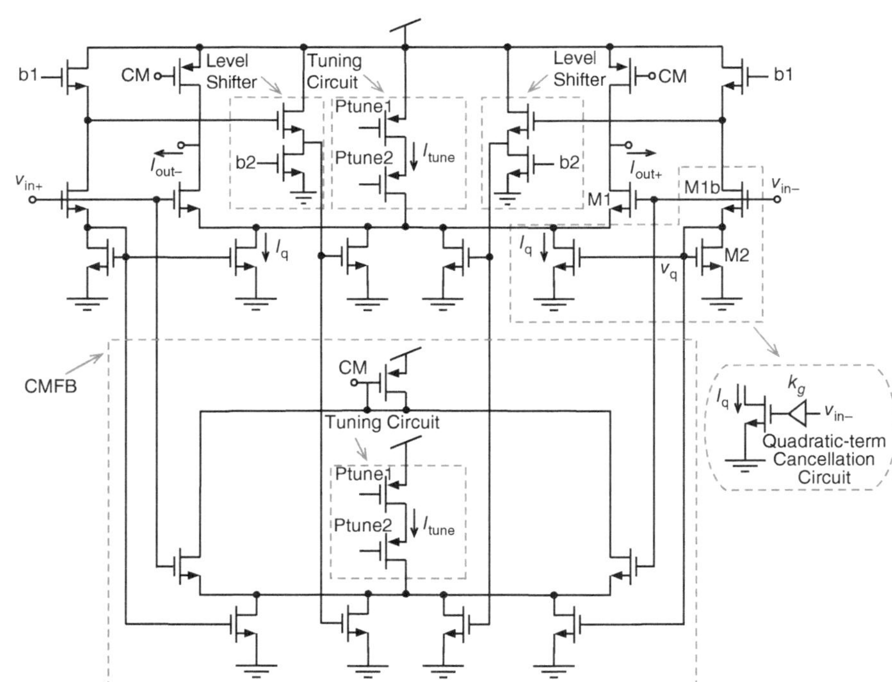

Figure 8.19 Example of ΣΔM loop-filter transconductor with quadratic-term cancellation technique [13].

Note that, for the tuning to be effective, each transconductance of the ΣΔM is usually implemented as the parallel connection of unit transconductors and a number of Monte Carlo simulations are required in order to guarantee that the performance of the transconductors – particularly the linearity – is not affected by the mismatch. In this example, a unit transconductance of 25 μA/V was used, while the values of I_{tune} ranged from 5 μA to 25 μA.

As stated in the previous section, the linearity and gain specifications are not so demanding as in this case of front-end integrators. For instance, considering the same modulator specifications described in previous section, the electrical specifications of the loop-filter transconductors will vary depending on their position in the modulator chain. The most restrictive specifications are the following:

- finite DC gain: >50 dB
- differential input/output swing: 0.3 V
- third-order nonlinearity: > −56 dB

As an illustration, Figure 8.20 shows some typical characteristics obtained by electrical simulation of transconductors using the test-bench schematic shown in Figure 8.21. Figure 8.20a shows the output current as a function of the input current for different values of I_{tune}, ranging from 5 μA to 23 μA in this example. The linearity of the transconductance can be calculated by computing the derivative of this I–V characteristic. This is represented in Figure 8.20b. Following this procedure, the maximum transconductance variation for each of the values of I_{tune} can be calculated. In this example, a deviation below 2% is obtained in all cases within the specified input voltage range (0.3 V).

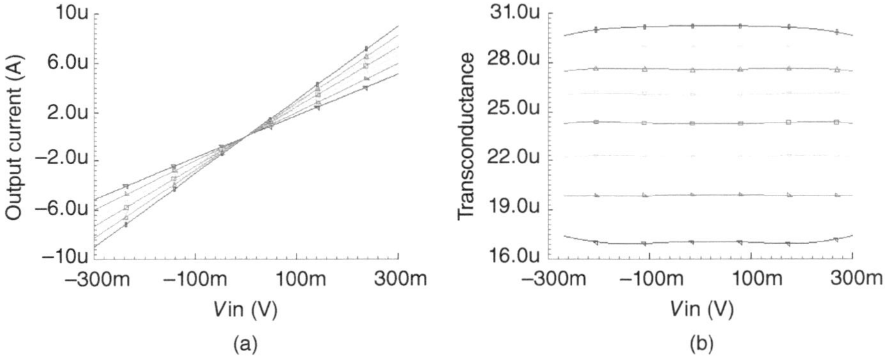

Figure 8.20 Illustrating the electrical performance of ΣΔM loop-filter transconductors: (a) output current versus input voltage for different values of the tuning current I_{tune}; (b) transconductance versus input voltage. I_{tune} ranges from 5 μA to 23 μA.

Figure 8.22 shows the frequency response of a Gm-C integrator made up of a unit transconductor (with a nominal transconductance of 25 μA/V) and the unit integration capacitor (3.65 pF in this example). Note that the phase error at unity gain of the integrator is below 1°, while the high-frequency poles are placed at a much higher frequency than the GB of the integrator. This is due to the fact that the transconductor in Figure 8.19 is composed of feedforward paths, facilitating a higher speed.

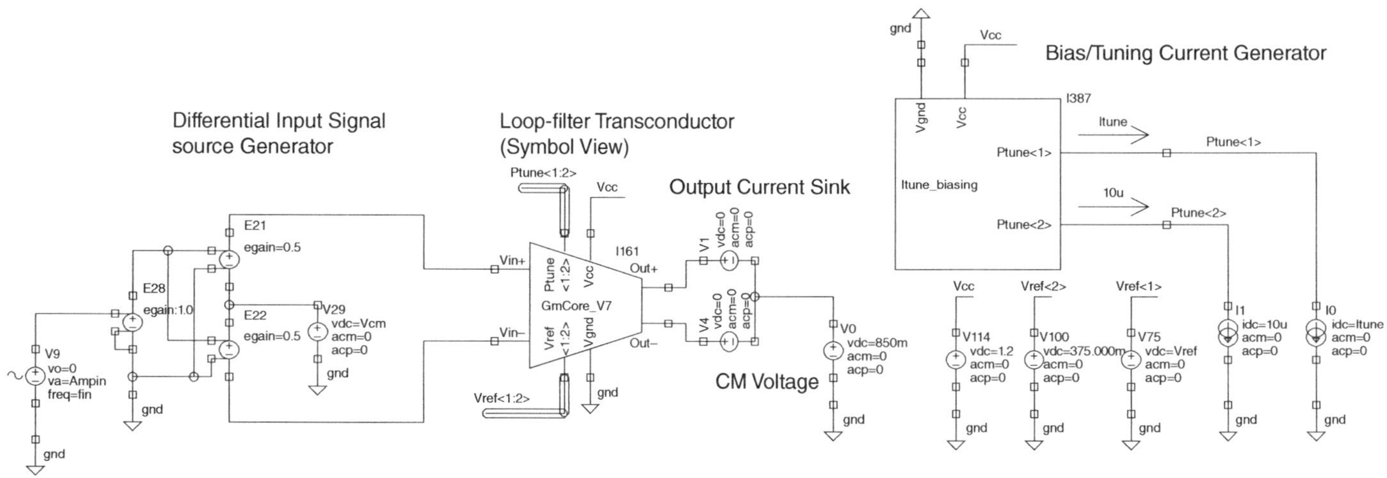

Figure 8.21 Cadence Virtuoso test-bench schematic used for characterizing the ΣΔM loop-filter transconductors. Note that this test bench can be used for DC, AC, and transient analysis. Alternatively, a load circuit can be connected at the output of the transconductor in order to emulate a circuit environment closer to the actual situation of loop-filter transcondutors embedded in a CT-ΣΔM.

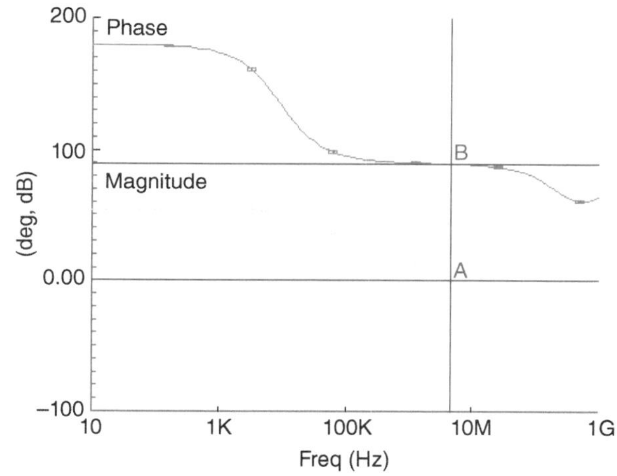

Figure 8.22 Frequency response of the loop-filter (unit) transconductor in Figure 8.19.

8.3.3 Widely Programmable Transconductors

There are many applications where reconfigurable, programmable transconductors are needed in order to adapt the performance of ΣΔM loop-filter to different ADC requirements. One of the most common ways to implement reconfigurable transconductors is based on connecting switchable unitary elements to get the required overall transconductance. A compact, *scaling-friendly* way to do this consists of using so-called inverter-based unitary transconductors.

Figure 8.23 shows an example of a switchable inverter-based transconductor [15], in which some unitary transconductors are always connected while others are switchable. The inverter-based unitary transconductors are very simple and modular, to increase the flexibility and programmability of the modulator, while minimizing the power consumption. Different values of unitary transconductances can be used in order to adjust the values of the ΣΔM loop-filter coefficients with a higher accuracy. For instance, one design presented in the literature [15], two unitary transconductances, g_{mu} and $1/2g_{mu}$, were considered, with g_{mu} = 100 μA/V. These values can be easily changed by appropriately scaling the value of the tail current, I_{bias}, of the unitary transconductors as well as the transistor sizing. This will be discussed in Chapter 9.

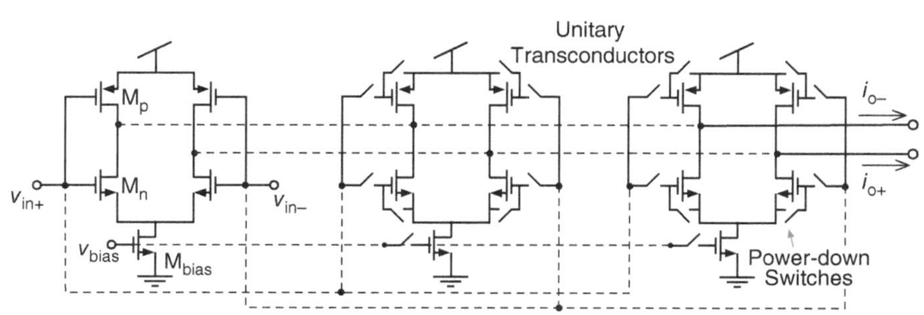

Figure 8.23 Switchable inverter-based transconductor used in reconfigurable CT-ΣΔMs.

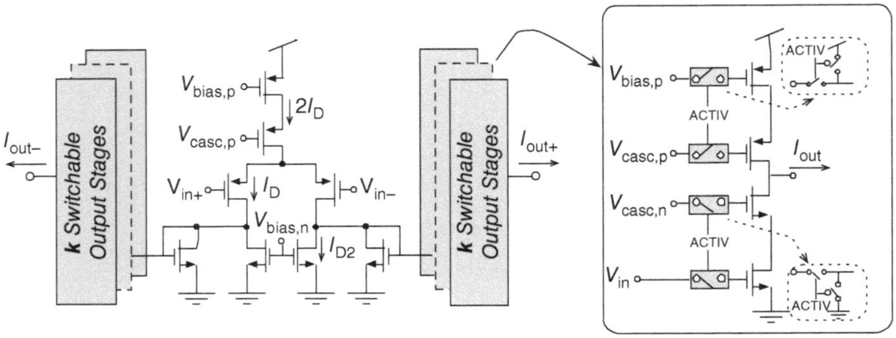

Figure 8.24 Widely programmable transconductor based on current starving and switchable output stages [17].

Another interesting way to implement highly programmable transconductors for ΣΔMs consists of combining so-called *current-starving* techniques with switchable output stages [16]. Figure 8.24 shows an example circuit that realizes this concept [17]. The transconductor topology uses a simple symmetric OTA in the first stage. The input differential pair uses pMOS transistors biased with a current I_D to exhibit a transconductance g_m at the quiescent point. Current starving is implemented by means of subtracting a DC current $I_{D2} = A \cdot I_D$ from that of the input transistors, with A being the *starving factor*. The remaining current at the quiescent point, $I_{D3} = (1 - A) \cdot I_D$, flows through a diode-connected nMOS transistor working as active load to the symmetric OTA. The second transconductor stage consists of the parallel connection of k output stages, which are electrically identical. Therefore, the overall current copy factor, N, equals the number of active output stages in this particular example. Each output stage consists of a simple common-source inverting amplifier that employs nMOS and pMOS cascode transistors to obtain a high output resistance.

The parallel connection of the output stages is programmable and their deactivation is independently controlled by digital signals (denoted as ACTIV in Figure 8.24), so that the value of N can be arbitrarily chosen according to the reconfigurability requirements in the transconductor. It can be shown that, for a given N, the effective transconductance-over-current ratio, $g_{m,\text{eff}}/I_{D,\text{eff}}$ of the transconductor is given by:

$$\alpha \equiv \frac{N}{1 + N \cdot (1 - A)} \tag{8.3}$$

where α stands for the energy efficiency factor of the overall transconductor compared to one of its input transistors. In this way, the output current range and granularity of the transconductor can be arbitrarily programmed with an energy efficiency that increases both with the current starving factor A in the first stage and the overall current copy factor N [17].

8.4 Design Considerations of Comparators

Comparators are essential building blocks of ΣΔ ADCs. These circuits are used for building the quantizer embedded in the modulator. Because of their position in the modulator loop, the comparator specifications are not very demanding in most practical cases, since circuit errors are attenuated by NTF in the same way as the quantization noise. However, its main performance limitations (offset, hysteresis, and comparison time) must be carefully taken into account in order to optimize the performance of ΣΔMs. Typical static specifications require an offset and a hysteresis of the order of tens of millivolts, and a maximum comparison time of a quarter of the clock period; that is, around half of the time interval corresponding to the strobe phase.[4] These specifications can be achieved by using the so-called regenerative latch topologies described below.

8.4.1 Regenerative Latch-based Comparators

The majority of comparator circuits used in ΣΔMs are based on a DT positive-feedback regenerative latch, which is built by cross-coupling a pair of inverters, as conceptually depicted in Figure 8.25a [18]. The inverters amplify the differential input voltage $v_{i+} - v_{i-}$ to obtain the saturated differential output $v_{o+} - v_{o-}$, according to the characteristics drawn in Figure 8.25b. During the so-called reset phase (ϕ_r high), the differential input is stored in the input sampling capacitors C and the circuit is driven to the central state Q_0 (see Figure 8.25b). During the comparison or *strobe* phase (ϕ_a high), the differential input is retrieved, forcing an initial state either on the right ($v_{i+} > v_{i-}$) or on the left ($v_{i+} < v_{i-}$) of Q_0. From this initial state, the action of positive feedback forces the output to evolve either towards $Q_{\rm H}$ (for $v_{i+} > v_{i-}$) or towards $Q_{\rm L}$ (for $v_{i+} < v_{i-}$). In both cases, the dynamic evolution around the central point is realized at a high speed due to the action of positive feedback.

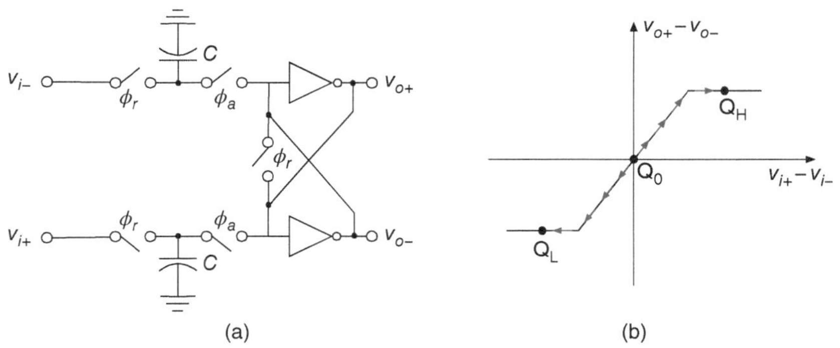

Figure 8.25 Positive-feedback regenerative latch: (a) conceptual schematic; (b) illustrating dynamic trajectories in the input–output characteristic.

Figure 8.26 shows some regenerative latch CMOS circuits [19–,22], which are commonly used for implementing comparators in state-of-the-art ΣΔMs. All of them are based on the conceptual model shown in Figure 8.25a. In practice, the static resolution of all these circuits is limited by dissymmetries between the positive and the negative branches of the fully-differential circuit, as well as other second-order circuit phenomena such as the kick-back noise on the integrator connected to the comparator. Thus, a preamplifier is usually placed at the input of the regenerative latch in order to improve the static resolution of these kinds of comparators [23].

Figure 8.27 shows a typical example of a CMOS regenerative latch comparator with preamplifier.[5] It consists of a pMOS differential input pair ($M_{1,2}$), a CMOS regenerative latch circuit, and a set-reset (SR) flip-flop to store the voltage provided by the latch. The latch circuit is composed of an nMOS

[4]The strobe phase is the clock phase in which the comparator is active: when it is comparing the input signals. This phase is sometimes referred to as the amplification phase or simply the comparison phase.

[5]Some comparators used in SC-ΣΔMs with feedforward paths include an SC network at the input of the preamplifier to merge with SC adders. These SC networks may also be used for improving the preamplifier performance and for implementing reconfiguration of the number of bits of the embedded quantizer [24].

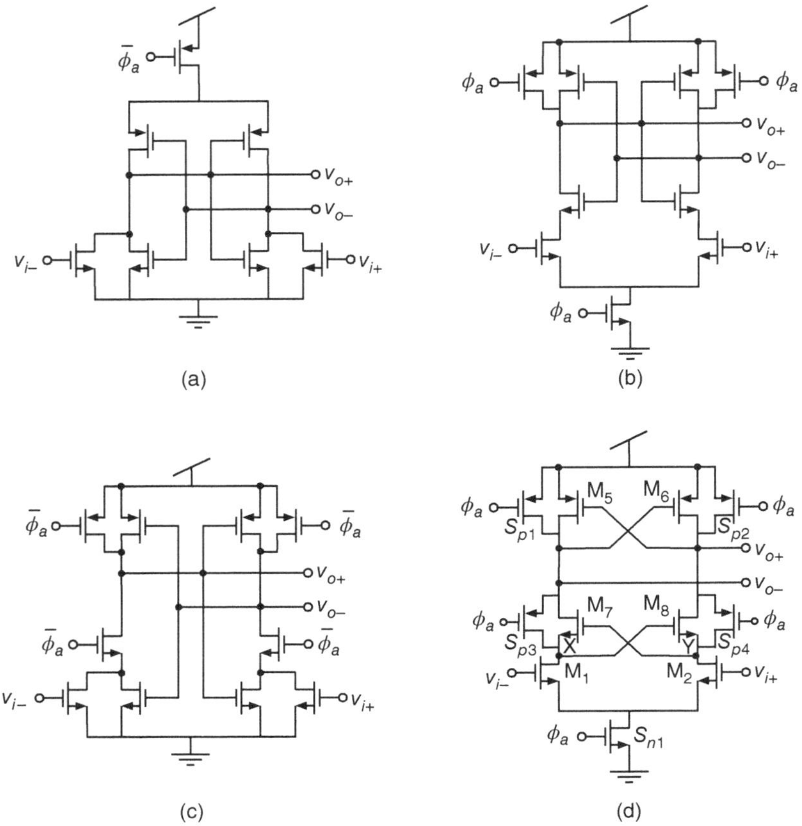

Figure 8.26 CMOS latches frequently used in ΣΔMs, designed by: (a) Nikoli *et al.* [19]; (b) Kobayashi *et al.* [20]; (c) Yukawa, *et al.* [21]; (d) Wang and Razavi [22].

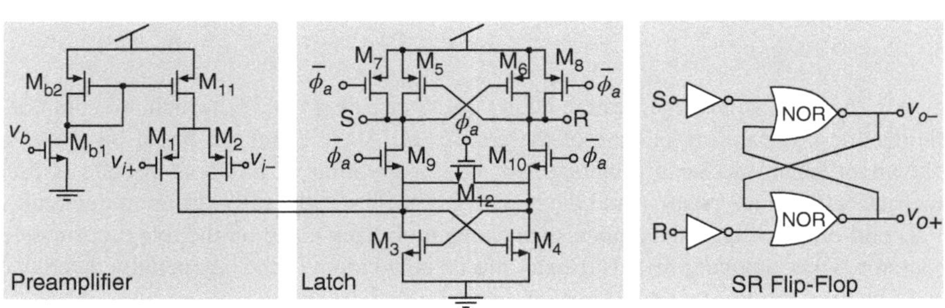

Figure 8.27 Example of a typical CMOS regenerative latch comparator with preamplifier.

flip-flop ($M_{3,4}$) with a pair of nMOS switches ($M_{9,10}$) for strobing and an nMOS switch (M_{12}) for resetting, and a pMOS flip-flop ($M_{5,6}$) with a pair of pMOS precharge switches ($M_{7,8}$).

The operation of the circuit in Figure 8.27 is as follows. During clock phase $\overline{\phi}_a$ (the reset phase), the latch is in reset mode and the input differential pair injects a current proportional to the differential input voltage ($v_{i+} - v_{i-}$). This generates an initial voltage imbalance across the on-resistance of switch M_{12}. The voltage difference across M_{12} is amplified when the latch becomes enabled during the comparison (amplification) phase ϕ_a, achieving a very fast comparison time due to the action of positive feedback.

8.4.2 Design Guidelines of Comparators

The design of the comparators used in ΣΔMs is carried out according to the high-level specifications extracted from behavioral simulations as described in Chapter 5. The main design parameters included in the behavioral models that degrade the performance of ΣΔMs are essentially two static parameters: offset and hysteresis. In addition, the transient response of the comparator must be fast enough to complete its operation within the comparison clock phase. Therefore, the analog sections[6] of the comparator – the preamplifier and the latch – must be carefully sized accordingly to these specifications.

The preamplifier aims to fulfill several goals: to obtain high DC gain in order to reduce the comparator input-referred offset, low kick-back noise, and high speed, while keeping low parasitic input capacitances. Therefore, increasing the preamplifier DC gain should not be done exclusively by increasing the size of the input differential pair transistor – in order to increase g_m – for a given bias current, because this would also increase the input parasitic capacitance. This design trade-off can be solved by using a preamplifier schematic with a high output resistance, like that shown in Figure 8.28. In this circuit, transistors M_4 and M_5 are biased in the ohmic region and are hence used as resistors to increase the output resistance of the preamplifier.

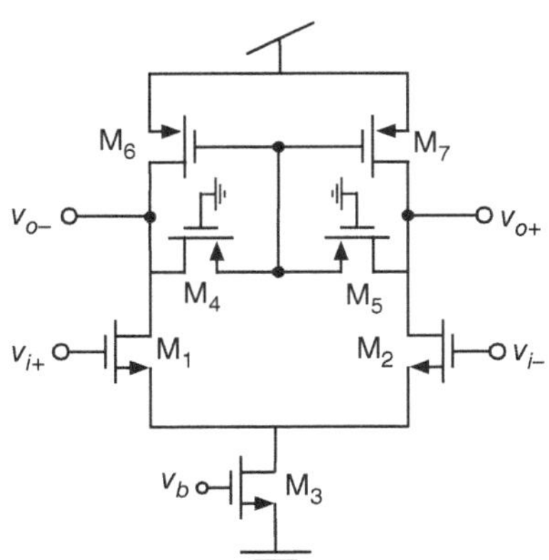

Figure 8.28 Single-stage preamplifier with ohmic-biased transistors (M_4 and M_5) to increase the output resistance.

Another important design trade-off that takes place in the latch involves the required mismatch specifications (which have a direct effect on the static resolution) and the comparator speed. Using minimum sizes for the inverter transistors (M_{3-6} in Figure 8.27) minimizes the MOS parasitic capacitances, thus benefiting the transient response. However, minimum-sized transistors increase the sensitivity to technology process variations.

A minimum length should be used for the switches involved (M_{7-10} in Figure 8.27) in order to reduce the switch on-resistance as well as the effect of charge injection. Note that the input voltage

[6]The SR flip-flop is a digital circuit and in the majority of cases can be designed using minimum sizes.

difference is sensed across switch M_{12}, and hence its on-resistance becomes critical for correct operation of the comparator. Indeed, the performance of the latch is very sensitive to the size of M_{12}, which has motivated some designers to use more robust topologies.

This is the case for the latch shown in Figure 8.26d [22]. The operation of this circuit is governed by one clock phase, denoted as ϕ_a in the figure. Thus, when the strobe signal (ϕ_a) triggers a logic one, switch S_{n1} turns on and transistors $M_{1,2}$ will process the differential output of the preamplifier. After a short time, one of the input differential pair transistors will turn off, depending on the sign of the input voltage imbalance, thus creating a differential voltage between nodes X and Y. Meanwhile, M_7 and M_8 will turn on and, later, the cross-coupled scheme will start working, leading to fast regeneration of the initial voltage imbalance [24].

Even though regenerative latch comparators are very fast, the comparison time and the dynamic resolution must be properly characterized at transistor level. The following subsections give some practical ways to do this.

8.4.3 Characterization of Offset and Hysteresis Based on the Input-ramp Method

Comparator offset and hysteresis can be characterized in electrical simulations using the test-bench circuit shown in Figure 8.29a. A slow ramp waveform input signal is applied to the comparator, so that the offset and hysteresis parameters can be extracted from the output voltage waveform, as conceptually depicted in Figure 8.29b. Note that in order to obtain a more precise value of both performance metrics (offset and hysteresis), the voltage limits of the input ramp should be approximately of the same order as the specified offset and hysteresis.

Let us consider an example in which the target specifications for the offset and the hysteresis are ± 10 mV and ± 20 mV, respectively. In order to take into account the impact of mismatch and technology parameter variations, a large number of simulations considering all technology corners as well as a Monte Carlo analysis should be carried out. As an illustration, Figure 8.30 shows some typical output waveforms obtained in HSPICE for the characterization of the offset and hysteresis.

The method based on the input ramp does not allow for memory or hysteresis issues in the comparator operation because the voltage value of the input signal is always increasing (or decreasing) in amplitude, which is not common in practice. A more realistic test bench uses a sinusoidal or triangular input signal. In this way, the input signal is forced to take alternating signs in order to allow for memory effects [24].

Another disadvantage of the ramp-waveform test-bench method is that a slower ramp is needed to obtain the required accuracy in the characterization of the offset and the hysteresis. The slower the input ramp is, the longer the transient simulation. This might lead to long CPU times, since typically 100–200 simulations are needed for an accurate Monte Carlo simulation.

8.4.4 Characterization of Offset and Hysteresis Based on the Bisectional Method

A more efficient way of characterizing the input offset of the comparator is based on the so-called *bisectional* method. In this method, graphically illustrated in Figure 8.31, a root-finding algorithm works by repeatedly halving an interval and then selecting the subinterval in which the root exists [25]. Given two points, a and b, such that $f(a)$ and $f(b)$ have opposite signs, the intermediate value theorem says that $f(x)$ must have at least one root in the interval $[a, b]$ as long as $f(x)$ is continuous. The bisectional algorithm is therefore applied to the subinterval in which the sign change occurs until a solution under a defined tolerance is obtained. It can be shown that the number of iterations required to reach convergence with a tolerance error ϵ is $n = \log_2[(b - a)/\epsilon]$ [25].

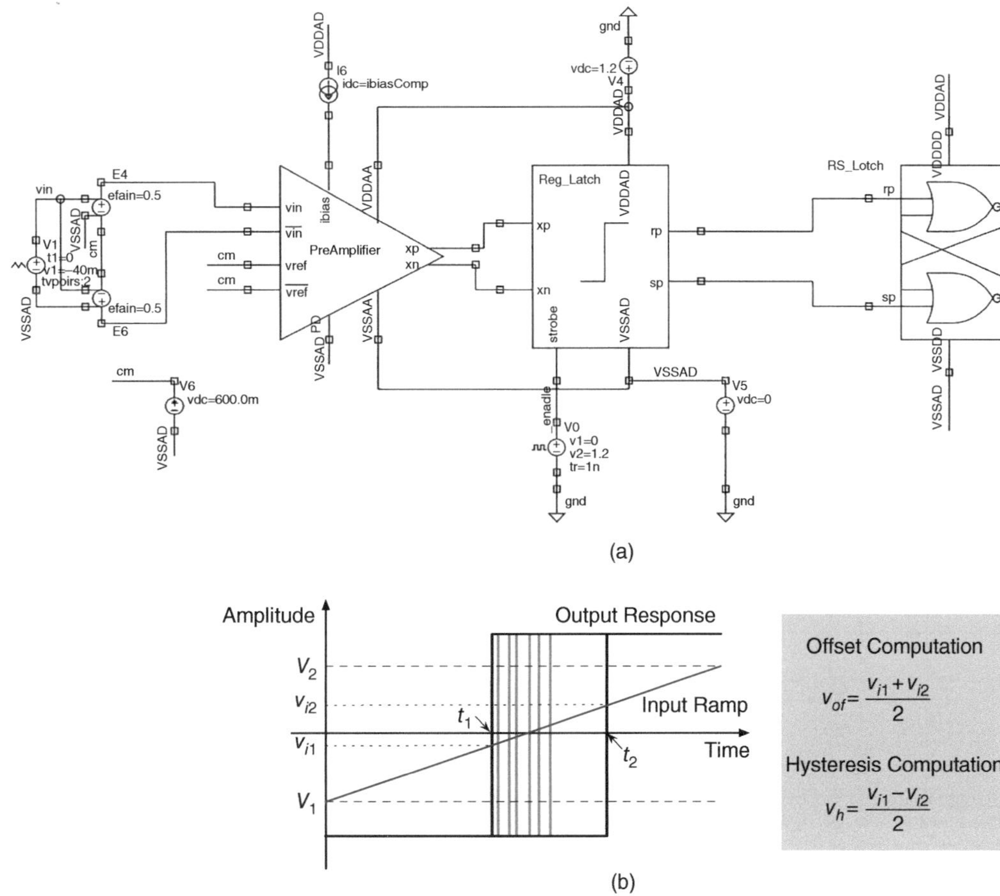

Figure 8.29 Characterization of offset and hysteresis based on the input-ramp method: (a) Cadence Virtuoso test-bench schematic; (b) conceptual output response and computation of offset and hysteresis parameters.

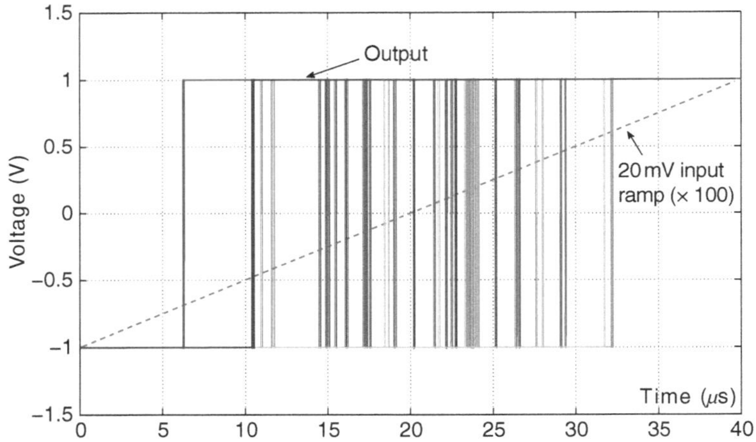

Figure 8.30 Illustrating the electrical characterization of hysteresis in regenerative latch comparators. Monte Carlo simulation in HSPICE.

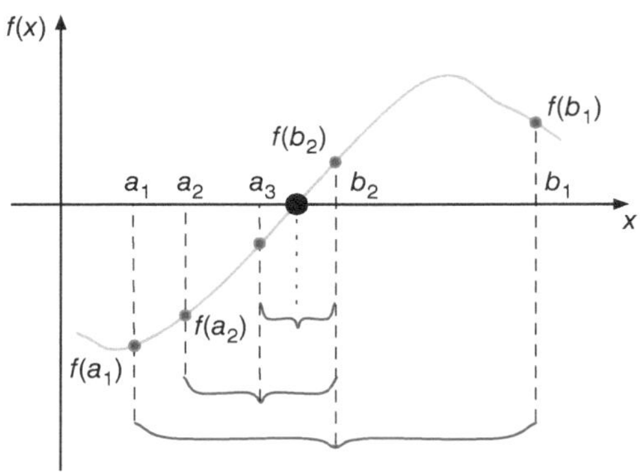

Figure 8.31 Graphical illustration of the bisectional method algorithm.

Figure 8.32a shows the test bench used in the Cadence Virtuoso schematic editor for the comparator offset computation. The preamplifier and the regenerative latch are placed inside a feedback loop. The preamplifier input signal is controlled by a block that implements the bisectional algorithm described above. The Verilog-A code shown in Figure 8.32b is used for this purpose [26]. The input of this block is the differential output of the regenerative latch, so the algorithm essentially selects the next input voltage based on the comparator output, following the procedure depicted in Figure 8.31. Note that a delayed version of the strobe clock signal triggers the Verilog-A block corresponding to the bisectional algorithm. This time delay must be longer than the response time of the regenerative latch, but lower than half of the cycle of the comparator strobe.

8.4.5 Characterizing the Comparison Time

Figure 8.33 illustrates the electrical simulation method, which is usually used to characterize the resolution speed and the comparison time of comparators. An input step of ΔV is applied, where ΔV

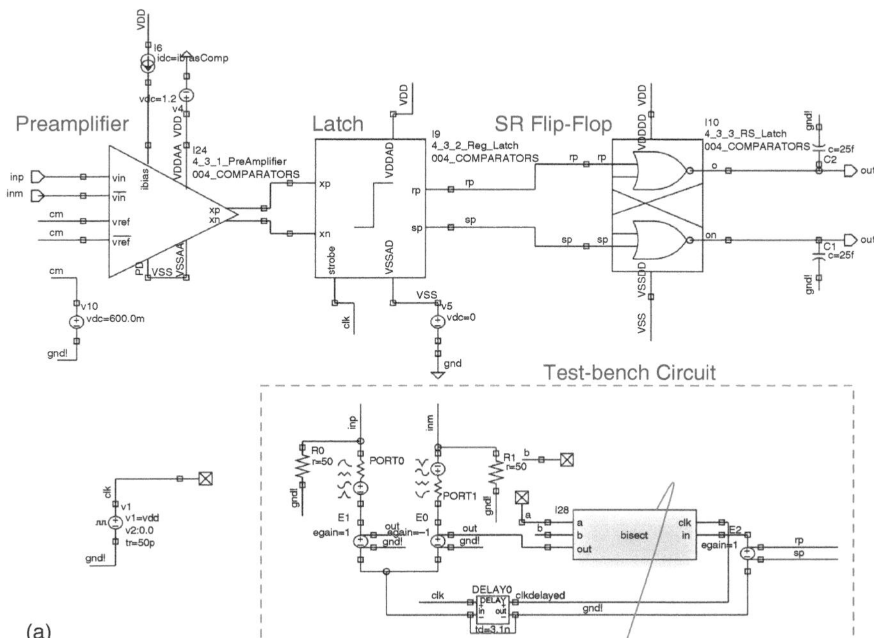

(a)

```
// VerilogA for GVdP_comparator, bisect, veriloga

`include "constants.vams"
`include "disciplines.vams"
module bisect(in, clk, a, b, out);

   input in; voltage in;
   input clk; voltage clk;
   output a; voltage a;
   output b; voltage b;
   output out; voltage out;

   parameter real va = -0.1;
   parameter real vb = 0.1;
   parameter real vth = 0.6;
   parameter integer dir = +1 from [-1:+1] exclude 0;
                 // if dir=+1, rising clock edge triggers
                 // if dir=-1, falling clock edge triggers
   parameter real td = 0 from [0:inf);  // output delay
   parameter real tt = 0 from [0:inf);  // output transition time

   real     val;
   real     vbl;
   real     vout;
   real     level;
analog begin
    @(cross(V(clk) - vth, dir)) begin
       val=V(a);
       vbl=V(b);
       vout=V(out);
       level = (V(in)-0);
       if (level<0)
        val=vout;
       else
        vbl=vout;
    end
    @(initial_step) begin
       val=va;
       vbl=vb;
    end
   vout=(val+vbl)/2;
   V(a) <+ transition( val, td, tt );
   V(b) <+ transition( vbl, td, tt );
   V(out) <+ transition( vout, td, tt );
end
endmodule
```

(b)

Figure 8.32 Cadence Virtuoso test bench used for the characterization of the comparator offset based on the bisectional algorithm method: (a) Cadence Virtuoso schematic; (b) Verilog-A code for the bisectional algorithm.

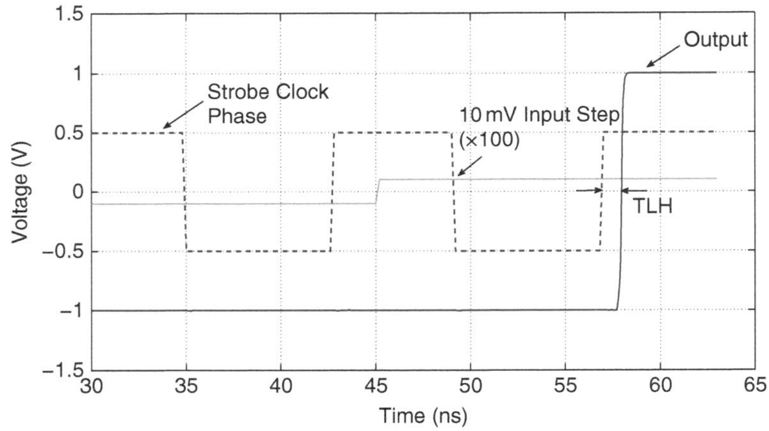

Figure 8.33 Characterization of comparison time in regenerative latch comparators.

is the specified static resolution of the comparator (10 mV in this example). The comparison time, also referred to as resolution time, is defined in this test bench as the interval between the time instant in which the strobe phase becomes low at the end of the comparison phase and the time instant when the output reaches the corresponding logic level; that is, V_{DD} for a positive input step or V_{SS} otherwise. Monte Carlo and technology corner variations need to be checked and the worst-case values are usually taken as design specifications.

8.5 Design Considerations of Current-Steering DACs

The DACs used in the feedback path of ΣΔMs are mostly implemented using SC and SI or current-steering circuit techniques.[7] The design of SC feedback DACs takes into account considerations of the design of switches and capacitors according to the design criteria discussed in Section 8.1. Indeed, the switches and capacitors used for building SC DACs are embedded in the SC integrators and quantizers (see for instance Figure 7.9), so SC DACs do not require special attention, apart from the design issues related to the switches.

In contrast, CS DACs – mostly used in CT-ΣΔMs – are essential building blocks affecting the performance of the modulator, due to several nonideal circuit phenomena already discussed in Chapters 3–5, namely: clock jitter error, transient response (and its effect on the excess loop delay error), and linearity (due to device mismatch of unit current sources). CS DACs are particularly suitable for wideband CT-ΣΔMs because of their potential benefits in terms of high-speed operation and the convenience of interfacing with both Gm-C and active-RC CT-ΣΔMs [27]. This is illustrated in Figure 8.34, which shows a conceptual schematic of the input summing node of a CT-ΣΔM. Note in Figure 8.34a that the output current of the feedback CS DAC is *naturally*[8] added with the output current of the loop-filter Gm-C transconductor. In the case of active-RC implementations,

[7]The difference between SI and CS techniques is discussed later.

[8]As will be discussed later, the high impedance output node of a Gm-C transconductor may be not suited to *inject* the feedback current provided by a CS DAC, for two main reasons. On the one hand, large signal swings reduce the voltage headroom required to keep the current cell transistors in the saturation region. On the other hand, the high impedance node causes small current errors to result in large voltage errors on the capacitors.

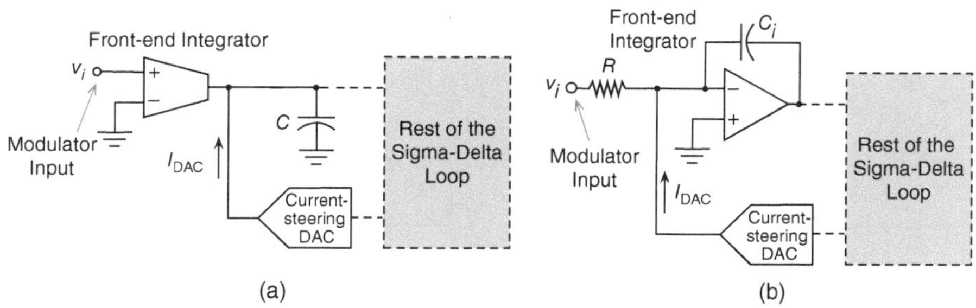

Figure 8.34 Illustrating the connection of CS DACs at the input summing node of: (a) Gm-C ΣΔMs; (b) active-RC ΣΔMs.

the current-mode adding operation takes place at the *virtual ground* input node of the amplifier, as illustrated in Figure 8.34b.

This section looks at the main criteria taken into account in the design of CS DACs – from the circuit-level perspective – focussing on those issues that must be considered during the electrical (transistor-level) design of ΣΔMs. Special emphasis is placed on the main performance metrics of CS DACs, showing how to characterize these metrics using practical test benches implemented in SPICE-like simulators.

8.5.1 Fundamentals and Basic Concepts of CS DACs

Figures 8.35a and 8.35b show two conceptual schemes of a CS DAC [28], where the output current is obtained by adding a number of switched unit current cells together. The switches are controlled by the DAC input bits, which are in turn the outputs of the quantizer embedded in the ΣΔM loop filter. Thus, if a binary code is used, N binary-weighted scaled current cells are required, as shown in Figure 8.35a. Note that although this solution requires the minimum number of elements, they are more sensitive to device element mismatch, mainly due to the very different values of the cell currents. For that reason, a thermometer-coded CS DAC like that shown in Figure 8.35b is usually chosen to implement the feedback DACs of CT-ΣΔMs. This approach *relaxes* the matching requirements at the price of exponentially increasing the number of unit elements; that is, the number of current cells and switches.

Note that the circuit schemes shown in Figures 8.35a and 8.35b cannot be implemented in practice due to the current glitches appearing when current cells are left open. For that reason, the scheme shown in Figure 8.35c – commonly referred to as *current-steering* – is usually preferred. In this approach the currents provided by the cells are *steered* or *redirected* either to the DAC output summing node or to a dummy low-impedance node, by means of complementary switches.

The scheme in Figure 8.35c is particularly useful in fully-differential implementations, which is one of the most common situation in practice. In this case, conceptually depicted in Figure 8.35d, unit current cells are steered either to the positive or to the negative output current, depending on the corresponding input digital code.

8.5.2 Practical Realization of CS DACs

Figure 8.36a shows an alternative implementation to Figure 8.35d, which provides a fully-balanced (complementary) differential output current with maximized power efficiency. As conceptually

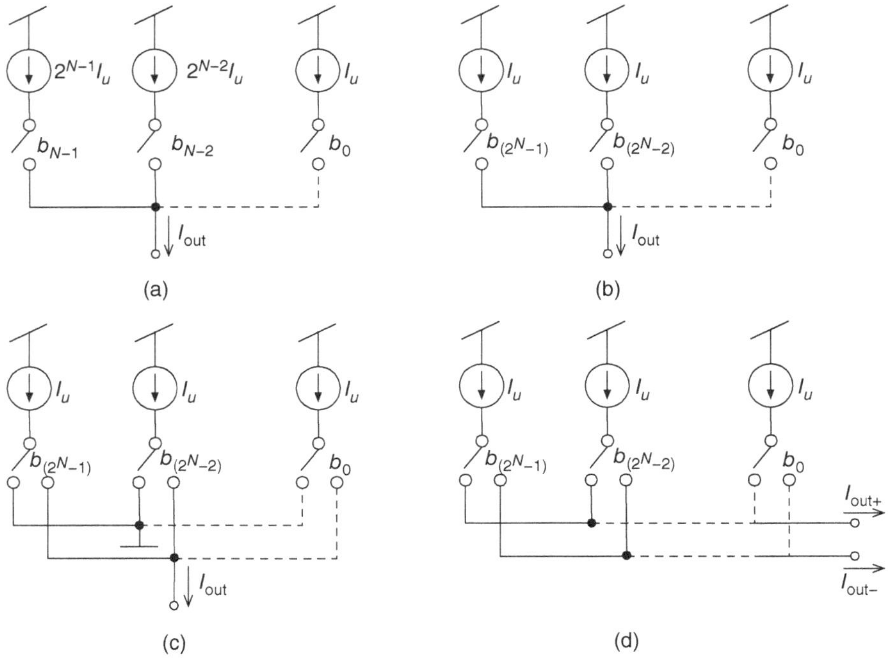

Figure 8.35 Conceptual schemes of switched-current DACs: (a) based on binary-weighted current cells; (b) based on thermometer-coded current cells;(c) single-ended current-steering DAC; (d) fully-differential current-steering DAC.

highlighted in the figure, two different kinds of current cells – p-type and n-type – are used for providing the current sources and sinks needed. The main difficulty with this DAC topology is to achieve the required voltage headroom for both types of current cell.

In order to palliate this limitation, many different DAC topologies have been proposed to relax the voltage headroom specification. The most common approach consists of keeping one of the current cell types fixed and the other switchable. This is conceptually illustrated in the CS DAC architectures depicted in Figures 8.36b and 8.36c. The former, which uses a fixed n-type current cell and switchable p-type current cells, allows for larger headroom for the (switchable) p-type current cells. The opposite situation is shown in Figure 8.36c. In both cases, the power efficiency is reduced to 50%. The choice of one of these two topologies will be conditioned by the voltage headroom requirement for each type of cell in a given design.

Figure 8.36d shows a CS DAC architecture that retains a power efficiency of 100% with relaxed specifications for the voltage headroom of both types of current cell. In this approach, p-type and n-type current cells are connected to different nodes, which are strategically chosen to provide the highest headroom voltage for each type of cell. This approach was proposed and successfully implemented by Crombez *et al.* [16]. In this case, illustrated in Figure 8.37, the p-type current cells are connected to the source node of the n-type cascode transistors of the Gm-C integrator output stage, while the p-type current cells are connected to the corresponding source terminal of the nMOS cascode transistors, thus maximizing the headroom for both types of cell. Note also that the source terminal of the cascode transistors provides a low-impedance node to the CS DAC output, thus reducing its output swing and making its design more robust [16].

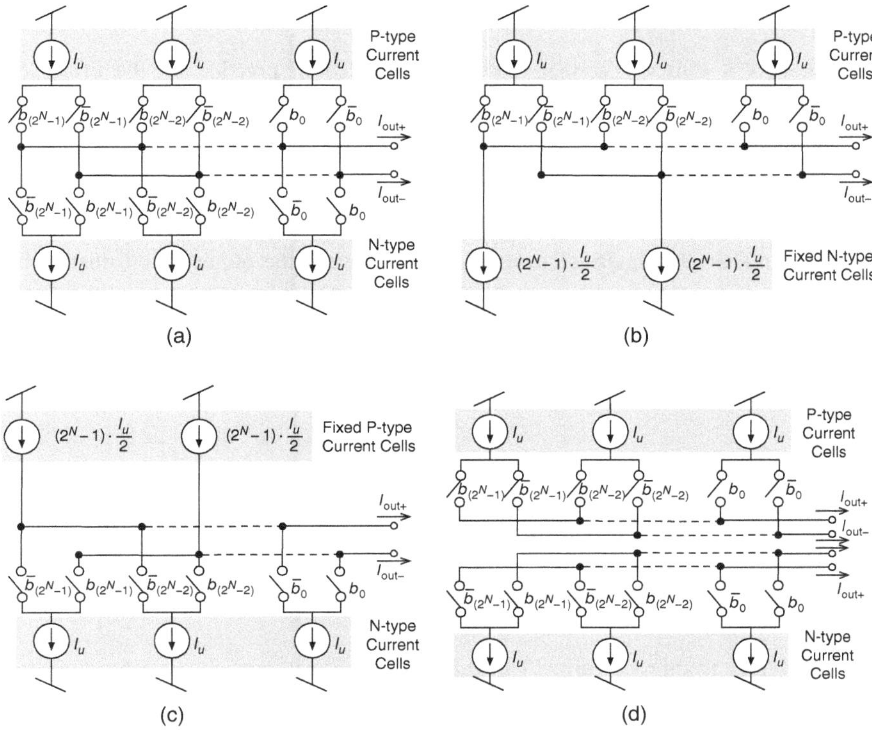

Figure 8.36 Conceptual schemes of fully-balanced (complementary) current-steering DACs considering: (a) switchable p-type and n-type current cells; (b) fixed n-type current cells; (c) fixed p-type current cells; (d) different output nodes for p-type and n-type current cells. P-type and n-type current cells are sometimes called current sources and current sinks, respectively.

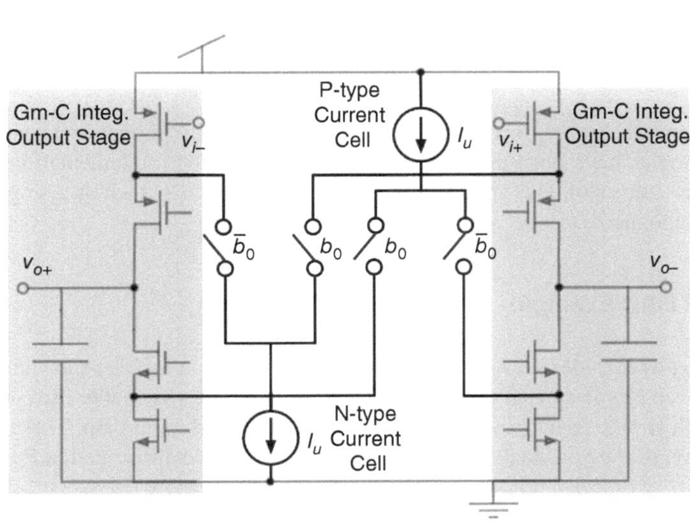

Figure 8.37 Illustrating the use of CS DACs with p-type and n-type current cells connected at different nodes of Gm-C integrators [16].

8.5.3 Current Cell Circuits, Error Limitations, and Design Criteria

The operation of CS DACs described above is degraded in practice by the effect of circuit implementation of switches and current cells. The main limiting factors[9] affecting the performance of CS DACs in ΣΔMs are caused by random errors due to device mismatch and systematic errors due to output impedance of the current cell, thermal gradients, layout edge effects, and incomplete settling.

Therefore, current cells are designed to satisfy a number of design specifications derived from behavioral simulations: the value of the output resistance, the transient response, mismatch, output swing, and so on. There is usually a strong trade-off between the required cell mismatch and the settling time. Thus, in order to relax the mismatch requirements without penalizing the linearity of the modulator, linearization techniques – such as DEM or digital calibration – are used (see Section 2.4).

Essentially, the following requirements should be taken into account when designing the current cells of CS DACs in CT-ΣΔMs:

- Reduce the mismatch among unit current cells.
- Increase the current cell output impedance, making use of cascode stages.
- Ensure that transistors operate in the saturation region for the specified integrator output swing. This is easier to achieve when the ΣΔM loop-filter integrators are active-RC, since the signal swing is limited at the virtual ground node of the OTA.
- Reduce the clock feedthrough error. This can be done by keeping parasitic capacitances as low as possible. An additional strategy to reduce this error consists of minimizing the voltage difference between the switch on-state and the switch off-state. This technique – often referred to as *soft driving*[10] – also reduces the overshoot currents due to the clock signal transitions and allows switches to operate in the saturation region as well, thus acting as cascode transistors to further boost the current cell output impedance.
- Reduce the CS DAC noise contribution to the required IBN specification of the modulator. Note that both thermal and flicker noise sources associated with current cells are summed at the ΣΔM input node, thus constituting an ultimate limiting factor.

In order to apply these design criteria to optimize power consumption, a huge number of different current cell topologies have been proposed in the literature on the implementation of CS DACs in CT-ΣΔMs. The circuit topologies span everything from basic cells – such as a single transistor – to simple cascode, regulated cascode, and so on.

8.5.4 CS 4-bit DAC Example

As a case study, let us consider a CS 4-bit DAC based on the circuit topology shown in Figure 8.36c. It consists of two fixed p-type current sources and 15 ($2^4 - 1$) n-type switchable current cells, which are controlled through simple nMOS switches by thermometer-coded input data. These data are entered into the DAC through a set of D-latches driving the switch gates, as illustrated in Figure 8.38.

[9]Design considerations of switches, as described in Section 8.1, must be also taken into account. In most practical cases, simple nMOS/pMOS switches satisfy the required specifications of CS DACs.

[10]One of the main disadvantages of using soft-driving techniques is that an additional level-shifter circuit is required to generate the switch control voltages.

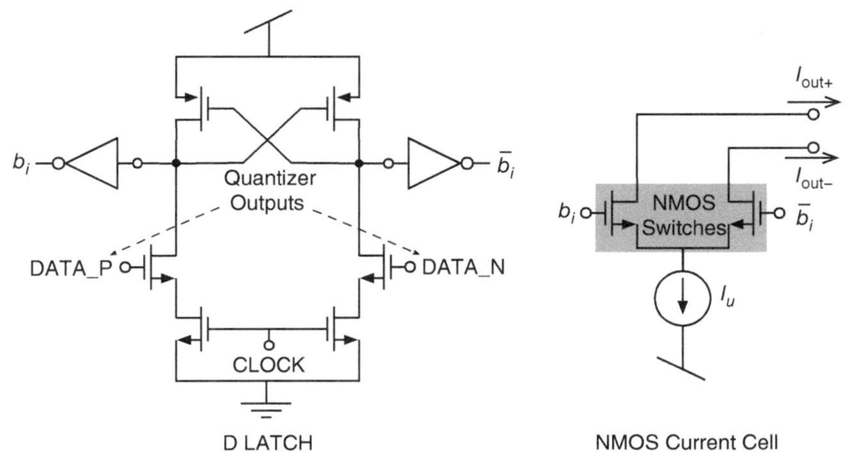

Figure 8.38 Illustrating the use of D-latches to connect the ΣΔM quantizer output and the nMOS switches of CS DACs.

The 4-bit CS DAC was used for implementing the NRZ feedback DAC of a fifth-order cascade 3-2 Gm-C CT-ΣΔM intended to digitize 20-MHz signals with 12-bit effective resolution, with a sampling frequency of 240 MHz [13]. These modulator specifications were mapped onto building-block specifications using SIMSIDES, as described in Section 5.6.2. As a result, the following specifications were defined for the CS DAC:

- output resistance: > 12 MΩ
- settling time: < 0.5 ns
- mismatch error: < 0.15% LSB

There is a strong design trade-off between the required mismatch error and the settling time error. This trade-off was relaxed in this design example by using DEM linearization techniques, which reduced the mismatch requirement to < 0.6% LSB. The unit current cell was chosen to be $I_u = 48$ μA, which leads to 360- μA (i.e., $7.5 I_u$) pMOS current sources.

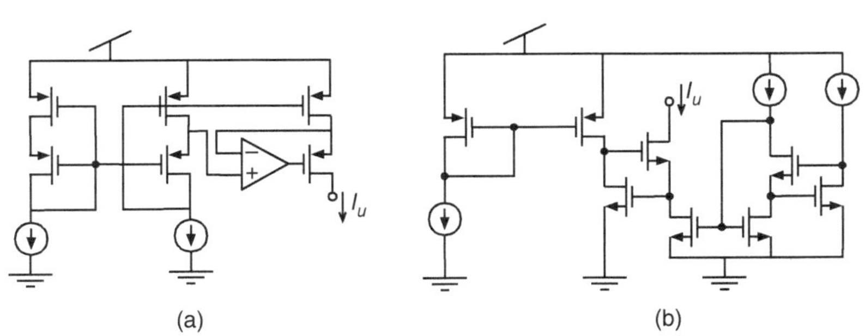

Figure 8.39 Examples of CS DAC current cells: (a) pMOS gain-boosted current cell; (b) nMOS regulated cascode current cell.

Another important limitation of the CS DAC in this design example was caused by the headroom voltage required for the pMOS current cells. As the common-mode voltage of the modulator was set to 0.75 V (forced by a design requirement of the loop-filter transconductors) and the FS voltage range was 0.3 V, then the resulting headroom voltages for the pMOS and the nMOS current sources were 0.3 V and 0.6 V, respectively. In order to satisfy all these specifications, a gain-boosted pMOS current source like that shown in Figure 8.39a was considered, while regulated cascode topologies, as depicted in Figure 8.39b, can be used for the nMOS current cells. These cells can be designed to satisfy the aforementioned specifications, while dissipating a reasonable power consumption of 0.49 mW and 0.1 mW, respectively for the pMOS and the nMOS cell [13].

8.6 Summary

This chapter presented a design guide with a collection of practical recipes to be taken into account in the electrical design and characterization of ΣΔMs. The most important design considerations of circuits used in ΣΔM building blocks were described, including amplifiers, transconductors, switches, comparators, and feedback DACs. Several design examples and simulation test benches were illustrated, putting emphasis on their implementation in Cadence IC Design.

The diverse considerations covered in this chapter are close to the end of the design flow of ΣΔMs discussed in this book, when going from ideal fundamentals and architecture considerations given in Chapter 1 and Chapter 2, to the impact of nonideal errors at system level discussed in Chapter 3 and Chapter 4, and to their application to the high-level synthesis and verification using behavioral models, simulation and optimization, as detailed in Chapter 5 and Chapter 6. This systematic top-down/bottom-up design methodology followed in this book will be completed in Chapter 9, going from circuit design to chip prototyping, measurement set-up and examples.

References

[1] R. del Río, F. Medeiro, B. Pérez-Verdú, J. M. de la Rosa, and A. Rodríguez-Vázquez, *CMOS Cascade ΣΔ Modulators for Sensors and Telecom: Error Analysis and Practical Design.* Springer, 2006.

[2] *The HSPICE Documentation Set.* Synopsys Inc., 2005.

[3] Y. Geerts, M. Steyaert, and W. Sansen, *Design of Multi-bit Delta-Sigma A/D Converters.* Kluwer Academic Publishers, 2002.

[4] S. Narayanan, "Application of Volterra series to intermodulation distortion of transistor feedback amplifier," *IEEE Trans. Circuit Theory*, pp. 518–527, November 1970.

[5] P. Wambacq, G. Gielen, P. Kinget, and W. Sansen, "High-Frequency Distortion Analysis of Analog Integrated Circuits," *IEEE Trans. on Circuits and Systems – II: Analog and Digital Signal Processing*, vol. 46, pp. 335–345, March 1999.

[6] W. Yu, S. Sen, and B. Leung, "Distortion Analysis of MOS Track-and-hold Sampling Mixers using Time-varying Volterra Series," *IEEE Trans. on Circuits and Systems – II: Analog and Digital Signal Processing*, vol. 46, pp. 101–113, February 1999.

[7] R. Schreier and G. C. Temes, *Understanding Delta-Sigma Data Converters.* IEEE Press, 2005.

[8] G. Mitteregger, C. Ebner, S. Mechnig, T. Blon, C. Holuigue, and E. Romani, "A 20-mW 640-MHz CMOS Continuous-Time ΣΔ ADC With 20-MHz Signal Bandwidth, 80-dB Dynamic Range and 12-bit ENOB," *IEEE J. of Solid-State Circuits*, vol. 41, pp. 2641–2649, December 2006.

[9] H. Shibata *et al.*, "A DC-to-1GHz Tunable RF ΔΣ ADC Achieving DR=74dB and BW=150MHz at f_0 =450MHz Using 550mW," *IEEE J. of Solid-State Circuits*, vol. 47, pp. 2888–2897, December 2012.

[10] T. C. Caldwell, D. Alldred, and Z. Li, "A Reconfigurable ΔΣ ADC With Up to 100MHz Bandwidth Using Flash Reference Shuffling," *IEEE Transactions on Circuits and Systems – I: Regular Papers*, vol. 61, pp. 2263–2271, August 2014.

[11] Y. Dong *et al.*, "A 72 dB-DR 465 MHz-BW Continuous-Time 1-2 MASH ADC in 28 nm CMOS," *IEEE J. of Solid-State Circuits*, vol. 51, pp. 2917–2927, August 2016.

[12] B. Thandri and J. Silva-Martinez, "A Robust Feedforward Compensation Scheme for Multistage Operational Transconductance Amplifiers With No Miller Capacitors," *IEEE J. of Solid-State Circuits*, vol. 38, pp. 237–243, February 2003.

[13] R. Tortosa, A. Aceituno, J. M. de la Rosa, A. Rodríguez-Vázquez, and F. V. Fernández, "A 12-bit, 40MS/s Gm-C Cascade 3-2 Continuous-Time Sigma-Delta Modulator," *Proc. of the IEEE Intl. Symp. on Circuits and Systems*, pp. 1–4, 2007.

[14] P. Allen and D. Holberg, *CMOS Analog Circuit Design*. Oxford University Press, 2nd ed., 2002.

[15] A. Morgado, R. del Río, and J. M. de la Rosa, "Design of a Power-efficient Widely-programmable Gm-LC Band-pass Sigma-Delta Modulator for SDR," *Proc. of the IEEE Intl. Symp. on Circuits and Systems (ISCAS)*, May 2016.

[16] P. Crombez *et al.*, "A Single-Bit 500 kHz-10 MHz Multimode Power-Performance Scalable 83-to-67 dB DR CT ΔΣ Modulator for SDR in 90 nm Digital CMOS," *IEEE J. of Solid-State Circuits*, vol. 45, pp. 1159–1171, June 2010.

[17] A. Morgado, R. del Río, and J. M. de la Rosa, "Energy Efficient Transconductor for Widely Programmable Analog Circuits and Systems," *Proc. of the IEEE Intl. Symp. on Circuits and Systems (ISCAS)*, May 2015.

[18] A. Rodríguez-Vázquez *et al.*, "Comparator Circuits," in *Wiley Encyclopedia of Electrical and Electronics Engineering*, pp. 577–600, John Wiley & Sons, 1999.

[19] B. Nikoli *et al.*, "Improved Sense-Amplifier-Based Flip-Flop: Design and Measurements," *IEEE J. of Solid-State Circuits*, vol. 35, pp. 876–884, June 2000.

[20] T. Kobayashi *et al.*, "A Current-controlled Latch Sense Amplifier and a Static Power-saving Input Buffer for Low-power Architecture," *IEEE J. of Solid-State Circuits*, vol. 28, pp. 523–527, April 1993.

[21] A. Yukawa, "A CMOS 8-bit High-Speed Converter IC," *IEEE J. of Solid-State Circuits*, vol. 20, pp. 775–779, June 1985.

[22] Y. Wang and B. Razavi, "An 8-Bit 150-MHz CMOS A/D Converter," *IEEE J. of Solid-State Circuits*, vol. 35, pp. 308–317, March 2000.

[23] G. Yin *et al.*, "A High-Speed CMOS Comparator with 8-b Resolution," *IEEE J. of Solid-State Circuits*, vol. 27, pp. 208–211, February 1992.

[24] A. Morgado, R. del Río, and J. M. de la Rosa, *Nanometer CMOS Sigma-Delta Modulators for Software Defined Radio*. Springer, 2012.

[25] W. Press *et al.*, *Numerical Recipes in C. The Art of Scientific Computing*. Cambridge University Press, 2nd ed., 1992.

[26] K. Kundert and O. Zinke, *The Designer's Guide to Verilog-AMS*. Kluwer Academic Publishers, 2004.

[27] S. Yan and E. Sánchez-Sinencio, "A Continuous-Time ΣΔ Modulator With 88-dB Dynamic Range and 1.1-MHz Signal Bandwidth," *IEEE J. of Solid-State Circuits*, vol. 39, pp. 75–86, January 2004.

[28] J. González and E. Alarcón, "Current-Steering High-Speed D/A Converters for Communications," in *CMOS Telecom Data Converters* (A. Rodríguez-Vázquez, F Medeiro and E Janssens, editors), Kluwer Academic Publishers, 2003.

[29] L. Breems, R. Rutten, and G. Wetzker, "A Cascaded Continuous-Time ΣΔ Modulator with 67-dB Dynamic Range in 10-MHz Bandwidth," *IEEE J. of Solid-State Circuits*, vol. 39, pp. 2152–2160, December 2004.

[30] M. Bolatkale *et al.*, "A 4GHz Continuous-Time ΔΣ ADC With 70dB DR and –74 dBFS THD in 125 MHz BW," *IEEE J. of Solid-State Circuits*, vol. 46, pp. 2857–2868, December 2011.

[31] K. Lee and R. Meyer, "A Current-controlled Latch Sense Amplifier and a Static Power-saving Input Buffer for Low-power Architecture," *IEEE J. of Solid-State Circuits*, vol. 20, pp. 1103–1113, December 1985.

[32] B. Razavi, *Design of Analog CMOS Integrated Circuits*. McGraw-Hill, 2000.

[33] A. Morgado, R. del Río, J. M. de la Rosa, L. Bos, J. Ryckaert, and G. van der Plas, "A 100kHz-10MHz BW, 78-to-52dB DR, 4.6-to-11mW Flexible SC ΣΔ Modulator in 1.2-V 90-nm CMOS," *Proc. of the IEEE European Solid-State Circuits Conf.*, pp. 418–421, September 2010.

[34] M. Felder and J. Ganger, "Analysis of Ground-Bounce Induced Substrate Noise Coupling in a Low Resistive Bulk Epitaxial Process: Design Strategies to Minimize Noise Effects on a Mixed-Signal Chip," *IEEE Trans. on Circuits and Systems – II: Analog and Digital Signal Processing*, vol. 46, pp. 1427–1436, October 1999.

[35] F. Maloberti, "Layout of Analog and Mixed Analog-Digital Circuits," in *Design of Analog-Digital VLSI Circuits for Telecommunications and Signal Processing* (J. Franca and Y. Tsividis, editors), Prentice-Hall, 1994.

[36] Y. Tsividis, *Mixed Analog-Digital VLSI Devices and Technology: An Introduction*. McGraw-Hill, 1996.

[37] S. Cao *et al.*, "ESD Design Strategies for High-Speed Digital and RF Circuits in Deeply Scaled Silicon Technologies," *IEEE Trans. on Circuits and Systems – I: Regular Papers*, vol. 57, pp. 2301–2311, September 2010.

[38] J. M. de la Rosa *et al.*, "A CMOS 110-dB@40-kS/s Programmable-Gain Chopper-Stabilized Third-Order 2-1 Cascade Sigma-Delta Modulator for Low-Power High-Linearity Automotive Sensor ASICs," *IEEE J. of Solid-State Circuits*, vol. 40, pp. 2246–2264, November 2005.

CHAPTER 9

PRACTICAL REALIZATION OF ΣΔMs: FROM CIRCUITS TO CHIPS

In addition to the circuits described in Chapter 8, additional circuitry is needed to complete a ΣΔM chip implementation. These additional or auxiliary circuits cover the generation of clock phases, common-mode voltage and bias currents, issues which are described in Section 9.1. The second part of the chapter deals with some of the most important design issues related to the layout (Section 9.2), chip prototyping (Section 9.3), and experimental verification (Section 9.4) of high-performance ΣΔM ICs.

The chapter concludes with Section 9.5, which includes some examples of ΣΔMs, considering both SC and CT circuit realizations, and covering different conversion regions of the resolution-versus-speed plane. The circuits described here are presented as case studies on the fundamentals and methods presented in earlier chapters, putting the emphasis on those design aspects that are critical for a particular application. The design criteria, modulator topologies, circuit limitations, and so on can be very different in ΣΔ ADCs intended for instrumentation, biomedical devices, communication systems, and so on. Therefore, the chapter points out relevant aspects of the circuit design rather than giving a detailed description.

9.1 Auxiliary ΣΔM Building Blocks

As stated earlier in this book, in addition to the basic building blocks used for implementing the loop filter and the quantizer of a ΣΔM, there are other subcircuits required to make an IC operative. This

Sigma-Delta Converters: Practical Design Guide, Second Edition. José M. de la Rosa.

section overviews the most important blocks, showing their fundamental schematics as well as some practical considerations that should be taken into account in their design.

9.1.1 Clock-phase Generators

As in any sampled-data system, the operation of ΣΔMs is governed by a clock signal. Usually, CT-ΣΔMs require one clock signal. In contrast, SC-ΣΔMs need to divide the clock period into several time intervals or *clock phases*, which are generated by a digital circuit commonly referred to as a *clock-phase generator*.

9.1.1.1 Phase Generation Figures 9.1a and 9.1b show two well-known digital circuits that are frequently used for generating the clock phases needed in SC-ΣΔMs. Essentially, the operation of both circuits is based on the use of bistable flip-flops to generate several periodic signals from an input signal, known as the master clock. The scheme in Figure 9.1a consists of a feedback loop made up of two NAND gates, each connected in series with a cascade of inverters. The number and sizes of these inverters provide the required clock-phase delays and nonoverlapping intervals.

Figure 9.1b shows an alternative implementation of the clock-phase generator, again based on the use of flip-flops [1]. In this case, there is not a *global* feedback loop and the different clock phases are generated by connecting D-type latches in cascade to provide the required delays and inversions.

In some high-frequency applications demanding a very high-precision, high-speed, low-jitter clock-phase scheme, the master clock signal used in Figures 9.1a and 9.1b is synthesized by an on-chip phase-locked loop (PLL) and an off-chip well-controlled crystal oscillator [2, 3], as conceptually depicted in Figure 9.1c.[1] Note that both clock drivers in Figures 9.1a and 9.1b generate two nonoverlapped clock phases, ϕ_1, ϕ_2, that control the sampling and integration operations of SC integrators. Delayed versions of the clock phases, ϕ_{1d}, ϕ_{2d}, are also generated in order to attenuate the error caused by signal-dependent charge injection produced during the turn-off process of input switches in SC integrators [5]. Complementary versions of the four clock phases, $\overline{\phi}_1, \overline{\phi}_2, \overline{\phi}_{1d}, \overline{\phi}_{2d}$ needed to control CMOS switches and some other ΣΔM subcircuits (such as latch-based comparators), are also synthesized by appropriately combining inverters and digital buffers, as illustrated in Figures 9.1a and 9.1b.

9.1.1.2 Phase Buffering All generated clock-phase signals need to be appropriately driven using a buffer tree like that conceptually shown in Figure 9.2. In this way, the differences in capacitive load among all phases become equalized. This is very important in practice because, if clock phases are not properly equalized, the different load capacitances connected at each clock phase will have a direct influence on the delays (and nonoverlapping intervals) among the phases, and may thus destroy the clock scheme generated by the circuits in Figures 9.1a and 9.1b, and consequently the operation of the SC-ΣΔM.

In order to design the clock-phase buffer tree, the parasitic capacitance loading each clock phase has to be accurately calculated. This piece of information can be extracted from electrical simulations by summing the input parasitic capacitances of all subcircuits (essentially CMOS switches and digital gates) to which the clock phases are connected.

[1] Some CT-ΣΔMs using clock signals in the gigahertz range use off-chip ultra-low-jitter signal sources and 180°-hybrid circuits to generate the master clock and its complementary signal [4].

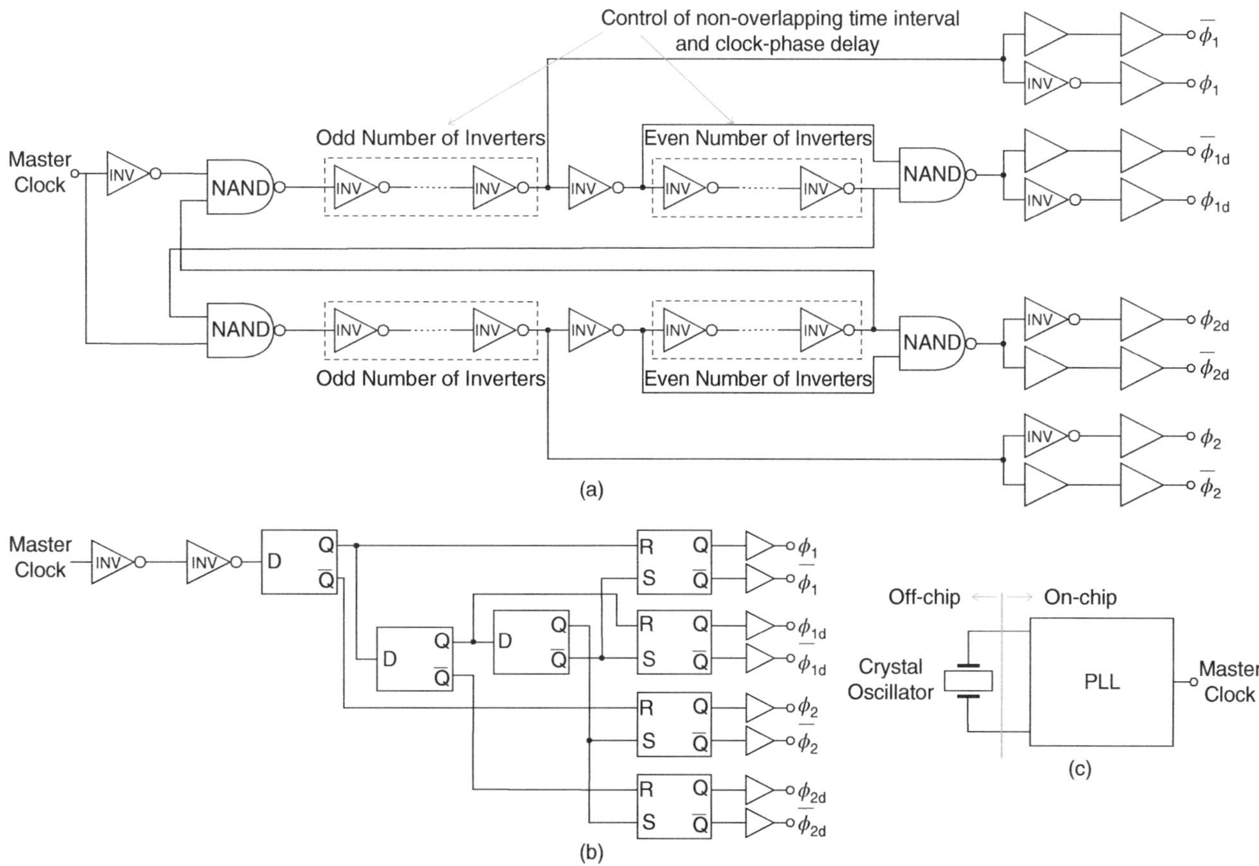

Figure 9.1 Conceptual schemes of clock-phase generators frequently used in SC-ΣΔMs: (a) feedback loop made up of two NANDs and a cascade of inverters; (b) D-latch based generator; (c) master clock generated by an (on-chip) PLL frequency synthesizer and an (off-chip) crystal oscillator. The use of on-chip PLLs allows generation of high-frequency, high-precision (low-jitter) clock signals, usually at the price of increased power consumption.

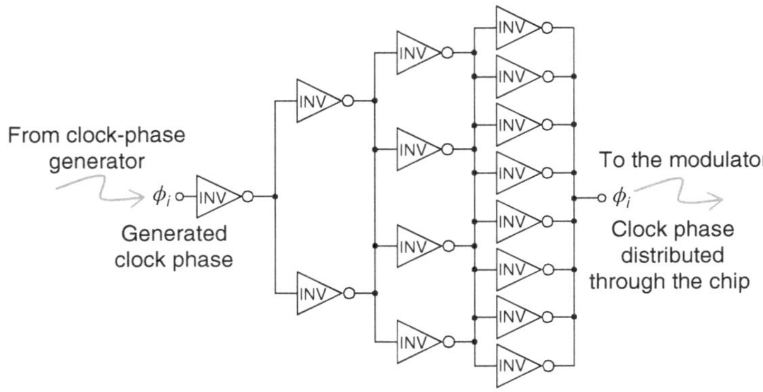

Figure 9.2 Example of buffer tree used for driving clock phases. The number and sizing of inverters depends on the fan-out requirements of each clock phase.

9.1.1.3 Phase Distribution

As clock phases are used in many different parts of SC-ΣΔMs, these signals need to be routed through the entire chip. To this end, a U-shaped bus – conceptually depicted in Figure 9.3a – is used. Each clock phase is isolated by implementing a Faraday cage with two ground walls at each side of the routed signals, as illustrated in Figure 9.3b. Note that each clock phase (ϕ_i) is closely routed with its complementary phase ($\overline{\phi}_i$). Both clock phases are surrounded by ground (GND in Figure 9.3b) strips of the same metal as that used for the clock phases. The whole bus is covered by the same ground above and below the routed phases, with plates implemented at the upper and lower metal layers, respectively.

Figure 9.3c shows a lumped LCR equivalent circuit that takes into account the circuit parasitics of the actual transmission line in Figure 9.3b. The values of L_T, C_T, and R_T can be extracted by using electromagnetic simulations that take into account the technological data of the different metal layers and dielectrics involved [6]. Note that post-layout simulations of the circuit in Figure 9.3c, together with the buffer tree described above, have to be carried out in order to optimize the design of the clock-phase generators in terms of silicon area and power consumption.

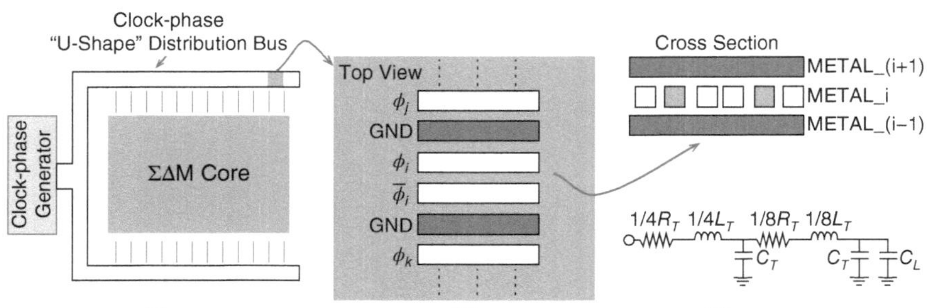

Figure 9.3 Clock-phase distribution and routing along a ΣΔM chip: (a) Conceptual U-shape bus distribution; (b) clock-phase signal isolation by using a Faraday cage; (c) equivalent circuit model extracted from electromagnetic simulations. In some practical situations, the inverted versions of clock signals are not globally routed, but locally generated by an inverter. Similarly, clock phases can be either globally or locally buffered depending on the fan-out requirements in each case [6].

9.1.2 Generation of Common-mode Voltage, Reference Voltage, and Bias Currents

The following reference and bias voltages are needed for correct operation of ΣΔMs:

- **reference voltage**, used in the embedded quantizer: the reference ladder of the flash ADC and the feedback DAC;
- **common-mode voltage**, extensively used by all ΣΔM subcircuits
- **bias currents**, required to bias all ΣΔM building blocks.

These reference and bias voltages and currents must have little dependence on the temperature, supply voltage, and technology process parameters in order for the ΣΔM to be robust. To this end, dedicated circuits have to be incorporated in ΣΔM ICs in order to generate these DC quantities.

9.1.2.1 Bandgap Circuit As in other analog ICs, the majority of ΣΔMs generate their internal reference voltages and bias currents from a DC temperature-independent voltage. This voltage is generated using well-known *bandgap*-reference generator circuits, often referred to as *bandgap circuits*.

Figure 9.4 shows an example of a bandgap circuit [7] used in ΣΔMs [6]. The circuit makes use of lateral bipolar transistors, which are usually available in standard CMOS technologies, to generate a DC voltage given by [7]:

$$V_{bg} = V_{EB1} + U_T \cdot \ln n \cdot \left(1 + \frac{R_2}{R_3}\right) \tag{9.1}$$

where V_{EB1} is the forward-biased emitter-base voltage of transistor Q_1, $U_T \equiv kT/q$, and $U_T \cdot \ln n$ stands for the difference between the emitter base voltages Q_1 and Q_2.

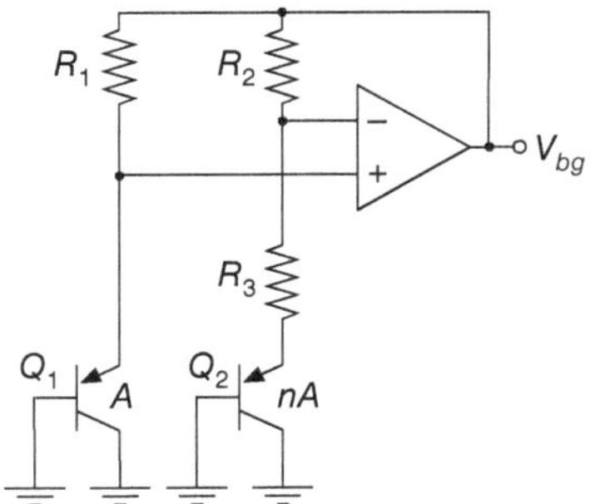

Figure 9.4 Example of bandgap circuit [7] used in some ΣΔMs [6].

Considering that $\frac{\partial V_{EB}}{\partial T} \simeq -1.5$ mV/K, and $\frac{\partial U_T}{\partial T} \simeq 0.087$ mV/K, a bandgap voltage (V_{bg}) with zero temperature coefficient can be achieved if $\ln n \cdot (1 + R_2/R_3) \simeq 17.2$. Taking this condition into account and assuming that $V_{EB1} \simeq 0.8$ V in Equation (9.1), a bandgap voltage of $V_{bg} \simeq 1.25$ V can be generated [7].

9.1.2.2 Reference Voltage Generator The reference voltage V_{ref} required for the modulator operation can be obtained as a linear function of the bandgap voltage,

$$V_{\text{ref}} = V_{\text{r+}} - V_{\text{r-}} = \alpha V_{bg} \tag{9.2}$$

where α is a proportionality factor. For instance, if $\alpha = 4/5$, then a reference voltage of $V_{\text{ref}} = 1$ V is obtained. This can be easily implemented[2] using a fully-differential amplifier in inverting configuration, like that shown in Figure 9.5. In this way, by simply choosing $R_2 = 4/5R_1$, a reference voltage of $V_{\text{ref}} = 1$ V is obtained. Note that a buffer – which can be implemented using a simple (asymmetric) OTA circuit – is used for driving the bandgap voltage V_{bg}.

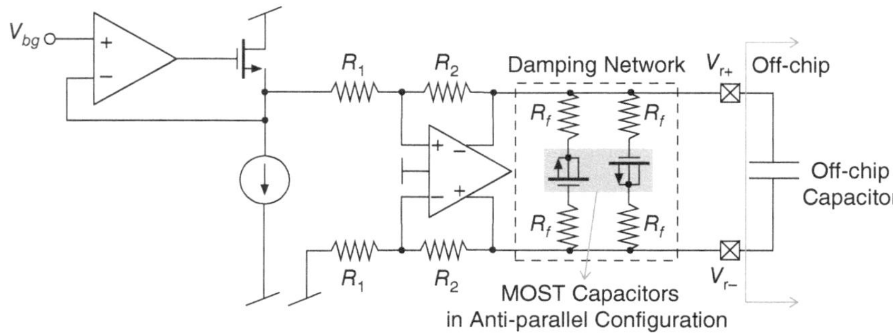

Figure 9.5 Example of reference voltage generator circuit [6].

The main design considerations that must be taken into account for the generation of reference voltages in ΣΔMs are a fast dynamic response (settling) as well as a low output impedance between the V_{r+} and V_{r-} lines, so that no dynamic distortion is introduced in the loop-filter integrators. To this end, the example shown in Figure 9.5 uses an off-chip capacitor connected between V_{r+} and V_{r-}. The value of this capacitor must be chosen according to the parasitic capacitances connected at these nodes (due to the bonding pads, bonding wires, lead frame, and package pin), so that the spurious components around half the sampling frequency are removed from the differential reference voltage. In addition, a damping network, which is made up of an RC circuit based on MOS capacitors connected in antiparallel configuration, is used for removing ringing voltages added to the reference voltages [6].

9.1.2.3 Master Bias Current Generator All current sources and sinks required to bias the ΣΔM subcircuits (opamps, comparators, and so on) need to be generated internally (on-chip) from a master bias current by a single circuit, commonly referred to as *master bias current generator*. Figure 9.6 shows an example of a master bias current generator where a single master current is generated from the bandgap voltage and an external (off-chip) resistor. This resistor can be also implemented on-chip using unsalicided poly resistors. The generated master bias current is mirrored and scaled so as to bias all amplifiers (used in integrators), the preamplifying stages of comparators, as well as other auxiliary (analog) ΣΔM building blocks such as the reference voltage generator and the common-mode voltage generator (described in Section 9.1.2.4).

In some applications, adaptive bias currents are implemented by means of programmable current mirrors, based on a combination of switchable transistors and unit resistors. In this way, the performance of the core amplifiers, and consequently the ΣΔM, can be adapted to a different set of specifications with optimized power consumption [8].

9.1.2.4 Common-mode Voltage Generator The common-mode voltage V_{CM} is usually defined as a half of the supply voltage; that is, $V_{CM} = (V_{DD} - V_{SS})/2$. This operation can be easily

[2]In a design prototype, ΣΔM reference voltages can be optionally provided by an off-chip circuit included in a test PCB, as will be discussed later. However, this solution is neither practical nor robust if the ΣΔM is embedded in a chip together with other circuit components which form a given electronic system.

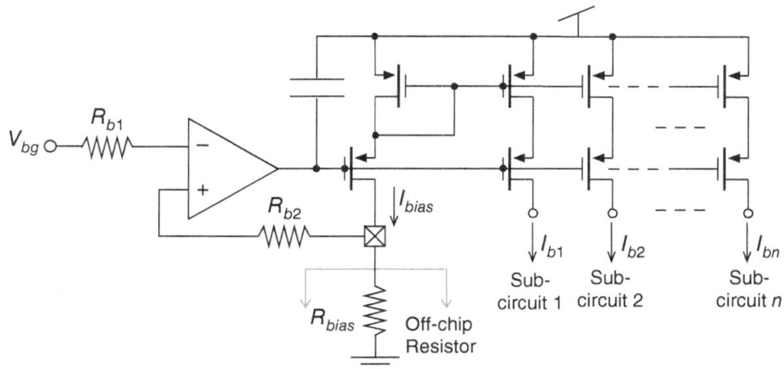

Figure 9.6 Example of master bias current generator. The resistor used for generating the master bias current R_{bias} can alternatively be implemented on chip using unsalicided poly resistors.

performed using a resistor divider and a buffer, as illustrated in the example shown in Figure 9.7. This circuit implements the required ratio of $1/2$ in a simple and robust way, using two identical resistors and a simple OTA configured as a buffer. Similar to the case of the reference voltage generators, a large (off-chip) capacitor may be used in combination with an on-chip damping network in order to "clean" the generated voltage and keep it constant and stable despite the switching-noise activity propagated across the substrate.

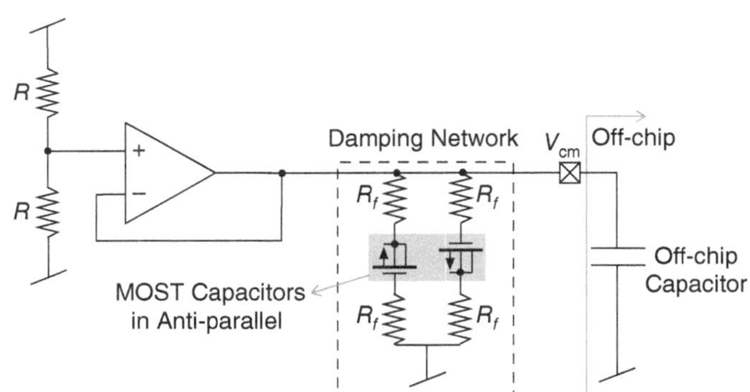

Figure 9.7 Example of common-mode voltage generator circuit [6].

9.1.3 Additional Digital Logic

Apart from the extra analog circuits described above, some ΣΔM ICs require the incorporation of additional digital circuits in order to carry out specific tasks. Among others, the following digital signal processes may be necessary in some applications:

- Output digital buffers driving the modulator outputs in order to drive the load capacitance due to either the bonding pads (in stand-alone IC implementations) or the decimation filter connected at the modulator output.

- Multi-bit DAC linearization techniques, for instance based on DEM algorithms such as DWA.
- Decoders used in quantizers, such as thermometric-to-binary decoders, frequently used in embedded flash ADCs.
- Serial-to-parallel registers, sometimes used for managing a large number of digital control signals, for instance in reconfigurable ΣΔMs.
- Digital power-down signals, used for turning off/on some parts of the circuits when needed in order to optimize the power consumption.

The practical implementation of these digital circuits will greatly depend on the specific purpose and application. The detailed description of the diverse techniques that can be used is beyond the scope of this book. The interested reader can find many examples of these logic circuits in a number of state-of-the-art references collected at the end of Chapter 10.

9.2 Layout Design, Floorplanning, and Practical Issues

As in any other mixed-signal IC, layout implementation is one of the most important steps in the design process of ΣΔMs. Indeed, as will be illustrated in this section, the performance of a ΣΔM can be completely destroyed if the layout is not carefully designed. To this end, a number of design strategies and practical tricks must be followed. Some of these layout techniques are based on general rules used in the design of analog IC circuits, while others are specific to the design of ΣΔMs. This section gives an overview of the most important and critical recommendations to give a high-quality ΣΔM layout.

9.2.1 Layout Floorplanning

At the very beginning of the layout design stage, it is essential to make an appropriate partitioning and placement or *floorplan* of the different parts that constitute the layout of the ΣΔM. This floorplan must take into account the recommendations given below.

9.2.1.1 Divide Layout into Different Parts or Regions The most common layout has three different regions corresponding to the analog, mixed-signal, and digital parts of the modulator:

- **The analog part** should include all analog subcircuits of the ΣΔM core, such as OTAs, the preamplifiers used in the latched comparators, capacitors, resistors, inductors, as well as any other auxiliary analog circuit; i.e., master bias current generator, reference voltage generator, and so on. This part of the layout must also include any critical parts that may affect the performance of the modulators – such as control circuits to implement reconfiguration techniques – power-down switches, and any other control (analog) circuits.
- **The mixed-signal part** usually includes CMOS switches (in SC-ΣΔMs), the latches used in the comparators, as well as any other ΣΔM subcircuit handling both analog and digital signals.
- **The digital part** includes the clock-phase generator, digital buffers, as well as any other digital logic circuit required for the operation of the modulator, such as DEM logic, decoders used in the quantizers, digital registers, and so on.

It is very important to keep in mind that there is not a direct correspondence between the schematic building blocks and their corresponding parts in the layout of the modulator. For instance, an SC integrator is made up of three essential circuit elements: opamps, capacitors, and CMOS switches. However, the first two parts (opamps and capacitors) are placed in the analog region of the layout, whereas CMOS switches are included in the mixed-signal region. So the routing between these parts is very important because their associated parasitics may severely degrade the performance of the modulator. Other examples include embedded multi-bit flash ADCs. These circuits are made up of a resistor ladder and a bank of comparators. The former are included in the analog part, while the latter are in turn subdivided into three different circuit blocks: preamplifiers, latches, and SR flip-flops (see Figure 8.27), respectively included in the analog, mixed-signal, and digital sections of the layout.

9.2.1.2 Shield Sensitive ΣΔM Analog Subcircuits from Switching Noise The placement of analog and digital parts should be carried out in such a way that there is an increased distance between the most sensitive analog blocks and the noisy digital parts. Nevertheless, given that the majority of standard CMOS technologies have a low-resistivity substrate, the switching activity of digital circuits may severely degrade the performance of the chip. Therefore, additional layout techniques are frequently used to attenuate the impact of noisy signals propagating across the substrate. Some of these techniques are:

- **Use guard rings** with dedicated bonding pads and pins surrounding each section of the circuit. Although these well-known techniques are not so effective in low-resistivity bulk epitaxial processes, they provide some attenuation, at least for the noisy surface currents propagating in the epitaxial layer [9].
- **Use separate voltage supplies** for the different parts of the modulator. This strategy implies using dedicated power supplies (V_{DD} and V_{SS}), each with its own bonding pad and chip package pin for the analog, mixed-signal, and digital parts, as well as for the guard rings.
- **Extensive use of decoupling capacitors** throughout the chip for each supply voltage (V_{DD}) and its corresponding ground (V_{SS}). This well-known technique allows to keep supply voltages clean.

9.2.1.3 Buses to Distribute Signals Shared by Different ΣΔM Parts There are a number of signals that are shared by a number of ΣΔM building blocks, and consequently must be distributed across the entire chip. These signals include, among others, the digital clock phases, DAC control signals, power-down signals, supply voltages, bias currents, reference voltages, and common-mode voltages. All these signals must be routed based on a U-shaped bus configuration similar to the one in Figure 9.3.

9.2.1.4 Be Obsessive about Layout Symmetry and Details of Analog Parts The design of the layout of the critical parts of the modulator, particularly the subcircuits forming the loop filter, must pay attention to all minor details and should follow useful layout rules for high-performance analog circuits, including:

- Use common-centroid layout structures with unit circuit elements (capacitors, resistors, transistors) in order to maximize the matching performance.
- Use fully-differential topologies to reduce common-mode interference.

- Use multiple contacts and vias in order to reduce the parasitic resistances associated with each connection and to avoid catastrophic failures derived from microfractures in the metal connections.
- Use single-finger transistors to build CMOS switches. This strategy helps to avoid crossings among digital signals (clock phases connected at the gate of transistors) and analog signals (connected at the drain/source terminals).[3]
- Optimize the width of metal connections, taking into account the maximum current density flowing through each metal path and their parasitic resistances and capacitances.
- Use stacked metal layers to reduce the parasitic resistances.

As an illustration of the recommendations above, Figure 9.8 shows the layout floorplanning of a cascade 2-1 SC-ΣΔM. A number of the aforementioned rules are highlighted. Note that, in addition to the previous recommendations, there are many other basic layout rules that should be followed in order to maximize the performance of analog circuits. The interested reader can find plenty of excellent papers and books dealing with this topic in the literature [7, 10, 11].

9.2.2 I/O Pad Ring

The design of the I/O pad ring enclosing the ΣΔM core layout is also very important to guarantee correct operation of the modulator. A number of practical rules must be followed in order to avoid any performance degradation caused by nonideal behavior associated with the design, placement, and/or routing of bonding pads. Among other things, the following guidelines should be taken into account:

- Divide the pad ring into different parts (analog, mixed-signal, digital, and so on) to further improve the isolation among the different ΣΔM regions and to avoid switching noise coupling. To this end, divide the power ring using power-cut cells – cells that provide a virtual cut through diodes in antiparallel configuration – as conceptually depicted in Figure 9.9 [12].
- Place the switching (noisy) pads as far as possible from the most sensitive analog pads.
- Use pads with ESD protection where needed, for instance for signals driving transistor gates.

As an illustration, Figure 9.10 shows the conceptual floorplanning of an I/O pad ring, highlighting some of the important strategies described above.

9.2.3 Importance of Layout Verification and Catastrophic Failure

As is well known, layout CAD tools, such as the Design Rule Checker (DRC) and Layout Versus Schematic (LVS) tool, are very useful for designers, ensuring that their layouts are free of errors. In addition, layout-extracted simulations including technology parasitics are also convenient in ensuring correct performance of the chip before sending it for fabrication.

[3]If single-finger transistors are used, the analog signals can be directly routed to the transistor diffusions, while the digital signals can be connected to the transistor gate either using polysilicon or metal layers.

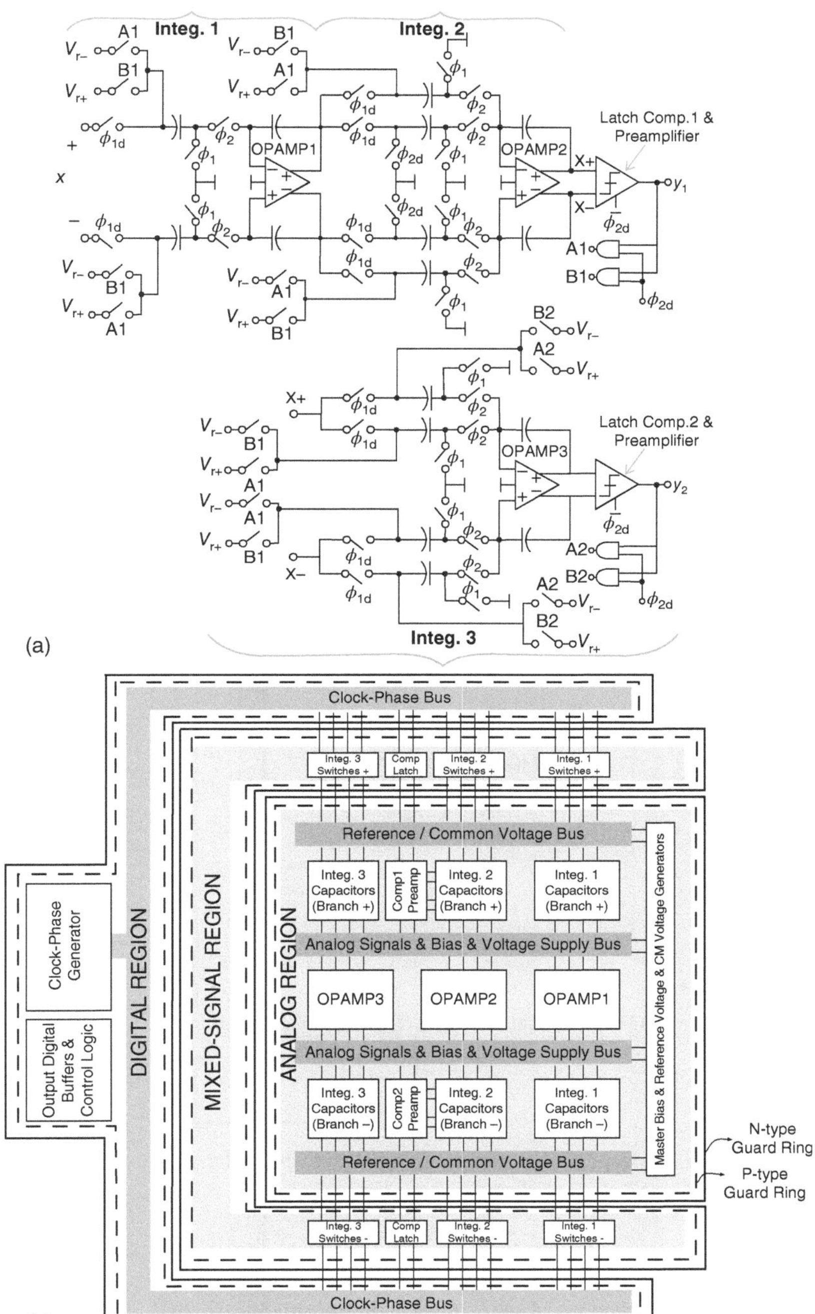

Figure 9.8 Example of layout floorplanning of ΣΔMs: (a) schematic of the modulator (cascade 2-1 SC-ΣΔM); (b) layout floorplanning. Each layout region as well as its corresponding guard rings have a dedicated supply voltage.

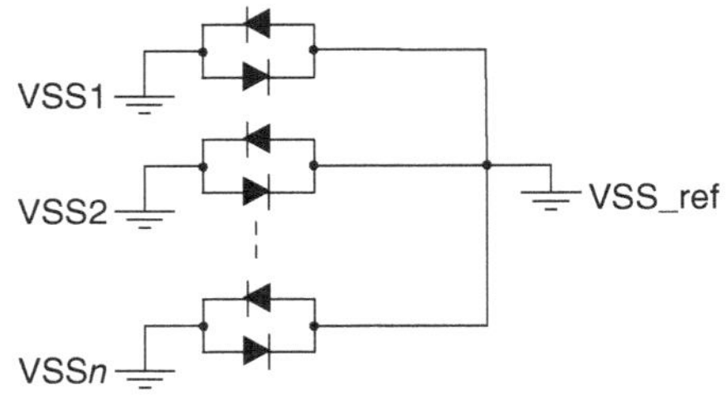

Figure 9.9 Conceptual illustration of the virtual connection among different grounds in the power ring using diodes in antiparallel configuration.

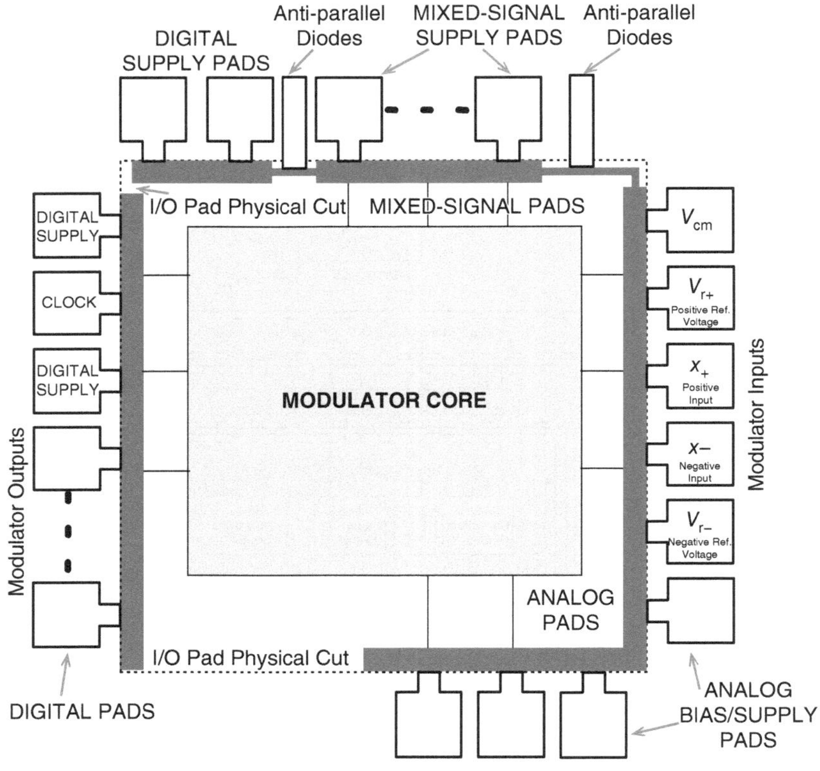

Figure 9.10 Conceptual illustration of an I/O pad ring that can be used in ΣΔMs.

In spite of the usefulness of these tools, in many practical situations – particularly in industrial first-silicon prototypes – it is very common to work with technologies that are still under development.[4] However, one of the consequences of using such new technologies is that the design kits (i.e., the technology files including electrical device models, layout rules, and so on) are also under development, which adds extra work for designers. For instance, it is relatively common that parasitic extraction

[4]Often, industrial parters working in a given project are interested in testing high-performance analog circuits like ΣΔMs in a cutting-edge technology.

tools such as Layout Parasitic Extractors (LPEs) are not available, and consequently, designers have to be very careful and conservative during the design process, and especially at the layout stage.

In this scenario, it is especially important to pay attention not only to the error messages provided by CAD tools (such as DRC/LVS) but also to the warning messages. The latter may look insignificant – particularly to novice designers – but these apparently minor problems may cause catastrophic failures. To illustrate this issue, which may be encountered in many practical scenarios, let us consider again the SC-ΣΔM shown in Figure 9.8a. The circuit implementation of this modulator involves over one thousand transistors, necessitating tedious and careful design and layout verification.

Let us assume that, due to a design error, the NWELL enclosing the pMOS transistors of the CMOS switches highlighted in Figure 9.11a are floating; that is, there is not a good contact. This error leads to generation of a warning message by the DRC/LVS tools. Assuming that there is no

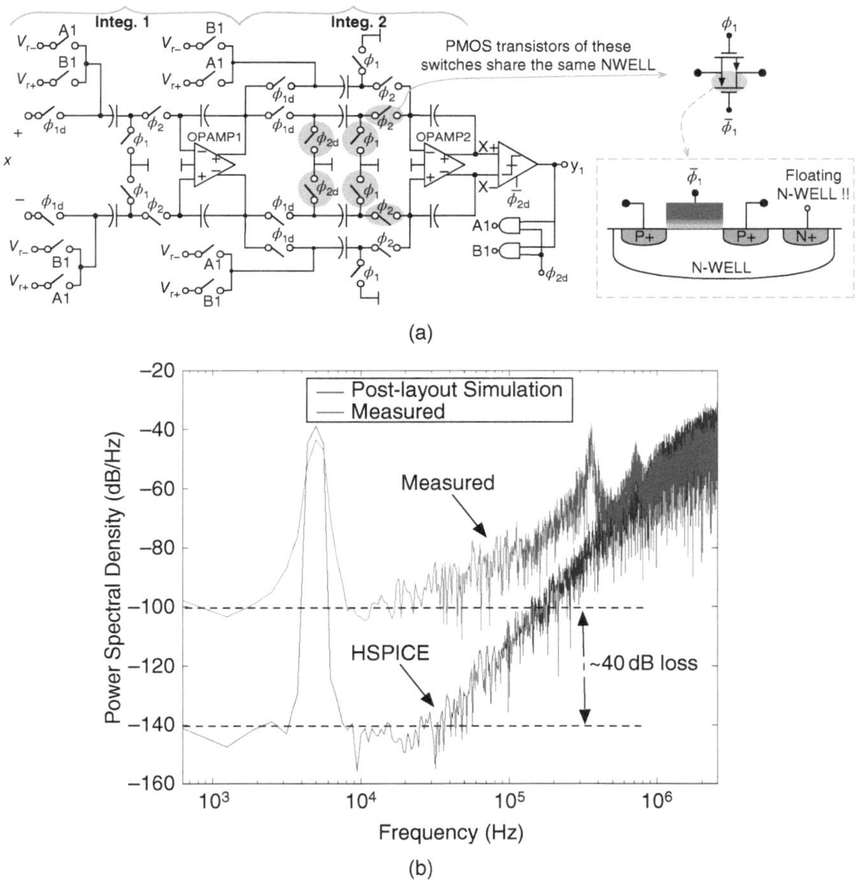

Figure 9.11 Illustrating the performance degradation in the SC-ΣΔM shown in Figure 9.8a caused by a floating NWELL: (a) SC schematic of the front-end stage of Figure 9.8a, highlighting the CMOS switches with floating NWELL; (b) experimental and simulated modulator output spectra, showing how the noise-shaping performance becomes severely degraded as a consequence of the floating NWELL.

LPE available, the layout-extracted netlist is essentially the same as that of the schematic. Therefore, transistor-level simulations will give good results, masking an error which obviously affects the signal transmitted from the first integrator to the second integrator. As a consequence, as illustrated in Figure 9.11b, experimental measurements reveal a severe performance degradation as compared to the transistor-level simulations. In other words, a single missing connection in an NWELL enclosing only six pMOS transistors may completely destroy the performance of a ΣΔM made up of more than one thousand transistors.

In conclusion, the layout phase of ΣΔMs is a critical stage of the design procedure and it must be carefully verified and checked using verification CAD tools, and without underestimating any warning message or apparently minor issue that may become the cause of a catastrophic failure.

9.3 Chip Package, Test PCB, and Experimental Setup

The last stage of the design phase of ΣΔMs deals with a number of tasks required to prepare the chip for the experimental testing of its performance in a lab. The most important considerations are the bonding diagram and chip package, the test PCB, and the experimental setup. Some of these issues must be addressed before sending the chip for fabrication. This section describes the most important practical issues to be considered regarding the packaging, prototyping, and testing of ΣΔMs, assuming that they are going to be measured as stand-alone ICs. Similar recommendations can be followed when the ΣΔM is a system-on-chip (SoC) implementation.

9.3.1 Bonding Diagram and Package

The bonding of the modulator chip has to be closely studied, with regards to the pad-ring placement described in previous section, the number of pads and pins assigned to supplies, the guard rings, reference voltages, and so on. Therefore, the selection of the most suitable chip package is critical because the parasitics associated with the package and its bonding connections to the chip may severely affect the performance of the modulator, particularly when high speed and/or high resolution are demanded.

The effect of package and bonding-wire parasitics can be analyzed in detail using dedicated CAD tools (such as Cadence Allegro), which allow simulation of the circuit while considering these parasitic effects. Some tools, like Cadence PKG, allow modeling of the package by an equivalent LCR circuit synthesized from the package's physical and electrical characteristics. The most important limitations come from the parasitic bonding inductances. It is recommended to use surface-mount devices such as ball grid arrays (BGAs) or quad flat packages (QFPs) – the latter being commonly used for testing ASIC prototypes in general and ΣΔ ICs in particular.

Some critical considerations must be also taken into account. On the one hand, bonding pads of the same type (analog, mixed-signal, digital, or digital I/Os) must be placed together and apart from the rest using *power-cut* diode cells like those shown in Figure 9.9. This allows power separation and avoids cross-talk. Moreover, supply pads and pins should be placed in parallel to reduce the total parasitic inductance due to the compensation given by the complementary mutual inductances [7]. On the other hand, double-bonding techniques and multiple pins are frequently used for the supplies of the different sections in order to reduce the inductance of the paths to the chip and decrease supply bounce. Note that if reference voltages are provided externally (off-chip), double bonding should be also used for the reference pads/pins in order to halve their bonding parasitic inductances.

9.3.2 Test PCB

In order to characterize the ΣΔM IC, an special-purpose PCB is needed in order to connect the chip to the different test instruments, providing the necessary signals, biasing, and supply voltages, as well as capturing the modulator output data for further processing in a computer. Due to the aforementioned practical limitations associated with the chip packages, other testing approaches are sometimes considered, particularly in very high-speed applications. For instance, the modulator die can be directly bonded to the test PCB, without using any package, in order to reduce the dimensions of bonding and consequently the parasitic inductances [6]. Alternatively, low-temperature co-fired ceramic substrates – commonly used in RF applications – can be used for reducing the effect of off-chip circuit element parasitics, since these elements can be embedded together with the modulator die into the same ceramic package.

Let us consider the most common approach, which consists of a chip package connected to the necessary circuits for testing in a multilayer PCB. Figure 9.12 shows the conceptual schematic of a PCB used for testing the ΣΔM of Figure 9.8 [13]. In this example, a 32-pin QFP package is used. Note that, in addition to the effect of circuit parasitics themselves, external electromagnetic interferences are injected into the ΣΔM chip through both inductive and capacitive coupling. In order to reduce the impact of off-chip circuit parasitics and to obtain a robust test circuit, the following circuit strategies are usually incorporated in the test PCB (some of them highlighted in Figure 9.12):

- **Separate the PCB into different areas or *planes*** corresponding to the analog, mixed-signal, and digital signals. Ground planes should be separated – usually with a gap larger than 1/8″ – and connected only at one point. In this way noisy return currents are minimized.
- **Use regulators to keep the values of the supply voltages stable**.
- **Use decoupling capacitors in the supply, biasing, and reference voltage lines**. It is common practice to combine a large tantalum capacitor with a small ceramic capacitor connected in parallel, in such a way that the ceramic capacitor is placed as close as as possible to the package pin. Both capacitors are usually connected together with one inductor in a π-filter configuration, as illustrated in Figure 9.12.
- **Keep digital signal paths as far as possible from the sensitive analog pins**.
- **Use termination resistors** for impedance coupling in the digital output lines.
- **Use ESD protection diodes** for sensitive input pins, particularly if they are not used in the corresponding bonding pads.

Apart from the aforementioned techniques, PCBs used for testing ΣΔMs should contain an AAF. A low-order (typically first- or second-order) RC filter is enough in the majority of practical situations.

9.4 Experimental Test Set-Up

The test set-up and instruments used for measuring the performance of ΣΔMs in the laboratory are also very important and need to be carefully planned. The number and type of lab equipment depends on different factors, including the nature of the signal to be tested (low-pass, band-pass, sinewave, modulated signal, and so on), the target modulator specifications (in-band noise power, linearity, and so on), the type of performance metrics to be measured (output spectrum, SNR/SNDR, HD3, IM3, INL, and so on), and so on.

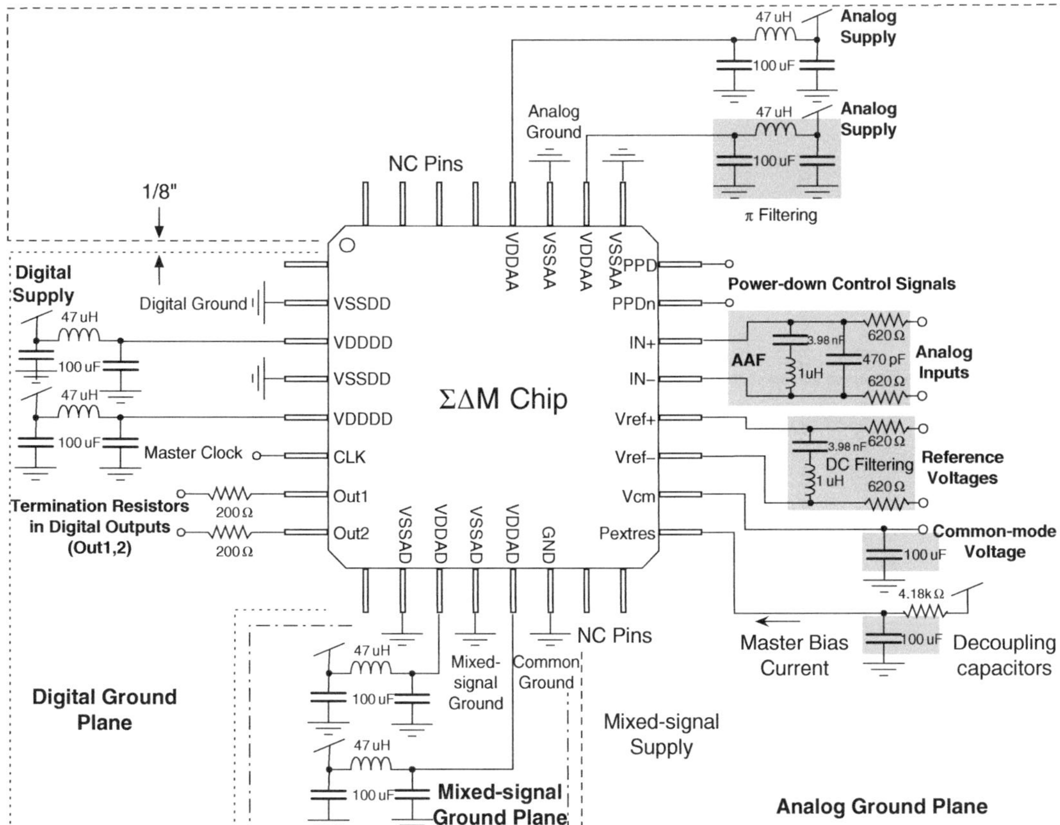

Figure 9.12 Conceptual schematic of a test PCB used for measuring the performance of the modulator in Figure 9.8. Note that this is a simplified version of the modulator presented by de la Rosa *et al.* [13], which included a programmable-gain front-end integrator and a preamplifier connected to the modulator. The additional pins required to include these circuits have been omitted for the sake of simplicity. Moreover, the values of circuit elements in this test chip correspond to the signal specifications reported in the same paper [13]: $B_w = 20$ kHz and DR = 110 dB.

9.4.1 Planning the Type and Number of Instruments Needed

It should be noted that every additional circuit element or lab instrument included in the measurement set-up is a potential source of error and interference that may degrade the modulator performance. For that reason, it is very important to think in advance – preferably during the design phase – about the kind of measurements that will be needed and the type and number of instruments that will be required. In many practical situations, the performance of the chip cannot be properly characterized experimentally due to the limitations of the lab equipment and instruments. Some examples of these limitations are the jitter error of the clock generators, the maximum frequency/linearity provided by the signal generator, and the maximum capture rate of the logic analyzer.

Generally speaking, and regardless of the specific measurements to be carried out, the following instruments at least are commonly needed for testing ΣΔMs:

- **Power supply generators** to generate the supply voltages, reference voltages, common-mode voltage as well as any other DC voltage or bias current required. Whenever possible, it is highly recommended to use a **voltage regulator circuit** – embedded in the test PCB – to generate all DC and bias signals from a single voltage supply to minimize the number of instruments and wires.
- **Analog (input) signal generators**, at least one sinewave generator, if possible providing balanced fully-differential signals with the required bandwidth and accuracy – in terms of noise and linearity. Note that the same common-mode voltage generator must be used for both the signal generator and the modulator chip. Otherwise, a systematic offset will be introduced at the modulator inputs, which may severely degrade the performance of the modulator.
- **Clock generator** with the required performance in terms of frequency, logic levels, and clock jitter error.
- **Data acquisition systems**, like logic analyzers or SoC test units (for instance Agilent 93000), are essential for capturing the modulator output bitstreams, which are transferred to a personal computer or to a workstation in order to process the data.[5]

Apart from these "essential" instruments, other lab equipment may be needed: spectrum analyzers to check the frequency spectrum of a given signal "on the fly", multimeters to measure the DC operating point, and so on.

9.4.2 Connecting Lab Instruments

The connection of the different instruments to the test PCB is critical, and must be implemented in such a way that the parasitics are minimized. The following recommendations must be followed:

- **Use the appropriate connectors** for each instrument in the PCB, as illustrated in Figure 9.13.
- **Reduce the length and number of cables** used for connecting the instruments.

[5]Many logic analyzers today have an embedded PC, so that the same instrument is used for capturing the ΣΔM output data and for processing this data, for instance using MATLAB. Alternatively, various logic analyzers and SoC test units can be used for generating the input signal waveforms, supply voltages, digital control signals, and so on.

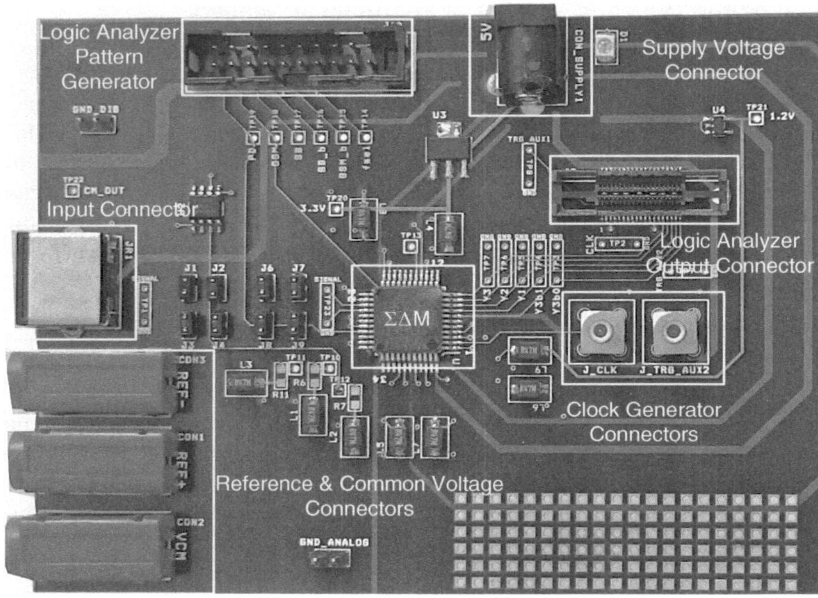

Figure 9.13 Illustrating some connectors of different instruments in a ΣΔM IC test PCB.

- Make sure that **the ground of each instrument is connected to its corresponding ground** in the PCB; that is, instruments providing analog signals should have their grounds connected to the analog ground in the PCB.
- Make sure that **all grounds are connected in a star configuration** as conceptually depicted in Figure 9.14a. Do not use the scheme shown in Figure 9.14b, where the ground of one instrument is connected to the ground of another instrument, and so on.

Last but not least, another important issue to take into account in the setup deals with the turn-on/off sequence that is followed to switch on/off the different instruments involved. The turn-on sequence should start with the supply voltage, followed by the clock signal generator, and finally the analog signal generator. The turning off should be carried out in the opposite order.

9.4.3 Measurement Set-Up Example

Figure 9.15 shows the conceptual diagram of a measurement set-up with a logic analyzer as the data acquisition system [6]. In this case, an Agilent A3631A unit generates the voltage supply, whereas the fully-differential input signal is generated by a Tektronix SG5010 audio oscillator. An optional SRS CG635 clock generator can be used for generating the clock signal, and an Agilent A16823B logic analyzer[6] is used for acquiring the modulator output bitstreams and also for generating the digital

[6]Depending on the electrical characteristics demanded for the clock signal, particularly the jitter error specification, it can be generated by a logic analyzer. If a very low-jitter clock signal is required, then an appropriate clock-signal generator should be used instead.

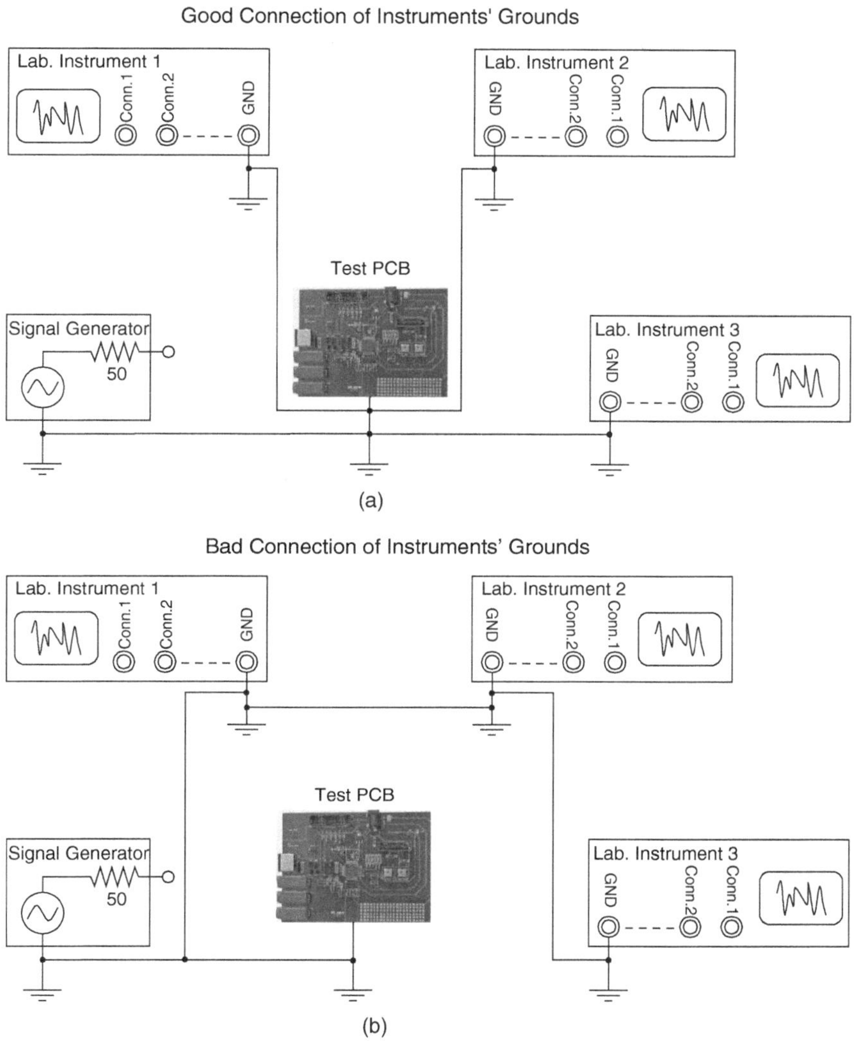

Figure 9.14 Illustrating the ground connection of different lab instruments to the test PCB.

control signals required to test the modulator. After bitstream acquisition, data is post-processed in MATLAB. Note that, in the case of cascade modulators in which the DCL is not implemented on-chip, a similar procedure to that described in Section 7.4 can be followed to process the output results.

9.5 ΣΔM Design Examples and Case Studies

In this section, some ΣΔM design examples and case studies are shown in order to illustrate some practical design considerations to take into acccount, depending on the target application in which the ΣΔ ADCs are to be used.

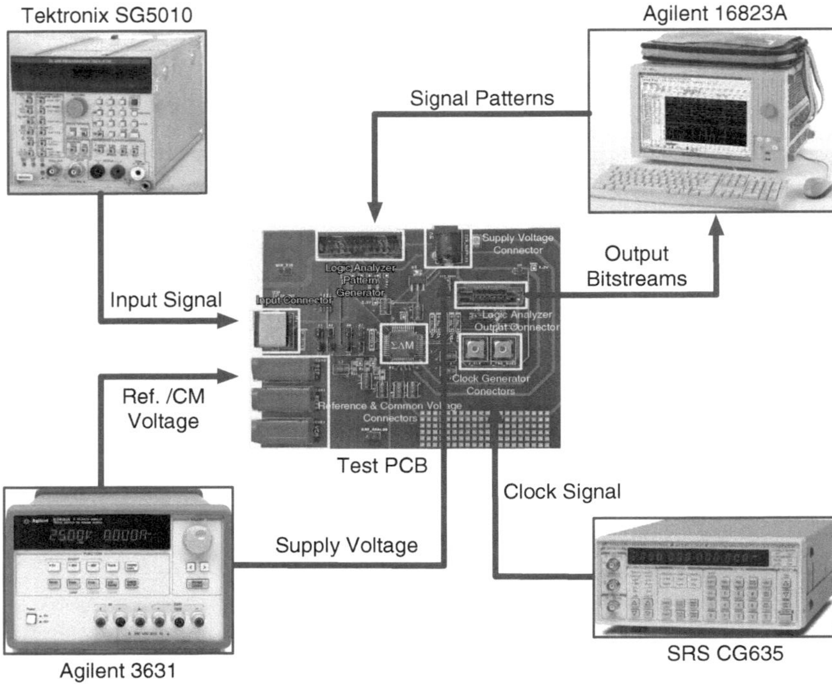

Figure 9.15 Diagram of a ΣΔM measurement set-up based on a logic analyzer.

9.5.1 Programmable-gain ΣΔMs for High Dynamic Range Sensor Interfaces

The first example under study is a ΣΔM intended for digitizing signals provided by a microelectromechanical transducer in a sensor interface system like that conceptually shown in Figure 9.16. A detailed description of the design, chip implementation and experimental measurements can be found in the literature [13]. This section focuses on some specific design considerations that need to be taken into account when targeting the digitization of high-resolution (> 16-bit) low-bandwidth (< 20 kHz) signals, by using ΣΔ ADCs. In this case, one of the most limiting factors is the electronic noise – including both thermal and flicker components – as well as the large dynamic range required to accommodate signals amplitudes ranging from microvolts to hundreds of millivolts, in hostile environmental conditions and in the presence of technology process variations – particularly due to a very large temperature range of [-40°C,175°C].

In the majority of sensor A/D interface systems, a programmable-gain preamplifier is placed between the sensor and the ADC in order to adjust the signal amplitude to the DR of the ADC. In this

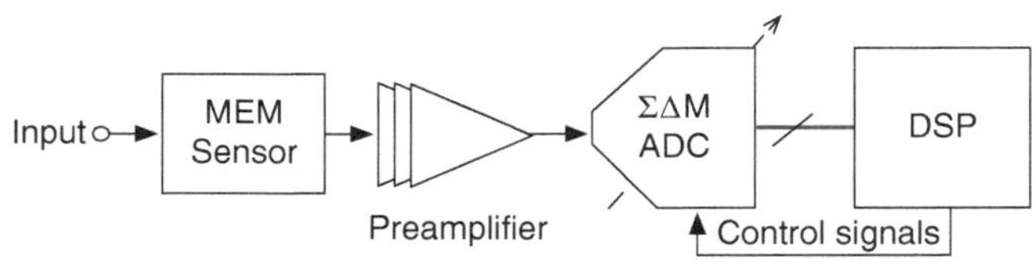

Figure 9.16 Block diagram of a sensor interface based on a programmable-gain ΣΔ ADC.

case study, however, a more compact solution that incorporates a programmable input-to-output gain to the ADC – referred to as ADC gain – is considered. This feature can be incorporated thanks to the versatility and robustness of ΣΔMs.

9.5.1.1 Main Design Criteria and Performance Limitations

Given that the resolution of the ADC must be > 16-bit, but the target signal bandwidth is low (< 20 kHz), it is expected that the IBN will be dominated by noise and harmonic distortion. In this example, a 0.35 μm CMOS process with 3.3-V supply is used, and hence there is plenty of voltage room for the amplifiers, so that highly linear operation can be achieved. Circuit noise, however, constitutes the ultimate physical limiting factor that has to be appropriately reduced in the design. Thus the first question that arises is: how much lower should the quantization in-band noise, IBN_Q, be as compared to the physical limit imposed by the circuit in-band noise, $\mathrm{IBN}_{\mathrm{noise}}$?

A well-known rule of thumb suggests that the most efficient design would be that in which $\mathrm{IBN}_{\mathrm{noise}} \simeq \mathrm{IBN}_Q$. The reasoning behind this rule is as follows. Making $\mathrm{IBN}_Q \ll \mathrm{IBN}_{\mathrm{noise}}$ would lead to an unnecessary increase in the power consumption and the silicon area, due to the increase of OSR, L and/or B, which in turn results in more demanding building-block dynamics, circuit complexity, and so on. The same reasoning applies if $\mathrm{IBN}_{\mathrm{noise}} \ll \mathrm{IBN}_Q$. In order to reduce thermal noise in SC circuit realizations, the sampling capacitor should be increased, with the subsequent penalty in power dissipation in order to reduce some circuit nonidealities, such as the incomplete setting error, and so on.

Based on this first design criterion, the next step is to select the appropriate ΣΔM architecture and noise-shaping parameters: L, B, and OSR. In this case study, a SC circuit realization is considered.[7] As discussed in Chapter 3, the main contributions to the circuit noise are the thermal and flicker noise of the front-end amplifier, the switches, the front-end DACs and the reference-voltage generator. Ignoring all thermal noise contributions except that of the switches, the expression of $\mathrm{IBN}_{\mathrm{noise}}$ given in Equation (3.41), can be simplified to:

$$\mathrm{IBN}_{\mathrm{noise}} \approx \frac{4kT}{C_{S1}\mathrm{OSR}}\left(1 + \frac{C_{S2}}{C_{S1}}\right) \tag{9.3}$$

where it has been assumed that the input SC branch and the DAC SC feedback branch (see Figure 3.15a), have different sampling capacitors, to implement a modulator gain defined as $\xi = C_{S1}/C_{S2}$.

As an initial estimate, let us consider the simple case of $\xi = 1$ and a single-bit quantizer ($B = 1$). Figure 9.17 represents the ENOB versus OSR for $C_{S1} = 2$ pF and different values of L, by considering IBN_Q and $\mathrm{IBN}_{\mathrm{noise}}$ as the only noise contributors. Note that for each curve, there is a breakpoint – highlighted in the curve – where $\mathrm{IBN}_{\mathrm{noise}} \simeq \mathrm{IBN}_Q$. Above this breakpoint, the resolution is dominated by thermal noise and hence doubling OSR gives rises to a mere 3-dB increase in ENOB. Below that breakpoint, ENOB increases with OSR at a different rate, depending on the value of L – according to Equation (1.12). Indeed, most state-of-the-art ΣΔMs are designed to operate close to, but at the right-hand side of, that breakpoint [13].

Based on the estimate of ENOB depicted in Figure 9.17, an exploration of different alternative architectures was carried out by considering the main circuit errors in a similar way as was described in Section 3.8. Based on this procedure – fine tuned by behavioral simulation – a single-bit cascade 2-1 ΣΔM with OSR = 128,256 was selected in this case study.

[7] A similar procedure to that shown here to select the values of L, B, and OSR can be followed if a CT-ΣΔM is considered, by using the general expressions of IBN, limited by quantization error (Equation (1.12)) and thermal noise (Equation (4.29)).

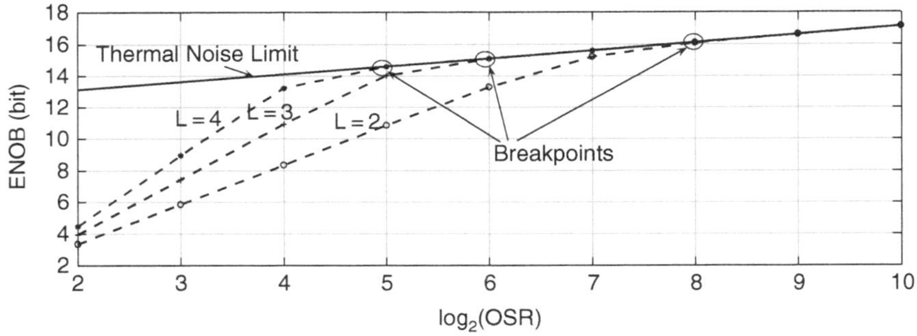

Figure 9.17 ENOB versus OSR for different values of L, considering both quantization noise and thermal noise contributions.

9.5.1.2 SC Realization with Programmable Gain and Double Sampling Figure 9.18 shows the block diagram and the conceptual schematic of the cascade 2-1 SC-ΣΔM under consideration. Double sampling is used in the input SC branch in order to achieve an extra gain of two, without penalizing the impact of circuit noise [14]. Note that different SC branches are used for the input signal and the feedback DAC so that a modulator gain can be implemented with the ratio $\xi = C_{S1}/C_{S2}$. An external digitally-controlled DC signal, $V_{\rm off}$, is also used to center the sensor signal in the modulator full-scale range [13]. In this design, a programmable gain, $\xi = 0.5, 1, 2, 4$ was implemented by using switchable capacitor arrays, each of them formed by a variable number of unitary capacitors, with $C_u = 1.5$ pF. In order to simultaneously keep the modulator gain and integrator weights, the total value of the capacitances has to be changed accordingly, by changing the number of unitary capacitors.

Figure 9.19 shows a detailed schematic of the front-end integrator. Capacitors are realized by using metal-insulator-metal (MiM) structures. All capacitances in the front-end integrator are made up of switchable capacitor arrays in which unitary capacitors are either connected or disconnected depending on the value required for the modulator gain. The arrangement of unitary instances in this programmable common-centroid structure is illustrated in Figure 9.20a for $\xi = 4$, together with the corresponding physical realization inside the ΣΔM chip (Figure 9.20b).

The modulator was designed by combining simulation and optimization at different hierarchical levels, following the design methodology described in previous chapters. Figure 9.20b shows a microphotograph of the chip, highlighting the details of the front-end integrator capacitor array used for the programmable gain. Figure 9.21a shows the in-band (20-kHz) output spectra corresponding to the four different cases of the modulator gain when an input signal of −20 dBV at 5 kHz is applied. The IBN is about −96 dB for each case of the modulator gain, corresponding to 16.2 bits with respect to the FS reference voltage, $V_{\rm ref} = 2$ V. This resolution can be markedly improved by the effect of ξ. This is illustrated in Figure 9.21b, where the measured SNDRs are represented against the input amplitude. Note that, thanks to the action of programmable gain, the input-referred DR is approximately −104 dBV; in other words, 110 dB below $V_{\rm ref}$.

9.5.1.3 Influence of Chopper Frequency on Flicker Noise As stated above, flicker noise is one of the physical limitations when considering high-resolution low-frequency applications like sensor A/D interfaces. In order to palliate the effect of this noise source, chopper stabilization techniques are usually included in the front-end integrators, as shown in Figure 9.18. It is interesting to see the effect of varying the chopper frequency, $f_{\rm ch}$, with respect to the clock (sampling) frequency, f_s,

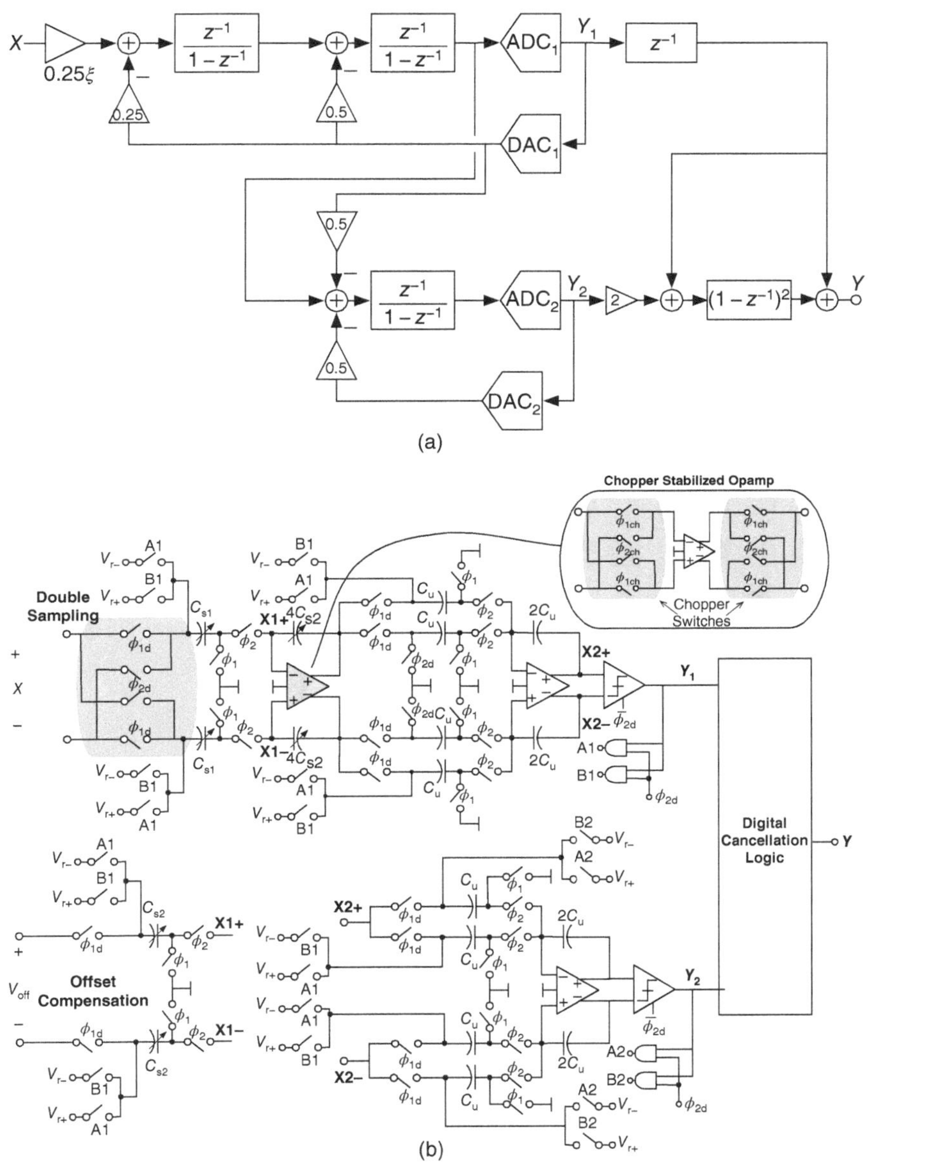

Figure 9.18 Chopper-stabilized cascade 2-1 SC-ΣΔM with programmable gain: (a) block diagram; (b) schematic.

in order to determine the value of f_{ch}/f_s that gives more effective noise suppression. This is illustrated in Figure 9.22, which shows several experimental output spectra for different values of f_{ch}/f_s. Note that the lower f_{ch}/f_s, the more flicker noise appears in the signal band.

The flexibility of ΣΔMs to adapt their performance metrics to different ADC requirements can be extended to other applications with higher-frequency signal bands, like communications, as illustrated in the circuit examples in the next sections.

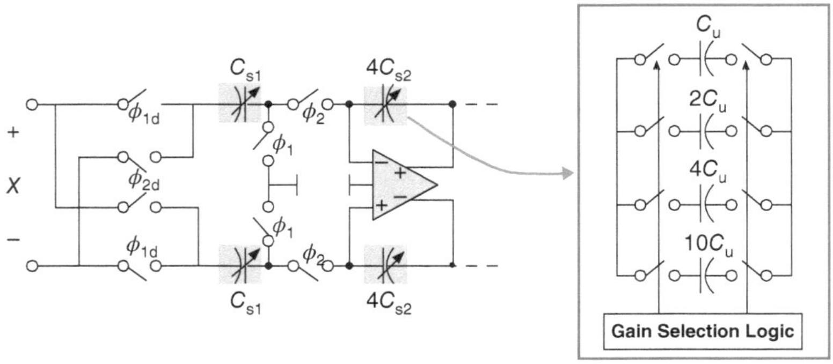

Figure 9.19 SC integrator with programmable capacitor arrays.

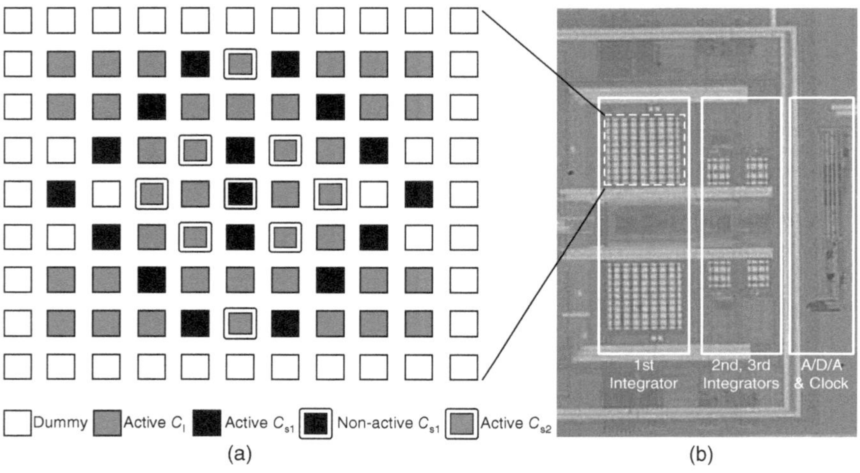

Figure 9.20 Programmable capacitor array: (a) conceptual layout for $\xi = 4$; (b) chip microphotograph [13].

9.5.2 Reconfigurable SC-ΣΔMs for Multi-standard Direct Conversion Receivers

As a second case study, let us consider the application of ΣΔMs for the design of reconfigurable ADCs in multi-standard wireless direct-conversion receivers intended for software-defined radio (SDR) systems. Figure 9.23 shows the block diagram of an SDR direct conversion receiver (DCR), in which the incoming RF signals, after being filtered and preamplified, are downconverted to baseband, where they are digitized by a reconfigurable SC LP-ΣΔ ADC. This receiver architecture is commonly used in multistandard applications because it eliminates the need for both IF and image-reject filtering and requires only a single oscillator and mixer. Such a receiver aims to cover the requirements of many wireless standards, including GSM, Bluetooth, GPS, UMTS, DVB-H, and WiMAX. These standards involve digitizing signals with B_w ranging from 100 kHz to 20 MHz, with ENOB ranging from 12- to 8-bit, respectively. These specifications were extracted from an iterative system-level simulation-based procedure considering the propagation of the different standard test signals through the receiver front-end. As will be shown by the experimental results, such requirements can be addressed by appropriately configuring the ΣΔM with $L = 2, 4, 6$, $B =$ 1–3 bits, and OSR $\in$(10,200).

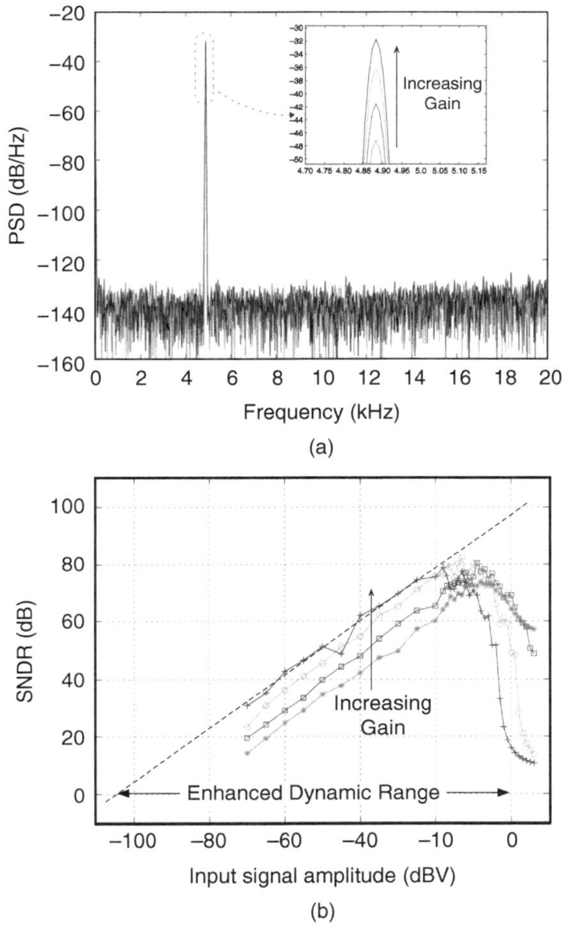

Figure 9.21 Measured performance of ΣΔM with programmable gain: (a) measured in-band (20-kHz) output spectra; (b) SNDR versus input signal [13].

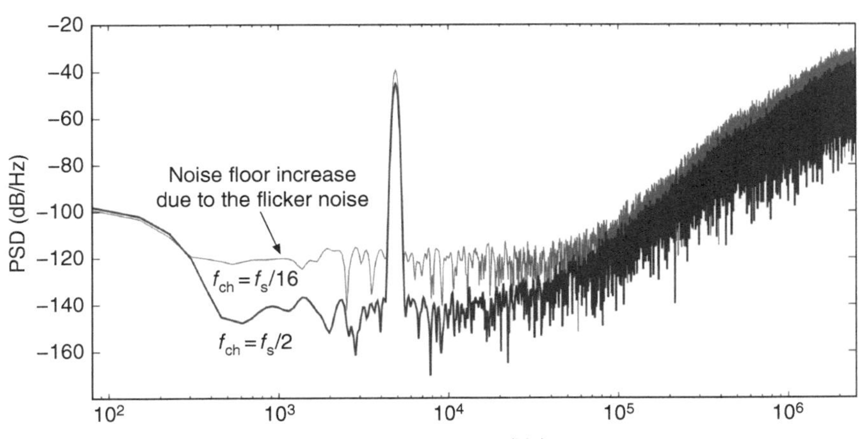

Figure 9.22 Influence of chopper frequency, f_{ch}, on the measured in-band noise [13].

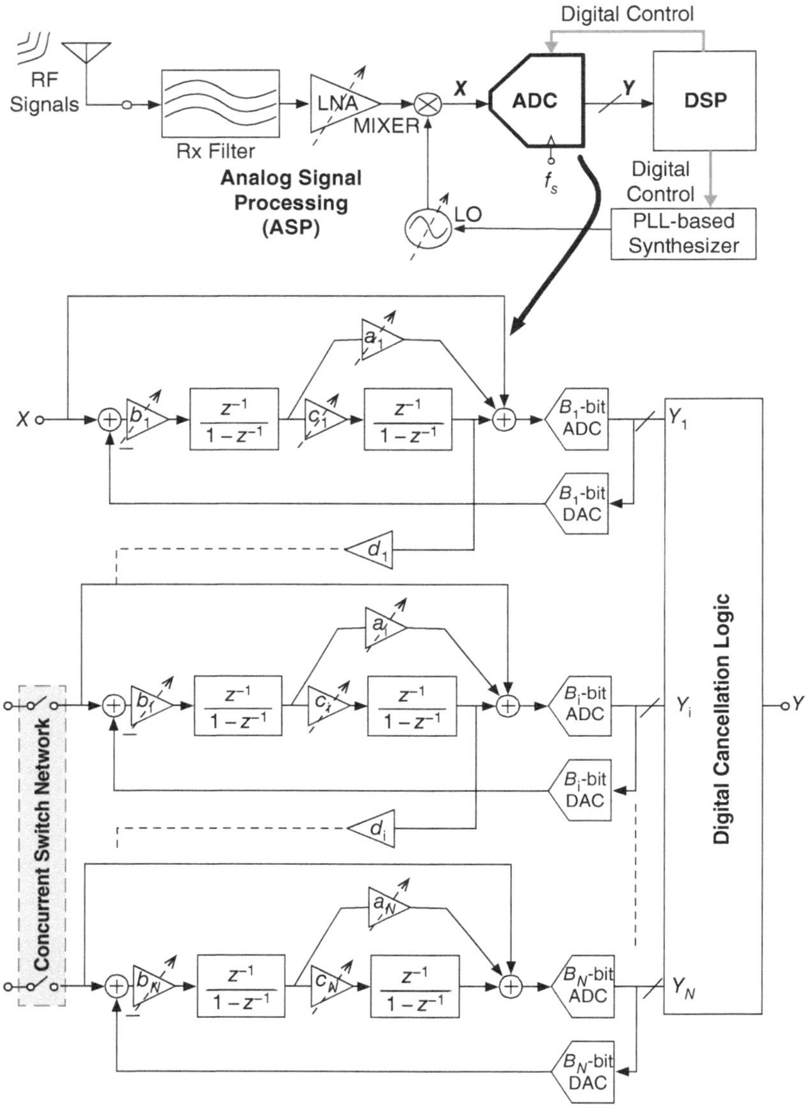

Figure 9.23 Expandable N-stage cascade SC-ΣΔM for wireless applications.

The ΣΔM considered in this example is also shown in Figure 9.23. It consists of an N-stage cascade topology, where all stages can be made independently switchable according to the desired quantization noise shaping, and the DCL can be programmed according to the value of L. If a stage is turned off, its building blocks can be powered down to save power. The number of bits of the internal quantizers, B_i, and/or the OSR can be also reconfigured in order to increase the flexibility of the ADC. In addition to reconfiguring its performance parameters – that is, OSR, L and B – a multimode ΣΔ ADC must digitize signals corresponding to different standards – for instance, GSM and Bluetooth signals and/or UMTS and WLAN signal – in a parallel or *concurrent* way. Concurrency can be also implemented in a cascade ΣΔM, as illustrated in Figure 9.23, where a switchable network is used to allow the ΣΔM to be configured as several *submodulators* working in parallel, each one processing a different input signal.

Different alternative implementations of the ΣΔM architecture shown in Figure 9.23 can be considered, by appropriately combining second-order stages and first-order stages in order to guarantee the stability of each sub-modulator [8, 15]. Figure 9.23 shows an example based on second-order stages, where a unity signal transfer function (USTF) is implemented in all stages in order to lower the sensitivity to circuit imperfections, especially in wideband standards, in which low OSR is required.

Another architectural strategy followed to improve the performance of the cascade topology involves including *resonators* in the loop filter. This technique allows the zeroes of NTF to be shifted from DC to optimal frequencies that maximize the effective modulator resolution without increasing the loop-filter order, with a subsequent saving in power consumption. As an illustration, the SC schematic of an intermediate stage with resonation is shown in Figure 9.24, where switchable capacitor arrays are used to implement the programmable loop-filter coefficients. In addition to the circuit elements shown in this figure, additional digital logic blocks – not shown for the sake of simplicity – are needed to implement the main reconfiguration functionalities by means of a set of control signals, which are also used to power up/down the different subcircuits according to the configuration used by the modulator.

9.5.2.1 Power-scaling Circuit Techniques The high-level sizing process described in Chapters 5 and 6 is carried out here by combining statistical optimization and behavioral simulation [16]. In this process, the worst cases for speed and for thermal noise are considered, so that

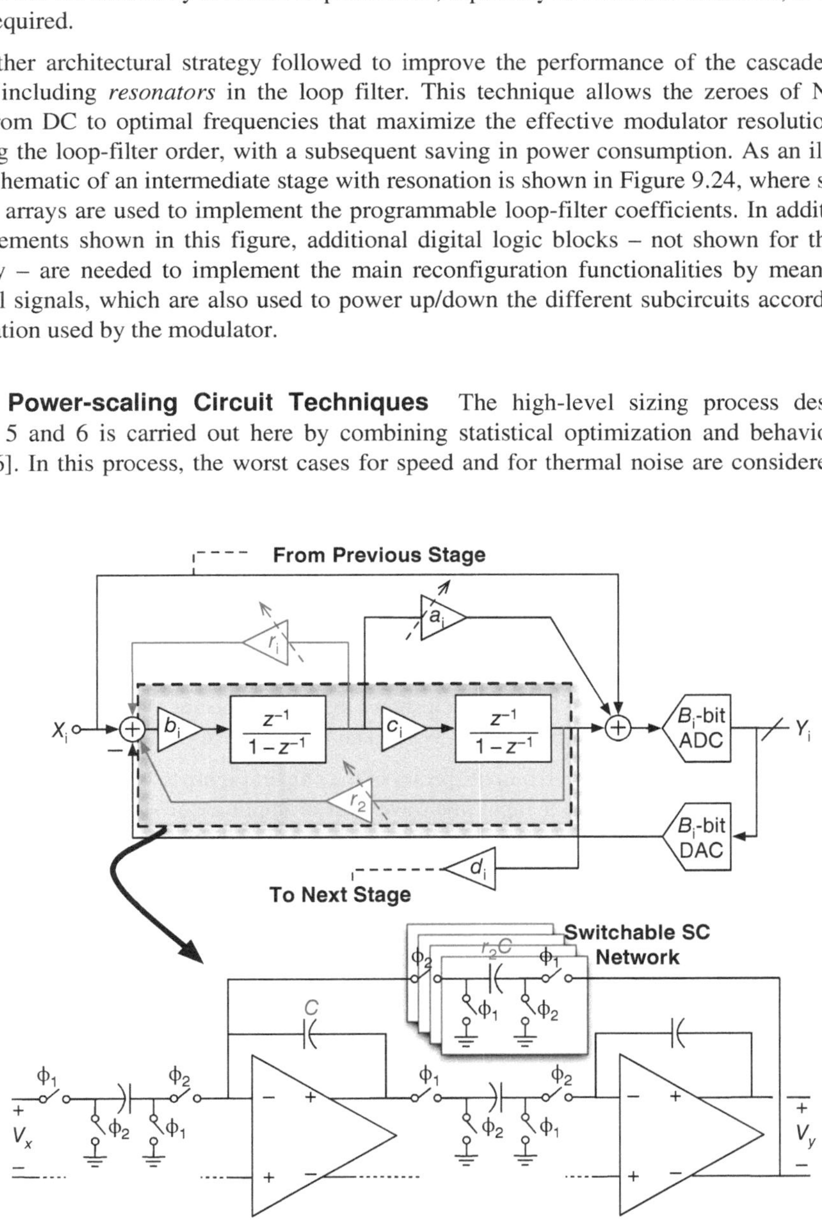

Figure 9.24 Conceptual schematic of programmable local resonation in SC-ΣΔMs.

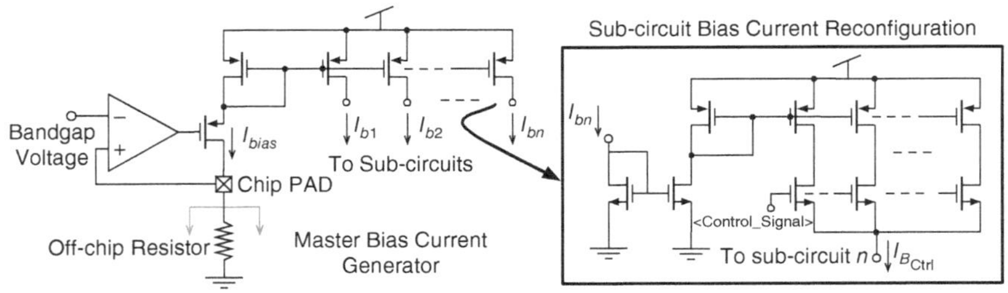

Figure 9.25 Example of programmable master bias current generation.

the ΣΔM subcircuit requirements are known and the transistor-level design can be completed. A key design issue in reconfigurable ADCs is to adapt the performance to the different specifications with optimized power consumption. To this end, the biasing of the main ΣΔM building blocks can be adjusted by using a programmable master bias current generator, like the one shown in Figure 9.25. In this example, reconfiguration is performed using binary-weighted pMOS current mirrors, and all mirrored currents are selected by control signals, which are applied to the gates of nMOS-based control switches. These mirrored currents go directly to all analog blocks, so that they have a dedicated control signals to adapt their bias current independently.

9.5.2.2 Experimental Results Figure 9.26 shows the microphotographs of the ΣΔM chip examples, highlighting their main parts. Both chips were designed for DCR in SDR applications and fabricated in a 90-nm CMOS technology. Figure 9.26a is a fourth-order two-stage (2-2) cascade with three-level quantization and USTF in both stages. Figure 9.26b is sixth-order and three-stage (2-2-2) with concurrency, programable local resonation and reconfigurable 3–5-level quantization. In both cases the OSR can be programmed by varying f_s from 40 to 240 MHz.

Reconfiguration can be implemented by varying OSR, L, or B. The latter is illustrated in Figure 9.27a, where the output spectrum of Figure 9.26b is shown for different levels of the last-stage quantizer (B). Another feature of this chip is its capability to concurrently handle different input signals. This is illustrated in Figure 9.27b, which shows the output spectra of the first and the second stages while processing different input sinewaves at 20 and 200 kHz, respectively.

Figure 9.28a shows the measured output spectra of the chip in Figure 9.26a corresponding to different operation modes, considering an input sine wave with an amplitude of −12.2 dB below full-scale (dBFS). The ENOB of the chip is better appreciated in Figure 9.28b, where the SNDR is depicted versus the input signal amplitude for the different standards under consideration. This chip featured a peak SNDR of 72.3/68.0/65.4/63.3/59.1/48.7 dB for 100kHz/500kHz/1MHz/2MHz/4MHz/10 MHz, while consuming 4.6/5.35/6.2/8/8/11 mW, respectively. This the state of the art of reconfigurable ADCs at the time of its publication [8].

9.5.3 Using Widely-programmable Gm-LC BP-ΣΔMs for RF Digitizers

The third example considered in this chapter is a fourth-order BP CT-ΣΔM intended for the digitization of RF signals in SDR applications. The modulator architecture – shown in Figure 6.10 – consists of two Gm-LC resonators with a tunable notch frequency and a 4-bit flash ADC in the forward path and a FIR-based NRZ DAC in the feedback path. Both system-level and circuit-level reconfiguration techniques need to be considered in order to allow the modulator to digitize signals placed at different carrier frequencies, from 450 to 950 MHz with a programmable 1.2/2 GHz clock rate. A suitable

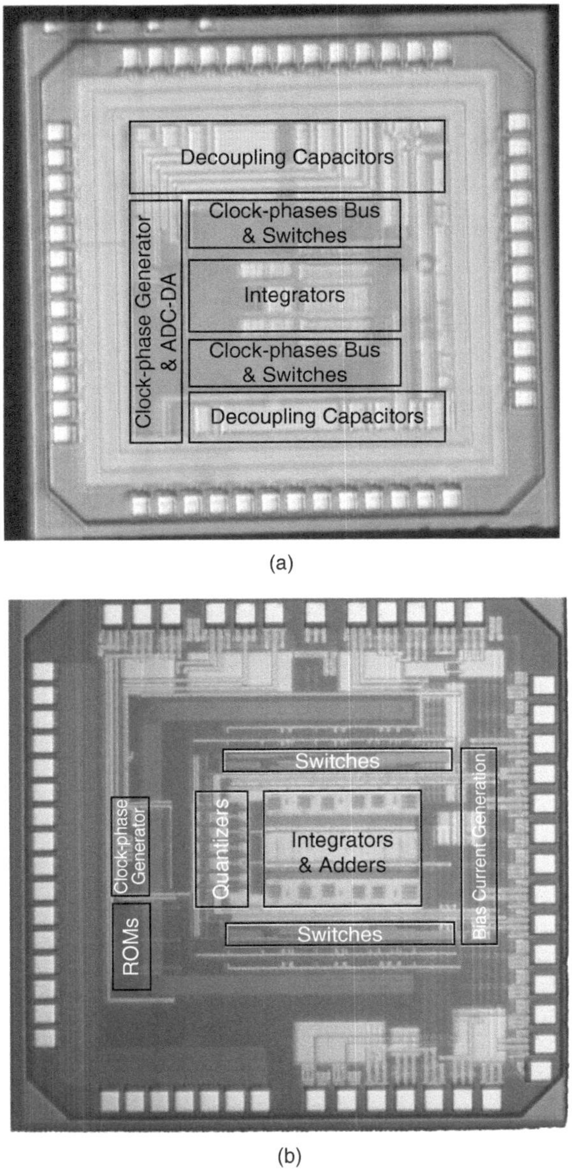

Figure 9.26 Chip examples: (a) fourth-order cascade 2-2 with three-level quantization [8]; (b) sixth-order 2-2-2 with concurrency and programmable quantization [17].

synthesis methodology for the loop-filter coefficients at system level and the use of inverter-based switchable transconductors allows the performance to be optimized in terms of robustness to circuit errors, stability, and power scalability. As a case study, a 65-nm CMOS technology with 1-V supply voltage is considered, and the design objectives are for digitizing signals with 40-MHz bandwidth and SNDR > 55 dB.

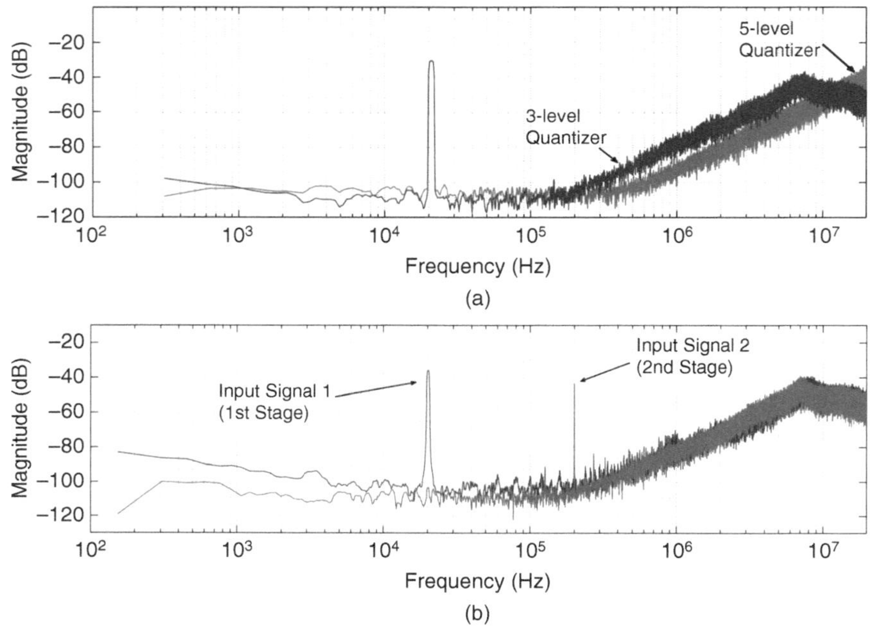

Figure 9.27 Illustrating reconfiguration and concurrency: (a) changing quantizer resolution; (b) handling two input signals concurrently.

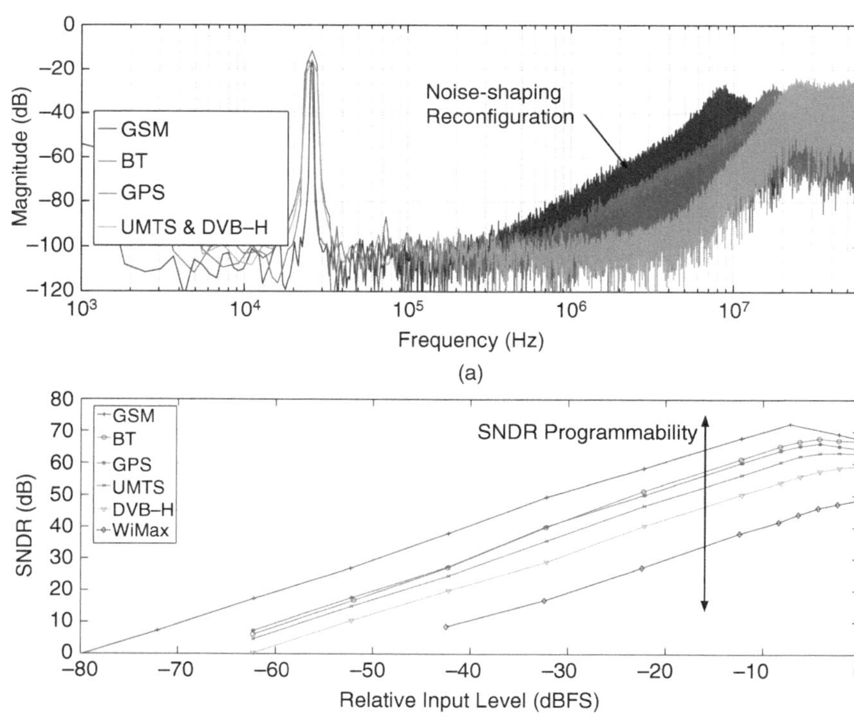

Figure 9.28 Experimental results: (a) output spectra; (b) SNDR vs input amplitude [8].

9.5.3.1 Application Scenario As stated previously in this book, CT-ΣΔMs have proved to be a competitive solution for the implementation of power-efficient ADCs operating in the gigahertz range, thus opening the door to their use in *digital-intensive* RF transceivers for SDRs and Internet-of-Things applications [4, 18–23]. One of the direct approaches to implementing such transceivers consists of placing a BP CT-ΣΔM as close as possible to the antenna, yielding the so-called *RF-to-digital converters* – conceptually depicted in Figure 9.29 [18, 20, 22, 24–26]. These ΣΔM-based RF digitizers use diverse strategies to reduce their demanding specifications in terms of the sampling frequency (f_s) and the dynamic range (DR), including, embedded out-of-band filtering [18], frequency-translating and *undersampling* or *subsampling* [22]. The latter approach allows digitization of RF signals placed at $f_{RF} > f_s/2$, while keeping high values of the OSR, since the signal bandwidth is typically $B_w \ll (f_{RF}, f_s)$. Other authors have proposed the use of a tunable center frequency or *notch frequency*, $f_n \equiv f_{RF}$, in order to simplify the design of the frequency synthesizer needed in the receiver [20, 22, 27].

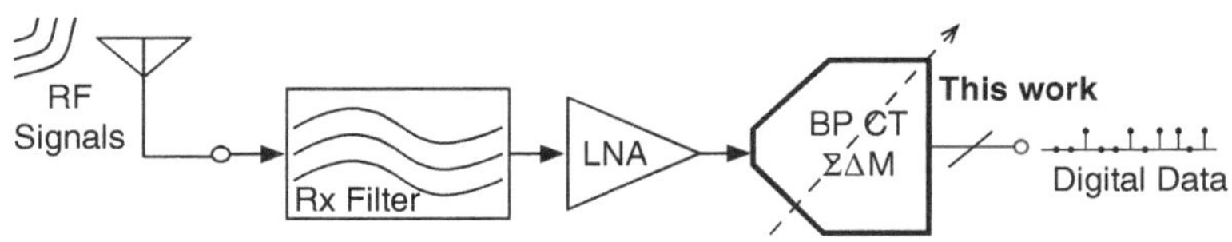

Figure 9.29 RF-to-digital conversion based on a programmable BP CT-ΣΔM.

In spite of their potential benefits, the performance of CT-ΣΔ RF-to-digital converters is still well short of what is needed for commercialization. The main challenge is associated with the design of the loop filter, so that high-quality and accurate resonance can be achieved, while keeping the necessary tuning range, robustness, and stability. As a result, LC tanks are great from a power and linearity perspective but typically support only an octave of range, whereas active-RC resonators can be widely tunable but require amplifiers with high gain at the ADC center (RF carrier) frequency.

As stated in Chapter 6, one of the key points in addressing these challenges is to optimize the design of BP CT-ΣΔMs by appropriately selecting the modulator loop-filter coefficients in terms of tunability, resonator input–output swing, and simplicity of the circuit implementation. In many cases, the design bottleneck is not as important at the circuit level as at the system level, where a suitable synthesis and sizing of the ΣΔM architecture can lead to a more efficient and robust design, while relaxing the required design margins of most performance metrics at transistor level. To this end, a systematic methodology is used to optimize the loop-filter implementation in order to reduce the power consumption to a level where it is competitive with conventional baseband digitization, while keeping the benefits of RF-to-digital conversion.

9.5.3.2 Gm-LC BP-ΣΔM High-level Sizing The BP CT-ΣΔM is realized by using Gm-LC resonators in order to increase the programmability of the loop-filter coefficients, which are realized as switchable multiples of a unitary transconductance element, g_{mu}, as will be detailed later. The quality factor, Q, of the Gm-LC resonators is enhanced by adding two extra transconductances, $g_{kq1,2}$, which can be calibrated in order to maximize the performance of the modulator. The 4-bit quantizer is made up of a flash ADC in the forward path and a NRZ current-steering FIR-DAC in the feedback loop.

The modulator has been synthesized by applying a CT-to-DT transformation to a BP DT-ΣΔM with an NTF, considering input signals with $B_w = 40$ MHz placed at a programable notch frequency $f_n = 450$–950 MHz. To this end, as discussed in Chapter 6, Schreier's toolbox can be used to synthesize the NTF for a given value of f_n and OBG, and the modulator loop-filter is derived from the impulse-invariant transformation. Once the ideal NTF and STF have been synthesized, the values of the loop-filter coefficients are selected in order to optimize the performance of the BP CT-ΣΔM in terms of robustness and stability, while maximizing DR and the SNDR with the minimum power consumption, P.

The values of DR and SNDR can be computed in a simulation using SIMSIDES [16], and P can be estimated as the sum of the power consumed by the resonators, P_{RES}, by the active adder, P_{ADD}, and by the embedded quantizer; that is, the ADC and the DAC. The latter is estimated from electrical simulations in Cadence Spectre, whereas P_{RES} and P_{ADD} are respectively derived as:

$$\begin{aligned} P_{\mathrm{RES}} &= V_{\mathrm{DD}} \cdot \left[\sum_{i=1}^{2} (I_{ci} + I_{cid}) + \frac{g_{m1} + g_{m2}}{g_{mID}} \right] \\ P_{\mathrm{ADD}} &= V_{\mathrm{DD}} \cdot \left[\frac{g_{m3}}{g_{mID}} + \frac{c_0 + c_{0d}}{R} \cdot V_{\mathrm{FS}} \right] \end{aligned} \tag{9.4}$$

where V_{DD} is the supply voltage and g_{mID} stands for the transconductance-versus-current efficiency. In this design, the same capacitances, $C_1 = C_2 = C$, inductances, $L_1 = L_2 = L$, and Q-enhancement gains, $g_{kq1} = g_{kq2} = g_{kq}$, are assumed for both resonators (see Figure 6.10).

Given a set of specifications defined in terms of SNDR, B_w, and P, the optimization procedure can be formulated as a design problem with five design variables: OBG, R_{gain}, $s_{\mathrm{r}1,2}$, and g_{kq}. Note that one of the design key strategies to minimize P involves reducing the values of the k_i coefficients by increasing the resonator weights $s_{\mathrm{r}1,2}$. Suitable selection of these weights might give a reduction of the power consumption of several orders of magnitude. However, this scaling procedure must be compensated by appropriately increasing the gain of the active adder in order to get the required loop-filter gain.

In order to optimize all involved high-level design parameters, the following step-by-step procedure is followed:

1) Parameters R_{gain} and OBG are optimized based on a parametric simulation analysis: these parameters are swept, and other parameters are fixed to ideal values

2) Parameters $s_{\mathrm{r}1}$ and $s_{\mathrm{r}2}$ are optimized. In this step, ideal values of $g_{kq1,2}$ are used, and R_{gain} and OBG are set to the values obtained in step 1

3) Parameter g_{kq} is optimized and the other parameters are fixed to the values determined in the previous steps.

Figure 9.30 illustrates this step-by-step optimization procedure, depicting some of the performance metrics obtained at each step for a test signal placed at 900 MHz. Note that in each optimization step, a SIMSIDES model of the modulator shown in Figure 6.10 is simulated, and initial ranges considered for the design variables are heuristically set. The design parameters are selected according to a trade-off between SNDR, P, and stability. Other ultimate limiting factors – such as thermal noise – need to be considered from the beginning of the design procedure. In this modulator example, the main thermal noise sources are the front-end transconductance and the feedback DAC, whose contribution to the overall noise floor must be reduced sufficiently to give the required performance. Clock jitter is another important limiting factor affecting the performance of BP CT-ΣΔMs operating in the gigahertz range. This is illustrated in Figure 9.31, where the influence of clock-jitter on the output spectra of the modulator is shown in Cadence Spectre for the different standards under consideration: CDMA, LTE700, and GSM900. Note that the degradation caused by the action of jitter is similar in

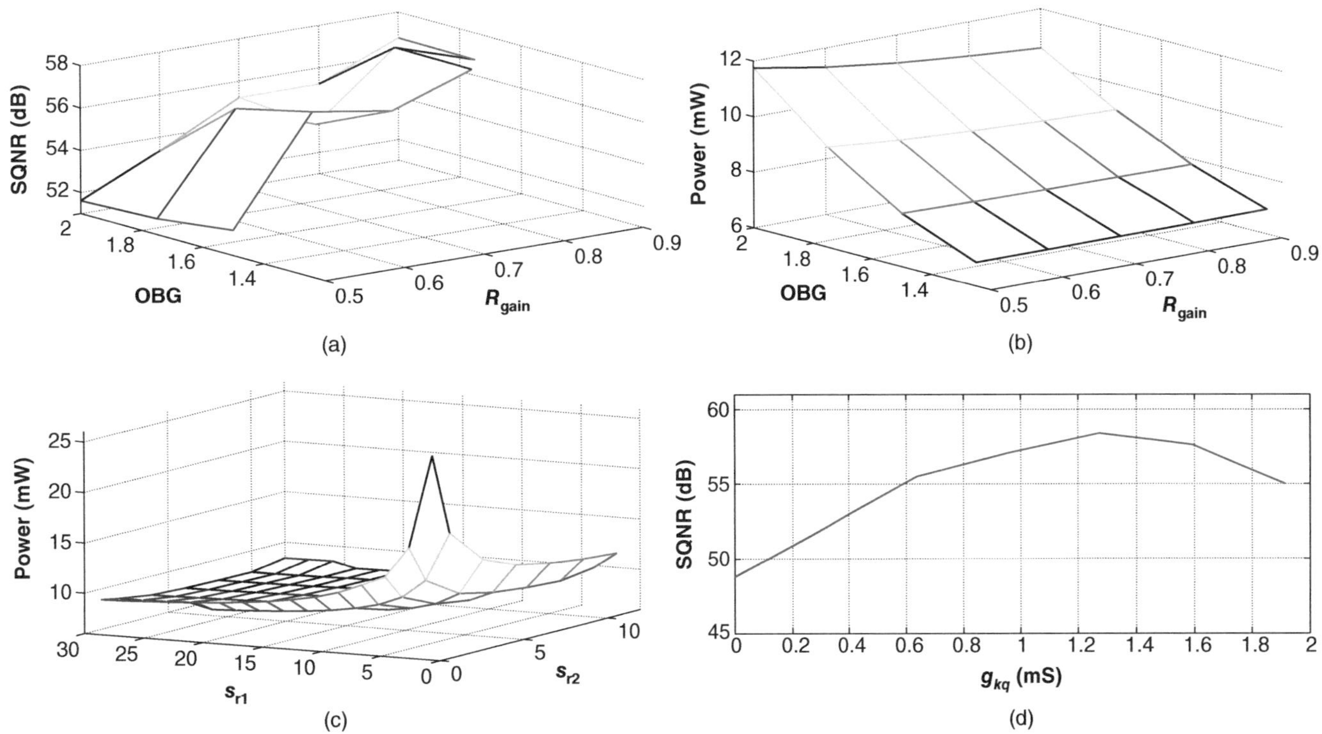

Figure 9.30 Illustrating system-level optimization procedure for $f_n = 900$ MHz and $B_w = 40$ MHz: (a) SQNR vs OBG and R_{gain} (Step 1); (b) P vs OBG and R_{gain} (Step 1); (c) P vs $s_{\text{r1,2}}$ (Step 2); (d) SQNR vs g_{kq} (Step 3).

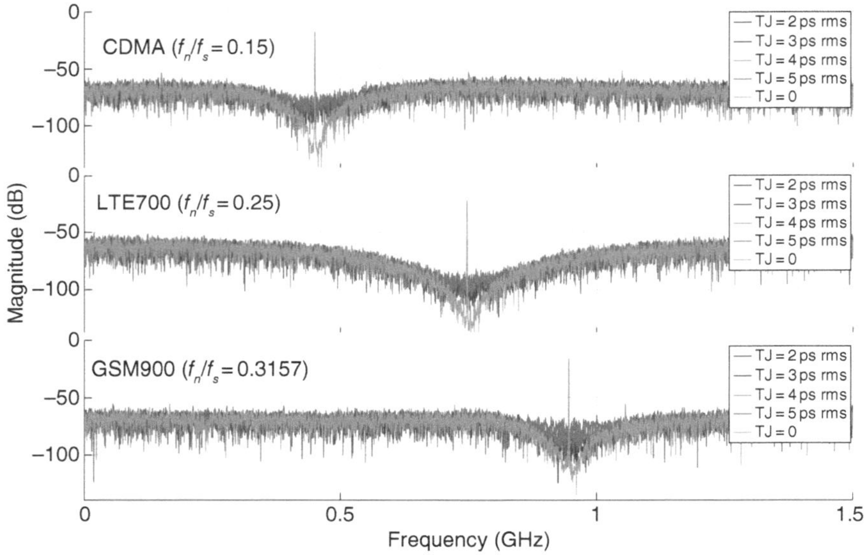

Figure 9.31 Effect of clock-jitter error (TJ) on the performance of a Gm-LC BP-ΣΔM for different standards and f_n/f_s.

all standards, regardless the value of f_n, which means that the value of the clock frequency, f_s, is more important than f_n [27].

Table 9.1 summarizes the results of this system-level optimization procedure, showing the values of selected design parameters for each standard and operation mode targeted by the modulator.

Table 9.1 Example of high-level synthesis of the BP CT-ΣΔM in Figure 6.10.

	Loop-filter coefficients					
	c_0	c_{0d}	c_1	c_{1d}	c_2	c_{2d}
CDMA-450	−0.62	1.22	−2.92	2.76	1.72	−1.28
LTE-700	−0.73	1.33	−3.32	2.10	1.48	−2.07
GSM-900	−0.46	0.80	0	−3.32	−3.12	−1.28
	System-level design parameters					
	OBG	s_{r1}	s_{r2}	R_{gain}	g_{kq1}(mS)	g_{kq2}(mS)
CDMA-450	2.5	16	4	1	5.8	5.8
LTE-700	2.5	16	4	1	2.0	2.0
GSM-900	2.5	16	4	0.8	1.6	1.6
	Signal and circuit parameters					
	f_n (GHz)	f_s (GHz)	L(nH)	C(pF)	R(kΩ)	Q
CDMA-450	0.45	1.25	6.7	18.6	1000	8.8
LTE-700	0.75	2	6.9	6.5	50	12.4
GSM-900	0.95	2	7.1	4.0	50	14.1

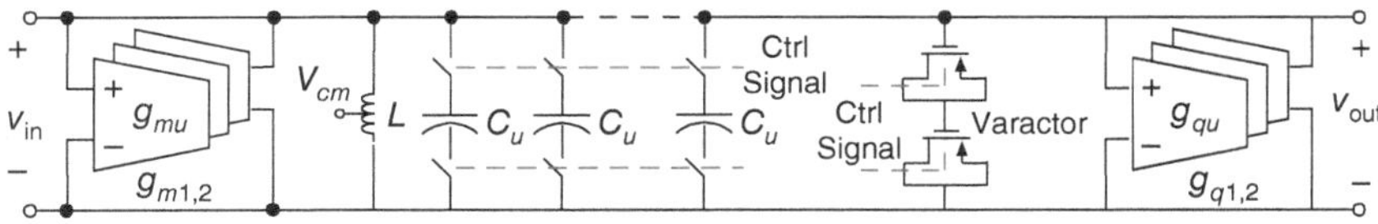

Figure 9.32 Conceptual schematic of Gm-LC resonators used in the BP CT-ΣΔM.

The values of the main circuit elements – transconductors, capacitors, and inductors – can be computed from the coefficient values shown in Table 9.1, and are used as the starting point for the design of the modulator subcircuits described in the next section.

9.5.3.3 BP CT-ΣΔM Loop-Filter Reconfiguration Techniques The main subcircuits of the BP CT-ΣΔM are the Gm-LC resonators, the active adder, and the 4-bit quantizer (see Figure 6.10). The latter is implemented using a conventional 4-bit flash ADC made up of a resistor ladder and regenerative-latch based comparators and a feedback current-steering DAC that uses cascode current cells to improve the output resistance. Special care is taken with the reconfiguration and design strategies applied to the resonators and the active adder, since they are key factors in the optimization of the performance in terms of reconfigurability, power consumption, and robustness to circuit variations.

The Gm-LC resonators are implemented using switchable inverter-based transconductors, as shown in Figure 8.23. The circuit is made up of an LC tank and reconfigurable transconductors. The forward-path transconductors – which correspond to $g_{m1,2}$ in Figure 6.10 – are used to process the resonator input signals, while FB transconductors implement $g_{kq1,2}$ coefficients in Figure 6.10. In both cases, the values of the required transconductances are obtained by connecting switchable unitary inverter-based transconductors, as illustrated in Figure 8.23. So, depending on the operation mode in which the modulator is working, these unitary elements are either switched on or off (and correspondingly powered up and down) in order to maximize the ADC performance with the minimum power consumption.

Note from Figure 8.23 that some unitary transconductors are always connected, while others are switchable. The inverter-based unitary transconductors are very simple and modular, to increase the flexibility and programmability of the modulator while minimizing the power consumption. Table 9.2 shows the sizes of all unitary transconductors. In order to adjust the values of the requested loop-filter coefficients with a higher accuracy, different values of the unitary transconductance, $g_{mu} = 100\ \mu$A/V are used, namely g_{mu}, $1/2g_{mu}$. The same strategy is followed for g_{qu}, where $1/2g_{qu}$, $1/4g_{mu}$ are also used. These values can be easily changed by appropriately scaling the value of the tail current, $I_{\text{bias}} =$ 12.5 μA, of the unitary transconductors as well as the sizing, as shown in Table 9.2. The common-mode voltage, $V_{\text{cm}} = 0.6$ V, is set by the inductor, so that no common-mode feedback circuit is required.

Another reconfiguration and tuning strategy included in the resonators involves programming the resonant frequency using switchable unitary MiM capacitors connected through CMOS switches, which are in turn digitally controlled in order to program the value of f_n required for the BP CT-ΣΔM

Table 9.2 Sizing example of unitary transconductors (see Figure 8.23).

W/L (μm/μm)	g_{mu}	$1/2g_{mu}$	g_{qu}	$1/2g_{qu}$	$1/4g_{qu}$
M_n	2.2/0.24	1.1/0.24	8.8/0.12	4.4/0.12	4.4/0.12
M_p	1.7/0.24	0.85/0.24	7.44/0.12	3.72/0.12	3.72/0.12
M_{bias}	1.2/0.24	1.2/0.24	4.8/0.24	2.4/0.24	1.2/0.24

to work in each operation mode. Different values of unitary capacitances, namely 125 fF, 150 fF, 300 fF, 400 fF, 800 fF and 1 pF, are used in order to increase the granularity of the overall capacitance value. Additionally, a pMOS-based varactor is also used to fine-tune the resonance frequency required in each case.

The control of the resonators' Q factor is also critical for correct operation of the modulator, especially considering the high number of programmability techniques embedded in the circuit. To this end, the FB transconductors implementing $g_{kq1,2}$ are also digitally programmable, as will be described later. Moreover, such digital programmability is also used in practice to calibrate the performance of the resonators. In order to implement these calibration techniques, the additional source-follower analog buffers depicted in Figure 9.33 can be used to access and test the output of both resonators.

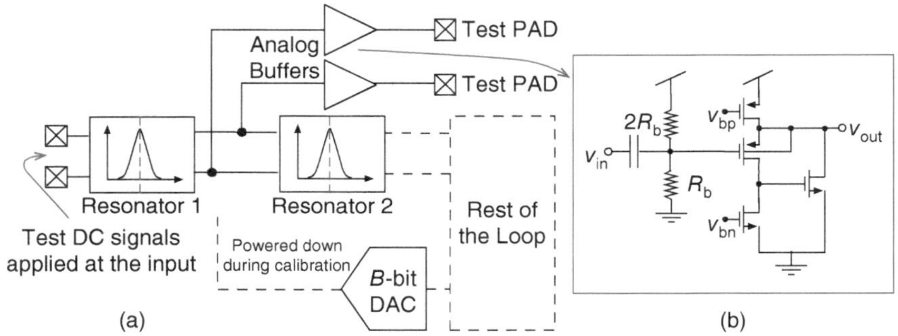

Figure 9.33 Conceptual scheme of resonators in calibration mode; (a) block diagram; (b) analog buffers used for testing purposes.

The different configuration modes implemented in both resonators are summarized in Table 9.3, where the main simulated performance metrics are shown for the different cases. The values and ranges of the digital control signals are also depicted:

- Ctrl-F stands for the digital signal setting the number of unitary g_{mu} to implement $g_{m1,2}$;
- Ctrl-Q sets the number of g_{qu}, $1/2g_{qu}$, and $1/4g_{qu}$ unitary transconductances connected to implement $g_{q1,2}$;

Table 9.3 Reconfiguration of resonators.

	Resonator 1				
	Ctrl-F	Ctrl-Q	Ctrl-SW	Ctrl-VAR (V)	f_n (MHz)
CDMA	11	53	10	0.4	449
LTE	9	20	2	0.3	759.6
GSM	14	6	1	0.4	955
	Resonator 2				
	Ctrl-F	Ctrl-Q	Ctrl-SW	Ctrl-VAR (V)	f_n (MHz)
CDMA	48	36	13	0.6	455
LTE	32	17	3	0.6	754
GSM	12	6	1	0.6	957

- Ctrl-SW sets the number of unitary MiM capacitors;
- Ctrl-VAR is the value of the varactor control voltage.

As an illustration of the reconfiguration of the resonators, Figure 9.34 shows the output frequency response of the first resonator, considering a 100-mV input tone placed at 450 MHz (CDMA), 750 MHz (LTE) and 950 MHz (GSM). It can be shown how both the Q-factor and f_n can be modified from the nominal value by changing the digital control signals. This is essential to adapt the performance of the modulator to the specifications of each standard.

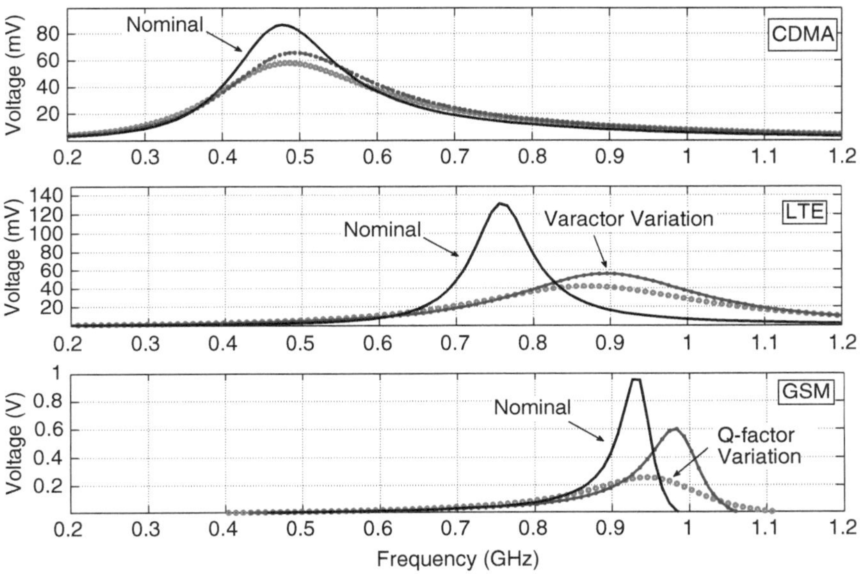

Figure 9.34 Effect of varying the digital control signals on the output frequency response of Gm-LC resonators like those shown in Figure 9.32.

Another critical subcircuit in these kinds of ΣΔM topologies is the analog active adder, since it has to sum and amplify signals in the gigahertz range. In this design example, the adder circuit is designed to have a nominal gain of 64 in order to reduce the rest of modulator loop-filter coefficients, with subsequent benefits in terms of input–output swing, power dissipation, and simplicity of design of the transconductors, as described above. However, the price to pay is that a more complex circuit is needed to implement the adder, thus consuming a significant part of the overall modulator power consumption. This is thus one of the modulator circuit design bottlenecks. In order to achieve the required gain, the multi-stage (four-stage) amplifier shown in Figure 9.35 has been considered. Each stage is made up of a programmable-gain transconductor and a transimpedance amplifier.

Figure 9.36 is a schematic of the transconductors used in the active adder, with the sizing and biasing as an inset in the figure. The circuit has an inverter-based topology similar to that used in the resonators, except for the input signal forward path, which includes an RC filter, made up of a MiM capacitor and an nMOS-based resistor. The latter can be adjusted by tuning the control signal Ctrl-Add – ranging from 0.6 to 0.9 V – which allows the gain, offset and delay of the adder to be controlled and calibrated for each operation mode of the modulator.

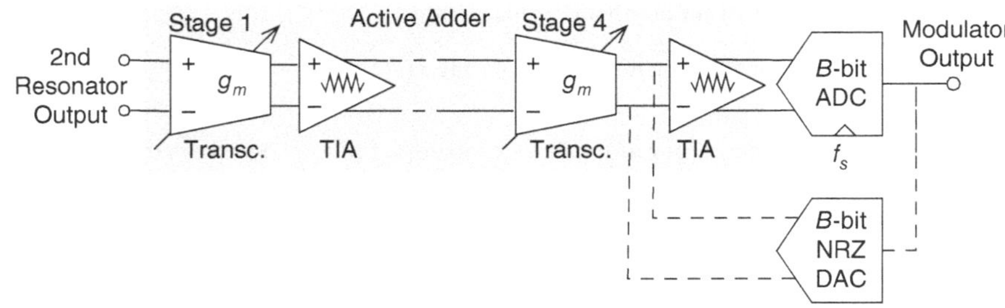

Figure 9.35 Fourth-order active loop-filter adder.

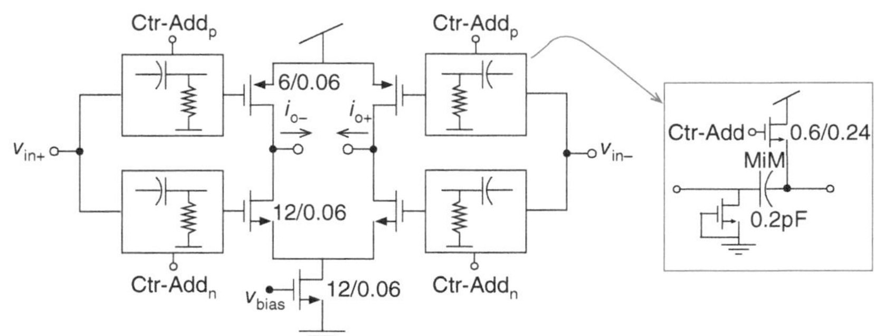

Figure 9.36 Programmable transconductors used in the loop-filter adder.

As an illustration, Figure 9.37 shows a 100-sample Monte Carlo simulation of the frequency response of the adder, with a worst-case gain of 45. As also happens with the resonators, the loop-filter adder includes a high degree of programmability in order to optimize its frequency response for the different wireless standards under study. To this end, the adder can also be calibrated experimentally using a scheme similar to that shown in Figure 9.33. However, in this case the same bonding pad used to test the output of the second resonator is used as an input pad to calibrate the adder. This scheme is implemented by a CMOS switch between the output buffer of the second resonator and the test pad, so that the latter can be used either as an output pad for testing the second resonator or as an input pad for testing the adder.

9.5.3.4 Embedded 4-bit Quantizer with Calibration Figure 9.38 shows the block diagram of the ADC used to implement the 4-bit quantizer of the modulator. The thermometer-code digital output of the flash ADC is binary encoded and stored in a ROM, which provides a 4-bit digital output clocked at f_s = 1.25 GHz or 2 GHz, depending on the operation mode. In order to simplify the testing process, a serial-to-parallel register (SPR) is used as a demultiplexer, so that the 4-bit output bitstream rated at f_s is transformed into a 16-bit signal clocked at $f_s/4$. As an illustration, Figure 9.39 shows the operation of a digital block that transforms a single 4-bit serial data input clocked at 2 GHz into four 1-bit parallel data outputs clocked at 500 MHz. A strobe signal is used to trigger and capture the parallel output data.

The analog output signal provided by the loop-filter adder is quantized by a flash ADC architecture – made up of a bank of 15 comparators and a resistor ladder that generates the internal

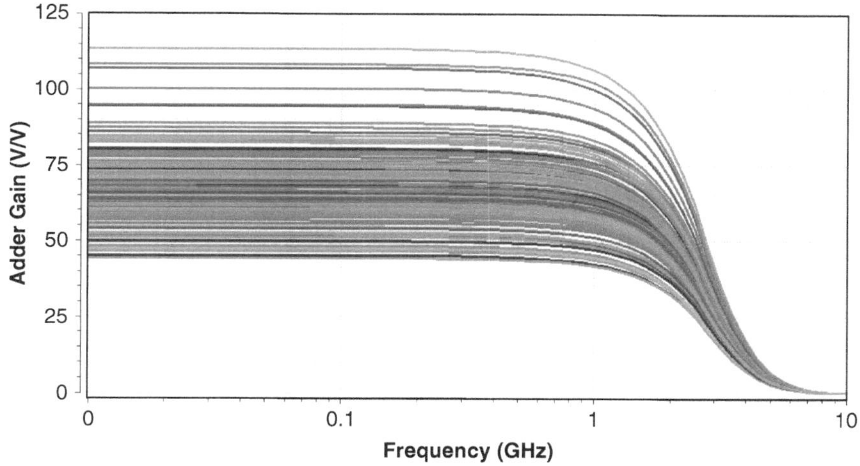

Figure 9.37 Influence of the process variability on the adder frequency response.

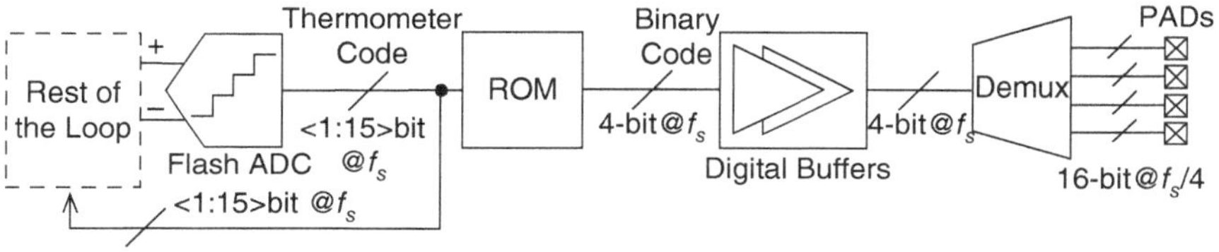

Figure 9.38 Block diagram of the 4-bit ADC embedded in the BP CT-ΣΔM.

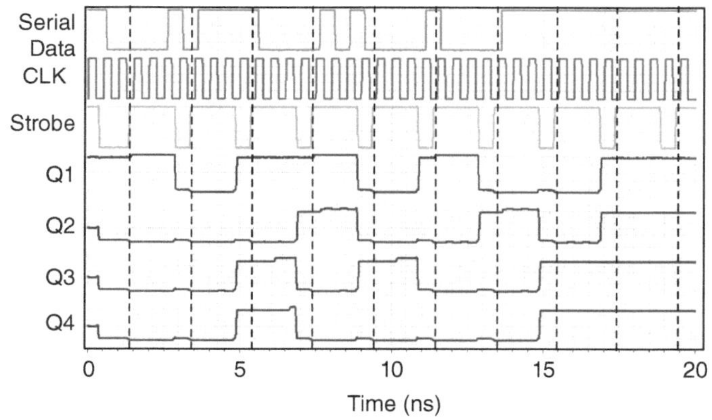

Figure 9.39 Illustrating the serial-to-parallel conversion of the output bitstream.

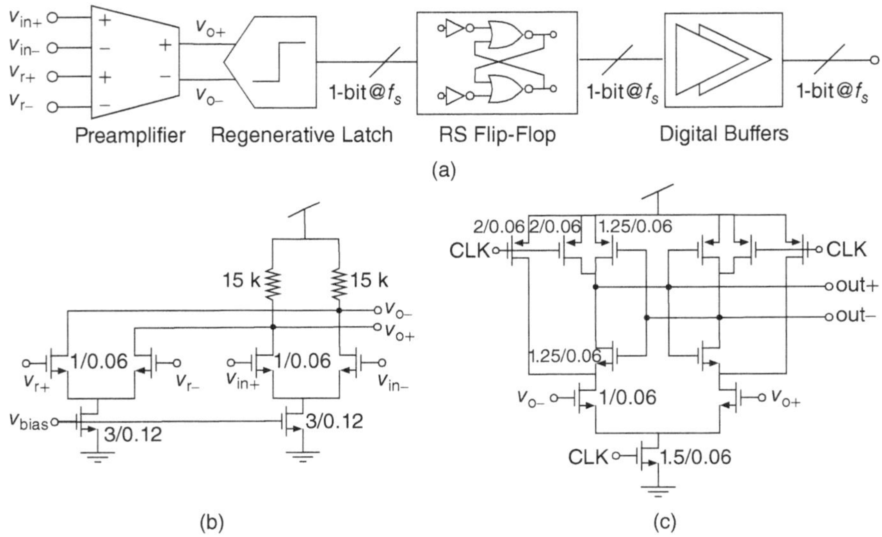

Figure 9.40 Regenerative-latch based comparator: (a) block diagram; (b) preamplifier; (c) regenerative latch.

references from the full-scale reference voltage, $V_{\text{ref}} = 1.2$ V. Figure 9.40 shows the comparator schematic, which is made up of:

- a simple OTA preamplifier to improve the static resolution: the offset and hysteresis;
- a conventional positive-feedback regenerative latch, built by cross-coupling a pair of CMOS inverters;
- a set-reset (SR) flip-flop to store the voltage provided by the latch.

During the comparison or strobe phase – when the clock signal, CLK, is high – the action of the positive feedback forces the output to evolve very quickly to the corresponding logic state according to the sign of the input signal. When CLK is a logic zero, the outputs of the latch are reset to logic one and the SR flip-flop keeps the comparator output value until the next strobing phase. As discussed in Section 8.4.2, Monte Carlo simulations and technology corner analyses need to be completed in order to appropriately size the comparators so that the time response is fast enough and the quantizer offset and hysteresis do not affect the performance of the modulator.

In addition, to reduce the offset by design, a calibration circuit – conceptually shown in Figure 9.41 – is included in the resistor ladder of the quantizer. This adjusts the offset of the comparators and reduces its influence on the modulator performance. The circuit is based on a switchable resistor network that allows each comparator offset to be changed by 40mV per control bit. Once the different comparator offsets are computed, they can be calibrated off-chip with 40-mV accuracy by means of the control signals CTRLi ($i = 1, 2 ... n$).

The feedback loop of the BP CT-ΣΔM is essentially made up of the D-latches to implement the required delays and the corresponding DACs (see Figure 6.10). Current-steering DACs have been used for their potential benefits in terms of high speed and the convenience to interface with Gm-C resonators. In this example, the nMOS and pMOS current cells shown in Figure 9.42 have been used

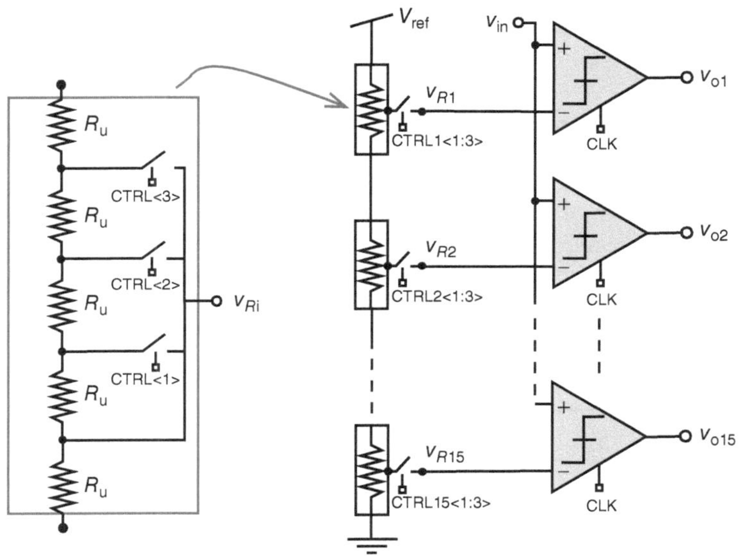

Figure 9.41 Conceptual (single-ended) schematic of the calibration circuit used in the embedded 4-bit quantizer.

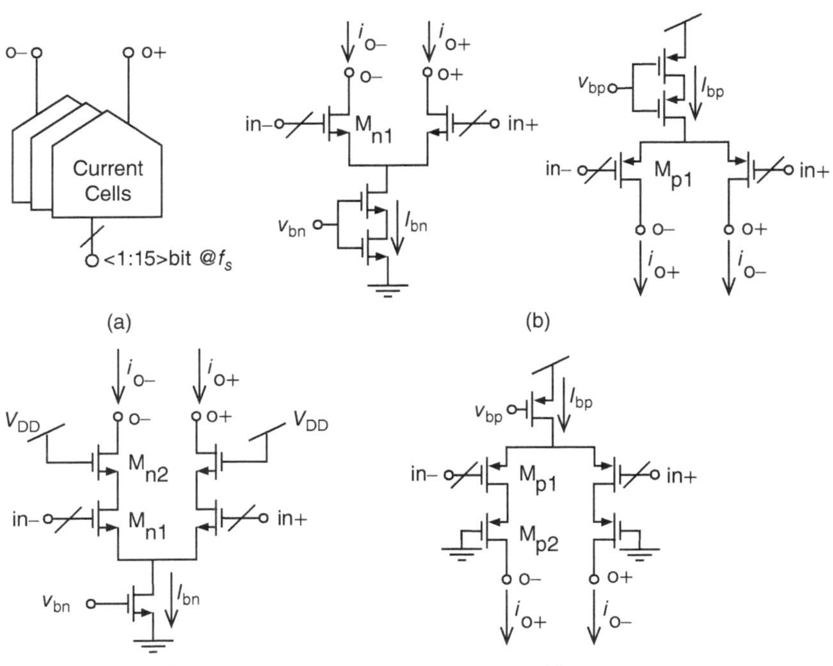

Figure 9.42 Current-steering DAC used in this design example: (a) diagram; (b) CMOS cells implementing feedback currents $I_{c0,c0d}$; (c) nMOS cell for $I_{c1,2}$; (d) pMOS cell for $I_{c1d,c2d}$ (see Figure 6.10).

in order to implement the loop-filter feedback coefficients of the modulator. Simple current cells have been used to implement the ELD compensation loop-filter coefficients, while the rest of feedback currents were provided by cascode cells.

9.5.3.5 Biasing, Digital Control Programmability and Testability In order to adapt the performance of the BP CT-ΣΔM to the different operation modes with optimized power consumption, the biasing of all building blocks is provided by using the master bias current generator, conceptually illustrated in Figure 9.43. The master bias current, I_{bias}, can be generated through an external resistor, say $R_{\text{bias}} = 100\ \text{k}\Omega$, and the mirrored currents biasing all modulator subcircuits are provided by either pMOS or nMOS current mirrors. All mirrored currents go directly to all analog blocks, so that they have a dedicated signal to control their bias current independently.

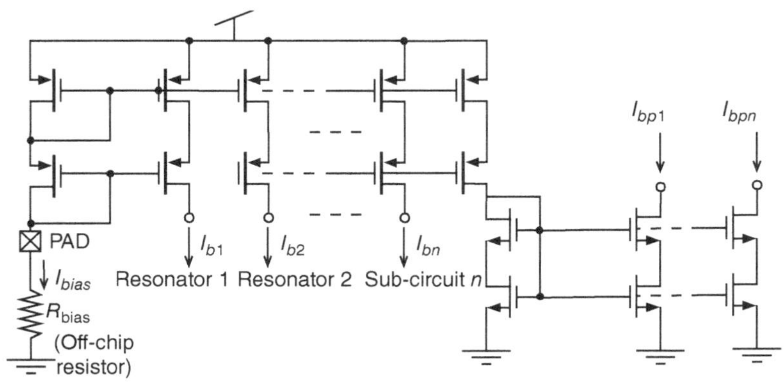

Figure 9.43 Master bias current circuit used in the BP CT-ΣΔM example.

Note that the diverse reconfiguration strategies included in the modulator used as the case study need more than 180 digital control signals. Such a large number of signals requires a serial-to-parallel solution to provide the necessary control signals to the chip, while keeping a reduced number of bonding pads to make the testing feasible. To this end, a SPR is used to collect serial input data and transform it into 180 parallel control bits. Figure 9.44 shows the block diagram of the SPR, which is based on the use of 2 × 182 D flip-flops, which can be integrated together with the modulator on the same chip. The top register (serial register) is used to shift the input data in serial mode, while the bottom register (parallel register), places all digital signals previously shifted in parallel. A low-rate (typically < 10 kHz) clock signal – different from that used in the modulator – is needed for the operations of the registers. An additional digital signal – called "load" – enables either the shifting operation in the top register or the loading in the bottom register. In this way, the BP CT-ΣΔM can be digitally programmed in an easy way, by loading the required control configuration for each operation mode.

Figure 9.45 shows the Arduino-based circuit diagram used to generate the required serial data for the SPR integrated with the BP CT-ΣΔM chip. An Arduino MEGA 2560 board was considered in this example, although the presented methodology can be extended to any other Arduino model [28]. In this configuration, the Arduino pins 48, 52 and 53 are respectively used to generate the serial data input (serial in), the enable signal, and clock signal of the SPR. As the bitstream provided by the Arduino were square waveforms of 5-V amplitude, the signal range was scaled down to 1.2 V using potentiometers, and the resulting pulses were transmitted to the chip through a set of digital

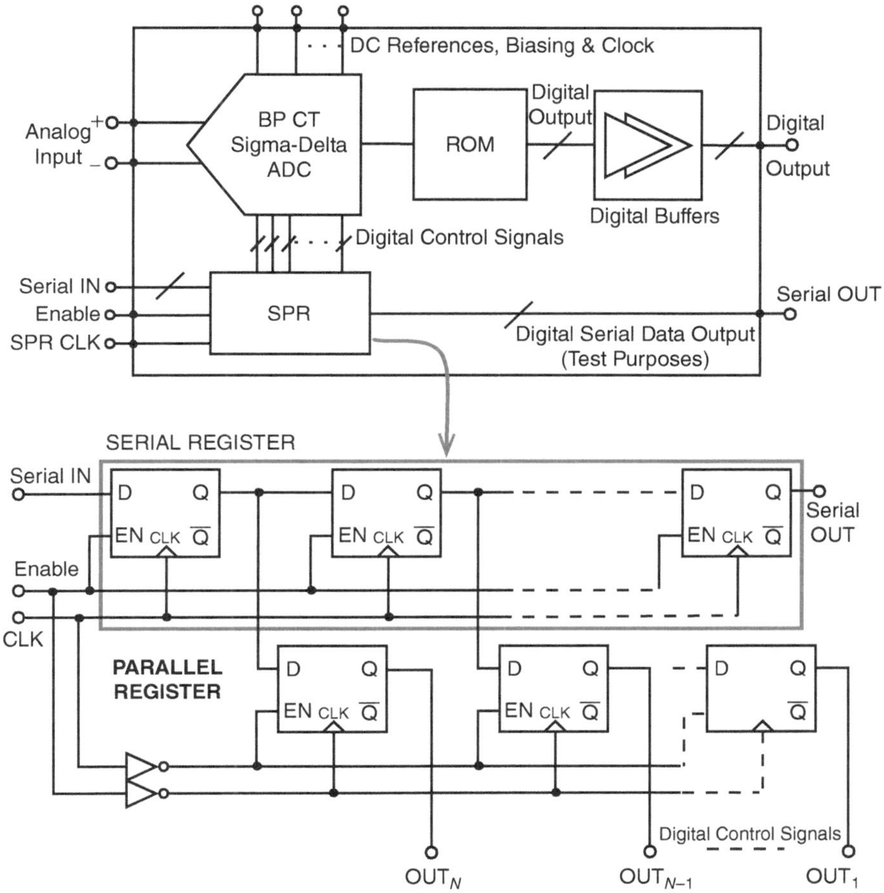

Figure 9.44 Block diagram of the serial-to-parallel register used to generate the required digital control signals of the Gm-LC BP CT-ΣΔM.

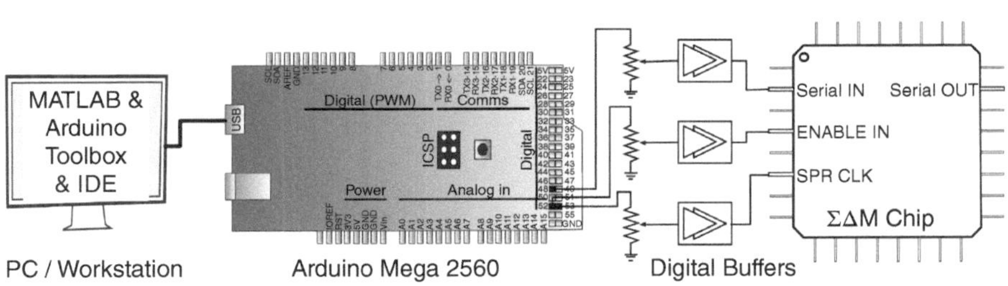

Figure 9.45 Circuit scheme of the Arduino-based signal pattern generator.

buffers. An Arduino prototyping shield board was used so that the whole system can be mounted into a soldered single module that was easily connected to the test PCB.

The MATLAB hardware support package for Arduino was used in this example to communicate and program the circuit in Figure 9.45 over a USB cable [29]. To this end, a very simple MATLAB script – listed in Figure 9.46 – was used. This script essentially loads the signal patterns, which can in turn be generated in MATLAB in CSV file format. Once the signal pattern is read, the commands `a.pinMode` and `a.digital.Write` are used to transmit the generated data to the Arduino board, which in turn generates the serial input data for the SPR included in the ΣΔM chip.

Figure 9.47 shows a photograph of the PCB used to test the BP CT-ΣΔM chip (highlighted in the figure). A general-purpose MICTOR connector is used to connect the Arduino-based hardware to the PCB. Figure 9.48 shows the serial input data provided by the Arduino and the corresponding data loaded on chip. It can be shown how both serial data match each other and hence, are properly loaded

```
clear all;
clc;
B=csvread('ControlSignalsCalRes1_1.csv');
a=arduino('COM5');
a.pinMode(53,'outPut');
a.pinMode(52,'outPut');
a.pinMode(48,'outPut');
for r=1:375
clk_in=B(r,1)
enable_in=B(r,2)
data_in=B(r,3)
a.digitalWrite(48,data_in);a.digitalWrite(52,enable_in);
a.digitalWrite(53,clk_in);
pause(0.1);
end
```

Figure 9.46 MATLAB script used to control the Arduino-based signal pattern generator.

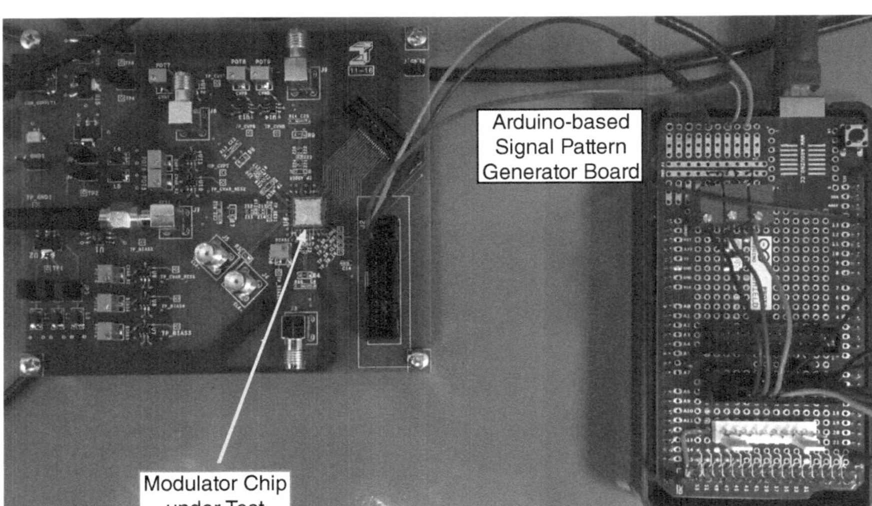

Figure 9.47 Set-up including the chip, test PCB and an Arduino-based pattern generator.

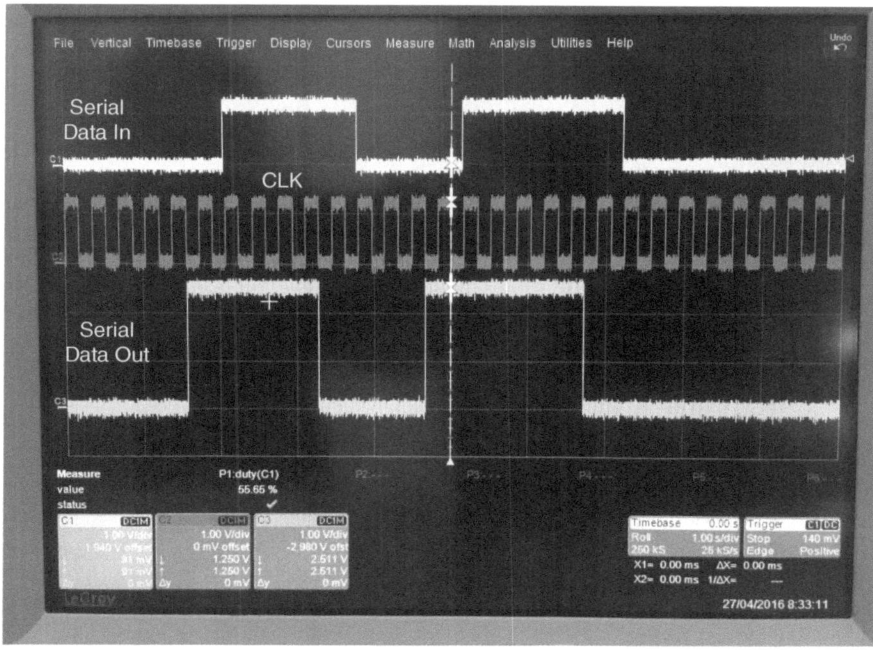

Figure 9.48 Illustrating the signal pattern generated by Arduino and the corresponding output data loaded and generated in the BP CT-ΣΔM chip.

in the modulator. The main benefit of this approach is that the generated signal pattern can be readily modified in MATLAB, using a simple script like that shown in Figure 9.46, and a new data series can be modified and loaded on the chip in order to reconfigure the operation mode in which the modulator chip is being characterized. This testing strategy can be useful in characterizing experimentally the performance of ΣΔ ADCs with a high degree of programmability, without requiring expensive test equipment in the lab.

9.6 Summary

This chapter discussed some practical design considerations to be taken into account to go from ΣΔM circuits to chip implementation and lab measurements. In addition to the required peripheral circuits – such as clock circuit generation, biasing circuitry, and digital control signal processing – some critical issues dealing with the layout, die prototyping and test setup were discussed. Some of these practical design considerations were exemplified with three ΣΔM case studies covering different specifications, ranging from high-resolution sensor interfaces to broadband and reconfigurable ADCs for wireless telecom systems.

The next chapter concludes this book with an exhaustive study of the state of the art of ΣΔMs, in order to help designers select the optimum ΣΔM architecture and circuit technique for their applications, and also to identify the trends, challenges, and practical solutions that have been proposed in each case by the ΣΔM design community, showing the path towards a next generation of ΣΔMs targeting new emerging digital applications.

References

[1] R. del Río, F. Medeiro, B. Pérez-Verdú, J. M. de la Rosa, and A. Rodríguez-Vázquez, *CMOS Cascade ΣΔ Modulators for Sensors and Telecom: Error Analysis and Practical Design*. Springer, 2006.

[2] L. Breems, R. Rutten, and G. Wetzker, "A Cascaded Continuous-Time ΣΔ Modulator with 67-dB Dynamic Range in 10-MHz Bandwidth," *IEEE J. of Solid-State Circuits*, vol. 39, pp. 2152–2160, December 2004.

[3] G. Mitteregger, C. Ebner, S. Mechnig, T. Blon, C. Holuigue, and E. Romani, "A 20-mW 640-MHz CMOS Continuous-Time ΣΔ ADC With 20-MHz Signal Bandwidth, 80-dB Dynamic Range and 12-bit ENOB," *IEEE J. of Solid-State Circuits*, vol. 41, pp. 2641–2649, December 2006.

[4] M. Bolatkale *et al.*, "A 4GHz Continuous-Time ΔΣ ADC With 70dB DR and –74 dBFS THD in 125 MHz BW," *IEEE J. of Solid-State Circuits*, vol. 46, pp. 2857–2868, December 2011.

[5] K. Lee and R. Meyer, "A Current-controlled Latch Sense Amplifier and a Static Power-saving Input Buffer for Low-power Architecture.," *IEEE J. of Solid-State Circuits*, vol. 20, pp. 1103–1113, December 1985.

[6] A. Morgado, R. del Río, and J. M. de la Rosa, *Nanometer CMOS Sigma-Delta Modulators for Software Defined Radio*. Springer, 2012.

[7] B. Razavi, *Design of Analog CMOS Integrated Circuits*. McGraw-Hill, 2000.

[8] A. Morgado, R. del Río, J. M. de la Rosa, L. Bos, J. Ryckaert, and G. van der Plas, "A 100kHz-10MHz BW, 78-to-52dB DR,4.6-to-11mW Flexible SC ΣΔ Modulator in 1.2-V 90-nm CMOS," *Proc. of the IEEE European Solid-State Circuits Conf.*, pp. 418–421, September 2010.

[9] M. Felder and J. Ganger, "Analysis of Ground-Bounce Induced Substrate Noise Coupling in a Low Resistive Bulk Epitaxial Process: Design Strategies to Minimize Noise Effects on a Mixed-Signal Chip," *IEEE Trans. on Circuits and Systems – II: Analog and Digital Signal Processing*, vol. 46, pp. 1427–1436, October 1999.

[10] F. Maloberti, "Layout of Analog and Mixed Analog-Digital Circuits," in *Design of Analog-Digital VLSI Circuits for Telecommunications and Signal Processing* (J. Franca and Y. Tsividis, editors), Prentice-Hall, 1994.

[11] Y. Tsividis, *Mixed Analog-Digital VLSI Devices and Technology: An Introduction*. McGraw-Hill, 1996.

[12] S. Cao *et al.*, "ESD Design Strategies for High-Speed Digital and RF Circuits in Deeply Scaled Silicon Technologies," *IEEE Trans. on Circuits and Systems – I: Regular Papers*, vol. 57, pp. 2301–2311, September 2010.

[13] J. M. de la Rosa *et al.*, "A CMOS 110-dB@40-kS/s Programmable-Gain Chopper-Stabilized Third-Order 2-1 Cascade Sigma-Delta Modulator for Low-Power High-Linearity Automotive Sensor ASICs," *IEEE J. of Solid-State Circuits*, vol. 40, pp. 2246–2264, November 2005.

[14] C. C. Enz and G. C. Temes, "Circuit Techniques for Reducing the Effects of Op-Amp Imperfections: Autozeroing, Correlated Double Sampling, and Chopper Stabilization," *Proc. of the IEEE*, vol. 84, pp. 1584–1614, November 1996.

[15] A. Morgado, R. del Río, and J. M. de la Rosa, "An Adaptive ΣΔ Modulator for Multi-Standard Hand-Held Wireless Devices," *Proc. of the IEEE Asian Solid-State Circuits Conf.*, pp. 232–235, 2007.

[16] J. Ruiz-Amaya *et al.*, "High-Level Synthesis of Switched-Capacitor, Switched-Current and Continuous-Time ΣΔ Modulators Using SIMULINK-based Time-Domain Behavioral Models," *IEEE Trans. on Circuits and Systems – I: Regular Papers*, vol. 52, pp. 1795–1810, Sep. 2005.

[17] A. Morgado *et al.*, "High-Efficiency Cascade ΣΔ Modulators for the Next Generation Software-Defined-Radio Mobile Systems," *IEEE Transactions on Instrumentation and Measurement*, vol. 61, pp. 2860–2869, Nov. 2012.

[18] K. Koli *et al.*, "A 900-MHz Direct Delta-Sigma Receiver in 65-nm CMOS," *IEEE J. of Solid-State Circuits*, vol. 45, pp. 2807–2818, Dec. 2010.

[19] P. Shettigar and S. Pavan, "Design Techniques for Wideband Single-Bit Continuous-Time ΔΣ Modulators With FIR Feedback DACs," *IEEE J. of Solid-State Circuits*, vol. 47, pp. 2865–2879, December 2012.

[20] H. Shibata *et al.*, "A DC-to-1GHz Tunable RF ΔΣ ADC Achieving DR=74dB and BW=150MHz at f_0 =450MHz Using 550mW," *IEEE ISSCC Digest of Technical Papers*, pp. 150–151, February 2012.

[21] E. Martens *et al.*, "RF-to-Baseband Digitization in 40nm CMOS With RF Bandpass ΔΣ Modulator and Polyphase Decimation Filter," *IEEE J. of Solid-State Circuits*, vol. 47, pp. 990–1002, April 2012.

[22] S. Gupta *et al.*, "A 0.8-2GHz Fully-Integrated QPLL-Timed Direct-RF-Sampling Bandpass ΣΔ ADC in 0.13μm CMOS," *IEEE J. of Solid-State Circuits*, vol. 47, pp. 1141–1153, May 2012.

[23] M. Englund *et al.*, "A Programmable 0.7–2.7 GHz Direct ΔΣ Receiver in 40nm CMOS," *IEEE J. of Solid-State Circuits*, vol. 50, pp. 644–655, March 2015.

[24] A. Ashry and H. Aboushady, "A 4th Order 3.6 GS/s RF ΣΔ ADC With a FoM of 1pJ/bit," *IEEE Trans. on Circuits and Systems – I: Regular Papers*, vol. 60, pp. 2606–2617, October 2013.

[25] H. Chae *et al.*, "A 12 mW Low Power Continuous-Time Bandpass ΔΣ Modulator With 58 dB SNDR and 24 MHz Bandwidth at 200 MHz IF," *IEEE J. of Solid-State Circuits*, vol. 49, pp. 405–415, February 2014.

[26] C. Wu *et al.*, "A Wideband 400 MHz-to-4 GHz Direct RF-to-Digital Multimode ΔΣ Receiver," *IEEE J. of Solid-State Circuits*, vol. 49, pp. 1639–1652, July 2014.

[27] G. Molina *et al.*, "LC-Based Bandpass Continuous-Time Sigma-Delta Modulators with Widely Tunable Notch Frequency," *IEEE Trans. on Circuits and Systems – I: Regular Papers*, vol. 61, pp. 1442–1455, May 2014.

[28] Arduino, *Arduino MEGA 2560.* [Online]. Available: http://www.arduino.cc/en/Main/arduinoBoardMega2560, 2016.

[29] Mathworks, *MATLAB Support Package for Arduino Hardware*. [Online]. Available: http://www.mathworks.com/help/supportpkg/arduinoio/, 2016.

CHAPTER 10

FRONTIERS, TRENDS AND CHALLENGES: TOWARDS NEXT-GENERATION ΣΔ MODULATORS

Since the invention of ΣΔ modulation techniques, a multitude of ICs has been developed, using diverse technology processes and architectural and circuit techniques, and targeting a huge number of applications – ranging from high-resolution A/D interfaces for sensors and instrumentation to ultra-low-power biomedical applications and broadband communications [1–4]. However, the original concepts and fundamental signal processing techniques behind ΣΔMs have evolved a lot in recent years. Moreover, new application scenarios require innovative solutions to meet increasingly demanding specifications in a power-efficient manner, thus leading to a new generation of ΣΔ converters. An example is RF digitizers for SDR-based receivers, in which part of the RF signal processing is embedded in the ΣΔ loop [5]. Others examples use *digitally-assisted* analog circuits in the loop filter [6, 7] or time-to-digital conversion in the embedded quantizer [8], thus taking advantage of digital signal processing integrated at the nanoscale.

An in-depth understanding of the performance of state-of-the-art ΣΔMs, their trends, challenges, and circuits and systems solutions, constitutes a powerful tool for designers, enabling them to select the optimum ΣΔM architecture, circuit implementation, technology process, and so on for a given set of specifications and/or a particular application. With this objective in mind, and following the practical approach considered throughout this book, this chapter gives an overview of reported cutting-edge ΣΔM ICs fabricated in CMOS technologies. The main purpose here is to give an exhaustive analysis of the evolving trends, design challenges, and reported solutions in order to extract practical conclusions and guidelines, which might be useful for designers in their own projects. Moreover, a survey of some promising new ΣΔ techniques is given, highlighting future directions and visions about next-generation ΣΔ converters as well as the implications of their integration in analog-hostile *deep-nanometer* CMOS processes.

Sigma-Delta Converters: Practical Design Guide, Second Edition. José M. de la Rosa.

This chapter is inspired by previous surveys on the state-of-the-art performance of ADCs [3, 9–16]. The majority of these covered many kinds of ADCs, and not only ΣΔMs. The surveys published by Walden [9], Murmann [10], and later by Jonsson [13, 14] and Manganaro [15], are exhaustive and particularly accurate. However, here we focus on ΣΔM ICs. In addition to analyzing the statistics and comparing data collected from publications and extracted from standard performance metrics, in this chapter we give an overview of ΣΔM circuit and system techniques which represent state-of-the-art performance, highlighting current trends and challenges as well as the proposed solutions.

The data used for this survey was mainly collected from the *IEEE Journal of Solid-State Circuits* as well as proceedings of major conferences sponsored by the IEEE Solid-State Circuits Society (SSCS): the International Solid-state Circuits Conference (ISSCC), the European Solid-state Circuits Conference (ESSCIRC), the Custom Integrated Circuits Conference (CICC), the Symposium on VLSI Circuits (VLSI), the Asian Solid-State Circuits Conference (ASSCC), and the Radio Frequency Integrated Circuits Symposium (RFIC). In addition to these SSCS conference proceedings, details of some other remarkable ΣΔM ICs were published in the *IEEE Transactions on Circuits and Systems (Parts I and II)* and the *IEEE International Symposium on Circuits and Systems (ISCAS)* and these are also collected in the survey. Overall, more than 500 ΣΔM ICs have been analyzed and considered in this review. Although the works under study include papers published from 1990 to 2017, data was more exhaustively collected for the last fifteen years.[1]

Following this introduction, the chapter is organized as follows. Section 10.1 gives an overview of the state of the art of ΣΔMs, comparing their performance with Nyquist-rate ADCs. The diverse families of ΣΔM architectures and circuit techniques are compared in Section 10.2 and exhaustively analyzed in Section 10.3 in order to extract practical and empirical design guidelines. Sections 10.4–10.9 review some cutting-edge ΣΔM techniques, analyzing the design trends and challenges in the frontiers of ΣΔMs. Finally, the chapter concludes with a classified description of state-of-the-art references in Section 10.10 and some conclusions are given in Section 10.11.

10.1 State-of-the-Art ADCs: Nyquist-rate versus ΣΔ Converters

In order to put ΣΔMs in the context of the state of the art of data converters, we first compare their performance with Nyquist-rate ADCs using flash, two-step, folding, pipeline and successive approximation register (SAR) techniques. Figure 10.1 represents the performance of cutting-edge ADCs in terms of their main design specifications: the resolution, characterized here by the ENOB and B_w. This plot – often referred to as an *aperture plot* – illustrates the *conversion region* covered by the different ADC architectures. Note that a wide conversion region is covered by ΣΔ ADCs, comparable with that achieved by SAR converters. The reader can also see that two different state-of-the-art (B_w-vs-ENOB) fronts can be distinguished, as highlighted in Figure 10.1. One front is mainly dominated by ΣΔ ADCs, and goes from low-frequency (tens/hundreds of kilohertz) and high-resolution (~22-bit), to medium–high frequency (tens/hundreds of megahertz) and medium–high resolution (~12-bit). The other front is dominated by SAR and pipeline ADCs,[2] ranging from

[1] All these data have been compiled in an Excel spreadsheet, which is available online at http://www.imse-cnm.csic.es/~jrosa/CMOS-SDMs-Survey-IMSE-JMdelaRosa.xlsx. The database in this spreadsheet file is updated every six months and aims to be a complement to the popular and highly cited Murmann's database [17], which includes state-of-the-art performance for ADCs as presented at the IEEE's ISSCC and VLSI conferences.

[2] Oversampling techniques can be also embedded in either SAR or pipeline ADCs. An example cutting-edge ADC is the one reported by Shibata *et al.* who proposed a 28-nm CMOS oversampling CT pipeline ADC, which digitizes signals with 1.125-GHz bandwidth with almost 12-bit ENOB when clocked at 9 GS/s [18].

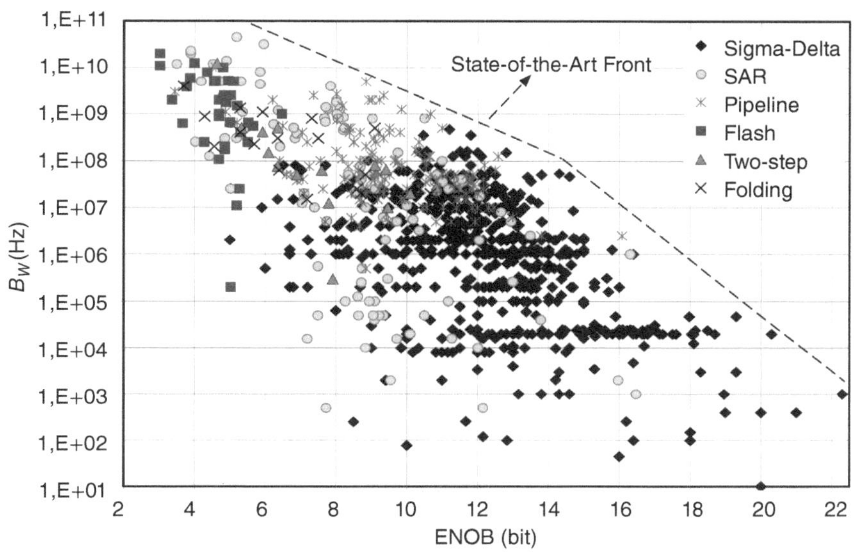

Figure 10.1 Aperture plot of state-of-the-art ADCs.

hundreds of megahertz and medium-resolution (of the order of 10–11-bit), to dozens of gigahertz and low-resolution (up to 5–6-bit). The ultimate limiting factors for both state-of-the-art fronts are thermal noise for low-frequency, high-resolution applications, and clock jitter for applications demanding the highest speeds.

10.1.1 Conversion Energy

It is clear from Figure 10.1 that the state of the art is dominated by ΣΔ, pipeline and SAR ADCs. Apart from their main specifications, ENOB and B_w, it is instructive to compare the efficiency of these ADCs. This can be quantified by measuring the amount of energy per converted sample – also referred to as the *conversion energy*, E, defined as [10, 14]:

$$E \equiv \frac{P(\mathrm{W})}{f_{\mathrm{snyq}}(\mathrm{Hz})} \tag{10.1}$$

where P stands for the power dissipated and $f_{\mathrm{snyq}} \equiv 2 \cdot B_w$, is the effective Nyquist rate – also referred to as the digital output rate – measured in samples per second (S/s).

Thus, it is common to analyze and graphically represent the efficiency of ADCs in the form of plots of *conversion energy* versus ENOB – also known as *energy plots* [10, 15]. As an illustration, Figure 10.2 plots E as a function of the ENOB achieved by state-of-the-art pipeline, SAR, and ΣΔ ADCs. This picture represents graphically the trade-off between resolution and conversion energy. As expected, there is a direct relationship between ENOB and E, so that the larger the resolution, the more conversion energy is needed. It is therefore convenient to express this trade-off in a single metric, which is often referred to as a figure of merit (FOM), and takes into account the main performance metrics of an ADC: the ENOB, B_w, and P.

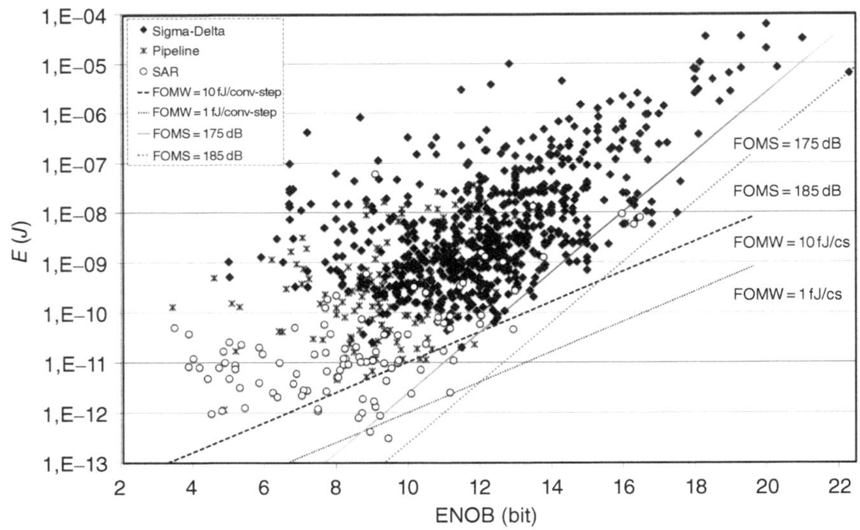

Figure 10.2 Energy plot of state-of-the-art ADCs.

10.1.2 Figures of Merit

Choosing the appropriate FOM is always a matter of debate and discussion. The following two FOMs will be considered here:

$$\text{FOMW} \equiv \frac{E(\text{J})}{2^{\text{ENOB(bit)}}}$$
$$\text{FOMS} \equiv \text{SNDR(dB)} + 10 \cdot \log_{10}[B_w(\text{Hz})/P(\text{W})] \tag{10.2}$$

where SNDR is related to ENOB as SNDR (dB) $= 10 \cdot \log_{10}[(3/2) \cdot 4^{\text{ENOB(bit)}}] = 6.02 \cdot \text{ENOB(bit)} + 1.76$ – according to Equation (1.18). FOMW, measured in joules per conversion step (J/cs), was proposed by Walden [9], whereas FOMS is based on the following FOM – originally proposed by Rabii and Wooley [19] – computed on a logarithmic scale, as suggested by Schreier and Temes [2]:

$$\text{FOMR} \equiv 2kT \cdot \frac{3 \cdot 2^{2 \cdot \text{ENOB(bit)}}}{E(\text{J})} = 2kT \cdot \frac{3 \cdot 2^{2 \cdot \text{ENOB(bit)}} \cdot f_{\text{snyq}}(\text{S/s})}{P(\text{W})} \tag{10.3}$$

with k being the Boltzmann constant and T the chip temperature (measured in kelvin). Note that FOMS can be also expressed in the following form:

$$\text{FOMS} \equiv P_{\text{sig}|_{\text{dBFS}}} - \text{NSD}|_{\text{dBFS/Hz}} - 10 \cdot \log_{10}[P(\text{W})] \tag{10.4}$$

where $P_{\text{sig}|_{\text{dBFS}}}$ denotes the input signal power, computed in decibels, referred to the FS range of the converter, and $\text{NSD}|_{\text{dBFS/Hz}} \equiv (P_{\text{nd}}/B_w)|_{\text{dBFS/Hz}}$ stands for the noise spectral density with P_{nd} being the noise plus distortion power referred to the FS. Therefore, FOMS essentially characterizes the efficiency of a converter according to its NSD extrapolated to a 1-W power consumption when an FS input ($P_{\text{sig}|_{\text{dBFS}}} = 0$) is considered. For example a converter with FOMS = 160 dB would have a NSD= −160 dBFS/Hz at $P = 1$ W, or NSD= −150 dBFS/Hz at $P = 0.1$ W [16]. Note that FOMW emphasizes power consumption, whereas FOMS emphasizes effective resolution. Therefore, the smaller the FOMW value and the larger the FOMS value, the "better" the ADC is.

In this chapter, FOMW and FOMS have been selected because they have been the most commonly used FOMs for the majority of the ΣΔM community. It should be noted that, according to the definition of ENOB given in Chapter 1, DR is used for computing ENOB and consequently for the value of FOMW shown in the tables at the end of this chapter. However, in those ΣΔM ICs exhibiting strongly nonlinear behavior, the effective resolution (the ENOB) is indeed limited by the peak SNDR, as stated in Chapter 1, and hence its corresponding FOMW is worse than that shown in the tables. However, the purpose in this chapter is to analyze the overall performance and trends of ΣΔ ADCs, instead of the specifications achieved by a particular design. From this perspective, and considering that not all authors reported all data required to compute FOMW, the analysis was based on expressions of FOMW derived from DR, rather than SNDR.

As a reference, FOMW and FOMS are also depicted in Figure 10.2, considering the following numerical values: FOMW = 1 and 10 fJ/conv-step; FOMS = 175 dB and 185 dB. The resulting lines of constant FOMW and FOMS in Figure 10.2 can be respectively determined from Equation (10.2) as:

$$\begin{aligned} E(\mathrm{J})|_{\mathrm{FOMW}} &= 2^{\mathrm{ENOB(bit)}} \cdot \mathrm{FOMW} \\ E(\mathrm{J})|_{\mathrm{FOMS}} &= \frac{3}{2} \cdot \frac{4^{\mathrm{ENOB(bit)}}}{10^{\frac{\mathrm{FOMS}}{10}}} \end{aligned} \tag{10.5}$$

Based on these metrics, it can be shown from Figure 10.2 that SAR and ΣΔ ADCs are the most efficient ADC techniques, each dominating different E-vs-ENOB regions. In terms of FOM, there are a number of SAR ADCs achieving a FOMW< 10 fJ/conv-step, and some have a FOMS slightly higher than 175 dB, while this value is exceeded by a number of ΣΔ ADCs with ENOB > 14bit. In general terms, the FOMS limit is more representative of the edge of the feasible region than FOMW. For this reason, FOMS should be preferred over FOMW in most cases.

10.2 Comparison of Different Categories of ΣΔ ADCs

In addition to comparing the performance of ΣΔMs with other kinds of ADCs, we can also compare various categories of ΣΔ ADCs;

- single-loop (single-quantizer) or cascade topologies;
- LP or BP;
- single-bit, multi-bit, or time-encoding quantization (TEQ);
- CT, either Gm-C or active-RC, or DT, mostly using SC circuit techniques.

Apart from these categories, *hybrid* ΣΔMs (H-ΣΔMs) include those topologies that combine diverse techniques: CT and DT, active and passive circuits; ΣΔ/Nyquist-rate ADCs, and so on.

Tables 10.1–10.13[3] sum up the performance of the state-of-the-art ΣΔM ICs considered in this survey. For the sake of clarity, the ICs included in the study have been classified according to their architecture/circuit characteristics into the following tables:

[3]A more detailed version of these tables is periodically updated and corrected at the book's companion website.

Table 10.1 State-of-the-art SC single-loop single-bit LP-ΣΔMs (sorted by FOMW).

Ref.	DR (bit)	B_w (Hz)	OSR	Architecture (order)	Tech./supply voltage (V)	Power (W)	FOMW (pJ/conv)
[6]	15.0	1.00E+05	60	4th	0.35 μm/1.5	1.40E-04	0.01
[408]	12.4	1.00E+05	64	4th	0.13 μm/0.25	3.56E-05	0.03
[105]	13.8	2.00E+04	100	3rd	0.18 μm/0.7	3.60E-05	0.06
[105]	12.3	8.00E+03	125	2nd	0.35 μm/1.2	5.60E-06	0.07
[36]	16.4	2.00E+04	152.5	3rd	0.18 μm/1.8	3.00E-04	0.09
[174]	13.5	2.00E+04	50	4th	0.13 μm/0.9	6.00E-05	0.13
[176]	14.4	2.00E+04	100	3rd	0.18 μm/1	1.30E-04	0.15
[177]	10.5	5.00E+04	65	3rd	90 nm/0.65	2.70E-05	0.19
[178]	16.3	2.50E+04	100	2nd	0.18 μm/0.7	8.70E-04	0.21
[179]	12.5	1.60E+04	48	3rd	0.5 μm/0.9	4.00E-05	0.22
[409]	12.3	2.00E+04	80	3rd	0.13 μm/0.4	6.30E-05	0.31
[105]	12.2	1.20E+02	42	3rd	0.35 μm/1.5	3.80E-07	0.35
[180]	13.0	8.00E+03	64	2nd	0.18 μm/0.65	4.55E-05	0.35
[181]	16.0	2.00E+04	64	4th	0.5 μm/1.5	1.00E-03	0.38
[410]	9.2	5.00E+05	500	1st	14 nm/1.2	2.32E-04	0.39
[182]	15.3	2.50E+05	96	5th	0.8 μm/3.3	4.30E-02	2.13
[183]	14.6	1.00E+03	64	5th	0.35 μm/1.5	2.00E-05	0.40
[106]	9.8	1.00E+04	70	3rd	0.13 μm/0.25	7.50E-06	0.41
[184]	13.7	1.25E+06	50	5th	0.25 μm/2.4	1.40E-02	0.42
[185]	14.4	1.35E+05	48	4th	0.25 μm/2.7	2.84E-03	0.50
[186]	12.2	1.00E+06	31	2nd	0.25 μm/2.4	5.00E-03	0.55
[187]	11.8	3.00E+05	166	2nd	0.13 μm/1.2	2.98E-04	0.14
[180]	11.0	1.60E+04	32	2nd	0.18 μm/0.65	4.55E-05	0.69
[188]	13.1	1.00E+05	520	2nd	0.13 μm/1.5	1.28E-03	0.73
[411]	12.0	2.00E+04	160	3rd	0.13 μm/0.4	1.40E-04	0.85
[189]	14.4	2.50E+04	100	3rd	0.35 μm/1	9.50E-04	0.88
[190]	10.0	2.00E+04	100	3rd	0.18 μm/0.7	3.60E-05	0.88
[191]	8.7	8.35E+05	30	2nd	45 nm/1.1	6.30E-04	0.89
[192]	12.2	8.00E+03	64	2nd	0.18 μm/0.7	8.00E-05	1.06
[193]	16.7	1.10E+04	64	4th	0.5 μm/2.5	2.50E-03	1.07
[21]	22.3	1.00E+03	320	4th	0.35 μm/5.4	1.27E-02	1.23
[188]	9.9	5.00E+05	104	2nd	0.13 μm/1.5	1.28E-03	1.33
[194]	11.5	7.80E+04	64	4th	0.18 μm/0.5	8.60E-04	1.90
[195]	16.4	4.80E+03	256	2nd	0.7 μm/5	1.71E-03	2.06
[188]	8.1	1.00E+06	52	2nd	0.13 μm/1.5	1.28E-03	2.32
[103]	11.5	2.00E+04	64	2nd	0.18 μm/1.8	4.20E-04	3.63
[196]	12.0	3.40E+03	74	2nd	0.7 μm/1.5	1.01E-04	3.63
[197]	16.0	2.50E+04	256	2nd	1 μm/5	1.38E-02	4.21
[198]	8.7	3.84E+06	24	3rd	0.25 μm/2.5	1.35E-02	4.23
[199]	14.3	3.00E+03	128	2nd	0.5 μm/1.5	5.50E-04	4.54
[200]	12.0	8.00E+03	64	3rd	1.2 μm/2	3.40E-04	5.19
[412]	9.9	1.00E+06	64	3rd(I/Q)	65 nm/1.2	1.27E-02	6.65
[201]	15.3	3.50E+03	286	2nd	0.6 μm/1.8	2.00E-03	7.08
[202]	13.0	8.00E+03	64	2nd	0.25 μm/1.8	1.00E-03	7.63
[203]	15.3	2.50E+05	64	5th	0.6 μm/5	2.10E-01	10.41
[204]	8.5	2.50E+02	16	3rd	0.35 μm/1.8	2.20E-06	12.15
[205]	13.4	9.77E+04	128	2nd	1.2 μm/5	2.59E-02	12.27
[175]	18.5	2.40E+04	—	6th	0.35 μm/5-3.3	2.30E-01	12.93

(*continued overleaf*)

Table 10.1 *(continued)*

Ref.	DR (bit)	B_w (Hz)	OSR	Architecture (order)	Tech./supply voltage (V)	Power (W)	FOMW (pJ/conv)
[206]	12.0	5.00E+04	102	2nd	0.35 μm/1	5.60E-03	13.67
[20]	21.0	4.00E+02	320	4th	3 μm/10	2.50E-02	14.90
[206]	13.0	2.00E+04	256	2nd	0.35 μm/1	5.60E-03	17.09
[198]	14.0	2.00E+04	192	3rd	0.25 μm/2.5	1.15E-02	17.55
[207]	20.0	4.00E+02	320	4th	0.6 μm/5	1.60E-02	19.07
[208]	13.5	2.50E+05	64	4th	1.5 μm/5	1.60E-01	27.62
[209]	10.3	1.00E+04	25–125	4th	0.18 μm/1.8	7.50E-04	29.13
[210]	14.5	8.00E+03	256	2nd	3 μm/5	1.20E-02	32.37
[211]	13.8	1.00E+03	250	3rd	2 μm/5	9.40E-04	32.95
[212]	14.2	9.77E+03	256	2nd	2 μm/5	1.30E-02	35.37

Table 10.2 State-of-the-art SC single-loop multi-bit LP-ΣΔMs.

Ref.	DR (bit)	B_w (Hz)	OSR	Architecture*	Tech./supply voltage (V)	Power (W)	FOMW (pJ/conv)
[213]	13.7	1.00E+05	16	2nd(9L)	0.18 μm/1.5	1.40E-04	0.05
[214]	13.8	2.00E+04	63	4th(3L)	0.13 μm/0.5	3.50E-05	0.06
[215]	15.0	2.40E+04	64	3rd(5b)	65 nm/0.6	1.33E-04	0.09
[156]	12.7	1.00E+06	24	2nd(–)	0.18 μm/1.5	1.35E-03	0.10
[216]	13.9	1.10E+06	60	2nd(5L)	0.18 μm/1.8	5.40E-03	0.16
[217]	15.0	1.60E+04	128	3rd(3b)	0.18 μm/1	1.70E-04	0.16
[218]	16.3	2.50E+04	100	2nd(18L)	0.18 μm/0.7	6.80E-04	0.17
[219]	13.2	4.20E+06	12	3rd(4b)	0.18 μm/1.5	1.30E-02	0.17
[413]	12.7	5.00E+06	12	2nd(9L)	0.13 um /1.2	1.21E-02	0.18
[220]	10.7	1.94E+06	20	2nd(5L)	90 nm/1.2	1.20E-03	0.19
[221]	14.0	2.20E+06	33	2nd(4b)	0.18 μm/1.8	1.38E-02	0.19
[222]	13.3	1.90E+06	16	2nd(4b)	0.18 μm/1.5	8.10E-03	0.21
[223]	15.0	2.40E+04	128	3rd(4b)	0.18 μm/1	3.50E-04	0.22
[414]	16.2	2.56E+02	128	2nd(4b)	0.18 μm/1	8.60E-06	0.22
[224]	12.3	1.25E+06	16	3rd(9L)	0.18 μm/1.2	3.30E-03	0.25
[225]	11.0	1.60E+04	32	2nd(–)	0.18 μm/1	1.70E-05	0.26
[415]	13.1	4.00E+06	15	3rd(4b)	0.18 μm/1.6	1.92E-02	0.27
[226]	13.4	1.10E+06	47	2nd(3b)	0.13 μm/1.5	7.00E-03	0.30
[227]	10.2	2.00E+07	8	3rd(4b)	0.18 μm/1.8	1.60E-02	0.35
[183]	14.6	1.00E+03	16	5th(1.5b)	0.35 μm/1.8	2.00E-05	0.40
[228]	13.0	2.50E+06	12	4th(4b)	0.18 μm/1.6	1.92E-02	0.47
[416]	14.6	1.00E+04	128	2nd(5L)	0.13 μm/1.2	2.41E-04	0.49
[229]	9.8	6.00E+06	8	3rd(5b)	0.18 μm/1.8	6.18E-03	0.58
[230]	14.0	1.25E+06	32	5th(1.5b)	0.25 μm/2.5	2.40E-02	0.59
[231]	11.3	4.00E+06	13	4th(4b)	90 nm/1.2–3	1.18E-02	0.60
[232]	13.7	1.25E+07	8	5th(4b)	0.18 μm/1.8	2.00E-01	0.60
[226]	14.4	3.00E+05	175	2nd(3b)	0.13 μm/1.5	8.00E-03	0.63
[233]	10.7	1.00E+06	20	2nd(4b)	90 nm/1.3	2.10E-03	0.63
[234]	12.8	2.00E+05	65	2nd(5b)	0.13 μm/1.5	2.40E-03	0.82
[234]	12.0	2.00E+05	65	2nd(5b)	0.13 μm/1.2	1.40E-03	0.85
[233]	12.5	2.00E+05	50	2nd(4b)	90 nm/1.3	2.10E-03	0.91

(continued overleaf)

Table 10.2 *(continued)*

Ref.	DR (bit)	B_w (Hz)	OSR	Architecture*	Tech./supply voltage (V)	Power (W)	FOMW (pJ/conv)
[235]	15.0	2.40E+04	128	3rd(3L)	0.13 μm/0.9	1.50E-03	0.95
[236]	17.2	2.00E+04	154	2nd(4b)	0.13 μm/3.3	9.90E-03	1.61
[237]	13.8	3.00E+05	96	2nd(3b)	0.18 μm/1.8	1.50E-02	1.75
[238]	17.0	3.13E+04	128	2nd(5b)	0.18 μm/3.3	1.47E-02	1.79
[239]	15.8	1.25E+06	24	3rd(4b)	0.65 μm/5	2.95E-01	2.07
[240]	12.7	2.20E+06	8	2nd(4b)	0.35u/3.3	6.20E-02	2.18
[241]	12.7	2.20E+06	8	5th(4b)	0.35 μm/3.3	6.20E-02	2.18
[234]	8.0	2.00E+06	12	2nd(5b)	0.13 μm/1.5	2.90E-03	2.81
[242]	12.8	6.25E+05	18	2nd(6b)	0.18 μm/2.7	3.00E-02	3.30
[243]	18.7	2.00E+04	128	5th(4b)	0.35 μm/5-3.3	6.80E-02	3.99
[242]	11.7	1.92E+06	12	2nd(6b)	0.18 μm/2.7	5.00E-02	4.00
[244]	19.0	4.00E+02	512	2nd(3b)	2 μm/5	2.18E-03	5.19
[245]	13.7	6.25E+05	12	4th(4b)	0.25 μm/2.5	1.00E-01	6.01
[28]	20.3	2.00E+04	128	5th(33L)	0.35 μm/5	3.30E-01	6.39
[245]	13.0	1.00E+06	12	4th(4b)	0.25 μm/2.5	1.05E-01	6.41
[242]	13.5	2.00E+05	58	2nd(6b)	0.18 μm/2.7	3.00E-02	6.47
[417]	13.2	1.25E+05	128	2-ch 3rd(4b)	0.18 μm/1.8	5.96E-02	0.26
[239]	12.0	6.25E+06	8	3rd(4b)	0.65 μm/5	3.80E-01	7.42
[246]	13.7	2.50E+05	16	4th(4b)	1.2 μm/5	5.80E-02	8.96
[247]	16.0	4.50E+01	512	1st(3b)	0.35 μm/2.6	6.00E-05	10.17
[248]	19.3	4.80E+04	64	7th(3L)	0.8 μm/5	7.60E-01	12.26
[249]	16.7	2.00E+04	64	2nd(5b)	0.5 μm/3.3	7.04E-02	16.53
[250]	16.2	2.40E+04	64	2nd(5b)	0.5 μm/3.3	6.86E-02	18.72
[242]	15.3	1.80E+04	639	2nd(6b)	0.18 μm/2.7	3.00E-02	20.37
[251]	18.0	2.00E+04	154	5th(17L)	0.35 μm/5	3.00E-01	27.83
[252]	15.7	2.00E+04	64	2nd(3b)	1.2 μm/5	6.75E-02	32.82
[253]	16.0	1.95E+04	128	3rd(1b–5b)	2 μm/5	8.50E-02	33.26
[254]	14.3	1.00E+03	16	5th(17L)	0.35 μm/1.8	9.00E-03	223.09

*Architecture is expressed in terms of the filter order and the number of bits(b)/levels(L) of the quantizer.

Table 10.3 State-of-the-art SC cascade single-bit LP-ΣΔMs.

Ref.	DR (bit)	B_w (Hz)	OSR	Architecture*	Tech./supply voltage (V)	Power (W)	FOMW (pJ/conv)
[255]	12.8	2.00E+04	64	2-2	0.13 μm/1	1.80E-05	0.06
[255]	11.5	4.00E+04	64	2-2	0.13 μm/1.2	1.80E-02	0.15
[144]	16.3	2.44E+04	128	2-1	90 nm/1	8.60E-04	0.22
[418]	16.1	2.00E+04	90	2-2	0.18 μm/1.8	9.80E-04	0.35
[180]	12.2	1.60E+04	32	2-1	0.18 μm/0.65	6.18E-05	0.42
[180]	13.0	8.00E+03	64	2-1	0.18 μm/0.65	6.18E-05	0.47
[256]	12.2	1.00E+06	31	2-2	0.25 μm/2.4	5.00E-03	0.53
[257]	13.6	1.00E+06	56	3-1	65 nm/1	1.70E-02	0.67
[19]	16.1	2.50E+04	80	2-1	0.8 μm/1.8	2.50E-03	0.71
[258]	14.3	2.00E+04	100	2-2	0.18 μm/1	6.60E-04	0.82
[259]	13.5	1.80E+05	36	2.2	0.4 μm/1.8	5.00E-03	1.20
[260]	18.1	2.00E+04	128	2-1	0.35 μm/3.3	1.47E-02	1.31

(continued overleaf)

Table 10.3 *(continued)*

Ref.	DR (bit)	B_w (Hz)	OSR	Architecture*	Tech./supply voltage (V)	Power (W)	FOMW (pJ/conv)
[261]	14.1	2.00E+05	100	2-1	0.13 μm/1.2	1.10E-02	1.57
[262]	15.0	1.10E+06	24	2-1-1	0.5 μm/3.3	2.00E-01	2.77
[263]	15.7	1.60E+05	64	2-1	1.2 μm/5	6.50E-02	3.82
[264]	14.8	1.00E+06	24	2-1-1	1 μm/5	2.30E-01	4.03
[265]	14.0	1.10E+06	24	2-2-2	0.35 μm/3.3	1.50E-01	4.16
[266]	16.7	2.20E+04	128	2-1	0.6 μm/3	2.20E-02	4.86
[267]	17.0	2.50E+04	128	2-1	1 μm/5	4.70E-02	7.17
[268]	18.1	1.25E+04	64	2-2	0.6 μm/5	7.50E-02	10.68
[269]	15.0	9.00E+04	64	1-1-1	1.5 μm/5	7.60E-02	12.89
[270]	18.1	2.40E+04	128	2-2	0.7 μm/5	5.00E-01	35.91
[271]	14.8	2.50E+04	64	2-2	3 μm/5	7.40E-02	51.03

*The cascade architecture is expressed in terms of the stage orders. For instance, the notation 2-2 is used for a fourth-order cascade ΣΔM with two second-order stages.

Table 10.4 State-of-the-art SC cascade multi-bit LP-ΣΔMs.

Ref.	DR (bit)	B_w (Hz)	OSR	Architecture*	Tech./dupply voltage (V)	Power (W)	FOMW (pJ/conv)
[272]	12.2	5.00E+06	13	2(4b)-2(4b)	0.13 μm/1.2	8.90E-03	0.19
[273]	11.4	2.00E+07	11	2-2(4b)	90 nm/1.2	2.79E-02	0.26
[274]	11.3	2.50E+06	8	1(4L)-1(4L)-1(4L)-1(4L)	65 nm/1.2	3.73E-03	0.29
[275]	12.5	6.25E+05	16	2(3b)-2(2b)	0.18 μm/1.2	2.10E-03	0.29
[419]	13.2	1.67E+07	33	2-2(4L)	65 nm/1	9.40E-02	0.31
[48]	12.2	5.00E+06	13	2(4b)-2(4b)	0.13 μm/1.2	1.60E-02	0.34
[91]	12.0	3.13E+06	8	0-3(17L)	0.18 μm/1.8	1.90E-02	0.74
[233]	9.4	2.00E+06	10	2(4b-dual)	90 nm/1.3	2.10E-03	0.79
[276]	10.4	1.92E+06	100	2-1(5L)	0.13 μm/1.2	4.30E-03	0.83
[277]	14.0	2.20E+06	16	2(3L)-1(3L)-1(3L)	0.25 μm/2.5	6.25E-02	0.87
[278]	10.8	2.00E+07	8	2-2(4b)	90 nm/1.4	7.80E-02	1.09
[276]	13.4	1.00E+05	195	2-1(5L)	0.13 μm/1.2	2.40E-03	1.11
[279]	15.0	2.00E+06	16	2(5b)-2(3b)-1(3b)	0.5 μm/2.5	1.50E-01	1.14
[280]	15.0	1.25E+06	8	2(4b)-1(4b)-1(4b)	0.5 μm/5	1.05E-01	1.28
[281]	13.0	2.20E+06	16	2-1-1(3b)	0.25 μm/2.5	7.17E-02	1.99
[281]	13.8	1.10E+06	32	2-1-1(3b)	0.25 μm/2.5	6.58E-02	2.10
[281]	12.7	2.20E+06	16	2-1-1(3b)	0.25 μm/2.5	6.58E-02	2.25
[282]	13.4	2.00E+04	64	2-2(1.5b)	0.35 μm/0.6	1.00E-03	2.31
[281]	13.7	1.10E+06	32	2-1-1(3b)	0.25 μm/2.5	7.17E-02	2.45
[27]	16.4	1.00E+06	8	2(5b)-2(3b)	0.25 μm/–	4.75E-01	2.75
[283]	13.0	1.10E+06	16	2-1-1(3b)	0.7 μm/5	5.50E-02	3.05
[284]	14.6	1.10E+06	29	2-1-1(2b)	0.35 μm/3.3	1.80E-01	3.29
[285]	13.0	7.81E+05	32	2-2(3b)	0.35 μm/2.5	5.00E-02	3.91
[281]	13.0	1.10E+06	16	2-1-1(4b)	0.35 μm/3.3	7.37E-02	4.09
[286]	12.0	1.05E+06	24	2-1(3b)	1 μm/5	4.10E-02	4.77
[281]	12.0	2.00E+06	16	2-1-1(4b)	0.35 μm/3.3	7.83E-02	4.78
[287]	10.2	8.33E+06	3	1-1-1-1-1-1-1-1(3L)	0.18. μm/1.8	9.50E-02	5.02
[265]	13.0	1.10E+06	24	2-2(5b)	0.35 μm/3.3	9.90E-02	5.49
[288]	8.5	4.00E+07	4	2(1.5b)-2(4b)	0.13 μm/1.2	1.75E-01	6.04
[289]	13.0	7.00E+05	16	2-2-2(3L)	0.7 μm/3.3	8.10E-02	7.06
[290]	14.3	1.00E+05	16	2(1.5b)-2(1.5b)-2(1.5b)	1.2 μm/5	4.00E-02	10.26

*The number of bits/levels of the quantizer is given in parentheses. Thus, 2-1(5L) denotes a cascade 2-1 ΣΔM topology with 5-level quantizer in the second stage. If no parentheses are shown, 1-bit quantization is assumed in a given modulator stage.

Table 10.5 State-of-the-art SC BP-ΣΔMs.

Ref.	DR (bit)	f_n (Hz)	B_w (Hz)	OSR	Architecture	Tech./supply voltage (V)	Power (W)	FOMW (pJ/conv)
[291]	14.4	4.00E+07	2.50E +06	12	4th	0.18 μm/1.8	1.50E-01	1.39
[292]	13.3	2.00E+07	1.25E+06	32	4-4	0.35 μm/3	3.70E-02	1.47
[293]	11.7	4.00E+07	1.00E+06	6	4th	0.18 μm/1.8	1.60E-02	2.46
[294]	13.2	1.00E+07	2.00E+05	33	2-0	0.25 μm/2.1	1.00E-02	2.66
[292]	11.7	2.00E+07	1.76E+06	23	4-4	0.35 μm/3	3.70E-02	3.16
[295]	15.7	1.26E+07	3.10E+05	65	4th	0.18 μm/1.8	1.15E-01	3.61
[296]	13.4	1.07E+07	6.00E+04	0	4th	0.25 μm/1	8.45E-03	6.65
[297]	12.0	1.60E+07	2.00E+06	16	6th	0.25 μm/2.5	1.10E-01	6.71
[296]	12.0	1.07E+07	1.00E+05	0	4th	0.25 μm/1	8.45E-03	10.03
[296]	12.7	1.07E+07	6.00E+04	0	4th	0.25 μm/1	8.45E-03	10.58
[298]	11.7	3.25E+06	2.00E+05	33	4-2	0.8 μm/3	1.44E-02	10.82
[299]	14.5	1.07E+07	4.00E+05	46	4th(4b)	0.15 μm/3.3	2.08E-01	11.22
[299]	15.3	1.07E+07	2.00E+05	93	4th(4b)	0.15 μm/3.3	2.08E-01	12.89
[296]	10.0	1.07E+07	2.00E+05	–	4th	0.25 μm/1	8.45E-03	20.63
[300]	12.6	5.66E+05	2.50E+05	20	2-2(3b)	0.25 μm/2.5	7.70E-02	24.81
[301]	9.0	2.00E+06	3.00E+04	133	4th	2 μm/3.3	8.00E-04	26.04
[302]	11.7	2.00E+07	2.70E+05	148	4th	0.35 μm/3	5.60E-02	31.17
[87]	12.2	2.00E+07	2.00E+05	200	4th	0.6 μm/3.3	7.20E-02	38.26
[303]	9.7	3.25E+06	2.00E+05	32	3rd	0.35 μm/3.3	1.87E-02	56.21
[302]	6.7	2.00E+07	3.84E+06	10	4th	0.35 μm/3	5.60E-02	70.13
[304]	12.0	1.07E+07	2.00E+05	93	6th	0.35 μm/3.3	1.16E-01	70.80
[299]	18.3	1.07E+07	3.00E+03	6167	4th(4b)	0.15 μm/3.3	2.08E-01	107.41
[305]	10.8	3.75E+06	2.00E+05	25	4th	0.8 μm/5	1.30E-01	182.29
[306]	6.8	1.07E+07	2.00E+05	107	2nd	0.35 μm/1	1.20E-02	269.23

Table 10.6 State-of-the-art CT single-loop single-bit LP-ΣΔMs.

Ref.	DR (bit)	B_w (Hz)	OSR	Architecture (order)	Tech./supply voltage (V)	Power (W)	FOMW (pJ/conv)
[420]	12.4	2.00E+06	80	3rd	65 nm/0.7	2.56E-04	0.018
[44]	13.5	3.60E+07	50	4th	90 nm/1.2	1.50E-02	0.018
[421]	12.4	6.00E+07	50	4th (1-bit+FIR)	65 nm/1.4	1.33E-02	0.021
[68]	15.0	2.00E+06	64	4th	0.13 μm/1.4	5.00E-03	0.038
[61]	14.5	1.00E+06	32	4th	0.18 μm/1.8	2.00E-03	0.04
[29]	15.6	2.40E+04	128	3rd	0.18 μm/1.8	1.10E-04	0.05
[307]	15.0	2.40E+04	127	3rd	0.18 μm/1.8	1.10E-04	0.07
[29]	14.8	2.40E+04	128	3rd	0.18 μm/1.8	1.22E-04	0.09
[33]	16.8	2.40E+04	128	3rd	0.18 μm/1.8	2.80E-04	0.05
[422]	16.8	2.40E+04	128	3rd	0.18 μm/1.8	2.80E-04	0.05

(continued overleaf)

Table 10.6 *(continued)*

Ref.	DR (bit)	B_w (Hz)	OSR	Architecture (order)	Tech./supply voltage (V)	Power (W)	FOMW (pJ/conv)
[57]	12.0	1.50E+07	80	3rd	65 nm/1.0	6.96E-03	0.06
[308]	14.6	1.00E+06	32	4th	0.18 μm/1.8	4.70E-03	0.09
[309]	12.8	2.00E+04	64	4th	0.13 μm/0.6	2.86E-05	0.10
[423]	14.8	2.00E+06	128	4th	0.18 μm/1.8	1.48E-02	0.13
[71]	12.5	2.50E+07	44.0	4th	65 nm/1.0	4.14E-02	0.14
[310]	9.6	1.56E+07	32	4th	0.13 μm/1.2	4.00E-03	0.16
[45]	9.9	6.00E+07	50	3rd	45 nm/1.4–1.8	2.00E-02	0.17
[311]	11.3	4.00E+06	18	3rd	65 nm/0.95–1.25	3.60E-03	0.18
[312]	10.8	1.00E+07	32	3rd	90 nm/1.2	6.80E-03	0.19
[313]	11.7	4.00E+06	64	3rd	90 nm/1.2	5.50E-03	0.21
[424]	14.2	2.00E+06	64	4th	0.18 μm/1.8	1.65E-02	0.22
[425]	14.2	2.00E+06	64	4th	0.18 μm/1.8	1.65E-02	0.22
[314]	11.8	4.00E+07	12	4th	90 nm/1.2	6.96E-02	0.24
[315]	10.1	1.00E+07	15	5th	0.11 μm/1.1	5.32E-03	0.25
[313]	12.7	1.92E+06	64	3rd	90 nm/1.2	6.44E-03	0.26
[316]	14.0	6.00E+05	213	4th	90 nm/1.3	5.40E-03	0.27
[317]	12.8	1.95E+06	32	5th	65 nm/2.5	8.55E-03	0.30
[318]	9.7	1.50E+07	50	5th	45 nm/1.1	9.00E-03	0.37
[313]	11.5	1.92E+06	64	3rd	90 nm/1.2	4.34E-03	0.39
[313]	13.5	5.00E+05	90	3rd	90 nm/1.2	5.00E-03	0.43
[170]	12.5	1.00E+06	140	2nd	0.18 μm/1.8	6.00E-03	0.52
[319]	11.8	6.40E+06	32	3rd	0.13 μm/1.2	2.52E-02	0.54
[319]	10.2	1.70E+07	12	3rd	0.13 μm/1.2	2.52E-02	0.61
[320]	10.7	1.25E+06	64	4th	0.13 μm/1.2	2.70E-03	0.65
[321]	11.3	2.00E+06	38	4th	0.18 μm/1.8	6.60E-03	0.65
[322]	11.8	2.50E+04	48	3rd	0.5 μm/1.5	1.35E-04	0.76
[323]	12.3	5.00E+05	64	5th	0.18 μm/1.8	4.40E-03	0.87
[60]	13.3	1.00E+05	65	4th	0.35 μm/2.5	1.80E-03	0.89
[324]	11.2	4.00E+06	50	3rd	0.18 μm/1.8	1.69E-02	0.92
[325]	10.4	1.30E+06	50	4th	0.11 μm/1.2	3.42E-03	0.99
[326]	9.4	5.00E+04	32	3rd	0.5 μm/1.5	7.50E-05	1.11
[327]	11.3	3.15E+06	81	2nd	90 nm/1.2	1.25E-02	1.26
[51]	12.4	1.23E+06	813	2nd	0.18 μm/1.8	1.80E-02	1.35
[325]	11.4	4.28E+05	150	4th	0.11 μm/1.2	3.42E-03	1.51
[328]	10.5	2.50E+06	16	2nd	1.2 μm/3	1.20E-02	1.66
[329]	12.0	2.50E+04	64	3rd	0.18 μm/0.5	3.70E-04	1.81
[330]	11.3	2.50E+04	48	3rd	0.5 μm/1.5	2.50E-04	1.98
[331]	10.0	3.10E+06	64	5th	0.6 μm/3.3	1.60E-02	2.52
[332]	13.0	4.00E+03	64	4th	0.5 μm/2.2	2.00E-04	3.05
[159]	10.0	1.20E+06	16	1st	0.35 μm/3.3	1.22E-02	4.96
[333]	11.7	2.56E+02	111	2nd	0.18 μm/1.4	1.33E-05	7.97
[334]	9.6	1.00E+06	25	2nd	2 μm/5	1.66E-02	10.70
[335]	10.0	2.50E+04	48	3rd	0.5 μm/1.5	7.50E-04	14.65
[426]	11.0	2.00E+03	80	3rd	0.15 μm/1.6	9.60E-05	14.70
[336]	12.8	1.00E+04	10000	3rd	0.18 μm/1.8	4.30E-03	31.21
[337]	12.8	1.00E+02	1500	2nd	0.7 μm/5	2.00E-03	1373.36

Table 10.7 State-of-the-art CT single-loop multi-bit LP-ΣΔMs.

Ref.	DR (bit)	B_w (Hz)	OSR	Architecture (order)	Tech./supply voltage (V)	Power (W)	FOMW (pJ/conv)
[427]	12.5	1.00E+07	32.5	3rd (1.5b)	65 nm/1.1	1.82E-03	0.016
[428]	12.1	2.00E+07	16	3rd(9L)	0.13 μm/1.2	5.10E-03	0.029
[429]	12.8	1.50E+07	40	3rd (2b)	65 nm/1	6.96E-03	0.032
[430]	12.3	2.00E+07	50	3rd (3b)	65 nm/1	6.80E-03	0.033
[431]	12.5	3.00E+07	14.0	3rd (4b)	55 nm/1.1–2.3	1.30E-02	0.037
[432]	11.7	1.60E+08	9.0	4th (5b)	16 nm	4.00E-02	0.037
[56]	11.8	8.00E+07	13.7	4th (3b)	20 nm/1	2.30E-02	0.043
[338]	15.2	2.40E+04	64	3rd(4b)	0.18 μm/1.8	9.00E-05	0.050
[339]	11.7	2.50E+07	16	3rd(4b)	90 nm/1.2	8.50E-03	0.050
[31]	12.7	1.80E+07	17.8	4th (9L)	28 nm/1.2–1.5	3.90E-03	0.050
[433]	11.3	1.00E+07	15.0	4th	40 nm	2.57E-03	0.051
[339]	11.7	2.50E+07	16.0	3rd(4b)	90 nm/1.2	8.50E-03	0.051
[169]	11.7	2.50E+07	10.0	3rd(4b)	90 nm/1.2	8.50E-03	0.051
[34]	17.5	2.00E+04	75.0	3rd(15L)	160 nm/1.6	3.90E-04	0.053
[434]	11.6	1.90E+07	26.0	TI(2ch) 3rd	16 nm	6.20E-03	0.061
[62]	13.0	2.00E+07	16	3rd(4b)	0.13 μm/1.2	2.00E-02	0.06
[435]	11.8	8.50E+06	17.6	3rd (3b)	90 nm/1.2	4.30E-03	0.071
[436]	12.2	1.00E+07	32.0	3rd (4b)	0.13 μm/1.2	7.20E-03	0.077
[437]	12.2	1.00E+07	32.0	4th (4b)	0.13 μm/1.2	7.20E-03	0.077
[438]	13.0	9.00E+06	16.0	3rd+2nd(3b)	65 nm/1.2	1.13E-02	0.077
[34]	16.8	2.00E+04	75.0	3rd (3b)	160 nm/1.6	3.90E-04	0.085
[433]	11.2	3.00E+06	31.0	3rd	65 nm	1.36E-03	0.096
[439]	10.6	7.50E+07	21.0	3rd (4b)	40 nm/1.1	2.28E-02	0.097
[440]	13.0	7.20E+06	12.8	4th (4b)	0.13 μm/1.2–1.4	1.37E-02	0.10
[441]	12.3	1.00E+07	25.0	5th (VCO)	0.18 μm/1.8V	5.80E-02	0.10
[340]	11.4	1.00E+07	15	5th(3b)	110 nm/1.1	5.32E-03	0.10
[442]	10.5	7.50E+07	21.3	3rd (4b)	40 nm/1.1	2.29E-02	0.11
[443]	11.2	8.50E+06	17.6	3rd(3b)	90 nm/1.2	4.30E-03	0.11
[66]	10.2	2.50E+07	10	3rd(4b)	90 nm/1.2	8.00E-03	0.14
[341]	13.0	2.00E+06	32	3rd(3b)	65 nm/1.2	4.52E-03	0.14
[444]	11.5	5.00E+06	24.0	3rd (9L)	90 nm/1	4.60E-03	0.15
[136]	12.0	2.00E+06	26	3rd(4b)	0.13 μm/1.5	3.00E-03	0.18
[342]	12.2	1.00E+06	32	3rd(4b)	0.13 μm/1.2	2.20E-03	0.23
[343]	11.3	5.00E+06	10	3rd(4b)	0.13 μm/1.2	6.00E-03	0.25
[70]	9.7	1.85E+07	16.0	4th (3b)	65 nm/1.2	8.20E-03	0.27
[344]	11.3	1.50E+07	10	3rd(4b)	0.18 μm/1.8	2.07E-02	0.27
[69]	11.2	1.00E+07	24.0	3rd (4b)	90 nm/1.2	1.31E-02	0.28
[64]	11.5	1.00E+07	13	3rd(3.5b)	0.13 μm/1.2	1.80E-02	0.31
[445]	12.2	1.60E+07	25.0	4th (4b)	0.18 μm/1.8	4.76E-02	0.32
[55]	11.2	5.00E+07	122.9	5th (17-L)	65 nm/1.3	8.80E-02	0.32
[345]	13.2	8.00E+06	16	4ther(4b)	65 nm/1.3	5.00E-02	0.33
[346]	12.7	2.00E+07	23	4th(4b)	0.13 μm/1.5	8.70E-02	0.34
[347]	13.0	1.20E+07	10	3rd(6b)	0.5 μm/2.5	7.50E-02	0.38
[67]	9.1	1.80E+07	15.0	5th (3b)	65 nm/1.2	7.90E-03	0.39
[23]	11.3	1.25E+08	16	3rd(4b)	45 nm/1.1,1.8	2.60E-01	0.41
[348]	9.7	2.50E+07	405	5th(4b)	0.18 μm/1.8	1.80E-02	0.44
[446]	10.6	8.90E+06	58.2	3rd (4b)	28 nm/1.1	1.25E-02	0.45
[152]	11.0	2.00E+07	24	3rd(3b)	0.13 μm/1.3	3.80E-02	0.46

(continued overleaf)

Table 10.7 *(continued)*

Ref.	DR (bit)	B_w (Hz)	OSR	Architecture (order)	Tech./supply voltage (V)	Power (W)	FOMW (pJ/conv)
[349]	13.3	1.00E+07	32	5th(3b)	0.18 μm/1.8	1.00E-01	0.50
[445]	10.3	3.20E+07	12.5	4th (4b)	0.18 μm/1.8	4.76E-02	0.59
[447]	12.5	1.00E+07	32.0	3rd (4b)	0.18 μm/1.8	6.00E-02	0.59
[350]	14.0	1.00E+05	130	2nd(3b)	65 nm/1	2.10E-03	0.64
[351]	13.9	2.50E+06	12	5th(4b)	0.25 μm/2.5	5.00E-02	0.65
[23]	10.5	1.25E+08	16	3rd(4b)	45 nm/1.1	2.60E-01	0.72
[350]	10.2	1.92E+06	16	2nd(3b)	65 nm/1	3.20E-03	0.72
[352]	10.0	2.00E+06	26	3rd(3b)	0.13 μm/1.2	3.00E-03	0.73
[448]	15.1	6.00E+05	48.0	3rd (5b)	0.18 μm/1.8	3.10E-02	0.73
[449]	12.2	1.00E+07	32.0	3rd (4b)	0.18 μm/1.8	7.00E-02	0.74
[353]	14.0	1.20E+05	54	4th(3b)	0.13 μm/1.25	3.00E-03	0.76
[354]	10.0	1.92E+06	48	1st(5b)	0.13 μm/1.2	3.10E-03	0.79
[116]	10.0	1.92E+06	48	1st(5b)	0.13 μm/1.2	3.10E-03	0.80
[355]	10.8	2.00E+07	16	3rd(4b)	0.13 μm/1.2	5.80E-02	0.82
[356]	12.5	6.00E+05	42	3rd(2b)	90 nm/1.5	6.00E-03	0.86
[357]	11.5	2.00E+06	32	5th(3L)	0.18 μm/1.8	1.10E-02	0.95
[358]	14.9	2.00E+04	300	2nd(4b)	45 nm/1.1	1.20E-03	0.98
[359]	11.0	1.50E+07	10	4th(4b)	0.13 μm/1.5	7.00E-02	1.14
[360]	11.0	1.50E+07	10	4th(4b)	0.13 μm/1.5	7.50E-02	1.22
[361]	10.0	2.00E+07	16	3rd(4b)	0.13 μm/1.2	5.80E-02	1.40
[362]	14.4	1.00E+06	16	3rd(5b)	0.5 μm/–	6.20E-02	1.43
[450]	10.8	2.00E+06	30.0	2nd (TI)	0.18 μm/1.5	1.27E-02	1.78
[363]	8.7	2.00E+07	16	2nd(3b)	0.25 μm/2.5	3.20E-02	1.92
[364]	8.9	2.00E+07	5	3rd(4b)	0.18 μm/1.8	1.03E-01	5.39

Table 10.8 State-of-the-art CT cascade LP-ΣΔMs.

Ref.	DR (bit)	B_w (Hz)	OSR	Architecture	Tech./supply voltage (V)	Power (W)	FOMW (pJ/conv)
[111]	12.5	1.00E+07	50	2-1(1b)	65 nm/1	1.57E-03	0.01
[25]	13.8	5.00E+07	18	3-1(1b)	28 nm/1.2–1.8	7.80E-02	0.05
[43]	12.5	5.03E+07	9.9	2-2 (4b)	40 nm/1.1	4.30E-02	0.07
[35]	12.8	1.50E+08	16.7	4-0 (7b)	0.13 μm/1.2	2.00E-02	0.09
[24]	14.6	4.57E+07	35	0-3(6b)	28 nm/0.9–1	2.35E-01	0.10
[366]	12.5	2.00E+07	8.5	2-2 I/Q	90 nm/1.2	5.60E-02	0.24
[22]	11.8	3.50E+08	8.6	1-2(17-L)	28 nm/1.0–1.8	7.65E-01	0.31
[115]	11.7	1.00E+07	5	1(3L)-1(1b)	0.18 μm/2	4.80E-02	0.72
[367]	11.0	1.80E+07	16	2-1-1(4b)	0.18 μm/1.8	1.83E-01	2.48
[368]	11.0	1.80E+07	10	2-1-1(4b)	0.18 μm/1.8	1.83E-01	2.48
[369]	10.9	2.00E+07	8	2-2(4b) I/Q	0.18 μm/1.8	2.16E-01	2.89
[369]	10.9	1.00E+07	8	2-2(4b)	0.18 μm/1.8	1.22E-01	3.27

I/Q denotes a quadrature topology.

Table 10.9 State-of-the-art CT BP-ΣΔMs.

Ref.	DR (bit)	f_n (Hz)	B_w (Hz)	OSR	Architecture (order)	Tech./supply voltage (V)	Power (W)	FOMW (pJ/conv)
[451]	12.0	6.00E+06	1.00E+07	17.3	2nd(4b)	65 nm/1.2	6.30E-03	0.08
[52]	11.3	2.4–2.5E+09	2.00E+07	80	6th	40 nm/–	2.00E-02	0.19
[63]	14.7	1.30E+05	2.00E+05	65	3rd	0.25 μm/1.8	2.70E-03	0.25
[127]	11.3	2.00E+08	2.50E+07	133.0	6th (4b)	65 nm/1	3.50E-02	0.28
[452]	11.3	1.80E+02	2.50E+07	16.0	6th (4b)	65 nm/1.2	3.50E-02	0.28
[80]	9.7	2.00E+08	2.40E+07	16.6	4th(9L)	65 nm/1.25	1.20E-02	0.30
[73]	10.0	2.40E+09	6.00E+07	25	6th	90 nm/1	4.00E-02	0.33
[370]	9.9	8.50E+01	5.00E+06	95	2nd	0.18 μm/1.8	6.00E-03	0.36
[371]	9.7	2.00E+08	2.40E+07	16	4th	65 nm/1.25	1.20E-02	0.39
[372]	12.0	–	3.84E+06	20	5th	0.18 μm/2.9	1.41E-02	0.45
[372]	13.5	–	1.23E+06	32	5th	0.18 μm/2.9	1.31E-02	0.46
[373]	11.0	2.00E+06	1.00E+06	24	2nd(4b)	0.18 μm/1.8	2.20E-03	0.54
[453]	10.3	1.70E+07	3.30E+07	7.0	4th(3b)	0.18 μm/1.8	5.44E-02	0.65
[26]	12.0	0–1E+09	1.50E+08	12	6th	65 nm/1	5.50E-01	0.80
[374]	13.3	1.00E+05	2.70E+05	24	4th	0.25 μm/2	4.60E-03	0.83
[375]	14.7	4.40E+07	8.50E+06	16	4th	0.18 μm/2.9	3.75E-01	0.83
[376]	8.3	2.44E+09	2.80E+07	64	4th	0.13 μm/1.2	1.50E-02	1.00
[454]	10.2	2.60E+08	2.00E+07	26.0	8×4th (5L)	65 nm/1.4	1.24E-01	2.63
[377]	9.3	2.25E+08	4.00E+06	13	4th	65 nm/1	1.30E-02	2.67
[378]	13.3	1.07E+07	2.00E+05	53	5th	0.25 μm/2.5	1.10E-02	2.73
[379]	12.7	1.00E+07	2.00E+04	250	2nd	1.2 μm/3.3	2.50E-04	3.42
[76]	6.9	2.44E+09	8.00E+07	38	4th	40 nm/1.1	5.28E-02	3.60
[380]	11.3	2.00E+08	1.00E+07	40	4th	0.18 μm/1.8	1.60E-01	3.72
[73]	10.0	2.44E+09	6.00E+07	60	6th	90 nm/1.2	4.00E-02	4.08
[376]	8.2	2.44E+09	2.50E+07	64	4th	0.13 μm/1.2	1.90E-02	4.65
[372]	15.0	1.00E+05	2.00E+04	650	5th	0.18 μm/2.9	9.10E-03	6.94
[381]	16.0	1.07E+07	2.00E+05	104	5th	0.18 μm/1.8	2.10E-01	8.01
[382]	9.3	4.09E+06	2.00E+06	31	2nd	0.25 μm/1.8	2.05E-02	8.36
[59]	9.0	9.00E+08	9.00E+06	55	4th	65 nm/1.2	8.00E-02	8.62
[382]	8.1	4.09E+06	4.00E+06	16	2nd	0.25 μm/1.8	2.05E-02	9.67
[53]	7.7	2.22E+09	8.00E+07	160	4th	40 nm/1.1	1.64E-01	9.86
[381]	14.0	1.07E+07	5.00E+05	42	5th	0.18 μm/1.8	2.10E-01	12.82
[383]	8.4	4.73E+07	3.84E+06	25	4th	0.35 μm/3.3	4.50E-02	16.99
[5]	6.8	2.7e9	1.50E+07	41.6	4th(1.5b)	40 nm/1.1	9.00E-02	26.92
[455]	6.3	2.50E+09	1.50E+07	41.6	3rd (3L)	40 nm/1.1	9.00E-02	37.03
[381]	19.3	1.07E+07	3.00E+03	6950	5th	0.18 μm/1.8	2.10E-01	54.22
[85]	8.3	7.97E+08	1.00E+06	1600	4th	0.13 μm/1.2	3.00E-02	58.06
[384]	10.8	1.07E+07	2.00E+05	100	6th	0.5 μm/5	6.00E-02	84.13
[74]	6.7	2.44E+09	1.00E+06	1628	2nd	0.13 μm/1.3	2.60E-02	126.40
[383]	9.2	4.73E+07	2.00E+05	473	2nd	0.35 μm/3.3	4.50E-02	195.30
[385]	6.7	7.00E+07	2.00E+05	700	2nd	0.5 μm/2.5	3.90E-02	937.79
[386]	7.2	1.00E+08	2.00E+05	1000	4th	0.35 μm/3.3	1.65E-01	2805.49
[387]	6.0	1.00E+09	5.00E+05	4000	4th	0.18 μm/1.8	2.90E-01	4531.25
[386]	8.7	1.00E+08	2.00E+05	1000	4th	0.35 μm/2.7–3.3	3.30E-01	5678.56

Table 10.10 State-of-the-art ΣΔMs with time-coded quantization.

Ref.	DR (bit)	B_w (Hz)	OSR	Architecture	Tech./supply voltage (V)	Power (W)	FOMW (pJ/conv)
[125]	12.0	2.00E+06	9	VCO 0-1 MASH	40 nm/1.1	3.50E-04	0.02
[168]	13.5	1.50E+07	21	3rd GRO	0.13 μm/1.2	1.14E-02	0.03
[148]	12.1	1.25E+08	8.6	3rd VCO	16 nm/1–1.5	5.40E-02	0.05
[456]	9.0	1.00E+04	60	VCO-based integrator	65 nm/0.3	5.10E-07	0.05
[139]	13.2	2.00E+07	45	1st VCO	0.13 μm/1.5	2.00E-02	0.05
[457]	9.4	2.00E+03	62.5	1st VCO	0.18 μm/1	1.50E-07	0.06
[146]	10.3	5.00E+07	10	VCO 0-1 MASH	65 nm/1	8.20E-03	0.07
[166]	12.1	5.00E+07	12.8	3rd VCO	65 nm/1.2–1.5	3.80E-02	0.09
[458]	11.4	1.67E+06	75	VCO-based integrator	0.13 μm/1.2	9.10E-04	0.10
[145]	12.0	5.00E+06	64	1st–VCO	90 nm/1.2	4.10E-03	0.10
[143]	12.8	1.00E+07	30	2nd–VCO	90 nm/1.2	1.60E-02	0.11
[143]	12.8	1.00E+07	30	2nd–VCO	90 nm/1.2	1.60E-02	0.11
[142]	12.5	4.00E+06	12.5-150	1(4b)–VCO	0.13 μm/1.2	6.10E-03	0.13
[140]	12.7	2.50E+07	8	5ther–VCO	0.18 μm/1.8	4.80E-02	0.14
[157]	11.5	2.81E+06	16	3rd TDC	0.13 μm/1.2	2.58E-03	0.16
[150]	10.5	3.50E+06	64	VCO-based	0.13 μm/1.2	1.65E-03	0.16
[459]	11.6	5.00E+07	120	VCO-based	65 nm/1	5.40E-02	0.17
[160]	9.8	2.00E+07	16	3rd–TEQ	65 nm/1	7.00E-03	0.19
[460]	11.3	1.25E+06	20	Two-step flash-(1-1-1 MASH)	40 nm/1.1	1.23E-03	0.19
[161]	11.2	1.80E+07	32	Ring oscillator	65 nm/1	1.70E-02	0.20
	11.8	9.00E+06	64			1.70E-02	0.26
[163]	12.2	5.00E+06	32-128	Ring oscillator	65 nm/0.9–1.2	1.15E-02	0.24
[461]	11.9	2.50E+06	10	VCO-based 0-1 MASH	0.18 μm/1.8	4.80E-03	0.25
[49]	11.9	2.50E+06	10.24	0-1 VCO	0.18 μm/1.8	4.80E-03	0.25
[165]	10.6	2.00E+06	75	VCO-based integrators	0.13 μm/1.2	1.80E-03	0.29
[149]	12.5	4.00E+06	150	2nd cascade–VCO	0.13 μm/1.4	1.38E-02	0.30
[138]	11.7	2.00E+07	25	2nd(5b)–VCO	0.13 μm/1.2	4.00E-02	0.30
	12.6	4.50E+06	128			1.70E-02	0.30
	11.5	4.50E+06	64			8.00E-03	0.31
[388]	9.7	2.00E+07	6	2nd–TDC	65 nm/1.3	1.05E-02	0.32
[389]	13.0	1.00E+05	25-250	1-1-1–TDC	0.13 μm/1.2	7.00E-04	0.43
[390]	9.7	1.00E+06	78	2nd–TDC	90 nm/1.2	1.30E-03	0.80
[158]	7.2	2.00E+06	36	PWM-based firster 3-b	0.18 μm/1.8	2.70E-03	4.59
[153]	8.2	1.00E+06	72	PWM-based	0.18 μm/1.8	9.50E-03	16.2

Table 10.11 State-of-the-art hybrid ΣΔMs.

Ref.	DR (bit)	B_w (Hz)	OSR	Hybrid technique	Tech./supply voltage (V)	Power (W)	FOMW (pJ/conv)
[462]	11.5	1.60E+08	6	ΣΔM-SAR	28 nm/1.15	6.39E-03	6.89E-03
[124]	13.4	4.50E+07	10	ΣΔM-SAR	65 nm/1.8	2.47E-02	2.54E-02
[38]	15.0	2.20E+06	32	ΣΔM-SAR	55 nm/1.2–1.8	4.50E-03	3.12E-02
[463]	13.0	5.00E+06	43.2	ΣΔM-SAR	28 nm/–	3.16E-03	3.86E-02
[50]	13.6	1.92E+06	16.92	ΣΔM-SAR	40 nm/1.2	1.91E-03	4.15E-02
[464]	12.2	1.00E+05	16	ΣΔM-SAR	65 nm/0.7	4.58E-05	5.00E-02
[465]	15.2	2.40E+04	64	ΣΔM-SAR	65 nm/1	9.40E-05	5.20E-02
[466]	8.8	1.10E+07	4	Noise-shaping-SAR	65 nm/1.2	8.06E-04	8.51E-02
[37]	16.7	2.42E+04	135	CT–DT	40 nm/1	5.00E-04	0.10
[467]	14.4	2.40E+04	64	ΣΔM-SAR	65 nm/1.1	1.21E-04	0.12
[40]	16.8	2.50E+04	128	ΣΔM-SAR	65 nm/1	8.00E-04	0.14

(continued overleaf)

Table 10.11 *(continued)*

Ref.	DR (bit)	B_w (Hz)	OSR	Hybrid technique	Tech./supply voltage (V)	Power (W)	FOMW (pJ/conv)
[391]	13.8	5.00E+06	8	ΣΔM-pipeline	0.18 μm/1.8	2.20E-02	0.15
[392]	13.0	5.00E+06	9	Active–Passive	55 nm/1.3	1.30E-02	0.16
[117]	12.2	5.00E+06	16	ΣΔM-cyclic	0.18 μm/1.8	7.90E-03	0.16
[118]	12.2	1.56E+06	8	ΣΔM/Two-step	0.18 μm/1.2	2.60E-03	0.18
[393]	11.2	1.20E+06	15000	Active–Passive	65 nm/1.4	1.16E-03	0.21
[39]	17.6	2.00E+04	–	ΣΔM-SAR	0.16 μm/1.8	1.65E-03	0.21
[468]	13.8	5.00E+06	24	ΣΔM-Pipeline	65 nm/ 1.25	3.70E-02	0.26
[123]	16.4	2.40E+04	500	ΣΔM-SAR	28 nm/1-3.3	1.13E-03	0.27
[469]	11.4	2.50E+06	8	Comparator-based	65 nm/1.2	3.73E-03	0.28
[122]	11.3	3.50E+06	5	VCO ΣΔM-SAR	0.18 μm/1.8	5.00E-03	0.28
[470]	11.4	5.00E+06	8	ΣΔM/Two-step ADC	0.13 μm/1.2	8.10E-03	0.30
[107]	10.9	2.00E+06	38	Active–Passive	0.28 μm/1.5	2.70E-03	0.35
[104]	12.5	2.00E+05	375	Digital-based	65 nm/1.3	9.50E-04	0.41
[115]	11.7	1.00E+07	5	ΣΔM-cyclic	0.18 μm/2	4.80E-02	0.72
[354]	10.0	1.92E+06	48	ΣΔM-SAR	0.13 μm/1.2	3.10E-03	0.79
[394]	8.9	5.00E+06	8	Digital-based	0.18 μm/1.8	4.00E-03	0.85
[395]	11.0	3.13E+06	16	CT–DT	65 nm/1.1	1.10E-02	0.86
[121]	10.8	1.50E+07	8	ΣΔM-pipeline	65 nm/1.25	4.60E-02	0.86
[396]	12.5	7.50E+06	16	CT–DT	0.18 μm/1.2	8.90E-02	1.02
[471]	10.3	2.00E+06	32	VCO-based/digital-based	0.18 μm/1.8	6.00E-03	1.19
[394]	12.4	2.00E+05	16	Digital-based	0.18 μm/1.8	4.00E-03	1.89
[114]	12.2	1.00E+07	4	ΣΔM-pipeline	0.18 μm/3.3	2.40E-01	2.55
[97]	15.9	4.80E+04	128	CT–DT	0.35 μm/3.3	1.80E-02	3.07
[99]	16.5	2.00E+04	256	CT–DT	65 nm/3.3	1.50E-02	4.10
[96]	16.7	2.00E+04	128	CT–DT	0.18 μm/3.3	3.73E-02	8.76
[65]	9.5	4.00E+06	500	Active–Passive	90 nm/1.2	5.40E-02	9.32
[397]	13.0	1.10E+04	64	CT–DT	0.5 μm/1.8	1.70E-03	9.43
[113]	14.5	1.25E+06	8	ΣΔM-pipeline	0.6 μm/5	5.50E-01	9.49

Table 10.12 State-of-the-art SC adaptive/reconfigurable ΣΔMs.

Ref.	Standard	DR (bit)	B_w (Hz)	OSR	Tech./supply voltage (V)	Power (W)	FOMW (pJ/conv)
[242]	AMPS	15.0	1.8E+04	639	0.18 μm/2.7	3.00E-02	22.1
	GSM	13.2	2.0E+05	58		3.00E-02	5.4
	CDMA	12.5	6.75E+05	17		3.00E-02	2.9
	UMTS	11.4	1.92E+06	12		5.00E-02	2.8
[398]	GSM/EDGE	14.3	1.0E+05	130	0.13 μm/1.2,3.3	2.39E-02	0.7
	UMTS	12.8	1.92E+06	12		2.45E-02	0.3
	WLAN	10.8	1.0E+07	6		4.45E-02	0.6
[89]	GSM	14.1	2.0E+05	100	0.13 μm/1.2,3.3	2.52E-02	8.5
	Bluetooth	13.2	1.0E+06	20		2.50E-02	4.1
	WCDMA	10.2	4.0E+06	10		4.45E-02	14.1
[399]	GSM	14.1	1.0E+05	130	0.18 μm/1.2	2.00E-03	0.6
	UMTS	12.5	1.92E+06	16		5.20E-03	0.2
		12.2	5.0E+06	16		1.36E-02	0.3
	LTE	11.6	1.0E+07	12		2.02E-02	0.3
		11.4	2.0E+07	5		3.47E-02	0.3
[400]	GSM	13.8	1.0E+05	250	90 nm/1.2	3.40E-03	1.2
	Bluetooth	12.5	5.0E+05	90		3.70E-03	0.6
	UMTS	10.7	2.0E+06	20		6.80E-03	1

(continued overleaf)

Table 10.12 *(continued)*

Ref.	Standard	DR (bit)	B_w (Hz)	OSR	Tech./supply voltage (V)	Power (W)	FOMW (pJ/conv)
[401]	GSM	12.7	1.0E+05	200	90 nm/1.2	4.60E-03	3.5
	Bluetooth	11.3	5.0E+05	80		5.30E-03	2.1
	GPS	11.6	1.0E+06	60		6.20E-03	1
	UMTS	10.7	2.0E+06	30		8.00E-03	1.2
	DVB-H	10.1	4.0E+06	15		8.00E-03	0.9
	WiMAX	8.5	5.0E+06	12		1.10E-02	1.6
	GSM	13.2	1.0E+05	256		4.90E-03	0.99

As reconfigurable ΣΔMs feature different performance metrics depending on their operating mode, these ICs have been sorted by date of publication.

Table 10.13 State-of-the-art CT reconfigurable ΣΔMs (sorted by date of publication).

Ref.	Standard	DR (bit)	B_w (Hz)	OSR	Tech./supply voltage (V)	Power (W)	FOMW (pJ/conv)
[403]	GSM	15	2.00E+05	64	0.18 μm/1.6,2.9	9.00E-03	0.69
	Bluetooth	13.5	1.23E+06	32		8.20E-03	0.29
	UMTS	12	3.84E+06	40		7.60E-03	0.24
[404]	GSM	13.3	2.00E+05	65	90 nm/1.1,1.3	1.40E-03	0.35
	Bluetooth	12.2	1.00E+06	100		3.40E-03	0.37
	UMTS	8.3	1.00E+07	20		7.00E-03	1.07
[98]	EDGE	14.3	1.35E+05	96	65 nm/2.5	2.60E-03	0.47
	CDMA	13.3	6.15E+05	62		3.10E-03	0.25
	UMTS	11.8	1.92E+06	40		3.70E-03	0.26
[405]	WLAN	10	5.00E+06	32	90 nm/1.2	6.80E-03	0.24
	DVB	11.3	4.00E+06	64		5.50E-03	0.27
	UMTS	11.2	1.92E+06	64		4.34E-03	0.5
	Bluetooth	12.5	1.00E+06	90		5.00E-03	0.85
	–	11.2	1.00E+06	32		1.70E-02	3.7
	–	10.83	2.30E+06	32		1.70E-02	1.94
[162]	–	10.88	5.00E+06	32	65 nm/1.2,2.5	1.70E-02	0.91
	–	11.83	1.00E+06	64		1.70E-02	2.32
	–	11.66	2.30E+06	64		1.70E-02	1.13
	–	12.63	1.00E+06	128		1.70E-02	1.34
	–	11.5	1.00E+06	64		8.00E-03	0.75
	–	8.84	1.50E+07	7		1.05E-02	0.76
	–	9.84	1.00E+07	10.4		1.05E-02	0.57
[365]	–	10.9	2.50E+06	20.8	65 nm/1.3	8.50E-03	0.74
	–	10.2	5.00E+06	41.6		8.50E-03	1.04
	GSM	13.32	2.00E+05	128	90 nm/1	2.80E-03	0.68
	Bluetooth	12.33	5.00E+05	96		2.60E-03	0.5
	UMTS	11.66	2.00E+06	32		3.60E-03	0.28
[406]	DVB-H	11.16	4.00E+06	24		4.90E-03	0.27
	WLAN	9.01	2.00E+07	16		8.50E-03	0.41
[407]	–	10.5	8.00E+06	25	0.18 μm/1.8	4.76E-02	0.81
	–	9.11	1.60E+07	12.5		4.76E-02	1.29
[402]	UMTS	9.7	1.92E+06	31	0.18 μm/1.8	8.90E-03	0.95
	DVB	9.8	4.0E+06	25		1.21E-02	0.84
[55]	–	11.1	1.00E+07	30.7	65 nm/1.3	5.50E-02	1.10
	–	10.7	2.00E+07	30.7		7.50E-02	1.10
	–	10.7	5.00E+07	12.3		8.80E-02	0.70
	–	10.1	1.00E+08	7.7		9.50E-02	0.80
[472]	–	11.2	1.00E+07	60	40 nm/–	1.94E-03	0.04
	–	11.4	2.00E+07			3.24E-03	0.03
	–	11.1	3.00E+07			4.30E-03	0.03
	–	11.1	4.00E+07			5.25E-03	0.03

- SC-ΣΔMs: single-loop single-bit LP-ΣΔMs (Table 10.1), single-loop multi-bit LP-ΣΔMs (Table 10.2), cascade single-bit LP-ΣΔMs (Table 10.3), cascade multi-bit LP-ΣΔMs (Table 10.4), and BP-ΣΔMs (Table 10.5),
- CT-ΣΔMs: single-loop single-bit LP-ΣΔMs (Table 10.6), single-loop multi-bit LP-ΣΔMs (Table 10.7), cascade LP-ΣΔMs (Table 10.8), and BP-ΣΔMs (Table 10.9),
- ΣΔMs with time-coded quantization (Table 10.10),
- Hybrid ΣΔMs: CT/DT ΣΔMs, active/passive ΣΔMs, digital-based ΣΔMs, and ΣΔM/Nyquist-rate ADCs (Table 10.11),
- Reconfigurable ΣΔMs, implemented using SC (Table 10.12) and CT circuits (Table 10.13).

In all cases, the main features of each reported IC are summarized in terms of the following performance metrics: DR (measured in bits), B_w, OSR, technology process, supply voltage, and power consumption.[4] In the case of BP-ΣΔMs (Tables 10.5 and 10.9), the notch frequency, f_n, is also given. A schematic description of each ΣΔM topology is sketched in the tables, highlighting the loop-filter order, the number of stages (in cascade ΣΔMs), the number of bits of the embedded quantizer, the type of quantization technique (in time-based quantizers), the operation mode and standard covered (in the case of reconfigurable ΣΔMs), and so on. A more complete description of the modulators – not shown in the tables of this chapter for the sake of simplicity – can be found in the spreadsheet available on the book's website.

10.2.1 Aperture Plot of ΣΔMs

Figure 10.3 represents the aperture plot of the state-of-the-art ΣΔ ADCs. It can be noted that a wide B_w-versus-ENOB *conversion region* is covered, ranging over seven orders of magnitude in frequency

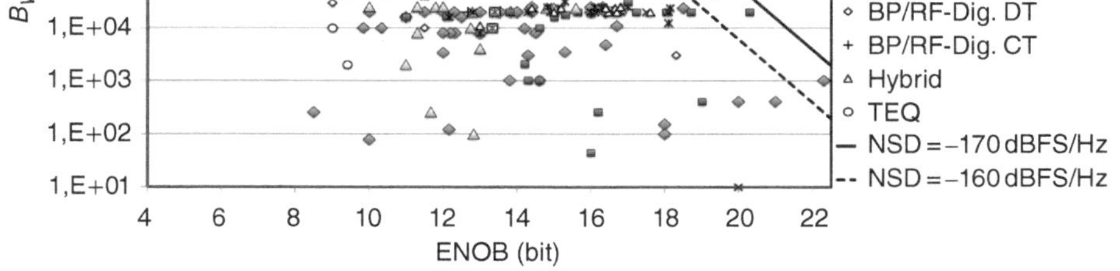

Figure 10.3 Aperture plot of state-of-the-art ΣΔMs: SL, single-loop; SB = single-bit; MB, multi-bit, RF-Dig., RF-to-digital.

[4]The data corresponding to the power consumption account only for the ΣΔM, excluding the power consumed by the decimation filter.

and 16 bits in the ENOB axis. One of the highest ENOBs achieved by ΣΔMs is 21 bits, which was reported in 1994 by Kerth *et al*, who designed a fourth-order single-loop single-bit SC-ΣΔM in a 3-μm CMOS technology, which digitized 400-Hz signals for seismic applications [20]. More recently, at ISSCC in February 2016, Steiner *et al.* presented a 0.35 μm CMOS fourth-order single-loop single-bit SC-ΣΔM implemented with *split steering amplifiers*, which exhibited 22.3-bit resolution within 1-kHz bandwidth, and featured FOMS > 185 dB [21]. At the other limit of the state-of-the-art front, the highest value of B_w considered in this survey was reported by Dong *et al.*, also at ISSCC 2016. This chip consisted of a cascade 1-2 CT-ΣΔM clocked at 8 GHz, implemented in a 28-nm CMOS technology, achieving 72-dB DR within a 465-MHz signal bandwidth [22].

As a reference, two state-of-the-art fronts are depicted in Figure 10.3, corresponding to NSD = −160 dBFS/Hz and NSD = −170 dBFS/Hz. Two different groups of cutting-edge designs can be distinguished among those achieving the best FOMS. The first covers a conversion region approximately limited by 500 kHz< B_w < 500 MHz and 10 bits <ENOB< 15 bits, and it is dominated by CT-ΣΔMs, considering different architectures at gigahertz-range sampling rates – where clock jitter is one of the physical limits – including single-loop [23], cascade [22, 24, 25], and BP/RF digitization [26]. The other front – mostly limited by thermal noise – covers high-resolution (ENOB > 16 bits) with low-bandwidth (B_w < 100 kHz) applications, where SC-ΣΔMs have the best metrics [20, 27, 28].

Analyzing the evolution over time of the the signal bandwidths handled by ΣΔMs, it can be observed that the tendency is towards extending the range of applications, by increasing the digitized signal band, although these designs are sometimes less power efficient. Figure 10.4 represents the digitized signal bandwidth (B_w) reported by ΣΔMs against year of publication for the years 1990–2017. It can be seen how ΣΔMs have evolved from low-frequency (high-resolution) to high-frequency (medium-to-low resolution) applications. Note also that the range of B_w has expanded over time from two orders of magnitude to almost seven orders of magnitude. This trend is a function of technology downscaling and it is expected to continue in the future.

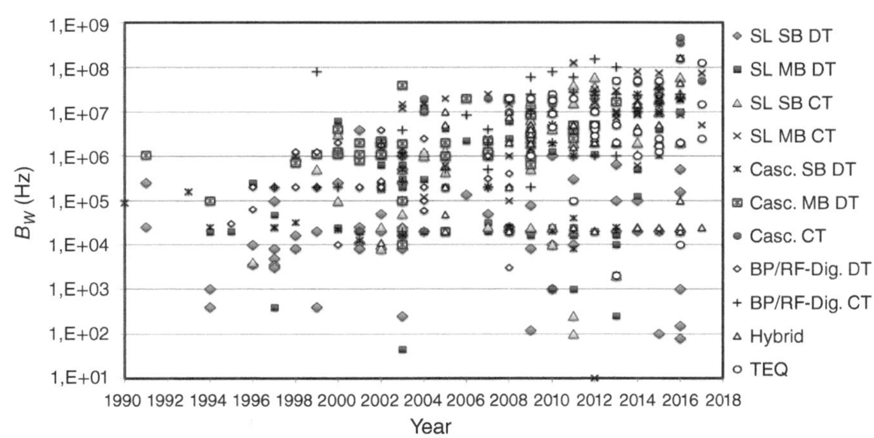

Figure 10.4 B_w spread of ΣΔMs over time for the years 1990–2017.

10.2.2 Energy Plot of ΣΔMs

Figure 10.5 depicts the energy plot of state-of-the-art ΣΔ ADCs. Note that the most efficient designs are mainly based on CT circuits, some of them featuring a FOMS between 175 and 185 dB

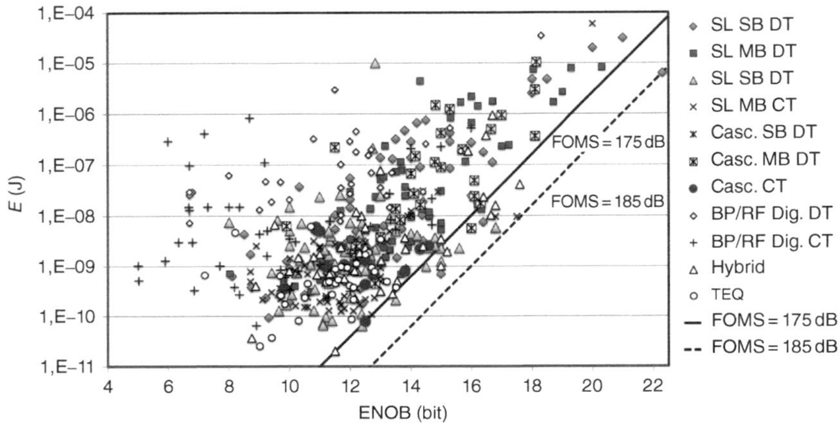

Figure 10.5 Energy plot of state-of-the-art ΣΔMs.

[24, 29–35]. However, as will be discussed later on, the use of inverter-based circuit techniques has been successfully demonstrated in some single-bit SC-ΣΔMs, including both single-loop [6, 36] and cascade [7] topologies. Moreover, the combined use of CT and SC techniques in so-called hybrid-ΣΔMs [37], as well as the combination with SAR ADCs [38–40], can also lead to energy efficient solutions, featuring FOMS > 175 dB in Figure 10.5.

The ΣΔM ICs corresponding to the best data points in Figures 10.3 and 10.5, define the *frontiers* of ΣΔ ADCs, showing the trends and challenges faced in diverse applications. The following circuits and systems techniques are included: digital-assisted analog circuits [6, 7, 32]; power-efficient analog filters and feedback DAC paths [30, 33, 39]; mostly-digital/scaling-friendly quantization [31, 41]; new cascade topologies [22, 24, 25]; RF/GHz-range [5, 23, 26], and so on. Some of these techniques will be discussed later in this chapter. Before that, let us take a closer look to the data in Tables 10.1–10.13. Analyzing these data in detail, a number of interesting conclusions can be drawn.

10.3 Empirical and Statistical Analysis of State-of-the-Art ΣΔMs

This section aims to identify practical pieces of information that can be used as guidelines for designers to make a better decision on the best ΣΔM architecture and circuit technique for a given set of specifications.

10.3.1 SC versus CT ΣΔMs

Although the SC circuit technique has been traditionally used in most reported ΣΔMs, more and more ICs are being implemented using CT circuits, especially in those applications targeting broadband signals and/or requiring low power consumption. This is illustrated in Figure 10.6a, which shows the evolution of FOMS of both SC and CT-ΣΔMs over time. Note that, although high values of FOMS (over 170 dB) have been achieved with both circuit implementations, the number of ΣΔMs using CT techniques has increased, especially over the last decade. This trend is expected to continue as this kind of ΣΔM will benefit from the technology downscaling process and the increase in transition frequency. Indeed the decrease in voltage supplies (and consequently the voltage headroom) and the

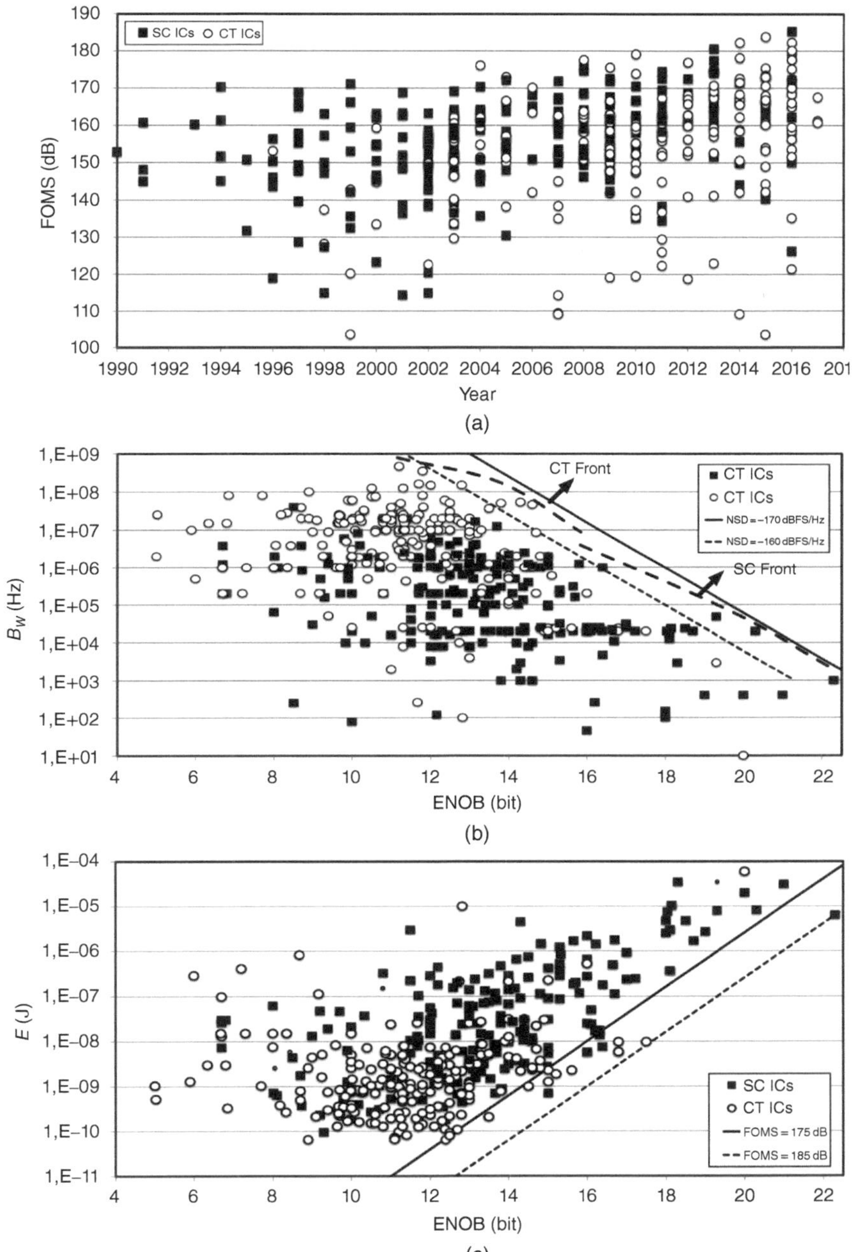

Figure 10.6 Comparison between state-of-the-art SC- and CT-ΣΔMs: (a) FOMS over time; (b) aperture plot; (c) energy plot. (Hybrid and TEQ ΣΔMs are not considered since they can use both SC and CT techniques).

increasing use of digital signal processing and calibration techniques are favouring the implementation of CT-ΣΔMs in nanometer CMOS, increasing their operating frequencies still further [15].

Figure 10.6b represents the *aperture plot* for state-of-the-art SC- and CT-ΣΔMs. Note that, as also happened in Figure 10.3, two state-of-the-art fronts (highlighted in the figure) can be clearly identified. One front – dominated by CT-ΣΔMs – goes from approximately 11-bit to 15-bit DR, covering a B_w range from 10 to 500 MHz. The other front – covered by SC-ΣΔMs – goes from 15-bit to 22-bit ENOB with B_w ranging from 1 kHz to 1 MHz. Therefore, as could be expected, it can be *empirically* deduced from these experimental data that CT-ΣΔMs are more suitable for medium-resolution (10–15 bits) and medium-high bandwidths (10–100 MHz), whereas SC-ΣΔMs are more appropriate for applications requiring high-resolution (16–21 bits) within low–medium bandwidths (100 Hz to 1 MHz).

The resolution-versus-bandwidth regions covered by the state-of-the-art fronts drawn in Figure 10.6b define the application areas where SC- and CT-ΣΔMs are more efficient. The efficiency is better illustrated in Figure 10.6c, where the energy plot is compared for both circuit techniques. Overall, it can be concluded that CT-ΣΔMs are more efficient than SC-ΣΔMs for ENOB$<$ 15 bits, although there are also a few designs featuring high efficiency for resolution between 16 and 17 bits.

As already discussed in this book, CT-ΣΔMs can be implemented using either active-RC or Gm-C integrators. Active-RC integrators have the advantages of better linearity and larger signal swing, whereas Gm-C integrators are potentially faster and have lower power consumption, at the price of reduced linearity compared to active-RC integrators [42]. In practice, the majority of state-of-the-art CT-ΣΔMs have been implemented using active-RC integrators. Sometimes an active-RC front-end integrator is chosen for its better linearity, with the other integrators of the ΣΔM loop filter being of Gm-C type. This is a common situation in LP-ΣΔMs, while most BP-ΣΔMs operating at notch frequencies of the order of hundreds of megahertz or in the gigahertz band, have been implemented using either Gm-C or Gm-LC integrators. In terms of FOM, active-RC CT-ΣΔMs have demonstrated to be more efficient than Gm-C implementations in a wider range of applications, ranging from hundreds of kilohertz to hundreds of megahertz.

10.3.2 Technology used in State-of-the-Art ΣΔMs

ΣΔ ADCs have demonstrated state-of-the-art performance in a wide variety of CMOS processes, taking advantage of technology downscaling to create the conditions to increase their conversion region for many different application scenarios. These technologies have been used in recent decades by ΣΔM designers – going from micrometer CMOS (3-μm technology node) down to *deep* nanometer processes (14–16 nm), employing tri-gate and Fin-FET transistors. Supply voltages have been correspondingly scaled down, with technology evolving from > 5V to 0.5 V and even less.

As an illustration, Figure 10.7 represents FOMS versus the technology node employed by state-of-the-art ΣΔMs considered in this survey. The most used technologies are 180-nm and 130-nm, although nodes below 100 nm – for example 90, 65, and even 40 nm – are becoming more and more popular. Some of these nanoscale technologies – including 14, 28, 40, 65, and 90 nm – have ΣΔM ICs featuring FOMS of over 170 dB and closer to 180 dB. This demonstrates that ΣΔ converters are benefiting from emerging fabrication processes, devices, and circuit techniques to achieve more efficient digitization in diverse applications.

10.3.3 Single-Loop versus Cascade ΣΔMs

Single-loop ΣΔ architectures have been more used than cascade topologies. Overall, approximately 85% of the state-of-the-art ΣΔM ICs considered in this survey are single-loop topologies, whereas only 15% are cascades. The latter have been mainly implemented using SC circuits, although recent

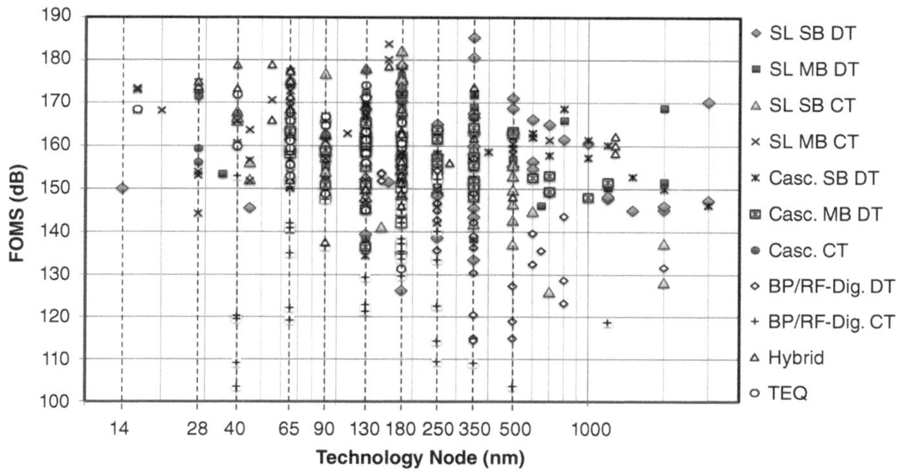

Figure 10.7 FOMS versus CMOS process nodes from 3μm to 14 nm.

synthesis methods as well as loop-filter design techniques are making the implementation of CT cascade ICs possible, successfully demonstrated in a number of ICs [22, 24, 25, 35, 43].

In terms of the loop-filter order of single-loop architectures, the second-order topology was the most used by SC-ΣΔMs (see Tables 10.1 and 10.2). In the case of CT implementations, third- and fourth-order topologies were preferred for single-bit topologies (see Table 10.6). In the case of multi-bit implementations, loop-filter orders higher than three were used in the vast majority of ICs, taking advantage of the better stability properties of high-order (> 3) implementations with multi-bit quantization. As far as the cascade topologies are concerned (see Tables 10.3, 10.4, and 10.8), the majority of implementations used a fourth-order loop filter with two stages in a 2-2 configuration or three stages in a 2-1-1 configuration. Third-order 2-1 implementations are also used by SC cascades, as shown in Tables 10.3 and 10.4.

Figure 10.8 compares the performance of single-loop and cascade state-of-the-art ΣΔMs in terms of their energy (Figure 10.8a) and aperture (Figure 10.8) plots. In terms of energy efficiency, Figure 10.8a shows that single-loop structures dominate the state of the art in a wide range of applications demanding ENOB from 12–22 bits. However, the highest values of digitized signal bandwidths are reached by cascade topologies, as shown in Figure 10.8b. Moreover, enhanced architectures such as SMASH topologies are improving the performance of cascade ΣΔMs. This approach may be particularly suited for the implemention of reconfigurable/adaptive ADCs, thus taking advantage of the very modular nature of their multi-stage topology.

10.3.4 Single-bit versus Multi-bit ΣΔMs

It is interesting to compare the performance of state-of-the-art ΣΔMs in terms of the type of quantizer embedded in the loop, either single-bit or multi-bit. Figure 10.9 compares single-bit and multi-bit ΣΔMs in terms of their energy plots (Figure 10.9a) and aperture plots (Figure 10.9b). Without loss of generality, only single-loop and cascade topologies have been included in these plots. The majority of cascade topologies include dual quantization with a multi-bit quantizer in the last stage in order to attenuate the impact of the nonlinearity of the feedback DAC. However, more and more cascade ICs are including internal quantizers with B up to 5 bits in all stages combined with appropriate linearization techniques. In addition, as stated earlier in this book, a number of ΣΔM ICs are including

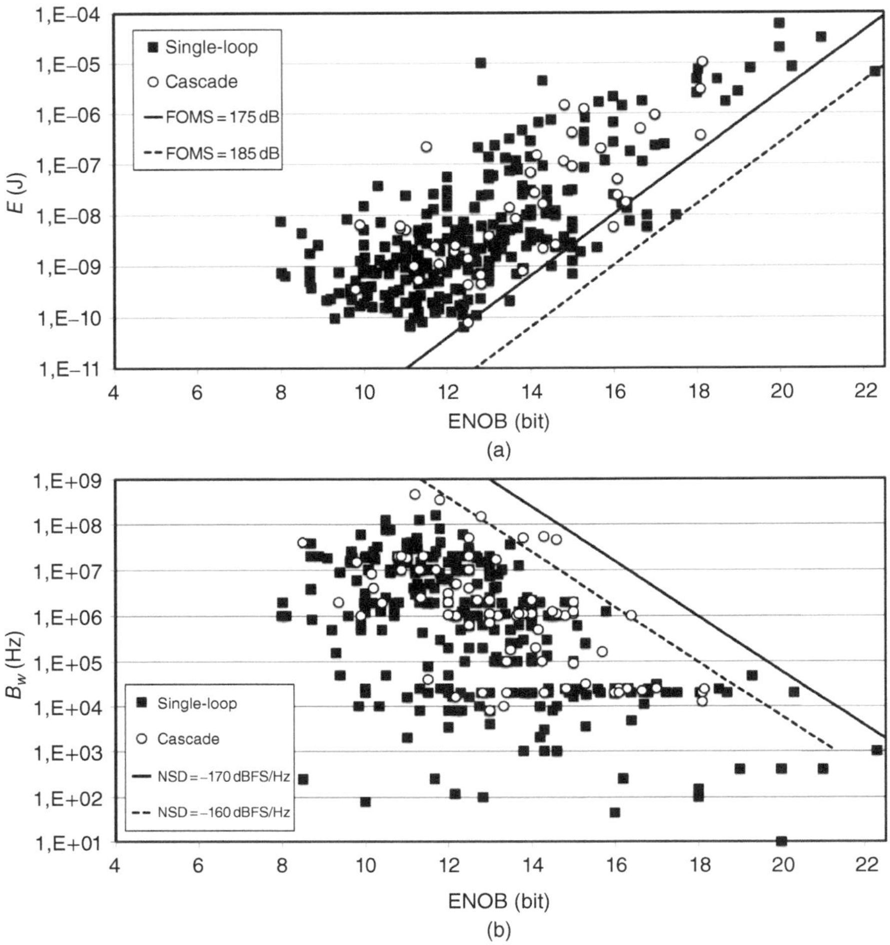

Figure 10.8 Single-loop versus cascade state-of-the-art ΣΔMs: (a) energy plot; (b) aperture plot. (Band-pass, hybrid, and TEQ ΣΔMs are not included in this plot).

tri-level quantizers to benefit from the extra level provided by fully-differential circuit implementation of the embedded flash ADCs, while keeping the inherently linear behavior of the feedback DAC.

In terms of conversion energy, there are more single-bit ΣΔMs than multi-bit ΣΔMs achieving a FOMS close to 185 dB, as illustrated in the energy plot shown in Figure 10.9a. Although multi-bit implementations can also achieve good efficiency for higher resolutions, of the order of 12–16 bits, there are more single-bit solutions reaching state-of-the-art performance within this range. However, for effective resolutions over 16-bit, there are more efficient ICs that employ multi-bit quantization, many of them approaching FOMS = 175 dB.

Figure 10.9b compares the aperture plots of single-bit and multi-bit ΣΔMs. It is seen that multi-bit modulators dominate the state-of-the-art fronts, except for those applications with resolutions over 20 bits within signal bandwidths below 1 kHz. A number of CT-ΣΔMs intended for broadband communications use multi-bit quantization. Apart from the obvious benefits in terms of the increased

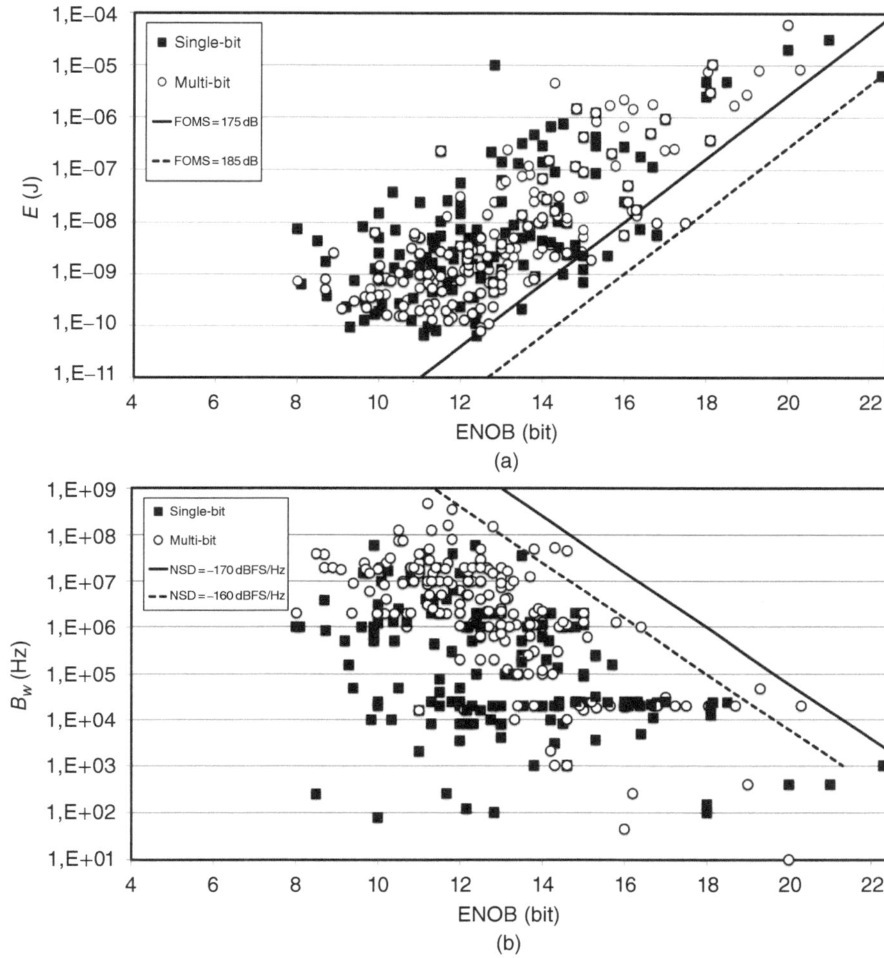

Figure 10.9 Single-bit versus multi-bit ΣΔMs: (a) energy plot; (b) aperture plot.

number of bits, multi-bit quantization is also appealing in practice for its lower sensitivity to clock jitter error. However, the price to pay for using multi-bit quantization is the inherent nonlinear operation of the multi-bit feedback DAC, which necessitates uses of linearization techniques, with a resultant penalty in power consumption, speed limitation, and circuit complexity. To address this problem, some authors have proposed the use of alternative implementations of the modulator feedback waveforms – such as DACs with finite impulsive response (FIR) [44, 45] – to reduce the sensitivity to clock jitter and to relax the ΣΔM loop-filter linearity specifications, giving rise to more competitive performance.

10.3.5 Low-pass versus Band-pass ΣΔMs

Figure 10.10 compares the performance achieved by LP-ΣΔMs and BP-ΣΔMs. It is clear from the energy plot shown in Figure 10.10a that LP-ΣΔMs give better performance in terms of conversion

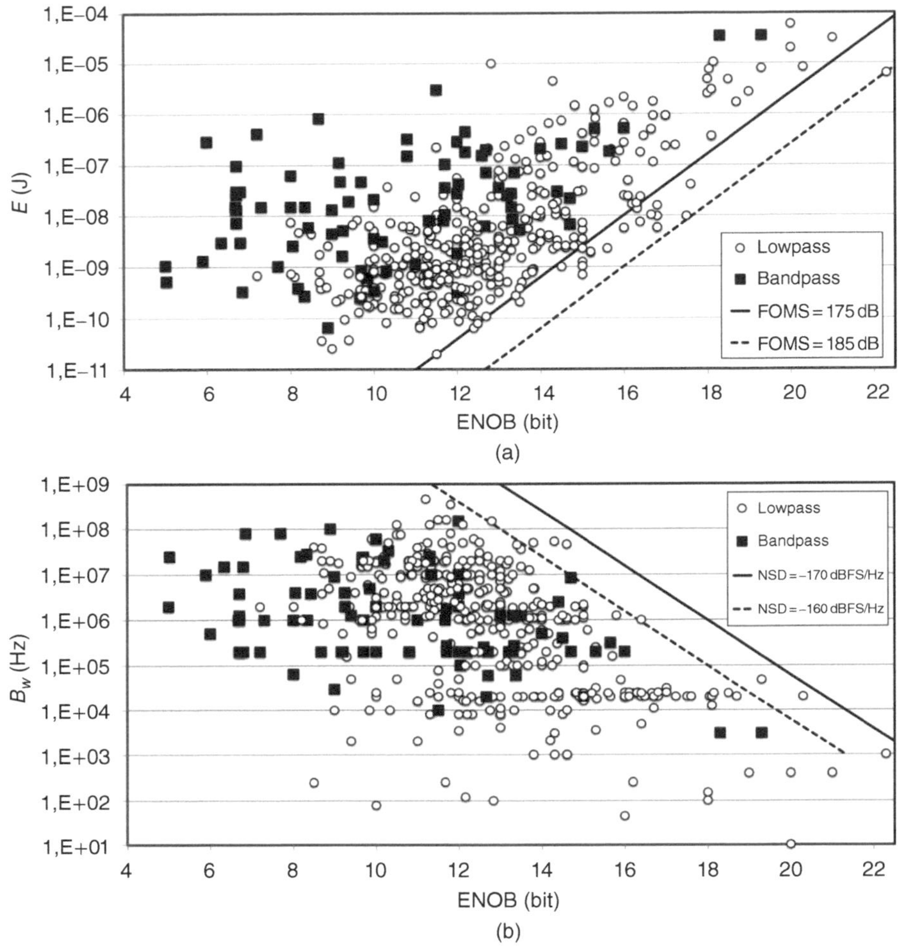

Figure 10.10 LP-ΣΔMs versus BP-ΣΔMs: (a) energy plot; (b) aperture plot.

energy than BP-ΣΔMs. However, it should be noted that the way in which the conversion energy is computed – based on B_w – might not be adequate for quantifying the efficiency of BP-ΣΔMs, because B_w is not always representative of the operating frequency of the modulator. For instance, looking at Table 10.9, it can be seen that there are several ICs that digitize signals with B_w in the range of tens or even hundreds of megahertz, although these signals are placed at center (notch) frequencies in the gigahertz range. For that reason some authors have proposed alternative FOMs, such as the following one [46]:

$$\mathrm{FOM_{BP}}|_{\mathrm{pJ/conv}} \equiv \frac{P_w(\mathrm{W})}{2^{\mathrm{ENOB(bit)}} \cdot (f_n + \frac{B_w}{2})} \cdot 10^{12} \tag{10.6}$$

This takes into account not only B_w, but also the notch frequency f_n to measure the conversion energy (E). The reader can see that the use of $\mathrm{FOM_{BP}}$ would increase the number of BP-ΣΔMs placed at the state of the art front, although the comparison might be not so fair in this case for LP-ΣΔMs.

Figure 10.10b shows the aperture plot of LP- and BP-ΣΔMs. Note that using this metric, the state-of-the-art front is also dominated by LP-ΣΔMs, although there are some BP-ΣΔMs approaching NSD = −160 dBFS/Hz, while digitizing signals with bandwidths over 100 MHz. As will be discussed in Section 10.4, new generations of BP-ΣΔMs are being developed that take advantage of technology downscaling to push ΣΔMs forward and make RF digitizers more feasible.

10.3.6 Emerging ΣΔM Techniques

As stated in previous sections, ΣΔMs have evolved in recent years towards an increasingly more efficient digitization. New circuits and systems techniques are continuously emerging, giving rise to new generations of ΣΔ architectures. These techniques can be grouped according to the ΣΔM building block in which they are applied, either the feedforward loop filter or quantizer, and the feedback (DAC) path.

Among others, the following techniques can be considered as emerging trends in the design of the ΣΔ loop filters:

- RF-to-digital and digital-assisted ΣΔ receivers [5]
- digital-inspired (inverter-based) loop filters [6]
- hybrid passive/active loop filters [47]
- amplifier/stage-shared loop-filter techniques [48].

Meanwhile, conventional ΣΔ quantization is being enhanced by the following circuits and systems techniques:

- time-based/frequency-based signal processing [49]
- hybrid SAR-ΣΔ ADCs [50]
- FIR-based feedback DAC techniques [44].

Some of these strategies are discussed in the next sections, highlighting their main characteristics and potential applications.

10.4 Gigahertz-range ΣΔMs for RF-to-digital Conversion

As stated already in this chapter, more and more CT-ΣΔMs are proving to be competitive solutions for the implementation of power-efficient ADCs operating in the gigahertz range [23, 26, 30, 45, 51–57]. This feature opens the door to a new category of digital-intensive RF transceivers [58] and SDRs, sometimes referred to as ΣΔ receivers [59]. One of the most common approaches in commercial mobile handsets is based on a direct conversion receiver, like the one conceptually depicted in Figure 10.11a. This receiver architecture is made up of analog signal conditioning circuitry (basically an LNA, a mixer, and a baseband filter) followed by an LP CT-ΣΔM. In addition to digitising the incoming signals, CT-ΣΔMs take advantage of their CT circuit nature to merge some RF functionalities, like out-of-band blocker/interfering-rejection filtering, frequency-mixing process,

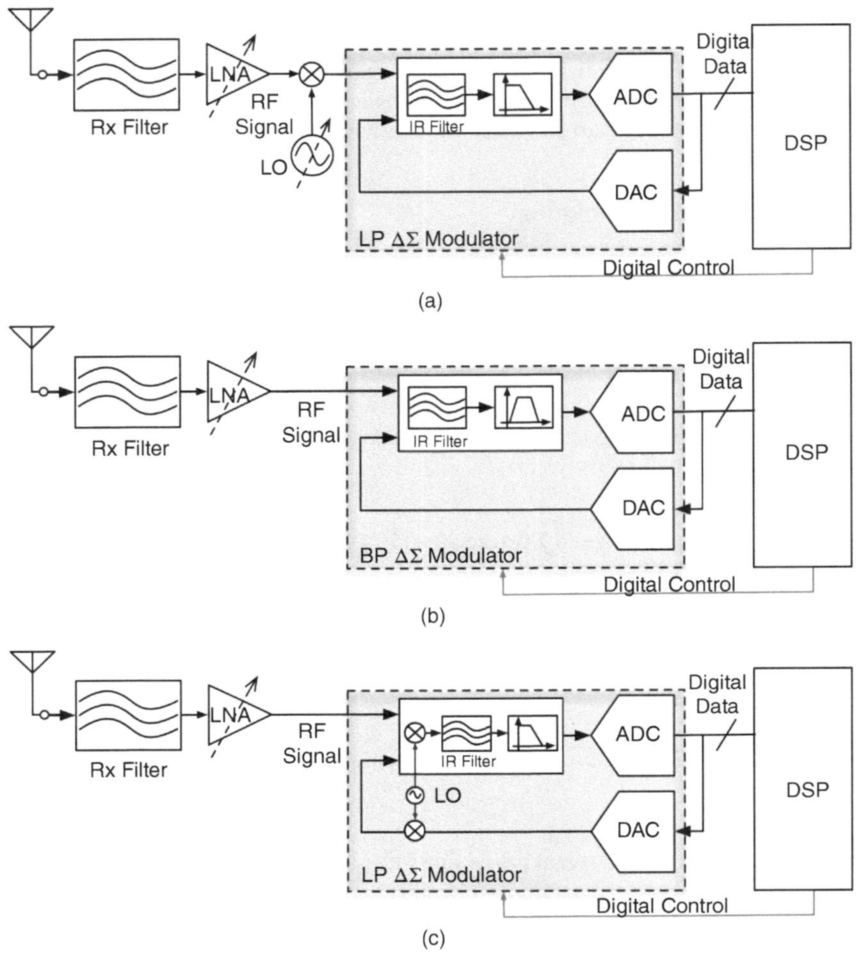

Figure 10.11 Conceptual diagrams of different CT-ΣΔ RF receivers: (a) direct conversion receiver based on a LP CT-ΣΔM with embedded interfering-rejection (IR) filter; (b) RF-to-digital converter based on a BP CT-ΣΔM; (c) direct ΣΔ receiver with embedded filtering and downconversion mixers.

channel selection, and antialiasing filtering [60–72]. All of these functionalities can be embedded within the ΣΔM feedback loop, as illustrated in Figure 10.11a, thus resulting in more compact RF receivers, with reduced complexity in the analog hardware and the subsequent benefits in terms of sensitivity to circuit imperfections and power/area scalability with technology.

An alternative trend towards the practical implementation of SDR receivers is based on placing gigahertz-range BP CT-ΣΔMs as close as possible to the antenna, as illustrated in Figure 10.11b, yielding the so-called *RF-to-baseband converters* or *RF-to-digital converters* [26, 54, 59, 73–80]. Although this idea is close to the ideal implementation of the SDR paradigm [81], such systems are still far from a deployment as a practical consumer product. They are mostly limited by the power-hungry nature of the ADC, particularly in terms of signal bandwidth, linearity, and dynamic range, supporting signal levels from the sensitivity to the full strength. In order to mitigate the demanding specifications of these ΣΔM-based RF digitizers, some authors have proposed strategies

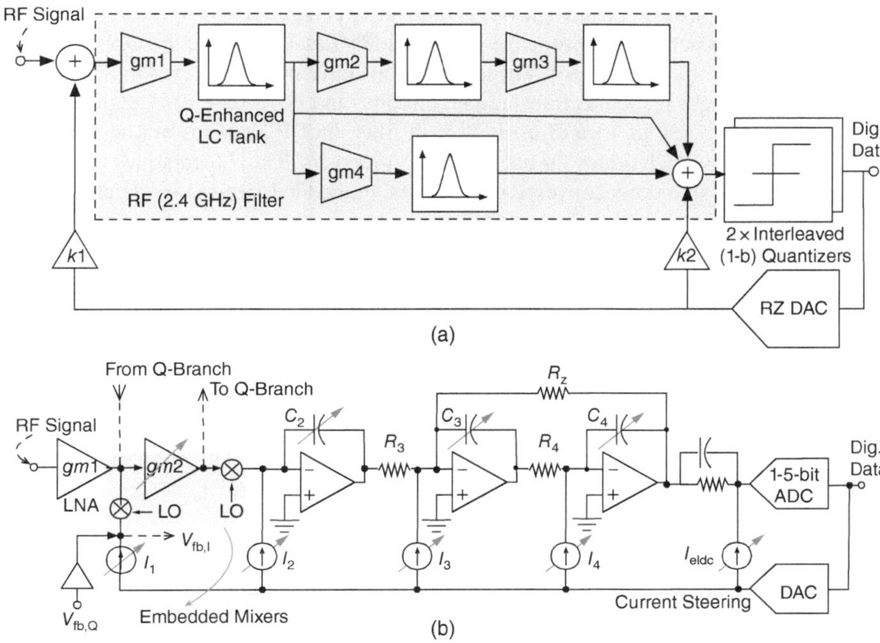

Figure 10.12 Exemplary conceptual schematics of RF-to-digital CT-ΣΔMs: (a) Gm-LC BP-ΣΔM [73]; (b) inductorless ΣΔ receiver (I-branch) [5].

such as embedded out-of-band filtering [59, 79], frequency-translating [75, 79] and undersampling or subsampling [54, 73, 74]. The latter approach allows digitization of RF signals placed at $f_{RF} > f_s/2$, while keeping high values of the OSR, since the signal bandwidth is typically $B_w \ll (f_{RF}, f_s)$ [82, 83].

One of the first successful approaches using subsampling techniques was proposed by Ryckaert *et al.* [73], who reported a sixth-order LC-based BP-ΣΔM – depicted in Figure 10.12a – for digitizing signals centered at 2.4 GHz with $f_s = 3$ GHz; a ratio of 4/5. To the best of the author's knowledge, Martens *et al.* have reported the highest sampling rate to date (8.8 GHz), using CMOS BP CT-ΣΔMs, digitising 2.2-GHz signals with 48-dB DR within 80-MHz bandwidth, while consuming 164 mW. This sampling rate is achieved by combining six time-interleaved quantizers with a polyphase decimation filter [53].

The majority of reported RF-to-digital BP CT-ΣΔMs use a fixed center frequency or *notch frequency*. This forces the RF receiver to use a programmable frequency synthesizer in order to place the signal band within the pass-band of the BP-ΣΔM. This issue has motivated interest in reconfigurable/programmable BP-ΣΔMs with wide tunable notch frequencies [84]. Shibata *et al.* reported a sixth-order reconfigurable LP/BP-ΣΔM, which features a DC-to-1 GHz tunable notch frequency, while consuming 550 mW [26]. Another interesting approach based on a reconfigurable 0.8–2 GHz notch frequency was reported by Gupta *et al.* [85]. A second-order CT BP-ΣΔM was integrated together with a quadrature phase-locked loop to allow quadrature phase synchronization between a raised-cosine feedback DAC and the embedded quantizer. The chip digitizes 1-MHz signals placed at 797 MHz with a DR of 8.3 bits, while consuming 41 mW [54].

The main challenge associated with the design of loop filters in BP-ΣΔ ADCs is achieving a high-quality and accurate resonance. For BP-ΣΔ ADCs intended for RF conversion, tuning range is also a concern. LC tanks are good from a power and linearity perspective but typically support only an octave of range, whereas active-RC resonators can be widely tunable but require amplifiers with high gain at the ADC's center frequency [16]. These limitations have motivated exploration of alternative signal processing techniques, such as translational circuits, and polyphase or N-path filters [86]. As stated in Chapter 2, although the idea of using N-path filters in BP ΣΔMs is not new [87], there has been a resurgence of these techniques for use in RF ΣΔ receivers. These techniques can be combined with either active or passive downconversion structures, embedded into the modulator loop filter as conceptually shown in Figure 10.11c, in order to achieve the required receiver linearity and sensitivity [5, 59, 88]. Based on this idea, Wu *et al.* reported a 400 MHz–to–4 GHz direct RF-to-digital ΣΔ receiver implemented in 65-nm CMOS, featuring SNDR > 68 dB within 4-MHz bandwidth across the carrier frequency band, while dissipating 40 mW [88]. More recently, Englund *et al.* presented an inductorless programmable 0.2–2.7 GHz ΣΔ receiver in 40-nm CMOS, as conceptually depicted in Figure 10.12b. This is able to digitize 20-MHz signal bandwidths with SNDR = 40 dB and 90 mW power consumption [5]. Although the performance of these converters is still well short of what is needed for comercialization, this trend is set to continue and it is expected that ΣΔ receivers will be one of the key blocks for practical implementations of SDRs and next-generation mobile systems.

10.5 Enhanced Cascade ΣΔMs

In spite of the higher sampling rates demonstrated by some CT-ΣΔMs discussed above, the increasing demand for digitizing signal bandwidths over 100 MHz, necessitates low values of the OSR to keep the required low noise density and power dissipation. This in turn requires an increase in the number of bits of the embedded quantizer, *B*, and/or the order of the loop filter, *L*. The former strategy is limited by the power consumption of the (flash) ADC embedded in the ΣΔM, while the latter strategy yields diminishing returns at low OSR. An alternative consists of using cascade ΣΔ topologies, although these can exhibit high sensitivity to mismatch.

10.5.1 SMASH CT-ΣΔMs

As stated in Chapter 2, SMASH ΣΔMs reduce the sensitivity of cascade ΣΔMs to noise leakage, thus benefiting from their higher noise-shaping capabilities. In addition, the performance of SMASH SC-ΣΔMs can be markedly improved if USTF is used in all stages of the cascade [89], but the gain error between the inter-stage paths computing the front-end stage quantization error, e_1, may degrade the modulator performance. Although SMASH ΣΔMs have been successfully implemented using SC circuits [90], the use of these architectures in the wideband applications, where the demand is, requires the use of CT circuits. However, practical implementation requires correct extraction of e_1, which means that the input and the output of the front-end stage quantizer must be delayed by the same amount – something which is difficult to control in CT circuits.

Yoon *et al.* [25] reported the successful implementation of a SMASH 3-1 CT-ΣΔM – conceptually depicted in Figure 10.13 – which circumvents these limitations by using a LP filter in the inter-stage path and a feedforward path around the integrator of the second stage in order to obtain a USTF. The chip, fabricated in 28-nm CMOS, is clocked at 1.8 GHz and achieves SNDR = 74.6 dB within a 50-MHz signal bandwidth, while consuming 78 mW. These metrics place this SMASH CT-ΣΔM on the state-of-the-art front.

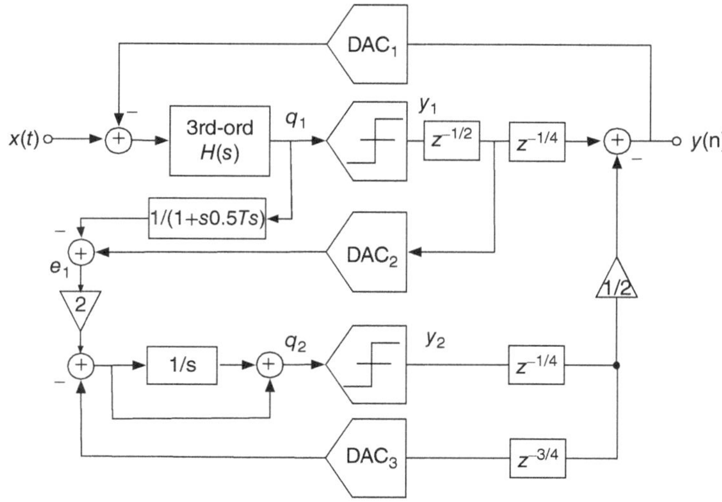

Figure 10.13 SMASH 3-1 CT-ΣΔM proposed by Yoon *et al.* [25].

10.5.2 Two-stage 0-L MASH

Another trend in the design of cascade ΣΔMs is based on a kind of two-step ADC made up of a front-end flash ADC followed by a conventional ΣΔM back-end stage, as conceptually depicted in Figure 10.14. The resulting architecture is called a 0-L MASH, since the front-end stage can be seen as a 0th order modulator – made up of a flash ADC, connected in cascade with a back-end Lth order loop that digitizes the residue of the front-end stage [91]. The output of the ADC is given by the sum of the outputs of both sub-converters, $v = v_0 + v_1 = u + \mathrm{NTF} \cdot (e_0 + e_1)$, where e_0 and e_1 are respectively the quantization errors of the front-end and back-end flash ADCs. Similarly to SMASH ΣΔMs, this topology does not need any DCL apart from a digital adder. Based on this strategy, Dong *et al.* [24] presented a 0-3 CT-ΣΔM, which consists of the cascade topology of a 16-level flash ADC followed by a third-order active-RC ΣΔM with an embedded 6-level flash ADC. The chip, integrated in 28-nm CMOS, does not need any additional DCL, and features SNDR= 71.4 dB within a 53.3-MHz signal band and consumes 235 mW.

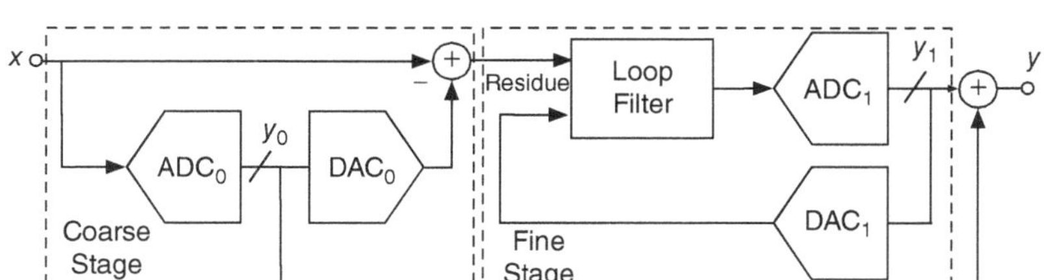

Figure 10.14 Conceptual block diagram of a 0-L MASH ΣΔ ADC [91].

10.5.3 Stage-sharing Cascade ΣΔMs

Regardless of the topology used, either single-loop or cascade, the increase of the modulator order implies addition of more active elements, with a resulting penalty in power dissipation. To reduce this problem, some authors have proposed sharing the active building blocks between the stages of the cascade. This strategy – known as *stage-sharing* and previously used in single-loop ΣΔMs [92] – was applied by Zanbaghi *et al.* to cascade topologies [48], as conceptually shown in Figure 10.15. In this example, the front-end integrators in both stages share the same opamp. Essentially the idea is to alternatively reuse the amplifiers in both integrators during the two nonoverlapping clock phases. The technique is limited by the number of stages, since the clock timing becomes more complicated. Stage-sharing is more robust than other similar techniques, such as double-sampling, since the latter can be severely degraded by folded noise and the timing requirements are more demanding. Obviously, this technique is only valid for DT-ΣΔMs, and it may result in a more efficient implementation of cascade topologies [48]; a 130-nm CMOS chip features 75-dB SNDR over a 5-MHz signal bandwidth with a power dissipation of 9 mW.

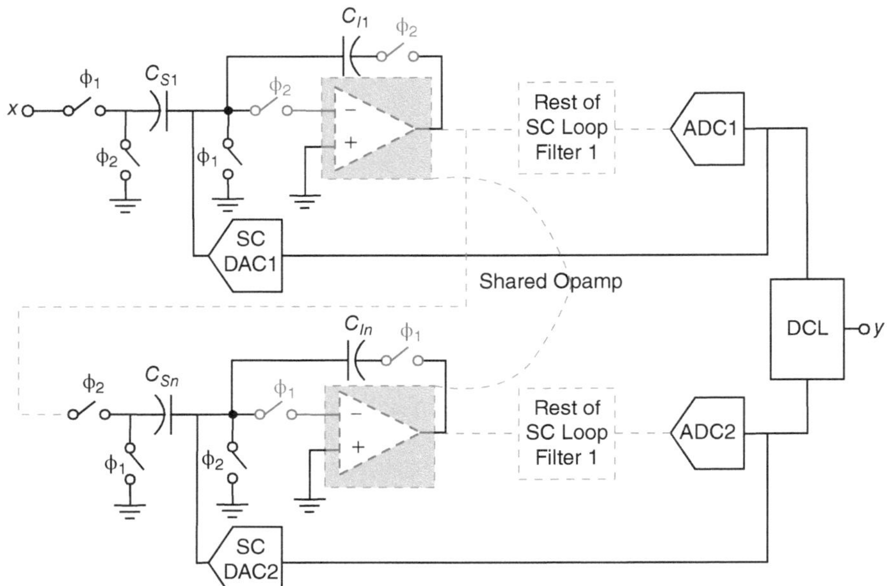

Figure 10.15 Stage-shared cascade SC-ΣΔM based ideas of Zanbaghi *et al.* [48].

10.5.4 Multi-rate and Hybrid CT/DT ΣΔMs

Multi-rate (MR) ΣΔMs are a particular kind of ΣΔMs in which a different sampling frequency is used in different parts of the modulator. The most common approach uses a lower sampling frequency in the building blocks placed at the front-end of the modulator – where most of the power is consumed – and a higher sampling rate in the subsequent subcircuits, where the dynamic requirements can be relaxed [93]. This concept can be applied to either single-loop or cascade architectures, and implemented using SC [93] and CT circuits [94].

10.5.4.1 Upsampling Cascade MR-ΣΔMs Figure 10.16a shows the conceptual implementation of a conventional cascade (two-stage) MR-ΣΔM.[5] For the sake of generality, multi-bit quantization will be assumed in all stages of the cascade, with B_i being the number of bits of the internal quantizer in the ith stage. The sampling frequency f_{si} of the different modulator blocks is depicted in the figure. The most common situation in conventional MR-ΣΔMs is that the front-end stage operates at f_{s1}, whereas the remaining i stages are sampled at $f_{si} > f_{s1}$. This approach – also referred to as upsampling MR-ΣΔM [95] – benefits from increasing values of OSR in the back-end stages, where the dynamic requirements are less demanding than in the front-end stages [90, 93, 94].

The operating principle behind the modulator in Figure 10.16a is conceptually the same as in a conventional *single-rate* cascade ΣΔM. All stage outputs are combined by the DCL transfer functions – clocked at f_{s2} – so that ideally only the quantization error of the last stage remains and it is shaped by an NTF whose order (L) is the sum of the orders of all stages in the cascade (L_i). Thus, assuming that $\text{STF}_i(z) = z^{-L_i}$ and $\text{NTF}_i(z) = (1 - z^{-1})^{L_i}$, and defining $\text{DCL}_i(z)$ so that $\text{DCL}_1(z) = \text{STF}_2(z)$ and $\text{DCL}_2(z) = \text{NTF}_1(z^r)$, it can be shown that the IBN power at the output of the modulator is approximately given by:[6]

$$\text{IBN}_{\text{USMR}} \simeq \frac{\Delta^2}{12} \frac{\pi^{2L}}{(2L+1)\text{OSR}_1^{2L+1} r^{2(L-L_1)+1}} \tag{10.7}$$

where $r \equiv f_{s2}/f_{s1}$, Δ stands for the quantization step of the last quantizer, and $\text{OSR}_1 \equiv f_{s1}/(2B_w)$ [95].

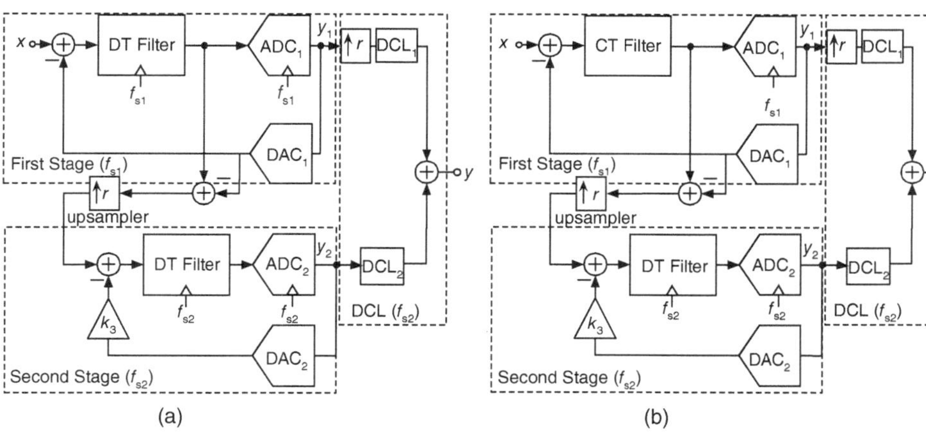

Figure 10.16 Conceptual block diagram of a conventional (upsampling) cascade MR-ΣΔM: (a) DT scheme; (b) hybrid CT/DT scheme.

The concept of MR-ΣΔMs can be extended to *hybrid* CT/DT implementations.[7] This is conceptually depicted in Figure 10.16b, which represents a cascade two-stage MR H-ΣΔM. The circuit

[5]Two-stage cascade architectures will be discussed in this section without loss of generality. The analysis can be extended to N-stage cascade MR-ΣΔMs [95].

[6]Subscript USMR is an acronym for "upsampling multi-rate", as opposed to the downsampling multi-rate (DSMR) concept discussed later in this section.

[7]Some authors have proposed combining the benefits of both SC and CT circuit techniques, giving rise to so-called hybrid CT/DT ΣΔMs (H-ΣΔMs). In these H-ΣΔMs, some parts of the loop filter – usually the front-end building blocks – are implemented using CT circuits, thus providing potentially faster operation and embedded antialiasing filtering with reduced power consumption [96–102]. H-ΣΔMs have been demonstrated using both single-loop [96, 97] and cascade ICs [101]. The most common situation in practice consists of using a CT front-end integrator; the remaining integrators in the modulator loop are implemented using SC circuit techniques.

nature (either CT or DT) of the different modulator blocks as well as their corresponding sampling frequencies are highlighted. The use of a complete CT stage (instead of just the front-end integrator, for instance) maximizes the embedded AAF, while taking advantage of the relaxed dynamic requirements of the front-end CT circuits as compared to their DT counterparts [101].

The analysis of Figure 10.16b can be carried out by applying a DT–CT transformation to the front-end stage of Figure 10.16b. Based on the CT–DT equivalence explained earlier in this book, the resulting MR H-ΣΔM is equivalent to the original MR DT-ΣΔM, and the IBN of the ΣΔM in Figure 10.16b is given by Equation (10.7).

10.5.4.2 Downsampling Hybrid CT/DT Cascade MR-ΣΔMs

Using lower values of the OSR in the front-end CT stage of Figure 10.16b may be beneficial when the subsequent SC stages can operate at high enough values of OSR. However, the use of high values of OSR in the back-end stages is not practical when digitizing wideband signals, for instance of the order of 10–100 MHz. Sampling rates of the order of gigahertz may be necessary to digitize such signals with low-to-medium effective resolutions (8–10-bit). Therefore, the use of back-end SC stages may be not feasible because of the prohibitive values of GB required for the operational amplifiers.

These problems can be alleviated if the concept of multirating is redefined in the opposite way to which it was conceived; in other words, going from an *upsampling* multi-rate system – in which the front-end stage rate is increased in the subsequent stages – to a *downsampling* multi-rate system – in which the front-end (CT) stage operates at the highest rate, thus benefiting from their potentially faster operation [95].

Figure 10.17 shows a conceptual block diagram of a *downsampling* (two-stage) cascade MR H-ΣΔM architecture [95]. In contrast to conventional (*upsampling*) MR H-ΣΔMs, the back-end (DT) stage operates at a rate lower than that of the front-end (CT) stage; that is, $f_{s2} < f_{s1}$. The main drawback of this approach is the aliasing caused by the downsampling processing, which requires using an interstage AAF. However, as shown by García-Sánchez and de la Rosa [95], the operation of the

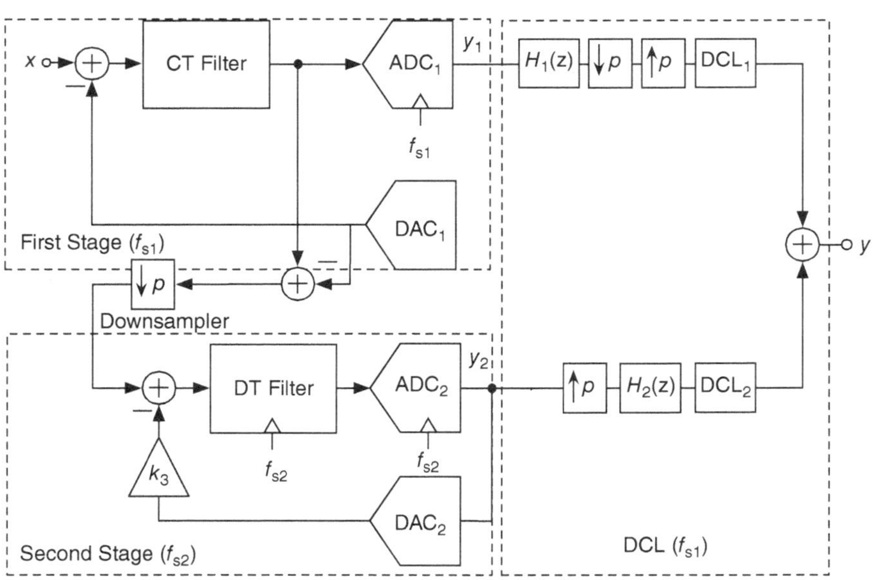

Figure 10.17 Block diagram of a downsampling cascade hybrid CT/DT MR-ΣΔM [95].

AAF can be completely translated to the digital domain, by using two additional digital blocks, whose transfer functions are named $H_1(z)$ and $H_2(z)$. In this way, Figure 10.17 has the same number of analog building blocks as a conventional cascade ΣΔM.

Therefore, the operation behind the modulator in Figure 10.17 is essentially the same as in conventional cascade ΣΔMs. The main difference is that the DCL transfer functions are designed so that they must remove not only the quantization error of the front-end stage $E_1(z)$, but also its aliased components. To this end, $H_1(z)$ and $H_2(z)$ must be reconfigurable and programmable according to the value of the downsampling ratio ($p \equiv f_{s1}/f_{s2} > 1$). These functions are completely implemented in the digital domain, without any extra analog hardware required, and can be easily synthesized for different values of p [95]. As a consequence, the resulting MR-ΣΔMs are potentially faster, less sensitive to circuit error mechanisms, and more power efficient than conventional upsampling MR architectures.

10.6 Power-efficient ΣΔM Loop-filter Techniques

The loop filter of ΣΔMs is becoming more and more efficient thanks to the inclusion of a number of emerging circuits and systems techniques, giving rise to new ΣΔM architectures and circuit realizations.

10.6.1 Inverter-based ΣΔMs

A simple and elegant way of implementing the loop filter of ΣΔMs consists of replacing their most power-hungry building blocks – the OTAs – by either comparators [103] or inverters [104, 105], which are more power- and area-efficient OTA topologies. This circuit technique was first applied by Chae and Han [105], who proposed an *inverter-based* second-order single-bit SC-ΣΔM, conceptually depicted in Figure 10.18. This chip achieved 63-dB peak-SNDR within 8-kHz bandwidth, while consuming only 5.6 μW from a 1.2-V supply (FOMS = 156 dB). Although the power supply rejection ration (PSRR) was not good, these results demonstrated for the first time that the use of *digital-inspired* analog circuit techniques may become energy efficient while keeping outstanding performance.

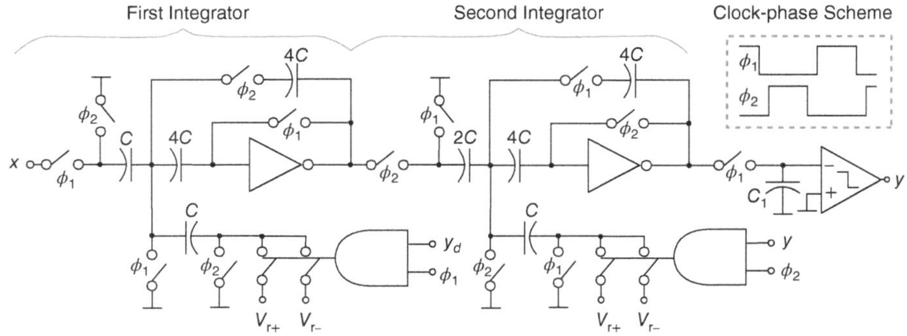

Figure 10.18 Inverter-based second-order single-loop ΣΔM [105].

In spite of these benefits, the reader should note that replacing conventional OTAs by simple CMOS inverters cannot be seen as a panacea in solving the analog design problems in ΣΔMs.

Inverter-based integrators give higher power efficiency than conventional integrators at the price of sacrificing other performance metrics, such as robustness over process-voltage-temperature (PVT) variations, impact of circuit parasitics, limited DC gain, and so on. These limitations have motivated some authors to propose improved versions of inverted-based integrators to implement the loop filter of ΣΔMs [6, 7, 32, 36, 106]. Thus, Christen proposed an inverter-based OTA where the operating point is controlled over the internal supply voltage by using a low-dropout (LDO) regulator, as conceptually depicted in Figure 10.19a [6]. This technique is applied to the design of a fourth-order feedforward single-loop single-bit SC-ΣΔM, which digitises 20-kHz signals with SNDR= 87.9 dB, while consuming 140 μW. Another cutting-edge inverter-based SC-ΣΔM was reported by Luo *et al.* [7], who proposed a gain-boosted class-C inverter, as depicted in Figure 10.19b. This technique allows an increase in the DC gain of the inverter-based OTA of over 80 dB, used to implement a cascade 2-1 SC-ΣΔM, featuring 91-dB SNDR within a 20-kHz signal bandwidth with 230 μW power dissipation.

Inverter-based OTAs have been also demonstrated in CT-ΣΔMs, as shown by Zeller *et al.* [32], who used an inverter-based single-opamp third-order filter – conceptually depicted in Figure 10.19c – in a pseudo-differential single-loop 1.5-bit CT-ΣΔM featuring 68.6-dB SNDR over a signal bandwidth of 10 MHz while consuming 1.82 mW. Thus, as demonstrated by these outstanding designs, the use of inverter-based loop filters is a very promising strategy for the implementation of high-performance ΣΔMs, and it is expected that the use of these circuit techniques might benefit from technology downscaling in the next few years.

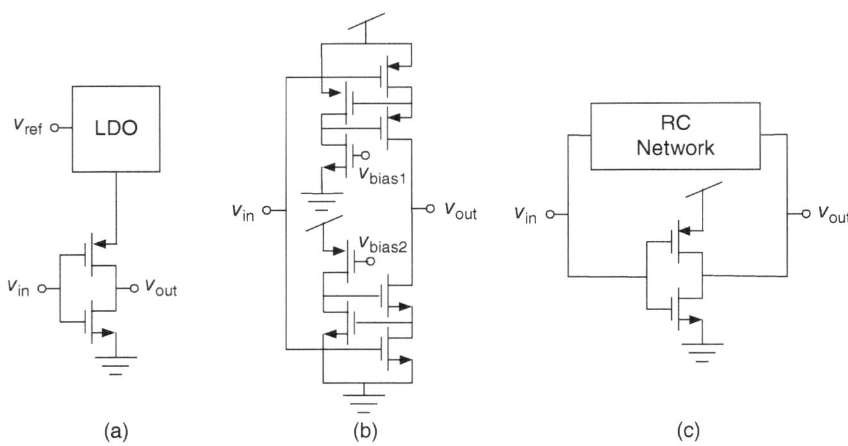

Figure 10.19 Inverter-based OTAs used in state-of-the-art ΣΔMs: (a) inverter with LDO-based biasing [6]; (b) gain-boosted class-C inverter [7]; (c) third-order inverter-based integrator used in the CT-ΣΔM reported by Zeller *et al.* [32].

10.6.2 Hybrid Active/Passive and Amplifier-less ΣΔMs

Another alternative way to reduce the number of OTAs in ΣΔMs consists of replacing some active building blocks with passive-RC network ΣΔMs [107]. In addition to reducing the power consumption, integrated passive networks exhibit potentially better linearity than their active counterparts, at the price of reducing the required gain to make the loop filter robust enough against thermal noise and other circuit nonidealities. These limitations can be partially mitigated by using a hybrid active/passive loop-filter, in which the front-end subcircuit (either an integrator or a resonator) is active, so that the effect of thermal noise of the passive circuit elements can be reduced by the preceding active blocks. Prompted by these potential benefits, passive RC networks are becoming a popular circuit solution for the implementation of both CT-ΣΔMs and SC-ΣΔMs [45, 47, 80, 107–110].

Figure 10.20 shows one of the first successful examples of this idea applied to CT-ΣΔMs, as proposed by Song *et al.* [107]. The modulator – consisting of a fifth-order single-loop architecture – uses five integrators, two of them implemented by passive RC networks. These passive integrators are placed in the second and fourth positions of the modulator chain, so that their nonideal effects can be attenuated by the gain of active integrators. Another interesting example was proposed by Srinivasan *et al.* [45], who reported a third-order CT-ΣΔM made up of a front-end passive-RC integrator followed by two Gm-C integrators. The chip – fabricated in a 45-nm CMOS technology – digitizes 60-MHz signals with 61-dB SNDR and 20-mW power consumption, while clocked at 6 GHz.

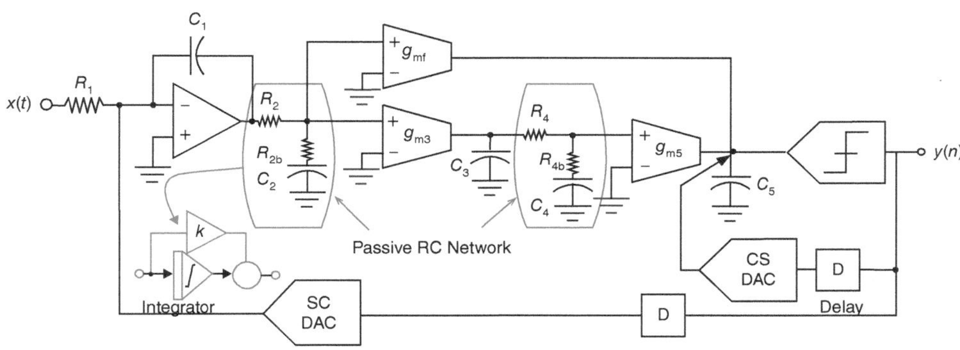

Figure 10.20 Example of the hybrid active-passive CT-ΣΔM reported by Song *et al.* [107].

Although hybrid active/passive loop filters have been mostly used in LP-ΣΔMs, these techniques can be also applied to BP-ΣΔMs. Due to their higher operating frequencies, these modulators present larger sensitivity to circuit element tolerances and technology process variations, as well as the degradation caused by parasitics. For instance, Chae *et al.* [80] reported a fourth-order BP CT-ΣΔM that digitizes 24-MHz signals located at 200 MHz with 58-dB SNDR while consuming 12 mW. The chip is implemented using single-opamp resonators, which – as discussed in Section 4.2 – incorporate positive feedback to achieve higher quality factors, and are implemented by combining an active LP filter and a passive high-pass filter, as illustrated in Figure 10.21.

Apart from being used in wideband CT-ΣΔMs, passive networks have also been applied to SC-ΣΔMs intended for ultra-low-power applications. A good example as reported by Yeknami *et al.* [47], who proposed several alternative (hybrid active/passive and fully-passive) implementations of a second-order single-bit SC-ΣΔM intended for medical implant devices. As an illustration, Figure 10.22 shows the conceptual schematic of a passive SC-ΣΔM, which digitises 500-Hz signals with 65-dB SNDR, while consuming just 0.43 μW.

This trend is expected to continue, being a very promising alternative technique in application scenarios where energy efficiency is a must. Indeed, hybrid active/passive circuits could be combined with other kinds of hybrid circuit techniques – for example in so-called CT/SC ΣΔMs – to take advantage of the different circuits involved: faster operation of CT circuits and lower sensitivity of SC circuits to errors [37, 96–101]. A good example of the combined use of these techniques is the CT-ΣΔM proposed by Nowacki *et al.* [111]. The modulator – conceptually depicted in Figure 10.23 – consists

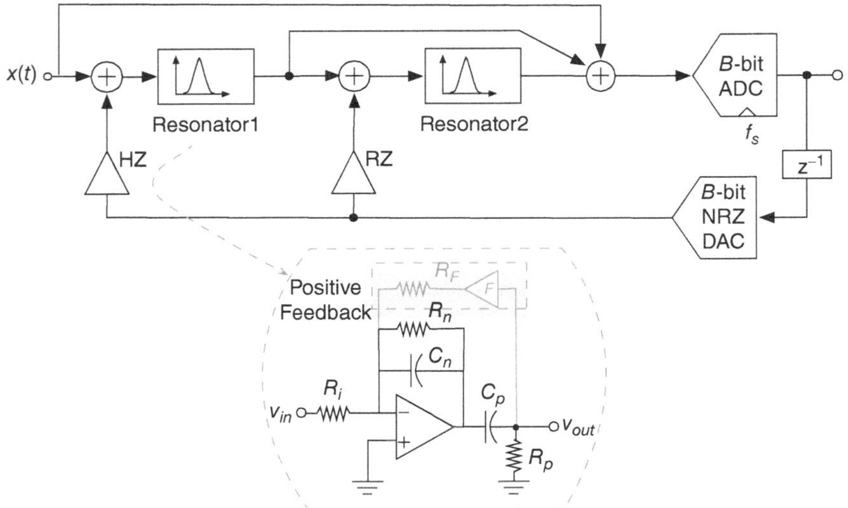

Figure 10.21 BP CT-ΣΔM proposed by Chae *et al.* [80], based on single-opamp resonators with Q-factor boosting.

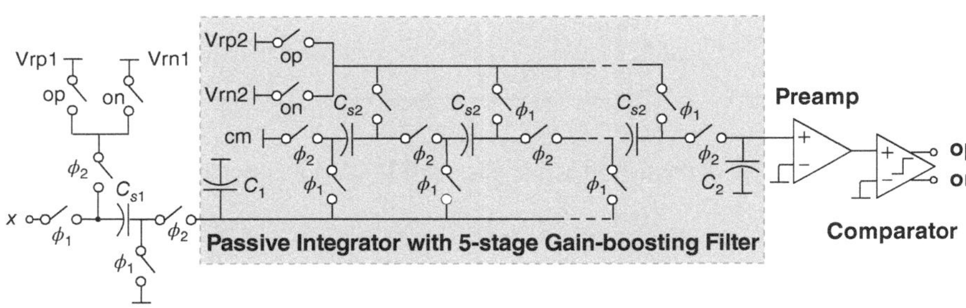

Figure 10.22 Passive second-order SC-ΣΔM proposed by Yeknami *et al.* [47] (single-ended version).

of a 1-bit 2-1 MASH topology made up of passive RC integrators with inter-stage (20-dB) gain stages and SC feedback DACs. The circuit consumes 1.57 mW and is able to digitize 10-MHz signals with 72.2-dB SNDR while clocked at 1 GHz, and featuring FOMS = 170.2 dB.

10.6.3 Power-efficient Amplifier Techniques

It is well known that the bandwidth requirements of loop-filter opamps in CT-ΣΔMs are much less than for their SC-ΣΔM counterparts, which is what makes CT loop-filter circuit techniques generally more efficient, as shown in the energy plot of Figure 10.5. As stated in Section 8.2.1, multi-stage feedforward compensated opamps have been used in some state-of-the-art CT-ΣΔMs [112], because they are more power-efficient than other topologies. This is because they do not waste current in charging and discharging the (explicit) compensation capacitor [30]. The main problem associated

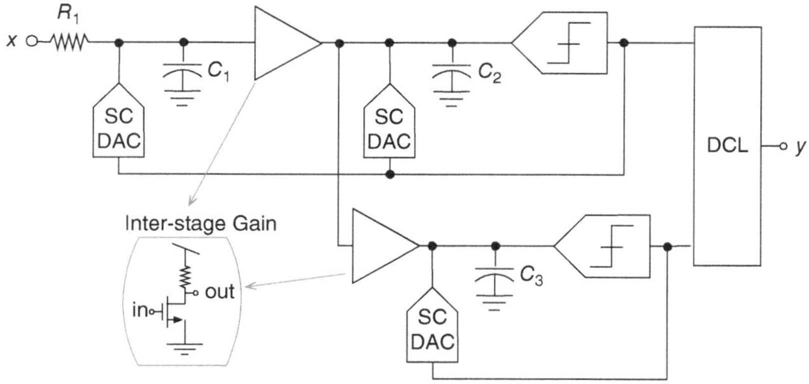

Figure 10.23 Block diagram of the passive/active cascade CT-ΣΔM proposed by Nowacki *et al.* [111].

with FF-compensated opamps is their long settling time, which is a limiting factor in SC-ΣΔMs. However, this is not relevant in CT-ΣΔMs, where the entire integrator output waveform is processed by the rest of the loop filter.

Another power-efficient technique is the so-called *assisted-opamp* integrator. This was proposed by Pavan *et al.* [29], and is conceptually illustrated in the single-bit CT-ΣΔM shown in Figure 10.24. Essentially, the problem is that the opamp in the front-end integrator cannot respond fast enough to the sharp steps of the 1-bit DAC feedback voltage. This effect causes large jumps in the virtual ground potential of the front-end opamp whenever the DAC switches – regardless of the DAC waveform used. The main consequence of these voltage jumps is nonlinearity, which severely degrades the performance of the modulator unless large bias currents are used to improve the dynamics of the opamp. Instead, an assisted-opamp integrator includes an additional (*assistant*) transconductor and a replica of the feedback DAC to pull current out of the OTA, thus reducing the virtual ground voltage excursions.

Assisted opamps have been successfully used in the front-end integrator of a CT-ΣΔM: this is usually the most power-hungry building block in a ΣΔM when a single-bit quantizer – that is, a comparator – is used. This technique can be applied to different OTA topologies [29, 30], including two-stage feedforward compensated and Miller-compensated OTAs, and giving state-of-the-art performance with SNDR= 88 dB within 24-kHz signal bandwidth with 110 μW.

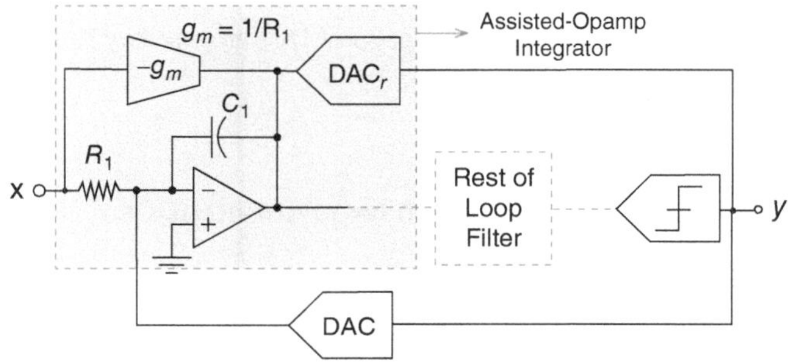

Figure 10.24 CT-ΣΔM with assisted-opamp integrator proposed by Pavan *et al.* [29].

10.7 Hybrid ΣΔM/Nyquist-rate ADCs

Another category of H-ΣΔM architectures combines a Nyquist-rate ADC – usually a pipeline, SAR, or cyclic ADC – and a ΣΔM [113–118].[8] In the majority of cases, the basic strategy followed in this kind of H-ΣΔM consists of replacing the embedded multi-bit flash ADC by another kind of Nyquist-rate ADC architecture. Essentially, the main advantage of this strategy is to provide a way of increasing the number of levels of the internal quantizer without the prohibitive exponential growth of the power consumption and silicon area of flash ADCs.

Hybrid ΣΔM-Nyquist ADCs have been implemented in both single-loop and cascade architectures, giving competitive performance in different application scenarios. Thus, based on the Leslie–Singh architecture [119], Brooks *et al.* [113] reported one of the first attempts to combine the benefits of Nyquist-rate and ΣΔMs in a single ADC, as illustrated in Figure 10.25. It involves cascading a 5-bit second-order ΣΔM and a 12-bit fourth-stage pipeline ADC. Since then, there have been a number of hybrid ΣΔM-Nyquist ADCs – considering both single-loop and cascade topologies – and giving competitive performance in different application scenarios [38, 50, 114–118, 120–123].

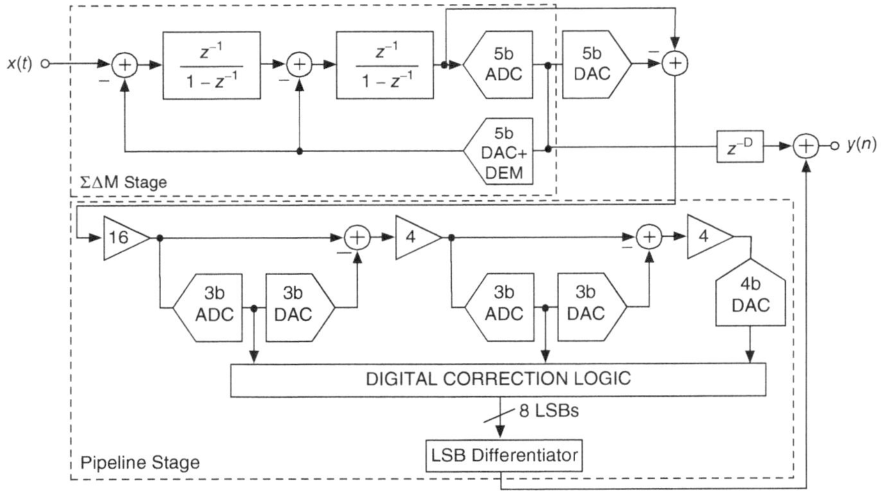

Figure 10.25 Example of hybrid ΣΔM-pipeline ADCs proposed by Brooks *et al.* [113].

10.7.1 Multi-bit ΣΔM Quantizers based on Nyquist-rate ADCs

In the majority of cases, the strategy followed by hybrid ΣΔM/Nyquist ADCs consists of replacing the multi-bit flash ADC embedded in the ΣΔM by another kind of Nyquist-rate ADC architecture: pipeline [113, 114, 118, 121], two-step flash [38, 50, 116, 122, 123], cyclic [115, 117] and integrating ADC [120]. The main advantage of this strategy is to provide a way of increasing the number of

[8] Alternatively, Nyquist-rate ADCs – such as SAR or pipeline – can include noise-shaping techniques to improve their performance [18].

levels of the internal quantizer without the prohibitive exponential growth of the power consumption and silicon area of conventional flash ADCs.

One of the main inconveniences of the combination of ΣΔMs and Nyquist-rate ADCs is that, as the number of bits of the Nyquist-rate sub-ADC increases, the circuit complexity, sensitivity to circuit errors, and the latency increase as well, with the risk of losing the benefits provided by ΣΔMs in terms of system simplicity and robustness against circuit nonidealities. Therefore, even though very competitive performance can be achieved by replacing multi-bit flash ADCs with alternative Nyquist-rate ADCs, the design trade-offs mentioned have motivated ΣΔM designers to explore alternatives to conventional multi-bit quantization, as will be discussed later.

Among other approaches, the use of an asynchronous SAR (ASAR) structure as the embedded quantizer ADC is gaining favor because of the potentially better efficiency [38, 50, 123, 124]. Figure 10.26 shows the conceptual diagram of the hybrid ASAR-based CT-ΣΔM proposed by Tsai *et al.* [50]. The chip, integrated in 40-nm CMOS, consists of a third-order loop-filter with a 6-bit ASAR quantizer followed by a digital ΣΔ truncator, which reduces the ASAR output level from $L_1 = 64$ to $L_2 = 9$ to achieve an optimum performance, and giving SNDR = 79.6 dB within a 1.92-MHz signal bandwidth with 1.91 mW. A similar approach is followed by Wang *et al.*[123], who embedded a 6-bit ASAR together with digital truncation and digital integration in a second-order CT-ΔM implemented in 28-nm CMOS, achieving FOMS= 173.8 dB. Later, Ho *et al.* [38] reported a fourth-order CT-ΣΔM with 4-bit ASAR and DWA to compensate for the nonlinearity of the multi-bit feedback DAC, obtaining SNDR = 90.4 dB, $B_w = 2.2$ MHz, and 4.5 mW.

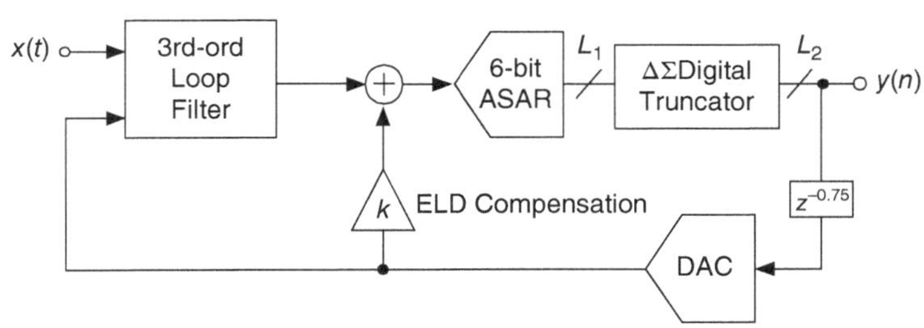

Figure 10.26 Hybrid ASAR-based third-order CT-ΣΔM proposed by Tsai *et al.* [50].

The combination of Nyquist-rate ADCs – especially SAR ADCs – and ΣΔMs is giving rise to new kinds of ADC architectures, which are in general more *digital-friendly* than conventional ADCs. A good example of this is the SAR-based ΣΔ ADC reported by Sanyal *et al.* [122, 125], conceptually depicted in Figure 10.27. The ADC can be seen as a two-stage 0-1 MASH, as discussed earlier, and uses a similar concept to the so-called *zoom*-ADC [39, 126, 127]. In the ADC of Figure 10.27, the first stage is a 5-bit SAR, which performs the coarse quantization of the input signal, with the residue digitized by a first-order ΣΔM mostly based on a voltage controlled oscillator (VCO), without needing OTAs for its implementation. Indeed, time/frequency-based quantization is another approach attempted by ΣΔM designers, as discussed in Section 10.8.

10.7.2 Incremental ΣΔ ADCs

Incremental ΣΔ ADCs (I-ΣΔ ADCs) can also be considered as hybrid ΣΔM/ Nyquist ADCs. These kinds of ADCs use oversampling and noise-shaping, but reset the loop-filter integrators and

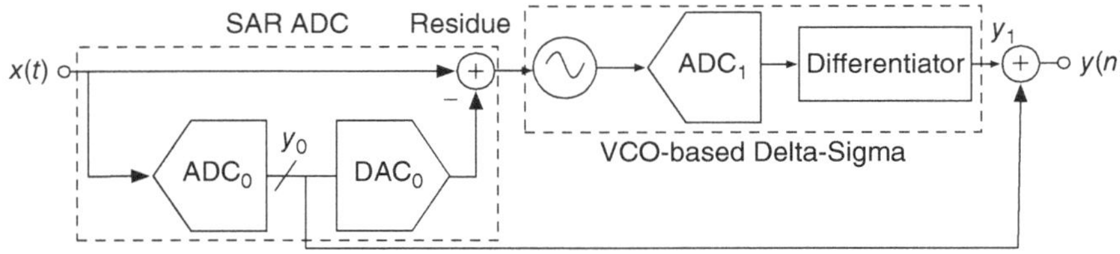

Figure 10.27 SAR-VCO ΣΔ ADC proposed in [122].

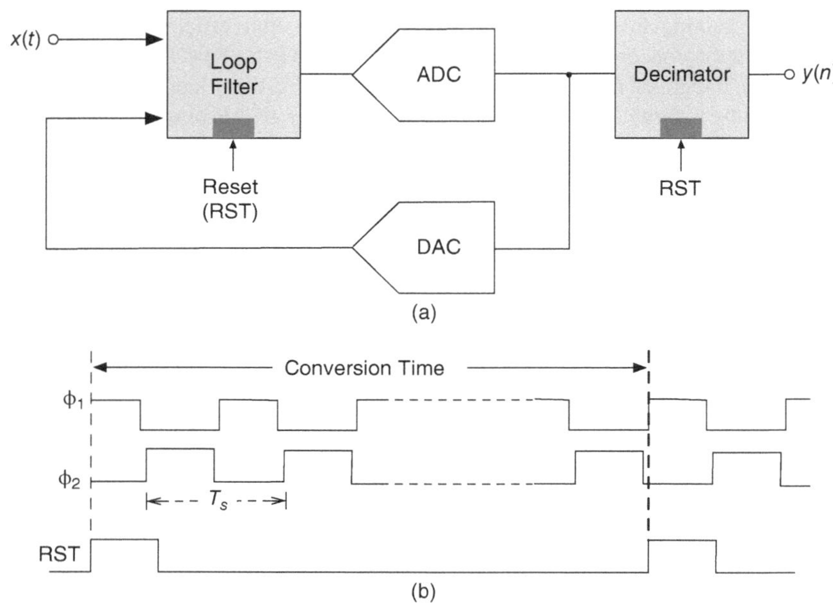

Figure 10.28 I-ΣΔ ADCs; (a) conceptual block diagram; (b) clock timing [131].

digital filters after each conversion [128], as conceptually depicted in Figure 10.28. These features allow these converters to be especially efficient in low-frequency high-resolution applications, like sensor interfaces using SC circuits [129] and other applications requiring to process multiplexed low-frequency signals with high-resolution [130].

The operation of I-ΣΔ ADCs is controlled by a reset pulse, which clears the memory of both the analog loop filter and the digital decimator after a number of clock periods according to the timing depicted in Figure 10.28. The number of oversampling clock periods is given by OSR, which is often referred to as the *conversion time period*. In this way, an I-ΣΔ ADC converts one sample of the analog input data at the end of every conversion period, thus operating as a Nyquist-rate ADC. Therefore, I-ΣΔ ADCs need 2^N clock periods to achieve N-bit ENOB, which leads to a trade off between latency (conversion time) and accuracy [131]. However, similar to ΣΔMs, the trade-off can be relaxed by increasing the loop-filter order, L, and the number of bits of the embedded quantizer, in order to obtain the required resolution without penalizing the conversion time.

It can be shown that the SQNR of an Lth order I-ΣΔ ADC with an embedded L_Q-level quantizer is approximately given by [132]:

$$\text{SQNR}_{\text{I}-\Sigma\Delta} \simeq L \cdot 20\log(\text{OSR}) + 20\log(L_Q - 1) - 20\log(L!) \tag{10.8}$$

Note that, in a similar way to conventional ΣΔMs, I-ΣΔ ADCs increase their resolution by combining different strategies based on three main design ingredients: OSR, L, and L_Q (directly related to B). However, as I-ΣΔ ADCs are indeed Nyquist-rate ADCs, this property has been exploited by some designers to get more efficient ADCs based on the combination of ΣΔ and Nyquist ADC techniques. Some cutting-edge I-ΣΔ ADCs are the *multi-step* or the *extended counting* architectures [133], which in turn can benefit from the hardware/stage-sharing technique discussed in Section 10.5.3.

Figure 10.29 illustrates the extended counting concept applied to I-ΣΔ ADCs [134]. The ADC consists of a DSMR cascade I-ΣΔ ADC with a front-end I-ΣΔ ADC and a back-end stage made up of a Nyquist-rate – which can be either a SAR or a cyclic ADC – to digitize the residue voltage of the front-end I-ΣΔ stage. In this way, the front-end stage performs the coarse quantization, resulting in the most significant bit (MSB) of the digitized signal, while the fine quantization (LSB) is carried out in the back-end stage. The outputs of both stages are digitally combined, giving a more accurate digitization [135].

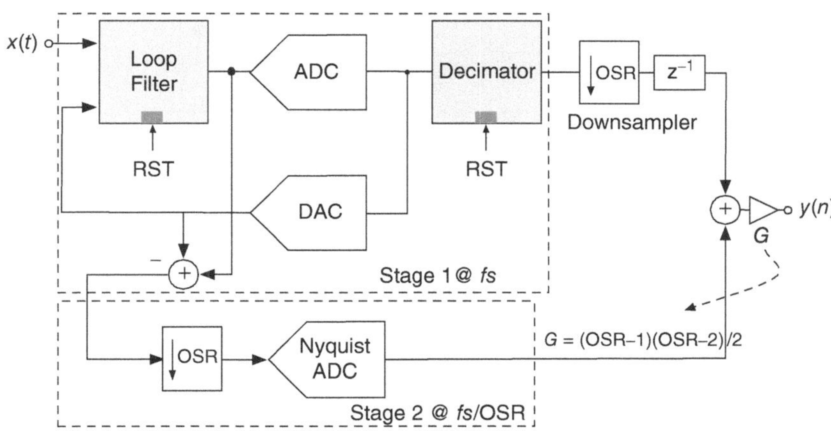

Figure 10.29 Illustrating the concept of *extending counting* technique applied to I-ΣΔ ADCs: cascade I-ΣΔ ADC using a back-end Nyquist-rate sub-ADC stage.

10.8 Time-based ΣΔ ADCs

Regardless of whether the ΣΔM quantizer is implemented by a conventional flash ADC, a tracking structure [31, 136], or any of the other Nyquist-rate ADCs above, the use of amplitude-based linearization becomes a design challenge due to the reduction in voltage supplies associated with CMOS technology downscaling. This problem has motivated the exploration of alternative quantizer implementations, based on time/frequency coding instead of amplitude coding. The benefit of this approach is that the resulting quantization methods are more suited for the implementation in low voltage (<1 V) nanometer CMOS processes, although special care must be taken in terms of those errors that are directly related to the timing, such as clock jitter error, in CT-ΣΔMs [137].

10.8.1 ΣΔMs with VCO/PWM-based Quantization

The use of VCO-based quantizers has been the most common way to replace conventional (amplitude-based) multi-bit quantizers [41, 58, 138–148]. The idea is based on translating the quantized information in the amplitude domain to the time domain by means of a voltage-to-frequency conversion. This can be performed by a VCO circuit, as conceptually depicted in Figure 10.30a, which shows one of the first implementations of this idea in ΣΔ ADCs [138]. The principle of operation behind this approach relies on the use of a ring oscillator to count the number of edges within a given

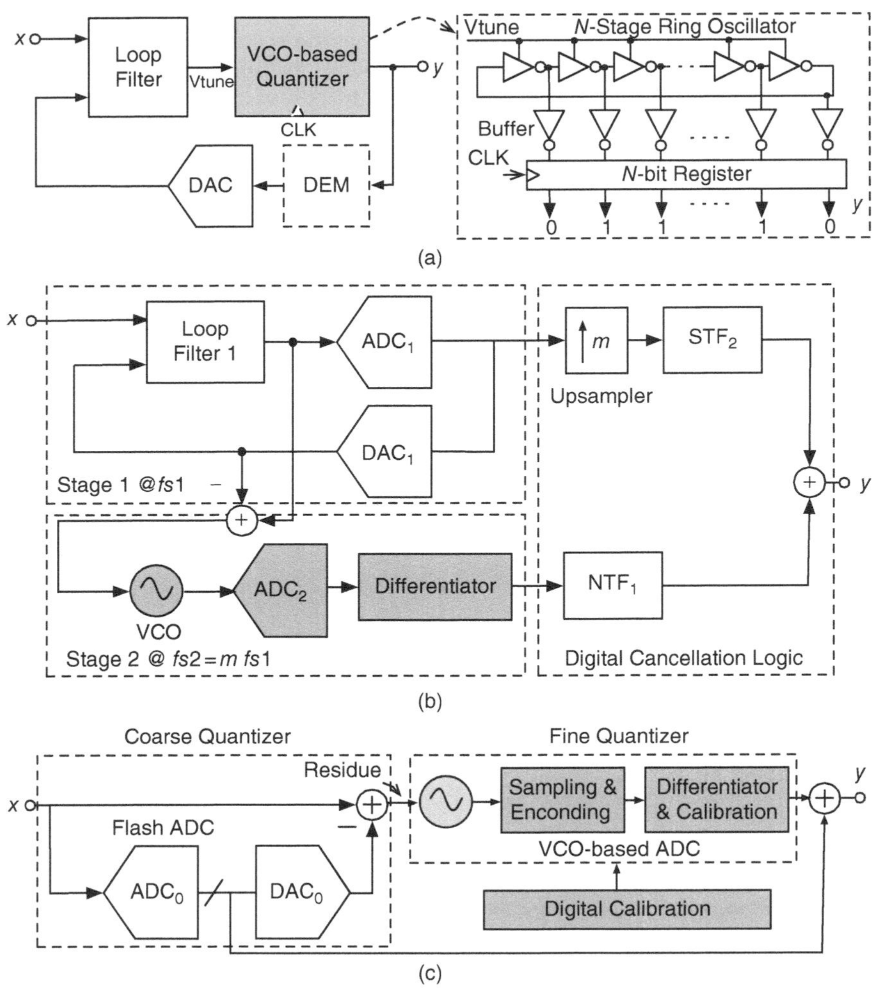

Figure 10.30 Conceptual diagram of VCO-based ΣΔMs: (a) single-loop topology [138]; (b) multi-rate cascade [149]; (c) cascade 0-1 with digital calibration [49].

time period. The result is directly related to the input signal, thus giving a digital representation of the amplitude. Moreover, VCO-based quantizers provide an implicit first-order noise-shaping filter due to their inherent differentiator operation, which is required to implement the frequency-to-voltage conversion in the digital domain. This feature is exploited by ΣΔMs to increase the noise-shaping filter order by one, without increasing the order of the analog loop filter [138].

The main limitation of VCO-based quantization is the inherent nonlinearity of the voltage-to-frequency conversion, requiring the use of either calibration or linearization techniques [144–146, 150, 151]. Some authors have suggested mitigating this limitation by using a phase detector instead of a frequency detector [139], which also provides a first-order noise shaping. The use of phase detectors has also been succesfully demonstrated for ELD compensation by Huang *et al.* [148], giving 71.9-dB SNDR within 125-MHz B_w in a chip integrated in 16-nm CMOS.

Another strategy involves embedding VCO-based quantizers in the back-end stage of a cascade ΣΔM, such that their nonlinearities can be attenuated by the noise shaping of previous stages in the cascade. One of the first implementations of this idea, proposed by Zaliasl *et al.* [149], consists in a two-stage cascade multi-rate ΣΔM, as illustrated in Figure 10.30b, in which the back-end stage (operating at a higher rate) is implemented by a VCO quantizer, thus benefiting from its potentially higher operation speed without being penalized by its intrinsic nonlinearity. This strategy is also behind the two-step hybrid Nyquist-rate/ΣΔ ADC proposed by Sanyal *et al.* [122]. Based on a similar approach, Ragab *et al.* [49] proposed a 0-1 MASH ΣΔ ADC – conceptually depicted in Figure 10.30c – made up of a front-end (coarse) 14-level flash ADC and a (fine) back-end 15-stage ring VCO, which includes digital background calibration. The chip achieves 12-bit ENOB within 2.5-MHz bandwidth, with a power consumption of 4.8 mW, giving FOMS = 161 dB.

An alternative to VCO-based quantizers is to use pulse-width modulation (PWM) [152, 153], as conceptually depicted in Figure 10.31. The PWM block converts the voltage at the output of the loop filter to a pulse $p(t)$ and a time-to-digital converter (TDC) generates digital codes that are discrete representations of the edges of the $p(t)$ signal; that is, a time-quantized representation. Embedded TDCs have been succesfully implemented by a number of recent ΣΔMs [154–158]. One of the main problems of TDCs and PWM-based quantizers is their limited dynamic range [137]. This issue can be addressed by using the *time-encoding quantizer* (TEQ) proposed by Hernandez *et al.* [159]. Essentially, the idea involves using a modulator whose loop filter is split into two loops, one of which is made to oscillate, thus providing an independent control of the oscillation, in contrast to PWM-based quantization schemes. Based on this idea, Prefasi *et al.* [160] reported a third-order CT-ΣΔM integrated in 65-nm CMOS and clocked at 2.56 GHz, giving SNDR = 61 dB within 20-MHz bandwidth.

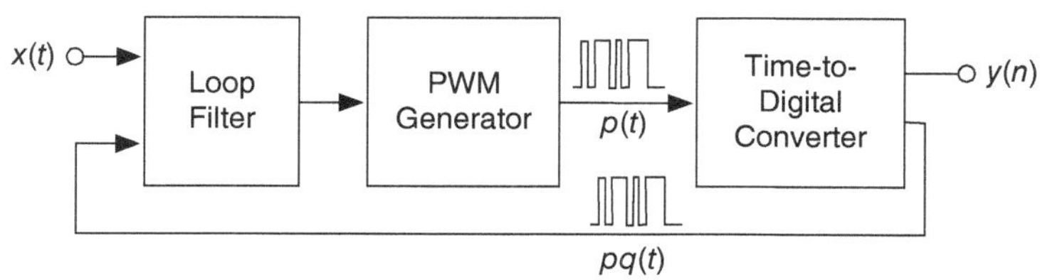

Figure 10.31 PWM-based ΣΔM concept reported by Dhanasekaran *et al.* [152].

10.8.2 Scaling-friendly Mostly-digital ΣΔMs

Time/frequency-based circuit techniques have not only been used for replacing quantizers, but also for other main building blocks of ΣΔMs. One of the first attempts to do so was by Taylor and Galton

[161], who used a voltage-controlled ring oscillator (VCRO) as the essential block of the ΣΔ ADC. The proposed modulator – mostly implemented with digital circuit techniques – consists of a digitally background calibrated VCRO, in which its inverters are sampled at the required output sample-rate, without containing any analog integrators, comparators, or feedback DACs. The main advantage of this approach is that, as the ΣΔM is essentially implemented by digital circuits, it may benefit from CMOS technology downscaling in terms of speed, with resulting benefits for low-voltage low-power wideband applications. Moreover, the operation of VCRO-based ΣΔMs essentially depends on the speed of its digital blocks, thus achieving high accuracy via digital signal processing. However, the main limitation of this approach is its higher sensitivity to timing errors, such as clock jitter error, and the aforementioned nonlinear behavior of VCOs. These problems have been overcome by using digital background calibration and self-canceling dither techniques [162], giving state-of-the-art performance results, with SNDR = 70–75 dB within B_w = 5–37.5 MHz with a tunable sampling rate of 1.3–2.4 GHz and FOMS = 160 dB [163].

Following the same philosophy of using VCOs as basic building blocks, Lee *et al.* [165] presented an OTA-less CT-ΣΔM implemented with both a VCO-based loop-filter and a quantizer, leading to a mostly-digital scaling-friendly ADC, which gives FOMS = 156 dB. Another innovative architecture, consisting of a third-order CT-ΣΔM with a loop filter based on VCO-based integrators, was presented by Young *et al.* [166]. The chip operates at 1.28 GS/s, achieving SNDR = 64 dB within 50-MHz bandwidth with 38 mW power dissipation. Therefore, although the performance of these VCO-based ΣΔMs is mainly limited by nonlinearity, these designs demonstrate that their metrics can be improved in *deeper* nanometer technologies.

10.8.3 GRO-based ΣΔMs

More recently, Yu *et al.* [8] proposed a 1-1 MASH ΣΔ TDC – conceptually depicted in Figure 10.32a – based on the use of the so-called gated switched-ring oscillator (GRO). The latter – conceptually depicted in Figure 10.32b – is controlled by an enable signal, so that it oscillates when such a signal is high, and *freezes* its phase otherwise. In this way, the GRO-based quantizer is in principle inherently linear since it operates with on/off states.[9] Moreover, the GROs introduce the required synchronization between both stages of the modulator, so that, a priori, no calibration is needed to compensate for the mismatch caused by the frequency differences between the two oscillators in the cascade. Based on the same approach, the same authors proposed a 1-3 MASH GRO-based TDC, which gave 0.22 ps_{rms} integrated noise within 15-MHz signal bandwidth, and FOMW = 0.19 pJ/conv [167].

Apart from using GRO-based ΣΔMs to implement TDCs, these kinds of quantizers can be used also to build ΣΔ ADCs that benefit from mostly-digital signal processing. An interesting approach based on this idea was proposed by Kim *et al.* [168]. Their ΣΔM architecture – conceptually depicted in Figure 10.33 – consists of a fourth-order single-loop architecture with a multi-bit time-based quantizer made up of two stages. The front-end stage of the quantizer is a 1.5-bit noise-shaped integrated quantizer (NSIQ) [156], and the back-end stage is a GRO-based quantizer. The NSIQ stage generates the quantization error in the time domain, which feeds to the back-end 3-bit GRO, which provides additional quantization levels as well as an extra order of noise-shaping. A digital back-end integrator is included in the quantizer so that it gives 6-bit quantization with second-order noise shaping. Overall, the modulator performs a fourth-order noise shaping with 6-bit embedded quantization, but using only a second-order analog loop filter – realized using active-RC integrators. The chip digitizes 15-MHz signals with 80-dB SNDR and a power consumption of 11.4 mW, giving FOMS = 174.1 dB.

[9] Multi-bit quantization can be implemented using the so-called multi-path GROs – rather than single-path GROs as shown in Figure 10.32b. In this way, the information in a multi-path GRO is quantized in the different phases provided by the stages (paths) of the GRO. For instance a seven-path GRO would be equivalent to a 3-bit quantizer [164].

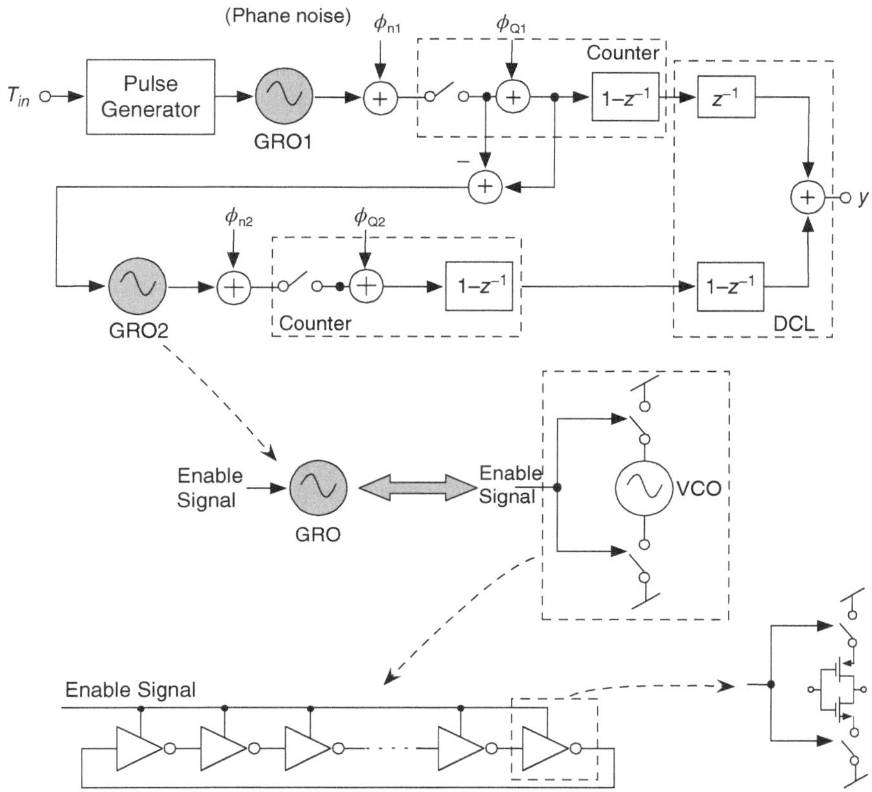

Figure 10.32 Mostly-digital GRO-based ΣΔ TDCs: (a) model of the GRO-based 1-1 MASH ΣΔ TDC proposed by Yu *et al.* [8] (b) GRO implemented with a ring VCO [164].

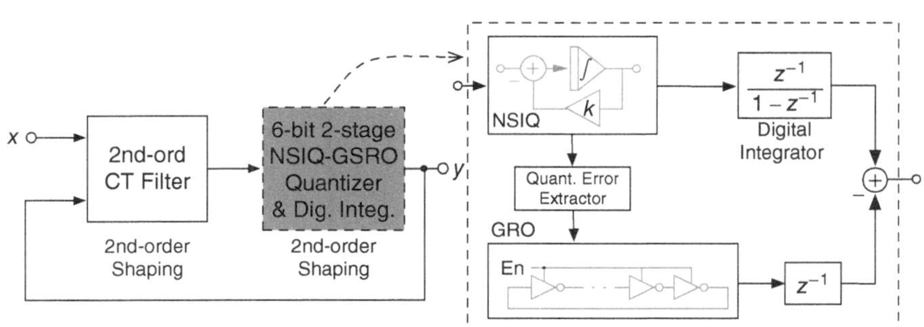

Figure 10.33 Conceptual diagram of a fourth-order CT-ΣΔM with 6-bit two-stage NSIQ-GRO quantizer [168].

10.9 DAC Techniques for High-performance CT-ΣΔMs

As discussed earlier, multi-bit quantization gives a number of inconveniences in terms of circuit complexity, area/power consumption, and the inherent nonlinearity caused by mismatches in the feedback DAC elements. Indeed, since a ΣΔ ADC is a feedback system, the performance of the ADC cannot be better than that of its internal feedback DAC. A number of linearization strategies have been proposed in the past to compensate for this error; for example, DEM, calibration, or the use of auxiliary DACs, which are based on a look-up-table computation of the INL [169]. In the majority of cases, these techniques introduce additional circuitry complexity, thus limiting the maximum resolution of the embedded quantizer to typically less than 4–5 bits.

To address this problem, some authors have proposed the use of alternative implementations of the modulator feedback waveforms – such as FIR DACs [30, 33, 45, 170], as discussed in Section 4.8.5. The reader should note that FIR DACs introduce an extra delay in the modulator loop filter. This delay has to be controlled so as to stabilise the system and guarantee correct performance. This problem can be solved by using the methods of moments, which allows perfect compensation of the FIR DAC's delay [171]. Another issue is related to the number of taps used in the FIR filter. This can be optimised to maximise the modulator performance and efficiency, as shown by Sukumaran and Pavan [33]. Theirs is one of the most efficient designs reported to date, according to the energy plot shown in Figure 10.5, featuring FOMS = 178 dB (B_w = 24 kHz, DR= 99 dB, $P = 0.3$ mW).

With the trend toward high-speed CT-ΣΔMs, nonidealities in the feedback DAC, especially dynamic errors, are becoming dominant. Although mismatch-shaping can be used to alleviate the errors associated with static element errors and static timing errors, mismatch-shaping increases the switching activity and therefore exacerbates the degradation caused by dynamic errors. Risbo *et al.* [172] have shown that it is possible to address this problem with an elaborate element selection logic that simultaneously shapes both mismatch-induced noise and the noise associated with the ISI dynamic error. The scheme was quite practical for a clock rate of 3 MHz, but would be difficult to implement at the gigahertz clock rates typical of current high-speed CT-ΣΔ ADCs. Consequently, other approaches, such as ISI calibration via analog adjustment or digital correction, are likely to be used in future high-speed CT-ΣΔ ADCs.

Current-mode DACs are typically used inside high-speed CT-ΣΔ ADCs, but resistor DACs [173] are preferred when noise is the most critical specification. Sukumaran and Pavan [33] describe a ΣΔ ADC using resistor DACs that achieves outstanding performance. Single-bit quantization is used to minimize power and an FIR DAC is used to counter the jitter sensitivity normally associated with single-bit quantization. Further minimization of DAC noise is possible by using tri-level DAC elements, or by using voltages outside the supply rails.

In addition to DACs built with MOS current sources or switched resistors, CT-ΣΔ ADCs can also make use of switched-capacitor DACs. These DACs were originally used to reduce the jitter sensitivity of LP ADCs, but more recently have been used by Harrison *et al.* [52] as voltage feedback to the Q-enhanced LC tanks of a BP ADC, as conceptually illustrated in Figure 10.34. In this case, inverters are used as voltage-mode DACs, driven by return-to-zero (RZ) latches. In the original usage scenario, the fact that the charge supplied by a switched capacitor is insensitive to the clock period gives the desired jitter-insensitivity. However, since the high peak current associated with the switched capacitor tends to require a more powerful amplifier than if the DAC current is steady, this technique has fallen out of favor. The more recent usage effectively implements voltage feedback to the LC tank, which is useful since current feedback alone provides only one degree of control over a second-order system.

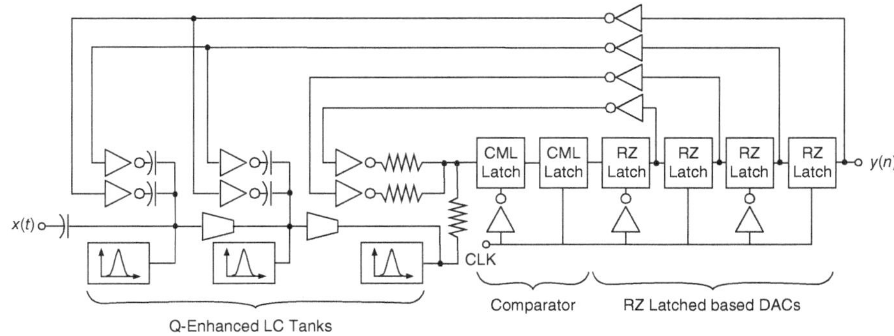

Figure 10.34 BP CT-ΣΔM based on Q-enhanced LC tanks and voltage-mode RZ latched DACs [52].

10.10 Classification of State-of-the-Art References

For the sake of clarity, this chapter is concluded by grouping the published references on the state-of-the-art ΣΔM ICs according to the same classification criteria followed in Tables 10.1–10.13:

- **SC single-loop single-bit LP-ΣΔMs**: [20, 103, 105, 106, 175–212].
- **SC single-loop multi-bit LP-ΣΔMs**: [28, 156, 183, 213, 214–254].
- **SC cascade single-bit LP-ΣΔMs**: [19, 180, 255–271].
- **SC cascade multi-bit LP-ΣΔMs**: [27, 91, 233, 265, 272–290].
- **SC BP-ΣΔMs**: [87, 291–306].
- **CT single-loop single-bit LP-ΣΔMs**: [29, 44, 45, 51, 60, 61, 159, 170, 307–337].
- **CT single-loop multi-bit LP-ΣΔMs**: [23, 62, 64, 66, 116, 136, 152, 338–364].
- **CT cascade LP-ΣΔMs**: [115, 365–369].
- **CT BP-ΣΔMs**: [26, 52, 53, 59, 63, 73, 74, 76, 85, 370–387].
- **ΣΔMs with time-coded quantization**: [138–140, 142, 143, 160, 161, 388–390].
- **Hybrid ΣΔMs**: [37, 65, 96, 97, 99, 104, 107, 113–115, 117, 118, 354, 391–397].
- **SC reconfigurable ΣΔMs**: [89, 242, 398–402].
- **CT reconfigurable ΣΔMs**: [98, 162, 365, 403–407].

10.11 Summary and Conclusions

This chapter has presented an analysis of the state of the art on CMOS ΣΔMs, giving a systematic classification of their architectures, circuit techniques, and target applications. Based on the practical approach followed in this book, the empirical data extracted from this statistical survey has been applied to identify current design trends and challenges in order to help designers to select the most suitable ΣΔM for a given application and/or range of specifications from among the different families. With this objective in mind, a number of design trade-offs have been discussed at different abstraction levels: architecture level (i.e., single-loop versus cascade, single-bit versus multi-bit, low-pass versus band-pass), circuit-level (i.e., SC versus CT, Gm-C versus active-RC), and technology process. The conclusions derived from the presented statistical and empirical analysis can be used as a practical guide to be incorporated in the top-down/bottom-up systematic design procedure described in this book.

Among others, design trends and circuits and systems approaches which are in the frontiers of data converters including among others, RF-to-digital and GHz- range ΣΔMs; robust and enhanced cascade architectures; digital-based, hybrid active/passive and power-efficient loop-filter circuit implementations; Nyquist- rate/ΣΔ ADCs, time-based quantization and high-performance feedback DACs have been outlined, identifying outstanding designs in each case. These trends are expected to continue and evolve, with the development and improvement of some of the architectural and circuit strategies discussed in this chapter, and especially those based on the use of continuous-time circuits, enhanced by an intensive use of digital-assisted analog techniques, which will push ΣΔ converters forward.

The increasing demand for more bandwidth and reduced power consumption as well as the pressures of nanometer-scale technologies and emerging MOS devices will drive the improvement of some architecture and circuit techniques – particularly using CT circuits, multi-bit and time-coded quantization, mismatch-shaping and feedforward amplifiers, as well as an intensive use of digital-assisted analog circuit techniques. Although single-stage multi-bit CT-ΣΔMs are dominant implementations, multi-stage (cascade) architectures are demonstrating further increases in bandwidth and are expected to be widely commercialized.

Apart from the application of ΣΔ techniques to digitization of baseband signals, the push for improved efficiency and the direct digitization of RF signals will motivate increased commercialization of systems employing BP architectures. Indeed, more and more analog signal conditioning functions will be embedded in the feedback loop of the modulators, thus benefiting from noise-shaping to make the overall system more robust, compact and easy to adapt to different standards and operation modes. Some good examples of this trend are the digital-intensive ΣΔ receivers outlined in this chapter, but similar strategies could be used for other application scenarios, such as wireless sensor networks and biomedical devices.

To fuel subsequent generations of ΣΔ ADCs, researchers will pursue more exotic implementations, including time/frequency-interleaving and polyphase ADCs, some of them discussed in this book. These new inventions and design strategies will continue raising the energy efficiency and bandwidth of ΣΔMs to unprecedented levels, enhancing their dominance over other types of ADC, particularly in those applications requiring resolutions greater than 12 bits, while pushing their figures of merit towards physical limits, such as the room-temperature thermal noise limit (in low-frequency ranges) and clock jitter (at high frequencies). Addressing these issues in the nanoscale will in turn result in more design innovations and new kinds of data converter in which ΣΔ techniques will play a key role for many years to come [16].

References

[1] P. M. Aziz *et al.*, "An Overview of Sigma-Delta Converters," *IEEE Signal Processing Magazine*, vol. 13, pp. 61–84, January 1996.

[2] R. Schreier and G. C. Temes, *Understanding Delta-Sigma Data Converters*. IEEE Press, 2005.

[3] J. M. de la Rosa, "Sigma-Delta Modulators: Tutorial Overview, Design Guide, and State-of-the-Art Survey," *IEEE Trans. on Circuits and Systems I: Regular Papers*, vol. 58, pp. 1–21, January 2011.

[4] S. Pavan, R. Schreier, and G. C. Temes, *Understanding Delta-Sigma Data Converters*. Wiley-IEEE Press, 2nd ed., 2017.

[5] M. Englund *et al.*, "A Programmable 0.7-2.7 GHz Direct ΔΣ receiver in 40nm CMOS," *IEEE J. of Solid-State Circuits*, vol. 50, pp. 644–655, March 2015.

[6] T. Christen, "A15-bit140μW Scalable-Bandwidth Inverter-Based Modulator for a MEMS Microphone With Digital Output," . *IEEE J. of Solid-State Circuits*, vol. 48, pp. 1605–1614, July 2013.

[7] H. Luo *et al.*, "A 0.8-V 230-μW 98-dB DR Inverter-Based ΣΔ Modulator for Audio Applications," *IEEE J. of Solid-State Circuits*, vol. 48, pp. 2430–2441, October 2013.

[8] W. Yu, K. Kim, and S. Cho, "A 148fs-rms Integrated Noise 4 MHz Bandwidth Second-Order ΔΣ Time-to-Digital Converter With Gated Switched-Ring Oscillator," *IEEE Trans. on Circuits and Systems – I: Regular Papers*, vol. 61, pp. 2281–2289, August 2014.

[9] R. H. Walden, "Analog-to-Digital Converter Survey and Analysis," *IEEE J. on Selected Areas in Communications*, vol. 17, pp. 539–550, April 1999.

[10] B. Murmann, "A/D Converter Trends: Power Dissipation, Scaling and Digitally Assisted Architectures," *Proc. of the IEEE Custom Integrated Circuits Conf.*, pp. 105–112, 2008.

[11] H. S. Lee and C. G. Sodini, "Analog-to-Digital Converters: Digitizing the Analog World," *Proceedings of the IEEE*, vol. 96, pp. 323–334, February 2008.

[12] B. E. Jonsson, "On CMOS Scaling and A/D-Converter Performance," *Proc. of the NORCHIP Conf.*, pp. 1–4, November 2010.

[13] B. E. Jonsson, "A Survey of A/D-Converter Performance Evolution," *Proc. of the IEEE Intl. Conf. on Electronics, Circuits, and Systems*, pp. 766–769, December 2010.

[14] B. E. Jonsson, "An Empirical Approach to Finding Energy Efficient ADC Architectures," *Proc. of the Intl. Workshop on ADC Modeling, Testing and Data Converter Analysis and Design and IEEE ADC Forum*, June 2011.

[15] G. Manganaro, *Advanced Data Converters*. Cambridge University Press, 2012.

[16] J. M. de la Rosa, R. Schreier, K.-P. Pun, and S. Pavan, "Next-Generation Delta-Sigma Converters: Trends and Perspectives," *IEEE J. on Emerging and Selected Topics in Circuits and Systems*, vol. 5, pp. 484–499, December 2015.

[17] B. Murmann, *ADC Performance Survey 1997–2018*. [Online]. Available: http://www.stanford.edu/~murmann/adcsurvey.html, 2018.

[18] H. Shibata *et al.*, "A 9-GS/s 1.125-GHz BW Oversampling Continuous-Time Pipeline ADC Achieving −164-dBFS/Hz NSD," *IEEE J. of Solid-State Circuits*, vol. 52, pp. 3219–3234, December 2017.

[19] S. Rabii and B. A. Wooley, "A 1.8V Digital-Audio Sigma-Delta Modulator in 0.8 μm CMOS," *IEEE J. of Solid-State Circuits*, vol. 32, pp. 783–796, June 1997.

[20] D. A. Kerth *et al.*, "A 120-dB Linear Switched-Capacitor Delta-Signal Modulator," *IEEE ISSCC Digest of Technical Papers*, pp. 196–197, February 1994.

[21] M. Steiner and N. Greer, "A 22.3b 1kHz 12.7mW Switched-Capacitor ΔΣ Modulator with Stacked Split-Steering Amplifiers," *IEEE ISSCC Digest of Technical Papers*, pp. 284–285, February 2016.

[22] Y. Dong *et al.*, "A 72 dB-DR 465 MHz-BW Continuous-Time 1-2 MASH ADC in 28 nm CMOS," *IEEE J. of Solid-State Circuits*, vol. 51, pp. 2917–2927, August 2016.

[23] M. Bolatkale *et al.*, "A 4GHz Continuous-Time ΔΣ ADC With 70dB DR and –74 dBFS THD in 125 MHz BW," *IEEE J. of Solid-State Circuits*, vol. 46, pp. 2857–2868, December 2011.

[24] Y. Dong *et al.*, "A Continuous-Time 0–3 MASH ADC Achieving 88 dB DR With 53 MHz BW in 28 nm CMOS," *IEEE J. of Solid-State Circuits*, vol. 49, pp. 2868–2877, December 2014.

[25] D.-Y. Yoon, S. Ho, and H.-S. Lee, "An 85dB-DR 74.6dB-SNDR 50MHz-BW CT MASH ΔΣ Modulator in 28nm CMOS," *IEEE ISSCC Digest of Technical Papers*, pp. 272–273, February 2015.

[26] H. Shibata *et al.*, "A DC-to-1GHz Tunable RF ΔΣ ADC Achieving DR=74dB and BW=150MHz at f_0 =450MHz Using 550mW," *IEEE ISSCC Digest of Technical Papers*, pp. 150–151, February 2012.

[27] R. Brewer *et al.*, "A 100dB SNR 2.5MS/s Output Data Rate ΔΣ ADC," *IEEE ISSCC Digest of Technical Papers*, pp. 172–173, February 2005.

[28] Y. Yang *et al.*, "A Single Die 124dB Stereo Audio Delta Sigma ADC With 111dB THD," *Proc. of the IEEE European Solid-State Circuits Conf.*, pp. 252–255, September 2007.

[29] S. Pavan and P. Sankar, "Power Reduction in Continuous-Time Delta-Sigma Modulators Using the Assisted Opamp Technique," *IEEE J. of Solid-State Circuits*, vol. 45, pp. 1365–1379, July 2010.

[30] P. Shettigar and S. Pavan, "Design Techniques for Wideband Single-Bit Continuous-Time ΔΣ Modulators With FIR Feedback DACs," *IEEE J. of Solid-State Circuits*, vol. 47, pp. 2865–2879, December 2012.

[31] Y.-S. Shu *et al.*, "A 28fJ/conv-step CT ΔΣ Modulator with 78dB DR and 18MHz BW in 28nm CMOS Using a Highly Digital Multibit Quantizer," *IEEE ISSCC Digest of Technical Papers*, pp. 268–269, February 2013.

[32] S. Zeller *et al.*, "A 0.039mm^2 Inverter-Based 1.82mW 68.6dB-SNDR 10MHz-BW CT-ΣΔ-ADC in 65nm CMOS Using Power- And Area-Efficient Design Techniques," *IEEE J. of Solid-State Circuits*, vol. 49, pp. 1548–1560, July 2014.

[33] A. Sukumaran and S. Pavan, "Low Power Design Techniques for Single-Bit Audio Continuous-Time Delta Sigma ADCs Using FIR Feedback," *IEEE J. of Solid-State Circuits*, vol. 49, pp. 2515–2525, November 2014.

[34] C. Berti *et al.*, "A 106 dB A-Weighted DR Low-Power Continuous-Time ΣΔ Modulator for MEMS Microphones," *IEEE J. of Solid-State Circuits*, vol. 51, pp. 1607–1618, July 2016.

[35] C. Briseno-Vidrios *et al.*, "A 4 Bit Continuous-Time ΣΔ Modulator With Fully Digital Quantization Noise Reduction Algorithm Employing a 7 Bit Quantizer," *IEEE J. of Solid-State Circuits*, vol. 51, pp. 1398–1409, June 2016.

[36] S. Lee *et al.*, "A 300-μW Audio ΔΣ Modulator With 100.5-dB DR Using Dynamic Bias Inverter," *IEEE Trans. on Circuits and Systems – I: Regular Papers*, vol. 63, pp. 1866–1875, November 2016.

[37] T.-Y. Lo, "A 102dB Dynamic Range Audio Sigma-Delta Modulator in 40nm CMOS," *Proc. of the IEEE Asian Solid-State Circuits Conf.*, pp. 257–260, 2011.

[38] C.-Y. Ho *et al.*, "A 4.5mW CT Self-Coupled ΔΣ Modulator with 2.2MHz BW and 90.4dB SNDR Using Residual ELD Compensation," *IEEE ISSCC Digest of Technical Papers*, pp. 274–275, February 2015.

[39] B. Gonen *et al.*, "A 1.65mW 0.16mm^2 Dynamic Zoom-ADC with 107.5dB DR in 20kHz BW," *IEEE ISSCC Digest of Technical Papers*, pp. 282–283, February 2016.

[40] Y. H. Leow *et al.*, "A 1 V 103 dB 3rd-Order Audio Continuous-Time ΔΣ ADC With Enhanced Noise Shaping in 65 nm CMOS," *IEEE J. of Solid-State Circuits*, vol. 51, pp. 2625–2638, November 2016.

[41] A. Ghosh and S. Pamarti, "Linearization Through Dithering: A 50 MHz Bandwidth, 10-b ENOB, 8.2 mW VCO-Based ADC," *IEEE J. of Solid-State Circuits*, vol. 50, pp. 2012–2024, September 2015.

[42] M. Ortmanns and F. Gerfers, *Continuous-Time Sigma-Delta A/D Conversion: Fundamentals, Performance Limits and Robust Implementations*. Springer, 2006.

[43] A. Edward *et al.*, "A 43-mW MASH 2-2 CT ΣΔ Modulator Attaining 74.4/75.8/76.8 dB of SNDR/SNR/DR and 50 MHz of BW in 40-nm CMOS," *IEEE J. of Solid-State Circuits*, vol. 52, pp. 448–459, February 2017.

[44] P. Shettigar and S. Pavan, "A 15mW 3.6GS/s CT-ΔΣ ADC with 36MHz Bandwidth and 83dB DR in 90nm CMOS," *IEEE ISSCC Digest of Technical Papers*, pp. 268–269, February 2012.

[45] V. Srinivasan *et al.*, "A 20mW 61dB SNDR (60MHz BW) 1b 3rd-Order Continuous-Time Delta-Sigma Modulator Clocked at 6GHz in 45nm CMOS," *IEEE ISSCC Dig. of Tech. Papers*, pp. 158–159, February 2012.

[46] A. Rodríguez-Vázquez, F. Medeiro, and E. Janssens, *CMOS Telecom Data Converters*. Kluwer Academic Publishers, 2003.

[47] A. Yeknami, F. Qazi, and A. Alvandpour, "Low-Power DT ΔΣ Modulators Using SC Passive Filters in 65nm CMOS," *IEEE Trans. on Circuits and Systems I – Regular Papers*, vol. 61, pp. 358–370, Feb. 2014.

[48] R. Zanbaghi *et al.*, "A 75-dB SNDR, 5-MHz Bandwidth Stage-Shared 2-2 MASH ΔΣ Modulator Dissipating 16 mW Power," *IEEE Trans. on Circuits and Systems – I: Regular Papers*, vol. 59, pp. 1614–1625, August 2012.

[49] K. Ragab and N. Sun, "A 12-b ENOB 2.5-MHz BW VCO-Based 0-1 MASH ADC With Direct Digital Background Calibration," *IEEE J. of Solid-State Circuits*, vol. 52, pp. 433–447, February 2017.

[50] H.-C. Tsai *et al.*, "A 64-fJ/Conv.-Step Continuous-Time Modulator in 40-nm CMOS Using Asynchronous SAR Quantizer and Digital ΔΣ Truncator," *IEEE J. of Solid-State Circuits*, vol. 48, pp. 2637–2648, November 2013.

[51] E. H. Dagher *et al.*, "A 2-GHz Analog-to-Digital Delta-Sigma Modulator for CDMA Receivers With 79-dB Signal-to-Noise Ratio in 1.23-MHz Bandwidth," *IEEE J. of Solid-State Circuits*, vol. 38, pp. 1819–1828, November 2004.

[52] J. Harrison *et al.*, "An LC Bandpass ΔΣ ADC with 70dB SNDR Over 20MHz Bandwidth Using CMOS DACs," *IEEE ISSCC Digest of Technical Papers*, pp. 146–147, February 2012.

[53] E. Martens *et al.*, "RF-to-Baseband Digitization in 40nm CMOS With RF Bandpass ΔΣ Modulator and Polyphase Decimation Filter," *IEEE J. of Solid-State Circuits*, vol. 47, pp. 990–1002, April 2012.

[54] S. Gupta *et al.*, "A 0.8-2GHz Fully-Integrated QPLL-Timed Direct-RF-Sampling Bandpass ΣΔ ADC in 0.13 μm CMOS," *IEEE J. of Solid-State Circuits*, vol. 47, pp. 1141–1153, May 2012.

[55] T. C. Caldwell, D. Alldred, and Z. Li, "A Reconfigurable ΔΣ ADC with up to 100MHz Bandwidth using Flash Reference Shuffling," *IEEE Trans. on Circuits and Systems – I: Regular Papers*, vol. 61, pp. 2263–2271, August 2014.

[56] S. Ho *et al.*, "A 23mW, 73dB Dynamic Range, 80MHz BW Continuous-Time Delta-Sigma Modulator in 20nm CMOS," *Proc. of the IEEE Symp. on VLSI Circuits*, 2014.

[57] Y. Zhang *et al.*, "A 1V 59fJ/Step 15MHz BW 74dB SNDR Continuous-Time ΔΣ Modulator with Digital ELD Compensation and Multi-Bit FIR Feedback," *Proc. of the IEEE Asian Solid-State Circuits Conf.*, pp. 321–324, 2014.

[58] F. Opteynde, "A Maximally-digital Radio Receiver Front-end," *IEEE ISSCC Digest of Technical Papers*, vol. 450-451, February 2010.

[59] K. Koli *et al.*, "A 900-MHz Direct Delta-Sigma Receiver in 65-nm CMOS," *IEEE J. of Solid-State Circuits*, pp. 2807–2818, Dec. 2010.

[60] L. Breems *et al.*, "A 1.8-mW CMOS ΣΔ Modulator with Integrated Mixer for A/D Conversion of IF Signals," *IEEE J. of Solid-State Circuits*, vol. 35, pp. 468–475, April 2000.

[61] K. Philips *et al.*, "A Continuous-Time ΣΔ ADC With Increased Immunity to Interferers," *IEEE J. of Solid-State Circuits*, vol. 39, pp. 2170–2177, December 2005.

[62] G. Mitteregger, C. Ebner, S. Mechnig, T. Blon, C. Holuigue, and E. Romani, "A 20-mW 640-MHz CMOS Continuous-Time ΣΔ ADC With 20-MHz Signal Bandwidth, 80-dB Dynamic Range and 12-bit ENOB," *IEEE J. of Solid-State Circuits*, vol. 41, pp. 2641–2649, December 2006.

[63] S. B. Kim *et al.*, "A 2.7 mW, 90.3 dB DR Continuous-Time Quadrature Bandpass Sigma-Delta Modulator for GSM/EDGE Low-IF Receiver in 0.25 μm CMOS," *IEEE J. of Solid-State Circuits*, vol. 44, pp. 891–900, March 2009.

[64] H. Kim *et al.*, "Adaptive Blocker Rejection Continuous-Time ΣΔ ADC for Mobile WiMAX Applications," *IEEE J. of Solid-State Circuits*, vol. 44, pp. 2766–2779, October 2009.

[65] R. Winoto and B. Nikolic, "A Highly Reconfigurable 400–1700MHz Receiver Using a Down-Converting Sigma-Delta A/D with 59-dB SNR and 57-dB SFDR over 4-MHz Bandwidth," *Proc. of the IEEE Symp. on VLSI Circuits*, pp. 142–143, 2009.

[66] J. G. Kauffman *et al.*, "An 8.5mW Continuous-Time ΔΣ Modulator With 25MHz Bandwidth Using Digital Background DAC Linearization to Achieve 63.5dB SNDR and 81dB SFDR," *IEEE J. of Solid-State Circuits*, vol. 46, pp. 2869–2881, December 2011.

[67] M. Anderson *et al.*, "A Filtering ΔΣ ADC for LTE and Beyond," *IEEE J. of Solid-State Circuits*, vol. 49, pp. 1535–1547, July 2014.

[68] R. S. Rajan *et al.*, "Design Techniques for Continuous-Time ΔΣ Modulators With Embedded Active Filtering," *IEEE J. of Solid-State Circuits*, vol. 49, pp. 2187–2198, October 2014.

[69] R. Ritter *et al.*, "A 10 MHz Bandwidth, 70 dB SNDR Continuous-Time Delta-Sigma Modulator With Digitally Improved Reconfigurable Blocker Rejection," *IEEE J. of Solid-State Circuits*, vol. 51, pp. 660–670, March 2016.

[70] X. Liu *et al.*, "A 65 nm CMOS Wideband Radio Receiver With ΔΣ-Based A/D-Converting Channel-Select Filters," *IEEE J. of Solid-State Circuits*, vol. 51, pp. 1566–1578, July 2016.

[71] L. Breems, R. Rutten, R. van Veldhoven, and G. van der Weide, "A 2.2GHz Continuous-Time ΔΣ ADC With −102dBc THD and 25 MHz Bandwidth," *IEEE J. of Solid-State Circuits*, vol. 42, pp. 2906–2916, December 2016.

[72] S. Loeda, F. Pourchet, and A. Adams, "A 10/20/30/40 MHz Feedforward FIR DAC Continuous-Time ΔΣ ADC With Robust Blocker Performance for Radio Receivers," *IEEE J. of Solid-State Circuits*, vol. 51, pp. 860–870, April 2016.

[73] J. Ryckaert, J. Borremans, B. Verbruggen, L. Bos, C. Armiento, J. Craninckx, and G. van der Plas, "A 2.4 GHz Low-Power Sixth-Order RF Bandpass ΔΣ Converter in CMOS," *IEEE J. of Solid-State Circuits*, vol. 44, pp. 2873–2880, November 2009.

[74] N. Beilleau, H. Aboushady, F. Montaudon, and A. Cathelin, "A 1.3V 26mW 3.2GS/s Undersampled LC Bandpass ΣΔ ADC for a SDR ISM-band Receiver in 130nm CMOS," *Proc. of the IEEE Radio Frequency Integrated Circuits Symp.*, 2009.

[75] C. Y. Lu *et al.*, "A Sixth-Order 200 MHz IF Bandpass Sigma-Delta Modulator With Over 68 dB SNDR in 10 MHz Bandwidth," *IEEE J. of Solid-State Circuits*, vol. 45, pp. 1122–1136, June 2010.

[76] J. Ryckaert *et al.*, "A 6.1GS/s 52.8mW 43dB DR 80MHz Bandwdith 2.4GHz RF Bandpass ΔΣ ADC in 40nm CMOS," *Proc. of the IEEE Radio Frequency Integrated Circuits Symp.*, pp. 443–446, June 2010.

[77] R. Schreier *et al.*, "An IF Digitizer IC Employing A Continuous-Time Bandpass Delta-Sigma ADC," *Proc. of the IEEE Radio Frequency Integrated Circuits Symp.*, pp. 323–326, 2012.

[78] A. Ashry and H. Aboushady, "A 4th Order 3.6 GS/s RF ΣΔ ADC With a FoM of 1 pJ/bit," *IEEE Trans. on Circuits and Systems – I: Regular Papers*, vol. 60, pp. 2606–2617, October 2013.

[79] P. M. Chopp and A. Hamoui, "A 1-V 13-mW Single-Path Frequency-Translating ΔΣ Modulator With 55-dB SNDR and 4-MHz Bandwidth at 225 MHz," *IEEE J. of Solid-State Circuits*, vol. 48, pp. 473–486, February 2013.

[80] H. Chae *et al.*, "A 12 mW Low Power Continuous-Time Bandpass ΔΣ Modulator With 58 dB SNDR and 24 MHz Bandwidth at 200 MHz IF," *IEEE J. of Solid-State Circuits*, vol. 49, pp. 405–415, February 2014.

[81] J. Mitola, "The Software Radio Architecture," *IEEE Communications Magazine*, pp. 26–38, May 1995.

[82] A. I. Hussein and W. Kuhn, "Bandpass ΣΔ Modulator Employing Undersampling of RF Signals for Wireless Communications," *IEEE Trans. on Circuits and Systems II: Analog and Digital Signal Processing*, vol. 47, pp. 614–620, July 2000.

[83] N. Beilleau, H. Aboushady, and M. Loureat, "Using Finite Impulse Response Feedback DACs to design ΣΔ Modulators based on LC Filters," *Proc. of the IEEE Intl. Midwest Symp. on Circuits and Systems (MWSCAS)*, pp. 696–699, August 2005.

[84] G. Molina *et al.*, "LC-Based Bandpass Continuous-Time Sigma-Delta Modulators with Widely Tunable Notch Frequency," *IEEE Trans. on Circuits and Systems – I: Regular Papers*, vol. 61, pp. 1442–1455, May 2014.

[85] S. Gupta *et al.*, "A QLL-Timed Direct-RF Sampling Band-Pass ΣΔ ADC with a 1.2 GHz Tuning Range in 0.13 μm CMOS," *Proc. of the IEEE Radio Frequency Integrated Circuits Symp.*, June 2011.

[86] B. Razavi, "The Role of Translational Circuits in RF Receiver Design," *Proc. of the IEEE Custom Integrated Circuits Conf.*, 2014.

[87] A. K. Ong and B. A. Wooley, "A Two-Path Bandpass ΣΔ Modulator for Digital IF Extraction at 20MHz," *IEEE J. of Solid-State Circuits*, vol. 32, pp. 1920–1934, December 1997.

[88] C. Wu *et al.*, "A Wideband 400 MHz-to-4 GHz Direct RF-to-Digital Multimode ΔΣ Receiver," *IEEE J. of Solid-State Circuits*, vol. 49, pp. 1639–1652, July 2014.

[89] A. Morgado, R. del Río, and J. M. de la Rosa, "A New Cascade ΣΔ Modulator for Low-Voltage Wideband Applications," *IET Electronics Letters*, vol. 43, pp. 910–911, August 2007.

[90] N. Maghari, S. Kwon, and U.-K. Moon, "74 dB SNDR Multi-Loop Sturdy-MASH Delta-Sigma Modulator Using 35 dB Open-Loop Opamp Gain," *IEEE J. of Solid-State Circuits*, vol. 44, pp. 2212–2221, August 2009.

[91] A. Gharbiya and D. A. Johns, "A 12-bit 3.125MHz Bandwidth 0-3 MASH Delta-Sigma Modulator," *IEEE J. of Solid-State Circuits*, vol. 44, pp. 2010–2018, July 2009.

[92] N. Waki *et al.*, "A Low-Distortion Two-Path Fourth-Order Bandpass Delta-Sigma Modulator Using Horizontal opamp Sharing," *IEEE Proc. of IEEE Midwest Symposium on Circuits and Systems*, pp. 1372–1375, August 2007.

[93] F. Colodro and A. Torralba., "Multirate ΣΔ Modulators," *IEEE Trans. on Circuits and Systems II: Analog and Digital Signal Processing*, vol. 49, pp. 170–176, March 2002.

[94] M. Ortmanns *et al.*, "Multirate Cascaded Continuous-Time ΣΔ Modulators," *Proc. of the IEEE Intl. Symp. on Circuits and Systems*, pp. 4225–4228, May 2002.

[95] J. G. García-Sánchez and J. M. de la Rosa, "Multirate Downsampling Hybrid CT/DT Cascade Sigma-Delta Modulators," *IEEE Trans. on Circuits and Systems I: Regular Papers*, vol. 59, pp. 285–294, February 2012.

[96] P. Morrow *et al.*, "A 0.18 μm 102dB-SNR Mixed CT SC Audio-band ΔΣ ADC," *IEEE ISSCC Digest of Technical Papers*, pp. 177–178, February 2005.

[97] K. Nguyen *et al.*, "A 106dB SNR Hybrid Oversampling ADC for Digital Audio," *IEEE ISSCC Digest of Technical Papers*, pp. 176–177, February 2005.

[98] B. Putter, "A 5th-Order CT/DT Multi-Mode ΣΔ Modulator," *IEEE ISSCC Digest of Technical Papers*, pp. 244–245, February 2007.

[99] M. Choi *et al.*, "A 101-dB SNR Hybrid Delta-Sigma Audio ADC Using Post Integration Time Control," *Proc. of the IEEE Custom Integrated Circuits Conf.*, pp. 89–92, 2008.

[100] H. Kwan *et al.*, "Design of Hybrid Continuous-Time Discrete-Time Delta-Sigma Modulators," *Proc. of the IEEE Intl. Symp. on Circuits and Systems*, pp. 1224–1227, May 2008.

[101] S. Kulchycki *et al.*, "A 77-dB Dynamic Range, 7.5-MHz Hybrid Continuous-Time/Discrete-Time Cascade ΣΔ Modulator," *IEEE J. of Solid-State Circuits*, vol. 43, pp. 796–804, April 2008.

[102] J. M. de la Rosa, A. Morgado, and R. del Río, "Hybrid Continuous-Time/Discrete-Time Cascade ΣΔ Modulators with Programmable Resonation.," *Proc. of the IEEE Intl. Symp. on Circuits and Systems*, pp. 2249–2252, May 2009.

[103] M. C. Huang and S. I. Lu, "A Fully Differential Comparator-Based Switched-Capacitor ΔΣ Modulator," *IEEE J. of Solid-State Circuits*, vol. 44, pp. 369–373, May 2009.

[104] R. H. M. Veldhoven *et al.*, "An Inverted-Based Hybrid ΣΔ Modulator," *IEEE ISSCC Digest of Technical Papers*, vol. 492-493, February 2008.

[105] Y. Chae and G. Han, "Low Voltage, Low Power, Inverter-Based Switched-Capacitor Delta-Sigma Modulator," *IEEE J. of Solid-State Circuits*, vol. 44, pp. 369–373, May 2009.

[106] F. Michel and M. Steyaert, "A 250mW 7.5μW 61dB SNDR CMOS SC ΔΣ Modulator Using a Near-Threshold-Voltage-Biased CMOS Inverter Technique," *IEEE ISSCC Digest of Technical Papers*, pp. 476–477, February 2011.

[107] T. Song *et al.*, "A 2.7-mW 2-MHz Continuous-Time ΣΔM With a Hybrid Active-Passive Loop Filter," *IEEE J. of Solid-State Circuits*, vol. 43, pp. 330–341, February 2008.

[108] G. K. Balachandran *et al.*, "A 1.16mW 69dB SNR (1.2MHz BW) Continuous Time ΣΔ ADC with Immunity to Clock Jitter," *Proc. of the IEEE Custom Integrated Circuits Conf.*, pp. 1–4, 2010.

[109] J. L. A. de Melo *et al.*, "A 0.4-V 410-nW Opamp-Less Continuous-Time ΣΔ Modulator for Biomedical Applications," *Proc. of the IEEE Intl. Symp. on Circuits and Systems (ISCAS)*, pp. 1340–1343, May 2014.

[110] A. Roy and R. J. Baker, "A Passive 2nd-Order Sigma-Delta Modulator for Low-Power Analog-to-Digital Conversion," *Proc. of the IEEE Intl. Midwest Symp. on Circuits and Systems (MWSCAS)*, pp. 595–598, August 2014.

[111] B. Nowacki *et al.*, "A 1V 77dB-DR 72dB-SNDR 10MHz-BW 2-1 MASH CT-ΣΔM," *IEEE ISSCC Digest of Technical Papers*, vol. 274-275, February 2016.

[112] C.-Y. Ho *et al.*, "A 75.1dB SNDR, 80.2dB DR, 4th-order Feed-forward Continuous-Time Sigma-Delta Modulator with Hybrid Integrator for Silicon TV-tuner Application," *Proc. of the IEEE Asian Solid-State Circuits Conf.*, pp. 261–264, 2011.

[113] T. L. Brooks *et al.*, "A Cascaded Sigma-Delta Pipeline A/D Converter with 1.25MHz Signal Bandwidth and 89dB SNR," *IEEE J. of Solid-State Circuits*, vol. 32, pp. 1896–1906, December 1997.

[114] A. Bosi *et al.*, "An 80MHz 4× Oversampled Cascaded ΔΣ-Pipelined ADC with 75dB DR and 87dB SFDR," *IEEE ISSCC Digest of Technical Papers*, vol. 54-55, February 2005.

[115] C. C. Lee and M. P. Flynn, "A 14b 23MS/s 48mW Resetting ΣΔ ADC with 87dB SFDR 11.7b ENOB & 0.5mm^2 area," *IEEE Symp. VLSI Circuits. Digest of Technical Papers*, pp. 182–183, 2008.

[116] M. Ranjbar *et al.*, "A 3.1 mW Continuous-Time ΔΣ Modulator With 5-Bit Successive Approximation Quantizer for WCDMA," *IEEE J. of Solid-State Circuits*, vol. 45, pp. 1479–1491, August 2010.

[117] C. C. Lee and M. P. Flynn, "A 14b 23MS/s 48mW Resetting ΣΔ ADC," *IEEE Trans. on Circuits and Systems I – Regular Papers*, vol. 58, pp. 1167–1177, June 2011.

[118] O. Rajaee *et al.*, "Low-OSR Over-Ranging Hybrid ADC Incorporating Noise-Shaped Two-Step Quantizer," *IEEE J. of Solid-State Circuits*, vol. 46, pp. 2458–2468, November 2011.

[119] T. C. Leslie and B. Singh, "An Improved Sigma-Delta Modular Architecture," *Proc. of the IEEE Intl. Symp. on Circuits and Systems*, pp. 372–375, May 1990.

[120] T. Oh *et al.*, "A Second-Order ΔΣ ADC Using Noise-Shaped Two-Step Integrating Quantizer," *IEEE J. of Solid-State Circuits*, vol. 48, pp. 1465–1474, June 2013.

[121] S.-C. Lee and Y. Chiu, "A 15-MHz Bandwidth 1-0 MASH ΣΔ ADC With Nonlinear Memory Error Calibration Achieving 85-dBc SFDR," *IEEE J. of Solid-State Circuits*, vol. 49, pp. 695–707, March 2014.

[122] A. Sanyal *et al.*, "A Hybrid SAR-VCO ΔΣ ADC with First-Order Noise Shaping," *Proc. of the IEEE Custom Integrated Circuits Conf.*, pp. 1–4, 2014.

[123] T.-C. Wang, Y.-H. Lin, and C.-C. Liu, "A 0.022mm^2 98.5dB SNDR Hybrid Audio Delta-Sigma Modulator with Digital ELD Compensation in 28nm CMOS," *Proc. of the IEEE Asian Solid-State Circuits Conf.*, pp. 317–320, 2014.

[124] B. Wu *et al.*, "A 24.7 mW 65 nm CMOS SAR-Assisted CT ΔΣ Modulator With Second-Order Noise Coupling Achieving 45 MHz Bandwidth and 75.3 dB SNDR," *IEEE J. of Solid-State Circuits*, vol. 51, pp. 2893–2905, December 2016.

[125] A. Sanyal and N. Sun, "A 18.5-fJ/step VCO-Based 0-1 MASH ΔΣ ADC with Digital Background Calibration," *IEEE Symposium VLSI Circuits. Digest of Technical Papers*, December 2016.

[126] K. Souri and K. A. Makinwa, "A 0.12mm^2 7.4μW Micropower Temperature Sensor With an Inaccuracy of ±0.2°C (3σ) from −30°C to 125°C," *IEEE J. of Solid-State Circuits*, vol. 46, pp. 1693–1700, July 2011.

[127] Y. Chae *et al.*, "A 6.3μW 20 bit Incremental Zoom-ADC with 6 ppm INL and 1μV Offset," *IEEE J. of Solid-State Circuits*, vol. 48, pp. 3019–3027, December 2013.

[128] R. van de Plassche, "A Sigma-Delta Modulator as an A/D Converter," *IEEE Trans. on Circuits and Systems I*, vol. 25, pp. 510–514, July 1978.

[129] C.-H. Chen *et al.*, "A Micro-Power Two-Step Incremental Analog-to-Digital Converter," *IEEE J. of Solid-State Circuits*, vol. 50, pp. 1796–1808, August 2015.

[130] S. Tao and A. Rusu, "A Power-Efficient Continuous-Time Incremental Sigma-Delta ADC for Neural Recording Systems," *IEEE Trans. on Circuits and Systems I: Regular Papers*, vol. 62, pp. 1489–1498, June 2015.

[131] S. Pavan, R. Schreier, and G. Temes, "Incremental Analog-to-Digital Converters," in *Understanding Delta-Sigma Data Converters*, Wiley-IEEE Press, 2nd ed., 2017.

[132] C.-H. Chen *et al.*, "Incremental Analog-to-Digital Converter for High-Resolution Energy-Efficient Sensor Interfaces," *IEEE J. of Emerging and Selected Topics in Circuits and Systems*, vol. 5, pp. 612–623, December 2015.

[133] Y. Zhang *et al.*, "A 16 b Multi-Step Incremental Analog-to-Digital Converter With Single-Opamp Multi-Slope Extended Counting," *IEEE J. of Solid-State Circuits*, vol. 52, pp. 1066–1076, April 2017.

[134] A. Agah *et al.*, "A High-Resolution Low-Power Oversampling ADC with Extended-Range for Bio-Sensor Arrays," *IEEE Symposium VLSI Circuits. Digest of Technical Papers*, pp. 244–245, December 2007.

[135] G. Temes, "Incremental and Extended-Range Data Converters," *in Design, Modeling and Testing of Data Converters*, pp. 143–159, Springer, 2014.

[136] L. Dorrer *et al.*, "A 3-mW 74-dB SNR 2-MHz Continuous-Time Delta-Sigma ADC With a Tracking ADC Quantizer in 0.13-µm CMOS," *IEEE J. of Solid-State Circuits*, vol. 40, pp. 2416–2427, December 2005.

[137] L. Hernández and E. Prefasi, "Analog-to-Digital Conversion Using Noise Shaping and Time Encoding," *IEEE Trans. on Circuits and Systems – I: Regular Papers*, vol. 55, pp. 2026–2037, August 2008.

[138] M. Z. Straayer *et al.*, "A 12-Bit, 10-MHz Bandwidth, Continuous-Time ΣΔ ADC With a 5-Bit, 950-MS/s VCO-Based Quantizer," *IEEE J. of Solid-State Circuits*, vol. 43, pp. 805–814, April 2008.

[139] M. Park and M. H. Perrott, "A 78 dB SNDR 87 mW 20 MHz Bandwidth Continuous-Time ΣΔADC With VCO-Based Integrator and Quantizer Implemented in 0.13µm CMOS," *IEEE J. of Solid-State Circuits*, vol. 44, pp. 3344–3358, December 2009.

[140] C.-Y. Lu *et al.*, "A 25MHz Bandwidth 5th-Order Continuous-Time Low-Pass Sigma-Delta Modulator With 67.7dB SNDR Using Time-Domain Quantization and Feedback," *IEEE J. of Solid-State Circuits*, vol. 45, pp. 1795–1808, September 2010.

[141] T. Watanabe and T. Terasawa, "An All-Digital A/D Converter TAD with 4-Shift-Clock Construction for Sensor Interface in 0.65-µm CMOS," *Proc. of the IEEE European Solid-State Circuits Conf.*, pp. 178–181, September 2010.

[142] S. Z. Asl *et al.*, "A 77dB SNDR, 4MHz MASH ΔΣ Modulator with a Second-Stage Multi-rate VCO-Based Quantizer," *Proc. of the IEEE Custom Integrated Circuits Conf.*, September 2011.

[143] K. Reddy and B. Haroun, "A 16mW 78dB-SNDR 10MHz-BW CT-ΔΣ ADC Using Residue-Cancelling VCO-Based Quantizer," *IEEE ISSCC Digest of Technical Papers*, pp. 152–153, February 2012.

[144] Y.-D. Chang *et al.*, "A 379nW 58.5dB SNDR VCO-Based ΔΣ Modulator for Bio-Potential Monitoring," *Proc. of the IEEE Symp. on VLSI Circuits*, pp. 66–67, 2013.

[145] S. Rao *et al.*, "A 4.1mW, 12-bit ENOB, 5MHz BW, VCO-based ADC with On-Chip Deterministic Digital Background Calibration in 90nm CMOS," *Proc. of the IEEE Symp. on VLSI Circuits*, pp. 68–69, 2013.

[146] A. Ghosh and S. Pamarti, "A 50MHz bandwidth, 10-b ENOB, 8.2mW VCO-based ADC enabled by filtered-dithering based linearization," *Proc. of the IEEE Custom Integrated Circuits Conf.*, pp. 1–4, 2013.

[147] P. Zhu, X. Xing, and G. Gielen, "A 40MHz-BW 35fJ/step-FoM nonlinearity-cancelling two-step ADC with dual-input VCO-based quantizer," *Proc. of the IEEE European Solid-State Circuits Conf.*, pp. 63–66, September 2014.

[148] S.-J. Huang *et al.*, "A 125MHz-BW 71.9dB-SNDR VCO-Based CT ΔΣ ADC with Segmented Phase-Domain ELD Compensation in 16nm CMOS," *IEEE ISSCC Digest of Technical Papers*, pp. 470–471, February 2017.

[149] S. Zaliasl *et al.*, "A 12.5-bit 4 MHz 13.8 mW MASH ΔΣ Modulator With Multirated VCO-Based ADC," *IEEE Trans. on Circuits and Systems I: Regular Papers*, vol. 59, pp. 1604–1613, August 2012.

[150] M. Gande *et al.*, "Blind Calibration Algorithm for Nonlinearity Correction Based on Selective Sampling," *IEEE J. of Solid-State Circuits*, vol. 49, pp. 1715–1724, August 2014.

[151] M. Amin and B. Leung, "Design Techniques for Linearity in Time-Based ΣΔ Analog-to-Digital Converter," *IEEE Trans. on Circuits and Systems – II: Express Briefs*, vol. 63, pp. 433–437, May 2016.

[152] V. Dhanasekaran *et al.*, "A 20MHz BW 68dB DR CT ΔΣ ADC Based on a Multi- Bit Time-Domain Quantizer and Feedback Element," *IEEE ISSCC Digest of Technical Papers*, pp. 174–175, February 2009.

[153] W. Jung *et al.*, "An All-Digital PWM-Based ΔΣ ADC with an Inherently Matched Multi-bit Quantizer," *Proc. of the IEEE Custom Integrated Circuits Conf.*, 2014.

[154] Y. Cao *et al.*, "A 1.7mW 11b 1-1-1 MASH ΔΣ Time-to-Digital Converter," *IEEE ISSCC Digest of Technical Papers*, vol. 480–481, February 2011.

[155] B. Young *et al.*, "A 2.4ps resolution 2.1mW Second-Order Noise-Shaped Time-to-Digital Converter with 3.2ns Range in 1-MHz Bandwidth," *Proc. of the IEEE Custom Integrated Circuits Conf.*, pp. 1–4, 2010.

[156] N. Maghari and U.-K. Moon, "A Third-Order DT ΔΣ Modulator Using Noise-Shaped Bi-Directional Single-Slope Quantizer," *IEEE J. of Solid-State Circuits*, vol. 46, pp. 2882–2891, December 2011.

[157] M. Gande *et al.*, "A 71dB Dynamic Range Third-Order ΔΣ TDC Using Charge-Pump," *Proc. of the IEEE Symp. on VLSI Circuits*, pp. 168–169, 2012.

[158] Y. Mortazavi *et al.*, "A Mostly Digital PWM-Based ΔΣ ADC With an Inherently Matched Multibit Quantizer/DAC," *IEEE Trans. on Circuits and Systems – II: Express Briefs*, vol. 63, pp. 1049–1053, November 2016.

[159] L. Hernández-Corporales *et al.*, "A 1.2-MHz 10-bit Continuous-Time Sigma-Delta ADC Using a Time Ecoding Quantizer," *IEEE Trans. on Circuits and Systems II: Express Briefs*, vol. 56, pp. 16–20, January 2009.

[160] E. Prefasi *et al.*, "A 7mW 20MHz BW Time-Encoding Oversampling Converter Implemented in a 0.08mm^2 65nm CMOS Circuit," *IEEE J. of Solid-State Circuits*, vol. 46, pp. 1562–1574, July 2011.

[161] G. Taylor and I. Galton, "A Mostly Digital Variable-Rate Continuous-Time ADC ΔΣ Modulator," *IEEE ISSCC Digest of Technical Papers*, pp. 298–299, February 2010.

[162] G. Taylor and I. Galton, "A Mostly-Digital Variable-Rate Continuous-Time Delta-Sigma Modulator ADC," *IEEE J. of Solid-State Circuits*, vol. 45, pp. 2634–2646, December 2010.

[163] G. Taylor and I. Galton, "A Reconfigurable Mostly-Digital ΔΣ ADC with a Worst-Case FOM of 160dB," *Proc. of the IEEE Symp. on VLSI Circuits*, pp. 166–167, 2012.

[164] M. Z. Straayer and M. H. Perrot, "A Multi-Path Gated Ring Oscillator TDC With First-Order Noise Shaping," *IEEE J. of Solid-State Circuits*, vol. 44, pp. 1089–1098, April 2009.

[165] K. Lee, Y. Yoon, and N. Sun, "A 1.8mW 2MHz-BW 66.5dB-SNDR ΔΣ ADC Using VCO-Based Integrators with Intrinsic CLA," *Proc. of the IEEE Custom Integrated Circuits Conf.*, 2013.

[166] B. Young *et al.*, "A 75dB DR 50MHz BW 3rd Order CT-ΔΣ Modulator Using VCO-Based Integrators," *Proc. of the IEEE Symp. on VLSI Circuits*, 2014.

[167] W. Yu, K. Kim, and S. Cho, "A 0.22ps$_{rms}$ Integrated Noise 15-MHz Bandwidth Fourth-Order ΔΣ Time-to-Digital Converter Using Time-Domain Error-Feedback Filter," *IEEE J. of Solid-State Circuits*, vol. 50, pp. 1251–1262, May 2015.

[168] T. Kim, C. Han, and N. Maghari, "An 11.4mW 80.4dB-SNDR 15 MHz-BW CT Delta-Sigma Modulator Using 6b Double-Noise-Shaped Quantizer," *IEEE ISSCC Digest of Technical Papers*, pp. 468–469, February 2017.

[169] J. G. Kauffman *et al.*, "A 72 dB DR, CT ΔΣ Modulator Using Digitally Estimated, Auxiliary DAC Linearization Achieving 88 fJ/conv-step in a 25 MHz BW," *IEEE J. of Solid-State Circuits*, vol. 49, pp. 392–404, February 2014.

[170] B. Putter, "ΣΔ ADC with Finite Impulse Response Feedback DAC," *IEEE ISSCC Digest of Technical Papers*, February 2004.

[171] S. Pavan, "Continuous-Time Delta-Sigma Modulator Design Using the Method of Moments," *IEEE Trans. on Circuits and Systems I – Regular Papers*, vol. 61, pp. 1629–1637, June 2014.

[172] L. Risbo, R. Hezar, B. Kelleci, H. Kiper, and M. Fares, "A 108dB-DR 120dB-THD and 0.5Vrms Output Audio DAC with Inter-symbol-interference-shaping Algorithm in 45nm CMOS," *IEEE ISSCC Digest of Technical Papers*, pp. 484–485, February 2011.

[173] C. B. Brahm, "Feedback Integrating System," *US Patent 3192371*, 1965.

[174] J. Roh *et al.*, "A 0.9-V 60-μW 1-Bit Fourth-Order Delta-Sigma Modulator With 83-dB Dynamic Range," *IEEE J. of Solid-State Circuits*, vol. 43, pp. 361–370, February 2008.

[175] C. B. Wang *et al.*, "A 113-dB DSD Audio ADC Using a Density-Modulated Dithering Scheme," *IEEE J. of Solid-State Circuits*, vol. 28, pp. 114–119, January 2003.

[176] L. Yao, M. Steyaert, and W. Sansen, "A 1-V 140-μW 88-dB Audio Sigma-Delta modulator in 90-nm CMOS," *IEEE J. of Solid-State Circuits*, vol. 39, pp. 1809–1818, November 2004.

[177] S. Gambini and J. Rabaey, "A 100kS/S 65dB DR ΣΔ ADC with 0.65V Supply Voltage," *Proc. of the IEEE European Solid-State Circuits Conf.*, pp. 202–205, September 2007.

[178] H. Park *et al.*, "A 0.7-V 100-dB 870-μW Digital Audio ΣΔ Modulator," *IEEE Symp. VLSI Circuits. Digest of Technical Papers*, pp. 178–179, 2008.

[179] V. Peluso *et al.*, "A 900-mV Low-Power ΔΣ A/D Converter with 77-dB Dynamic Range," *IEEE J. of Solid-State Circuits*, vol. 12, pp. 1887–1897, December 1998.

[180] J. Sauerbrey *et al.*, "A 0.65V Sigma-Delta Modulators," *Proc. of the IEEE Intl. Symp. on Circuits and Systems*, pp. 1021–1024, May 2003.

[181] A. L. Coban and P. E. Allen, "A 1.5V 1.0mW Audio ΔΣ Modulator with 98dB Dynamic Range," *IEEE ISSCC Digest of Technical Papers*, pp. 50–51, February 1999.

[182] P. Rombouts *et al.*, "A 250-kHz 94-dB Double-Sampling ΣΔ Modulation A/D Converter With a Modified Noise Transfer Function," *IEEE J. of Solid-State Circuits*, vol. 10, pp. 1657–1662, October 2003.

[183] J. Xu *et al.*, "Power Optimization of High Performance ΔΣ Modulators for Portable Measurement Applications," *Proc. of the IEEE Asian Solid-State Circuits Conf.*, 2010.

[184] Z. Cao *et al.*, "A 14mW 2.5 MS/s 14 bit ΣΔ Modulator Using Split-Path Pseudo-Differential Amplifiers," *IEEE J. of Solid-State Circuits*, vol. 42, pp. 2169–2179, October 2007.

[185] N. Klemmer and E. Hegazi, "A DLL-Biased, 14-Bit DS Analog-to-Digital Converter for GSM/GPRS/EDGE Handsets," *IEEE J. of Solid-State Circuits*, vol. 41, pp. 330–338, February 2006.

[186] K. A. Donoghue *et al.*, "A Digitally Calibrated 5-mW 2-MS/s 4th-Order ΔΣ ADC in 0.25-μm CMOS with 94dB SFDR," *Proc. of the IEEE European Solid-State Circuits Conf.*, pp. 422–425, September 2010.

[187] B. Nowacki *et al.*, "A 1.2V 300μW Second-Order Switched-Capacitor ΔΣ Modulator Using Ultra Incomplete Settling with 73dB SNDR and 300kHz BW in 130nm CMOS," *Proc. of the IEEE European Solid-State Circuits Conf.*, pp. 271–274, September 2011.

[188] F. Chen, S. Ramaswamy, and B. Bakkaloglu, "A 1.5V 1mA 80dB Passive ΣΔ ADC in 0.13μm Digital CMOS Process," *IEEE ISSCC Digest of Technical Papers*, vol. 54-55, pp. 244–245, February 2003.

[189] M. Dessouky and A. Kaiser, "Very Low-Voltage Digital-Audio ΔΣ Modulator with 88-dB Dynamic Range Using Local Switch Bootstrapping," *IEEE J. of Solid-State Circuits*, vol. 36, pp. 349–355, March 2001.

[190] Y. Chae *et al.*, "A 0.7V 36μW 85dB-DR Audio ΔΣ Modulator Using Class-C Inverter," *IEEE ISSCC Digest of Technical Papers*, pp. 490–491, February 2008.

[191] T. Musah *et al.*, "A 630μW Zero-Crossing-Based ΔΣ ADC Using Switched-Resistor Current Sources in 45nm CMOS," *Proc. of the IEEE Custom Integrated Circuits Conf.*, September 2009.

[192] J. Sauerbrey *et al.*, "A 0.7-V MOSFET-Only Switched-Opamp ΣΔ Modulator in Standard Digital CMOS Technology," *IEEE J. of Solid-State Circuits*, vol. 37, pp. 1662–1669, December 2002.

[193] M. Snoeij *et al.*, "A 4th-Order Switched-Capacitor Sigma-Delta A/D Converter Using a High-Ripple Chebyshev Loop Filter," *Proc. of the IEEE Intl. Symp. on Circuits and Systems*, pp. 615–618, May 2001.

[194] J. Wang *et al.*, "A 0.5V Feedforward Delta-Sigma Modulator with Inverter-Based Integrator," *Proc. of the IEEE European Solid-State Circuits Conf.*, September 2009.

[195] F. Medeiro *et al.*, "Quick Design of High-Performance ΣΔ Modulators using CAD Tools: A 16.4b 1.71mW CMOS ΣΔM for 9.6ksample/s A/D Conversion," *Proc. of 2nd IEEE-CAS Region 8 Workshop on Analog and Mixed IC Design)*, pp. 22–27, 1997.

[196] V. Peluso *et al.*, "A 1.5-V 100-μW ΔΣ Modulator with 12-b Dynamic Range Using the Swtiched-Opamp Technique," *IEEE J. of Solid-State Circuits*, vol. 32, pp. 943–952, July 1997.

[197] B. P. Brandt and B. A. Wooley, "A 50-MHz Multibit Sigma-Delta Modulator for 12-b 2-MHz A/D Conversion," *IEEE J. of Solid-State Circuits*, vol. 26, pp. 1746–1756, December 1991.

[198] T. Burger and Q. Huang, "A 13.5-mW 185-MSample/s ΔΣ Modulator for UMTS/GSM Dual-Standard IF Reception," *IEEE J. of Solid-State Circuits*, vol. 36, pp. 1868–1878, December 2001.

[199] D. Senderowicz *et al.*, "Low-Voltage Double-Sampled ΣΔ Converters," *IEEE J. of Solid-State Circuits*, vol. 32, pp. 1907–1919, December 1997.

[200] S. Au and B. H. Leung, "A 1.95-V, 0.34-mW, 12-b Sigma-Delta Modulator Stabilized by Local Feedback Loops," *IEEE J. of Solid-State Circuits*, vol. 32, pp. 321–328, March 1997.

[201] J. Gilo *et al.*, "A 1.8V 94dB Dynamic Range ΔΣ Modulator for Voice Applications," *IEEE ISSCC Digest of Technical Papers*, pp. 230–231, February 1996.

[202] T. Tille and R. Meyer, "A 1.8-V MOSFET-Only ΣΔ Modulator Using Substrate Biased Depletion-Mode MOS Capacitors in Series Compensation," *IEEE J. of Solid-State Circuits*, vol. 36, pp. 1041–1047, July 2001.

[203] P. Maulik *et al.*, "A 16-Bit 250-kHz Delta-Sigma Modulator and Decimation Filter," *IEEE J. of Solid-State Circuits*, vol. 35, pp. 458–467, April 2000.

[204] A. Gerosa *et al.*, "A Fully-Integrated Two-Channel A/D Interface for hate Acquisition of Cardiac Signals in Implantable Pacemakers," *Proc. of the IEEE European Solid-State Circuits Conf.*, pp. 157–160, September 2003.

[205] C. Thanh *et al.*, "A Second-Order Double-Sampled Delta-Sigma Modulator Using Individual-Level Averaging," *IEEE J. of Solid-State Circuits*, vol. 32, pp. 1269–1273, August 1997.

[206] M. Keskin *et al.*, "A 1-V 10-MHz Clock-Rate 13-Bit CMOS ΔΣ Modulator Using Unity-Gain-Reset Opamps," *IEEE J. of Solid-State Circuits*, vol. 37, pp. 817–824, July 2002.

[207] D. B. Kasha *et al.*, "A 16-mW, 120-dB Linear Switched-Capacitor Delta-Sigma Modulator With Dynamic Biasing," *IEEE J. of Solid-State Circuits*, vol. 34, pp. 921–926, July 1999.

[208] F. Op'T-Eynde *et al.*, "A CMOS Fourth-Order 14b 500k-Sample/s Sigma-Delta ADC Converter," *IEEE ISSCC Digest of Technical Papers*, pp. 62–63, February 1991.

[209] H.-B. Le *et al.*, "A Regulator-Free 84dB DR Audio-Band ADC for Compact Digital Microphones," *Proc. of the IEEE Asian Solid-State Circuits Conf.*, 2010.

[210] B. E. Boser and B. A. Wooley, "The Design of Sigma-Delta Modulation Analog-to-Digital Converters," *IEEE J. of Solid-State Circuits*, vol. 23, pp. 1298–1308, December 1988.

[211] S. Nadeem *et al.*, "16-Channel Oversampled Analog-to-Digital Converter," *IEEE J. of Solid-State Circuits*, vol. 29, pp. 1077–1085, September 1994.

[212] T. V. Burmas *et al.*, "A Second-Order Double-Sampled Delta-Sigma Modulator Using Additive-Error Switching," *IEEE J. of Solid-State Circuits*, vol. 31, pp. 284–293, March 1996.

[213] A. Pena-Perez *et al.*, "A 84dB SNDR 100kHz Bandwidth Low-Power Single Op-Amp Third-Order ΔΣ Modulator Consuming 140μW," *IEEE ISSCC Digest of Technical Papers*, pp. 478–479, February 2011.

[214] Z. Yang *et al.*, "A 0.5-V 35-μW 85-dB DR Double-Sampled ΔΣ Modulator for Audio Applications," *IEEE J. of Solid-State Circuits*, vol. 47, pp. 722–735, March 2012.

[215] L. Liu *et al.*, "A 95dB SNDR Audio ΔΣ Modulator in 65nm CMOS," *Proc. of the IEEE Custom Integrated Circuits Conf.*, September 2011.

[216] K. Lee *et al.*, "A Power-Efficient Two-Channel Time-Interleaved ΣΔ Modulator for Broadband Applications," *IEEE J. of Solid-State Circuits*, vol. 42, pp. 1206–1215, June 2007.

[217] L. Liu *et al.*, "A 1-V 15-Bit Audio ΔΣ-ADC in 0.18μm CMOS," *IEEE Trans. on Circuits and Systems I – Regular Papers*, vol. 59, pp. 915–925, May 2012.

[218] H. Park *et al.*, "A 0.7-V 870-μW Digital-Audio CMOS Sigma-Delta Modulator," *IEEE J. of Solid-State Circuits*, vol. 44, pp. 1078–1088, April 2009.

[219] K. Lee *et al.*, "A Noise-Coupled Time-Interleaved ΔΣ ADC with 4.2MHZ BW, -98dB THD, and 79dB SNDR," *IEEE ISSCC Digest of Technical Papers*, pp. 494–495, February 2008.

[220] J. Koh, Y. Choi, and G. Gómez, "A 66dB DR 1.2V 1.2mW Single-Amplifier Double-Sampling 2nd-order ΔΣ ADC for WCDMA in 90nm CMOS," *IEEE ISSCC Digest of Technical Papers*, pp. 170–171, February 2005.

[221] S. Kwon and F. Maloberti, "A 14mW Multi-bit ΔΣ Modulator with 82dB SNR and 86dB DR for ADSL2+," *IEEE ISSCC Digest of Technical Papers*, pp. 161–162, February 2006.

[222] K. Lee *et al.*, "An 8.1-mW, 82dB Delta-Sigma ADC with 1.9-MHz BW and –98dB THD," *IEEE J. of Solid-State Circuits*, vol. 44, pp. 2202–2211, August 2009.

[223] L. Liu *et al.*, "A 1V 350μW 92dB SNDR 24kHz ΔΣ Modulator in 0.18μm CMOS," *Proc. of the IEEE Asian Solid-State Circuits Conf.*, 2010.

[224] N. Maghari *et al.*, "A +5dBFS Third-Order Extended Dynamic Range Single-Loop ΔΣ Modulator," *Proc. of the IEEE Custom Integrated Circuits Conf.*, September 2010.

[225] K. Tiew *et al.*, "A 0.06-mm^2 Double-Sampling Single-OTA 2nd-Order ΔΣ Modulator in 0.18-μm CMOS Technology," *Proc. of the IEEE Asian Solid-State Circuits Conf.*, 2011.

[226] R. Gaggl, M. Inversi, and A. Wiesbauer, "A Power Optimized 14-Bit SC ΔΣ Modulator for ADSL CO Applications," *IEEE ISSCC Digest of Technical Papers*, pp. 82–83, February 2004.

[227] J. Chae *et al.*, "A 63dB 16mW 20MHz BW Double-Sampled ΔΣ Analog-to-Digital Converter with an Embedded-Adder Quantizer," *Proc. of the IEEE Custom Integrated Circuits Conf.*, September 2010.

[228] Y. Wang *et al.*, "A 2.5MHz BW and 78dB SNDR Delta-Sigma Modulator Using Dynamically Biased Amplifiers," *Proc. of the IEEE Custom Integrated Circuits Conf.*, September 2008.

[229] E. Bonizzoni *et al.*, "Third-Order ΣΔ Modulator with 61-dB SNR and 6-MHz Bandwidth Consuming 6 mW," *Proc. of the IEEE European Solid-State Circuits Conf.*, pp. 218–221, September 2008.

[230] R. Reutemann, P. Balmelli, and Q. Huang, "A 33mW 14b 2.5MSample/s ΣΔ A/D Converter in 0.25μm Digital CMOS," *IEEE ISSCC Digest of Technical Papers*, vol. 1, p. 316, 2002.

[231] Y. Fujimoto, Y. Kanazawa, P. Re, and K. Iizuka, "A 100 MS/s 4 MHz Bandwidth 70 dB SNR ΔΣ ADC in 90 nm CMOS," *IEEE J. of Solid-State Circuits*, vol. 44, pp. 1697–1708, June 2009.

[232] P. Balmelli and Q. Huang, "A 25-MS/s 14-b 200-mW ΣΔ Modulator in 0.18-μm CMOS," *IEEE J. of Solid-State Circuits*, vol. 39, pp. 2161–2169, December 2004.

[233] J. Yu and F. Maloberti, "A Low-Power Multi-Bit ΣΔ Modulator in 90-nm Digital CMOS Without DEM," *IEEE J. of Solid-State Circuits*, vol. 40, pp. 2428–2436, December 2005.

[234] G. Gomez and B. Haroun, "A 1.5V 2.4/2.9mW 79/50dB DR ΣΔ Modulator for GSM/WCDMA in a 0.14μm Digital Process," *IEEE ISSCC Digest of Technical Papers*, February 2002.

[235] M. G. Kim *et al.*, "A 0.9 V 92 dB Double-Sampled Switched-RC Delta-Sigma Audio ADC," *IEEE J. of Solid-State Circuits*, vol. 43, pp. 1195–1206, May 2008.

[236] Y. Kim *et al.*, "A 105.5dB, 0.49mm^2 Audio ΣΔ Modulator Using Chopper Stabilization and Fully Randomized DWA," *Proc. of the IEEE Custom Integrated Circuits Conf.*, September 2008.

[237] R. Gaggl *et al.*, "A 85-dB Dynamic Range Multibit Delta-Sigma ADC for ADSL-CO Applications in 0.18-μm CMOS," *IEEE J. of Solid-State Circuits*, vol. 38, pp. 1105–1114, July 2003.

[238] J. Wu *et al.*, "Multi-bit Sigma Delta ADC with Reduced Feedback Levels, Extended Dynamic Range and Increased Tolerance for Analog Imperfections," *Proc. of the IEEE Custom Integrated Circuits Conf.*, September 2007.

[239] Y. Geerts *et al.*, "A High-Performance Multibit ΔΣ CMOS ADC," *IEEE J. of Solid-State Circuits*, vol. 35, pp. 1829–1840, December 1999.

[240] J. Chen *et al.*, "A 94dB SFDR 78dB DR 2.2MHz BW Multi-bit Delta-Sigma Modulator with Noise Shaping DAC," *Proc. of the IEEE Custom Integrated Circuits Conf.*, September 2007.

[241] A. J. Chen and Y. P. Xu, "Multibit Delta-Sigma Modulator With Noise-Shaping Dynamic Element Matching," *IEEE J. of Solid-State Circuits*, vol. 44, pp. 1125–1133, June 2009.

[242] M. R. Miller and C. S. Petrie, “A Multibit Sigma-Delta ADC for Multimode Receivers,” *IEEE J. of Solid-State Circuits*, vol. 38, pp. 475–482, March 2003.

[243] Y. Yang *et al.*, “A 114-dB 68-mW Chopper-Stabilized Stereo Multibit Audio ADC in 5.62mm^2,” *IEEE J. of Solid-State Circuits*, vol. 38, pp. 2061–2068, December 2003.

[244] O. Nys and R. K. Henderson, “A 19-bit Low-Power Multibit Sigma-Delta ADC Based on Data Weighted Averaging,” *IEEE J. of Solid-State Circuits*, vol. 32, pp. 933–942, July 1997.

[245] T. Kuo *et al.*, “A Wideband CMOS Sigma-Delta Modulator With Incremental Data Weighted Averaging,” *IEEE J. of Solid-State Circuits*, vol. 37, pp. 11–17, January 2002.

[246] R. T. Baird *et al.*, “A Low Oversampling Ratio 14-b 500-kHz ΔΣ ADC With a Self-Calibrated Multibit DAC,” *IEEE J. of Solid-State Circuits*, vol. 31, pp. 312–320, March 1996.

[247] J. Johansson *et al.*, “A 16-bit 60μW Multi-Bit ΣΔ Modulator for Portable ECG Applications,” *Proc. of the IEEE European Solid-State Circuits Conf.*, September 2003.

[248] J. Gilo *et al.*, “A 5V, 118dB ΔΣ Analog-to-Digital Converter for Wideband Digital Audio,” *IEEE ISSCC Digest of Technical Papers*, pp. 218–219, February 1997.

[249] E. Fogleman and I. Galton, “A Dynamic Element Matching Technique for Reduced-distortion Multibit Quantization in Delta-Sigma ADCs,” *IEEE Trans. on Circuits and Systems – II: Analog and Digital Signal Processing*, vol. 48, pp. 158–170, February 2001.

[250] E. Fogleman *et al.*, “A 3.3-V Single-Poly CMOS Audio ADC Delta-Sigma Modulator with 98-dB Peak SINAD and 105-dB Peak SFDR,” *IEEE J. of Solid-State Circuits*, vol. 35, pp. 297–307, March 2000.

[251] A. Prasad *et al.*, “A 120dB 300mW Stereio Audio A/D Converter with 110dB THD+N,” *Proc. of the IEEE European Solid-State Circuits Conf.*, pp. 191–194, September 2004.

[252] F. Chen and B. H. Leung, “A High Resolution Multibit Sigma-Delta Modulator with Individual Level Averaging,” *IEEE J. of Solid-State Circuits*, vol. 30, pp. 453–460, April 1995.

[253] J. Sauerbrey *et al.*, “A Dual-Quantization Multi-Bit Sigma Delta Analog/Digital Converter,” *Proc. of the IEEE Intl. Symp. on Circuits and Systems*, pp. 437–440, May 1994.

[254] J. Xu *et al.*, “A 9μW 88dB DR Fully-Clocked Switched-Opamp ΔΣ Modulator with Novel Power and Area Efficient Resonator,” *Proc. of the IEEE Custom Integrated Circuits Conf.*, September 2011.

[255] L. Wang and L. Theogarajan, “An 18μW 79dB-DR 20kHz-BW MASH ΔΣ Modulator Utilizing Self-Biased Amplifiers for Biomedical Applications,” *Proc. of the IEEE Custom Integrated Circuits Conf.*, September 2011.

[256] K. A. Donoghue *et al.*, “A Digitally Corrected 5-mW 2-MS/s SC ΔΣ ADC in 0.25-μm CMOS With 94-dB SFDR,” *IEEE J. of Solid-State Circuits*, vol. 46, pp. 2673–2684, November 2011.

[257] K. Cornelissens and M. Steyaert, “A 1-V 84-dB DR 1-MHz Bandwidth Cascade 3-1 Delta-Sigma ADC in 65-nm CMOS,” *Proc. of the IEEE European Solid-State Circuits Conf.*, pp. 332–335, 2009.

[258] C. Kuo *et al.*, “A Low-Voltage Fourth-Order Cascade Delta-Sigma Modulator in 0.18-μm CMOS,” *IEEE J. of Solid-State Circuits*, vol. 57, pp. 2450–2461, September 2010.

[259] O. Oliaei *et al.*, “A 5-mW Sigma-Delta Modulator With 84-dB Dynamic Range for GSM/EDGE,” *IEEE J. of Solid-State Circuits*, vol. 37, pp. 2–10, January 2002.

[260] J. M. de la Rosa *et al.*, “A CMOS 110-dB@40-kS/s Programmable-Gain Chopper-Stabilized Third-Order 2-1 Cascade Sigma-Delta Modulator for Low-Power High-Linearity Automotive Sensor ASICs,” *IEEE J. of Solid-State Circuits*, vol. 40, pp. 2246–2264, November 2005.

[261] A. Morgado, R. del Río, and J. M. de la Rosa, “An Adaptive ΣΔ Modulator for Multi-Standard Hand-Held Wireless Devices,” *Proc. of the IEEE Asian Solid-State Circuits Conf.*, pp. 232–235, 2007.

[262] Y. Geerts *et al.*, “A 3.3-V, 15-bit, Delta-Sigma ADC with a Signal Bandwidth of 1.1 MHz for ADSL Applications,” *IEEE J. of Solid-State Circuits*, vol. 34, pp. 1829–1840, July 1999.

[263] G. Yin *et al.*, “A 16-b 320-kHz CMOS A/D Converter Using Two-Stage Third-Order ΣΔ Noise Shaping,” *IEEE J. of Solid-State Circuits*, vol. 28, pp. 640–647, June 1993.

[264] A. M. Marques, V. Peluso, M. S. J. Steyaert, and W. M. Sansen, "A 15-b Resolution 2-MHz Nyquist Rate ΔΣ ADC in a 1-μm CMOS Technology," *IEEE J. of Solid-State Circuits*, vol. 33, pp. 1065–1075, July 1998.

[265] J. C. Morizio *et al.*, "14-bit 2.2-MS/s Sigma-Delta ADC's," *IEEE J. of Solid-State Circuits*, vol. 35, pp. 968–976, July 2000.

[266] G. J. Gomez, "A 102-dB Spurious-Free DR ΣΔ ADC Using a Dynamic Dither Scheme," *IEEE J. of Solid-State Circuits*, vol. 47, pp. 531–535, June 2000.

[267] L. A. Williams and B. A. Wooley, "A Third-Order Sigma-Delta Modulator with Extended Dynamic Range," *IEEE J. of Solid-State Circuits*, vol. 35, pp. 193–202, March 1994.

[268] C. B. Wang, "A 20-bit 25-kHz Delta-Sigma A/D Converter Utilizing a Frequency-Shaped Chopper Stabilization Scheme," *IEEE J. of Solid-State Circuits*, vol. 36, pp. 566–569, March 2001.

[269] M. Rebeschini, N. R. van Bavel, P. Rakers, R. Greene, J. Caldwell, and J. R. Haug, "A 16-b 160 kHz CMOS A/D Converter Using Sigma-Delta Modulation," *IEEE J. of Solid-State Circuits*, vol. 25, pp. 431–440, April 1990.

[270] I. Fujimori *et al.*, "A 5-V Single-Chip Delta-Sigma Audio A/D Converter with 111dB Dynamic Range," *IEEE J. of Solid-State Circuits*, vol. 32, pp. 329–336, March 1997.

[271] G. Miao *et al.*, "An Oversampled A/D Converter Using Cascaded Fourth Order Sigma-Delta Modulation and Current Steering Logic," *Proc. of the IEEE Intl. Symp. on Circuits and Systems*, pp. 412–415, May 1998.

[272] R. Zanbaghi *et al.*, "A 75dB SNDR, 10MHz Conversion Bandwidth Stage-Shared 2-2 MASH ΔΣ Modulator Dissipating 9mW," *Proc. of the IEEE Custom Integrated Circuits Conf.*, September 2011.

[273] P. Malla *et al.*, "A 28mW Spectrum-Sensing Reconfigurable 20MHz 72dB-SNR 70dB-SNDR DT ΔΣ ADC for 802.11n/WiMAX Receivers," *IEEE ISSCC Digest of Technical Papers*, February 2008.

[274] K. Yamamoto and A. C. Carusone, "A 1-1-1-1 MASH Delta-Sigma Modulator Using Dynamic Comparator-Based OTAs," *Proc. of the IEEE Custom Integrated Circuits Conf.*, September 2011.

[275] N. Maghari *et al.*, "74dB SNDR Multi-Loop Sturdy-MASH Delta-Sigma Modulator Using 35dB Open-Loop Opamp Gain," *IEEE J. of Solid-State Circuits*, vol. 44, pp. 2212–2221, August 2009.

[276] A. Dezzani and E. Andre, "A 1.2-V Dual-Mode WCDMA/GPRS ΣΔ Modulator," *IEEE ISSCC Digest of Technical Papers*, February 2003.

[277] T.-H. Chang *et al.*, "A 2.5-V 14-bit, 180-mW Cascaded ΣΔ ADC for ADSL2+ Application," *IEEE J. of Solid-State Circuits*, vol. 42, pp. 2357–2368, November 2007.

[278] J. Paramesh *et al.*, "An 11-Bit 330MHz 8× OSR Σ – Δ Modulator for Next-Generation WLAN," *IEEE Symposium VLSI Circuits. Digest of Technical Papers*, December 2006.

[279] K. Vleugels, S. Rabii, and B. Wooley, "A 2.5-V Sigma-Delta Modulator for Broadband Communications Applications," *IEEE J. of Solid-State Circuits*, vol. 36, pp. 1887–1899, December 2001.

[280] I. Fujimori, L. Longo, A. Hairapetian, K. Seiyama, S. Kosic, J. Cao, and S.-L. Chan, "A 90-dB SNR 2.5-MHz Output-Rate ADC Using Cascaded Multibit Delta-Sigma Modulation at 8× Oversampling Ratio," *IEEE J. Solid-State Circuits*, vol. 35, pp. 1820–1828, December 2000.

[281] R. del Río, F. Medeiro, B. Pérez-Verdú, J. M. de la Rosa, and A. Rodríguez-Vázquez, *CMOS Cascade ΣΔ Modulators for Sensors and Telecom: Error Analysis and Practical Design*. Springer, 2006.

[282] G.-C. Ahn *et al.*, "A 0.6-V 82-dB Delta-Sigma Audio ADC Using Switched-RC Integrators," *IEEE J. of Solid-State Circuits*, vol. 40, pp. 1868–1878, December 2005.

[283] F. Medeiro *et al.*, "A 13-bit, 2.2-MS/s, 55-mW Multibit Cascade ΣΔ Modulator in CMOS 0.7-μm Single-Poly Technology," *IEEE J. of Solid-State Circuits*, vol. 34, pp. 748–760, June 1999.

[284] S. K. Gupta and V. Fong, "A 64-MHz Clock-Rate ΣΔ ADC With 88-dB SNDR and –105-dB IM3 Distortion at a 1.5-MHz Signal Frequency," *IEEE J. of Solid-State Circuits*, vol. 37, pp. 1653–1661, December 2002.

[285] H. Lampinen and O. Vinio, "An Optimization Approach to Designing OTAs for Low-Voltage Sigma-Delta Modulators," *IEEE J. of Solid-State Circuits*, vol. 50, pp. 1665–1671, December 2001.

[286] B. P. Brandt and B. A. Wooley, "A 50-MHz Multibit Sigma-Delta Modulator for 12-b 2-MHz A/D Conversion," *IEEE J. of Solid-State Circuits*, vol. 26, pp. 1746–1756, December 1991.

[287] T. C. Caldwell and D. A. Johns, "An 8th-Order MASH Delta-Sigma With an OSR of 3," *Proc. of the IEEE European Solid-State Circuits Conf.*, September 2009.

[288] A. Tabatabaei *et al.*, "A Dual Channel ΣΔ ADC with 40MHz Aggregate Signal Bandwidth," *IEEE ISSCC Digest of Technical Papers*, February 2003.

[289] A. R. Feldman *et al.*, "A 13-Bit, 1.4-MS/s Sigma-Delta Modulator for RF Baseband Channel Applications," *IEEE J. of Solid-State Circuits*, vol. 33, pp. 1462–1469, October 1998.

[290] I. Dedic, "A Sixth-Order Triple-Loop ΣΔ CMOS ADC with 90dB SNR and 100kHz Bandwidth," *IEEE ISSCC Digest of Technical Papers*, pp. 188–189, 1994.

[291] F. Ying and F. Maloberti, "A Mirror Image Free Two-path Bandpass ΣΔ Modulator with 72 dB SNR and 86 dB SFDR," *IEEE ISSCC Digest of Technical Papers*, pp. 84–85, February 2004.

[292] T. O. Salo *et al.*, "80-MHz Bandpass ΔΣ Modulators for Multimode Digital IF Receivers," *IEEE J. of Solid-State Circuits*, vol. 38, pp. 464–474, March 2003.

[293] J. Galdi *et al.*, "Two-Path Band-Pass ΣΔ Modulator with 40-MHz IF 72-dB DR at 1-MHz Bandwidth Consuming 16mW," *Proc. of the IEEE European Solid-State Circuits Conf.*, pp. 248–251, September 2007.

[294] R. Maurino and C. Papavassiliou, "A 10mW 81dB Cascaded Multibit Quadrature ΣΔ ADC with a Dynamic Element Matching Scheme," *Proc. of the IEEE European Solid-State Circuits Conf.*, pp. 451–454, September 2005.

[295] K. Yamamoto *et al.*, "A Delta-Sigma Modulator with a Widely Programmable Center Frequency and 82-dB Peak SNDR," *Proc. of the IEEE Custom Integrated Circuits Conf.*, pp. 65–69, September 2007.

[296] C.-H. Kuo and S.-I. Liu, "A 1-V 10.7-MHz Fourth-Order Bandpass ΔΣ Modulators Using Two Switched Opamps," *IEEE J. of Solid-State Circuits*, vol. 39, pp. 2041–2045, November 2004.

[297] A. Tabatabaei and A. Wooley, "A Two-Path Bandpass Sigma-Delta Modulator with Extended Noise Shaping," *IEEE J. of Solid-State Circuits*, vol. 35, pp. 1799–1809, December 2000.

[298] A. Hairapetian, "An 81 MHz IF receiver in CMOS," *IEEE J. of Solid-State Circuits*, vol. 31, pp. 1981–1986, December 1996.

[299] T. Yamamoto, M. Kasahara, and T. Matsuura, "A 63 mA 112/94 dB DR IF Bandpass ΔΣ Modulator With Direct Feed-Forward Compensation and Double Sampling," *IEEE J. of Solid-State Circuits*, vol. 43, pp. 1783–1794, August 2008.

[300] T. Ueno *et al.*, "A Fourth-Order Bandpass $\Delta - \Sigma$ Modulator Using Second-Order Bandpass Noise-Shaping Dynamic Element Matching," *IEEE J. of Solid-State Circuits*, vol. 37, pp. 809–816, July 2002.

[301] B.-S. Song, "A Fourth-Order Bandpass Delta-Sigma Modulator with Reduced Number of Op Amps," *IEEE J. of Solid-State Circuits*, vol. 30, pp. 1309–1315, December 1995.

[302] T. O. Salo *et al.*, "A 80-MHz Bandpass ΣΔ Modulator for a 100-MHz IF Receiver," *IEEE J. of Solid-State Circuits*, vol. 37, pp. 798–808, July 2002.

[303] W.-T. Cheng *et al.*, "A 75dB Image Rejection IF-Input Quadrature Sampling SC ΣΔ Modulator," *Proc. of the IEEE European Solid-State Circuits Conf.*, pp. 455–458, September 2005.

[304] P. Cusinato *et al.*, "A 3.3-V CMOS 10.7-MHz Sixth-Order Bandpass ΣΔ Modulator with 74-dB Dynamic Range," *IEEE J. of Solid-State Circuits*, vol. 36, pp. 629–638, April 2001.

[305] S. A. Jantzi *et al.*, "Quadrature Bandpass ΔΣ Modulation for Digital Radio," *IEEE J. of Solid-State Circuits*, vol. 32, pp. 1935–1950, December 1997.

[306] V. S. Cheung *et al.*, "A 1-V 10.7-MHz Switched-Opamp Bandpass ΣΔ Modulator Using Double-Sampling Finite-Gain-Compensation Technique," *IEEE J. of Solid-State Circuits*, vol. 37, pp. 1215–1225, October 2002.

[307] S. Pavan and P. Sankar, "A 110μW Single Bit Audio Continuous-Time Oversampled Converter with 92.5-dB Dynamic Range," *Proc. of the IEEE European Solid-State Circuits Conf.*, September 2009.

[308] F. Munoz, K. Philips, and A. Torralba, "A 4.7mW 89.5dB DR CT Complex ΔΣ ADC with Built-In LPF," *IEEE ISSCC Digest of Technical Papers*, February 2005.

[309] J. Zhan *et al.*, "A 0.6-V 82-dB 28.6-μW Continuous-Time Audio Delta-Sigma Modulator," *IEEE J. of Solid-State Circuits*, vol. 46, pp. 2326–2335, October 2011.

[310] A. Jain and S. Pavan, "A 4mW 1GS/s Continuous-Time ΔΣ Modulator with 15.6MHz Bandwidth and 67dB Dynamic Range," *Proc. of the IEEE European Solid-State Circuits Conf.*, pp. 259–262, September 2011.

[311] K. Matsukawa *et al.*, "A 69.8dB SNDR 3rd-order Continuous-Time Delta-Sigma Modulator with an Ultimate Low-Power Tuning System for a Worldwide Digital TV-Receiver," *Proc. of the IEEE Custom Integrated Circuits Conf.*, September 2010.

[312] P. Crombez *et al.*, "A Single-Bit 6.8mW 10MHz Power-Optimized Continuous-Time ΔΣ with 67dB DR in 90nm CMOS," *Proc. of the IEEE European Solid-State Circuits Conf.*, September 2009.

[313] P. Crombez *et al.*, "A 500kHz-10MHz Multimode Power-Performance Scalable 83-to-67dB DR CTΣΔ in 90nm Digital CMOS with Flexible Analog Core Circuitry," *IEEE Symp. VLSI Circuits. Digest of Technical Papers*, pp. 70–71, 2009.

[314] X. Xing *et al.*, "A 40MHz 12bit 84.2dB-SFDR Continuous-Time Delta-Sigma Modulator in 90nm CMOS," *Proc. of the IEEE Asian Solid-State Circuits Conf.*, 2011.

[315] K. Matsukawa *et al.*, "A Fifth-Order Continuous-Time Delta-Sigma Modulator With Single-Opamp Resonator," *IEEE J. of Solid-State Circuits*, vol. 45, pp. 697–706, April 2010.

[316] A. Das *et al.*, "A 4th-order 86dB CT ΔΣ ADC with Two Amplifiers in 90nm CMOS," *IEEE ISSCC Digest of Technical Papers*, February 2005.

[317] D. Kim *et al.*, "A Continuous-Time, Jitter Insensitive ΣΔ Modulator Using a Digitally Linearized Gm-C Integrator with Embedded SC Feedback DAC," *IEEE Symp. VLSI Circuits. Digest of Technical Papers*, pp. 38–39, 2011.

[318] R. H. Veldhoven *et al.*, "Technology portable, $0.04mm^2$, GHz-rate ΣΔ Modulators in 65nm and 45nm CMOS," *IEEE Symp. VLSI Circuits. Digest of Technical Papers*, pp. 72–73, 2009.

[319] E. Prefasi *et al.*, "A $0.1mm^2$, Wide Bandwidth Continuous-Time ΣΔ ADC Based on a Time Encoding Quantizer in 0.13μm CMOS," *IEEE J. of Solid-State Circuits*, vol. 44, pp. 2745–2754, October 2009.

[320] J. Zhang *et al.*, "A 1.2-V 2.7-mW 160MHz Continuous-Time Delta-Sigma Modulator with Input-Feedforward Structure," *Proc. of the IEEE Custom Integrated Circuits Conf.*, pp. 475–478, September 2009.

[321] R. van Veldhoven, "A 3.3mW ΣΔ Modulator for UMTS in 0.18μm CMOS with 70dB Dynamic Range in 2MHz Bandwidth," *IEEE ISSCC Digest of Technical Papers*, February 2002.

[322] F. Gerfers, M. Ortmanns, and Y. Manoli, "A 1.5-V 12-bit Power-Efficient Continuous-Time Third-Order ΣΔ Modulator," *IEEE J. of Solid-State Circuits*, vol. 38, pp. 1343–1352, August 2003.

[323] K. Philips, "A 4.4mW 76dB Complex ΣΔ ADC for Bluetooth Receivers," *IEEE ISSCC Digest of Technical Papers*, February 2003.

[324] Y.-C. Chang *et al.*, "A 4MHz BW 69dB SNDR Continuous-Time Delta-Sigma Modulator with Reduced Sensitivity to Clock Jitter," *Proc. of the IEEE Asian Solid-State Circuits Conf.*, pp. 265–268, 2011.

[325] T. Nagai *et al.*, "A 1.2V 3.5mW ΔΣ Modulator with a Passive Current Summing Network and a Variable Gain Function," *IEEE ISSCC Digest of Technical Papers*, February 2005.

[326] L. Samid and Y. Manoli, "A Micro Power Continuous-Time ΣΔ Modulator," *Proc. of the IEEE European Solid-State Circuits Conf.*, September 2003.

[327] L. S. M Anderson, "Design and Measurement of a CT ΔΣ ADC With Switched-Capacitor Switched-Resistor Feedback," *IEEE J. of Solid-State Circuits*, vol. 44, pp. 473–483, February 2009.

[328] C.-H. Lin *et al.*, "A 5MHz Nyquist Rate Continuous-Time Sigma-Delta Modulator for Wideband Wireless Communication," *Proc. of the IEEE Intl. Symp. on Circuits and Systems*, pp. 368–371, May 1999.

[329] K.-P. Pun, S. Chatterjee, and P. Kinget, "A 0.5-V 74-dB SNDR 25-kHz Continuous-Time Delta-Sigma Modulator With a Return-to-Open DAC," *IEEE J. of Solid-State Circuits*, vol. 42, pp. 496–507, March 2007.

[330] F. Gerfers *et al.*, "Implementation of a 1.5V Low-Power Clock-Jitter Insensitive Continuous-Time ΣΔ Modulator," *Proc. of the IEEE Intl. Symp. on Circuits and Systems*, pp. 652–655, May 2002.

[331] L. Luh *et al.*, "A 400MHz 5th-Order CMOS Continuous-Time Switched-Current ΣΔ Modulator," *Proc. of the IEEE European Solid-State Circuits Conf.*, September 2000.

[332] E. van der Zwan and E. Dijkmans, "A 0.2mW CMOS ΣΔ Modulator for Speech Coding with 80dB Dynamic Range," *IEEE J. of Solid-State Circuits*, vol. 31, pp. 1873–1880, December 1996.

[333] F. Cannillo *et al.*, "1.4V 13μW 83dB DR CT-ΣΔ Modulator with Dual-Slope Quantizer and PWM DAC for Biopotential Signal Acquisition," *Proc. of the IEEE European Solid-State Circuits Conf.*, pp. 267–270, September 2011.

[334] L. Luh *et al.*, "A 50-MHz Continuous-Time Switched-Current ΣΔ Modulator," *Proc. of the IEEE Intl. Symp. on Circuits and Systems*, pp. 579–582, May 1998.

[335] M. Ortmanns *et al.*, "A Continuous-Time Sigma-Delta Modulator with Switched Capacitor Controlled Current Mode Feedback," *Proc. of the IEEE European Solid-State Circuits Conf.*, September 2003.

[336] H. Zare-Hoseini *et al.*, "A Low-Power Continuous-Time ΔΣ Modulator for Electret Microphone Applications," *Proc. of the IEEE Asian Solid-State Circuits Conf.*, 2010.

[337] N. Sarhangnejad *et al.*, "A Continuous-Time ΣΔ Modulator with a Gm-C Input Stage, 120-dB CMRR and -87dB THD," *Proc. of the IEEE Asian Solid-State Circuits Conf.*, 2011.

[338] S. Pavan, "Excess Loop Delay Compensation in Continuous-Time Delta-Sigma Modulators," *IEEE Trans. on Circuits and Systems II: Express Briefs*, vol. 55, pp. 1119–1123, November 2008.

[339] P. Witte *et al.*, "A 72dB-DR ΔΣ CT Modulator Using Digitally Estimated Auxiliary DAC Linearization Achieving 88fJ/conv in a 25MHz BW," *IEEE ISSCC Digest of Technical Papers*, pp. 154–155, February 2012.

[340] K. Matsukawa *et al.*, "A 5th-Order Delta-Sigma Modulator with Single-Opamp Resonator," *IEEE Symp. VLSI Circuits. Digest of Technical Papers*, pp. 68–69, 2009.

[341] S.-J. Huang and Y.-Y. Li, "A 1.2V 2MHz BW 0.084mm^2 CT ΔΣ ADC with -97.7dBc THD and 80dB DR Using Low-Latency DEM," *IEEE ISSCC Digest of Technical Papers*, pp. 172–173, February 2009.

[342] J.-C. Tsai *et al.*, "A Continuous-Time ΔΣ ADC with Clock Timing Calibration," *Proc. of the IEEE Asian Solid-State Circuits Conf.*, pp. 369–370, 2008.

[343] M. Ranjbar *et al.*, "A Robust STF 6mW CT ΔΣ Modulator with 76dB Dynamic Range and 5MHz Bandwidth," *Proc. of the IEEE Custom Integrated Circuits Conf.*, September 2010.

[344] K. Reddy and S. Pavan, "A 20.7mW Continuous-Time ΔΣ Modulator with 15MHz Bandwidth and 70dB Dynamic Range," *Proc. of the IEEE European Solid-State Circuits Conf.*, pp. 210–213, September 2008.

[345] Y.S. Shu, B. S. Song, and K. Bacrania, "A 65nm CMOS CT ΔΣ Modulator with 81dB DR and 8MHz BW Auto-Tuned by Pulse Injection," *IEEE ISSCC Digest of Technical Papers*, February 2008.

[346] J. G. Kauffman *et al.*, "A 78dB SNDR 87mW 20MHz Bandwidth Continuous-Time ΔΣ ADC With VCO-Based Integrator and Quantizer Implemented in 0.13μm CMOS," *IEEE J. of Solid-State Circuits*, vol. 44, pp. 3344–3358, December 2009.

[347] M. Moyal *et al.*, "A 700/900mW/Channel CMOS Dual Analog Front-End IC for VDSL with Integrated 11.5/14.5dBm Line Drivers," *IEEE ISSCC Digest of Technical Papers*, February 2003.

[348] X. Chenand *et al.*, "A 18mW CT ΔΣ Modulator with 25MHz Bandwidth for Next Generation Wireless Applications," *Proc. of the IEEE Custom Integrated Circuits Conf.*, pp. 73–76, September 2007.

[349] W. Yang *et al.*, "A 100mW 10MHz-BW CT ΔΣ Modulator with 87dB DR and 91dBc IMD," *IEEE ISSCC Digest of Technical Papers*, February 2008.

[350] M. Vadipour *et al.*, "A 2.1mW/3.2mW Delay-Compensated GSM/WCDMA ΣΔ Analog-Digital Converter," *IEEE Symp. VLSI Circuits. Digest of Technical Papers*, pp. 180–181, 2008.

[351] Z. Li and T. S. Fiez, "A 14 Bit Continuous-Time Delta-Sigma A/D Modulator With 25MHz Signal Bandwidth," *IEEE J. of Solid-State Circuits*, vol. 42, pp. 1873–1883, September 2007.

[352] L. Dorrer *et al.*, "10-Bit, 3 mW Continuous-Time Sigma-Delta ADC for UMTS in a 0.12 μm CMOS process," *Proc. of the European Solid-State Circuits Conf.*, September 2003.

[353] M. Schimper *et al.*, "A 3mW continuous-time ΣΔ modulator for EDGE/GSM with high adjacent channel tolerance," *Proc. of the European Solid-State Circuits Conf.*, pp. 183–186, September 2004.

[354] M. Ranjbar *et al.*, "A Low-Power 1.92MHz CT ΔΣ Modulator With 5-bit Successive Approximation Quantizer," *Proc. of the IEEE Custom Integrated Circuits Conf.*, pp. 5–8, September 2009.

[355] J.-G. Jo *et al.*, "A 20MHz Bandwidth Continuous-Time ΣΔ Modulator with Jitter Immunity Improved Full-Clock Period SCR (FSCR) DAC and High Speed DWA," *Proc. of the IEEE Asian Solid-State Circuits Conf.*, 2010.

[356] P. Fontaine, A. N. Mohieldin, and A. Bellaouar, "A Low-Noise Low-Voltage CT ΔΣ Modulator with Digital Compensation of Excess Loop Delay," *IEEE ISSCC Digest of Technical Papers*, pp. 498–499, 2005.

[357] Y. Aiba *et al.*, "A Fifth-Order Gm-C Continuous-Time ΔΣ Modulator With Process-Insensitive Input Linear Range," *IEEE J. of Solid-State Circuits*, vol. 44, pp. 2381–2391, September 2009.

[358] L. Dorrer *et al.*, "A Continuous-Time ΔΣ ADC for Voice Coding with 92dB DR in 45nm CMOS," *IEEE ISSCC Digest of Technical Papers*, pp. 502–503, February 2008.

[359] S. Patón, A. di Giandomenico, L. Hernández, A. Wiesbauer, T. Poetscher, and M. Clara, "A 70-mW 300-MHz CMOS Continuous-Time ΣΔ ADC With 15-MHz Bandwidth and 11 Bits of Resolution," *IEEE J. of Solid-State Circuits*, vol. 39, pp. 1056–1063, July 2004.

[360] A. D. Giandomenico *et al.*, "A 15MHz Bandwidth Sigma-Delta ADC with 11 Bits of Resolution in 0.13μm CMOS," *Proc. of the IEEE European Solid-State Circuits Conf.*, September 2003.

[361] J.-G. Jo *et al.*, "A 20-MHz Bandwidth Continuous-Time Sigma-Delta Modulator With Jitter Immunity Improved Full Clock Period SCR (FSCR) DAC and High-Speed DWA," *IEEE J. of Solid-State Circuits*, vol. 46, pp. 2469–2477, November 2011.

[362] S. Yan and E. Sánchez-Sinencio, "A Continuous-Time ΣΔ Modulator With 88-dB Dynamic Range and 1.1-MHz Signal Bandwidth," *IEEE J. of Solid-State Circuits*, vol. 39, pp. 75–86, January 2004.

[363] J. Arias *et al.*, "A 32-mW 320-MHz Continuous-Time Complex Delta-Sigma ADC for Multi-Mode Wireless-LAN Receivers," *IEEE J. of Solid-State Circuits*, vol. 41, pp. 339–351, February 2006.

[364] T. C. Caldwell and D. A. Johns, "A Time-Interleaved Continuous-Time ΔΣ Modulator with 20MHz Signal Bandwidth," *Proc. of the IEEE European Solid-State Circuits Conf.*, pp. 447–450, September 2005.

[365] J. Sauerbrey *et al.*, "A Configurable Cascaded Continuous-Time ΔΣ Modulator with up to 15MHz Bandwidth," *Proc. of the IEEE European Solid-State Circuits Conf.*, pp. 426–429, September 2010.

[366] L. Breems, R. Rutten, R. van Veldhoven, and G. van der Weide, "A 56 mW Continuous-Time Quadrature Cascaded ΣΔ Modulator With 77 dB DR in a Near Zero-IF 20 MHz Band," *IEEE J. of Solid-State Circuits*, vol. 42, pp. 2696–2705, December 2007.

[367] J. Kamiishi *et al.*, "A Self-Calibrated 2-1-1 Cascaded Continuous-Time ΔΣ Modulator," *Proc. of the IEEE Custom Integrated Circuits Conf*, pp. 9–12, September 2009.

[368] Y.-S. Shu, J. Kamiishi, K. Tomioka, K. Hamashita, and B.-S. Song, "LMS-Based Noise Leakage Calibration of Cascaded Continuous-Time ΣΔ Modulators," *IEEE J. of Solid-State Circuits*, vol. 45, pp. 368–379, February 2010.

[369] L. Breems, R. Rutten, and G. Wetzker, "A Cascaded Continuous-Time ΣΔ Modulator with 67-dB Dynamic Range in 10-MHz Bandwidth," *IEEE J. of Solid-State Circuits*, vol. 39, pp. 2152–2160, December 2004.

[370] B. D. Vuyst and P. Rombouts, "A 5-MHz 11-bit Delay-Based Self-Oscillating ΣΔ Modulator in $0.025mm^2$," *Proc. of the IEEE Custom Integrated Circuits Conf.*, September 2010.

[371] H. Chae *et al.*, "A 12mW Low-Power Continuous-Time Bandpass ΔΣ Modulator with 58dB SNDR and 24MHz Bandwidth at 200MHz IF," *IEEE ISSCC Digest of Technical Papers*, pp. 148–149, February 2012.

[372] R. H. M. Veldhoven, "A Triple-Mode Continuous-Time ΣΔ Modulator With Switched-Capacitor Feedback DAC for a GSM-EDGE/CDMA2000/UMTS Receiver," *IEEE J. of Solid-State Circuits*, vol. 38, pp. 2069–2076, December 2003.

[373] M. S. Kappes, "A 2.2-mW CMOS Bandpass Continuous-Time Multibit Δ-Σ ADC With 68 dB of Dynamic Range and 1-MHz Bandwidth for Wireless Applications," *IEEE J. of Solid-State Circuits*, vol. 38, pp. 1098–1104, July 2003.

[374] F. Esfahani *et al.*, "An 82dB CMOS Continuous-Time Complex Bandpass Sigma-Delta ADC for GSM/EDGE," *Proc. of the IEEE Intl. Symp. on Circuits and Systems*, pp. 1049–1052, May 2003.

[375] R. Schreier *et al.*, "A 375-mW Quadrature Bandpass ΔΣ ADC With 8.5-MHz BW and 90-dB DR at 44 MHz," *IEEE J. of Solid-State Circuits*, vol. 41, pp. 2632–2640, December 2006.

[376] A. Ashry *et al.*, "A 3.6GS/s, 15mW, 50dB SNDR, 28MHz Bandwidth RF ΣΔ ADC with FoM of 1pJ/bit in 130nm CMOS," *Proc. of the IEEE Custom Integrated Circuits Conf.*, September 2011.

[377] P. M. Chopp and A. A. Hamoui, "A 1V 13mW Frequency-Translating ΔΣ ADC with 55dB SNDR for a 4MHz Band at 225MHz," *Proc. of the IEEE Custom Integrated Circuits Conf.*, September 2011.

[378] E. J. van der Zwan *et al.*, "A 10.7-MHz IF-to-Baseband ΣΔ A/D Conversion System for AM/FM Radio Receivers," *IEEE J. of Solid-State Circuits*, vol. 35, pp. 1810–1819, December 2000.

[379] F. Chen and B. Leung, "A 0.25-mW Low-Pass Passive Sigma-Delta Modulator with Built-In Mixer for a 10-MHz IF Input," *IEEE J. of Solid-State Circuits*, vol. 32, pp. 774–782, June 1997.

[380] C.-Y. Lu *et al.*, "A Sixth-Order 200MHz IF Bandpass Sigma-Delta Modulator With Over 68dB SNDR in 10MHz Bandwidth," *IEEE J. of Solid-State Circuits*, vol. 45, pp. 1122–1136, June 2010.

[381] P. G. R. Silva *et al.*, "An IF-to-Baseband ΣΔ Modulator for AM/FM/IBOC Radio Receivers With a 118 dB Dynamic Range," *IEEE J. of Solid-State Circuits*, vol. 42, pp. 1076–1089, May 2007.

[382] S.-B. Kim *et al.*, "Continuous-Time Quadrature Bandpass Sigma-Delta Modulator for GPS/Galileo Low-IF Receiver," *Proc. of the IEEE Radio Frequency Integrated Circuits Symp.*, pp. 127–130, June 2007.

[383] R. Yu and Y. P. Xu, "Bandpass Sigma-Delta Modulator Employing SAW Resonator as Loop Filter," *IEEE Trans. on Circuits and Systems I – Regular Papers*, vol. 54, pp. 723–735, April 2007.

[384] J. V. Engelen and R. van de Plassche, *BandPass Sigma-Delta Modulators: Stability Analysis, Performance and Design Aspects*. Kluwer Academic Publishers, 1999.

[385] I. Hsu and H. C. Luong, "A 70-MHz Continuous-Time CMOS Band-pass ΣΔ Modulator for GPS Receivers," *Proc. of the IEEE Intl. Symp. on Circuits and Systems*, pp. 750–753, May 2000.

[386] H. Tao and J. M. Khoury, "A 400-MS/s Frequency Translating Bandpass Sigma-Delta Modulator," *IEEE J. of Solid-State Circuits*, vol. 34, pp. 1741–1752, December 1999.

[387] K. Thomas *et al.*, "A 1GHz CMOS Fourth-Order Continuous-Time Bandpass Sigma-Delta Modulator for RF Receiver Front-End A/D Conversion," *Proc. of the IEEE Asia and South Pacific Design Automation Conf.*, pp. 665–670, 2005.

[388] V. Dhanasekaran *et al.*, "A Continuous-Time Multi-bit ΔΣ ADC Using Time Domain Quantizer and Feedback Element," *IEEE J. of Solid-State Circuits*, vol. 46, pp. 639–650, March 2011.

[389] Y. Cao *et al.*, "A 0.7mW 13b Temperature-Stable MASH ΔΣ TDC with Delay-Line Assisted Calibration," *Proc. of the IEEE Asian Solid-State Circuits Conf.*, pp. 361–364, 2011.

[390] B. Young *et al.*, "A 2.4ps Resolution 2.1mW Second-Order Noise-Shaped Time-to-Digital Converter with 3.2ns Range in 1MHz Bandwidth," *Proc. of the IEEE Custom Integrated Circuits Conf.*, September 2009.

[391] O. Rajaee *et al.*, "A 79dB 80MHz 8×-OSR Hybrid Delta-Sigma/Pipeline ADC," *IEEE Symp. VLSI Circuits. Digest of Technical Papers*, pp. 74–77, 2009.

[392] C.-Y. Ho *et al.*, "A 75.1dB SNDR, 80.2dB DR, 4th-Order Feed-Forward Continuous-Time Sigma-Delta Modulator with Hybrid Integrator for Silicon TV-tuner Applications," *Proc. of the IEEE Asian Solid-State Circuits Conf.*, pp. 261–264, 2011.

[393] G. K. Balachandran *et al.*, "A 1.16mW 69dB SNR (1.2MHz BW) Continuous-Time ΣΔ ADC with Immunity to Clock Jitter," *Proc. of the IEEE Custom Integrated Circuits Conf.*, September 2009.

[394] J. H. Shim, I.-C. Park, and B. Kim, "A Third-Order ΣΔ Modulator in 0.18μm CMOS with Calibrated Mixed-Mode Integrators.," *IEEE J. of Solid-State Circuits*, vol. 40, pp. 918–925, April 2005.

[395] Y. Kim *et al.*, "An 11mW 100MHz 16X-OSR 64dB-SNDR Hybrid CT/DT ΔΣ ADC with Relaxed DEM Timing," *Proc. of the IEEE Custom Integrated Circuits Conf.*, September 2009.

[396] S. Kulchycki *et al.*, "A 77-dB Dynamic Range, 7.5-MHz Hybrid Continuous-Time/Discrete-Time Cascade ΣΔ Modulator," *IEEE J. of Solid-State Circuits*, vol. 43, pp. 796–804, April 2008.

[397] O. Bajdechi *et al.*, "A 1.8-V ΔΣ Modulator Interface for an Electret Microphone With On-Chip Reference," *IEEE J. of Solid-State Circuits*, vol. 37, pp. 279–285, March 2002.

[398] T. Christen, T. Burger, and Q. Huang, "A 0.13μm CMOS EDGE/UMTS/WLAN Tri-Mode ΣΔ ADC with -92dB THD," *IEEE ISSCC Digest of Technical Papers*, pp. 240–241, February 2007.

[399] T. Christen and Q. Huang, "A 0.13 μm CMOS 0.1-20 MHz Bandwidth 86-70 dB DR Multi-Mode DT ΔΣ ADC for IMT-Advanced," *Proc. of the IEEE European Solid-State Circuits Conf.*, pp. 414–417, September 2010.

[400] L. Bos *et al.*, "Multirate Cascaded Discrete-Time Low-Pass ΔΣ Modulator for GSM/Bluetooth/UMTS," *IEEE J. of Solid-State Circuits*, vol. 45, pp. 1198–1208, June 2010.

[401] A. Morgado, R. del Río, J. M. de la Rosa, L. Bos, J. Ryckaert, and G. van der Plas, "A 100kHz-10MHz BW, 78-to-52dB DR,4.6-to-11mW Flexible SC ΣΔ Modulator in 1.2-V 90-nm CMOS," *Proc. of the IEEE European Solid-State Circuits Conf.*, pp. 418–421, September 2010.

[402] C.-Y. Ho *et al.*, "A Quadrature Bandpass Continuous-Time Delta-Sigma Modulator for a Tri-Mode GSM-EDGE/UMTS/DVB-T Receiver," *IEEE J. of Solid-State Circuits*, vol. 46, pp. 2571–2582, November 2011.

[403] R. van Veldhoven, "A Triple-mode Continuous-time ΣΔ Modulator with Switched-capacitor Feedback DAC for a GSM-EDGE/CDMA2000/UMTS Receiver," *IEEE J. of Solid-State Circuits*, vol. 38, pp. 2069–2076, December 2003.

[404] S. Ouzounov *et al.*, "A 1.2V 121-Mode CT Delta-Sigma Modulator for Wireless Receivers in 90nm CMOS," *IEEE ISSCC Digest of Technical Papers*, pp. 242–243, February 2007.

[405] P. Crombez *et al.*, "A Single-Bit 500 kHz-10 MHz Multimode Power-Performance Scalable 83-to-67 dB DR CT ΔΣ Modulator for SDR in 90 nm Digital CMOS," *IEEE J. of Solid-State Circuits*, vol. 45, pp. 1159–1171, June 2010.

[406] Y. Ke *et al.*, "A 2.8-to-8.5mW GSM/Bluetooth/UMTS/DVB-H/WLAN Fully Reconfigurable CT ΔΣ with 200kHz to 20MHz BW for 4G Radios in 90nm Digital CMOS," *IEEE Symp. VLSI Circuits. Digest of Technical Papers*, pp. 153–154, 2010.

[407] V. Singh *et al.*, "A 16MHz BW 75dB DR CT ΔΣ ADC Compensated for More than One Cycle Excess Loop Delay," *Proc. of the IEEE Custom Integrated Circuits Conf.*, September 2011.

[408] Z. Qiao, Z. Zhou, and Q. Li, "A 250mV 77dB DR 10kHz BW SC ΔΣ Modulator Exploiting Subthreshold OTAs," *Proc. of the IEEE European Solid-State Circuits Conf.*, pp. 419–422, September 2014.

[409] Y. Yoon, D. Choi, and J. Roh, "A 0.4 V 63 W 76.1 dB SNDR 20 kHz Bandwidth Delta-Sigma Modulator Using a Hybrid Switching Integrator," *IEEE J. of Solid-State Circuits*, vol. 50, pp. 2342–2352, October 2015.

[410] T. Oshita *et al.*, "A Compact First-Order ΣΔ Modulator for Analog High-Volume Testing of Complex System-on-Chips in a 14 nm Tri-Gate Digital CMOS Process," *IEEE J. of Solid-State Circuits*, vol. 51, pp. 378–390, February 2016.

[411] Y. Yoon, H. Roh, and J. Roh, "A True 0.4-V Delta-Sigma Modulator Using a Mixed DDA Integrator Without Clock Boosted Switches," *IEEE Trans. on Circuits and Systems II: Express Briefs*, vol. 61, pp. 229–233, April 2014.

[412] M. Yonekura and H. Ishikuro, "I/Q Mismatch Compensation ΔΣ Modulator Using Ternary Capacitor Rotation Technique," *Proc. of the IEEE European Solid-State Circuits Conf.*, pp. 229–232, September 2015.

[413] M. D. Bock and P. Rombouts, "A Double-Sampling Cross Noise-coupled Sigma Delta Modulator with a Reduced Amount of Opamps," *Proc. of the IEEE Custom Integrated Circuits Conf.*, 2013.

[414] S. Porrazzo *et al.*, "A 1-V 99-to-75dB SNDR, 256Hz–16kHz bandwidth, 8.6-to-39μW Reconfigurable SC ΔΣ Modulator for Autonomous Biomedical Applications," *Proc. of the IEEE European Solid-State Circuits Conf.*, pp. 367–370, September 2013.

[415] X. Meng *et al.*, "A 19.2-mW, 81.6-dB SNDR, 4-MHz Bandwidth Delta-Sigma Modulator With Shifted Loop Delays," *Proc. of the IEEE European Solid-State Circuits Conf.*, pp. 221–224, September 2015.

[416] A. Nilchi *et al.*, "A Low-Power Delta-Sigma Modulator Using a Charge-Pump Integrator," *IEEE Trans. on Circuits and Systems – I: Regular Papers*, vol. 60, pp. 1310–1321, May 2013.

[417] D. Behera and N. Krishnapura, "A 2-Channel 1 MHz BW, 80.5 dB DR ADC using a ΔΣ Modulator and Zero-ISI Filter," *Proc. of the IEEE European Solid-State Circuits Conf.*, pp. 415–418, September 2014.

[418] M. Grassi *et al.*, "A Multi-Mode SC Audio ΣΔ Modulator for MEMS Microphones with Reconfigurable Power Consumption, Noise-Shaping Order, and DR," *Proc. of the IEEE European Solid-State Circuits Conf.*, pp. 245–248, September 2016.

[419] S. Wu and J. Wu, "A 81-dB Dynamic Range 16-MHz Bandwidth ΔΣ Modulator Using Background Calibration," *IEEE J. of Solid-State Circuits*, vol. 48, pp. 2170–2179, September 2013.

[420] J. Melo, J. Goes, and N. Paulino, "A 0.7 V 256 μW ΔΣ Modulator With Passive RC Integrators Achieving 76 dB DR in 2 MHz BW," *IEEE Symposium VLSI Circuits. Digest of Technical Papers*, pp. C290–C291, December 2015.

[421] A. Jain and S. Pavan, "A 13.3mW 60MHz Bandwidth, 76dB DR 6GS/s CTΔΣM with Time Interleaved FIR Feedback," *IEEE Symposium VLSI Circuits. Digest of Technical Papers*, December 2016.

[422] S. Billa, A. Sukumaran, and S. Pavan, "A 280μW 24kHz-BW 98.5dB-SNDR Chopped Single-Bit CT ΔΣM Achieving <10Hz 1/f Noise Corner Without Chopping Artifacts," *IEEE ISSCC Digest of Technical Papers*, pp. 276–277, February 2016.

[423] A. Sukumaran and S. Pavan, "Design of Continuous-Time ΔΣ Modulators With Dual Switched-Capacitor Return-to-Zero DACs," *IEEE J. of Solid-State Circuits*, vol. 51, pp. 1619–1629, July 2016.

[424] T. Nandi, K. Boominathan, and S. Pavan, "A Continuous-time ΔΣ Modulator with 87 dB Dynamic Range in a 2 MHz Signal Bandwidth Using a Switched-Capacitor Return-to-Zero DAC," *Proc. of the IEEE Custom Integrated Circuits Conf.*, 2012.

[425] T. Nandi, K. Boominathan, and S. Pavan, "Continuous-Time Modulators With Improved Linearity and Reduced Clock Jitter Sensitivity Using the Switched-Capacitor Return-to-Zero DAC," *IEEE J. of Solid-State Circuits*, vol. 48, pp. 1795–1805, August 2013.

[426] J. Garcia, S. Rodriguez, and A. Rusu, "A Low-Power CT Incremental 3rd Order ΣΔ ADC for Biosensor Applications," *IEEE Trans. on Circuits and Systems – I: Regular Papers*, vol. 60, pp. 25–36, January 2013.

[427] S. Zeller, C. Muenker, and R. Weigel, "A 0.039mm^2 Inverter-Based 1.82mW 68.6dB-SNDR 10MHz-BW CT-ΣΔ-ADC in 65nm CMOS," *Proc. of the IEEE European Solid-State Circuits Conf.*, pp. 319–322, September 2013.

[428] C. Ding, Y. Manoli, and M. Keller, "A 5.1mW 74dB DR CT ΔΣ Modulator with Quantizer Intrinsic ELD Compensation Achieving 75fJ/conv.-step in a 20MHz BW," *Proc. of the IEEE European Solid-State Circuits Conf.*, pp. 213–216, September 2015.

[429] Y. Zhang *et al.*, "A Continuous-Time Delta-Sigma Modulator for Biomedical Ultrasound Beamformer Using Digital ELD Compensation and FIR Feedback," *IEEE Trans. on Circuits and Systems I: Regular Papers*, vol. 62, pp. 1689–1698, July 2015.

[430] T. Kaneko *et al.*, "A 76-dB-DR 6.8-mW 20-MHz Bandwidth CT ΔΣ ADC with a High-linearity Gm-C Filter," *Proc. of the IEEE European Solid-State Circuits Conf.*, pp. 253–256, September 2016.

[431] C. Lo *et al.*, "A 75.1dB SNDR 840MS/s CT ΔΣ Modulator with 30MHz Bandwidth and 46.4fJ/conv FOM in 55nm CMOS," *IEEE Symp. VLSI Circuits. Digest of Technical Papers*, pp. C60–C61, December 2013.

[432] S. Wu *et al.*, "A 160MHz-BW 72dB-DR 40mW Continuous-Time ΔΣ Modulator in 16nm CMOS with Analog ISI-Reduction Technique," *IEEE ISSCC Digest of Technical Papers*, vol. 280-281, February 2016.

[433] K. Matsukawa *et al.*, "A 10 MHz BW 50 fJ/conv. Continuous Time ΔΣ Modulator with High-order Single Opamp Integrator using Optimization-based Design Method," *IEEE Symposium VLSI Circuits. Digest of Technical Papers*, pp. 160–161, December 2012.

[434] T. Kao *et al.*, "A 16nm FinFet 19/39MHz 78/72dB DR Noise-Injected Aggregated CTSDM ADC for Configurable LTE Advanced CCA/NCCA Application," *IEEE Symposium VLSI Circuits. Digest of Technical Papers*, pp. C260–C261, December 2015.

[435] C. Weng *et al.*, "A Continuous-Time Delta-Sigma Modulator Using ELD-Compensation-Embedded SAB and DWA-Inherent Time-Domain Quantizer," *IEEE J. of Solid-State Circuits*, vol. 51, pp. 1235–1245, May 2016.

[436] K. Kim, "Silicon Technologies and Solutions for the Data-Driven World," *IEEE ISSCC Digest of Technical Papers*, pp. 8–14, February 2015.

[437] T. Kim, C. Han, and N. Maghari, "A 7.2 mW 75.3 dB SNDR 10 MHz BW CT Delta-Sigma Modulator Using Gm-C-Based Noise-Shaped Quantizer and Digital Integrator," *IEEE J. of Solid-State Circuits*, vol. 51, pp. 1840–1850, August 2016.

[438] M. Anderson *et al.*, "A 9 MHz Filtering ADC with Additional 2nd-order ΔΣ Modulator Noise Suppression," *Proc. of the IEEE European Solid-State Circuits Conf.*, pp. 323–326, September 2013.

[439] C. Briseno-Vidrios *et al.*, "A 75-MHz Continuous-Time Sigma-Delta Modulator Employing a Broadband Low-Power Highly Efficient Common-Gate Summing Stage," *IEEE J. of Solid-State Circuits*, vol. 52, pp. 657–668, March 2017.

[440] R. Zanbaghi *et al.*, "An 80-dB DR, 7.2-MHz Bandwidth Single Opamp Biquad Based CT ΔΣ Modulator Dissipating 13.7-mW," . *IEEE J. of Solid-State Circuits*, vol. 48, pp. 487–501, February 2013.

[441] R. Kaald *et al.*, "A 500 MS/s 76dB SNDR Continuous Time Delta Sigma Modulator with 10MHz Signal Bandwidth in 0.18μm CMOS," *Proc. of the IEEE Custom Integrated Circuits Conf.*, 2013.

[442] C. Briseno-Vidrios *et al.*, "A 75 MHz BW 68dB DR CT-ΣΔ Modulator with Single Amplifier Biquad Filter and A Broadband Low-power Common-gate Summing Technique," *IEEE Symposium VLSI Circuits. Digest of Technical Papers*, pp. C254–C255, December 2015.

[443] C. Weng *et al.*, "An 8.5MHz 67.2dB SNDR CTDSM with ELD Compensation Embedded Twin-T SAB and Circular TDC-based Quantizer in 90nm CMOS," *IEEE Symposium VLSI Circuits. Digest of Technical Papers*, December 2014.

[444] C.-H. Weng, Y.-Y. Lin, and T.-H. Lin, "A 1-V 5-MHz Bandwidth 68.3-dB SNDR Continuous-Time Delta-Sigma Modulator With a Feedback-Assisted Quantizer," *IEEE Trans. on Circuits and Systems I: Regular Papers*, vol. 64, pp. 1085–1093, May 2017.

[445] V. Singh *et al.*, "A 16 MHz BW 75 dB DR CT ΔΣ ADC Compensated for More Than One Cycle Excess Loop Delay," *IEEE J. of Solid-State Circuits*, vol. 47, pp. 1884–1895, August 2012.

[446] R. Ritter *et al.*, "A Multimode CT ΔΣ-Modulator with a Reconfigurable Digital Feedback Filter for Semi-Digital Blocker/Interferer Rejection," *Proc. of the IEEE European Solid-State Circuits Conf.*, pp. 225–228, September 2015.

[447] J. Huang *et al.*, "A 10-MHz Bandwidth 70-dB SNDR 640MS/s Continuous-Time ΣΔ ADC Using Gm-C Filter with Nonlinear Feedback DAC Calibration," *Proc. of the IEEE Custom Integrated Circuits Conf.*, 2013.

[448] A. Bandyopadhyay *et al.*, "A 97.3 dB SNR, 600 kHz BW, 31mW Multibit Continuous Time ΔΣ ADC," *IEEE Symposium VLSI Circuits. Digest of Technical Papers*, December 2014.

[449] J. Huang, S. Yang, and J. Yuan, "A 75 dB SNDR 10-MHz Signal Bandwidth Gm-C-Based Sigma-Delta Modulator With a Nonlinear Feedback Compensation Technique," *IEEE Trans. on Circuits and Systems I: Regular Papers*, vol. 62, pp. 2216–2226, September 2015.

[450] Y. Hu *et al.*, "A Continuous-Time ΔΣ ADC Utilizing TimeInformation for Two Cycles of Excess Loop Delay Compensation," *IEEE Trans. on Circuits and Systems – II: Express Briefs*, vol. 62, pp. 1063–1067, November 2015.

[451] Y. Xu *et al.*, "Dual-mode 10MHz BW 4.8/6.3mW Reconfigurable Lowpass/ Complex Bandpass CT ΣΔ Modulator with 65.8/74.2dB DR for a Zero/Low-IF SDR Receiver," *Proc. of the IEEE Radio Frequency Integrated Circuits Symp.*, pp. 313–316, 2014.

[452] H. Chae and M. Flynn, "A 69 dB SNDR, 25 MHz BW, 800 MS/s Continuous-Time Bandpass ΔΣ Modulator Using a Duty-Cycle-Controlled DAC for Low Power and Reconfigurability," *IEEE J. of Solid-State Circuits*, vol. 51, pp. 649–659, March 2016.

[453] Y. Zhang *et al.*, "A 54.4-mW 4th-order Quadrature Bandpass CT ΣΔ Modulator with 33-MHz BW and 10-bit ENOB for a GNSS receiver," *Proc. of the IEEE Radio Frequency Integrated Circuits Symp.*, pp. 343–346, 2015.

[454] J. Jeong, N. Collins, and M. Flynn, "A 260 MHz IF Sampling Bit-Stream Processing Digital Beamformer With an Integrated Array of Continuous-Time Band-Pass ΔΣ Modulators," *IEEE J. of Solid-State Circuits*, vol. 51, pp. 1168–1176, May 2016.

[455] M. Englund *et al.*, "A 2.5-GHz 4.2-dB NF Direct ΔΣ Receiver with a Frequency-Translating Integrator," *Proc. of the IEEE European Solid-State Circuits Conf.*, pp. 371–374, September 2014.

[456] N. Narasimman and T. Kim, "A 0.3 V, 49 fJ/Conv.-step VCO-based Delta Sigma Modulator with Self-compensated Current Reference for Variation Tolerance," *Proc. of the IEEE European Solid-State Circuits Conf.*, pp. 237–240, September 2016.

[457] Y. Chang *et al.*, "A 379nW 58.5dB SNDR VCO-Based ΔΣ Modulator for Bio-Potential Monitoring," *IEEE Symposium VLSI Circuits. Digest of Technical Papers*, pp. C66–C67, December 2013.

[458] Y. Yoon *et al.*, "A 0.04-mm^2 0.9-mW 71-dB SNDR Distributed Modular ΔΣ ADC with VCO-based Integrator and Digital DAC Calibration," *Proc. of the IEEE Custom Integrated Circuits Conf.*, 2015.

[459] K. Reddy *et al.*, "A 54mW 1.2GS/s 71.5dB SNDR 50MHz BW VCO-Based CT ΔΣ ADC Using Dual Phase/Frequency Feedback in 65nm CMOS," *IEEE Symposium VLSI Circuits. Digest of Technical Papers*, pp. C256–C257, December 2015.

[460] Y. Wu *et al.*, "A 103fs-rms 1.32mW 50MS/s 1.25MHz Bandwidth Two-Step Flash-ΔΣ Time-to-Digital Converter for ADPLL," *Proc. of the IEEE Radio Frequency Integrated Circuits Symp.*, pp. 95–98, 2015.

[461] K. Ragab and N. Sun, "A 12b ENOB, 2.5MHz-BW, 4.8mW VCO-Based 0-1 MASH ADC with Direct Digital Background Nonlinearity Calibration," *Proc. of the IEEE Custom Integrated Circuits Conf.*, 2015.

[462] C. Wang *et al.*, "A Mode-Configurable Analog Baseband for Wi-Fi 11ac Direct- Conversion Receiver Utilizing a Single Filtering ΔΣ ADC," *Proc. of the IEEE Radio Frequency Integrated Circuits Symp.*, pp. 170–173, 2016.

[463] G. Wei *et al.*, "A 13-ENOB, 5 MHz BW, 3.16 mW Multi-Bit Continuous-Time ΔΣ ADC in 28 nm CMOS with Excess-Loop-Delay Compensation Embedded in SAR Quantizer," *IEEE Symposium VLSI Circuits. Digest of Technical Papers*, pp. C292–C293, December 2015.

[464] Z. Chen, M. Miyahara, and A. Matsuzawa, "A Stability-Improved Single-Opamp Third-Order ΣΔ Modulator by Using a Fully-Passive Noise-Shaping SAR ADC and Passive Adder," *Proc. of the IEEE European Solid-State Circuits Conf.*, pp. 249–252, September 2016.

[465] B. Zhan *et al.*, "A 91.2dB SNDR 66.2fJ/Conv. dynamic amplifier based 24kHz ΔΣ Modulator," *Proc. of the IEEE Asian Solid-State Circuits Conf.*, pp. 317–320, 2016.

[466] J. Fredenburg and M. Flynn, "A 90-MS/s 11-MHz-Bandwidth 62-dB SNDR Noise-Shaping SAR ADC," *IEEE J. of Solid-State Circuits*, vol. 47, pp. 2898–2904, December 2012.

[467] I. Ahmed *et al.*, "A Low-Power Gm-C-Based CT-ΔΣ Audio-Band ADC in 1.1V 65nm CMOS," *IEEE Symposium VLSI Circuits. Digest of Technical Papers*, pp. C294–C295, December 2015.

[468] S. Lee, B. Elies, and Y. Chiu, "An 85dB SFDR 67dB SNDR 8OSR 240MS/s ΣΔ ADC," *IEEE Symposium VLSI Circuits. Digest of Technical Papers*, pp. 164–165, December 2012.

[469] K. Yamamoto and A. Carusone, "A 1-1-1-1 MASH Delta-Sigma Modulator With Dynamic Comparator-Based OTAs," *IEEE J. of Solid-State Circuits*, vol. 47, pp. 1866–1883, August 2012.

[470] T. Oh, N. Maghari, and U.-K. Moon, "A 5MHz BW 70.7dB SNDR Noise-Shaped Two-Step Quantizer Based ΔΣ ADC," *IEEE Symposium VLSI Circuits. Digest of Technical Papers*, pp. 162–163, December 2012.

[471] J. Hamilton, S. Yan, and T. Viswanathan, "An Uncalibrated 2MHz, 6mW, 63.5dB SNDR Discrete-Time Input VCO-Based ΔΣ ADC," *Proc. of the IEEE Custom Integrated Circuits Conf.*, 2012.

[472] S. Loeda *et al.*, "A 10/20/30/40 MHz Feed-Forward FIR DAC Continuous-Time ΔΣ ADC with Robust Blocker Performance for Radio Receivers," *IEEE Symposium VLSI Circuits. Digest of Technical Papers*, pp. C262–C263, December 2015.

APPENDIX A

STATE-SPACE ANALYSIS OF CLOCK JITTER IN CT-ΣΔMs

This appendix presents a summary of the state-space formulation proposed by Oliaei [1] as a way to analyze the effect of clock-jitter error on CT-ΣΔMs. Some expressions derived from this analysis are used in Chapter 4 to explain the lingering effect of clock jitter.[1]

A.1 State-space Representation of NTF(z)

Let us consider the conceptual block diagram of a CT-ΣΔM like that shown in Figure A.1. Using the impulse-invariant transformation [3], the DT equivalent of the loop filter transfer function, $H(z)$, can be written as:

$$H(z) = \frac{n_{(L-1)}z^{L-1} + \ldots + n_2 z^2 + n_1 z + n_0}{d_L z^L + d_{(L-1)}z^{L-1} + \ldots + d_2 z^2 + d_1 z + d_0} \tag{A.1}$$

[1]The interested reader can see the complete analysis in the literature [2].

Sigma-Delta Converters: Practical Design Guide, Second Edition. José M. de la Rosa.

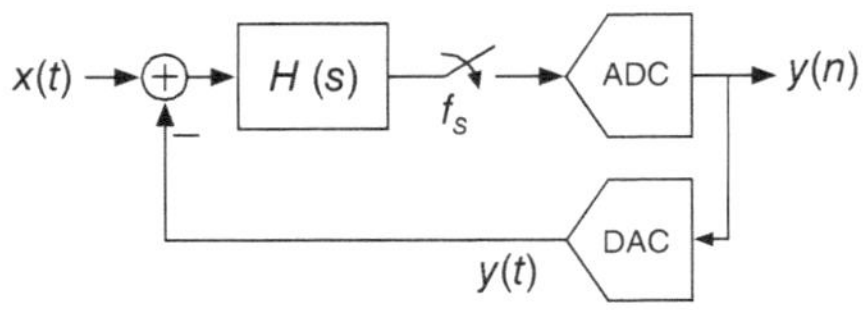

Figure A.1 Conceptual diagram of a CT-ΣΔM.

where n_i and d_i are function of the coefficients of the original CT modulator loop filter. Therefore, NTF(z) can be derived from Equation (A.1) as:

$$\begin{aligned} \text{NTF}(z) &= \frac{1}{1+H(z)} = \\ &= \frac{d_L + d_{(L-1)}z^{-1} + \ldots + d_1 z^{-(L-1)} + d_0 z^{-L}}{d_L + (d_{(L-1)} + n_{(L-1)})z^{-1} + \ldots(d_1 + n_1)z^{-(L-1)} + (d_0 + n_0)z^{-L}} \end{aligned} \tag{A.2}$$

Figure A.2 shows the transpose of the Direct II implementation of Equation (A.2) [4]. This block diagram does not have any direct correspondence with the physical implementation of the modulator and it is only used to represent the relationship between the input, $e(n)$, the output, $q(n)$, and the state variables of NTF(z). The so-called state-space representation of NTF(z) can be implemented by the block diagram in Figure A.3, which is described by the following finite difference equations [4]:

$$\begin{aligned} \overline{v(n+1)_0} &= \overline{F_0} \cdot \overline{v(n)_0} + \overline{p_0} \cdot e(n) \\ q(n) &= \overline{g_0}^T \cdot \overline{v(n)_0} + e(n) \end{aligned} \tag{A.3}$$

where $\overline{F_0}$ is the state matrix, $\overline{v(n)_0}$ is an $L \times 1$ state vector, $\overline{p_0}$ and $\overline{g_0}$ are $L \times 1$ vectors, respectively given by:

$$\overline{F_0} = \begin{bmatrix} 0 & 0 & 0 & \ldots & 0 & -\dfrac{n_0+d_0}{d_L} \\ 1 & 0 & 0 & \ldots & 0 & -\dfrac{n_1+d_1}{d_L} \\ 0 & 1 & 0 & \ldots & 0 & -\dfrac{n_2+d_2}{d_L} \\ 0 & 0 & 1 & \ldots & 0 & -\dfrac{n_3+d_3}{d_L} \\ \ldots & \ldots & \ldots & \ldots & \ldots & \ldots \\ 0 & 0 & 0 & \ldots & 1 & -\dfrac{n_{(L-1)}+d_{(L-1)}}{d_L} \end{bmatrix} \tag{A.4}$$

$$\overline{p_0} = \begin{bmatrix} \dfrac{n_0}{d_L} \\ \dfrac{n_1}{d_L} \\ \ldots \\ \dfrac{n_{(L-1)}}{d_L} \end{bmatrix}, \overline{g_0} = \begin{bmatrix} 0 \\ 0 \\ \ldots \\ 1 \end{bmatrix} \tag{A.5}$$

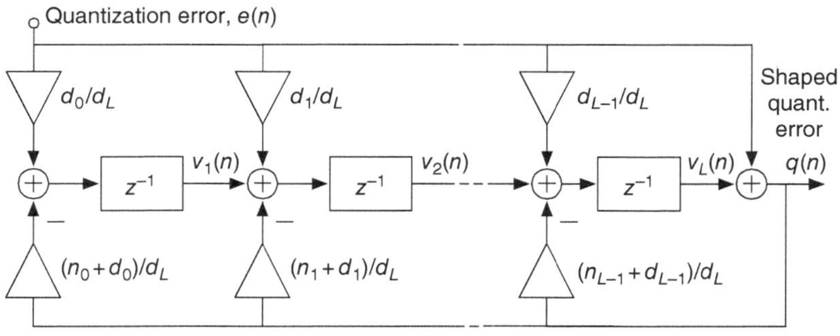

Figure A.2 Transpose of the Direct II implementation of NTF(z).

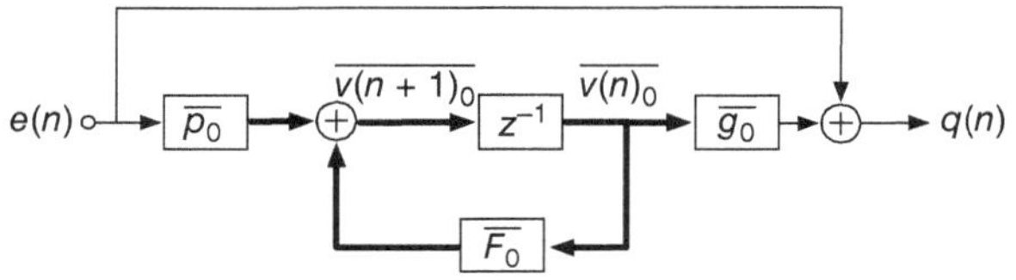

Figure A.3 State-space representation of NTF(z).

Equation system (A.3) can be solved recursively to find the relation between the initial state $\overline{v(n)}_0$, previous inputs $e(k)$, current input $e(n)$, and output $q(n)$ of the system [4]. This gives:

$$q(n) = e(n) + \overline{g_0}^T \cdot \overline{F_0}^n \cdot \overline{v(n)}_0 + \sum_{k=0}^{n-1} [\overline{g_0}^T \cdot \overline{F_0}^{(n-1-k)} \cdot \overline{p_0} \cdot e(k)] \tag{A.6}$$

A.2 Expectation Value of $(\Delta q_n)^2$

In order to compute the power of the jitter error analyzed in Chapter 4, it is necessary to derive the mathematical expectation of $(\Delta q_n)^2$, which is given by:

$$\mathcal{E}\{\Delta q_n^2\} \equiv \mathcal{E}\{[q(n) - q(n-1)]^2\} = 2 \cdot [\mathcal{E}\{q(n)^2\} - \mathcal{E}\{q(n) \cdot q(n-1)\}] \tag{A.7}$$

where the fact that $\mathcal{E}\{q(n)^2\} = \mathcal{E}\{q(n-1)^2\}$ has been taken into account.

Assuming $\overline{v(n)}_0 = 0$ in Equation (A.6) and considering that $\mathcal{E}\{e(k)e(j)\} = 0$ for $k \neq j$, it can be shown that [2]:

$$\mathcal{E}\{q^2(n)\} = \mathcal{E}\{e^2(n)\} \cdot [1 + \sum_{k=0}^{n-1} (\overline{g_0}^T \cdot \overline{F_0}^{(n-1-k)} \cdot \overline{p_0})^2] \tag{A.8}$$

$$\mathcal{E}\{q(n)q(n-1)\} = \mathcal{E}\{e^2(n)\} \cdot [\overline{g_0}^T \cdot \overline{p_0} + \sum_{k=0}^{n-1} (\overline{g_0}^T \cdot \overline{F_0}^{(n-1-k)} \cdot \overline{p_0} \cdot \overline{g_0}^T \cdot \overline{F_0}^{(n-2-k)} \cdot \overline{p_0})] \tag{A.9}$$

Applying diagonalization to the state-space representation of the NTF, the system equation in (A.3) transforms into:

$$\begin{aligned}\overline{v(n+1)} &= \overline{F} \cdot \overline{v(n)} + \overline{p} \cdot e(n)\\ q(n) &= \overline{g}^T \cdot \overline{v(n)} + e(n)\end{aligned} \tag{A.10}$$

where $\overline{v(n)}$ is the state vector of the diagonalized system and $\overline{F}$, $\overline{p}$, and $\overline{g}$ are respectively given by:

$$\overline{F} = \overline{T}^{-1} \cdot \overline{F_0} \cdot \overline{T} = \begin{bmatrix} \lambda_1 & 0 & 0 & \dots & 0\\ 0 & \lambda_2 & 0 & \dots & 0\\ 0 & 0 & \lambda_3 & \dots & 0\\ \dots & \dots & \dots & \dots & \dots\\ 0 & 0 & 0 & \dots & \lambda_L \end{bmatrix} \tag{A.11}$$

$$\overline{p} = \overline{T}^{-1} \cdot \overline{p_0} = \begin{bmatrix} p_1\\ p_2\\ \dots\\ p_L \end{bmatrix}, \overline{g}^T = \overline{g_0}^T \cdot \overline{T} = \begin{bmatrix} g_1 & g_2 & \dots & g_L \end{bmatrix} \tag{A.12}$$

where $\overline{T}$ is the matrix formed by the eigenvectors of $\overline{F}$, λ_j are the eigenvalues, and L is the order of the modulator loop filter.

Taking into account the diagonalized versions of $\overline{F_0}$, $\overline{p}$, and $\overline{g}$, the expressions in Equations (A.8) and (A.9) can be written as:

$$\mathcal{E}\{q^2(n)\} = \mathcal{E}\{e^2(n)\} \cdot \left(1 - \sum_{k=1}^{L}\sum_{j=1}^{L} g_k \cdot p_k \cdot g_j \cdot p_j \cdot \frac{\lambda_k{}^{-1} \cdot \lambda_j{}^{-1}}{1 - \lambda_k{}^{-1} \cdot \lambda_j{}^{-1}}\right) \tag{A.13}$$

$$\mathcal{E}\{q(n)q(n-1)\} = \mathcal{E}\{e^2(n)\} \cdot \left(\overline{g}^T \cdot \overline{p} - \sum_{k=1}^{L}\sum_{j=1}^{L} g_k \cdot p_k \cdot g_j \cdot p_j \cdot \frac{\lambda_k{}^{-1}}{1 - \lambda_k{}^{-1} \cdot \lambda_j{}^{-1}}\right) \tag{A.14}$$

A.3 In-band Noise Power due to Clock Jitter

Replacing the above expressions in Equation (A.7), and considering that

$$\mathcal{E}\{e^2(n)\} = \frac{X_{\mathrm{FS}}{}^2}{12(2^B - 1)^2} \tag{A.15}$$

the expectation of Δq_n^2 can be written as:

$$\mathcal{E}\{\Delta q_n^2\} = \frac{X_{\mathrm{FS}}{}^2}{6(2^B - 1)^2} \cdot \overline{\Psi}(\overline{g}, \overline{p}, \overline{\lambda}, L) \tag{A.16}$$

where

$$\overline{\Psi}(\overline{g}, \overline{p}, \overline{\lambda}, L) \equiv 1 - \overline{g}^T \cdot \overline{p} + \sum_{k=1}^{L}\sum_{j=1}^{L} g_k \cdot p_k \cdot g_j \cdot p_j \cdot \frac{\lambda_k{}^{-1} \cdot (1 - \lambda_j{}^{-1})}{1 - \lambda_k{}^{-1} \cdot \lambda_j{}^{-1}} \tag{A.17}$$

The expression above provides a method to compute $\mathcal{E}\{\Delta q_n^2\}$ and hence, the quantization IBN from the eigenvalues of a $L \times L$ matrix, which involves the summation of L^2 terms, usually with $L < 5$ to keep the ΣΔM loop filter stable.

In this way, the IBN degraded by clock jitter can be computed by applying the above formulation in Equations (4.52) and (4.53), yielding:

$$\mathrm{IBN}_{\epsilon_Q} = \sigma_j{}^2 B_w \left[\frac{(2\pi A_{\mathrm{in}} f_{\mathrm{in}})^2}{f_s} + \frac{f_s X_{\mathrm{FS}}^2}{3(2^B - 1)^2} \overline{\Psi}(\overline{g}, \overline{p}, \overline{\lambda}, L) \right] \tag{A.18}$$

which, as stated in Chapter 4, includes two main terms: a signal-dependent term and a modulator-dependent term.

References

[1] O. Oliaei, "State-Space Analysis of Clock Jitter in Continuous-Time Oversampling Data Converters," *IEEE Trans. on Circuits and Systems II: Analog and Digital Signal Processing*, vol. 50, pp. 31–37, January 2003.

[2] R. Tortosa, J. M. de la Rosa, F. V. Fernández, and A. Rodríguez-Vázquez, "Clock Jitter in Multi-bit Continuous-time ΣΔ Modulators with Non-Return-to-Zero Feedback Waveform," *Elsevier Microelectronics Journal*, vol. 39, pp. 137–151, January 2008.

[3] O. Shoaei, *Continuous-Time Delta-Sigma A/D Converters for High Speed Applications*. PhD dissertation, Carleton University, 1995.

[4] J. G. Proakis and D. G. Manolakis, *Digital Signal Processing. Principles, Algorithmics and Applications*. Prentice-Hall, 1998.

APPENDIX B

SIMSIDES USER GUIDE

SIMSIDES (SIMulink-based SIgma-DElta Simulator) is a time-domain behavioral simulator for ΣΔMs that has been developed as a toolbox in the MATLAB/SIMULINK environment. SIMSIDES can be used for simulating any arbitrary ΣΔM architecture, implemented with both DT and CT circuit techniques. To this end, a complete list of ΣΔM building blocks (integrators, resonators, quantizers, embedded DACs, and so on) is included in the toolbox. The behavioral models of these building blocks take into account the most important error mechanisms of various different circuit techniques, including SC, SI, and CT circuits. These models, validated through transistor-level electrical simulations and by experimental measurements taken from a number of silicon prototypes, have been incorporated into the SIMULINK environment as C-MEX S-functions. As detailed in Chapter 5, this approach drastically increases the computational efficiency in terms of CPU time and accuracy of the simulation results.

The behavioral models included in SIMSIDES have been compiled and tested in a number of operating systems, including Mac OS X, UNIX (Solaris), Linux, and Microsoft Windows. Both 32-bit and 64-bit system platforms have been successfully tested for the majority of them. The toolbox has been updated and successfully used in a number of MATLAB/SIMULINK versions over recent years. This appendix is a user guide to SIMSIDES, giving an overview of the most significant features of the simulator.

B.1 Getting Started: Installing and Running SIMSIDES

A free copy of SIMSIDES can be downloaded from http://www.imse-cnm.csic.es/simsides. After completing the online registration form and accepting the terms and conditions, a zip file called simsides.zip is downloaded. The following steps must be followed to install the toolbox:

1) Uncompress the simsides.zip file to a directory of your computer hard disk. Let us assume that the directory is called SIMSIDES.

2) Start MATLAB.

3) Set the MATLAB search path in order to add the SIMSIDES directory. To do this, go to File menu in MATLAB and select Set Path. The Set Path dialog box opens, listing all folders on the search path. From this dialog box, click the button Add with Subfolders and select the SIMSIDES directory to add to the search path. In order to reuse the newly modified search path including the SIMSIDES directory and subdirectories, click Save, and finally click Close. This procedure–illustrated in Figure B.1a – must be done only the first time SIMSIDES is installed on the hard disk.

In order to start SIMSIDES, type simsides at the MATLAB prompt and the SIMSIDES main window is displayed, as illustrated in Figure B.1b.

B.2 Building and Editing ΣΔM Architectures in SIMSIDES

To create a new ΣΔM architecture in SIMSIDES, select File and then New Architecture in the main menu, and a new SIMULINK model window is displayed. Alternatively, an existing ΣΔM architecture can be opened by selecting File -> Open Architecture, as illustrated in Figure B.2.

In order to define a ΣΔM block diagram in SIMSIDES, the required building blocks can be incorporated from the Edit menu, as shown in Figure B.3. Both SIMULINK and SIMSIDES library models can be included by selecting Edit -> SIMULINK Library or Edit -> Add block, respectively. The latter option allows users to browse through all SIMSIDES library models. In this way, clicking on Edit -> Add block displays a new window where the user can select either ideal or real building blocks, by choosing either the Add Ideal Block or Add Real Block menus, respectively. In both cases, building-block models are organized in a set of sublibraries covering integrators, quantizers and comparators, D/A converters, resonators, and auxiliary blocks. The latter are only available in real libraries.

Some model libraries are grouped in sublibraries that contain models corresponding to different kinds of circuit implementations. For instance, if library Real Integrators is selected, a new window is displayed where the user can select the circuit technique (CT, SC, or SI) as well as the type of integrator (either FE or LD in the case of SC and SI integrators, and Gm-C, Gm-MC, active-RC, MOSFET-C in the case of CT integrators). As an illustration, Figure B.3 shows different sublibraries contained in the Real Integrators library. A complete list of model libraries and sublibraries, as well as a description of their models, is given in Appendix C.

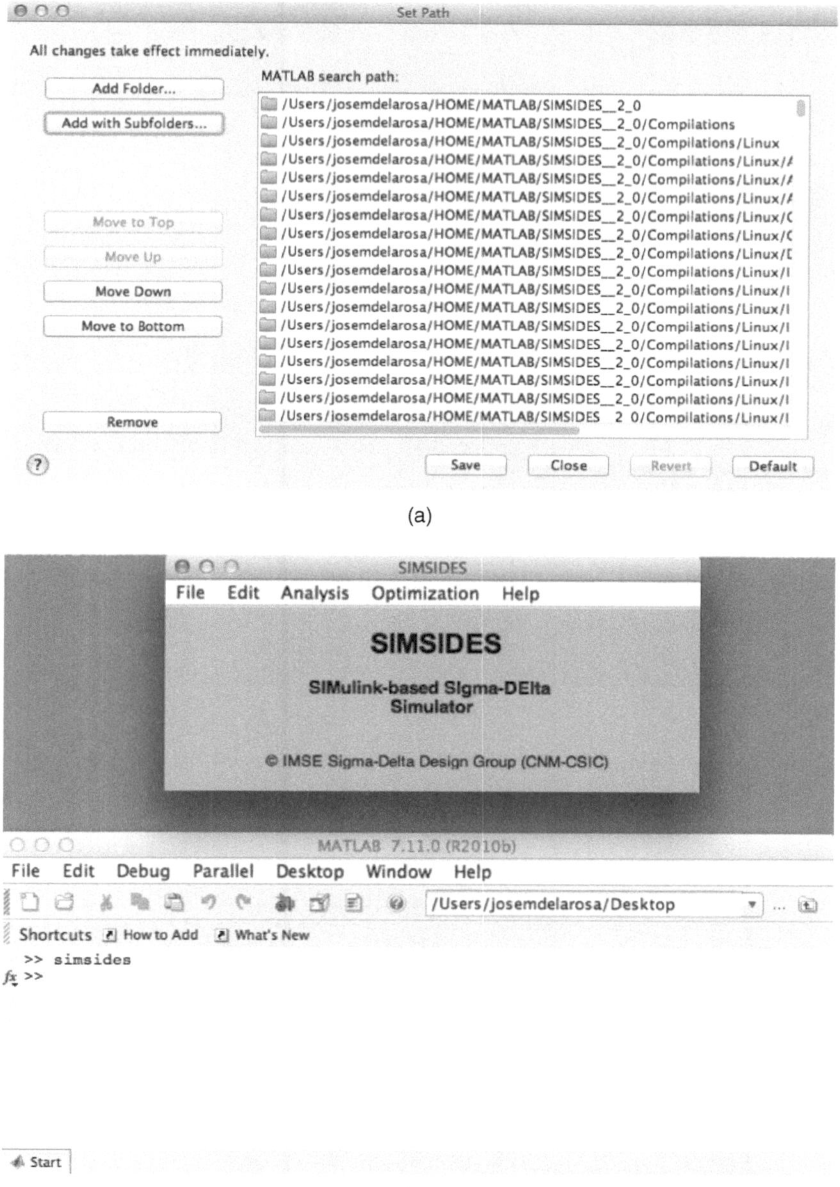

Figure B.1 Installing and starting SIMSIDES: (a) setting the MATLAB path; (b) starting SIMSIDES at the MATLAB prompt.

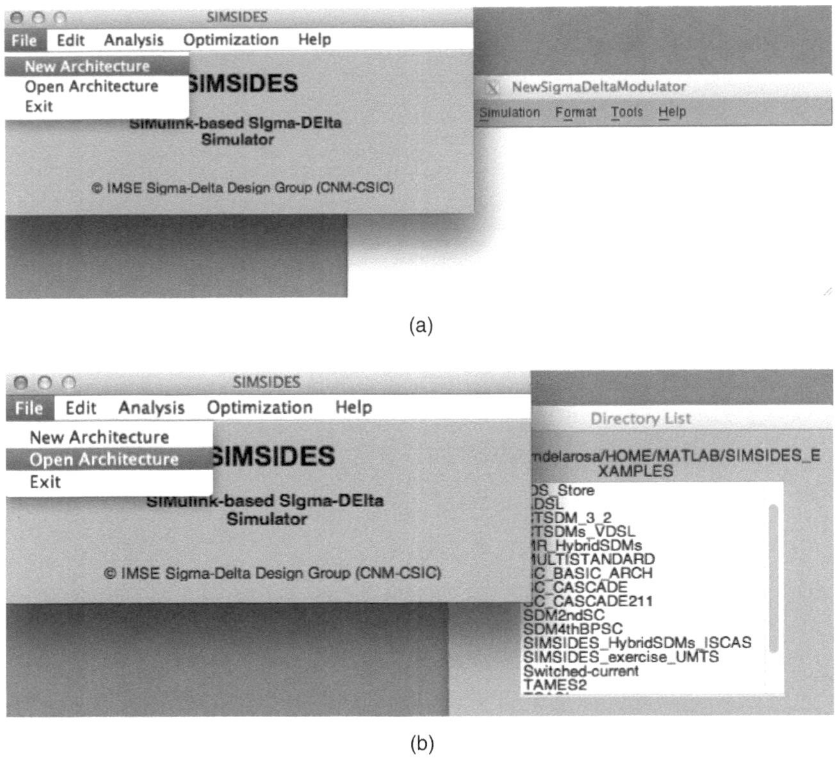

Figure B.2 Building and editing ΣΔMs in SIMSIDES: (a) creating a new ΣΔM architecture; (b) opening an existing model.

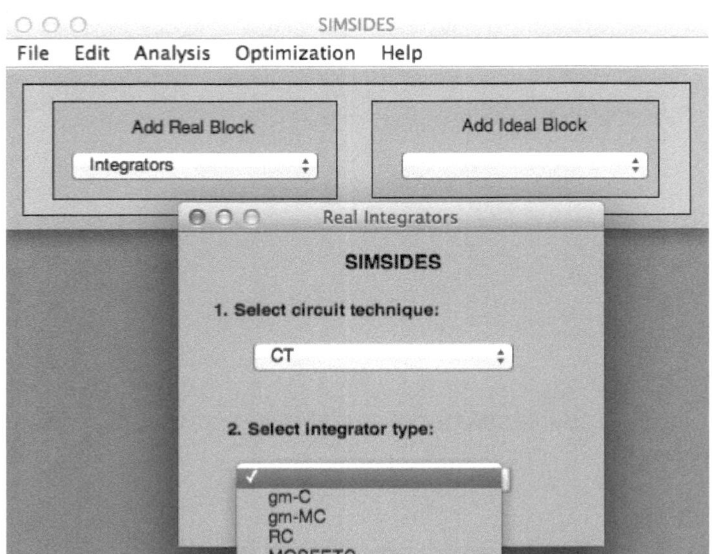

Figure B.3 Illustrating different sublibraries included in the `Real Integrators` library.

Once the ΣΔM block diagram is completed and the different building-block model parameters have been defined, the modulator can be simulated in SIMULINK following the same procedure as for the simulation of an arbitrary model in SIMULINK; that is, choosing the `Simulation -> Start` menu in the SIMULINK model window.

B.3 Analyzing ΣΔMs in SIMSIDES

Simulation output data can be post-processed in SIMULINK using the `Analysis` menu. As illustrated in Figure B.4, the `Analysis` menu includes the following submenus:

- `Node Spectrum Analysis`, which computes and plots the FFT magnitude spectrum of a given signal.
- `Integrated Power Noise`, used for calculating and graphically representing the IBN within a given signal bandwidth.
- `SNR/SNDR`, which computes the SNR and/or SNDR within the band of interest, considering both LP- and BP-ΣΔMs.
- `Harmonic Distortion`, which computes dynamic harmonic distortion figures, such as the THD and intermodulation distortion figures.
- `Histogram`, used for representing histograms and analyzing the input/output swing in ΣΔM building blocks.
- `INL/DNL`, which calculates static harmonic distortion.
- `MTPR`, used for computing the multi-tone power ratio (MTPR).
- `Parametric Analysis`, which is for simulation of the impact of a given model parameter on the performance of ΣΔMs.
- `Monte Carlo Analysis`, to do Monte Carlo simulations.

The required parameters and details of these analysis menus are described below.

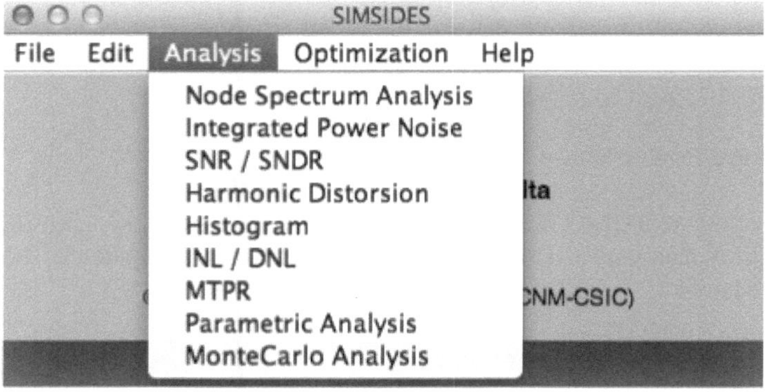

Figure B.4 Analysis menu in SIMSIDES.

B.3.1 Node Spectrum Analysis

Figure B.5a shows the SIMSIDES `Node Spectrum Analysis` window. The following parameters are required to compute the FFT magnitude spectrum:

- `Name of the signal(s) to process`, where different variable names can be introduced, separated by commas. These variables can be output data generated in the simulations (for instance, the modulator output data stream), which have been previously saved in the MATLAB workspace by using the `To Workspace` SIMULINK block.
- `Sampling frequency`: the sampling frequency in hertz.
- `Window`, which defines the window function used for computing the FFT. The main window functions available in MATLAB can be selected: Kaiser, Barlett, Blackman, Hamming, Hanning, Chebyshev, Boxcar, and triangular.
- `Number of Points`: the number of points (`N` in Figure B.5) for the selected window function and for FFT computation.
- `Window Parameters`, where other parameters required to define the window function are defined (like the `Beta` parameter used in Kaiser windows).

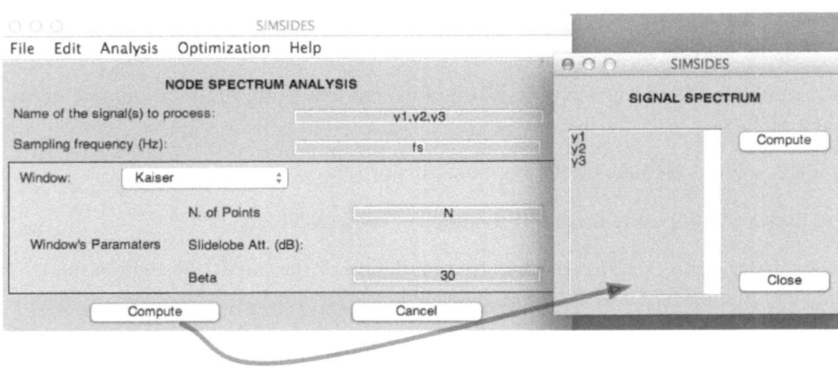

(a) (b)

Figure B.5 Node spectrum analysis menu.

Once these parameters have been defined, the output spectrum can be computed by clicking on the `Compute` button and then selecting the signal to be processed from the new window that is displayed (`Signal Spectrum` window shown in Figure B.5b).

B.3.2 Integrated Power Noise

Figure B.6 shows the SIMSIDES `Integrated Power Node` window, used for computing the IBN of any arbitrary data sequence obtained from simulations. To compute the IBN, the following parameters are required:

- `Name of the signal(s) to process.`
- `Sampling frequency`: the sampling frequency in hertz.

- `Oversampling ratio`: the value of OSR that defines the signal bandwidth in which the IBN is computed.
- `Input frequency`, where it is assumed that a single-tone input signal is applied.
- `Window Parameters`: the parameters required to define the window function used for computing the IBN.
- `Kind of Spectrum`, which specifies the signal nature: low-pass (LP) or band-pass (BP).

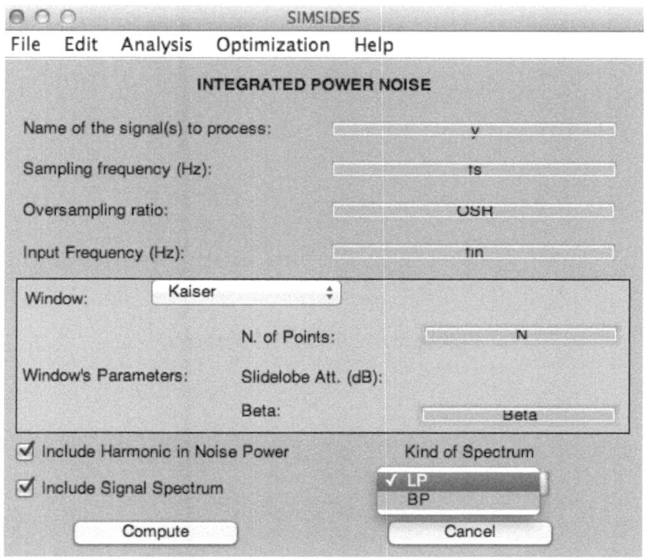

Figure B.6 Integrated power noise menu.

After defining all parameters described above, the IBN is computed by clicking on the `Compute` button. Harmonic distortion can be also taken into account in the calculation of the IBN by clicking the `Include Harmonic in Noise Power` button. The signal spectrum can be also plotted together with the IBN by choosing the `Include Signal Spectrum` option.

B.3.3 SNR/SNDR

Figure B.7 shows the SIMSIDES `SNR/SNDR` window. The parameters required to calculate the SNR/SNDR of a given signal are essentially the same as those used for computing IBN, as described in the previous section. In this case, either the SNR or the SNDR is computed depending on the `Figure of merit` selected. Note that this kind of analysis calculates the SNR/SNDR for a given value of the input signal amplitude. If a SNR-versus-amplitude curve is required, a parametric analysis should be chosen, as will be described in Section B.3.8.

B.3.4 Harmonic Distortion

Figure B.8 shows the SIMSIDES `Harmonic Distortion` window, which is used for computing the harmonic distortion power. Two different figures of merit can be calculated: the THD and

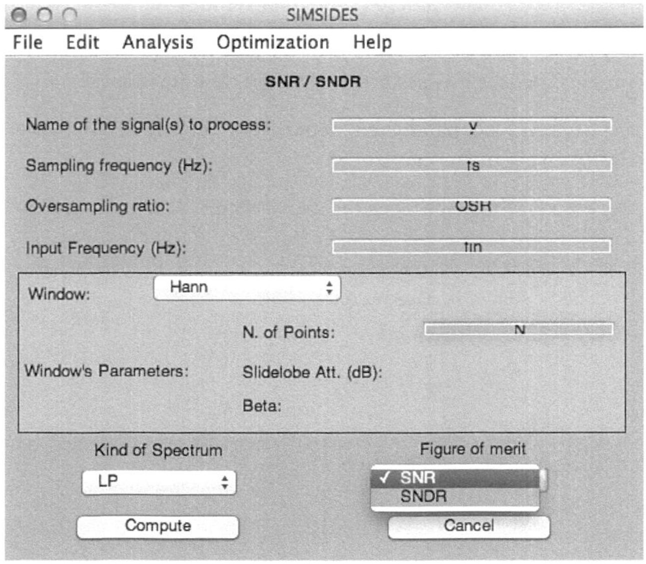

Figure B.7 SNR/SNDR menu.

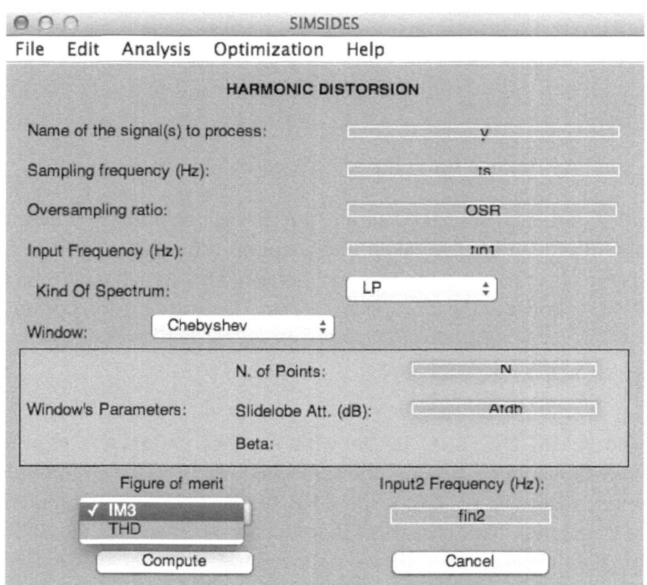

Figure B.8 Harmonic distortion analysis menu.

third-order intermodulation distortion (IM3 in Figure B.8). The latter requires a two-tone input signal. For that reason, there is an additional parameter called Input2 Frequency, which defines the frequency of the second input tone.

B.3.5 Integral and Differential Non-Linearity

The INTEGRAL AND DIFFERENTIAL NON-LINEARITY menu, illustrated in Figure B.9, is used for characterizing the static linearity in SIMSIDES. The analysis is based on either Histograms or Input Ramp Waveform, as selected by the user. Other parameters required to do this analysis are the Input Amplitude and the Number of bits, which specifies the ideal resolution of the A/D conversion, expressed in bits.

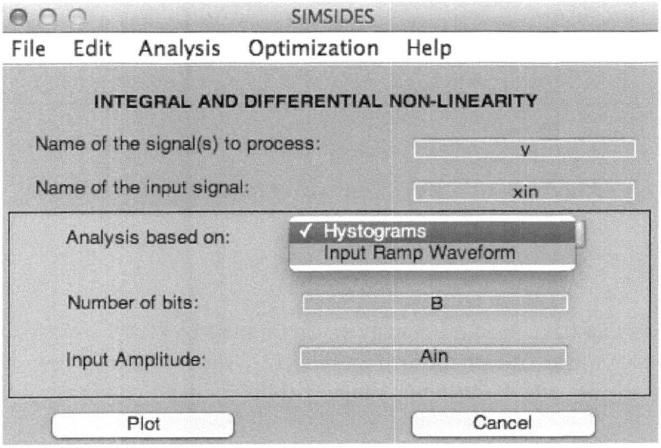

Figure B.9 Integral and differential non-linearity analysis menu.

B.3.6 Multi-tone Power Ratio

SIMSIDES can also analyze the harmonic distortion in those telecom applications such as ADSL, where a discrete multi-tone (DMT) signal is used. In this case, the linearity of the system is measured by a figure called the multi-tone power ratio (MTPR). The corresponding SIMSIDES menu – shown in Figure B.10 – allows the user to compute MTPR for DMT input signals of different types:

- Suppressing 1 carrier of each 16: 1 out of 16 carrier channels are suppressed.
- Suppressing 8 carrier of each 128: 8 out of 128 carrier channels are suppressed.
- Suppressing 16 carrier of each 256: 16 out of 256 carrier channels are suppressed.

In addition, the following parameters are also needed to compute MTPR:

- Number of carriers, which stands for the number of carrier channels into which the DMT signal is divided.
- Bins by carrier: the number of bins assigned to each carrier channel in the FFT.

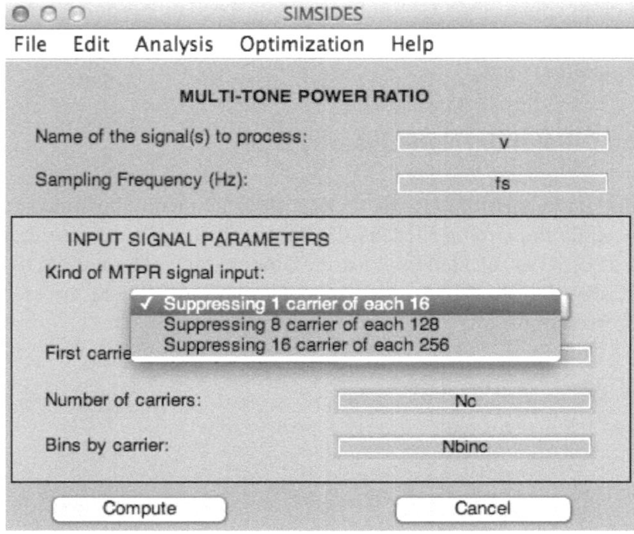

Figure B.10 Multi-tone power ratio analysis menu.

B.3.7 Histogram

Histograms of signals that have been previously saved on the MATLAB workspace can be computed using the `HISTOGRAM` menu (illustrated in Figure B.11), where the `Number of bins` specifies the number of intervals into which the signal range will be divided to compute the histogram.

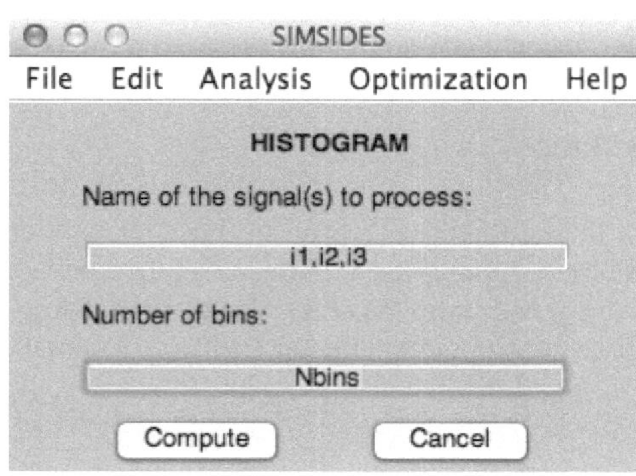

Figure B.11 Histogram analysis menu.

B.3.8 Parametric Analysis

Figure B.12 shows the SIMSIDES `PARAMETRIC ANALYSIS` menu. This menu is used for analyzing the impact of varying a model parameter on the performance of ΣΔMs. Either one parameter or two parameters can be varied simultaneously by selecting the `Second Parameter` option. For each parameter, the following data must be specified:

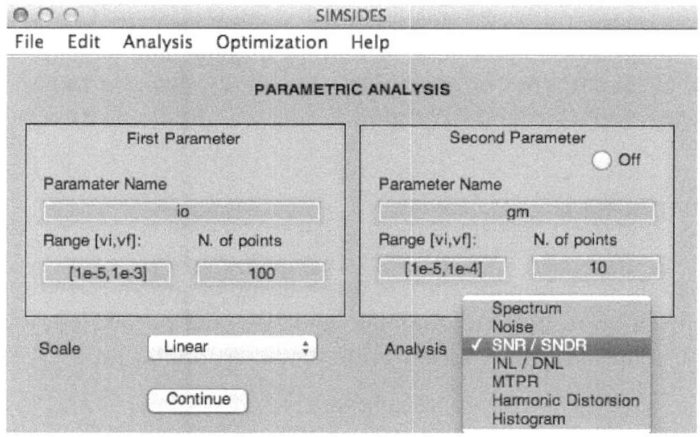

Figure B.12 Parametric analysis menu.

- `Parameter Name`: the name of the model parameter to be varied. This model parameter can be a variable used in a ΣΔM building block model (like for instance I_o, g_m, etc) or a simulation parameter, like the input signal amplitude, sampling frequency, and so on.
- `Range [vi,vf]`, which defines the variation range, defined by an interval, with a lower value given by `vi` and a upper value by `vf`.
- `No. of points`: the number of points into which the variation interval is divided.
- `Scale`, which specifies if the variation range is either linear or logarithmic.
- `Analysis`, which specifies the type of analysis to be carried out, including output spectrum, IBN, SNR/SNDR, INL, MTPR, harmonic distortion, histograms, and so on.

B.3.9 Monte Carlo Analysis

Figure B.13 shows the SIMSIDES menu to run a Monte Carlo analysis. This is a particular case of parametric analysis, which has essentially the same functionalities and model parameters. The only

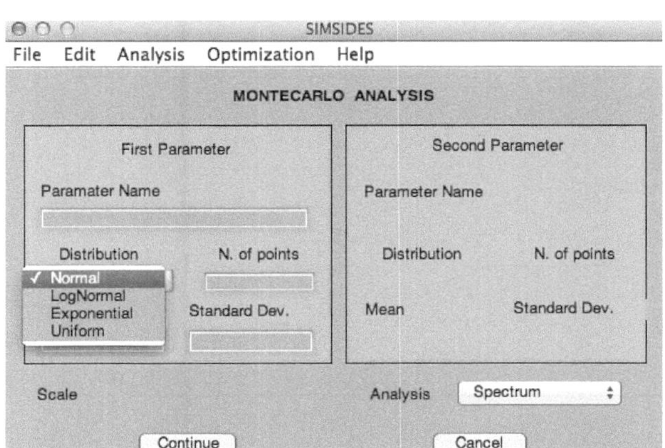

Figure B.13 Monte Carlo analysis menu.

difference is that the variation of the parameters involved in the Monte Carlo analysis are randomly varied according to a probability distribution with a mean value and a standard deviation specified in the analysis menu. Different types of probability distributions can be chosen, including `Normal`, `Log-Normal`, `Exponential`, and `Uniform`.

B.4 Optimization Interface

SIMSIDES includes an optimization interface to assist designers to combine time-domain behavioral simulation with optimization in order to automate and systematize the high-level design of ΣΔMs. Figure B.14a shows a snapshot of the optimization menu. A number of solvers such as `Fmincon`,

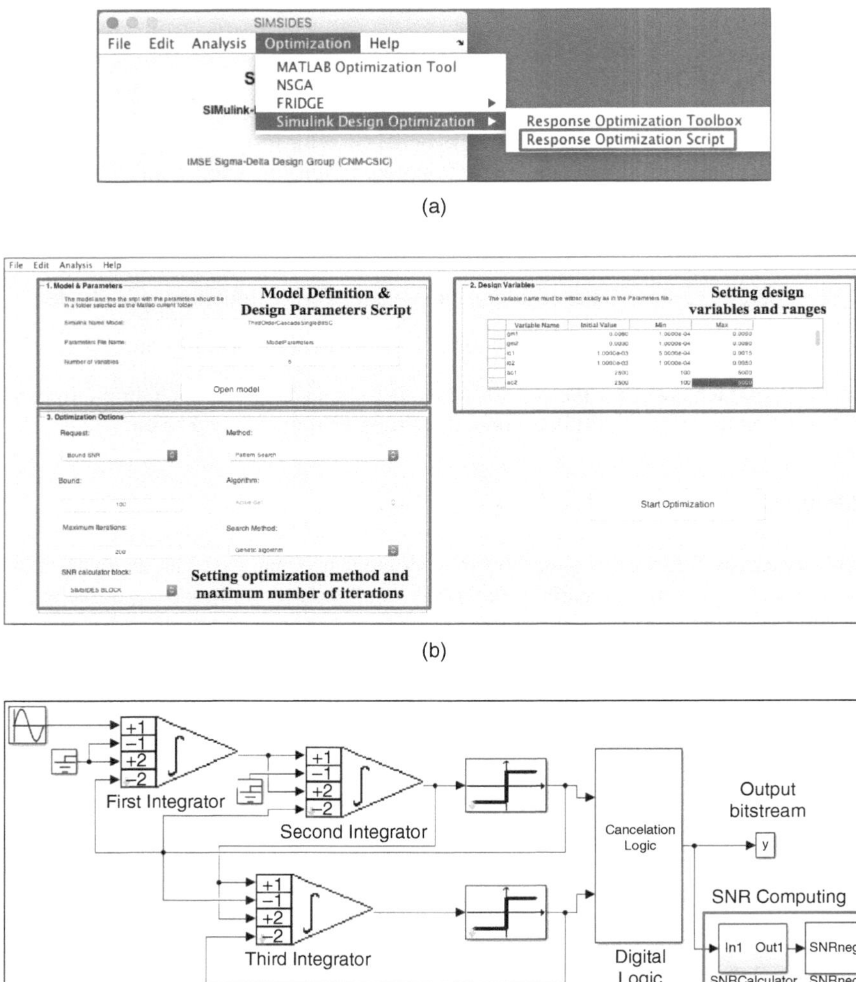

(c)

Figure B.14 SIMSIDES optimization menu: (a) main window; (b) GUI for setting up the optimization; (c) model including the additional block to compute SNR while running the optimization.

`Fminsearch`, and `Patternsearch`, search methods like `Gradient Descent`, `Simplex Search`, `Pattern`, and so on, as well as diverse optimization algorithms: `Neider-Mead`, `Genetic`. In addition, optimization engines – including multi-objective evolutionary algorithms (MOEAs) such as NSGA-II – are also included.

The `SIMULINK Design Optimization` toolbox can be used with any arbitrary SIMULINK model, and so SIMSIDES models can be also combined with this toolbox. However, although a powerful and user friendly GUI is provided, the formulation of the optimization problem is not an easy task when applied to the design of ADCs, and particularly to ΣΔMs. The reason is that a suitable performance metric – like for instance the SNR – cannot be directly set from the optimization toolbox, which is mostly focused on optimizing the frequency response of filters based on fine tuning their Bode diagrams.

In order to address these problems, a dedicated interface has been developed and embedded in SIMSIDES in order to help designers combine the benefits of its time-domain behavioral models for ΣΔMs with the different optimization methods and algorithms available in MATLAB. Figure B.14b is an excerpt of the SIMSIDES optimization interface. This optimization menu allows designers to customize their optimization problem by defining all the required pieces of information: the name of the SIMSIDES model, the MATLAB script including main simulation parameters, the number of design variables, the initial values and ranges of variables, and the optimization method. To this end, designers need to follow these three basic steps:

- Build a model of the ΣΔM in SIMSIDES
- Create a MATLAB script with all design variables to be optimized, as well as the parameters required to simulate the ΣΔM model
- Enter the information corresponding to the SIMSIDES model and the MATLAB script in the SIMSIDES optimization interface.

Different performance metrics can be used for optimization purposes. Without loss of generality, the SNR has been considered. Therefore, the design objective is to maximize the SNR while optimizing the ΣΔM building-block design variables in order to minimize the power consumption. However, the optimization solvers and algorithms available in MATLAB are intended to minimize a given function, rather than to maximize it, as is the case here. Therefore, in order to overcome this limitation, an additional block is added to the SIMSIDES model – depicted in Figure B.14c – which calculates the SNR at the output of the ΣΔM output bitstream and obtains the negative value of the computed SNR, so that the optimization problem can be formulated as:

$$\text{maximize } [f(\overline{x})] = \text{minimize } [-f(\overline{x})] \tag{B.1}$$

where $f(\overline{x})$ is the performance metric to be optimized – SNR in this case – and $\overline{x}$ denotes the vector of design variables involved in the optimization; that is, the ΣΔM building-block specifications to be minimized/maximized in order to get the maximum SNR with the minimum power consumption. For the design variables, initial values and ranges need to be entered as depicted in Figure B.14b. Also, the algorithm solver and search method used in the optimization are set, as is the maximum number of iterations to be considered in order to limit the CPU time in case the synthesis process does not converge to any solution. During the optimization procedure, the toolbox gives information about both the number of iterations and the number of simulations. The former refers to the number of times the optimization algorithm has been run whereas the later stands for the number of times the ΣΔM has been simulated.

B.5 Tutorial Example: Using SIMSIDES to Model and Analyze ΣΔMs

This section illustrates the use of SIMSIDES through a simple example in which several kinds of analysis will be carried out to show the main features of the toolbox. Figure B.15 shows the block diagram of the modulator under study, which consists of a third-order cascade 2-1 DT-ΣΔM with single-bit quantization in both stages.

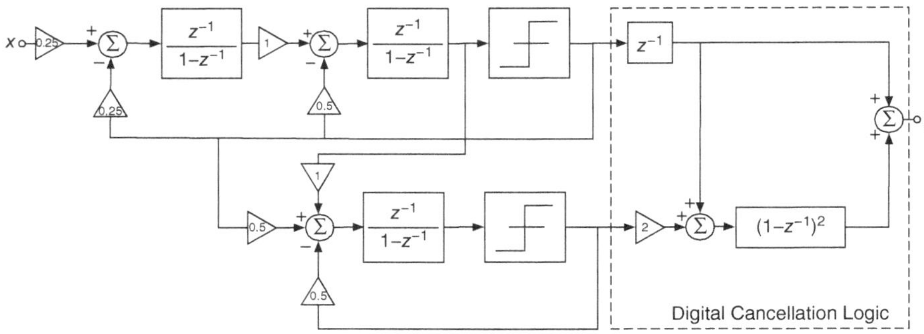

Figure B.15 *Z*-domain block diagram of a cascade 2-1 DT-ΣΔM.

B.5.1 Creating the Cascade 2-1 ΣΔM Block Diagram in SIMSIDES

The modulator block diagram shown in Figure B.15 can be implemented using the model libraries available in SIMSIDES. To this end, the same procedure as described in Section B.2 is followed:

- Go to SIMSIDES main menu, select `File -> New Architecture` and introduce a name for the new ΣΔM architecture.
- Include the integrators and comparators from the SIMSIDES model libraries. To do this, select `Edit -> Add Block`. In this example, the FE integrators in Figure B.15 are implemented by using the `SC_FE_Integrator_All_Effects` blocks from the `Real Integrators` library, whereas single-bit quantizers are modeled by the `Real_Comparator_Offset&Hysteresis` comparator block available in the `Quantizers&Comparators` library. These building blocks can be incorporated in the new architecture by simply dragging and dropping the models from their corresponding SIMSIDES libraries, as illustrated in Figure B.16a.
- Incorporate the remaining building blocks from the SIMULINK model library. To do this, go to `Edit -> Simulink Library` and drag the required models. In this example the following blocks are required: `Sine Wave` and `Ground` blocks from the `Sources` library, `Unit Delay` and `Discrete Filter` block from the `Discrete` library, and `To Workspace` from the `Sinks` library.
- Finally, once all required blocks have been included in the new architecture, they are connected to implement the required ΣΔM architecture, as shown in Figure B.16b.

B.5.2 Setting Model Parameters

The modulator parameters and model parameters required to simulate the block diagram of Figure B.16 can be either set up in the MATLAB command window or they can be saved in an M-file that is

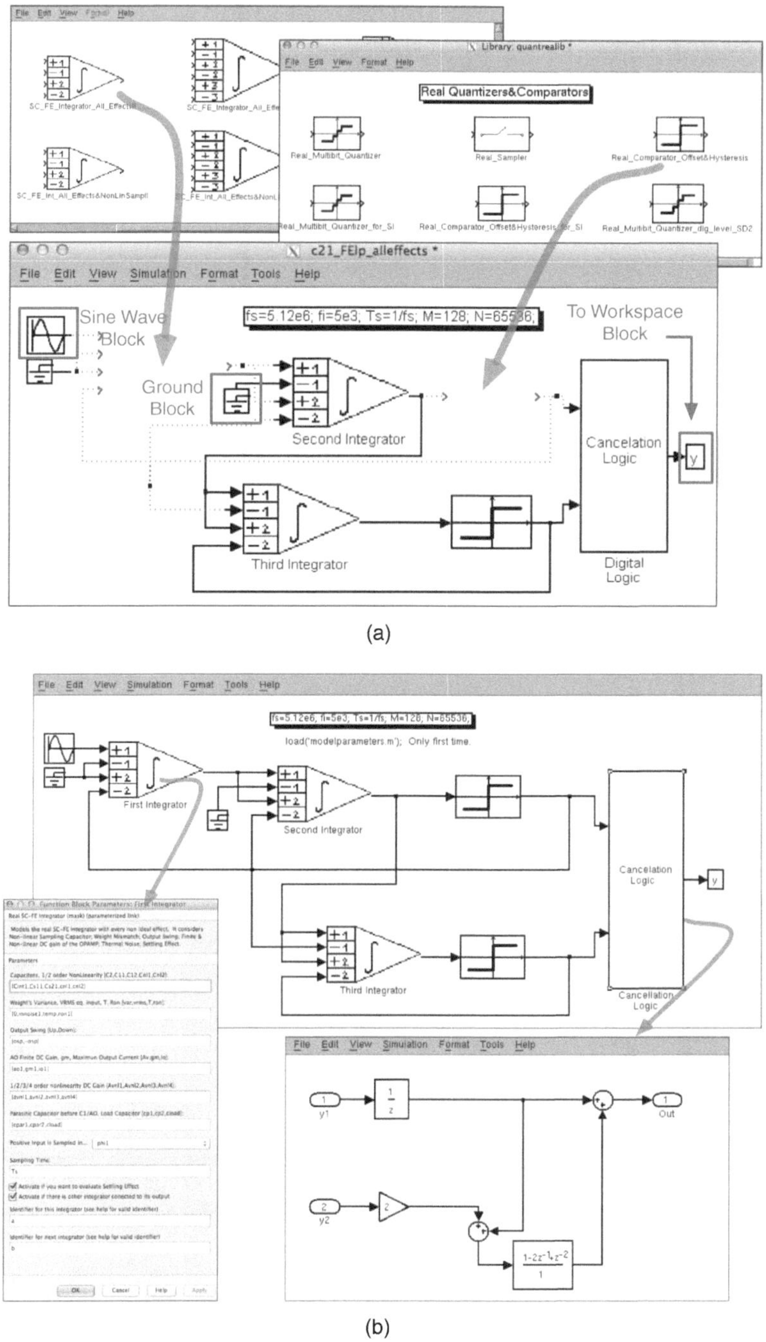

Figure B.16 SIMSIDES block diagram of the ΣΔM shown in Figure B.15: (a) building and editing the block diagram; (b) complete modulator block diagram in SIMSIDES.

loaded when needed. As an illustration, Figure B.17 shows the M-file used for setting up all the model parameters of Figure B.16. It also includes a brief description of the different parameters and variables included. For the sake of completeness, Table B.1 includes the values of all building-block parameters as described in the SIMSIDES user masks, as well as other auxiliary block parameters (such as those used in the `Sine Wave` and `To Workspace` blocks) that are required during simulation. In addition to these model parameters, simulation parameters must be set up to run a simulation. To do this, go to the `Simulation -> Simulation Parameters` menu and define the following parameters:

- Simulation time: `Start Time: 0.0`; `Stop Time: (N-1)*Ts`
- Solver options: `Type: Variable Step`; `Max Step Size: Auto`

```
% SDM parameters:
% Sampling Frequency(fs), Input Frequency (fi), Sampling Time (Ts)
% OverSampling Ratio (OSR=M); Number of points (N)
fs=5.12e6; fi=5e3; Ts=1/fs; M=128; N=65536;
% Model parameters
kt=0.026*1.6e-19; % Boltzmann constant

% First Integrator's parameters
Cint1=24e-12; % integration capacitor For gain=1
Cs11=6e-12; % sampling capacitor (branch 1)
Cs21=6e-12; % sampling capacitor (branch 2)
innoise1=0; % rms value of the input equivalent noise
ao1=2.63e3; % open-loop OTA DC gain
gm1=4.5e-3; % transconductance
io1=0.977e-3; % maximum OTA output current
ron1=60; % sampling switch-on resistance

% Second- and Third- Integrators
Cint2=3e-12;
Cs12=1.5e-12;
Cs22=1.5e-12;
innoise2=0;
ao2=1.38e3;
gm2=0.87e-3;
io2=0.25e-3;
ron2=650;

% Common integrator parameters
temp=175; % temperature
osp=2.7; % output swing
cnl1=0; % capacitor first-order non-linear coef.
cnl2=25e-6; % capacitor second-order non-linear coef.
avnl1=0; % DC gain first-order non-linear coef.
avnl2=15e-2; % DC gain second-order non-linear coef.
avnl3=0; % DC gain third-order non-linear coef.
avnl4=0; % DC gain fourth-order non-linear coef.
cpar1=0.6e-12; % parasitic (opamp) input capacitance
cpar2=0.6e-12;
cload=2.28e-12; % opamp (intrinsic) load capacitance
% Comparators
vref=2; % DAC reference voltage
hys=30e-3; % comparator hysteresis
```

Figure B.17 M-file including all model parameters required to simulate the ΣΔM in Figure B.16b.

Note that integrator building blocks are identified in order to properly compute the equivalent load capacitances required for the incomplete settling error model.

B.5.3 Computing the Output Spectrum

The output spectrum of the ΣΔM can be computed in SIMSIDES as follows:

Table B.1 Building-block model parameters used for simulating the ΣΔM in Figure B.16b.

Building block	Parameter description	Value/variable
Input sine wave	Sine type	Time based
	Amplitude	0.5
	Bias	0
	Frequency (rad/s)	2*pi*fi
	Phase (rad)	0
	Sample time	0
	Interpret vector parameters	Selected
First integrator	Integration and sampling capacitors (Branch 1, Branch 2)	[Cint1,Cs11,Cs21]
	Capacitor nonlinear coefficients	[cnl1,cnl2]
	Weights variance, rms eq. input noise, temperature	[0,innoise1,temp]
	OTA DC gain, transconductance, max. output current	[ao1,gm1,io1]
	Positive/negative output swing	[osp,-osp]
	Switch on-resistance	ron1
	OTA DC gain nonlinear coefs.	[avnl1,2,3,4]
	Parasitic capacitances before the OTA	[cpar1,cpar2]
	Load capacitance	cload
	Positive input 1 is sampled at …	phi1
	Sampling time	Ts
	Identifier for this integrator	a
	Identifier for the next integrator	b
Second, third integrators	Integration and sampling capacitors (Branch 1, Branch 2)	[Cint2,Cs12,Cs22]
	Capacitor nonlinear coefficients	[cnl1,cnl2]
	Weights variance, rms eq. input noise, temperature	[0,innoise2,temp]
	OTA DC gain, transconductance, max. output current	[ao2,gm2,io2]
	Positive/negative output swing	[osp,-osp]
	Switch on-resistance	ron2
	OTA DC gain nonlinear coefs.	[avnl1,2,3,4]
	Parasitic capacitances before the OTA	[cpar1,cpar2]
	Load capacitance	cload
	Positive input 1 is sampled at …	phi1
	Sampling time	Ts
	Identifier for this integrator (second integrator)	b
	Identifier for this integrator (third integrator)	c
	Identifier for the next integrator	c
Comparators	Vhigh, Vlow	[vref, -vref]
	Offset, hysteresis	[0,hys]
	Phase ON	phi1
	Sampling time	Ts
	Identifier for this quantizer	quant1
To workspace (*y*)	Variable name	y
	Limit data points to last	N
	Decimation	1
	Sample time	Ts
	Save format	Array

- Set up model parameters using the M-file shown in Figure B.17.
- Simulate the modulator in Figure B.16b from the menu `Simulation -> Start`.
- Once the simulation has finished, go to `Analysis -> Node Spectrum Analysis` menu in SIMSIDES.
- Define the parameters requested in that menu. In this example, the sampling frequency is defined as `fs` and a Kaiser window function is used with a number of points `N` and `Beta = 20`.
- Click on `Compute` and then `Plot`, and the output spectrum shown in Figure B.18 is displayed.

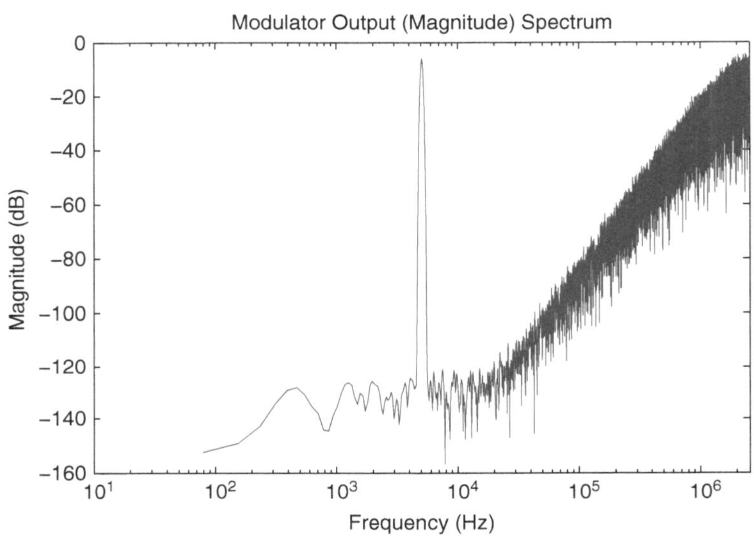

Figure B.18 Output spectrum (magnitude) of the ΣΔM in Figure B.16b.

B.5.4 SNR versus Input Amplitude Level

Figure B.19 shows the SNDR versus input amplitude level (or SNDR curve) of the ΣΔM in Figure B.16b. This figure has been obtained by using the `Analysis` menu and choosing `SNR/SNDR` analysis. In this example, the following parameters are used:

- `Parameter Name: Ain`, where `Ain` is the `Amplitude` parameter defined in the `Input Sine Wave` block in Figure B.16b.
- `Range [vi,vf]: [1e-6,2]`
- `N. of points: 50`
- `Scale: Logarithmic`
- `Analysis: SNR/SNDR`
- `Second Parameter: Off`

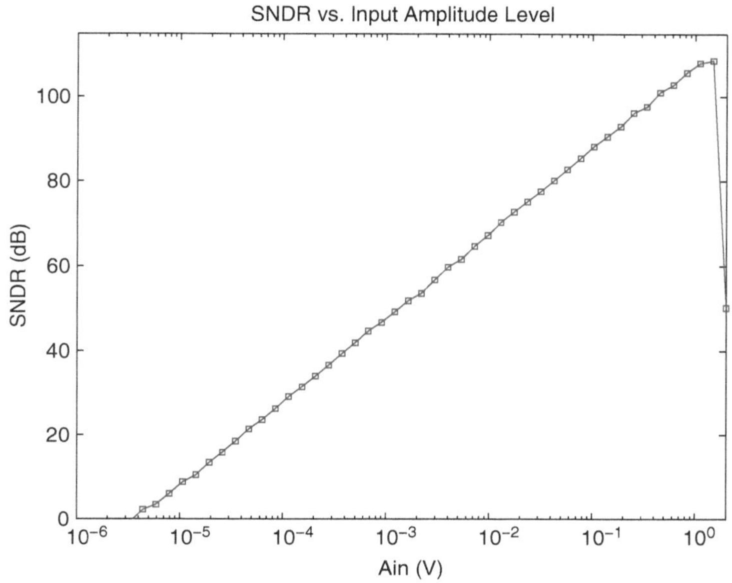

Figure B.19 SNDR versus input amplitude level of the ΣΔM in Figure B.16b.

Once the aforementioned parameters are set up, click on `Continue` and the `SNR/SNDR` window menu shown in Figure B.7 is displayed. The requested parameters (sampling frequency, oversampling ratio, and so on) are set up according to the values given in Figure B.17, namely:

- `Name of the signal(s) to process: y`
- `Sampling frequency (Hz): fs`
- `Oversampling ratio: M`
- `Input Frequency (Hz): fi`
- `Window: Kaiser`
- `N. of Points: N`
- `Beta: 20`
- `Kind of Spectrum: LP`
- `Figure of merit: SNDR`

After setting up these parameters, click on `Compute` and then `Plot` to obtain the curve given in Figure B.19.

B.5.5 Parametric Analysis Considering Only One Parameter

The `Parametric Analysis` menu can be used for studying the effect of a given model parameter on the modulator performance. For instance, let us consider the effect of the OTA transconductance

g_m of the front-end integrator in Figure B.16b. In order to analyze the impact of this parameter on the effective resolution of the modulator, go to the `Parametric Analysis` menu and set up the following parameters:

- `Parameter name: gm1`, which stands for g_m of the front-end integrator block in Figure B.16b.
- `Range [vi,vf]: [1e-5,1e-3]`
- `N. of points: 50`
- `Scale: Linear`
- `Analysis: SNR/SNDR`
- `Second Parameter: Off`

Once these parameters have been defined, click on `Continue` and proceed in a similar way to the previous examples in order to compute the SNDR. Figure B.20 shows the results of this analysis, depicting the SNDR versus `gm1`.

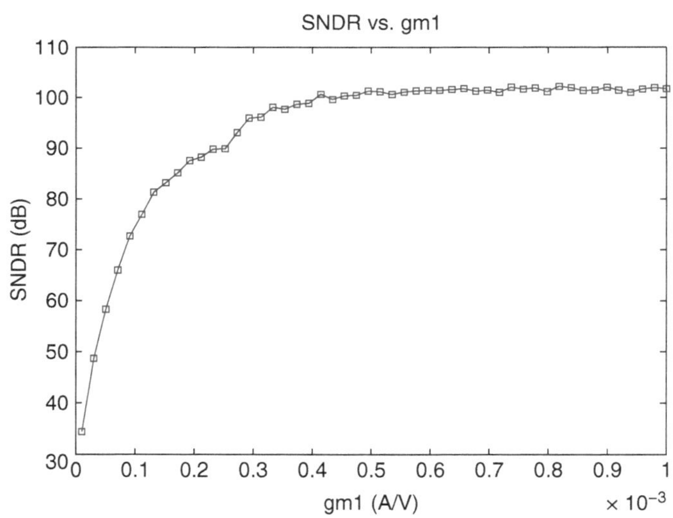

Figure B.20 Using parametric analysis to study the effect of a single model parameter: SNDR versus transconductance of the front-end amplifier for the ΣΔM in Figure B.16b.

B.5.6 Parametric Analysis Considering Two Parameters

The `Parametric Analysis` menu can be used also for implementing parametric analyses considering the variation of two different parameters. As an example, Figure B.21 shows the effect of both the OTA transconductance g_{m1} and the maximum output current I_{o1} of the front-end amplifier on the SNDR of the ΣΔM in Figure B.16b.

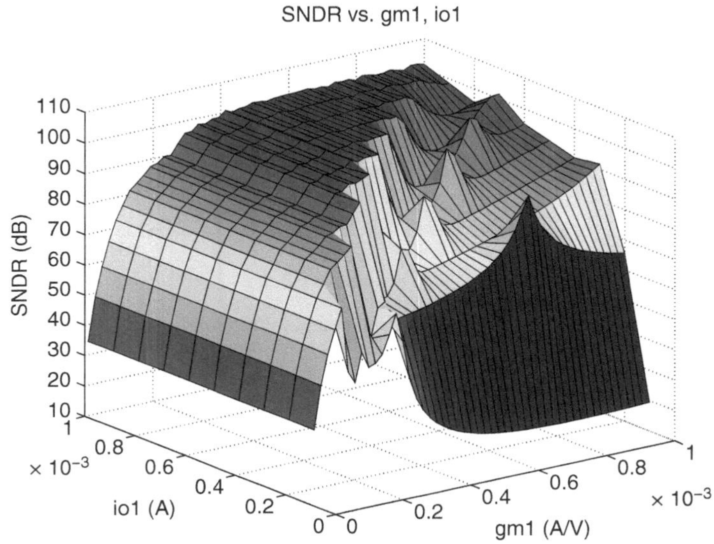

Figure B.21 Parametric analysis considering the effect of two parameters (g_{m1} and I_{o1}) on the SNDR.

In order to obtain the graph in Figure B.21, the following parameters are set up in the `Parametric Analysis` menu:

- `Parameter name: io1`, which stands for the maximum output current I_o of the front-end integrator.
- `Range [vi,vf]: [1e-4,1e-3]`
- `N. of points: 10`

B.5.7 Computing Histograms

Finally, to conclude this example, Figure B.22 illustrates the histograms of the integrators' outputs in the front-end stage of the modulator in Figure B.16b.

These histograms have been obtained by using the `Analysis -> Histograms` menu from SIMSIDES and setting up the following model parameters:

- `Name of the signal(s) to process: y1,y2`, which are the names given to the output of the integrators saved into the MATLAB workspace using `To Workspace` blocks from the SIMULINK elementary library.
- `Number of bins: 100`

B.6 Getting Help

SIMSIDES includes a help menu (illustrated in Figure B.23). This user guide can be opened by selecting `Help -> User Manual` in the SIMSIDES main window. In addition, a complete list

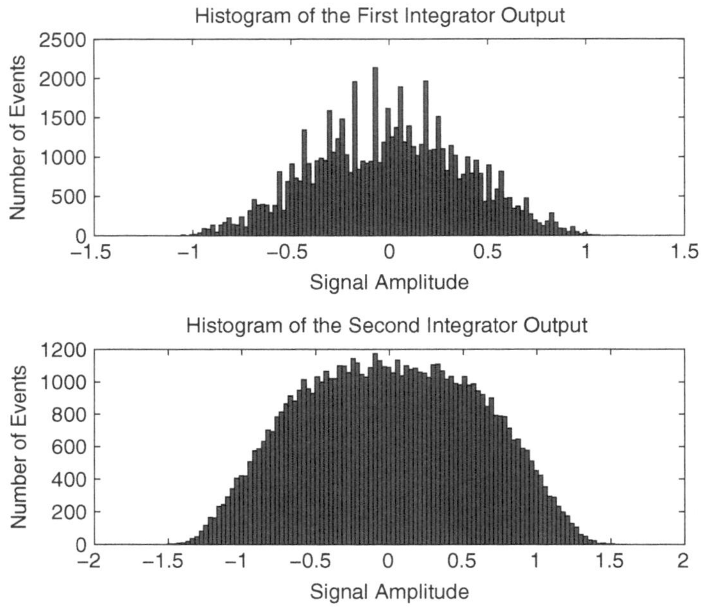

Figure B.22 Illustrating the use of histograms of the modulator in Figure B.16b.

of all behavioral models (and their corresponding parameters) included in SIMSIDES – described in Appendix C – can be also obtained from this menu by selecting `Help -> Libraries and Models`.

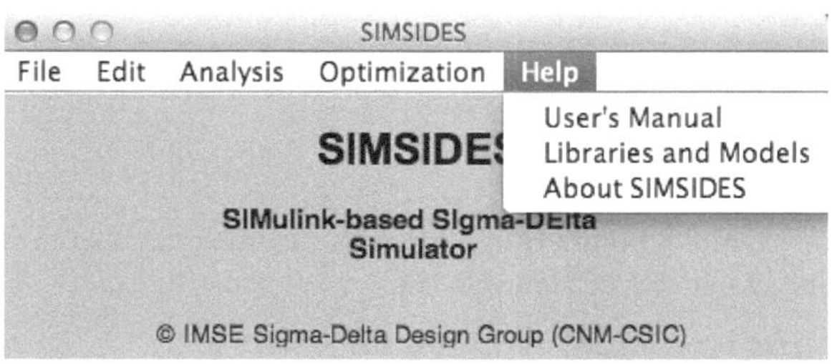

Figure B.23 Help menu.

APPENDIX C

SIMSIDES BLOCK LIBRARIES AND MODELS

This appendix contains a compilation of most of the ΣΔM building blocks and libraries included in SIMSIDES. A brief description of their purpose and functionality, as well as their main model parameters, is provided. Revised versions of this appendix are periodically updated and provided together with the user manuals included in the SIMSIDES website.

C.1 Overview of SIMSIDES Libraries

Table C.1 compiles all libraries included in SIMSIDES together with a brief description of their contents. These libraries are divided into two main categories: ideal libraries and real libraries. The former contains ideal building blocks, whereas the latter includes behavioral models that incorporate circuit-level nonidealities. The libraries containing integrators and resonators are subdivided into several specific sublibraries, which include in turn building-block models corresponding to different circuit-level implementations. For instance, SC integrators are subdivided into FE and LD integrators; CT integrators are subdivided into Gm-C, active-RC, and so on.

Sigma-Delta Converters: Practical Design Guide, Second Edition. José M. de la Rosa.

Table C.1 Overview of SIMSIDES libraries.

Ideal libraries	**Sublibraries**	**Building blocks**
Integrators	–	Ideal DT/CT integrators
Resonators	–	Ideal resonators
Quantizers and comparators	–	Ideal quantizers
D/A converters	–	Ideal DACs
Real libraries	**Sublibraries**	**Building blocks**
Integrators	SC FE integrators	Forward-Euler SC integrators
	SC LD integrators	Lossless-direct SC integrators
	SI FE integrators	Forward-Euler SI integrators
	SI LD integrators	Lossless-direct SI integrators
	gm-C integrators	Gm-C integrators
	gm-MC integrators	Miller OTA integrators
	RC integrators	Active-RC integrators
	MOSFET-C integrators	MOSFET-C integrators
Resonators	SC FE resonators	Resonators based on FE SC integrators
	SC LD resonators	Resonators based on LD SC integratorss
	SI FE resonators	Resonators based on FE SI integrators
	SI LD resonators	Resonators based on LD SI integrators
	gm-C resonators	Resonators based on Gm-C integrators
	gm-LC resonators	Resonators based on Gm-LC integrators
Quantizers and comparators	–	Nonideal single-bit and multi-bit quantizers
D/A converters	–	Nonideal single-bit and multi-bit DACs
Auxiliary blocks	–	Adders, latches, DEM blocks, etc

C.2 Ideal Libraries

As shown in Table C.1, SIMSIDES includes four ideal libraries: integrators, resonators, quantizers, and DACs. The building blocks contained in these libraries are described in the following sections.

C.2.1 Ideal Integrators

There are three kinds of ideal integrators in this library: `Ideal_CT_Integrator`, `Ideal_FE_Integrator`, and `Ideal_LD_Integrator`.

C.2.1.1 Building-block Model Purpose and Description The `Ideal_CT_Integrator` block models the ideal S-domain transfer function of a CT integrator, given by,

$$\mathrm{ITF}(s) = g \cdot \frac{1}{s} \tag{C.1}$$

where g stands for the integrator gain (also referred to as integrator weight).

The `Ideal_FE_Integrator` and `Ideal_LD_Integrator` blocks model the ideal Z-domain transfer functions of DT FE and LD integrators, respectively given by:

$$
\begin{aligned}
\mathrm{ITF_{FE}}(z) &= g \cdot \frac{z^{-1}}{1 - z^{-1}} \\
\mathrm{ITF_{LD}}(z) &= g \cdot \frac{z^{-1/2}}{1 - z^{-1}}
\end{aligned}
\tag{C.2}
$$

C.2.1.2 Model Parameters The following model parameters are included in the dialog box of the models above:

- `Gain`, which defines the integrator weight g.
- `Sampling Time`: the sampling period of both the `Ideal_FE_Integrator` and `Ideal_LD_Integrator` blocks.

C.2.2 Ideal Resonators

This library includes the various ideal resonator blocks described below.

C.2.2.1 Ideal_LD_Resonator This block consists of a DT resonator made up of two LD integrators connected in a feedback loop, as illustrated in Figure C.1a. Its model parameters are the following:

- `Af`: the forward loop gain.
- `Afb`, which stands for the feedback loop gain.
- `Sampling Time`.

C.2.2.2 Ideal_FE_Resonator This block models a DT resonator made up of two FE integrators connected in a feedback loop, as shown in Figure C.1b. In this case, the model parameters are the following:

- `Affe`, which stands for the feedforward loop gain.
- `Afb1`: the global feedback loop gain.
- `Afb2`: the local feedback loop gain.
- `Sampling Time`.

C.2.2.3 Ideal_CT_Resonator This model corresponds to a CT (biquad) resonator with an S-domain transfer function given by:

$$
\mathrm{RTF}(s) = \frac{(\pi/2) \cdot s}{s^2 + (\pi/2)^2} \tag{C.3}
$$

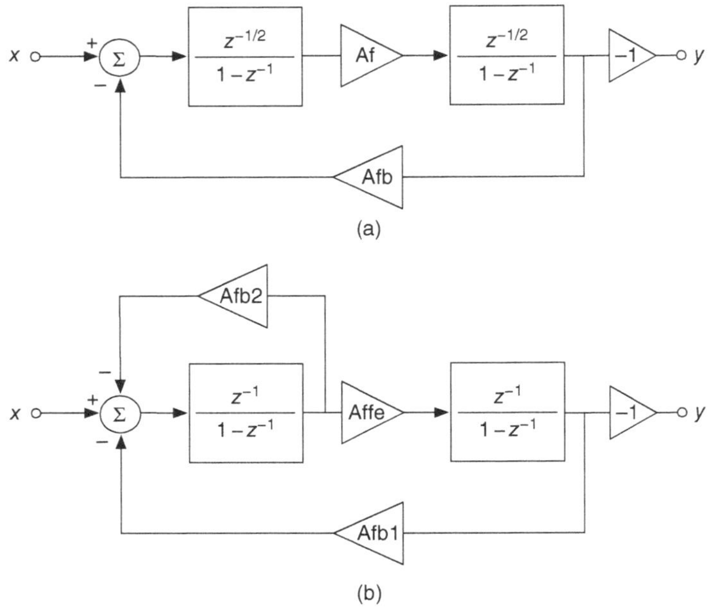

Figure C.1 *Z*-domain block diagram of: (a) Ideal_LD_Resonator. (b) Ideal_FE_Resonator.

The above transfer function is implemented using the `Transfer Fcn` block from the `Continuous` SIMULINK library.

C.2.3 Ideal Quantizers

This library includes several building blocks that model single-bit and multi-bit/multi-level quantizers intended for both voltage-mode (SC/CT) and current-mode (SI) ΣΔMs.

C.2.3.1 Ideal_Comparator This block models the input/output DC characteristic of an ideal comparator, given by a sign function as,

$$v_o = \begin{cases} V_{\text{high}} & v_i \geq 0 \\ V_{\text{low}} & v_i < 0 \end{cases} \tag{C.4}$$

where v_i and v_o stand for the input and output voltages, respectively, and V_{high} and V_{low} represent the analog values of the logic one and logic zero, respectively.

The model parameters of the `Ideal_Comparator` block are the following:

- `Vhigh,Vlow`, which stand respectively for V_{high} and V_{low}.
- `Phase`, which models the clock phase in which the input signal is sampled. Two nonoverlapping clock phases, denoted as `phi1,2` are assumed.
- `Sampling Time`: the clock signal period.

- `Identifier for this Quantizer`, which defines an identification name for the block (used by some dynamic errors like the incomplete settling error in SC circuits).

C.2.3.2 Ideal_Comparator_for_SI This block models a current-mode comparator used in SI-ΣΔMs. The behavioral model is exactly the same as that used in the `Ideal_Comparator` block, except that in this case the input signal is a current-mode signal that is modeled as a matrix made up of two vector elements: the current signal itself and the output conductance of the current-mode building block (i.e, integrator, resonator, and so on) connected at the input of the comparator.

In this way, the information provided to the model at each sampling time is a vector of two elements, as illustrated in Figure C.2, where i_i is the input current, v_o is the output voltage, and g_{oi} is the output conductance of the building block connected at the comparator input.

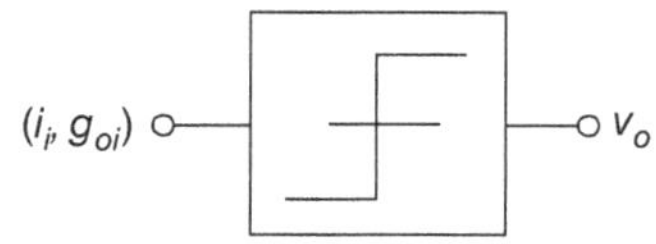

Figure C.2 Input and output signals in an Ideal_Comparator_for_SI model.

C.2.3.3 Ideal_Multibit_Quantizer This building block includes the ideal behavioral model of a multi-bit quantizer with a DC input/output *midrise* characteristic, as illustrated in Figure C.3a.

The model parameters of this block are the following:

- `Vhigh, Vlow`, which stand respectively for the maximum and minimum values of the quantizer output FS range; see Figure C.3a.
- `Number of bits` of the quantizer.

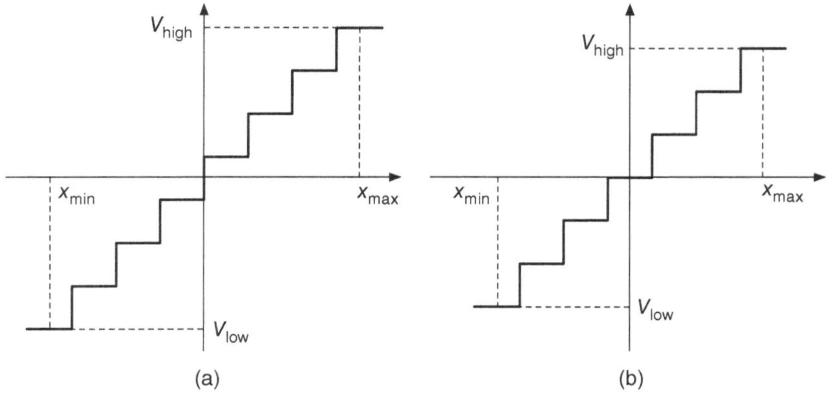

Figure C.3 Illustrating the input/output DC characteristic of: (a) multi-bit (3-bit) *midrise* quantizer; (b) multi-level (7-level) *midtreat* quantizer.

- Xmax-Xmin, which defines the input FS range of the quantizer.
- Phase ON, which is the clock phase in which the input signal is sampled, considering a two-phase clock signal generator.
- Sampling Time, which denotes the clock period.

C.2.3.4 Ideal_Multibit_Quantizer_for_SI This block models a current-mode multi-bit quantizer with the same input/output characteristic as an Ideal_Multibit_Quantizer, but considering that the input signal has two components – i_i and g_{oi} – in the same way as in Figure C.2.

C.2.3.5 Ideal_Multibit_Quantizer_levels This building block models a multi-level quantizer in which the input/output characteristic is defined as a function of the number of levels, instead of the number of bits, as in the Ideal_Multibit_Quantizer block. Thus the same parameters are used, except for the number of bits, which is replaced by Number of levels. If this parameter is even, a *midrise* quantization characteristic, like that shown in Figure C.3a, is implemented. Otherwise, a *midtreat* characteristic, like that depicted in Figure C.3b, is provided.

C.2.3.6 Ideal_Multibit_Quantizer_levels_SD2 This building block implements the same behavioral model as the previous one, but the output is a thermometric-coded bit array. This is used in combination with multi-level DACs with DEM techniques, as will be illustrated in Section C.8.

C.2.3.7 Ideal_Sampler This building block models an ideal S&H circuit that can be used in CT-ΣΔMs, in which the signal is sampled at the input of internal quantizers.

The model parameters are the following:

- Sampling Time, which denotes the clock period.
- Input clock phase, which defines the sampling clock phase.

C.2.4 Ideal D/A Converters

The blocks included in this library model different kinds of ideal DACs:

- Ideal_DAC, which is used for modeling ideal single-bit DACs for either SC- or CT-ΣΔMs. They are simply modeled as a voltage gain, called Gain in the model.
- Ideal_DAC_for_SI, which models a single-bit DAC for SI-ΣΔMs.
- Ideal_DAC_dig_level_SD2, which consists of a multi-level ideal DAC.

The last two models and their associated parameters are described below.

C.2.4.1 Ideal_DAC_for_SI Figure C.4 shows the equivalent circuit of the Ideal_DAC_for_SI block. It consists of a voltage-controlled current source in parallel with a finite output

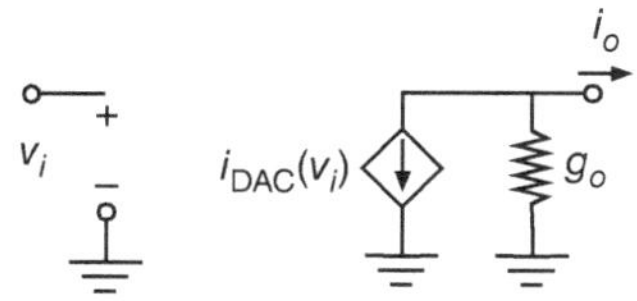

Figure C.4 Equivalent circuit of Ideal_DAC_for_SI model.

conductance g_o. The current source $i_{\rm DAC}$ is a sign function of the input voltage v_i given by,

$$i_{\rm DAC}(v_i) = \begin{cases} +I_{\rm ref} & v_i = +v_{\rm ref} \\ -I_{\rm ref} & v_i = -v_{\rm ref} \end{cases} \tag{C.5}$$

where $I_{\rm ref}$ and $v_{\rm ref}$ stand for the modulator FS reference current and voltage, respectively.

The model parameters included in the `Ideal_DAC_for_SI` model are the following:

- `Gain`, which models the DAC gain: $I_{\rm ref}/v_{\rm ref}$.
- `Gout`, which stands for g_o.
- `Sampling Time`, which models the clock period.
- `Input Clock Phase`: the clock phase at which the DAC input is sampled.

C.2.4.2 Ideal_DAC_dig_level_SD2 This block transforms a thermometric-coded digital input into its corresponding analog level. The model parameters used in this behavioral model are:

- `Vhigh`, which defines the upper limit of the quantization FS range.
- `Vlow`, which corresponds to the lower limit of the quantization FS range.
- `Number of levels` of the embedded quantizer.

C.3 Real SC Building-Block Libraries

SIMSIDES includes two libraries of SC integrators and two libraries of SC resonators. These libraries are described below.

C.3.1 Real SC Integrators

There are two SC integrator model libraries in SIMSIDES: one for FE SC integrator models and the other for LD SC integrators. In both cases, integrator models are classified according to the nonideal effects that are included in the model and the number of SC branches connected at the integrator input. In this way, for each model there are four building blocks using the same behavioral model except for the number of input SC branches.

As an illustration, Figure C.5 shows the symbol used in SIMSIDES for one-branch SC FE integrators (Figure C.5a) and two-branch SC FE integrators (Figure C.5b), together with their

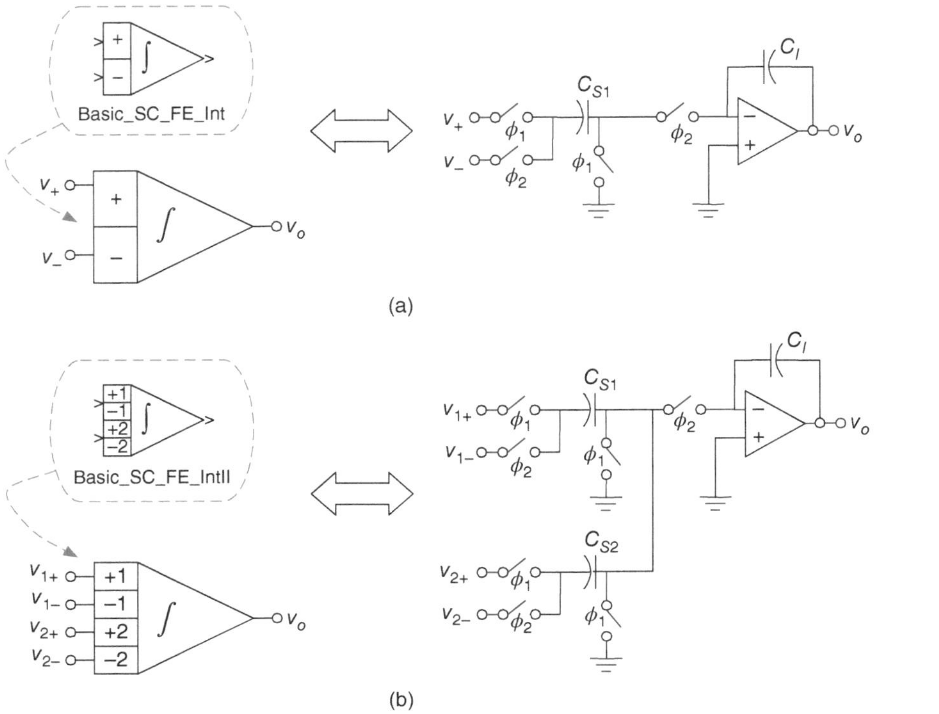

Figure C.5 SC integrator symbol in SIMSIDES: (a) one-branch integrator; (b) two-branch integrator.

equivalent SC circuits. Note that, although single-ended conceptual schematics are shown in this figure, fully-differential circuits are assumed in the behavioral models.

Both integrators in Figure C.5 use the same behavioral model, which consists of an ideal SC FE integrator with output swing limitation. The behavioral model corresponding to a one-branch SC FE integrator is named `Basic_SC_FE_Int`, while the model of the two-branch SC FE integrator is named `Basic_SC_FE_IntII`. Following this nomenclature, `Basic_SC_FE_IntIII` and `Basic_SC_FE_IntIV` models are used for three- and four-branch SC FE integrators, respectively.

Table C.2 lists all SC integrator models available in SIMSIDES, including brief descriptions of the nonidealities included in each. Note that the model names included in Table C.2 correspond to one-branch integrators. The same models are available for integrators with up to four input branches.

The model named `SC_FE_Int_1b_SD2` represents one-branch SC FE integrators with all circuit nonidealideal effects, including the degradation caused by the switch on-resistance on the integrator GB and SR. In this model, notation `1b` is used for denoting one input SC branch. Similarly, other models whose names include `nb` denote n input SC branches.

The model `SC_FE_Int_1b_DEM_SD2` includes the same nonideal effects as the `SC_FE_Int_1b_SD2` model, but it also allows modeling of the sampling capacitor as an array of unit capacitors. This array is used in combination with multi-level quantizers and DACs that incorporate DWA/DEM algorithms, as described in Section C.8. In this way, the number of unit capacitors in the array must be exactly the same as the number of DAC levels. This array should include the DAC mismatch error, which can be modeled as a Gaussian distribution. Both

Table C.2 Library of SC (FE/LD) integrators included in SIMSIDES.

Model name	Circuit effects included
`Basic_SC_FE_Int` `Basic_SC_LD_Int`	Output swing limitation
`SC_FE_Int_Non_linear_C` `SC_LD_Int_Non_linear_C`	Output swing limitation, capacitor nonlinearity.
`SC_FE_Int_Weight_Mismatch` `SC_LD_Int_Weight_Mismatch`	Output swing limitation, capacitor mismatch.
`SC_FE_Int_Non_Linear_Sampling` `SC_LD_Int_Non_Linear_Sampling`	Output swing limitation, nonlinear switch on-resistance.
`SC_FE_Int_FiniteDCgain` `SC_LD_Int_FiniteDCgain`	Finite OTA DC gain, output swing limitation, parasitic OTA caps.
`SC_FE_Int_Finite&Non_LinearDCGain` `SC_LD_Int_Finite&Non_LinearDCGain`	Finite nonlinear OTA DC gain, output swing limitation parasitic OTA caps.
`SC_FE_Int_Noise` `SC_LD_Int_Noise`	OTA thermal noise, output swing limitation, parasitic/load OTA caps.
`SC_FE_Int_Settling` `SC_LD_Int_Settling`	Incomplete settling error, output swing limitation, parasitic/load OTA caps.
`SC_FE_Integrator_All_Effects` `SC_LD_Integrator_All_Effects`	switch on-resistance, capacitor nonlinearity and mismatch, settling error, finite (nonlinear) DC gain, thermal noise, parasitic/load capacitors, output swing limitation.
`SC_FE_Integrator_All_Effects&NonLinSamp` `SC_LD_Integrator_All_Effects&NonLinSamp`	switch nonlinear on-resistance, capacitor nonlinearity and mismatch, settling error, finite (nonlinear) DC gain, thermal noise, parasitic/load capacitors, output swing limitation.
`SC_FE_Int_1b_SD2`	switch on-resistance and its effect on GB and SR, capacitor nonlinearity and mismatch, settling error, finite (nonlinear) DC gain, thermal noise, parasitic/load capacitors, output swing limitation.
`SC_FE_Int_1b_DEM_SD2`	switch on-resistance and its effect on GB and SR, array of unit sampling capacitors, capacitor nonlinearity and mismatch, settling error, finite (nonlinear) DC gain, thermal noise, parasitic/load capacitors, output swing limitation.

Table C.3 Model parameters used in SIMSIDES SC (FE/LD) integrators.

Parameter name (in alphabetical order)	Brief description
`Array of sampling capacitors for DEM branch`	Array of unit capacitors used with multi-level DACs with DEM
`B (switch parameters)`	MOS large-signal transconductance (analytic model)
`Bandwidth (BW)`	Input signal bandwidth
`Capacitor (first/second)-order nonlinearity`	Capacitor (first/second) order nonlinearity
`Finite and Linear Ron`	Switch on-resistance, linear model
`Finite DC Gain of the AO`	Finite OTA DC gain
`g (switch parameters)`	Finite switch on-conductance (analytic model)
`Identifier for this integrator`	Identifier used for settling error model
`Input Equivalent Thermal Noise`	OTA input-referred thermal noise
`Input parameters [A,fi,ph] (switch)`	Amplitude, frequency, and phase of the sinewave input (table look-up model)
`Integration/Sampling Capacitor`	Integration/sampling capacitors
`Integration additional load`	Additional load capacitance at the integration phase
`Load Capacitor (cload)`	Integrator load capacitance
`Maximum output current (Io)`	OTA maximum output current
`Nonlinearity of the DC Gain`	OTA DC gain nonlinear coefficients
`Output Swing Up/Down`	Maximum/minimum output swing limits
`Parasitic Capacitor before the AO (Cp)`	Parasitic capacitance at the OTA input
`pcoef (switch parameters)`	Nonlinear coefficients of the switch on-resistance (table look-up model)
`Positive Input is Sampled in...`	Input-switch clock phase
`Ron`	Switch on-resistance
`Sampling additional load`	Additional load capacitance at the sampling phase
`Sampling Time`	Clock signal period
`Switch on-resistance (Ron)`	Switch on-resistance
`Temp`	Temperature (K)
`Transconductance of the AO (gm)`	OTA transconductance
`Variance`	Variance of the capacitor mismatch error

SC_FE_Int_1b_SD2 and SC_FE_Int_1b_DEM_SD2 models provide detailed information about the integrator equivalent input-referred thermal noise, equivalent load capacitances, and transient response model parameters. All these pieces of information are displayed in the MATLAB command window after simulation.

Table C.3 lists the most important parameters used by the SC integrator behavioral models in SIMSIDES, as well as brief descriptions of each.

C.3.2 Real SC Resonators

SIMSIDES has two SC resonator model libraries corresponding to FEI-based resonators and LDI-based resonators. All building-block models correspond to the Z-domain block diagrams shown in Figure C.1, but they are implemented using the SC integrator models described in the previous section. As an illustration, Figure C.6 shows the SIMSIDES block diagram of an SC LDI-based resonator (Figure C.6a) and an SC FEI-based resonator (Figure C.6b), which correspond to Figures C.1a and C.1b, respectively.

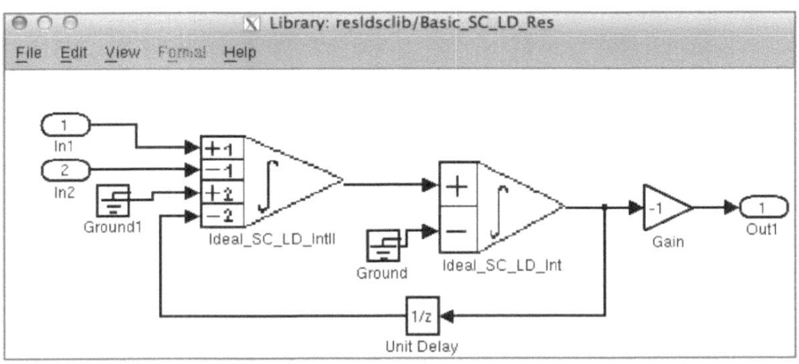

(a)

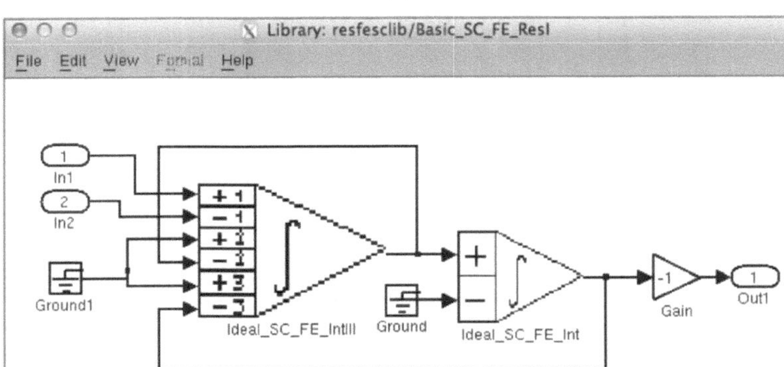

(b)

Figure C.6 Block diagram of SC resonators in SIMSIDES: (a) LDI-based resonator; (b) FEI-based resonator.

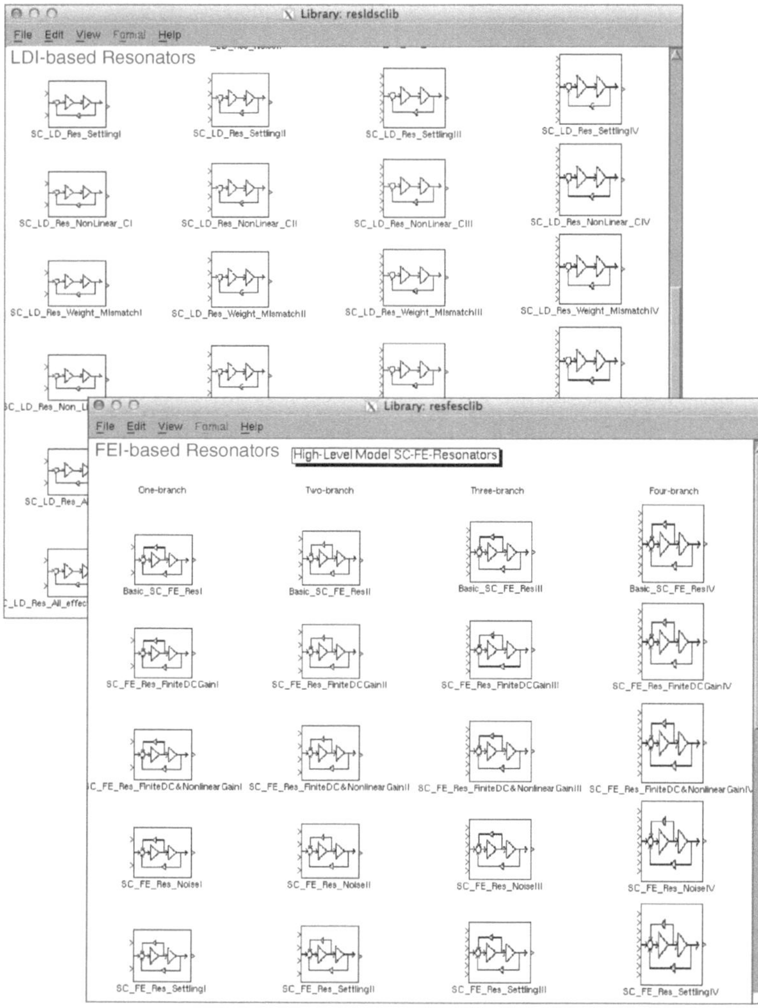

Figure C.7 Excerpt of SC resonator libraries in SIMSIDES.

Following the same philosophy as used for SC integrators, the behavioral models of SC resonators in SIMSIDES are classified attending to the number of input SC branches and the circuit nonideal effects included in the models. As an illustration, Figure C.7 shows an excerpt of both SC resonator libraries. Note that each row of blocks includes the same circuit nonideal effects, with the only difference being the number of input branches.

Table C.4 lists all the SC resonator models available in SIMSIDES, including a brief description of the nonidealities considered in each. The parameters used in these models are the same as those included in SC integrator models – listed in Table C.3. In addition to these parameters, the resonator gain can also be defined by the user by setting a parameter named `Gain` (see Figure C.6), which can be defined in the model dialogue box.

Table C.4 Library of SC (FE/LD) resonators included in SIMSIDES.

Model name	Circuit effects included
`Basic_SC_FE_Res` `Basic_SC_LD_Res`	Output swing limitation
`SC_FE_Res_NonLinear_C` `SC_LD_Res_NonLinear_C`	Output swing limitation, capacitor nonlinearity.
`SC_FE_Res_Weight_Mismatch` `SC_LD_Res_Weight_Mismatch`	Output swing limitation, capacitor mismatch.
`SC_FE_Res_Non_Linear_Sampling` `SC_LD_Res_Non_Linear_Sampling`	Output swing limitation, nonlinear switch on-resistance.
`SC_FE_Res_FiniteDCgain` `SC_LD_Res_FiniteDCgain`	Finite OTA DC gain, output swing limitation, parasitic OTA caps.
`SC_FE_Res_FiniteDC&NonLinearGain` `SC_LD_Res_FiniteDC&NonLinearGain`	Finite nonlinear OTA DC gain, output swing limitation, parasitic OTA caps.
`SC_FE_Res_Noise` `SC_LD_Res_Noise`	OTA thermal noise, output swing limitation, parasitic/load OTA caps.
`SC_FE_Res_Settling` `SC_LD_Res_Settling`	Incomplete settling error, output swing limitation, parasitic/load OTA caps.
`SC_FE_Res_All_effects` `SC_LD_Res_All_effects`	Switch on-resistance, capacitor nonlinearity and mismatch, settling error, finite (nonlinear) DC gain, thermal noise, parasitic/load capacitors.
`SC_FE_Res_All_effects&NonLinSamp` `SC_LD_Res_All_effects&NonLinSamp`	Switch nonlinear on-resistance, capacitor nonlinearity and mismatch, settling error, finite (nonlinear) DC gain, thermal noise, parasitic/load capacitors, output swing limitation.

C.4 Real SI Building-Block Libraries

SIMSIDES includes all necessary building blocks for the simulation of SI-$\Sigma\Delta$Ms. This section describes SI integrators and resonators, as well as their main model parameters.

C.4.1 Real SI Integrators

Following the same classification criteria as used for SC building-block models, there are two libraries of SI integrators in SIMSIDES: one for FE SI integrators and the other for LD SI integrators. Figure C.8 shows the symbol used in SIMSIDES for SI integrators together with a conceptual schematic of an FE SI integrator and an LD SI integrator. In both cases, different models are included in SIMSIDES, classified according to the number of nonideal effects that are taken into account, as detailed in Table C.5.

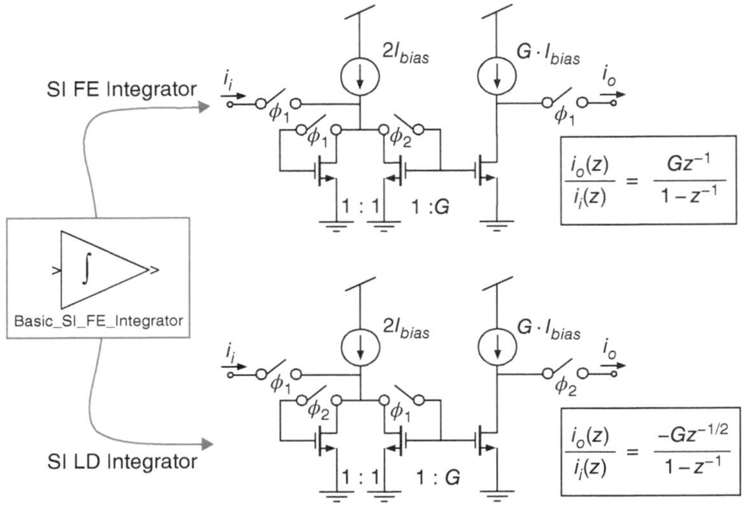

Figure C.8 SI integrator symbol used in SIMSIDES and its corresponding conceptual schematics.

Table C.5 Library of SI (FE/LD) integrators included in SIMSIDES.

Model name	Circuit effects included
`Ideal_SI_FE_Integrator` `Ideal_SI_LD_Integrator`	Ideal SI (FE/LD) integrator.
`Basic_SI_FE_Integrator` `Basic_SI_LD_Integrator`	Output current limits, nonlinear gain.
`SI_FE_Int_Finite_Conductance` `SI_LDI_Finite_Conductance`	Output current limits, finite nonlinear output conductance, input voltage limits, thermal noise.
`SI_FE_Int_Finite_Conductance&Settling` `SI_LDI_Finite_Conductance&Settling`	Output current limits, finite nonlinear input/output conductance, input voltage limits, settling error, thermal noise.

`SI_FE_Int_Finite_`...`&Settling&`
`Charge_Injection`
`SI_LDI_Finite_`...`&Settling&`
`Charge_Injection`
	Output current limits, finite nonlinear
	input/output conductance,
	input voltage limits,
	settling error, thermal noise,
	charge injection error.

Apart from the building blocks in Table C.5, a current-mode *buffer* block, `Buffer`, is included in all SIMSIDES SI libraries. This block transforms an input current vector into an output matrix made up of two vector elements: the input current signal itself in parallel with an output conductance, called `Gout of the source` in the model. Figure C.9 illustrates the operation of the `Buffer` block, showing the SIMSIDES block symbol and its equivalent circuit.

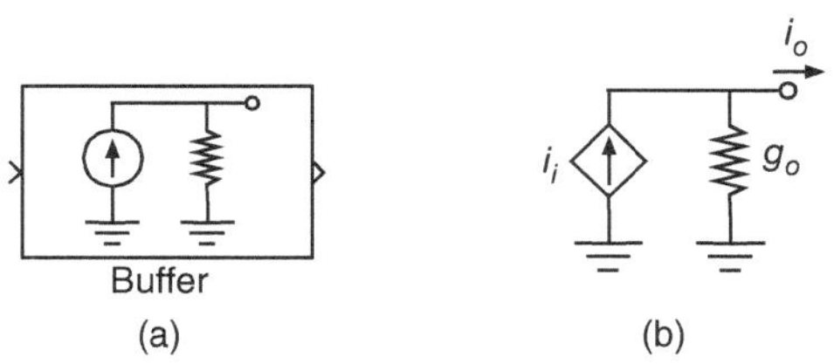

Figure C.9 SI buffer used in SIMSIDES: (a) symbol; (b) equivalent circuit.

C.4.2 Real SI Resonators

As in the case of SC circuits, two different types of SI resonators are modeled in SIMSIDES: SI FEI-based resonators and LDI-based resonators. Both block diagrams are depicted in Figure C.10. In both cases, integrator gain parameters – denoted as `Affe, Afb1,` and `Afb2` – can be defined by the user in the block dialogue window, as well as their associated gain errors – respectively denoted in the model as `MU1, MU2, MU3`.

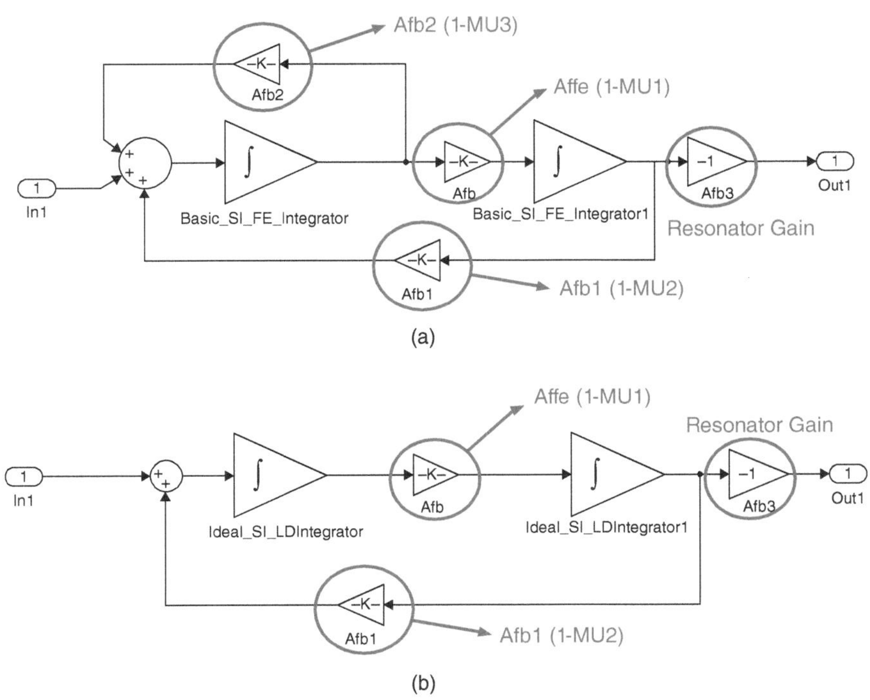

Figure C.10 Block diagram of SI resonators in SIMSIDES: (a) FEI-loop resonator; (b) LDI-loop resonator.

Table C.6 Library of SI (FE/LD) resonators included in SIMSIDES.

Model name	Circuit effects included
`Ideal_SI_FE_Resonator` `Ideal_SI_LD_Resonator`	Ideal SI (FE/LD) resonator.
`Basic_SI_FE_Resonator` `Basic_SI_LD_Resonator`	Output current limits, nonlinear gain.
`SI_FE_Res_Finite_Conductance` `SI_LD_Resonator_Finite_Conductance`	Output current limits, finite nonlinear output conductance, input voltage limits, thermal noise.
`SI_FE_Res_Finite_Conductance&Settling` `SI_LD_Res_Finite_Conductance&Settling`	Output current limits, finite nonlinear input/output conductance, input voltage limits, settling error, thermal noise.
`SI_FE_Res_Finite_`... `&Settling&Charge_Injection` `SI_LD_Res_Finite_`... `&Settling&Charge_Injection`	Output current limits, finite nonlinear input/output conductance, input voltage limits, settling error, thermal noise, charge injection error.

Table C.6 lists all SI resonator blocks included in SIMSIDES, together with a brief description of the error mechanisms taken into account in each model. These errors and their associated model parameters, which are the same as those used in SI integrators, are described below.

C.4.3 SI Errors and Model Parameters

This section gives a brief description of the main errors and model parameters included in the SI model libraries of SIMSIDES. For the sake of clarity, these errors are linked to some of the building blocks in which they are included.

C.4.3.1 Basic_SI_FE(LD)_Integrator and Basic_SI_FE(LD)_Resonator

The model parameters included in these blocks are:

- `Integrator Gain`: integrator weight (see Figure C.8).
- `Iomax(=-Iomin)`: maximum/minimum integrator output current.
- `MU1,MU3`: first-order and third-order coefficients of the gain nonlinearity.

The Z-domain transfer function of this integrator block is given by the following expression,

$$i_o(z) = (1 - \mathrm{MU1}) \cdot i_{oi}(z) + \mathrm{MU3} \cdot i_{oi}^3(z) \tag{C.6}$$

where $i_{oi}(z)$ is the Z-transform of the ideal output current.

The above expression is also used for both LDI- and FEI-based SI resonators, although in this case the nonlinear coefficients are called `SHI1, SHI3` instead of `MU1, MU3`.

C.4.3.2 SI_FE(LD)_Int_Finite_Conductance These blocks include the effect of finite input–output conductance ratio error. To this end, the equivalent circuit shown in Figure C.11 is solved during sampling phase. This circuit corresponds to the SI LD integrator[1] of Figure C.8 on clock phase ϕ_1. In this circuit, the input signal is modeled as an ideal current in parallel with a finite conductance, denoted as g_{oi}. Memory-cell transistor 1 (M1) is configured in the hold phase and is modeled by its drain current i_{d1} in parallel with an output conductance g_o. In contrast, memory-cell transistor 2 (M2) is operating in sampling phase, and it is modeled by the parallel connection of its output conductance with an input resistor. This resistor is a nonlinear function of the drain current i_{d2} of M2, given by,

$$v_i(i_{d2}) \simeq \mathrm{A1} \cdot i_{d2} + \mathrm{A3} \cdot i_{d2}^3 \tag{C.7}$$

where `A1, A3` stand for the first-order and third-order nonlinear coefficients of the input resistance. These parameters are defined in the models together with:

- `Gout`: output conductance g_o of memory cells.
- `Vmax/Vmin`: maximum/minimum values of v_i in Equation (C.7).
- `Ibias`: integrator bias current (see Figure C.8).

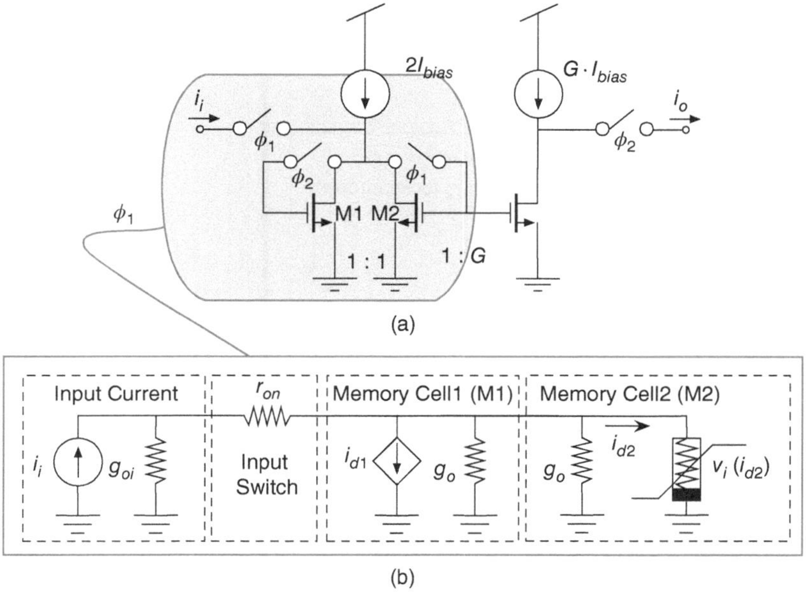

Figure C.11 Modeling finite input-output conductance ratio error in SIMSIDES: (a) SI LD integrator; (b) equivalent circuit during sampling phase (ϕ_1).

[1] A similar circuit is used for SI FE integrators.

C.4.3.3 SI_FE(LD)_Int_Finite_Conductance & Settling & ChargeInjection Those SI building blocks including nonlinear incomplete settling and charge injection require the following additional model parameters:

- `Gmo`: operating-point small-signal transconductance of memory transistors.
- `Cgs`: gate-to-source capacitance of memory transistors.
- `Eq`: charge injection error.

The charge injection error `Eq` is defined as a relative error on the voltage stored in the gate-to-source capacitance, given by:

$$v_{gs,\text{nonideal}} = (1 - \text{Eq}) \cdot v_{gs,\text{ideal}} \tag{C.8}$$

C.5 Real CT Building-Block Libraries

Figure C.12 shows the CT building-block model libraries included in SIMSIDES. There are four libraries of CT integrators and two libraries of CT resonators, which are classified according to the circuit nature of the building blocks: Gm-C, Gm-MC, Gm-LC, active-RC, and MOSFET-C.

C.5.1 Real CT Integrators

Tables C.7–C.10 list all the models[2] included in the CT integrator libraries shown in Figure C.12, together with a brief description of the nonideal effects included. An explanation of the most important blocks and their associated model parameters is given in the following subsections.

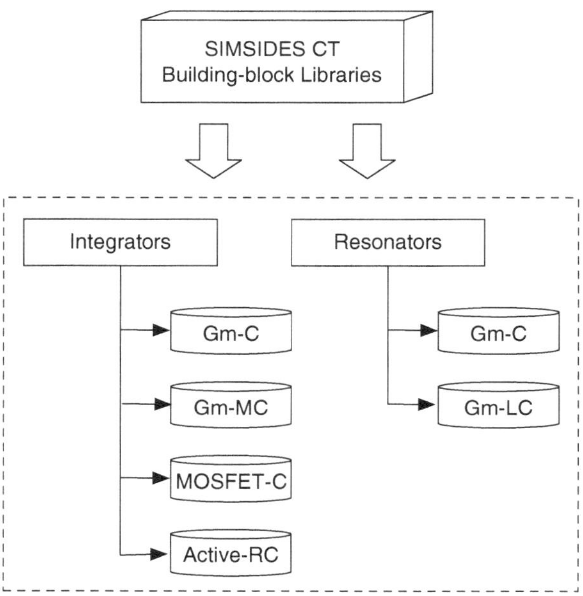

Figure C.12 Classification of SIMSIDES CT model libraries.

[2]RC_Int_1,2,3 in models allow setting of transistor-level parameters, such as channel-length modulation, gate-to-source overdrive voltage, saturation voltage, supply voltage, and so on.

Table C.7 Gm-C integrator library models in SIMSIDES.

Model name	Circuit effects included
Ideal_OTA_C_CTint Transconductor	Ideal Gm-C integrator. Input saturation voltage, nonlinear transconductance.
gm_no_noise_new	Output saturation voltage, third-order intercept point.
1pole_gm	Gm-C output impedance.
OTA_C_CT_1pole	Input/output saturation voltage, finite OTA DC gain, nonlinear transconductance, one-pole dynamic, time-constant error, nonlinear transconductance, thermal noise.
OTA_C_CT_2poles OTA_C_CT_2polesb	Input/output saturation voltage, finite OTA DC gain, nonlinear transconductance, two-pole dynamic, time-constant error, nonlinear transconductance, thermal noise.

Table C.8 Gm-MC integrator library models in SIMSIDES.

Model name	Circuit effects included
Gm_MC_CTInt_1pole	Input/output saturation voltage, finite OTA DC gain, parasitic capacitances, one-pole dynamic, thermal noise.
Gm_MC_CTInt_2poles	Input/output saturation voltage, finite OTA DC gain, parasitic capacitances, two-pole dynamic, thermal noise.
Gm...1pole&Large_signal_distortion	Input/output saturation voltage, output current limit, finite OTA DC gain, parasitic capacitances, one-pole dynamic.
Gm...2poles&Large_signal_distortion	Input/output saturation voltage, output current limit, finite OTA DC gain, parasitic capacitances, two-pole dynamic.
Gm...1pole&Small_Signal_Distortion	Input/output saturation voltage, output current limit, finite OTA DC gain, nonlinear transconductance, parasitic capacitances, one-pole dynamic.
Gm...2poles&Small_Signal_Distortion	Input/output saturation voltage, output current limit, finite OTA DC gain, nonlinear transconductance, parasitic capacitances, two-pole dynamic.

Table C.9 Active-RC integrator library models in SIMSIDES.

Model name	Circuit effects included
`RC_CTInt_1pole`	OTA output swing limitation, finite OTA DC gain, parasitic capacitances, capacitance voltage coefficient, one-pole dynamic, thermal noise.
`RC_CTInt_2poles`	OTA output swing limitation, finite OTA DC gain, parasitic capacitances, capacitance voltage coefficient, two-pole dynamic, thermal noise.
`RC...1pole&Large_signal_distortion`	OTA output swing limitation, output current limit, finite OTA DC gain, parasitic capacitances, capacitance voltage coefficient, one-pole dynamic, thermal noise.
`RC...2poles&Large_signal_distortion`	OTA output swing limitation, output current limit, finite OTA DC gain, parasitic capacitances, capacitance voltage coefficient, two-pole dynamic, thermal noise.

`RC_Int_1in`
`RC_Int_2in`
`RC_Int_3in`
	OTA output swing limitation, finite OTA DC gain, nonlinear trans., slew rate, parasitic capacitances, one-pole dynamic, thermal noise.

Table C.10 MOSFET-C integrator library models in SIMSIDES.

Model name	Circuit effects included
`MOSFET_C_CTInt_1pole`	OTA output swing limitation, finite OTA DC gain, parasitic capacitances, capacitance voltage coefficient, one-pole dynamic, thermal noise.
`MOSFET_C_CTInt_2poles`	OTA output swing limitation, finite OTA DC gain, parasitic capacitances, capacitance voltage coefficient, two-pole dynamic, thermal noise.
`MOS...1pole&Large_signal_distortion`	OTA output swing limitation, output current limit, finite OTA DC gain, parasitic capacitances one-pole dynamic, thermal noise.
`MOS...2poles&Large_signal_distortion`	OTA output swing limitation, output current limit, finite OTA DC gain, parasitic capacitances two-pole dynamic, thermal noise.

C.5.1.1 Model Parameters used in Transconductors and Gm-C Integrator Building Blocks The building blocks listed in Table C.7 are used in SIMSIDES to model transconductors and Gm-C integrators, considering the effect of different circuit-level nonideal effects. This section gives a description of the most significant model parameters included in these blocks.

- `Transconductance of the OTA` and `Integration Capacitor`, define respectively the transconductance g_m and the integration capacitor C of the Gm-C integrator, whose ITF is obtained by replacing $g = g_m/C$ in Equation (C.1).
- `[Upper,Lower] bound saturation voltage`, defines the maximum and minimum values of the output voltage v_o, as illustrated in Figure C.13.
- `Input Voltage [Upper, Lower] saturation limit`, defines the maximum and minimum values of the input voltage, respectively denoted as v_{imax} and v_{imin} in Figure C.13.
- `[Second, Third] order distortion coefficient`, stand for the second- and third-order nonlinear transconductance coefficients $g_{m(1,2)}$, where it is assumed that the transconductance depends on the Gm-C integrator input voltage v_i as,

$$g_m \simeq g_{mo} \cdot (1 + g_{m1} \cdot v_i + g_{m2} \cdot v_i^2) \tag{C.9}$$

  with g_{mo} being the nominal value of the transconductance. The maximum current provided by the transconductor is defined as I_o.
- `DC voltage gain`stands for the finite OTA DC gain, defined as g_m/g_o, with g_o being the output conductance of the Gm-C integrator.
- `Integration constant time error`, defined as C_p/C, with C_p being the parasitic capacitance at the output of the Gm-C integrator.
- `High frequency pole`, which defines the value of the high-frequency pole when a two-pole dynamic model is considered.

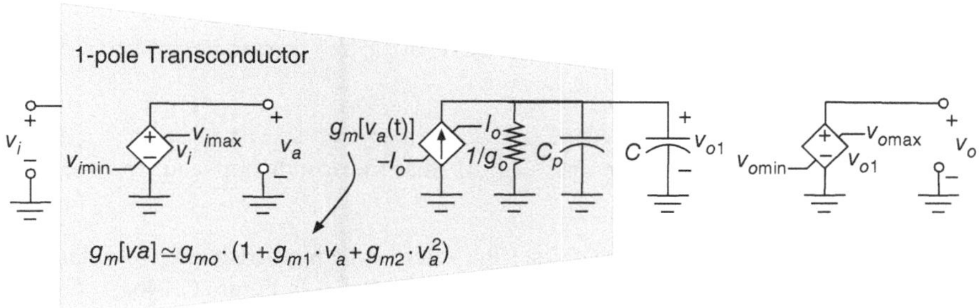

Figure C.13 One-pole Gm-C integrator model used in SIMSIDES.

C.5.1.2 Gm-MC Integrators SIMSIDES contains a library of Gm-C integrators based on the connection of a transconductance element and a Miller capacitance, also referred to as Gm-MC integrators. Figure C.14a shows the conceptual schematic of the Gm-MC integrators modeled in SIM-

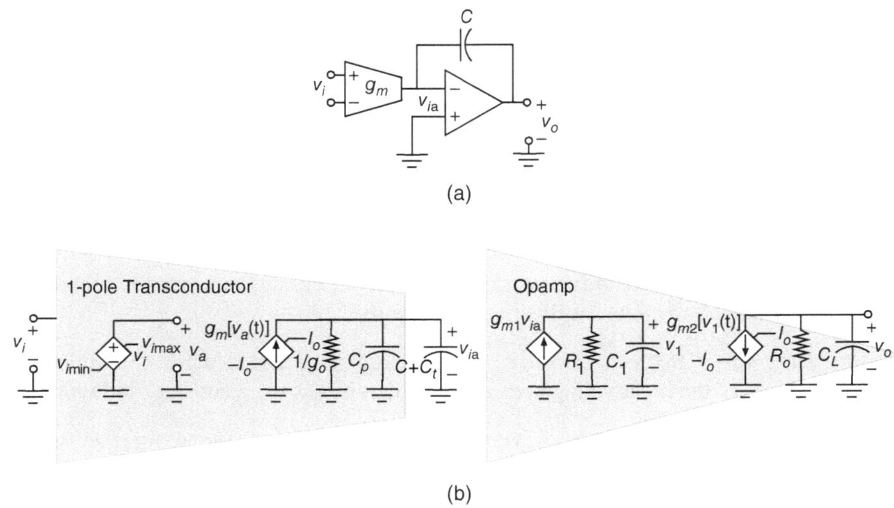

Figure C.14 Two-pole Gm-MC integrator model used in SIMSIDES: (a) conceptual schematic; (b) equivalent circuit of the transconductor and the opamp.

SIDES. Several models that account for the effect of different circuit nonidealities are included in this library. The most accurate one, named `Gm_MC_CTInt_1pole&Small_Signal_Distortion`, is modeled by the equivalent circuit shown in Figure C.14b, which involves a two-pole dynamic model. In addition to the model parameters included in Gm-C integrators, Gm-MC integrator models use the additional parameters described below.

- `Output transconductor parasitic, Integration capacitor ratio (Cp/C)`, which stands for C_P/C (see Figure C.14b).
- `Output operational parasitic, Integration capacitor ratio(CI/C)`: C_L/C.
- `Operational parasitic Output, OTA output capacitor ratio (Ct/C)`: C_t/C.
- `Transconductor and Op. Amplifier Unity gain frequency (Hz) [Gb1, Gb2]`, which stand for the GB of the transconductor and for the opamp in Figure C.14b, respectively.
- `High Frequency pole`, which defines the value of the high-frequency (nondominant) pole when a two-pole dynamic is considered – given by $1/(R_1C_1)$ in Figure C.14b.
- `Origin transconductance`, which refers to the operating-point transconductance g_{mo} of the nonlinear characteristic given in Equation (C.9).

C.5.1.3 Active-RC Integrators

Among all active-RC integrator models listed in Table C.9, the most accurate and complete one is named `RC_CTInt_2poles&Large_signal_distortio`. Figure C.15 shows the conceptual schematic (Figure C.15a) and its corresponding equivalent model (Figure C.15b). There are two versions of this model: one based on a linear input resistance R and

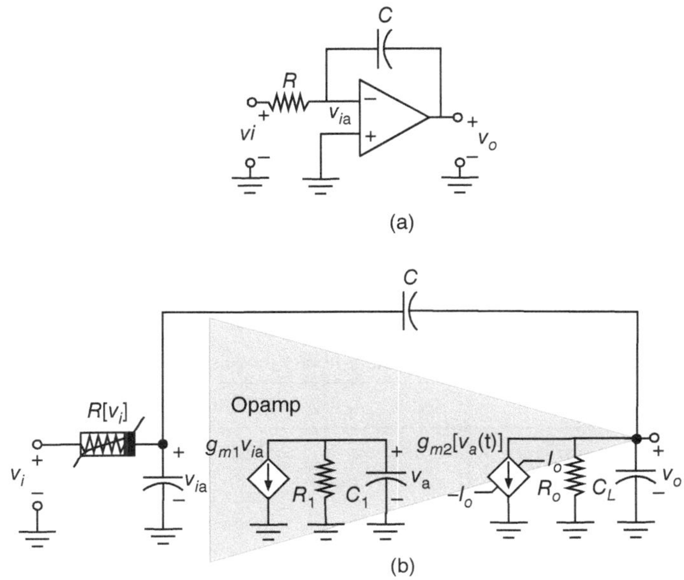

Figure C.15 Two-pole active-RC integrator model used in SIMSIDES: (a) conceptual schematic; (b) equivalent circuit.

another in which R is a nonlinear function of the input voltage given by,

$$R(v_i) \simeq R \cdot (1 + R_1 \cdot v_i + R_2 \cdot v_i^2) \tag{C.10}$$

where $R_{(1,2)}$ stand for the first- and second-order nonlinear coefficients.

Apart from the model parameters used by the CT building blocks described in earlier sections, the following parameters are used in active-RC integrator models:

- `Output resistance - integration resistance ratio`, which stands for R_o/R in Figure C.15b.
- `Integrator Ideal Unity gain frequency (RC)`, defined as $1/(RC)$.
- `Opamp ideal Unity gain frequency`: the GB of the opamp in Figure C.15b.
- `High frequency pole (Hz)`, defined as $1/(R_1C_1)$ (see Figure C.15b).

C.5.1.4 MOSFET-C Integrators In addition to active-RC integrator models, SIMSIDES includes also a library of MOSFET-C integrators whose conceptual schematic is shown in Figure C.16. Essentially, these building blocks are the same as those used for modeling active-RC integrators, except that the integrator resistance R is replaced by a MOSFET transistor.

C.5.2 Real CT Resonators

Tables C.11 and C.12 list all the models included in the CT resonator libraries shown in Figure C.12, together with a brief description of their nonideal effects. These libraries include various building

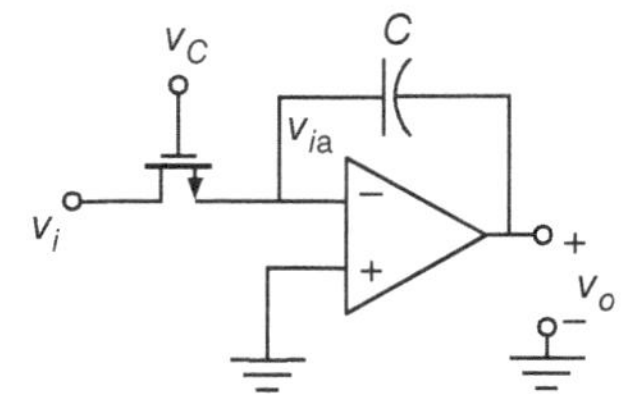

Figure C.16 Conceptual schematic of a MOSFET-C integrator like that modeled in SIMSIDES by the building blocks listed in Table C.10.

Table C.11 Gm-C resonator library models in SIMSIDES.

Model name	Circuit effects included
`Ideal_gmC_CT_Resonator`	Ideal Gm-C resonator.
`gmC_CT_Res_1pole`	Finite OTA DC gain, time-constant error, one-pole dynamic, thermal noise.
`gmC_CT_Res_2poles` `gmC_CT_Res_2polesfull`	Finite OTA DC gain, time-constant error, two-pole dynamic, thermal noise.
`gmC_CT_Res_1pole_larged`	Input/output saturation voltage, output current limit, finite OTA DC gain, time-constant error, one-pole dynamic.
`gmC_CT_Res_2poles_larged`	Input/output saturation voltage, output current limit, finite OTA DC gain, nonlinear transconductance, time-constant error, two-pole dynamic.
`gmC_CT_Res_1pole_small&larged`	Input/output saturation voltage, output current limit, finite OTA DC gain, nonlinear transconductance, time-constant error, one-pole dynamic.
`gmC_CT_Res_2poles_small&larged`	Input/output saturation voltage, output current limit, finite OTA DC gain, nonlinear transconductance, time-constant error, two-pole dynamic.

blocks that are classified according to the accuracy of their models as well as the circuit nonidealities that are taken into account. As an illustration, Figure C.17 depicts both the CT resonator libraries included in SIMSIDES: Gm-C resonators and Gm-LC resonators. Figure C.18 shows the conceptual schematic of both kinds of CT resonators. The model parameters involved in the building blocks shown in Figure C.17 are detailed below.

C.5.2.1 Gm-C Resonators Apart from those parameters used in Gm-C integrators, the following model parameters are used in the Gm-C resonator model library:

Table C.12 Gm-LC resonator library models in SIMSIDES.

Model name	Circuit effects included
`Ideal_gmLC_CT_Resonator`	Ideal Gm-LC resonator.
`gmLC_CT_Res_1pole`	Input/output saturation voltage, inductance quality factor and series parasitic resistance, finite OTA DC gain, time-constant error, one-pole dynamic, thermal noise.
`gmLC_CT_Res_2poles`	Input/output saturation voltage, inductance quality factor and series parasitic resistance, finite OTA DC gain, time-constant error, two-pole dynamic, thermal noise.
`gmLC...1pole_large_dist`	Input/output saturation voltage, inductance quality factor and series parasitic resistance, output current limitation, finite OTA DC gain, time-constant error, one-pole dynamic, thermal noise.
`gmLC...2poles_large_dist`	Input/output saturation voltage, inductance quality factor and series parasitic resistance, output current limitation, finite OTA DC gain, time-constant error, two-pole dynamic, thermal noise.
`gm...1pole_small&large_dist`	Input/output saturation voltage, inductance quality factor and series parasitic resistance, output current limitation, nonlinear transconductance, finite OTA DC gain, time-constant error, one-pole dynamic, thermal noise.
`gmLC...2poles_small&larged`	Input/output saturation voltage, inductance quality factor and series parasitic resistance, output current limitation, nonlinear transconductance, finite OTA DC gain, time-constant error, two-pole dynamic, thermal noise.

- `Transconductance of the first, second, and third OTA (gm1,gm2, gm3)`: g_{m1}, g_{m2}, and g_{m3} in Figure C.18a.
- `Capacitors`: C_1, C_2 in Figure C.18a.
- `Nonlinear transconductance coefficients [gmnl1, gmnl2]`, which stand for nonlinear coefficients $g_{m(1,2)}$ in Equation (C.9).
- `DC gain of OTAs [Av1,Av2,Av3]`, which defines the finite OTA DC gain of transconductors in Figure C.18a.
- `Percentual integration constant time error [et1, et2]`: the time constant errors associated to both feedforward transconductances in Figure C.18a, given by,

$$\epsilon_{t1} = \frac{C_{p1} + C_{p3}}{C_1} \quad , \quad \epsilon_{t2} = \frac{C_{p2}}{C_2} \tag{C.11}$$

where C_{pi} stands for the parasitic capacitance of the ith transconductance in Figure C.18a.

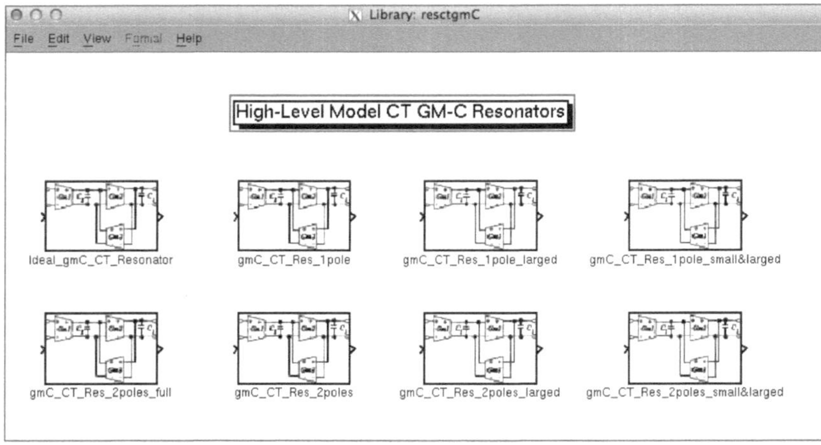

(a)

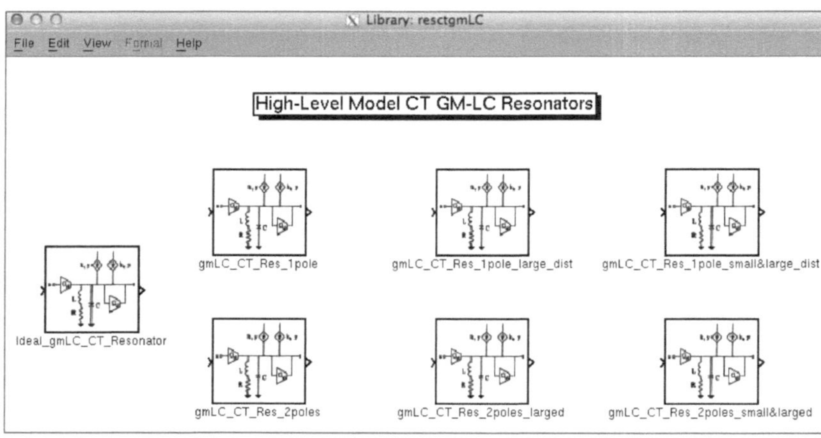

(b)

Figure C.17 CT resonator libraries included in SIMSIDES: (a) Gm-C resonators; (b) Gm-LC resonators.

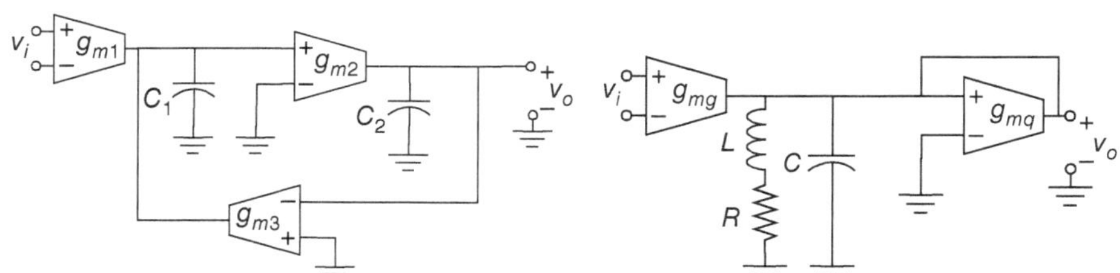

(a) (b)

Figure C.18 Conceptual schematics of the CT resonators modeled in SIMSIDES: (a) Gm-C resonator; (b) Gm-LC resonator.

C.5.2.2 Gm-LC Resonators

The main model parameters used by Gm-LC resonator blocks are described in this section. In essence, these parameters deal with the inductor element in Figure C.18b and its associated resonant frequency.

- `Frequency resonance`: the resonant frequency of the Gm-LC resonator.
- `Inductor Q`: the Q-factor of the inductor in Figure C.18b.
- `Series resistance`: the parasitic resistance R of the inductor.

The rest of the model parameters used in Gm-C resonators have the same meaning as those used in Gm-C integrators and resonators.

C.6 Real Quantizers & Comparators

Table C.13 lists the blocks included in the real `Quantizers&Comparators` SIMSIDES library, together with a brief description of their operation and main circuit nonidealities. In addition to the ideal parameters described in Section C.2.3, additional model parameters are required to model the different circuit nonidealities. These error parameters are listed in Table C.14.

Note that, as well as blocks corresponding to comparators and quantizers, there is a building block called `Real_Sampler`, which is used for modeling the S&H circuits that are connected at the input of embedded quantizers in CT-ΣΔMs. One of the most important errors associated with this building block is the clock jitter, which is modeled as an uncertainty in the sampling time δt corresponding to a stationary process with zero mean and standard deviation defined by the user (see Table C.14).

Table C.13 Real quantizers and comparator models included in SIMSIDES.

Model name	Circuit effects included
`Real_Comparator_Offset&Hysteresis`	Voltage-mode comparator with offset, (random and deterministic) hysteresis.
`Real_Comparator_Offset&Hysteresis_for_SI`	Current-mode comparator with offset and nonlinearity (INL).
`Real_Multibit_Quantizer`	Voltage-mode multi-bit quantizer with gain error, offset (random and deterministic) hysteresis.
`Real_Multibit_Quantizer_for_SI`	Current-mode multi-bit quantizer with gain error, offset, INL, (random and deterministic) hysteresis.
`Real_Multibit_Quantizer_dig_level_SD2`	Voltage-mode multi-level quantizer with gain error, offset, INL, (random and deterministic) hysteresis.
`Real_Sampler`	Sampling and hold circuit with clock jitter error.

Table C.14 Error model parameters used in SIMSIDES real quantizers.

Parameter name (in alphabetical order)	Brief description
`Gain error in LSB`	Gain error measured in LSB.
`Jitter typical deviation`	Standard deviation of clock jitter error.
`Kind of hysteresis`	Comparator hysteresis. It may be either deterministic or random hysteresis.
`INL in LSB`	Integral nonlinearity error measured in LSB.
`Number of levels`	Number of quantizer levels.
`Offset`	Offset error.
`Offset error in LSB`	Offset error measured in LSB.
`Seed for random jitter generation`	Seed number used for generating random jitter error.

C.7 Real D/A Converters

Table C.15 lists the different building blocks included in the real `D/A Converters` SIMSIDES library, together with a brief description of their operation and main circuit errors.

Table C.15 Real DAC models included in SIMSIDES.

Model name	Circuit effects included
`Real_DAC_Multibit`	Voltage-mode multi-bit DAC with offset, gain error, and INL error.
`Real_DAC_Multibit_SI`	Current-mode multi-bit DAC with offset, gain error, and INL error.
`Real_DAC_Multibit_delay_Jitter`	Voltage-mode multi-bit DAC with offset error, gain error, INL error, delay error, and clock jitter error.
`Real_DAC_Multibit_delay_Jitter_SI`	Current-mode multi-bit DAC with finite output conductance, offset error, gain error, INL error, delay error, and clock jitter error.
`Real_DAC_pulse_types`	Voltage-mode multi-bit DAC with selectable NRZ/RZ/HRZ output waveform.
`Real_DAC_Multibit_pulse_types`	Voltage-mode multi-bit DAC with selectable NRZ/RZ/HRZ output waveform, gain error, offset error, and INL error.
`Real_DAC..._delay_jitter`	Voltage-mode multi-bit DAC with selectable NRZ/RZ/HRZ output waveform, gain error, offset error, INL error, delay error, and clock jitter error.

Error parameters associated with the models listed in Table C.15 have the same meaning as those used in multi-bit quantizers, except for the selectable NRZ/RZ/HRZ DAC waveform and the delay error. The latter can be chosen to be either a constant delay or a signal-dependent delay, given by,

$$\text{delay}(v_i) = \text{d0} + \frac{\text{d1}}{\text{x1} \cdot |v_i|} < \text{dmax} \tag{C.12}$$

where `d0, d1, x1, dmax` are model parameters set by the user.

C.8 Auxiliary Blocks

In addition to the building blocks described in previous sections, SIMSIDES includes a library named `Auxiliary blocks` that contains some other blocks (like adders, DEM algorithms, and digital

Table C.16 Auxiliary building-block models used in SIMSIDES.

Analog adders	
Model name	**Brief description**
`Analog_Adder_Ideal_SD2`	Ideal SC passive adder with parasitic input capacitance and load capacitance.
`Analog_Adder_real_SD2`	Real SC passive adder with parasitic input capacitance and load capacitance, switch on-resistance, settling error, capacitor nonlinearity, and thermal noise.
Digital adders	
Model name	**Brief description**

`Dig_add_generic_2outs`
`Dig_add_3L_5L_13L`
`Dig_add_3L_3L_5L_2outs`
`Dig_add_3L_3L_7L_2outs`
`Dig_add_3L_5L_9L_2outs`
`Dig_add_3L_5L_13L_2outs`
	Digital subtraction of a M_1-level thermometric-coded signal and a M_2-level thermometric-coded signal, which is scaled by a factor of `d`. The result is a $(M_1 + M_2/\text{d})$-level thermometric-coded digital output.

Digital latches	
Model name	**Brief description**
`D_latch_simplest` `D_latch`	Digital "D" latches.
DAC with DEM algorithms	
Model name	**Brief description**
`DEM_id_SD2`	Ideal DEM algorithm.
`DAC - DEM - V04`	DAC block with a selectable DEM algorithm. There are three options: No DEM, DWA, Pseudo-DWA.
`Mux_SD2`	Building block used for sampling an input (analog) signal by a number of different branches corresponding to the number of DAC unit capacitors.

Table C.17 Error model parameters used in SIMSIDES auxiliary blocks.

Parameter name (in alphabetical order)	Brief description
`Comparator input capacitor (C)`	Parasitic capacitance at the comparator/quantizer input.
`DEM type`	DEM algorithm: (1) No DEM, (2) DWA, (3) Pseudo-DWA. (Default = 1)
`Input capacitor (C)`	Input capacitance of the analog adder.
`Nonlinearities of the capacitors`	Capacitance nonlinear coefficients in an analog adder.
`Number of elements`	Number of DAC unit elements.
`Output type`	Digital output code: (1) Binary output, (2) Trilevel output including common mode (Default = 1).
`Time interval between sampling and comparison (delta)`	Delay between the time instant when the adding operation is performed and the time instant when comparison time takes place.

latches) also needed to simulate ΣΔMs. Table C.16 lists the models included in the library, together with brief descriptions of their operation. The most significant parameters used by these models are listed in Table C.17.

As an illustration on the use of some of the auxiliary blocks, Figure C.19a shows the SIMSIDES block diagram of a second-order feedforward SC-ΣΔM, which includes an embedded 16-level quantizer and DAC with a selectable DEM algorithm.

The block diagram in Figure C.19a includes the following building blocks:

- `Mux_SD2`. This block samples the modulator input signal in a number of unit capacitors, which corresponds to the number of unit elements used in the multi-level DAC minus one. As shown in Figure C.19b, the only model parameter of this block is called `Number of elements`, and equals 15 in this example.
- `SC_FE_Int_All_Effects_DEM_SD2`. This block is used for modeling the front-end integrator and includes all error mechanisms with values defined in the block dialogue box in Figure C.19c. This model also includes a parameter `Array of sampling capacitors for DEM branch`. The number of capacitors in the array must be exactly the same as the number of DAC levels: the number of quantization levels. As an illustration, Figure C.20 shows the MATLAB code used for generating different alternative capacitor arrays, together with other parameters used for simulating the block diagram in Figure C.19a. Note that this capacitor array must also include the DAC element mismatch, defined as a Gaussian probability distribution.
- `Basic_SC_FE_Int`. This block is used for modeling the second integrator, considering only ideal values of the output swing and capacitors, as shown in Figure C.19d.
- `Real_Multibit_Quantizer_dig_level_SD2`. This block, called `Q1` in the example, is used for modeling the quantizer while considering the nonideal effects listed in Table C.13. In

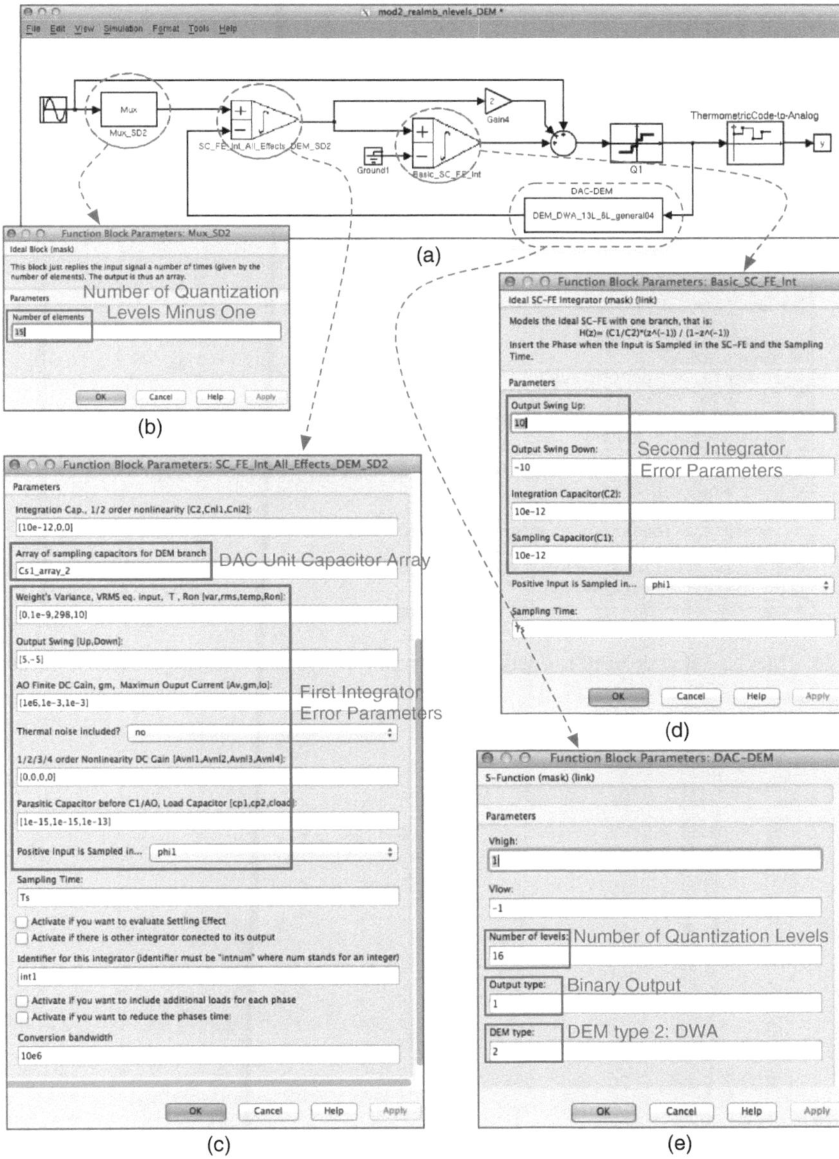

Figure C.19 Example of a second-order feedforward SC-ΣΔM with 16-level quantization and DEM: (a) SIMSIDES block diagram; (b) Mux_SD2 dialogue box; (c) first integrator block dialogue box; (d) second integrator block dialogue box; (e) DAC-DEM block dialogue box.

this example, these nonideal effects have not been taken into account. The output of this block is a thermometric-coded bit array. This array is transformed into an analog signal for further processing using the block named `ThermometricCode-to-Analog`.

- `DAC-DEM`. This block uses the `DAC - DEM - V04` model (see Table C.16). The most important model parameters of these block are highlighted in Figure C.19e.

```
%%%%%%%%%% Modulator: simulation parameters %%%%%

fs=35.2e6; % Sampling frequency
Ts=1/fs; % Sampling time
fi=137.5e3; % Input frequency
OSR=64; % OverSampling Ratio
N=65536; % Number of Points

%%%%%%%%%% DEM Parameters and Capacitor Feedback Arrays %%%%%%%%%%%

nlevels=16; % Number of quantization levels
Cs1=10e-12; % Sampling Capacitor
sigma=[1/100]; % Capacitor mismatch standard deviation

%% Capacitor array

n_elements_1=(nlevels-1);
% Binary array
n_elements_2=floor((nlevels)/2);
% Tri-level array
Cs1_array_1=Cs1/n_elements_1*(ones(1,n_elements_1));
% No mismatch - binary
Cs1_array_2=Cs1/
n_elements_1*(ones(1,n_elements_1)+sigma*randn(1,n_elements_1));
% With mismatch - binary
Cs1_array_3=Cs1/
n_elements_2*(ones(1,n_elements_2)+sigma*randn(1,n_elements_2));
% With mismatch - tri_level
Cs1_array_4=Cs1/n_elements_2*(ones(1,n_elements_2));
% No mismatch - tri_level
```

Figure C.20 MATLAB code used for defining capacitor arrays and DEM parameters of Figure C.19e.

Index

Note: Page numbers in *italics* refer to figures, those in **bold** refer to tables.